Organosilicon Chemistry

Organosilicon Chemistry

Novel Approaches and Reactions

Edited by
Tamejiro Hiyama
Martin Oestreich

WILEY-VCH

Editors

Dr. Tamejiro Hiyama
Chuo University
Research & Development Initiative
1-313-27 Kasuga, Bunkyo-ku
112-8551 Tokyo
Japan

Dr. Martin Oestreich
Technische Universität Berlin
Institut für Chemie; Sekr. C3
Straße des 17. Juni 115
10623 Berlin
Germany

Library of Congress Card No.: applied for

British Library Cataloguing-in-Publication Data
A catalogue record for this book is available from the British Library.

Bibliographic information published by the Deutsche Nationalbibliothek
The Deutsche Nationalbibliothek lists this publication in the Deutsche Nationalbibliografie; detailed bibliographic data are available on the Internet at <http://dnb.d-nb.de>.

Print ISBN: 978-3-527-34453-6
ePDF ISBN: 978-3-527-81475-6
ePub ISBN: 978-3-527-81477-0
oBook ISBN: 978-3-527-81478-7

Typesetting SPi Global, Chennai, India
Printing and Binding Markono Print Media Pte Ltd, Singapore

Printed on acid-free paper

10 9 8 7 6 5 4 3 2 1

Contents

Foreword

When I was a student, sixty years ago, there was essentially no silicon chemistry in our course. We were perhaps lucky to hear about silicone in the first year, but only because our lecturer was F. S. Kipping's son, Barry. Had we had any industrial chemistry, we might have learned of the 'direct' synthesis of dichlorodimethylsilane, and been introduced to catalytic hydrosilylation. At some stage someone must have mentioned tetramethylsilane. Today it would be a deficient undergraduate course that didn't mention these topics and some at least of the many more aspects of organosilicon chemistry that have sprung up in the meantime. This book is a testament to how different the subject is today.

My own introduction took place in the middle of my graduate work, in two lectures we had from visiting speakers: Henry Gilman and Colin Eaborn. I remember nothing of Gilman's lecture beyond the interminable introduction he was given by F. G. Mann. They were both beyond the end of their careers, but Eaborn, who was at the beginning of his, gave us a memorable lecture about aromatic electrophilic substitution. He was using protodesilylation as a convenient reaction with which to study substituent effects, and in due course it so caught his fancy that he became more and more an organosilicon chemist and less and less a physical organic chemist using kinetics to study the mechanism of organic reactions. But he was not focused on synthesis as such, and in that lecture he did not point out, nor did I appreciate, how significant it was that the electrophile attacked the benzene ring at the carbon atom to which the silicon was attached. That came later, in 1968, in Eaborn's book-length review with Bott of the chemistry of the silicon-carbon bond, where he made explicit his understanding that a silyl group stabilised a β-carbocation, and that a silyl group attached to carbon was a better electrofuge than a proton.

Through the 1960s, silicon inched its way into the minds of synthetic organic chemists. Adrian Brook worked on his eponymous reaction, which had started life in Gilman's laboratory. Gilbert Stork and Paul Hudrlik introduced the t-butyldimethylsilyl group in 1968 as a protecting group, stable enough to be used to separate the regioisomers of silyl enol ethers. In the same year, Peterson developed another of Gilman's observations: that β-silyl alcohols underwent elimination to give alkenes in the presence of acid and base, and, shortly after, Bill Chan showed how useful it was in the methylenation of carbonyl compounds. Three years later, Stork and Ernie Colvin showed that a silylated epoxide opened in acid regioselectively to place the carbonyl group on the carbon atom that had carried the silyl group. All this work stimulated me into reading Eaborn and Bott's chapter in 1972, giving me the sense that here was a largely untapped resource, with some striking possibilities revealed by the work of Calas and Dunoguès in Bordeaux on allylsilanes. That very year E. J. Corey and Venkateswarlu showed that t-butyldimethylsilyl protecting groups could be removed selectively with fluoride ion, and the scene was set for organosilicon chemistry to become an essential part of synthetic organic chemistry.

Organosilicon chemistry dominated my life in chemistry for the next thirty years as my research group found more and more ways of using the properties of carbon-bound silyl groups to control carbocation reactions. We had to learn new ways with which to introduce silyl groups into organic structures, in order that its presence could control the reactivity in its neighbourhood. Several chapters in this book extend those methods in ways unimaginable when we met the problem. Better still, my research group found more and more new subjects of study as unexpected reactions took place and new opportunities presented themselves. It was an expanding area, and has continued to expand to this day.

I was not alone, as scores of chemists saw the potential for organosilicon chemistry in synthesis, especially attractive because the metallic properties of silicon, muted to be sure, but tunable, appear to be accompanied by a much lower risk of toxicity in its compounds. It would be invidious to select from the large number of colleagues, friends and acquaintances I met over those years as the subject developed so excitingly. Perhaps I may name just one: Teruaki Mukaiyama, for his version of the aldol, Mannich and Michael reactions has to have been one of the major developments in organic synthesis.

Organosilicon chemistry is now a mainstream subject. Silicon turns up time and again in organic chemistry lectures and research papers, sometimes without comment, as though it had always been there. This is symbolic of a mature subject, but not of course a subject that has nowhere else to go. In this book we see how embedded it now is in the chemistry of today.

October 2018

Ian Fleming
University of Cambridge

Preface

Over the years, we had been approached and encouraged several times by the publisher to produce a book on modern silicon chemistry. We declined in the beginning and later kept postponing our decision to accept. Our concern was to get the right timing as both organic and inorganic silicon chemistry have witnessed tremendous development particularly in the recent decade. We eventually agreed on editing a book dedicated to silicon chemistry relevant to homogeneous catalysis as we do feel that the timing is now perfect. The outcome of our efforts or, more precisely, those of the contributors intends to capture the exciting advances in catalytic processes where silicon is involved in one way or another. Silicon fulfills a plethora of roles in synthetic chemistry, be it as a linchpin for carbon–carbon bond formation or as an independent catalyst or even as a ligand for transition metal catalysts. Of course, progress in silicon–carbon bond formation and new techniques making use of silicon-based reagents are also covered.

We consider ourselves fortunate that the leaders in the field responded positively to our invitation ... and we are grateful that essentially every author stayed within the timeline. It was a joy to work with this group of contributors, and we think that the quality of the chapters is impressive. We do not claim to be comprehensive but believe that these timely summaries are a testament to silicon chemistry being an active and strong area of synthetic chemistry. We sincerely hope that you find the book a good read and a useful reference.

Long live silicon!

December 2018

Tamejiro Hiyama
Chuo University

Martin Oestreich
Technische Universität Berlin

1

Catalytic Generation of Silicon Nucleophiles

Koji Kubota and Hajime Ito

Institute for Chemical Reaction Design and Discovery (WPI-ICReDD), Hokkaido University, Sapporo, Hokkaido 060-8628, Japan
Faculty of Engineering, Division of Applied Chemistry and Frontier Chemistry Center, Hokkaido University, Kita 13 Nishi 8, Kita-ku, Sapporo, Hokkaido, 060-8628, Japan

1.1 Introduction

Silicon nucleophiles represent a class of important organometallic species for silicon–carbon, silicon–silicon, and silicon–boron bond formation reactions in synthetic chemistry [1]. Conventionally, the generation of silicon nucleophiles is accomplished by reactions of chlorosilanes with alkali metal (K, Na, Li), reactions of hydrosilanes with alkali metal hydride, cleavage of the silicon–silicon bond in disilanes or the silicon–boron bond in silylboron reagents by organometallic carbon nucleophiles, and transmetallation from other silicon-metal compounds [2]. However, these stoichiometric methods have significant limitations such as low functional-group compatibility due to the high reactivity of hard silyl anions with an alkali metal countercation. In this context, silicon-based organocuprates are widely used as soft silyl anion equivalents for silicon–carbon bond formation reactions, even though this method requires stoichiometric organometallic compounds and copper salt [3]. Recently, catalytic nucleophilic silylation reactions have attracted considerable attention because of their mild reaction conditions and unique selectivity and reactivity. This chapter mainly focuses on two types of activation modes for catalytic generation of silicon nucleophiles (Figure 1.1). First, transmetalation between silicon compounds containing a Si—X bond (X = Si, B, and Zn) and metal catalysts generates nucleophilic silyl metal intermediates (Figure 1.1a). Second, a catalytic amount of Lewis bases (Nu) activates the silicon–boron bond of silylboron reagents to form nucleophilic silyl species (Figure 1.1b). This chapter provides the recent advancements in the catalytic generation of silicon nucleophiles through these activation pathways and their applications in organic synthesis.

Organosilicon Chemistry: Novel Approaches and Reactions, First Edition.
Edited by Tamejiro Hiyama and Martin Oestreich

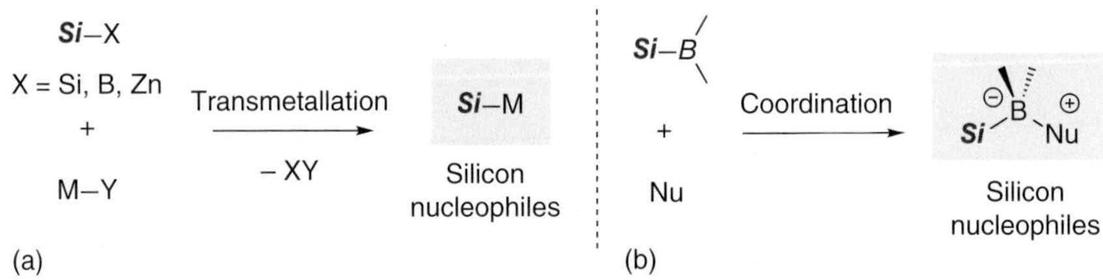

Figure 1.1 Representative pathways for catalytic generation of silicon nucleophiles. (a) Metal-catalyzed method. (b) Lewis base-catalyzed method.

1.2 Silicon Nucleophiles with Copper Catalysts

1.2.1 Copper-Catalyzed Nucleophilic Silylation with Disilanes

1.2.1.1 Silylation of α,β-Unsaturated Carbonyl Compounds

In 1998, the first example of copper-catalyzed nucleophilic 1,4-silylation of α,β-unsaturated carbonyl compounds with disilanes was reported by Ito et al. (Scheme 1.1) [4]. The reaction of cyclohexanone with a disilane in the presence of a copper salt and Bu_3P as a ligand proceeded to give the corresponding 1,4-silyl addition product in high yield. The silylation presumably goes through the σ-bond metathesis between a copper salt and a disilane to form the silylcopper intermediate, followed by its 1,4-addition to cyclohexanone. The copper catalyst is regenerated by the reaction between the resultant copper enolate and silyl triflate, which is formed at the first stage of this cycle. This mild protocol can be applied to a variety of substrates such as α,β-unsaturated cyclic and linear ketones and aldehydes to form the β-silyl carbonyl compounds in high yields.

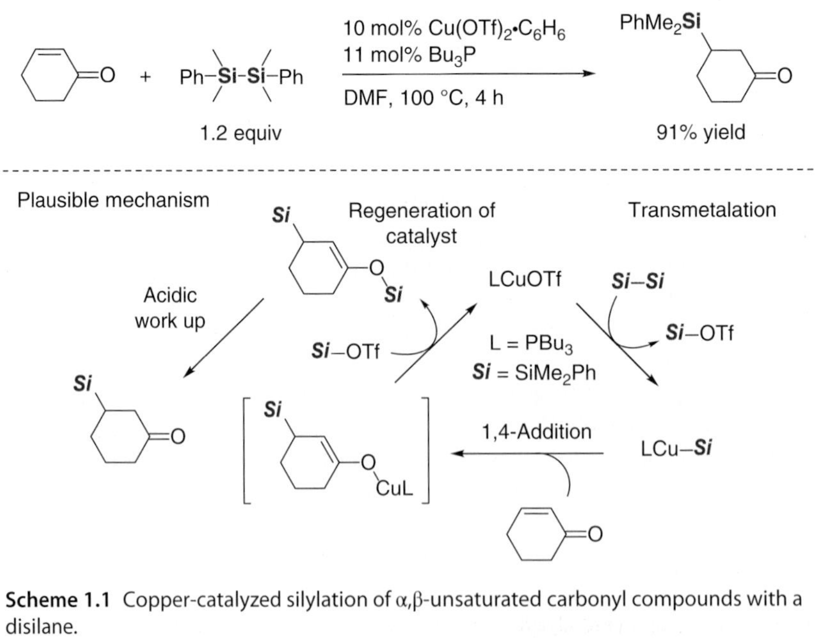

Scheme 1.1 Copper-catalyzed silylation of α,β-unsaturated carbonyl compounds with a disilane.

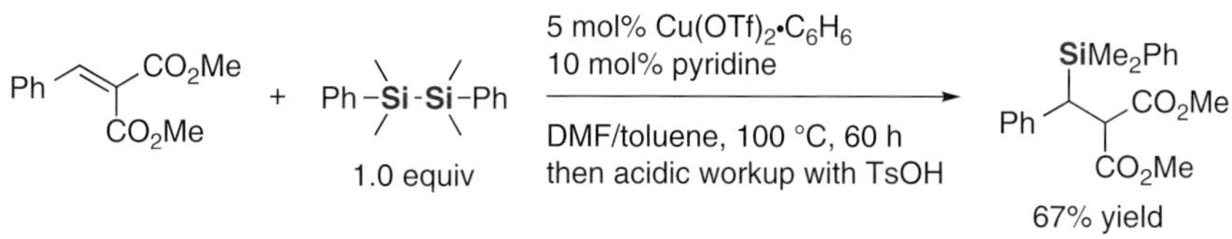

Scheme 1.2 Copper-catalyzed silylation of alkylidene malonates with a disilane.

1.2.1.2 Silylation of Alkylidene Malonates

Scheidt and coworkers reported the copper-catalyzed nucleophilic silylation of alkylidene malonates with disilanes in 2004 (Scheme 1.2) [5]. They found pyridine to be an effective ligand rather than phosphines for this reaction.

1.2.1.3 Silylation of Allylic Carbamates

In 2012, Ito et al. developed the copper-catalyzed allylic substitution with silicon nucleophiles (Scheme 1.3) [6]. This is the first example of a copper-catalyzed reaction between a disilane and allylic carbonates to produce allylsilanes, which are particularly useful reagents for stereoselective allylation of aldehydes in the presence of Lewis acids [7]. The regioselectivity of this reaction depends on the structure of substrates, suggesting that this reaction would proceed through the formation of a π-allyl copper(III) intermediate.

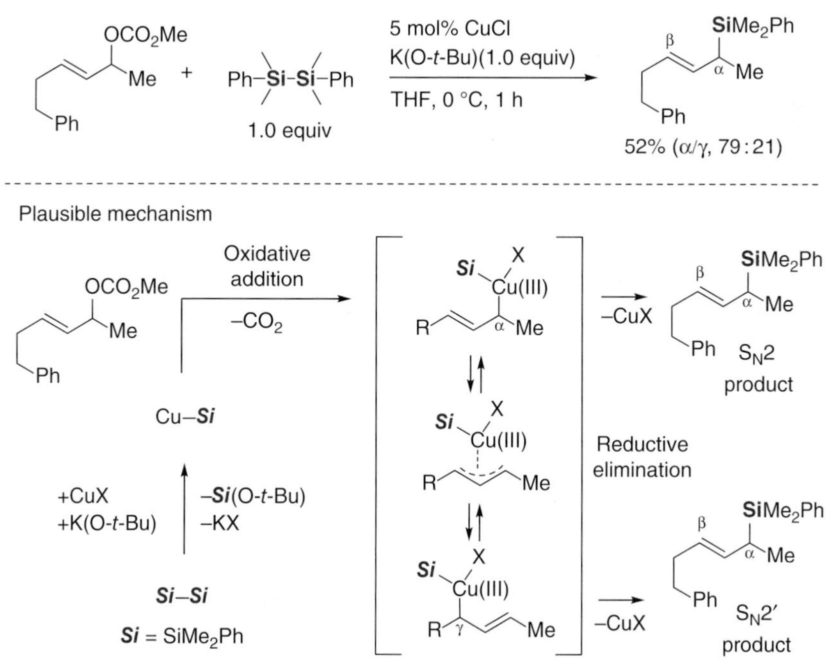

Scheme 1.3 Copper-catalyzed silylation of allylic carbamates with a disilane.

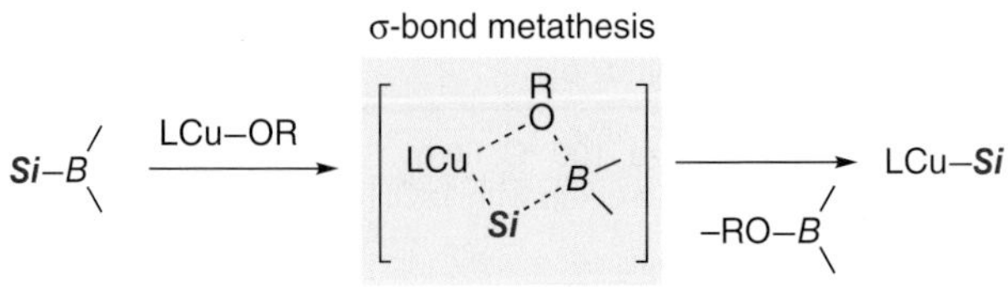

Figure 1.2 Activation of silicon–boron bond by σ-bond metathesis with a copper alkoxide.

1.2.2 Copper-Catalyzed Nucleophilic Silylation with Silylboronate

1.2.2.1 Silicon–Boron Bond Activation with Copper Alkoxide

Although disilanes are powerful sources for the generation of silicon nucleophiles as described, recent attention has focused on exploring the unique reactivity of silylboronates [8]. The difference in Lewis acidity between silicon and boron of silylboronates is exploited for facile boron–metal exchange at the silicon atom to generate silicon nucleophiles. The silicon–boron bond activation by σ-bond metathesis with copper alkoxide catalysts has become a general technique to access copper-stabilized silicon nucleophiles due to its mild reaction conditions (Figure 1.2) [8].

1.2.2.2 Silylation of α,β-Unsaturated Carbonyl Compounds

In 2010, the first example of copper-catalyzed nucleophilic silylation of α,β-unsaturated carbonyl compounds with silylboronates was reported by Hoveyda and a coworker (Scheme 1.4) [9]. They found that the copper-based chiral N-heterocyclic carbene (NHC) complex, generated in situ from the reaction of the corresponding silver-based carbene precursor with CuCl and Na(O-*t*-Bu), efficiently catalyzed conjugate silyl addition to cyclohexenone to form chiral β-silyl ketone in high yield with excellent enantioselectivity. Acyclic unsaturated ketones and cyclic dienones also undergo the enantioselective silyl conjugate addition with good to high enantioselectivity. Later, they also developed regio- and enantioselective conjugate silyl additions of acyclic and cyclic dienones and dienoates with a chiral NHC/copper complex catalyst [10].

Soon after, Riant and coworkers reported the copper-catalyzed asymmetric silylative aldol reaction between α,β-unsaturated carbonyl compounds and aromatic aldehydes (Scheme 1.5) [11]. The in situ–generated silylcopper intermediate reacts with methacryloxazolidinones to form the β-silylcopper enolate, followed by a diastereoselective aldol reaction to give the products bearing a chiral quaternary carbon center with excellent stereoselectivity.

In 2011, Córdova and coworkers reported that a copper-catalyzed silylation method can be combined with a chiral amine cocatalyst for iminium activation, as exemplified by the catalytic enantioselective silyl addition to α,β-unsaturated aldehydes (Scheme 1.6) [12].

The known copper-catalyzed silylation methods are generally sensitive to moisture and need to be carried out under an inert atmosphere. However, in 2012, Santos and a coworker discovered that the conjugative silylation of carbonyl compounds in water under air was efficiently catalyzed by a copper salt and 4-picoline (Scheme 1.7) [13]. Both copper and pyridine are required in this

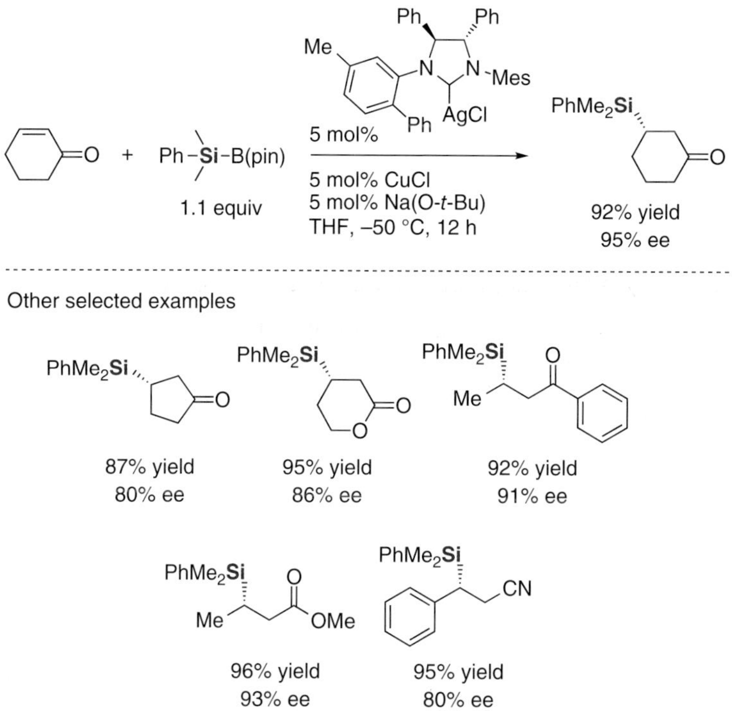

Scheme 1.4 Copper-catalyzed enantioselective conjugate silyl addition with a silylboronate.

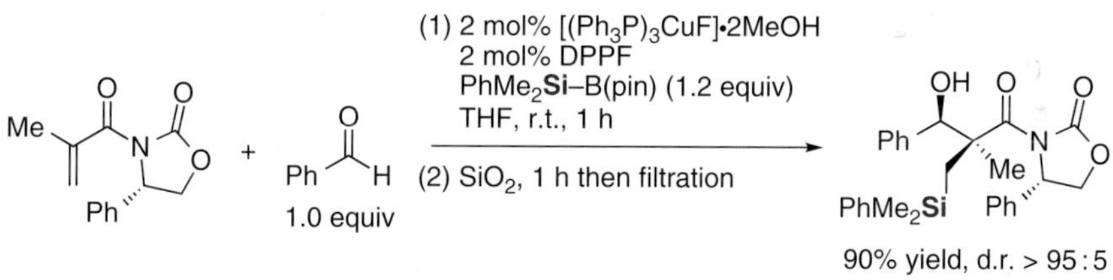

Scheme 1.5 Copper-catalyzed silylative aldol reaction with a silylboronate.

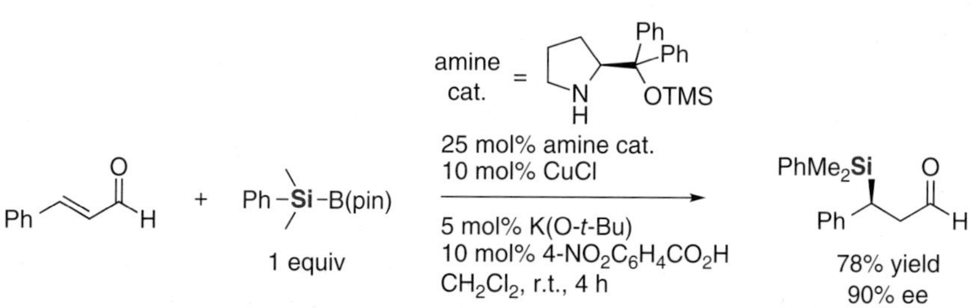

Scheme 1.6 Enantioselective silylation of α,β-unsaturated aldehydes by a copper/chiral amine cooperative catalyst.

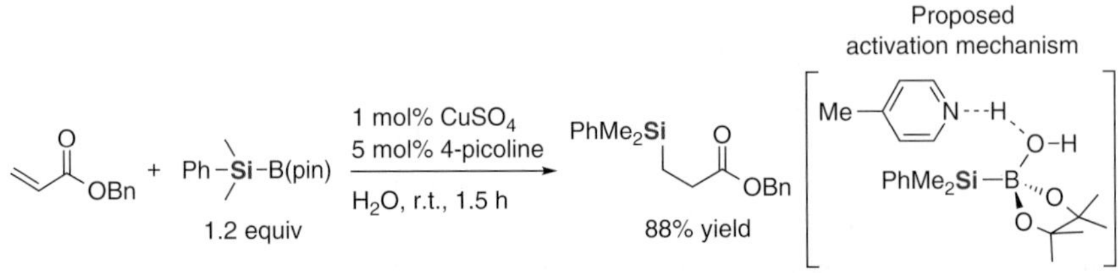

Scheme 1.7 Copper-catalyzed silylation of α,β-unsaturated carbonyl compounds in water at room temperature.

reaction. The role of the pyridine is proposed to deprotonate a nucleophilic water molecule to form an sp^3-hybridized silylboronate, followed by transmetalation with copper to generate a silylcopper active species.

The incorporation of silicon atoms into amino acids and peptides has attracted significant attention due to the large number of applications in chemical biology, and even as therapeutic agents [14]. The known methods for the preparation of silicon-containing amino acids have some limitations such as functional-group incompatibility due to the use of highly reactive carbon nucleophiles [15]. In 2015, Piersanti and coworkers developed the mild copper-based catalytic method for the regioselective silyl addition of dehydroalanine derivatives (Scheme 1.8) [16]. This transformation would serve as a useful platform for the preparation of silicon-containing peptide mimetics.

para-Quinone methides are useful intermediates for the synthesis of highly functionalized phenol derivatives. In 2015, Tortosa and coworkers found that the copper-catalyzed silylation–aromatization sequence of *para*-quinone methides afforded mono- and dibenzylic alkylsilanes in high yields (Scheme 1.9) [17]. The products, which react with electrophiles such as aldehydes, can be used as stable dibenzylic carbanion equivalents.

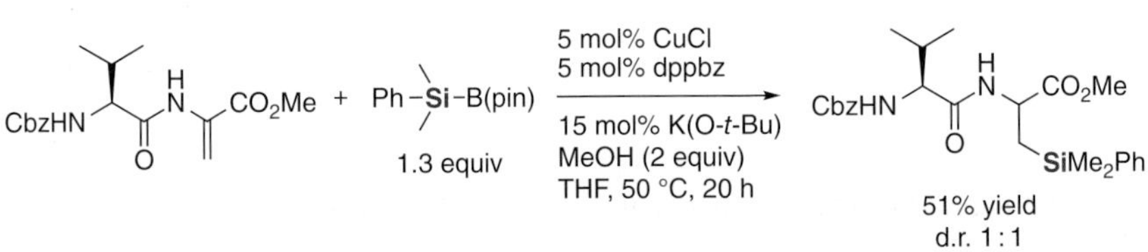

Scheme 1.8 Copper-catalyzed silylation of dehydroalanine derivatives for the synthesis of silicon-containing peptides.

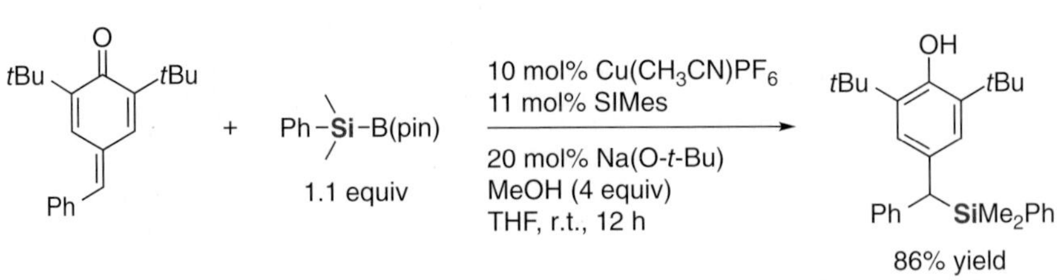

Scheme 1.9 Copper-catalyzed silylation of *para*-quinone methides.

1.2.2.3 Catalytic Allylic Silylation

In 2010, the first example of copper-catalyzed allylic silylation with a silylboronate was reported by Oestreich and a coworker (Scheme 1.10) [18]. The reaction of allyl chlorides with a silylboronate in the presence of the copper catalyst and a stoichiometric amount of NaOMe in THF afforded the corresponding allylsilanes with excellent γ-selectivity. The regioselectivity of this transformation was significantly influenced by the nature of leaving groups; γ-selectivity for halides and phosphates and α-selectivity for oxygen leaving groups such as carbonates, carbamates, and carboxylates. Explanations for the effects of the leaving groups on the regioselectivity have remained unclear.

Afterwards, the same group reported the enantioselective allylic silylation using a chiral NHC/copper complex catalyst, providing synthetically useful chiral allylsilanes (Scheme 1.11) [19]. The six-membered NHC ligand, which was originally introduced by McQuade and a coworker [20], efficiently promoted the reaction with high enantioselectivity.

At almost the same time, Shintani and coworkers also reported the copper-catalyzed enantioselective allylic substitution of allyl phosphates with a silylboronate to provide chiral allylsilanes (Scheme 1.12) [21]. The use of a chiral NHC ligand bearing an alcohol functional group is a key to achieving high enantioselectivity of this reaction. Notably, this powerful catalytic system can be applied to the synthesis of chiral allylsilanes having a tetra-substituted carbon center with excellent regio- and enantioselectivity by employing γ,γ-disubstituted allyl phosphates as substrates.

In 2015, Oestreich and coworkers found that McQuade's NHC/copper-catalyzed allylic silylation of racemic cyclic allyl phosphates underwent a

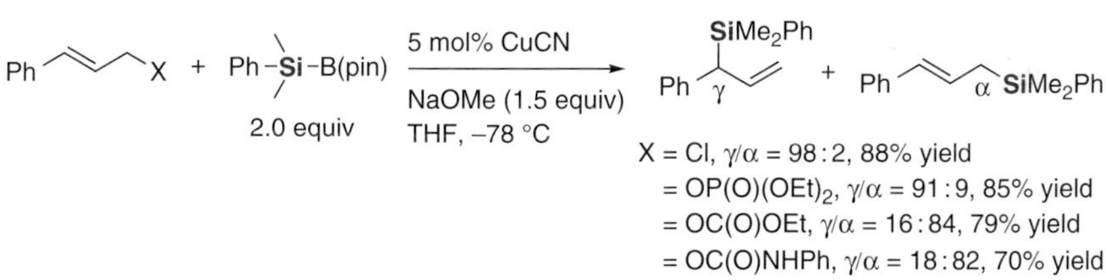

Scheme 1.10 Copper-catalyzed non-enantioselective allylic silylation.

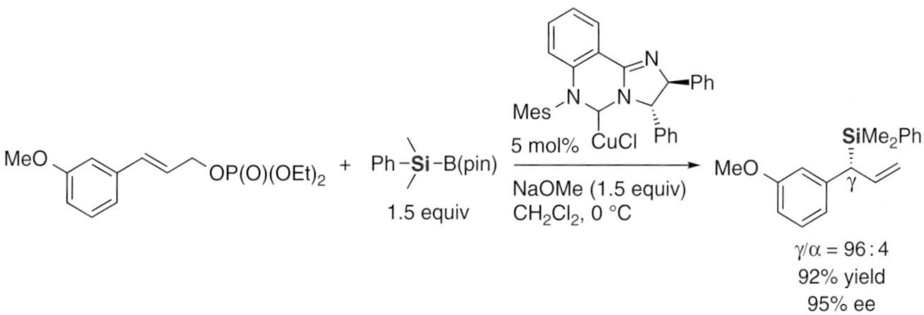

Scheme 1.11 Chiral NHC/copper-catalyzed enantioselective allylic silylation reported by Oestreich and coworkers.

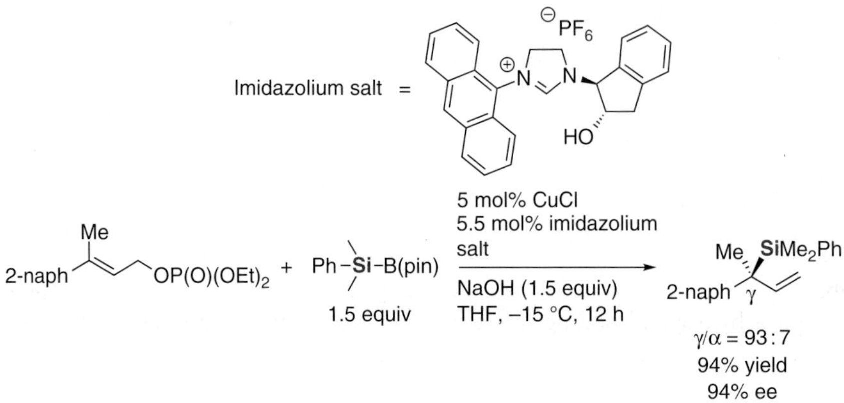

Scheme 1.12 Chiral NHC/copper-catalyzed enantioselective allylic silylation reported by Shintani and coworkers.

direct enantioconvergent transformation [22], where the enantiomeric allylic phosphates converged to the same enantiomer product by two distinctive S_N2' pathways with opposite diastereofacial selectivity (Scheme 1.13) [23].

For synthetic efficiency, it would be much more desirable to generate allylsilanes from allyl alcohols as allylic group precursors. In 2013, Li and coworkers reported that the allylic silylation of Morita–Baylis–Hillman alcohols with a silylboronate in the presence of the $Cu(OTf)_2$/pyridine catalyst in methanol proceeded smoothly to provide the corresponding allylsilanes in good yields with excellent

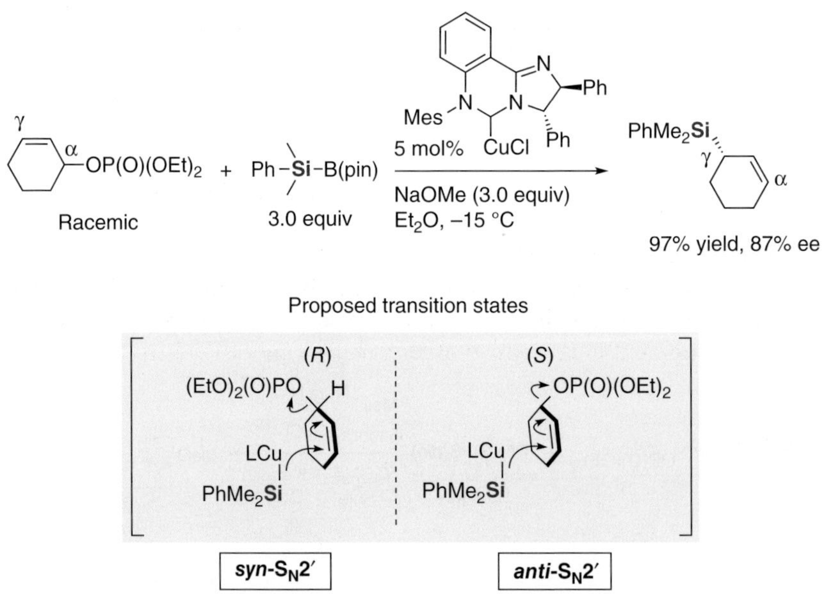

Scheme 1.13 Direct enantioconvergent transformation in the copper-catalyzed allylic silylation of cyclic allyl phosphates.

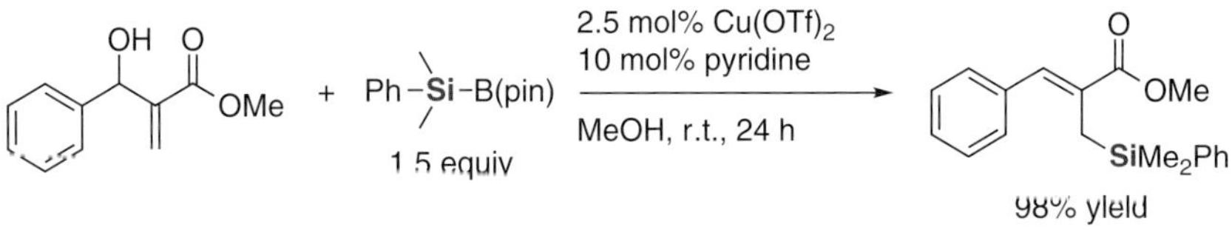

Scheme 1.14 Copper-catalyzed direct allylic silylation of allyl alcohols.

regioselectivity (Scheme 1.14) [24]. They proposed that the hydroxyl group of the substrate could be activated by the in situ–generated TfOH–pyridine complex through a hydrogen-bonding interaction. However, no experimental evidence for this activation mechanism were described.

1.2.2.4 Catalytic Silylation of Imines

Due to the considerable interest in silicon-containing peptides and amino acids, a copper-catalyzed addition of silicon nucleophiles to imines is an important transformation to form α-aminosilanes bearing sensitive functional groups. In 2011, the first example of copper-catalyzed nucleophilic silylation of imines with a silylboronate was reported by Oestreich and coworkers (Scheme 1.15) [25]. A series of aldimines and ketimines were reacted with a silylboronate in the presence of CuCN as a catalyst precursor and a catalytic amount of NaOMe to afford the corresponding racemic α-aminosilanes in good to high yields.

An enantioselective addition of silicon nucleophiles to aldimines was also developed by Oestreich and coworkers in 2014 (Scheme 1.16a) [26]. They found that the NHC/copper complex efficiently catalyzed the enantioselective silylation of aromatic and aliphatic aldimines with good to high enantioselectivity. At the almost same time, Mita et al. as well as He independently reported a similar copper-catalyzed method for the enantioselective addition of silicon nucleophiles to aldimines (Scheme 1.16b,c). The Mita and Sato group found that a copper/chiral diamine complex–catalyzed silylation of *N-tert*-butylsulfonylimines provided the optically active α-aminosilanes with high enantioselectivity (Scheme 1.16b) [27]. The He group used the C_1-symmetric chiral NHC ligand for this transformation, providing good yields and high enantioselectivity (Scheme 1.16c) [28].

1.2.2.5 Catalytic Silylation of Aldehydes

α-Hydroxylsilanes, useful intermediates in synthesis, have been used for several stereocontrolled rearrangements to access structurally complex molecules. Thus, the development of efficient synthetic methods for α-hydroxylsilanes has

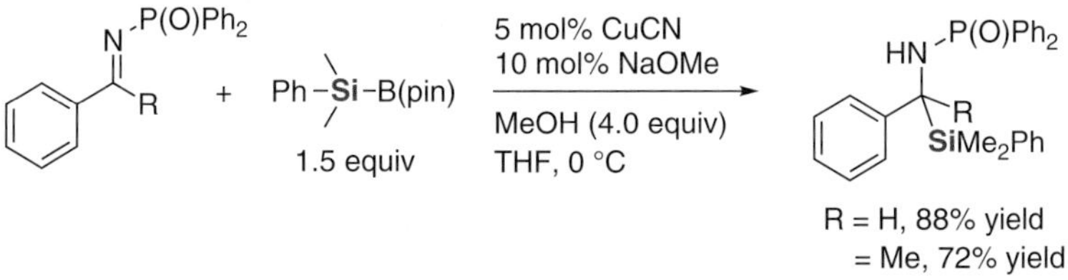

Scheme 1.15 Copper-catalyzed nucleophilic silylation of imines.

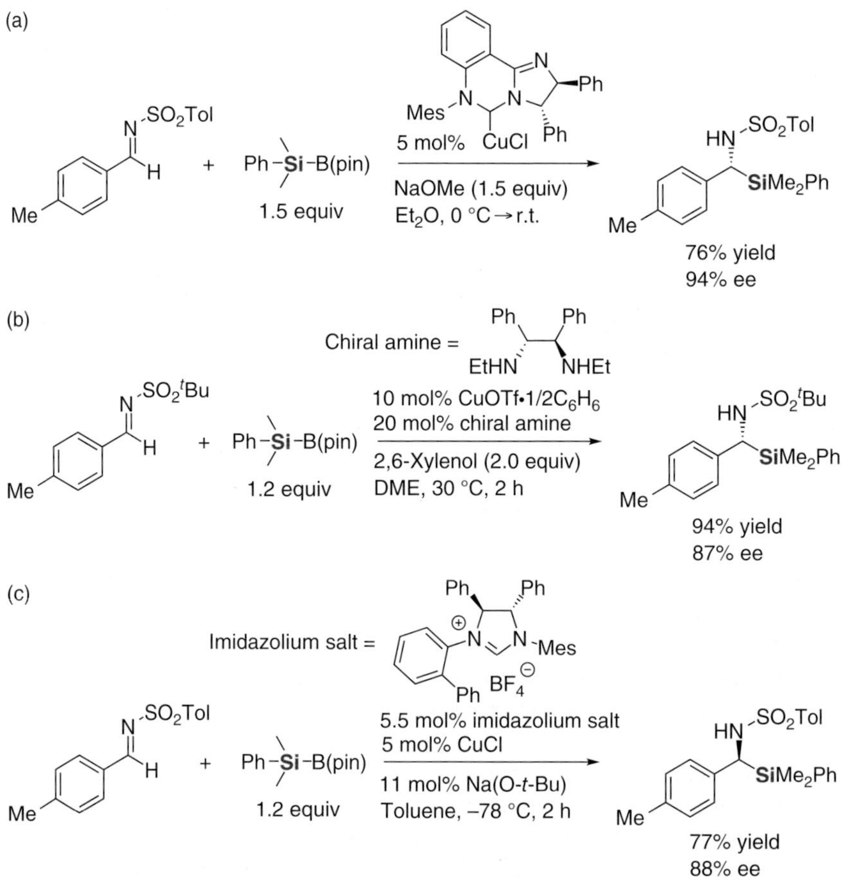

Scheme 1.16 Copper-catalyzed enantioselective nucleophilic silylation of aldimines. (a) Oestreich, (b) Mita and Sato, (c) He.

received increased attention in recent years. In 2011, the first example of copper-catalyzed nucleophilic silylation of aldehydes with a silylboronate was reported by Kleeberg et al. (Scheme 1.17) [29]. Both aromatic and aliphatic aldehydes reacted with a silylboronate to give the corresponding racemic α-hydroxylsilanes in high yields. The mechanism of this transformation was investigated in detail by stoichiometric and catalytic experiments as well as NMR spectroscopic measurements.

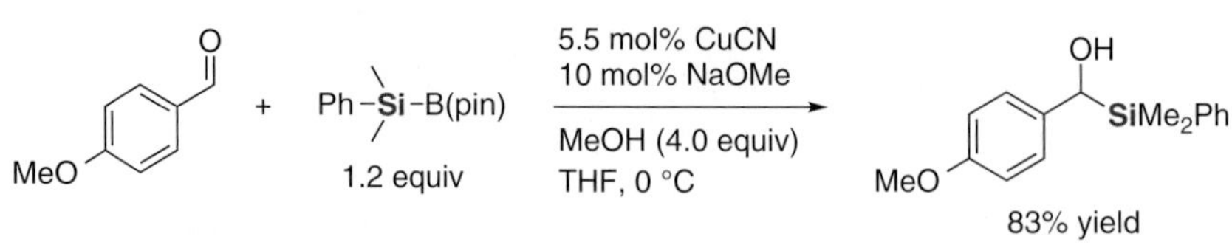

Scheme 1.17 Copper-catalyzed nucleophilic silylation of aldehydes.

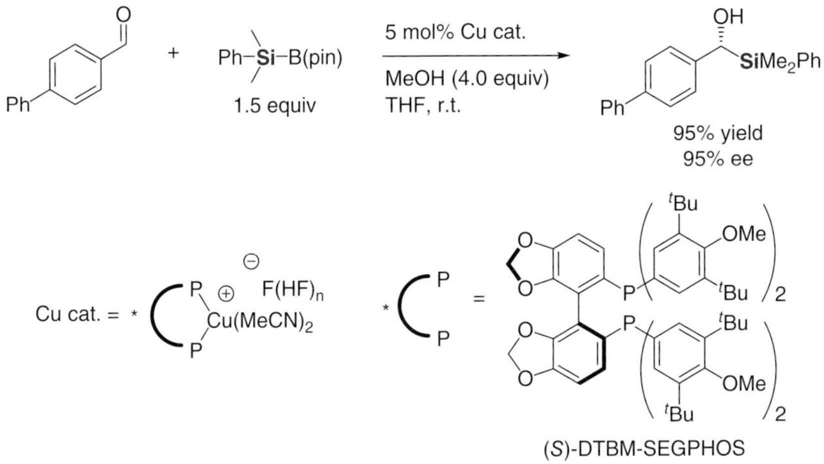

Scheme 1.18 Copper-catalyzed enantioselective nucleophilic silylation of aldehydes.

In 2013, Riant and coworkers reported the first enantioselective nucleophilic silylation of aldehydes by employing a newly developed copper/DTBM-SEGPHOS complex catalyst (Scheme 1.18) [30]. A series of aromatic and aliphatic aldehydes were converted into the corresponding chiral α-hydroxylsilanes in good yields with excellent enantioselectivity.

1.2.2.6 Catalytic Synthesis of Acylsilanes

Acylsilanes are stable compounds in which a silyl fragment is directly attached to a carbonyl group. Recently, the use of acylsilanes in organic synthesis has attracted significant attention due to the discovery of valuable new transformations [31]. Therefore, the development of efficient, direct synthetic methods for acylsilanes is highly desirable. In 2013, Riant and coworkers reported the copper-catalyzed silylation of anhydrides with a silylboronate to form the corresponding acylsilanes in high yields (Scheme 1.19) [32]. Notably, this process can be carried out in a one-pot procedure starting from easily available carboxylic acids.

1.2.2.7 Silylative Carboxylation with CO_2

Carbon dioxide (CO_2) is a nontoxic, abundant, and renewable carbon source. Therefore, the utilization of CO_2 in carbon–carbon bond-forming reaction is one of the most important subjects in organic synthesis. In 2012, Fujihara et al. developed the copper-catalyzed silacarboxylation of internal alkynes with CO_2

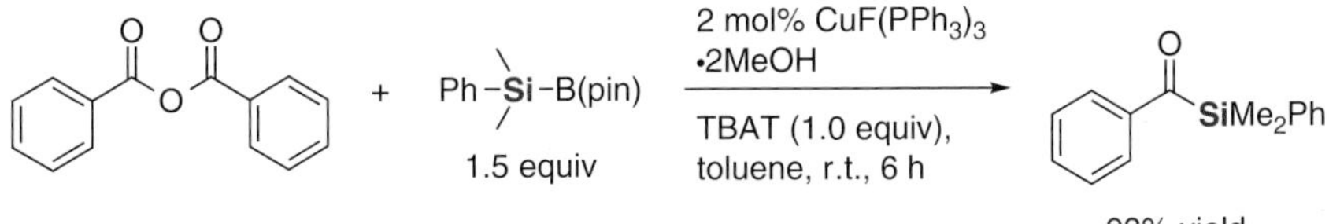

Scheme 1.19 Copper-catalyzed nucleophilic silylation of anhydrides to form acylsilanes.

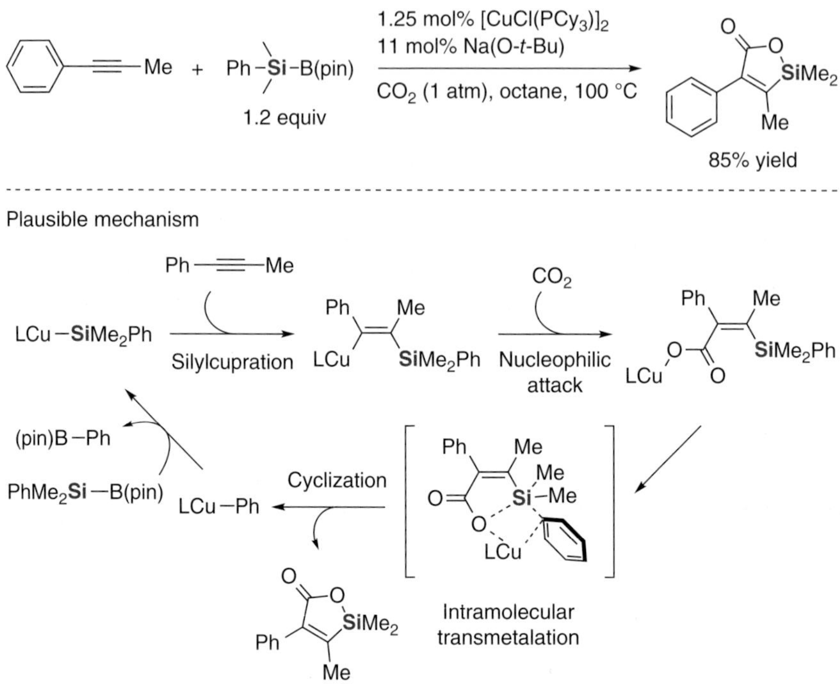

Scheme 1.20 Copper-catalyzed silacarboxylation of alkynes.

to give the silalactone derivatives (Scheme 1.20) [33]. The reaction presumably goes through the silylcupration of alkynes to form the silylated alkenylcopper intermediate, followed by the reaction with CO_2 to afford the copper carboxylate species. Then, intramolecular cyclization provides the phenylcopper complex by extrusion of the silalactone product. The catalyst is recovered by the σ-bond metathesis between the phenylcopper and a silylboronate.

The same group also reported the copper-catalyzed regiodivergent silacarboxylation of allenes with CO_2 (Scheme 1.21) [34]. The regioselectivity of the reaction is reversed by the proper choice of ligand; carboxylated vinylsilanes are obtained with *rac*-Me-Duphos as the ligand, whereas the use of PCy_3 provides carboxylated allylsilanes with high selectivity. The origin of the regioselectivity

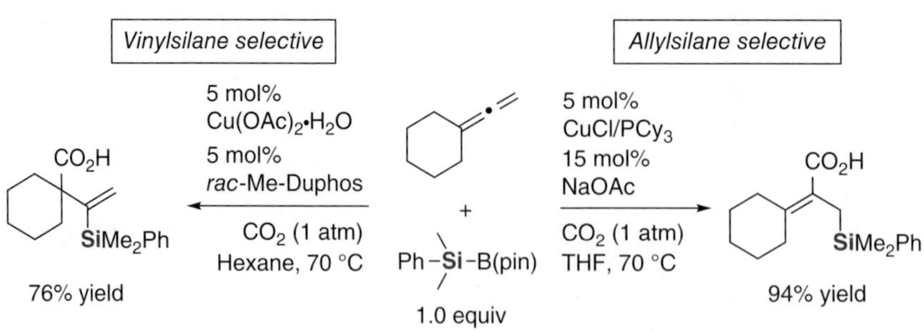

Scheme 1.21 Copper-catalyzed regiodivergent silacarboxylation of allenes.

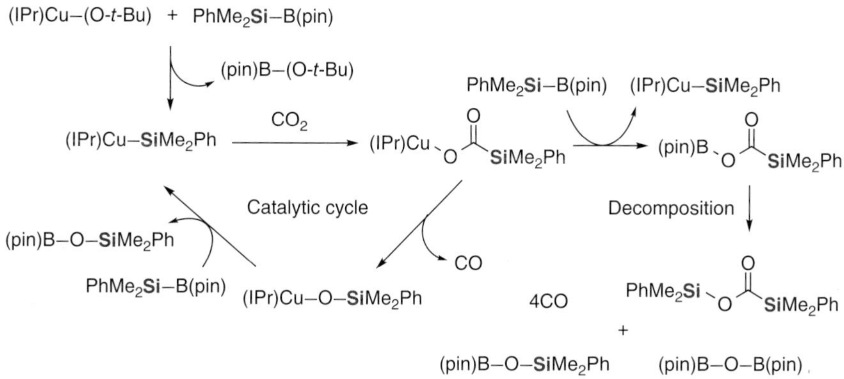

Scheme 1.22 CO_2 reduction with a silylboronate.

might be attributed to the difference in relative steric bulk of the copper/phosphine complex and a silyl group.

1.2.2.8 CO_2 Reduction via Silylation

In 2011, Kleeberg et al. studied the reaction of the [1,3-bis(diisopropylphenyl)imidazole-2-ylidene(IPr)]copper–silyl complex with CO_2 to form CO in detail experimentally (Scheme 1.22) [35]. The (IPr)copper-silyl complex reacted with CO_2 to provide the silanecarboxylic acid complex, followed by the formation of CO and a (IPr)Cu–O–$SiMe_2Ph$ linkage. This complex reacts with a silylboronate, regenerating the silyl complex and producing the (pin)B–O–$SiMe_2Ph$. It is also possible that the silanecarboxylic acid complex undergoes transmetalation with a silylboronate to form the silyl complex as well as decomposition byproducts, which were detected in the reaction mixture.

1.2.2.9 Silyl Substitution of Alkyl Electrophiles

Reactions for the carbon–silicon bond formation through a substitution of unactivated alkyl electrophiles with silicon nucleophiles have been practically unexplored until very recently. In 2016, the first example of copper-catalyzed nucleophilic silylation of alkyl triflates with a silylboronate was reported by Oestreich and a coworker (Scheme 1.23) [36]. Notably, this silylation is a stereospecific process, providing chiral silanes from the optically active substrates in a stereoinversion manner [37].

Afterwards, Oestreich's group extended their method to alkyl iodides (Scheme 1.24) [38]. Interestingly, in this case, the carbon–silicon bond-forming process proceeded through a radical mechanism. The catalytic cycle was investigated computationally in detail, leading to a full mechanistic picture that includes a single-electron-transfer (SET) process between the silylcopper intermediate and alkyl iodide to generate an alkyl radical intermediate, followed by a radical coupling to form the silylation product. They also demonstrated the silylative radical cyclization of alkenyl iodides to give the corresponding cyclic silylation compounds with high setereoselectivity.

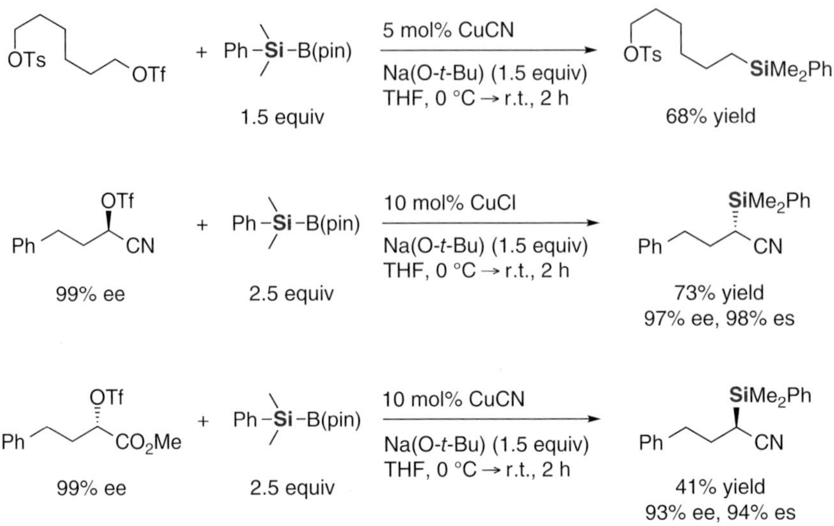

Scheme 1.23 Stereospecific copper-catalyzed silyl substitution of alkyl triflates.

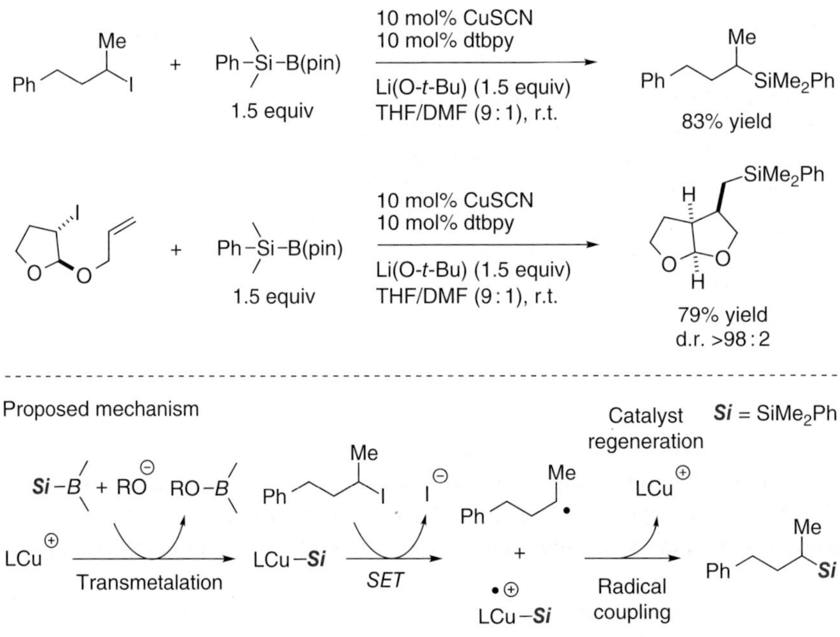

Scheme 1.24 Copper-catalyzed silyl substitution of alkyl iodides via a radical mechanism.

1.2.2.10 Decarboxylative Silylation

In addition to the carbon–silicon bond-forming cross-coupling of alkyl triflates and halides, Oestreich and a coworker disclosed the copper-catalyzed decarboxylative radical silylation of aliphatic carboxylic acid derivatives (Scheme 1.25) [39]. The reaction of aliphatic *N*-hydroxyphthalimide (NHPI) esters with a silylboronate

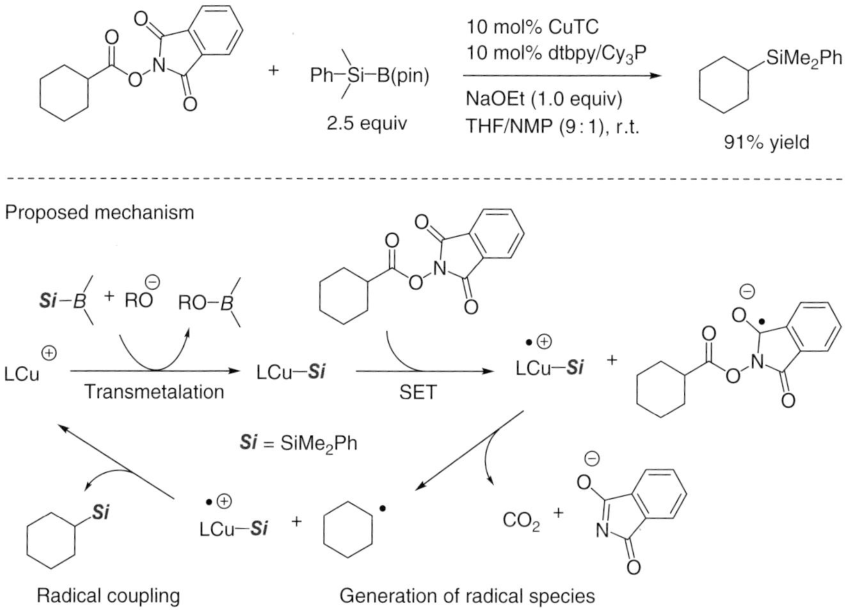

Scheme 1.25 Copper-catalyzed decarboxylative silylation of NHPI esters.

produced primary, secondary, and tertiary alkylsilanes in good to high yields. The radical-trapping and racemization experiments were consistent with a radical mechanism. They proposed the catalytic cycle involving an SET process from the silylcopper complex to the electron-accepting NHPI ester, followed by the radical coupling to give the silylation product, as shown in Scheme 1.25.

1.2.2.11 Silylative Cyclization

A silylation of carbon–carbon double bond with concomitant carbon–carbon bond formation would be highly beneficial for the efficient construction of structurally complex alkylsilanes. Tian and coworkers developed the copper-catalyzed enantioselective silylative cyclization with a silicon nucleophile to form three consecutive chiral carbon centers in one step (Scheme 1.26) [40]. This tandem reaction presumably goes through the regio- and enantioselective addition of a silylcopper intermediate to the allene and subsequent enantioselective 1,4-addition of a cyclohexadienone moiety to give the bicyclic silylation product with excellent stereoselectivity.

1.2.2.12 Silylative Allylation of Ketones

The addition of transition-metal species across allenes is a useful strategy for in situ generation of allylic nucleophiles. In 2015, Fujihara and Tsuji's group reported that the copper-catalyzed silylative allylation of ketones with allenes and silylboronates through in situ–generated β-silylated allylcopper intermediate (Scheme 1.27) [41]. Various ketones and allenes were converted into the corresponding homoallylic tertiary alcohols containing internal vinylsilane moieties in high yields.

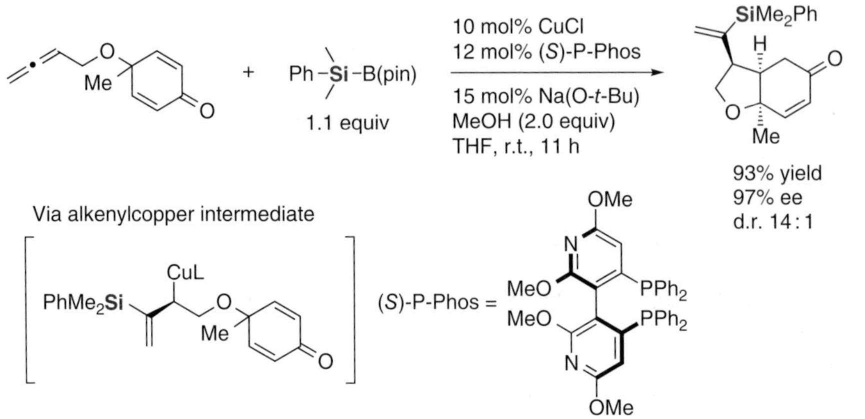

Scheme 1.26 Copper-catalyzed silylative cyclization of allenes.

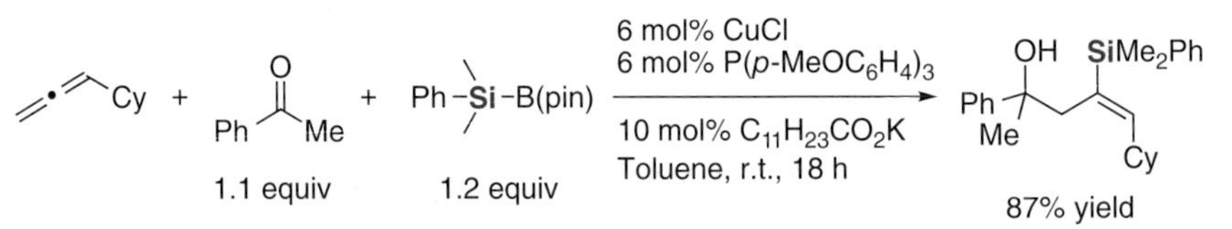

Scheme 1.27 Copper-catalyzed silylative allylation of ketones and allenes.

1.2.2.13 Silylation of Alkynes

A regioselective addition of silicon nucleophiles across alkynes provides an efficient synthetic route to various vinylsilanes. In 1986, a pioneering study on the silylation of alkynes with silicon nucleophiles was reported by Nozaki et al. (Scheme 1.28) [42]. They discovered that the reaction of terminal alkynes with $PhMe_2SiBEt_3Li$ in the presence of a catalytic amount of CuCN and methanol as a proton source produced the corresponding β-vinylsilanes with moderate to high regioselectivity.

More than 20 years after the report of Nozaki et al., Loh and coworkers discovered that the monophosphine/copper complex–catalyzed regioselective protosilylation of terminal alkynes proceeded to form α-vinylsilanes (Scheme 1.29) [43]. The choice of the ligand significantly influenced the

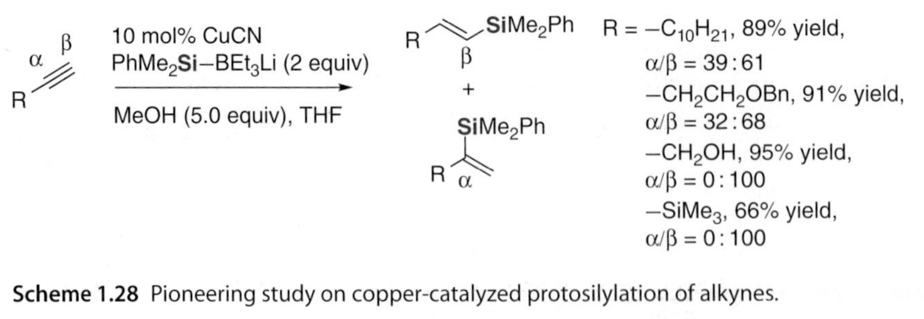

Scheme 1.28 Pioneering study on copper-catalyzed protosilylation of alkynes.

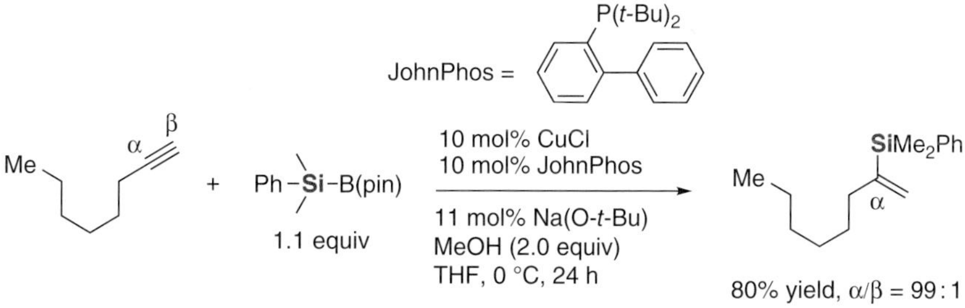

Scheme 1.29 Monophosphine/copper complex-catalyzed regioselective protosilylation of alkynes.

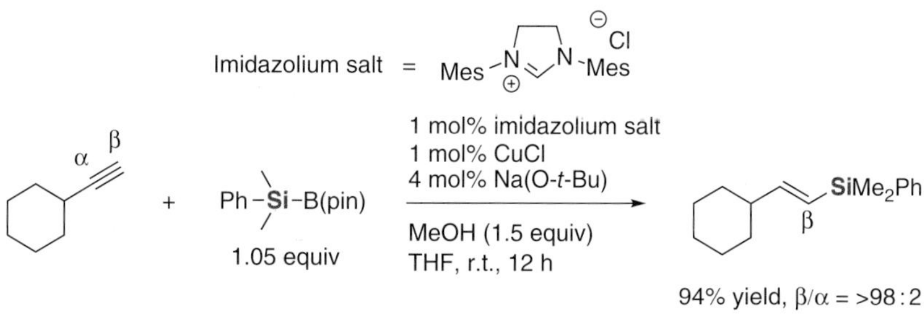

Scheme 1.30 NHC/copper complex-catalyzed regioselective protosilylation of alkynes.

regioselectivity in this reaction. It was found that the use of a bulky monophosphine ligand such as JohnPhos provided a high degree of Markovnikov selectivity.

In 2013, Hoveyda and coworkers reported that the NHC/copper complex-catalyzed protosilylation of terminal alkynes proceeded to form β-vinylsilanes with excellent regioselectivity (Scheme 1.30) [44]. Exclusive generation of the β-isomer, regardless of the electronic properties of the alkyne substituent, suggests that the regioselectivity can be controlled by steric factors, where the relatively larger dimethylphenylsilyl group is placed at the terminal carbon.

Later, Oestreich and coworkers reported a general method for the β-selective silylation without adding a ligand (Scheme 1.31) [45]. After fine-tuning the reaction parameters such as a copper precatalyst, base, and solvent, they found

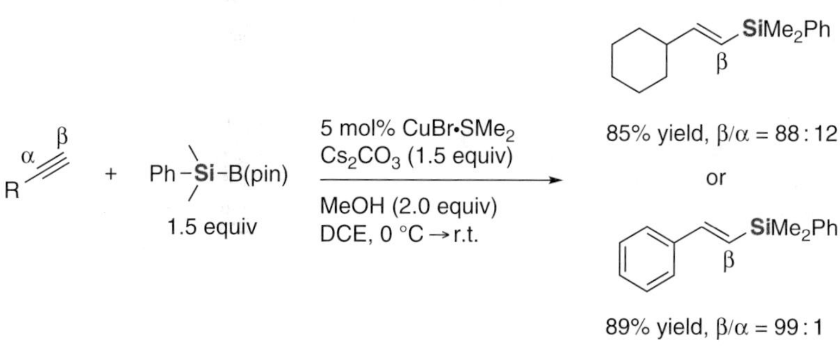

Scheme 1.31 Ligand-free copper-catalyzed protosilylation of alkynes.

that the use of $CuBr \cdot SMe_2$, Cs_2CO_3 and dichloroethane as a solvent afforded the desired β-selective silylation products in high yields with excellent regioselectivity. Both aromatic and aliphatic terminal alkynes were converted into the products with high β-selectivity.

In 2014, new protocols for the silylation of alkynes in water were independently reported by Santos and Lipshutz' groups (Scheme 1.32) [46, 47]. Santos and a coworker found that the silylation of activated alkynes with a silylboronate in the presence of $CuSO_4$ and 4-picoline as a ligand in water proceeded smoothly to give the corresponding silyl conjugate addition products in high yields (Scheme 1.32a) [46]. Lipshutz's group discovered that the use of the commercially available surfactant TPG-750-M served as the reaction medium and efficiently promoted the copper-catalyzed conjugate silylation of alkynes in water (Scheme 1.32b) [47].

Cooperative catalysts can promote complex, challenging transformations that could not be realized by a single catalyst operation. Riant and coworkers developed a tunable and stereoselective dual copper/palladium catalytic system for the silylative allylation of activated alkynes in 2014 (Scheme 1.33) [48]. This reaction presumably proceeds through the silylcupration of alkynes, followed by transmetalation between the alkenylcopper species and the π-allylpalladium intermediate to form trisubstituted alkenylsilanes. Notably, fine-tuning the reaction conditions allows selective access to both *Z* and *E* isomers.

The precise control of the regioselectivity in copper-catalyzed silylation of alkynes still remains a challenging subject. In 2015, Carretero and coworkers developed the regiodivergent silylation of internal alkynes with a traceless

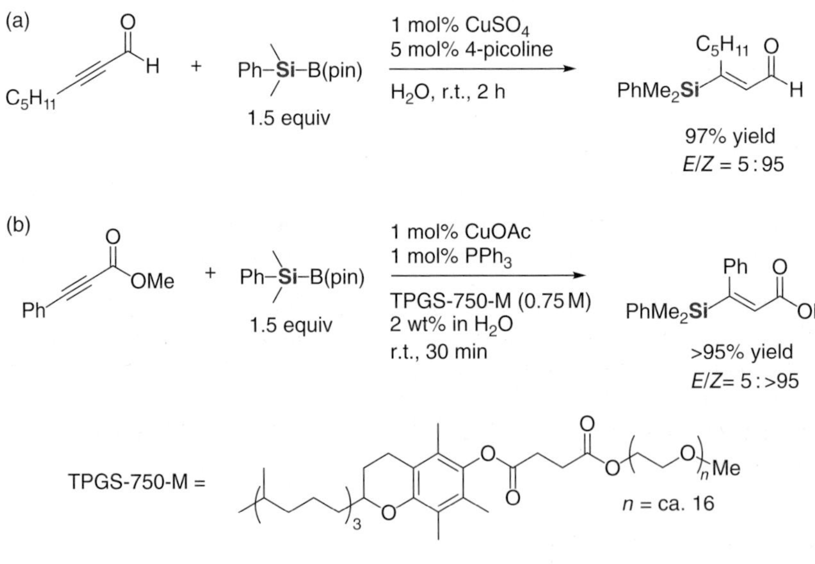

Scheme 1.32 Copper-catalyzed protosilylation of alkynes in water at room temperature. (a) Santos, (b) Lipshutz.

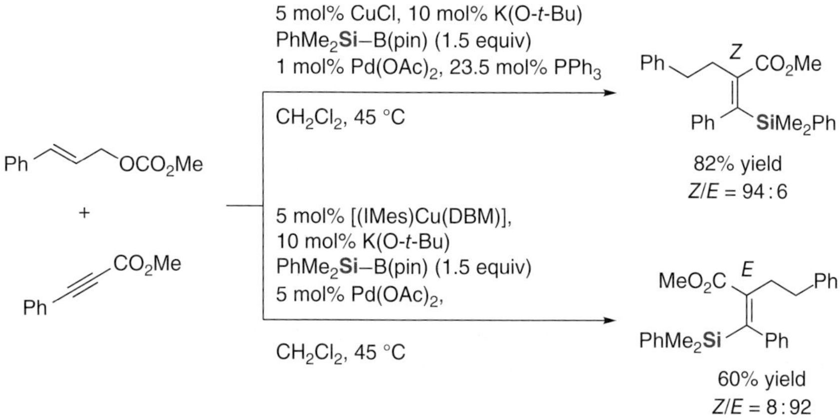

Scheme 1.33 Copper/palladium cooperative catalytic system for silylative allylation of alkynes.

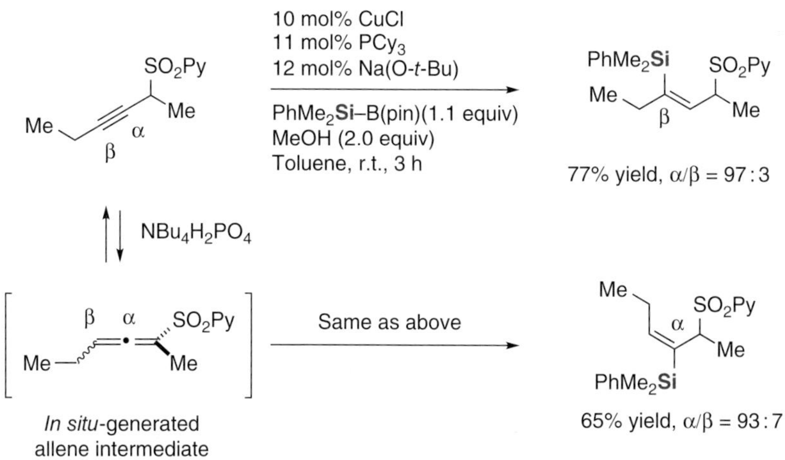

Scheme 1.34 Regiodivergent copper-catalyzed protosilylation of alkynes.

2-pyridylsulfonyl group as a directing group (Scheme 1.34) [49]. Either regioisomer could be obtained without modification of the starting substrates by virtue of an *in situ* base-promoted alkyne to allene equilibration which takes place prior to the addition of a silylcopper intermediate. In addition, this directing group could be removed after the silylation by the addition of carbon nucleophiles.

1.2.2.14 Propargylic Substitution

A γ-selective propargylic substitution is an efficient synthetic route to multi-substituted allenes. In 2011, the first copper-catalyzed propargylic substitution

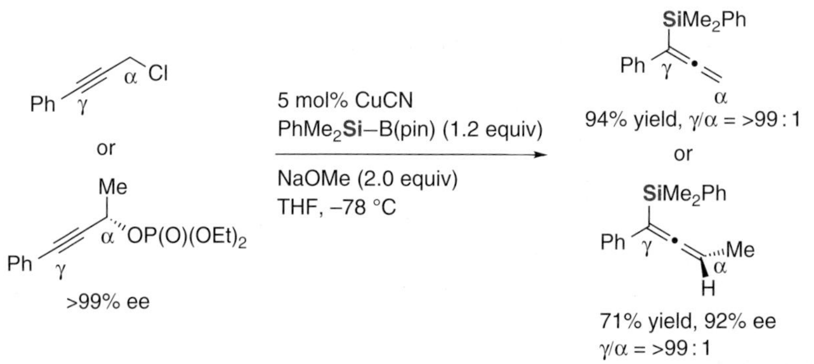

Scheme 1.35 Stereospecific γ-selective propargylic silyl substitution.

with silicon nucleophiles was reported by Oestreich and coworkers (Scheme 1.35) [50]. The reaction of a variety of propargylic chlorides in the presence of CuCN and NaOMe with a silylboronate afforded the corresponding allenylsilanes with excellent γ-selectivity. Notably, the central-to-axial chirality transfer was also observed in the silylation of the α-chiral propargylic phosphates to give the chiral allenylsilanes with high stereospecificity.

1.2.3 Copper-Catalyzed Nucleophilic Silylation with Silylzincs

In 2004, Oestreich and Weiner first introduced soft bis(triorganosilyl)zinc species, such as $(PhMe_2Si)_2Zn$, which can be used as a source for the generation of silicon nucleophiles in copper-catalyzed reactions [51]. In this context, Oestreich and coworkers developed a copper-catalyzed conjugate addition, silylation of allylic and propargylic electrophiles with $(PhMe_2Si)_2Zn$ reagent (Scheme 1.36) [52–54].

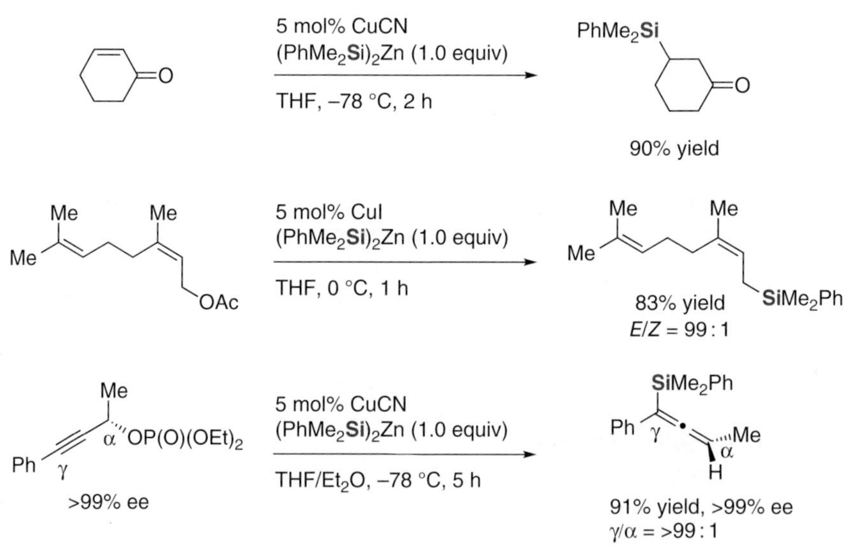

Scheme 1.36 Copper-catalyzed nucleophilic silylation with $(PhMe_2Si)_2Zn$ reagent.

1.3 Silicon Nucleophiles with Rhodium Catalysts

1.3.1 Rhodium-Catalyzed Nucleophilic Silylation with Disilanes

In 2006, the first example of catalytic generation of nucleophilic rhodium–silicon species with disilanes and its application to catalytic silylation of aryl and alkenyl cyanides was reported by Tobisu et al. (Scheme 1.37) [55]. The reaction mechanism was proposed to proceed via the silylmetalation of a cyano group by a silylrhodium species, which can be generated through reaction of rhodium chloride with a disilane, followed by the carbon–carbon bond-cleavage process.

1.3.2 Rhodium-Catalyzed Nucleophilic Silylation with Silylboronates

1.3.2.1 Conjugate Silylation

In 2006, the first example of rhodium-catalyzed conjugate silylation with a silylboronate was reported by Oestreich and coworkers (Scheme 1.38) [56]. The reaction of α,β-unsaturated carbonyl compounds with a silylboronate in the presence of a rhodium catalyst and Et_3N provided the 1,4-addition products in good to high yields. The same group also discovered that the use of (*S*)-BINAP as a chiral ligand in this reaction resulted in the formation of chiral silanes with excellent enantioselectivity [56–60]. This reaction presumably goes through the formation of a silylrhodium species and subsequent 1,4-silyl addition to α,β-unsaturated carbonyl compounds. The synthetic utility of this protocol was

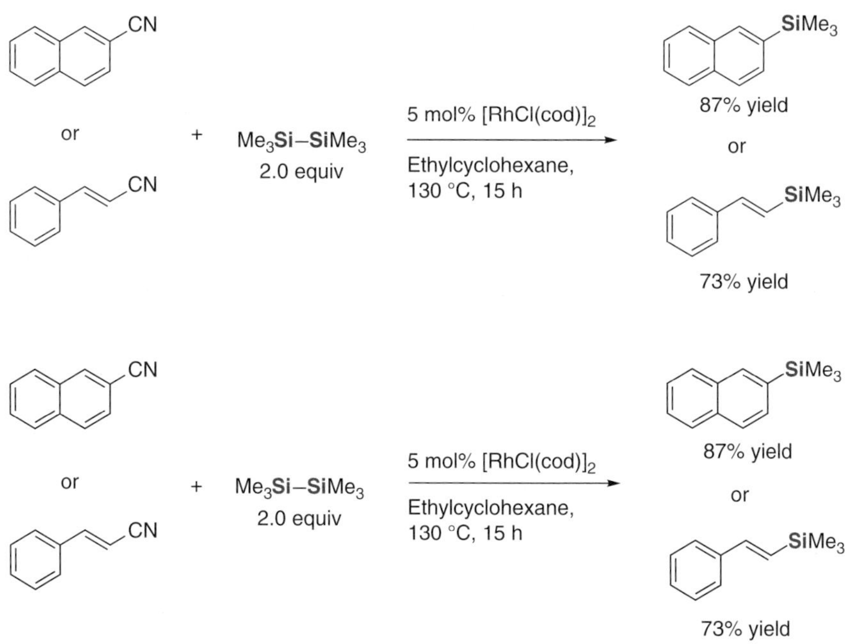

Scheme 1.37 Rhodium-catalyzed silylation of aryl and alkenyl cyanides.

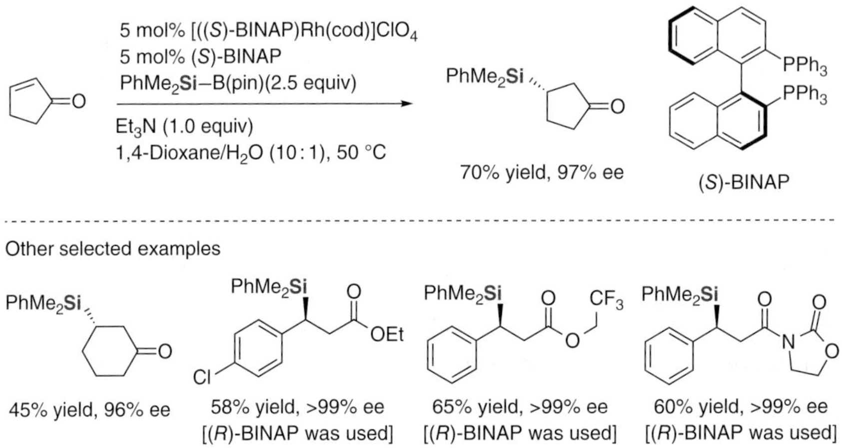

Scheme 1.38 Rhodium-catalyzed conjugate silylation with a silylboronate.

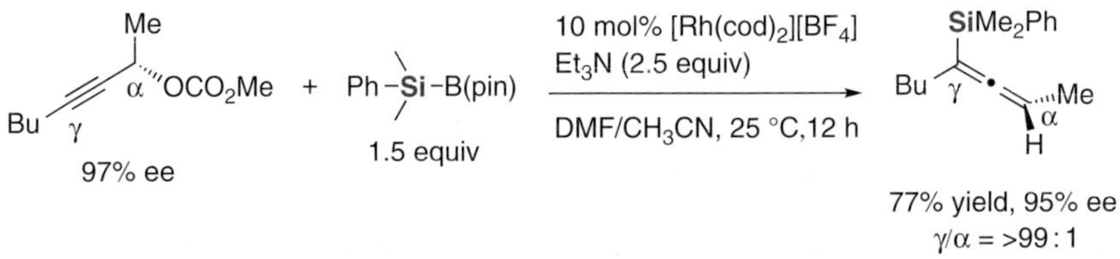

Scheme 1.39 Rhodium-catalyzed stereospecific silylation of propargylic carbonates.

successfully demonstrated by the synthesis of the C7–C16 fragment of (+)-Neopeltolide [59].

1.3.2.2 Coupling between Propargylic Carbonates to Form Allenylsilanes

In 2009, Sawamura and coworkers reported that the rhodium-catalyzed coupling reaction between propargylic carbonates and a silylboronate provided the corresponding allenylsilanes in high yields (Scheme 1.39) [61]. The reaction of an optically active substrate proceeded with excellent chirality transfer to form an axially chiral allenylsilane.

1.4 Silicon Nucleophiles with Nickel Catalysts

1.4.1 Nickel-Catalyzed Nucleophilic Silylation with Alkyl Electrophiles

The first metal-catalyzed cross-coupling reactions of unactivated secondary and tertiary alkyl electrophiles to form carbon–silicon bonds was reported by Fu and coworkers (Scheme 1.40) [62]. In the presence of a nickel/pybox catalyst, the cross-coupling reactions between secondary and tertiary alkyl bromides and a silylzinc reagent afforded a variety of alkyl silicon compounds. Stereochemical

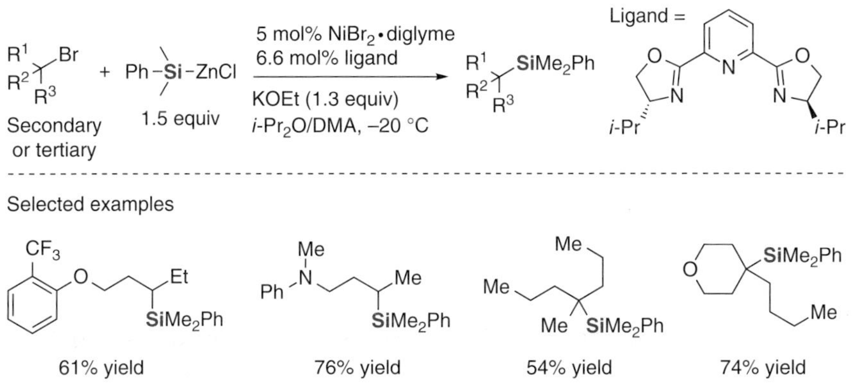

Scheme 1.40 Nickel-catalyzed silylation of unactivated alkyl electrophiles.

and radical-trap experiments are consistent with a radical mechanism involving a homolytic pathway for carbon–bromine bond cleavage.

1.5 Silicon Nucleophiles with Lewis Base Catalysts

1.5.1 N-Heterocyclic Carbene-Catalyzed Nucleophilic 1,4-Silylation

In 2011, Hoveyda and a coworker first discovered that NHCs, in the absence of a metal salt, activate the silicon–boron bond of a silylboronate to generate silicon nucleophiles (Scheme 1.41) [63]. The reaction of α,β-unsaturated carbonyl

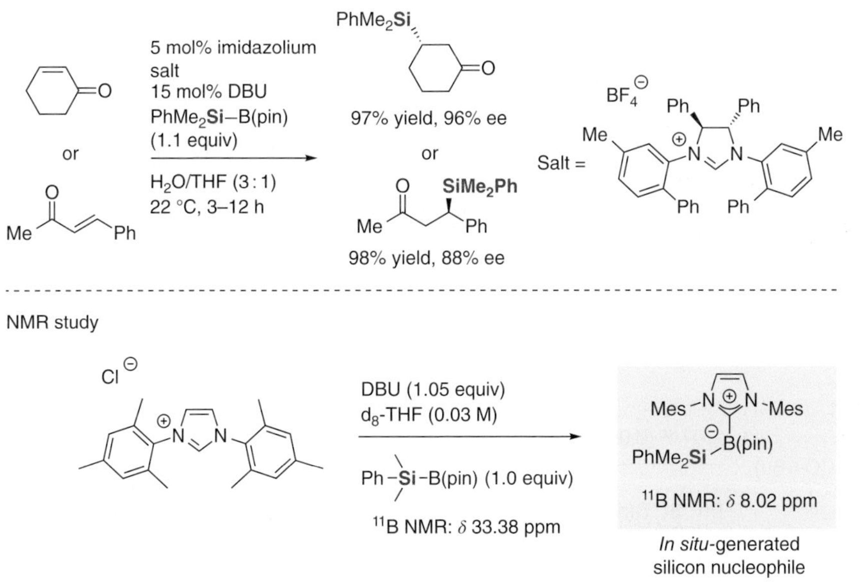

Scheme 1.41 Chiral NHC-catalyzed enantioselective conjugate silylation.

compounds with a silylboronate in the presence of catalytic amount of a imidazolium salt and 1,8-diazabicyclo(5.4.0)undec-7-ene (DBU) afforded the conjugate silyl addition products in high yields. Their *in situ* NMR studies suggested the formation of a NHC → B–Si ate complex in the reaction mixture. They also developed the enantioselective version of this process with a chiral NHC catalyst.

1.5.2 Alkoxide Base–Catalyzed 1,2-Silaboration

Alkoxide bases also activate the silicon–boron bond of a silylboronate to generate silicon nucleophiles. In 2012, the first alkoxide base–catalyzed reaction of a silylboronate was reported by Ito et al. (Scheme 1.42) [64]. In the presence of 10 mol% K(O-*t*-Bu), silylboration of aromatic alkenes with a silylboronate proceeded with excellent regioselectivity. Their *in situ* NMR studies revealed the formation of a *t*-BuO→B–Si ate complex in the reaction mixture. A related silylative cyclopropanation reaction of allyl phosphates with a stoichiometric amount of the alkoxide base was also reported by Shinatani et al. in 2014 [65].

1.5.3 Phosphine-Catalyzed 1,2-Silaboration

In 2015, Ohmiya and coworkers reported that trialkylphosphine-catalyzed *anti*-selective silaboration of alkynoates with a silylboronate produced β-boryl-α-silyl acrylates in high yields (Scheme 1.43) [66]. They proposed the reaction mechanism that includes the conjugate addition of Bu_3P to the alkynoate to form a zwitterionic allenolate intermediate, followed by nucleophilic silyl transfer from the activated silylboronate to give the ylide intermediates. Next, the nucleophilic attack of the ylide carbon to the boron atom bound to the enolate oxygen provides the cyclic borate and the subsequent elimination of Bu_3P associated with the boron–oxygen bond cleavage affords the *anti*-silaboration product.

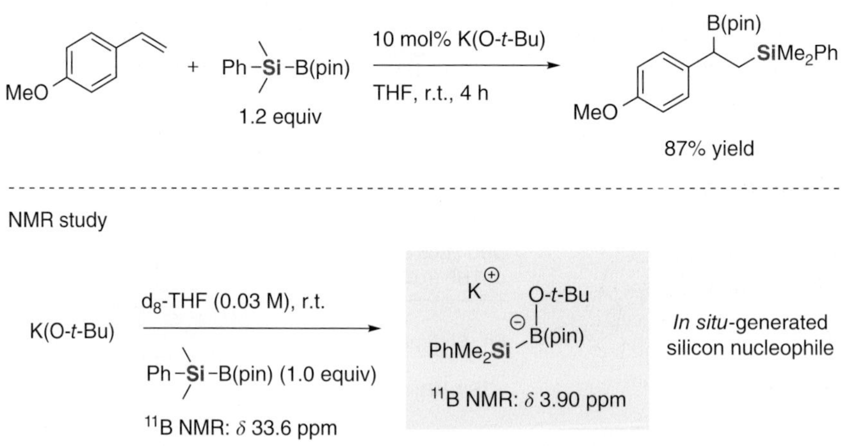

Scheme 1.42 Alkoxide base–catalyzed silaboration of aromatic alkenes.

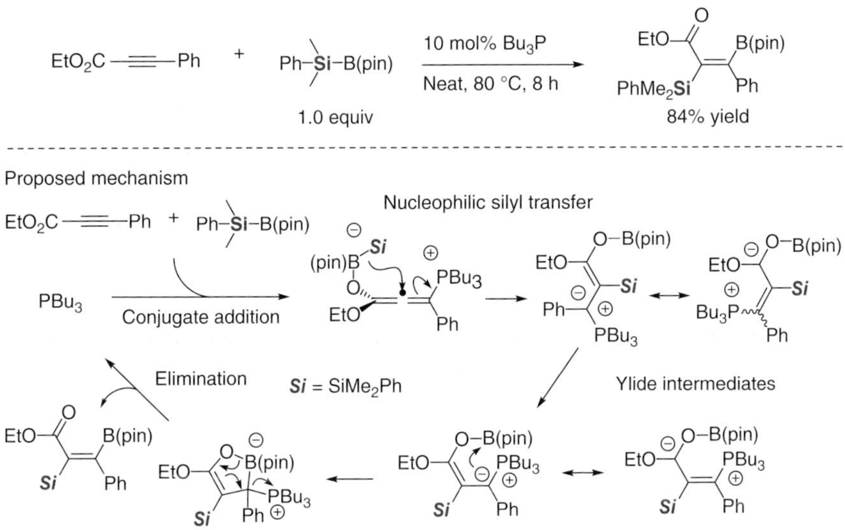

Scheme 1.43 Phosphine-catalyzed 1,2-silaboration of alkynyl esters.

1.6 Closing Remarks

A number of transition-metal-catalyzed reactions with *in situ*–generated silicon nucleophiles have been reported as alternative synthetic pathways for the preparation of a sensitive functional group containing sterically hindered silicon compounds. Different transition metals have specific reactivities in these silicon–carbon bond formation reactions. In particular, the combination of a copper catalyst and a silylboronate has been widely used for the catalytic generation of a soft silicon nucleophile, a silylcopper intermediate, which reacts with a broad range of electrophiles to form the corresponding silylation products with high selectivity. In addition to metal-catalyzed methods, less toxic metal-free protocols such as NHC- or alkoxide base–catalyzed nucleophilic silylations have also been developed in recent years. These significant achievements discussed in this chapter will find a wide range of applications in organic synthesis, pharmaceutical drug discovery, and materials science.

Abbreviations

THF	tetrahydrofuran
DMF	dimethylformamide
DME	1,2-dimethoxyethane
DMA	*N,N*-dimethylacetamide
DCE	dichloroethane
Ts	tosyl
Tf	trifluoromethanesulfonyl
2-naph	2-naphthyl

NMP	*N*-methyl-2-pyrrolidone
cod	1,5-cyclooctadiene
NHC	*N*-heterocyclic carbene
DBU	1,8-diazabicyclo(5.4.0)undec-7-ene
dppf	1,1′-bis(diphenylphosphino)ferrocene
TBAT	tetrabutylammonium difluorotriphenylsilicate
IMes	1,3-dimesitylimidazol-2-ylidene, 1,3-bis(2,4,6-trimethylphenyl)-imidazolium, 1,3-bis(2,4,6-trimethylphenyl)imidazol-2-ylidene
SIMes	1,3-bis(2,4,6-trimethylphenyl)-4,5-dihydroimidazol-2-ylidene
Cbz	benzyloxycarbonyl
Mes	mesityl
TMS	trimethylsilyl
dtbpy	4,4′-di-*tert*-butyl-2,2′-dipyridyl
Bn	benzyl
dppbz	1,2-bis(diphenylphosphino)benzene
Py	pyridine
Tol	*p*-tolyl
DBM	dibenzoylmethane
diglyme	1-methoxy-2-(2-methoxyethoxy)ethane
TC	thiophene-2-carboxylate

References

1 Brook, M.A. (2000). *Silicon in Organic, Organometallic, and Polymer Chemistry*. Weinheim: Wiley-Interscience.

2 Kawachi, A., Minamimoto, T., and Tamao, K. (2001). Boron–metal exchange reaction of silylboranes with organometallic reagents: a new route to arylsilyl anions. *Chem. Lett.* 30: 1216–1217.

3 Dieter, R.K. (2001). *Modern Organocopper Chemistry* (ed. N. Krause). Weinheim: Wiley-VCH.

4 Ito, H., Ishizuka, T., Tateiwa, J. et al. (1998). New method for introduction of a silyl group into α,β-enones using a disilane catalyzed by a copper(I) salt. *J. Am. Chem. Soc.* 120: 11196–11197.

5 Clark, C.T., Lake, J.F., and Scheidt, K.A. (2004). Copper(I)-catalyzed disilylation of alkylidene malonates employing a Lewis base activation strategy. *J. Am. Chem. Soc.* 126: 84–85.

6 Ito, H., Horita, Y., and Sawamura, M. (2012). Copper(I)-catalyzed allylic substitution of silyl nucleophiles through Si—Si bond activation. *Adv. Synth. Catal.* 354: 813–817.

7 (a) Hosomi, A. and Sakurai, H. (1976). Syntheses of γ,δ-unsaturated alcohols from allylsilanes and carbonyl-compounds in presence of titanium tetrachloride. *Tetrahedron Lett.* 17: 1295–1298. (b) Hosomi, A. and Sakurai, H. (1977). Chemistry of organosilicon compounds 99. Conjugate addition of allylsilanes to α,β-enones new method of stereoselective introduction of angular allyl group in fused cyclic α,β-enones. *J. Am. Chem. Soc.* 99: 1673–1675. (c) Fleming, I. and Paterson, I. (1979). Allylsilanes in organic synthesis: a method for the

introduction of two carbon substituents in place of carbonyl oxygen. *Synthesis* 1979: 446–448.

8 For a recent review, see: (a) Oestreich, M., Hartmann, E., and Mewald, M. (2013). Activation of the Si—B interelement bond: mechanism, catalysis, and synthesis. *Chem. Rev.* 113: 402–441. (b) Ohmura, T. and Suginome, M. (2009). Silylboranes as new tools in organic synthesis. *Bull. Chem. Soc. Jpn.* 82: 29–49.

9 Lee, K. and Hoveyda, A.H. (2010). Enantioselective conjugate silyl additions to cyclic and acyclic unsaturated carbonyls catalyzed by Cu complexes of chiral *N*-heterocyclic carbenes. *J. Am. Chem. Soc.* 132: 2898–2900.

10 Lee, K., Wu, H., Haeffner, F., and Hoveyda, A.H. (2012). NHC-Cu-catalyzed silyl conjugate additions to acyclic and cyclic dienones and dienoates. Efficient site-, diastereo- and enantioselective synthesis of carbonyl-containing allylsilanes. *Organometallics* 31: 7823–7826.

11 Welle, A., Petrignet, J., Tinant, B. et al. (2010). Copper-catalyzed domino silylative aldol reaction leading to stereocontrolled chiral quaternary carbons. *Chem. Eur. J.* 16: 10980–10983.

12 Ibrahem, I., Santoro, S., Himo, F., and Córdova, A. (2011). Enantioselective conjugate silyl additions to α,β-unsaturated aldehydes catalyzed by combination of transition metal and chiral amine catalysts. *Adv. Synth. Catal.* 353: 245–252.

13 Calderone, J.A. and Santos, W.L. (2012). Copper(II)-catalyzed silyl conjugate addition to α,β-unsaturated conjugated compounds: Brønsted base-assisted activation of Si—B bond in water. *Org. Lett.* 14: 2090–2093.

14 Franz, A.K. and Wilson, S.O. (2013). Organosilicon molecules with medicinal applications. *J. Med. Chem.* 56: 388–405.

15 (a) Kenworthy, M.N., Kilburn, J.P., and Taylor, R.J.K. (2004). Highly functionalized organolithium reagents for enantiomerically pure α-amino acid synthesis. *Org. Lett.* 6: 19–22. (b) Barfoot, C.W., Harvey, J.E., Kenworthy, M.N. et al. (2005). Highly functionalized organolithium and organoboron reagents for the preparation of enantiomerically pure α-amino acids. *Tetrahedron* 61: 3403–3417.

16 Bartoccini, F., Bartolucci, S., Lucarini, S., and Piersanti, G. (2015). Synthesis of boron- and silicon-containing amino acids through copper-catalyzed conjugate additions to dehydroalanine derivatives. *Eur. J. Org. Chem.* 3352–3360.

17 López, A., Parra, A., Jarava-Barrera, C., and Tortosa, M. (2015). Copper-catalyzed silylation of *p*-quinone methides: new entry to dibenzylic silanes. *Chem. Commun.* 51: 17684–17687.

18 Vyas, D.J. and Oestreich, M. (2010). Copper-catalyzed Si—B bond activation in branched-selective allylic substitution of liner allylic chlorides. *Angew. Chem. Int. Ed.* 49: 8513–8515.

19 Delvos, L.B., Vyas, D.J., and Oestreich, M. (2013). Asymmetric synthesis of α-chiral allylic silanes by enantioconvergent γ-selective copper-catalyzed allylic silylation. *Angew. Chem. Int. Ed.* 52: 4650–4653.

20 Park, J.K. and McQuade, D.T. (2012). Chiral 6-NHC ligand and copper complex: properties, application, and mechanism. *Synthesis* 44: 1485–1490.

21 Takeda, M., Shintani, R., and Hayashi, T. (2013). Enantioselective synthesis of α-tri- and α-tetrasubstituted allylsilanes by copper-catalyzed asymmetric allylic

substitution of allyl phosphates with silylboronates. *J. Org. Chem.* 78: 5007–5017.

22 (a) Ito, H., Kunii, S., and Sawamura, M. (2010). Direct enantio-convergent transformation of racemic substrates without racemization or symmetrization. *Nat. Chem.* 2: 972–976. (b) Langlois, J.-B., Emry, D., Mareda, J., and Alexakis, A. (2012). Mechanistic identification and improvement of a direct enantioconvergent transformation in copper-catalyzed asymmetric allylic alkylation. *Chem. Sci.* 3: 1062–1069.

23 Delvos, L.B. and Oestreich, M. (2015). Temperature-dependent direct enantioconvergent silylation of a racemic cyclic allylic phosphate by copper-catalyzed allylic substitution. *Synthesis* 47: 924–933.

24 Xuan, Q.-Q., Zhong, N.-J., Ren, C.-L. et al. (2013). Cu(II)-catalyzed allylic silylation of Morita–Baylis–Hillman alcohols via dual activation of Si—B bond and hydroxyl group. *J. Org. Chem.* 78: 11076–11081.

25 Vyas, D.J., Fröhlich, R., and Oestreich, M. (2011). Activation of the Si—B linkage: copper-catalyzed addition of nucleophilic silicon to imines. *Org. Lett.* 13: 2094–2097.

26 Hensel, A., Nagure, K., Delvos, L.B., and Oestreich, M. (2014). Enantioselective addition of silicon nucleophiles to aldimines using a preformed NHC-copper(I) complex as the catalyst. *Angew. Chem. Int. Ed.* 53: 4964–4967.

27 Mita, T., Sunagawa, M., Saito, K., and Sato, Y. (2014). Catalytic enantioselective silylation of *N*-sulfonylimines: asymmetric synthesis of α-amino acids from CO_2 via stereospecific carboxylation of α-amino silanes. *Org. Lett.* 16: 3028–3031.

28 Zhao, C., Jiang, C., Wang, J. et al. (2014). Eantioselective syntheses of α-silyl amines via a copper-*N*-heterocyclic carbene catalyzed nucleophilic silicon transfer to imines. *Asian J. Org. Chem.* 3: 851–855.

29 Kleeberg, C., Feldmann, E., Hartmann, E. et al. (2011). Copper-catalyzed 1,2-addition of nucleophilic silicon to aldehydes: mechanistic insight and catalytic systems. *Chem. Eur. J.* 17: 13538–13543.

30 Cirriez, V., Rasson, C., Hermant, T. et al. (2013). Copper-catalyzed addition of nucleophilic silicon to aldehydes. *Angew. Chem. Int. Ed.* 52: 1785–1788.

31 The selected examples, see: (a) Nicewicz, D.A., Yates, C.M., and Johnson, J.S. (2004). Catalytic asymmetric acylation of (silyloxy)nitrile anions. *Angew. Chem. Int. Ed.* 43: 2652–2655. (b) Sasaki, M., Kondo, Y., Kawahata, M. et al. (2011). Eantioselective synthesis of siloxyallenes from alkynoylsilanes by reduction and a brook rearrangement and their subsequent trapping in a [4+2] cycloaddition. *Angew. Chem. Int. Ed.* 50: 6375–6378.(c) Smirnov, P., Mathew, J., Nijs, J. et al. (2013). One-pot-promoted asymmetric alkynylation/brook-type rearrangement/ene-allene cyclization: highly selective formation of three new bonds and two stereocenters in acyclic systems. *Angew. Chem. Int. Ed.* 52: 13717–13721.

32 Cirriez, V., Rasson, C., and Riant, O. (2013). Synthesis of acylsilanes by copper(I)-catalyzed addition of silicon nucleophiles onto acid derivatives. *Adv. Synth. Catal.* 355: 3137–3140.

33 Fujihara, T., Tani, Y., Semba, K. et al. (2012). Copper-catalyzed silacarboxylation of internal alkynes by employing carbon dioxide and silylboranes. *Angew. Chem. Int. Ed.* 51: 11487–11490.

34 Tani, Y., Fujihara, T., Terao, J., and Tsuji, Y. (2014). Copper-catalyzed regiodivergent silacarboxylation of allenes with carbon dioxide and a silylborane. *J. Am. Chem. Soc.* 136: 17706–17709.

35 Kleeberg, C., Cheung, M.S., Lin, Z., and Marder, T.B. (2011). Copper-mediated reduction of CO_2 with pinB-SiMe$_2$Ph via CO_2 insertion into a copper–silicon bond. *J. Am. Chem. Soc.* 133: 19060–19063.

36 Scharfbier, J. and Oestreich, M. (2016). Copper-catalyzed Si—B bond activation in the nucleophilic substitution of primary aliphatic triflates. *Synlett* 27: 1274–1276.

37 Scharfbier, J., Hazrati, H., Irran, E., and Oestreich, M. (2017). Copper-catalyzed substitution of α-triflyloxy nitriles and esters with silicon nucleophiles under inversion of the configuration. *Org. Lett.* 19: 6562–6565.

38 Xue, W., Qu, Z.-W., Grimme, S., and Oestreich, M. (2016). Copper-catalyzed cross-coupling of silicon pronucleophiles with unactivated alkyl electrophiles coupled with radical cyclization. *J. Am. Chem. Soc.* 138: 14222–14225.

39 Xue, W. and Oestreich, M. (2017). Copper-catalyzed decarboxylative radical silylation of redox-active aliphatic carboxylic acid derivatives. *Angew. Chem. Int. Ed.* 56: 11649–11652.

40 He, Z.-T., Tang, X.-Q., Xie, L.-B. et al. (2015). Efficient access to Bicyclo[4.3.0] nonanes: copper-catalyzed asymmetric silylative cyclization of cyclohexadienone-tethered allenes. *Angew. Chem. Int. Ed.* 54: 14815–14818.

41 Tani, Y., Yamaguchi, T., Fujihara, T. et al. (2015). Copper-catalyzed silylative allylation of ketones and aldehydes employing allens and silylboranes. *Chem. Lett.* 44: 271–273.

42 Nozaki, K., Wakamatsu, K., Nonaka, T. et al. (1986). Transition-metal catalyzed regioselective addition of PhMe$_2$SiBEt$_3$Li and Bu$_3$SnBEt$_3$Li to acetylenic compounds in the presence of methanol. *Tetrahedron Lett.* 27: 2007–2010.

43 Wang, P., Yeo, W.-L., and Loh, T.-P. (2011). Copper-catalyzed highly regioselective silylcupration of terminal alkynes to form α-vinylsilanes. *J. Am. Chem. Soc.* 133: 1254–1256.

44 Meng, F., Jang, H., and Hoveyda, A.H. (2013). Exceptionally E- *and* β-selective NHC-Cu-catalyzed proto-silyl additions to terminal alkynes and site- and enantioselective proto-boryl additions to the resulting vinylsilanes: synthesis of enantiomerically enriched vicinal and geminal borosilanes. *Chem. Eur. J.* 19: 3204–3214.

45 Hazra, C.K., Fopp, C., and Oestreich, M. (2014). Copper(I)-catalyzed regioselective addition of nucleophilic silicon across terminal and internal carbon-carbon triple bonds. *Chem. Asian J.* 9: 3005–3010.

46 Calderone, J.A. and Santos, W.L. (2014). Copper(II)-catalyzed silylation of activated alkynes in water: diastereodivergent access to E- or Z-β-silyl-α,β-unsaturated carbonyl and carboxyl compounds. *Angew. Chem. Int. Ed.* 53: 4154–4158.

47 Linstadt, R.T.H., Peterson, C.A., Lippincott, D.J. et al. (2014). Stereoselective silylcupration of conjugated alkynes in water at room temperature. *Angew. Chem. Int. Ed.* 53: 4159–4163.

48 Vercruysse, S., Cornelissen, L., Nahra, F. et al. (2014). Cu/Pd cooperative dual catalysis: tunable stereoselective construction of tetra-substituted alkenes. *Chem. Eur. J.* 20: 1834–1838.

49 García-Rubia, A., Romero-Revilla, J.A., Manuleón, P. et al. (2015). Cu-catalyzed silylation of alkynes: a traceless 2-pyridylsulfonyl controller allows access to either regioisomer on demand. *J. Am. Chem. Soc.* 137: 6857–6865.

50 Vyas, D.J., Hazra, C.K., and Oestreich, M. (2011). Copper(I)-catalyzed regioselective propargylic substitution involving Si—B bond activation. *Org. Lett.* 13: 4462–4465.

51 Oestreich, M. and Weiner, B. (2004). Copper-catalyzed conjugate addition of a bis(triorganosilyl) zinc and a methyl(triorganosilyl) magnesium. *Synlett* (12): 2139–2142.

52 Oestreich, M. and Auer, G. (2005). Practical synthesis of allylic silanes from allylic esters and carbamates by stereoselective copper-catalyzed allylic substitution reactions. *Adv. Synth. Catal.* 347: 637–640.

53 Vyas, D.J. and Oestreich, M. (2010). Expedient access to branched allylic silanes by copper-catalyzed allylic substitution of liner allylic halides. *Chem. Commun.* 46: 568–570.

54 Hazra, C.K. and Oestreich, M. (2012). Copper(I)-catalyzed regio- and chemoselective single and double addition of nucleophilic silicon to propargylic chlorides and phosphates. *Org. Lett.* 14: 4010–4013.

55 Tobisu, M., Kita, Y., and Chatani, N. (2006). Rh(I)-catalyzed silylation of aryl and alkenyl cyanides involving the cleavage of C—C and Si—Si bonds. *J. Am. Chem. Soc.* 128: 8152–8153.

56 Walter, C., Auer, G., and Oestreich, M. (2006). Rhodium-catalyzed enantioselective conjugate silyl transfer: 1,4-Addition of silyl boronic esters to cyclic enones and lactones. *Angew. Chem. Int. Ed.* 45: 5675–5677.

57 Walter, C. and Oestreich, M. (2008). Catalytic asymmetric C—Si bond formation to acyclic α,β-unsaturated acceptors by Rh(I)-catalyzed conjugate silyl transfer using a Si—B linkage. *Angew. Chem. Int. Ed.* 47: 3818–3820.

58 Walter, C., Fröhlich, R., and Oestreich, M. (2009). Rhodium-catalyzed enantioselective 1,4-addition of nucleophilic silicon. *Tetrahedron* 65: 5513–5520.

59 Hartmann, E. and Oestreich, M. (2010). Asymmetric conjugate silyl transfer in iterative catalytic sequences: synthesis of the C7—C16 fragment of (+)-neopeltolide. *Angew. Chem. Int. Ed.* 49: 6195–6198.

60 Hartmann, E. and Oestreich, M. (2012). Two-directional desymmetrization by double 1,4-addition of silicon and boron nucleophiles. *Org. Lett.* 14: 2406–2409.

61 Ohmiya, H., Ito, H., and Sawamura, M. (2009). General and functional group-tolerable approach to allenylsilanes by rhodium-catalyzed coupling between propargylic carbonates and a silylboronate. *Org. Lett.* 11: 5618–5620.

62 Chu, C.K., Liang, Y., and Fu, G.C. (2016). Silicon–carbon bond formation via nickel-catalyzed cross-coupling of silicon nucleophiles with unactivated secondary and tertiary alkyl electrophiles. *J. Am. Chem. Soc.* 138: 6404–6407.

63 O'Brien, J.M. and Hoveyda, A.H. (2011). Metal-free catalytic C—Si bond formation in an aqueous medium. Enantioselective NHC-catalyzed silyl conjugate additions to cyclic and acyclic α,β-unsaturated carbonyls. *J. Am. Chem. Soc.* 133: 7712–7715.

64 Ito, H., Horita, Y., and Yamamoto, E. (2012). Potassium *tert*-butoxide-mediated regioselective silaboration of aromatic alkenes. *Chem. Commun.* 48: 8006–8008.
65 Shintani, R., Fujie, R., Takeda, M., and Nozaki, K. (2014). Silylative cyclopropanation of allyl phosphates with silylboronates. *Angew. Chem. Int. Ed.* 53: 6546–6549.
66 Nagao, K., Ohmiya, H., and Sawamura, M. (2015). *Anti*-selective vicinal silaboration and diboration of alkynoates through phosphine organocatalysis. *Org. Lett.* 17: 1304–1307.

2

Si—H Bond Activation by Main-Group Lewis Acids

Dieter Weber[1] *and Michel R. Gagné*[2]

[1] *Appalachian State University, A.R. Smith Department of Chemistry and Fermentation Sciences, 525 River Street, Boone, NC 28608, USA*
[2] *University of North Carolina at Chapel Hill, Department of Chemistry, Caudill Labs, 131 South Rd, Chapel Hill, NC 27514, USA*

2.1 Introduction to Silanes and the Si—H bond

2.1.1 Overview of the Discovery and the History of Silanes

Compounds containing Si—H bonds are called hydrosilicons, silicon hydrides, and/or silanes. The first silane was discovered by Heinrich Buff and Friedrich Wöhler in 1857. While conducting electrochemical experiments with silicon-containing aluminum [1], they observed the formation of an in-air flammable gas, which they named "Siliciumwasserstoffgas" (German for siliciuretted hydrogen) [2]. They could generate the same gas chemically by reacting hydrochloric acid with the impure aluminum. The following year, Wöhler highlighted further the spontaneous flammability of silanes in air by dropping magnesium silicide into diluted hydrochloric acid [3]. The primary component of Wöhler's gas was monosilane SiH_4. It is a colorless gas with a repulsive smell that produces headaches [4]. Wöhler attempted to draw analogies between hydrocarbon and hydrosilicon chemistry [5], which were later passionately discussed by the chemical community. For example, Mendeleev [6] and Reynolds [7, 8] favored Wöhler's opinion, but Henri Moissan [9] disagreed heavily in 1904 by making the following statement: "*the number of hydrocarbons is enormous, very few hydrosilicons have been made; the former are usually very stable, the latter unstable. How far does the analogy between silicon and carbon go? Is it only a question of time and trouble for the development of a system of silicon chemistry analogues to that now established in carbon chemistry?*" Soon thereafter, hydrocarbons and hydrosilicon compounds were deemed distinctly dissimilar by Stock, Barlow, and Pope as reported by Mellor [10]. This opinion still predominates today [11].

Organosilicon Chemistry: Novel Approaches and Reactions, First Edition.
Edited by Tamejiro Hiyama and Martin Oestreich

2.1.2 A Comparison of Hydrocarbons and Hydrosilicons

Why are hydrosilicons so different from hydrocarbons? There are several answers to this question that are not independent of one another and are worth discussing.

When looking at the electronegativities (ENs) of silicon and hydrogen versus carbon and hydrogen, an inverse trend can be noted. Carbon has an EN of 2.5 and hydrogen an EN of 2.1, which are both higher than that of silicon (EN of 1.8). As a result, the C—H bonds of alkanes have a small dipole moment with carbon δ^- and hydrogen δ^+. In contrast, the Si—H bonds found in silanes are polarized such that silicon has a δ^+ character and hydrogen has consequently a hydridic character. Silanes therefore undergo more facile heterolytic reactions such as hydrolysis than do alkanes due to their significantly higher affinity to oxygen- and other EN atom-containing compounds [12]. Because of the relationship between ENs and bond dissociation energies (BDEs) [13], the BDEs of most silicon compounds are considerably different from those of their carbon counterparts, as reported by Walsh [14, 15]. Generally, Si—H and Si—Si bonds are weaker than analog C—H and C—C bonds, and Si—X bonds (X = N, O, S, F, Cl, Br, I) are *significantly stronger* than the corresponding C—X bonds. Both energy factors together generate strong enthalpic driving forces for the oxidation of silanes. On the other hand, oxidation products of silanes exhibit mostly a lower entropy than do comparable products from hydrocarbons. For example, the combustion of silanes yields primarily SiO_2, a solid product (low entropy), while hydrocarbons yield gaseous CO_2 (high entropy). However, the strong enthalpic preference typically outweighs the entropic factor. The polarized nature of bonds to silicon additionally leads to lower activation energies than hydrocarbon analogs. For example, perchlorinated hydrocarbons are indifferent to water and amines at room temperature, while perchlorinated silanes undergo instant hydrolysis and disproportionation reactions. Thermodynamic and kinetic considerations thus provide an explanation for the instability of silanes when compared to hydrocarbons.

Another factor of note when comparing silanes with hydrocarbons was recently highlighted by Semenov [11]. Electronic absorption spectroscopy (EAS) and photoelectron spectroscopy (PES) data suggest that Si—Si single bonds exhibit delocalized σ-electrons comparable to the delocalization of π-electrons in conjugated polyenes. This delocalization allows Si—Si single bonds to interact with carbon π-systems, which can affect highest occupied molecular orbital (HOMO) and lowest unoccupied molecular orbital (LUMO) energies. This property of Si—Si single bonds explains their resemblance to C=C double bond reactivity, for example, the bromination reaction with Br_2 to yield 2 equiv of bromosilanes, the Prilezhaev epoxidation with *meta*-chloroperbenzoic acid (mCPBA) forming silyl ethers, or the bond cleavage under UV irradiation of cyclosilanes leading to the extrusion of silylenes (silicon analogs of carbene).

The diffuse character of silicon σ-electrons is also revealed in the β-silicon effect, which stabilizes a β-carbocation via hyperconjugation [16], but to a much larger extent than in their carbon analogs. The β-silicon stabilizing effect is similar in magnitude to the allylic stabilization of carbocations [17].

The stability of hydrosilicons can be significantly increased by replacing hydrogens with organic alkyl, alkenyl, phenyl, or alkynyl groups. In fact, according to Walsh, the BDEs of Si—C bonds are generally higher than those of the corresponding C—C analogs. As a result, most studies involve organosilanes, which are much easier to handle than air-sensitive and flammable silanes. However, due to recent advances by Oestreich and coworkers, monosilane (SiH_4) surrogates can be used to access the synthetic utility of this compound without exposing oneself to known hazards [18–23].

2.1.3 Stability of the Silicon–Hydrogen Bond

Generally, a BDE of 90 kcal/mol is reported for the Si—H bond [24], which renders this bond weaker than the H—H bond (103 kcal/mol) [25] or the C—H bond (98 kcal/mol for ethane) [14, 15]. An interesting aspect, however, is how the stability of Si—H bonds is affected by the electronic environment of the silicon atom, especially when compared to the C—H bond. BDE values reported by Walsh suggest that the Si—H bond strength is only marginally affected when hydrogens are replaced with alkyl groups. For example, the BDE of monosilane $H_3Si–H$ (90.3 kcal/mol) is basically the same as those of methylsilane $MeSiH_2–H$ (89.6 kcal/mol), dimethylsilane $Me_2SiH–H$ (89.4 kcal/mol), and trimethylsilane $Me_3Si–H$ (90.3 kcal/mol). This is counterintuitive to organic chemists, who are used to a decreasing BDE trend with more substituted C—H bonds: methane $H_3C—H$ (104.8 kcal/mol), primary C—H bonds as in ethane $MeCH_2—H$ (98 kcal/mol), secondary C—H bonds as in propane $Me_2CH—H$ (95 kcal/mol), and tertiary C—H bonds as in 2-methylpropane $Me_3C—H$ (92 kcal/mol); the latter being comparable to the stability of Si—H bonds.

The Si—H bond becomes slightly weaker in disilane $H_3SiSiH_2—H$ (86.3 kcal/mol) or phenylsilane $PhSiH_2—H$ (88.2 kcal/mol) and stronger with halogenation as in trichloro- $Cl_3Si—H$ (91.3 kcal/mol) or in trifluorosilane $F_3Si—H$ (100.1 kcal/mol), matching the trend of analogous carbon compounds: toluene $PhCH_2—H$ (87.9 kcal/mol), chloro- $Cl_3C—H$ (96 kcal/mol), and fluoroform $F_3C—H$ (106 kcal/mol). BDEs of Si—C bonds range from 88 to 89 kcal/mol and appear slightly stronger than the corresponding C—C bond energies (82–88 kcal/mol).

2.1.4 The Silylium Ion

Lewis acids that interact with the Si—H bond directly by binding to the hydride activate Si—H bonds by generating an electrophilic silicon (typically a tri-coordinate, tetravalent silylium ion) and a nucleophilic hydrogen. Free silylium ions are among the strongest known Lewis acids [26], as they coordinate to the majority of bonds that exhibit the slightest nucleophilicity, including solvents and counteranions; they are thus never truly "free" (see Section 2.4.3). As a result, this main-group Lewis acid is highly attractive for catalytic transformations that break strong σ and π bonds. However, to harness the high reactivity of silylium ions, its electronic (electrophilicity) and steric (accessibility) parameters must be tuned to achieve selectivity. For example, the electrophilicity of an intermediary silylium species can be altered by the extent to which the Si—H bond is activated by a Lewis acid, which ranges from minor polarization (lower reactivity) to full hydride abstraction (higher reactivity). Modification of the reactivity of silylium cations through sterics is often achieved by varying the size of the silane consumed in the reaction (less steric crowding around the Si atom results in higher reactivity of the corresponding silylium ion). Lessons learned from the application of the main-group Lewis acid tris(pentafluorophenyl)borane (BCF) in reactions involving the activation of silanes are extremely useful to highlight this point. These studies are described in greater detail in the following sections.

2.2 The Activation of Si—H Bonds by Boron Lewis Acids

2.2.1 Tris(pentafluorophenyl)borane (BCF)

Tris(pentafluorophenyl)borane $B(C_6F_5)_3$, commonly abbreviated as BCF, is an air-stable and water-tolerant Lewis acid [27]. This white solid is commercially available and can be prepared from boron halides (BX_3, X = F, Cl, Br) and pentafluorophenyllithium (F_5C_6Li) [28, 29], pentafluorophenylmagnesium bromide (F_5C_6MgBr) [30], or pentafluorophenylcopper $[Cu(F_5C_6)]_4(\eta^2\text{-toluene})_2$ [31]. BCF is a powerful Lewis acid with a Lewis acidity similar to that of BF_3 due to the electron-withdrawing nature of the three pentafluorophenyl rings [32]. The incorporation of B—C bonds, however, renders BCF less prone to hydrolysis than does BF_3 [33]. The steric bulk of the perfluorinated phenyl rings additionally makes BCF an ideal Lewis acid for frustrated Lewis pair (FLP) chemistry [34]. Initially, BCF was used only as a cocatalyst in Ziegler–Natta chemistry [35], but it now dominates the field of main-group Lewis acid and metal-free catalysis [36]. While BCF comes close to the definition of an ideal Lewis acid, it has several drawbacks, including the *para*-fluorine atom of the pentafluorophenyl rings being susceptible to nucleophilic aromatic substitution [37]. Less Lewis acidic compounds (according to the Gutmann–Beckett [38] and Childs et al. [39] scales) such as $B(p\text{-}C_6F_4H)_3$ [40] or more Lewis acidic boranes, such as the Lewis superacid tris(*p*-perfluorotolyl)borane (BTolF) [41], can be employed to circumvent this side reactivity. Another consideration in using BCF is its interaction with water. It has been reported that the Brønsted acidity of the water-BCF adduct (pK_a of 8.4 in MeCN) is comparable to the pK_a of HCl (8.5 in MeCN) [42], with the irreversibly deprotonated BCF hydroxide adduct perceived to be catalytically inactive as a Lewis acid [43]. For this reason, calls to develop more water-tolerant Lewis acids have been voiced [44]. Due to its symmetry, BCF is also not suitable for chiral Lewis acid catalysis, which stimulated the development of numerous chiral BCF analogs [45–48].

2.2.2 The Catalytic Activation of Si—H Bonds by BCF and Other Boranes

2.2.2.1 The Mechanism of Borane-Catalyzed Si—H Bond Activation

In a seminal study by Piers and a coworker, the ability of BCF to activate the Si—H bond of Ph_3SiH was demonstrated in hydrosilylation reactions [49]. Aromatic aldehydes, ketones, and esters were reduced to silyl ethers and silyl acetals, the latter being converted to aldehydes upon acidic workup (Scheme 2.1). It is noteworthy that when Piers employed more than 1 equiv of silane, further reduction of the silylated products was observed, implying the potential of silane/BCF systems to cleave C—O bonds. These reactions are discussed later.

Based on kinetic data, it was proposed that BCF activated the Si—H bond (via an unobserved η^1-[B]–H–[Si] adduct), leading to an electrophilic Si atom rather than the more traditional mechanism of activators of the carbonyl group by binding to oxygen (i.e. via the [B]—O=CR_2 adduct), which increases the electrophilicity of the carbonyl carbon atom. The latter was described previously for the more oxophilic (when compared to BCF) $BF_3{\cdot}OEt_2$/silane-mediated reductions of aldheydes and ketones to alcohols and symmetric ethers [50] or alcohols and alkanes [51] (Scheme 2.2).

Piers provided additional mechanistic findings while reporting the expansion of his new methodology to aliphatic carbonyl compounds [52]. The stoichiometric reaction of Ph_3SiH or Et_3SiH and BCF (in the absence of substrate) led to the conclusion that a fast equilibrium between these reagents and their η^1-adduct [R_3Si–H–$B(C_6F_5)_3$] existed, which pointed to a BCF-catalyzed Si—H bond cleavage mechanism that yields a borohydride and an electrophilic silicon species (Scheme 2.3).

Piers proposed that BCF coordination to the hydrogen atom of the silane in an η^1-manner enhances the electrophilicity of the silicon moiety, which upon addition (at Si) by the nucleophilic carbonyl oxygen generates a silyloxonium/borohydride ion pair that recombines to yield the reduced product and BCF catalyst. Additional experimental and computational studies supported this three-step

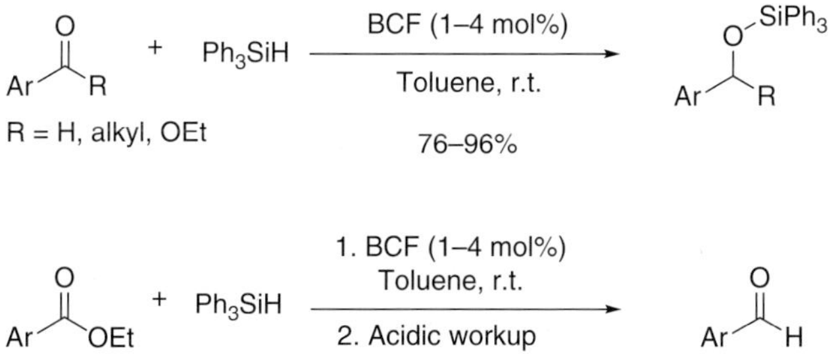

Scheme 2.1 Piers' seminal hydrosilylative reductions of carbonyl compounds using a BCF/silane system.

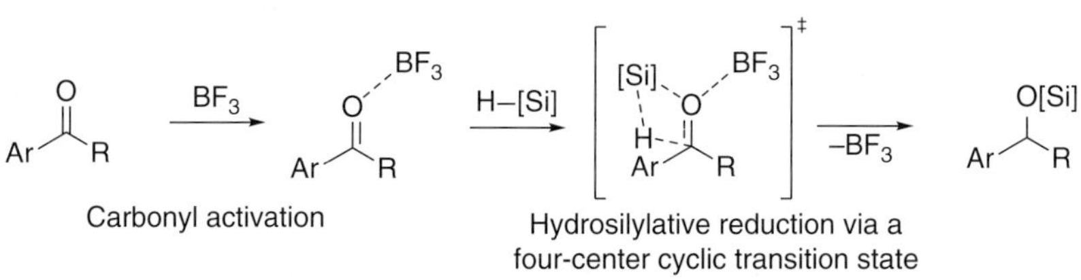

Scheme 2.2 Proposed carbonyl activation pathway for the BF_3-catalyzed hydrosilylation of carbonyl compounds.

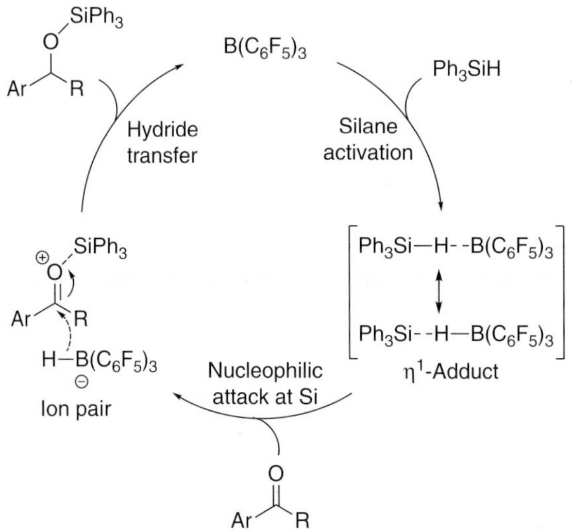

Scheme 2.3 Mechanism for the BCF-catalyzed hydrosilylative reduction of carbonyl compounds.

mechanism and pointed to an S_N2-type nucleophilic substitution around the silicon atom [53–56]. This basic three-step mechanism for the activation of silanes by BCF has been widely accepted, but more complex scenarios may arise depending on the specific reaction [57].

2.2.2.2 Additional Mechanistic Aspects

This simple mechanistic picture provides a foundation for the development of BCF-catalyzed reductions using silanes, with a central role for the η^1-[B]–H–[Si] adduct. The existence of an η^1-coordinated adduct between a strong boron Lewis acid and triethylsilane was spectroscopically (^{19}F NMR, ^{1}H NMR, and IR data) and structurally verified by Piers and coworkers [58]. The fluorinated boraindene in Scheme 2.4 is more Lewis acidic than BCF and shifts the equilibrium of the borane/silane mixture to the η^1-[B]–H–[Si] adduct at −78 °C in toluene. This cooling is accompanied by a characteristic color change from red to yellow, which could be reversed by warming the reaction mixture to room temperature. The crystal structure of the η^1-[B]–H–[Si] adduct revealed a slightly bent Si–H–B

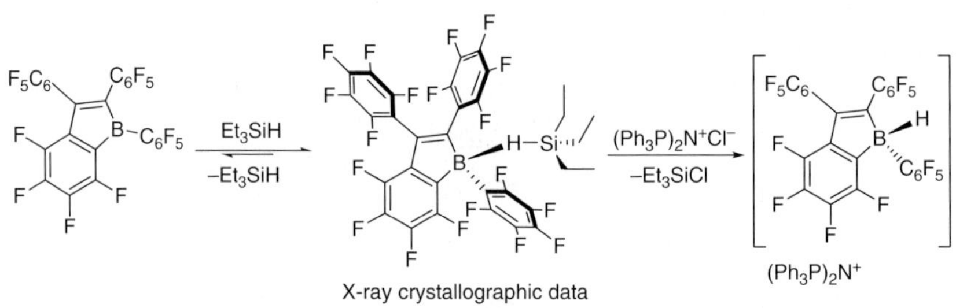

Scheme 2.4 Synthesis and reactivity of an isolated η^1-[B]–H–[Si] adduct.

bridge (157°). The Si—H bond of the η^1-adduct was 1.51(2) Å, only marginally elongated from the typical Si—H bond length of ≈1.48 Å. The B—H bond length in the borane–silane complex is significantly elongated (1.46(2) Å) relative to other borohydrides (1.14 Å). In addition, secondary C—F interactions between the *ortho*-fluorine atoms and the coordinated triethylsilane were highlighted. The structurally verified boraindene–triethylsilane complex reacted instantly with bis(triphenylphosphine)iminium chloride to form the expected borohydride and chlorotriethylsilane.

This mechanistic work experimentally supports the polarization of the silane for nucleophilic attack in the η^1-[B]–H–[Si] adduct formation. In Piers' original example, the nucleophilic oxygen of the carbonyl group cleaves the η^1-adduct, leading to the silyloxonium cation/borohydride ion pair. The recombination of those two species in a tight ion pair transfers the hydride from $[B–H]^-$ to the electrophilic carbonyl carbon and completes the reduction of the polar C=O double bond to the silylated C—O single bond. Computational studies by Sakata and Fujimoto supported the three-step mechanistic proposal of Piers for BCF-catalyzed reactions and confirmed carbonyl activation by BF_3 [55].

The fact that Piers observed over-reduction to alkanes when an excess of silane was employed showed that even silyl ethers, which have reduced basicity, can be suitable nucleophiles to attack the η^1-[B]–H–[Si] complex and initiate further reduction. Since numerous heteroatom nucleophiles could be envisioned to react with the electrophilic silicon of the η^1-adduct, many different functional groups could become activated by a silylium and be reduced by the generated $[B–H]^-$ species. The focus of the following sections is on how the η^1-[B]–H–[Si] complex enables the reduction of a number of π and σ bonds.

One aspect to consider when designing BCF-catalyzed silylative reductions with silanes involving the η^1-[B]–H–[Si] adduct is the steric bulk of the silane. Depending on the turnover limiting step in Piers' three-step mechanism (formation of the η^1-adduct, ion pair formation, or hydride transfer), large R-groups on the silane may become inhibitive at any step along the way. On the other hand, small groups or even multiple hydrogen atoms may increase the reducing power of the BCF/silane system. For R_3SiH-type silanes, Tolman's cone angles θ for the corresponding phosphines have been used in previous studies [59, 60] to estimate their steric bulk. And in cases when the cone angles of phosphines were unavailable (i.e. mixed phosphines of $PR'R''_2$), the cone angles of silanes were calculated from (PR'_3 and PR''_3) as follows: $\theta = 1/3\theta' + 2/3\theta''$ [61]. Although Si—C bond lengths differ from P—C bond lengths, the principle should transfer in a straightforward manner.

Another aspect to consider when designing catalytic reductive hydrosilylations with a borane/silane system are the stereoelectronic attributes of the borane catalyst. Even though similar in Lewis acidity, the electrophilic nature of the boron center in BCF allows this catalyst to interact disproportionally stronger with small (versus large) Lewis bases (such as hydrosilicons) than does BF_3. In addition, attractive interactions of the *ortho*-fluorine atoms of BCF with the alkyl groups of silanes provide energy for mechanistically required deformations of the BCF molecule [55].

It also should be noted that there are numerous examples involving BCF-type Lewis acid catalysts acting on silanes as the stoichiometric reductant, wherein Lewis bases are deliberately added to tune the activity and reactivity of the reducing system. These processes are related to Piers-type reductions; but in these cases, the η^1-[$(C_6F_5)_3B$]–H–[Si] adduct is cleaved by the Lewis base (LB) to form a $[R_3Si–LB]^+[(C_6F_5)_3B–H]^-$ ion pair. These reaction systems have significant parallels to key concepts in the FLP field and exhibit their own characteristic chemistry [62]. Traditional Lewis pairs combine to form adducts that consume both the Lewis base and the acid. However, when a beneficial bonding interaction is inhibited, most often due to steric congestion, these Lewis pairs may react cooperatively, for example, in the cleavage of nonpolar (i.e. H—H) bonds. The traditional definition of FLPs (no dative bonding between pairs) still stands, but it has been proposed to categorize FLP systems more by their reactivity (i.e. exhibiting FLP chemistry) instead of the presence of partial or complete dative bonding between pairs, since partially stabilized ion pairs can still exhibit FLP chemistry. For this reason, the deliberate addition of a Lewis base to enhance Si—H bond activation is considered an FLP approach in this chapter whether dative bonding occurred or not and some examples are highlighted. It should be noted that the term FLP chemistry is more broadly defined in some literature. For example, the hydrosilylation of carbonyl compounds developed by Piers could be viewed as FLP chemistry when defining the carbonyl oxygen of the substrate as the Lewis base. But independent of the definition, the formation of an η^1-[B]–H–[Si] adduct resembles a fundamentally different mode of Si—H bond activation when compared to transition metals [63]. These catalysts activate the Si—H bond by oxidative addition or σ bond metathesis [64]. An illustrative example is the recent report by Gagné and coworkers in which a phosphine was added to the reducing BCF/silane mixture [65]. Detailed mechanistic studies showed a change of the catalyst resting state from the Lewis acidic BCF to the milder electrophilic silyl phosphonium species ion paired with BCF hydride, which alters not only the acidity of the reaction solution but also allows further tunability of silylative reductions by modification of the phosphine.

2.2.3 Categorizing Reduction Types of π and σ Bonds Involving the η^1-[B]–H–[Si] Adduct

As outlined earlier, the reaction of the η^1-[B]–H–[Si] adduct is central to borane-catalyzed reductions of bonds with silanes. Since several bond types may be reduced under the same reaction conditions, it is helpful to categorize reduction types with representative examples.

2.2.3.1 Type I: The Reduction of Polar π Bonds (El=Nu/El≡Nu)

Polar π bonds contain a nucleophilic and an electrophilic atom El=Nu/El≡Nu and can react with the *in situ*-generated η^1-[B]–H–[Si] adduct (Scheme 2.5). The nucleophilic atom will be silylated and the electrophilic atom will accept the hydride from the borohydride with concomitant breakage of the π bond. The regioselectivity is determined by the polarity of the multiple bond. Since each π bond may potentially add one silane, double addition for triple bonds is

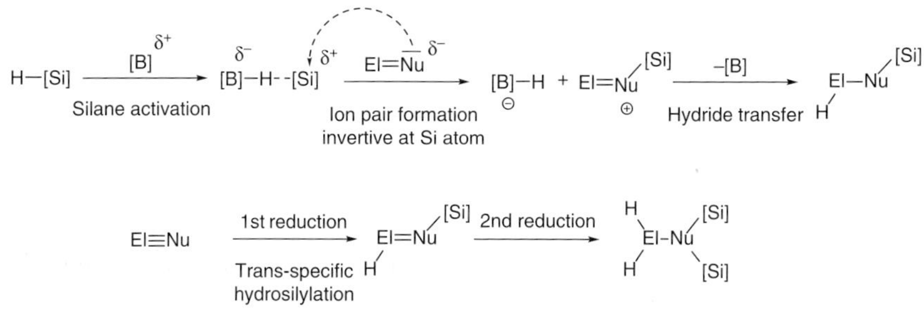

Scheme 2.5 The mechanistic steps and characteristics of type I reductions.

possible. A representative example for this reduction type is the above-discussed BCF-catalyzed reduction of aldehydes and ketones to silyl ethers.

Piers and coworkers also showed that his BCF/silane system reduced protected aromatic aldimines and ketimines to amines (Scheme 2.6) [66]. The basic three-step mechanism described earlier is in its essence still valid for this conversion, but competing mechanistic pathways involving simultaneously operating silylated imines, enamines, and amines were identified by Oestreich and coworkers [67–69]. In addition, Klankermayer and coworkers and Du and a coworker reported enantioselective variations more recently (not shown) [70–72].

In 2016, Ingleson and coworkers reported the reductive amination of aldehydes and ketones using primary and secondary arylamines (Scheme 2.7) [73]. Excess water and arylamine tolerance of BCF is crucial for this reaction because water is produced and amines are a starting reagent. Earlier perceptions that irreversible deprotonation of the BCF-water adduct to $[BCF–OH]^-$ would permanently

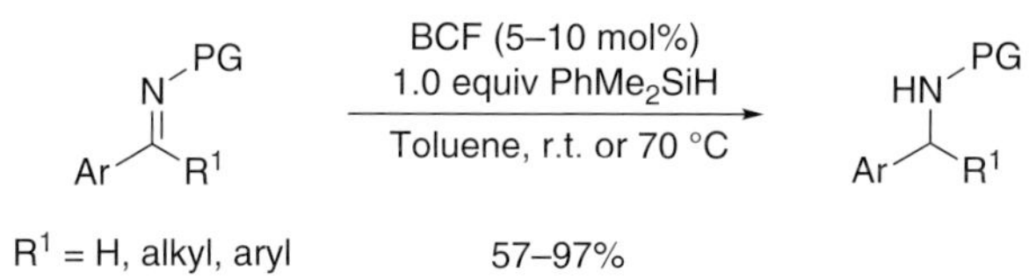

Scheme 2.6 Hydrosilylative reduction of imines.

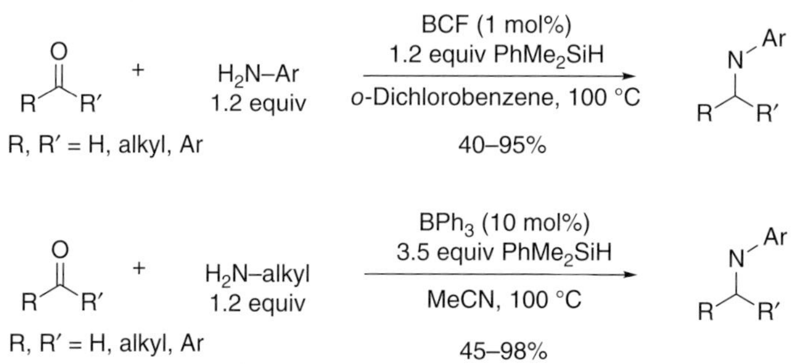

Scheme 2.7 Reductive amination of aldehydes and ketones.

poison the catalyst were shown not to apply in this case as the hydroamination proceeded efficiently at elevated temperatures [74, 75]. In a more recent study, Ingleson and a coworker demonstrated that the less acidic but also less oxophilic BPh_3 and $B(3,5\text{-}Cl_2C_6H_3)_3$ catalyzed the reaction efficiently and allowed even the more basic aliphatic amines to be employed as substrates [76].

A double reductive amination approach was used by Xu and coworkers to synthesize tetrahydroquinoxalines from 1,2-diaminobenzenes and α-keto-esters [77]. The optimal conditions were comprised of 5 mol% BCF and 4 equiv polymethylhydrosiloxane (PMHS) in toluene at 110 °C (not shown).

A clean and high-yielding BCF-catalyzed sila-reduction of thioketones to silyl thioethers was reported by Rosenberg and coworkers in 2005 [78]. Mono- and disilanes (not shown) were efficiently utilized in this reaction protocol and catalyst loading could be very low (Scheme 2.8).

Chang and coworkers reported the chemoselective reduction of nitriles to either primary amines or imines depending on the bulk of the reducing silane (Scheme 2.9) [79]. Primary amines were obtained with smaller Et_2SiH_2 resulting from a double reduction, but the bulkier Et_3SiH only hydrosilylated one π bond and led to imines.

In 2010, Rosenberg and coworkers demonstrated that the diastereoselectivity of the BCF-catalyzed reduction of α-diketones was dependent on the silane used in the reaction (Scheme 2.10) [80]. When benzil was reduced with Me_3SiH, the *anti*-product was obtained in a 96% *de*, but when the bulkier Ph_3SiH was employed, the *syn*-product was formed (92% *de*). In this reaction, two sequential type I reductions occur. The stereocontrol was explained by a stereochemistry-determining second hydride transfer to the partially reduced diketone, producing either the Felkin or the *anti*-Felkin product depending on the steric bulk of the silyl groups.

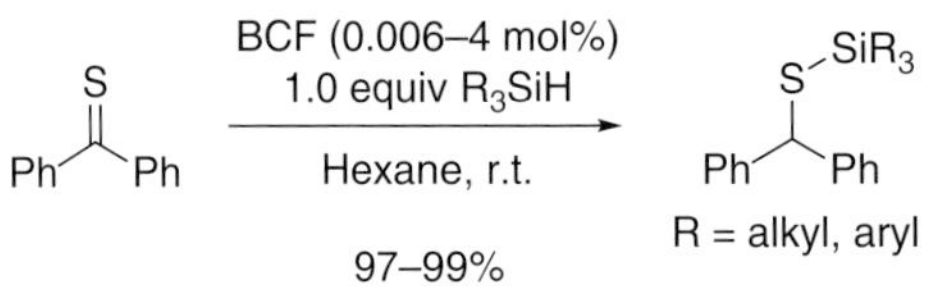

Scheme 2.8 BCF-catalyzed hydrosilylative reduction of thioketones.

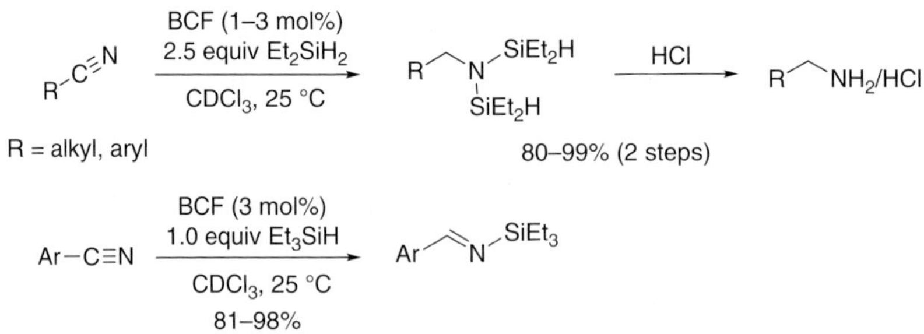

Scheme 2.9 Tunable hydrosilylative reduction of nitriles.

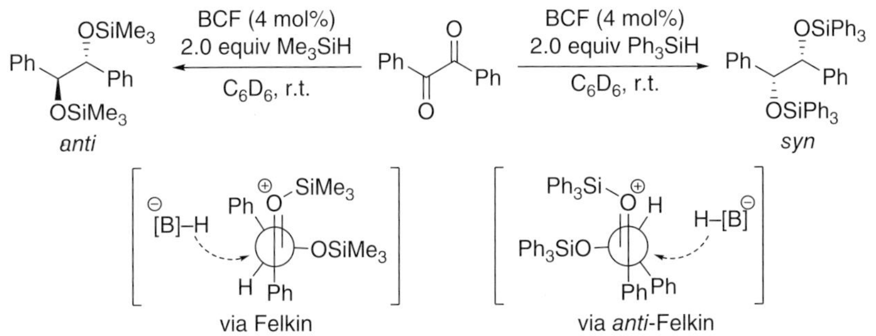

Scheme 2.10 Stereocontrolled hydrosilylative reduction of benzyl.

Yamamoto and coworkers investigated the stereoselective hydrosilylation of α- and β-substituted ketones [81]. The incorporation of alkynes in the γ-position of the carbonyl was proposed to create a σ-π-chelate between the silylcarboxonium ion and the alkyne, which provides the means to control the diastereoselectivity and yield the predominant *syn*-products for α-substituted ketones and *anti*-products for β-substituted ketones (Scheme 2.11). The implied diastereocontrol suggests in these reactions an affinity between intermediary silyloxonium cations and π-nucleophiles, but the mechanism remains to be investigated.

In 2016, Oestreich and coworkers reported an asymmetric version of a type I reduction. Acetophenones were reduced by a borane catalyst bearing only one perfluorophenyl ring and a 3,3′-substituted axially chiral binaphthyl backbone (Scheme 2.12) [82]. Key for effective stereo induction was the employment of an excess of the sterically less hindered $PhSiH_3$ as the reducing agent. The stereoselectivity of the reaction was catalyst loading dependent, leading the authors to speculate on the possibility of two (or more) competing mechanistic pathways. The substrate scope, reaction rate, yield, and enantioselectivity was further improved by Ryu and coworkers using a chiral oxazaborolidinium ion as the borane catalyst [83]. Aromatic and aliphatic ketones, and also epoxides and acetals, were efficiently converted to the chiral secondary alcohols. Oxazaborolidines are known chiral catalysts for Itsuno–Corey reductions (also

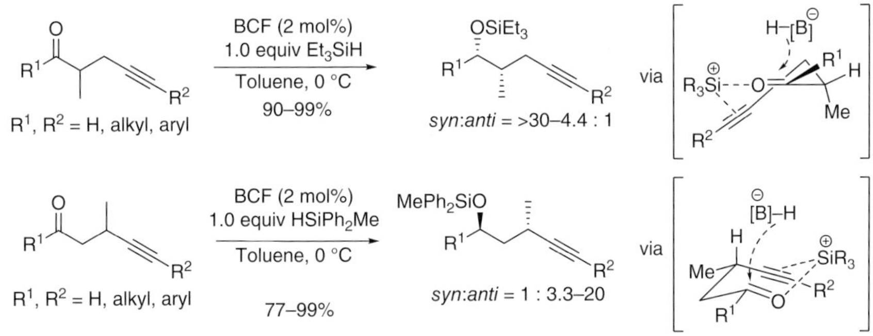

Scheme 2.11 Stereoselective hydrosilylation of alkynylic ketones via σ-π-chelation.

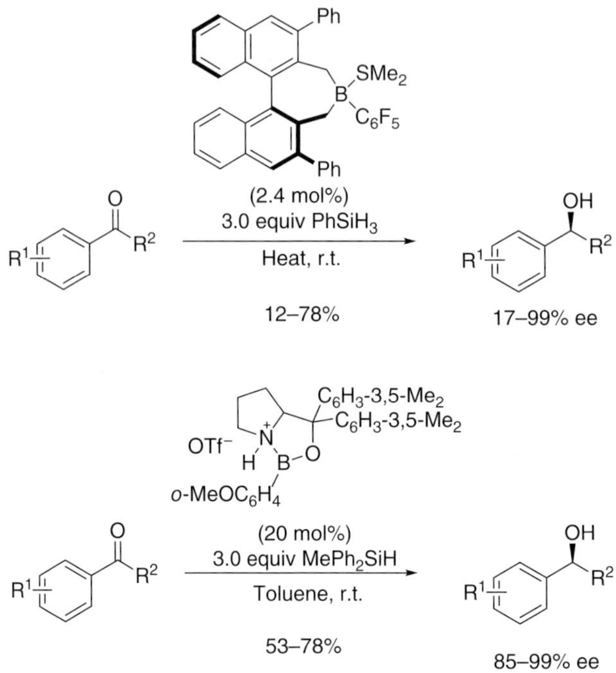

Scheme 2.12 Asymmetric hydrosilylative reduction of ketones.

known as the Corey–Bakshi–Shibata reduction), which operates mechanistically via carbonyl activation [84, 85]. While the authors did not rule out the Piers-type mechanism, a direct hydride transfer from the silane to the carbonyl carbon was proposed to rationalize the observed enantioinduction. Other approaches in enantioselective main-group catalysis were recently reviewed [86].

In 2015, Huang and coworkers reported an interesting example highlighting how the deliberate addition of a Lewis base can change the activation mode of silanes by Lewis acidic boranes. It was demonstrated that the less Lewis acidic and less electrophilic BEt_3 (in comparison to BCF) efficiently catalyzes the reduction of aldehydes, ketones, and esters with PMHS, an inexpensive waste product of the silicon industry, but only if catalytic amounts of NaOMe were present (Scheme 2.13) [87].

It was proposed that $NaBEt_3OMe$ reacts with the Si–H bonds of PMHS to form $NaBEt_3H$ [88] and a silyl ether by-product. After reduction of the carbonyl group by $NaBEt_3H$, the resulting Na alkoxide is formed along with BEt_3, which can reenter the cycle as a $NaBEt_3$-alkoxide adduct to activate more silane. Although no additional mechanistic studies have been reported, it is clear that an η^1-[B]–H–[Si] adduct plays no role in this reduction and that it is an alkoxyborate that activates the silane.

Laali and coworkers compared BCF versus metal triflates in catalytic reductions of aldehydes and ketones using silanes. It was demonstrated that the substrate scope could be expanded using metal triflates (not shown) [89]; and the

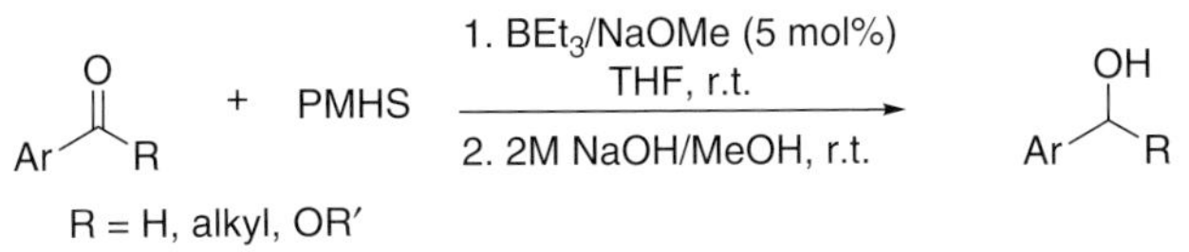

Scheme 2.13 Catalytic reductive hydrosilylation using BEt_3 and NaOMe as a Lewis acid/base pair.

application of this type I reduction was highlighted in the synthesis toward 1,4-dihydronaphthalenes by Adiwidjaja and coworkers [90].

2.2.3.2 Type II: The Reduction of Polar σ Bonds (El–Nu)

The reactivity of polar σ bonds (El–Nu) is analogous to that of polar π bonds with the nucleophilic end attacking the electrophilic silicon of the η^1-[B]–H–[Si] adduct (Scheme 2.14). In this case, however, the recombination of the borohydride with the silylated product in the ion pair now cleaves the σ bond to form two products, a hydrogenated electrophile (El—H) and a silylated nucleophile (Nu—Si). If an electrophile is bonded to several nucleophiles, all polar σ bonds may be sequentially cleaved. There are several different reaction classes that fall into this category.

2.2.3.2.1 Dehydrogenative Silylations (Dehydrogenative Coupling or Dehydrocoupling of Silanes)

This reaction class may occurs with any compound that contains a hydrogen that is acidic enough to react with $[BCF–H]^-$; and, as a result, is commonly observed as side reactions in BCF-catalyzed reactions using silanes. As part of his pioneering work, Piers and coworkers demonstrated that BCF efficiently catalyzed the dehydrogenative silylation of a variety of alcohols using numerous silanes [91]. Primary, secondary, tertiary, allylic, benzylic, propargylic, and aromatic alcohols were cleanly silylated in high yields with an excellent functional group tolerance (Scheme 2.15).

In this reaction, the oxygen acts as the nucleophile and the hydrogen as the electrophile. A plausible mechanism has the lone pair of the alcohol oxygen attacking the electrophilic silicon of the η^1-[B]–H–[Si] adduct yielding a silylated oxonium ion and borohydride ion pair, which reacts with the acidic proton of the silylated oxonium ion producing the silyl ether and H_2 gas.

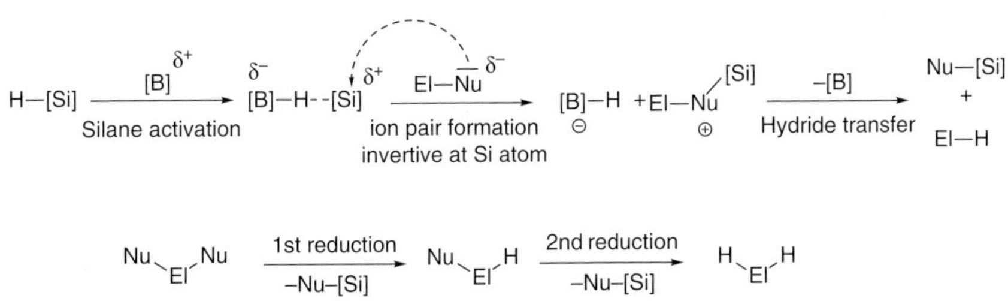

Scheme 2.14 The mechanistic steps and characteristics of type II reductions.

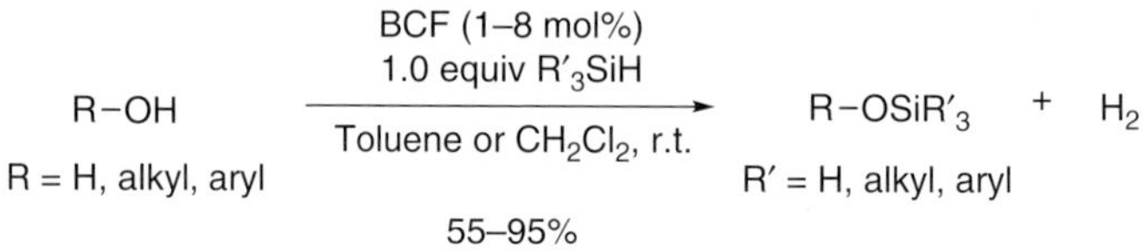

Scheme 2.15 Dehydrogenative silylation of alcohols.

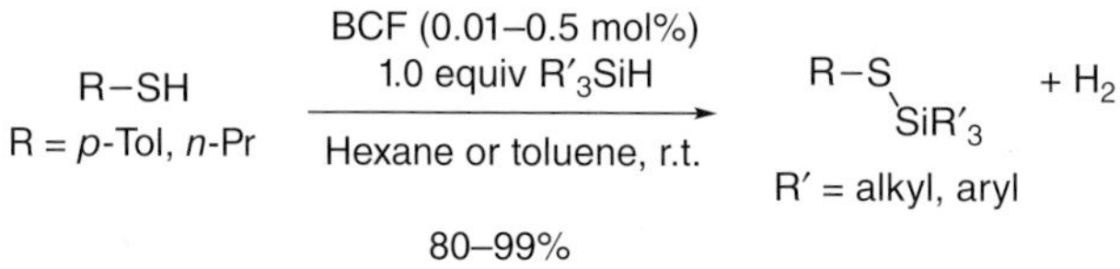

Scheme 2.16 Dehydrogenative silylation of thiols.

Another representative example was demonstrated by Rosenberg and coworkers who reported the dehydrogenative silylation of thiols [92] and the thiolation of polyphenylsilanes (Scheme 2.16) [93]. In all examples, the steric bulk of the silane strongly affected reaction rates and could even shut down the transformation. These reactions were used by Rosenberg and a coworker to selectively derivatize $(Ph_2SiH)_2$ adding to the synthetic repertoire of incorporating Si—Heteroatom bonds in silanes [94]; such methods were recently reviewed [95].

As part of his mechanistic work, Oestreich and coworkers reported that dehdydrogenative Si—N couplings were reversibly catalyzed by BCF at room temperature [69]. Shortly after Paradies and coworkers showed that primary and secondary aryl amines were well suited for this reaction, especially when the reaction temperature was elevated to 70 °C (Scheme 2.17) [96].

Very recently, Dobrovetsky and a coworker reported the chlorination of silanes using BCF or $Et_2O{\cdot}B(C_6F_5)_3$ as the catalyst and HCl as the chlorine source (Scheme 2.18) [97]. This reaction can be viewed as a dehydrogenative silylation of HCl.

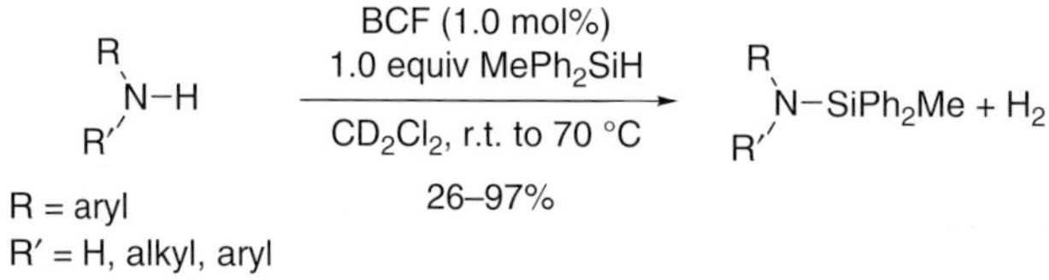

Scheme 2.17 Dehydrogenative silylation of amines.

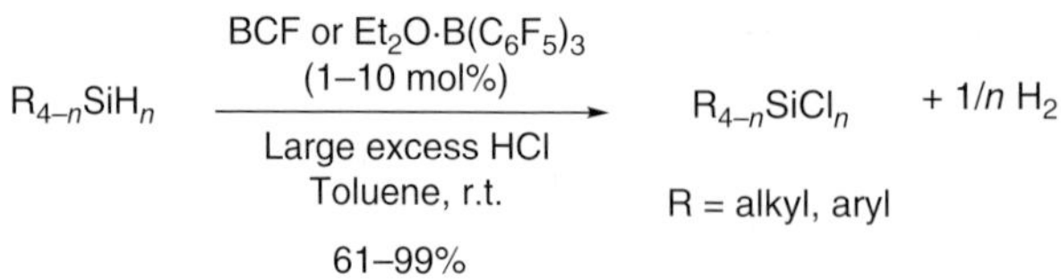

Scheme 2.18 Dehydrogenative silylation of hydrogen chloride.

Interestingly, when $Et_2O{\cdot}B(C_6F_5)_3$ was employed, mono- or dichlorinated products could be obtained in a controlled manner from di- and trihydrosilanes, explainable by a difference in mechanism. It was proposed that the BCF-catalyzed reaction proceeded via direct η^1-activation of the silane according to Piers' mechanism (η^1-[B]–H–[Si] adduct), while the $Et_2O{\cdot}B(C_6F_5)_3$ catalyzed the chlorination stepwise by first forming *in situ* a silyl ether species ([Si]–OEt), which was cleaved by the acid to the product ([Si]–Cl) and ethanol.

2.2.3.2.2 Silylative Deoxygenation of Ethers

In these reactions, C—O single bonds of dialkyl- (R_3C—OR′) or alkyl silyl ethers (alkoxysilanes, R_3C—O[Si]) are cleaved with the ether oxygen being the attacking nucleophile that reacts with the Si electrophile of the η^1-adduct. Depending on the starting ether, either a monosilyl (R_3C—O[Si]R'^{+}) or a disilyl oxonium (R_3C—O$[Si]_2^{+}$) is formed as the key positively charged intermediate. The electrophilic carbon atom reacts further with the borohydride to cleave the C—O bond and generate an alkane and a silyl ether (R_3CH + [Si]OR′) or an alkane and a disiloxane (R_3CH + [Si]O[Si]), respectively. The latter reaction is called the Piers–Rubinsztajn reaction and is used in the synthesis of functionalized silicones [98–101]. Because of their mechanistic similarity to the dehydrogenative silylation of alcohols, this reaction type is also referred to as the dehydrocarbonative silylation of ethers forming hydrocarbons (instead of H_2 gas) as (by) products.

Shortly after Piers initial reports on BCF-catalyzed carbonyl reductions, Gevorgyan et al. produced a protocol for the reduction of alcohols and ethers to silyl ethers or alkanes using BCF and Et_3SiH (Scheme 2.19) [102, 103].

Primary alcohols could either be converted to silyl ethers with an equimolar amount of silane (dehydrogenative silylation) or further reduced to alkanes with an excess of silane (3.0–6.0 equiv). Secondary, tertiary, and aromatic alcohols stopped at their corresponding silyl ethers independent of the amount of silane used in the reaction. Similarly, α-unsubstituted ethers cleaved to silyl ethers with 1.1 equiv of silane and they were fully reduced to alkanes with 3.0 equiv, and α-substituted or aromatic ethers cleaved only to the silyl ether stage. The latter reaction was used by Piers to access (perfluoroaryl)borane-functionalized carbosilane dendrimers (not shown) [104]. This study not only demonstrated the ability of BCF-catalyzed reductions using silanes to defunctionalize alcohols and ethers, which was later expanded by McRae using either *n*-butylsilane or diethylsilane and by Njardarson and coworkers reducing cyclic allylic ethers [105–107] but it also illustrated the potential to control the chemoselectivity of a reduction by tuning reaction conditions.

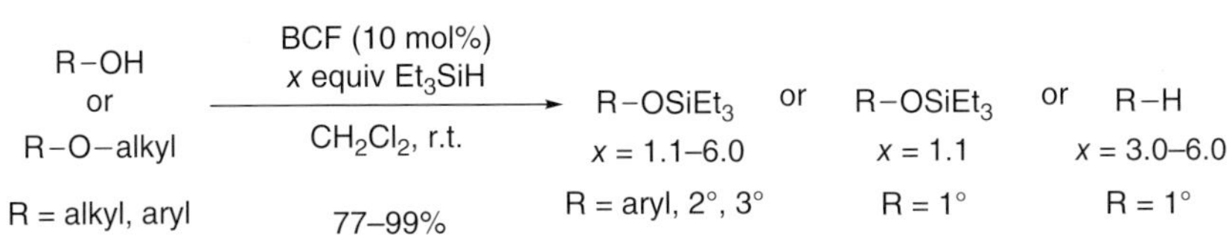

Scheme 2.19 Silylative deoxygenation of alcohols and ethers.

The silylative C—O bond cleavage was applied in natural product synthesis and biomass conversion [108]. This reduction was used in natural product synthesis to demethylate aromatic OMe protecting groups by cleaving the O—CH_3 bond under mild conditions [109–112]. The BCF-catalyzed silylative deoxygenation of carbohydrates was subsequently reported by Gagné and coworkers [113]. Overall, in the latter reaction, the rate and product distribution was sensitive to substrate, solvent, silane, protecting group, and catalyst (BCF versus $[Ir(III)–H]^+$ species) [114]. The complete deoxygenation of a variety of biomass-derived sugars (to C6 hydrocarbons) was achieved with Et_2SiH_2, but employment of Me_2EtSiH enabled partial reduction when BCF was used as the catalyst. Careful NMR monitoring of reactions of silylated mono- and disaccharides gave an insight into the stereoelectronic factors and mechanistic routes governing the chemo- and regioselective outcome of partial carbohydrate deoxygenation reactions [115]. The general chemoselective trend for C—O bond cleavage in carbohydrates is as follows: anomeric C—O bonds were identified as the most reactive sites for bond cleavage likely due to their steric accessibility and the stability of the oxocarbenium ions that result from alkoxide abstraction, followed by primary silyl ethers, which generally reacted faster than did the secondary functionalities. Oxygen Lewis basicity was also important and nuanced the reaction pathways. For example, Et_3Si-protected glucose yielded sorbitol as the major product after two hours by first facilitating an endocyclic cleavage of the anomeric C—O bond (Scheme 2.20). The product distribution shifted to 1,6-deoxysorbitol after 16 hours, which is formed by cleaving both primary silyl ether functionalities of sorbitol. The reduction of secondary silyl ethers was observed as a slower background reaction, leading to 2,3,4-hexanetriol via a silylated tetrahydrofuran (THF) intermediate [116].

Interestingly, the regioselectivity of anomeric C—O bond cleavages could be tuned by the Lewis basicity of the *exo*-oxygen substituent for monosaccharides [117]. The sterically less crowded and more Lewis basic 1-MeO-glucose demethoxylates selectively to 1-deoxyglucose, whereas the persilylated glucose ring

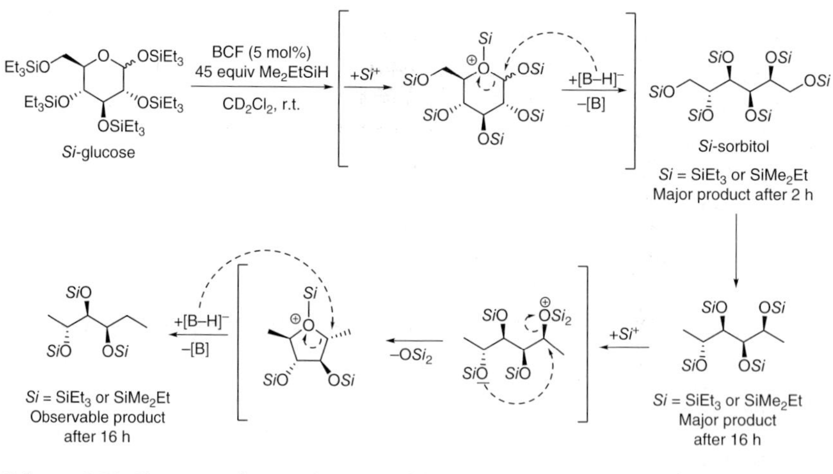

Scheme 2.20 Chemo- and regioselectivity of the silylative reduction of monosaccharides.

opens exclusively to sorbitol (Scheme 2.21). In these cases, the site of activation (where Si^+ resides) is rationalized as being heavily influenced by the Lewis basicity of the ether ($R_2O > ROSi$).

The regioselective cleavage of competing anomeric centers in persilylated disaccharides was also investigated [115]. Preferential cleavage of the glycosidic linkage was observed instead of ring-opening reactions for maltose (α-1,4′-glycosidic linkage), cellobiose (β-1,4′-glycosidic linkage), and trehalose (α,α′-1,1′-glycosidic linkage). But the regioselectivty was reversed for isomaltose with the α-1,6-glycosidic linkage yielding primarily ring-opened isomaltitol (Scheme 2.22).

As outlined in Scheme 2.22, the reduction of secondary silyl ether groups of 1,6-deoxysorbitol to 2,3,4-hexanetriol involved an intramolecular S_N2 cyclization of a neighboring silyloxy group forming a silylated THF intermediate. The formation of a cyclic intermediate enables a chemoselective C—O bond reduction of C6 sugars to yield chiral synthons by accessing deoxygenation pathways that are lower in energy than if no neighboring group were present [116]. Park and coworkers also reported a selective C—O bond cleavage of carbohydrates leading to similarly deoxygenated polyols using Piers' borane [$(C_6F_5)_2BH$], which

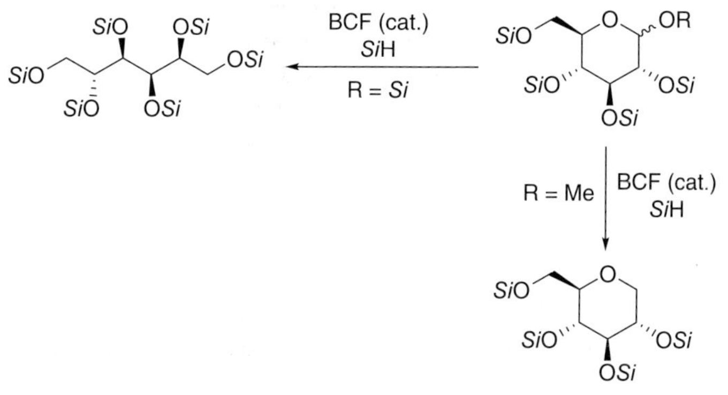

Scheme 2.21 Regioselectivity dependent on *exo*-oxygen substituent.

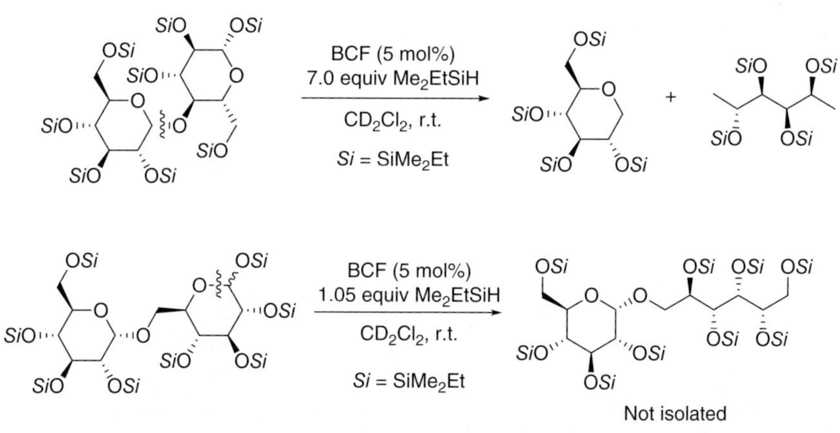

Scheme 2.22 Regioselectivity of competing anomeric centers in disaccharides.

was *in situ* generated from [$(C_6F_5)_2BOH$] in the presence of tetramethyldisiloxane (TMDS) and other silanes [118].

A fundamentally different neighboring group effect was utilized by Morandi and coworker, who reported an elegant regioselective monodehydroxylation of vicinal terminal diols at the primary position (Scheme 2.23) [119]. Key was the sequential use of two different silanes.

First, 1 equiv of Ph_2SiH_2 was used to convert the 1,2-diol to a cyclic siloxane via dehydrogenative silylation. The subsequent addition of 1.1 equiv of Et_3SiH completed the reduction of the primary silyl ether via the silyloxonium ion, as shown in Scheme 2.23. This reaction tolerated the C—X bond and C=C double bonds and allowed for low BCF catalyst loadings (0.1 mol%). A 1,3-diol was also highlighted in the substrate scope, presumably following the same cyclic siloxane mechanism, but this substrate required a higher catalyst loading (5 mol%). The synthetic utility of this methodology was demonstrated in the four-step synthesis of (*R*)-lisofylline; it found further use in natural product synthesis and in biomass conversion, which was recently highlighted [120].

Morandi and coworkers expanded his 1,2-diol protocol to internal vicinal diols [121], which preferentially underwent a stereoselective pinacol-type rearrangement through a concerted mechanism (Scheme 2.24). In this formulation, the secondary silane Ph_2SiH_2 (1.0 equiv) is first added to form the 5-membered cyclic siloxane. Upon addition of the tertiary silane (Et_3SiH), the cyclic siloxane

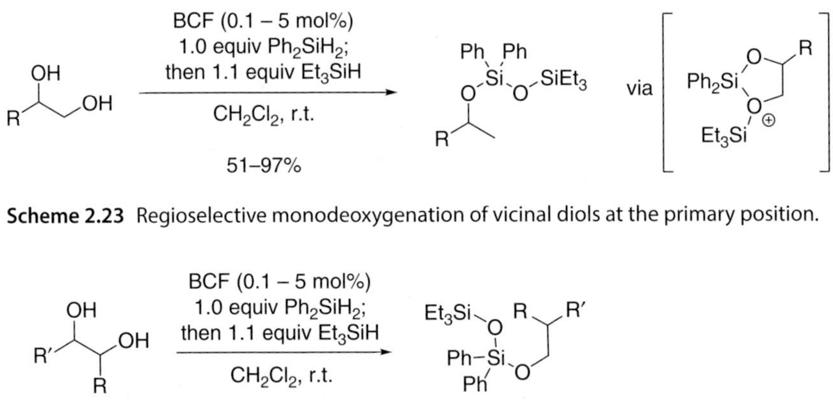

Scheme 2.23 Regioselective monodeoxygenation of vicinal diols at the primary position.

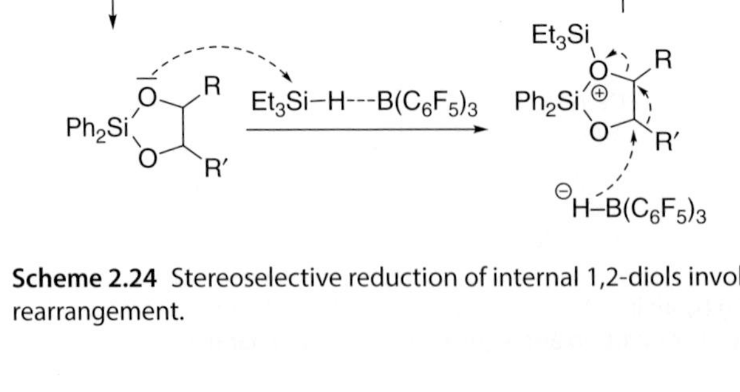

Scheme 2.24 Stereoselective reduction of internal 1,2-diols involving a pinacol-type rearrangement.

becomes silylated by the η[1]-[B]–H–[Si] adduct, which initiates an ionic rearrangement prior to cation reduction by the borohydride species. This illustrates an example where a C—C σ bond traps a silyloxonium ion via a concerted stereoinvertive carbocation migration/hydride transfer. A variety of internal diols were stereoselectively reduced to secondary and primary silylated alcohols and cyclic 1,2-diols had the ability to undergo stereoselective ring expansions and contractions due to the concerted nature of the alkyl migration.

The strong electrophilic character of intermediate silyloxonium ions was evident in the abovementioned cyclizations involving neighboring silyl ether groups or in pinacol-type alkyl migrations. Intramolecular addition of alkenes has also enabled the diastereoselective reductive carbocyclization of unsaturated carbohydrates, yielding cyclopropanes with polyhydroxylated side chains (Scheme 2.25) and vicinal dihydroxylated cyclopentanes (Scheme 2.26) [122].

The C—C bond-forming step is facilitated by an initial condensative THF cyclization. Silylation of this now more basic cyclic oxygen sets up a favorable alkene trapping path that leads to carbocyclization prior to reduction of the benzyl cation by the hydride counterion.

The formation of cyclopentane products requires a different mechanistic rationale since a methyl group is generated. After condensative THF formation, an invertive C—C bond migration leads to an *O*-silyloxocarbenium ion intermediate, which is further reduced by the BCF/silane system to the observed methyl group; deuterium labeling experiments with Et_3SiD were consistent with this proposal. Once this branched triol has been formed, electrophilic silylation of the primary silyl ether group initiates carbocyclization with the alkene tether prior to hydride reduction of the benzyl cation.

Exposing the same silylated alkenyl pentaol exclusively to Lewis acidic conditions (10 mol% BCF without any silane) leads to vinyl-tetrahydropyran (THP), which could be further reduced by a BCF/silane system to an alkenic tetraol (Scheme 2.27).

This reactivity could also be achieved by sequential addition of catalytic amounts of BCF followed by 1.1 equiv of Et_3SiH, and highlights the different roles BCF can fulfill depending on reaction conditions: (i) BCF acts as a Lewis

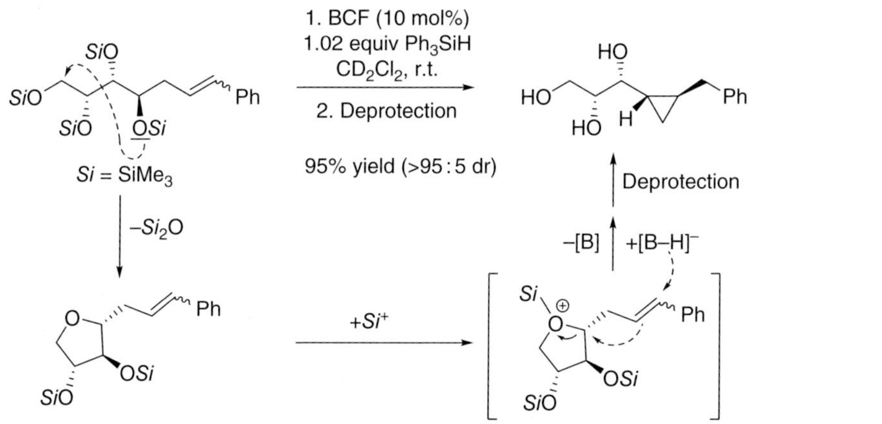

Scheme 2.25 Alkene trapping by an intermediate silaoxonium ion yielding cyclopropanes.

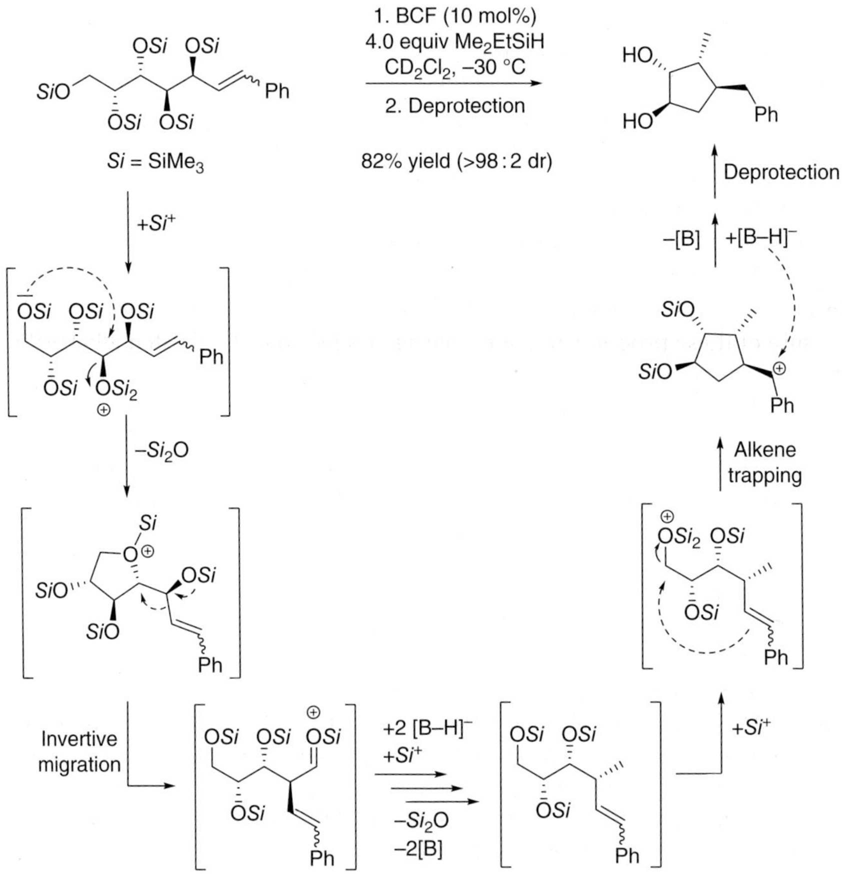

Scheme 2.26 Mechanism involving C—C migration and alkene trapping by a silaoxonium ion yielding cyclopentanes.

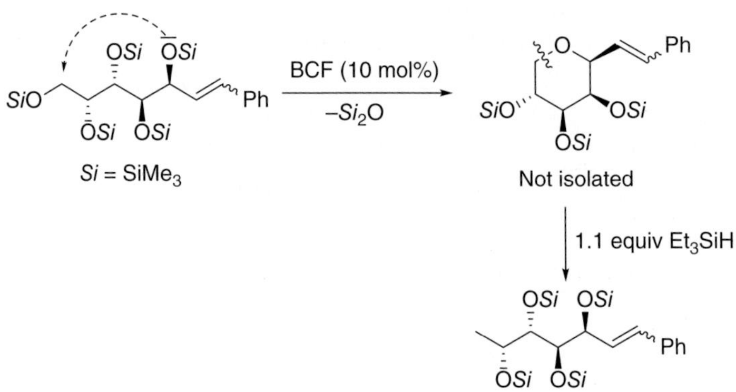

Scheme 2.27 Condensative vinyl-THP under BCF Lewis acidic conditions.

acid during THP formation; (ii) upon addition of the silane, BCF acts as a silane activator; and (iii) upon formation of its conjugate hydride $[BCF–H]^-$, it then becomes the reducing agent. With an increased understanding of the tunability of type I and type II reduction sequences and of the multiple roles of BCF in BCF/silane systems, Gagné achieved selective and controllable functional group modifications in a variety of bioactive natural compounds [123].

The ability of BCF/silane systems to mediate reductive C—C bond-forming ring closures of alkenes was further highlighted in reports by Chang and coworkers [124]. Furans were first catalytically ring opened by combining 2 mol% BCF and $PhMe_2SiH$ (2.05 equiv) to yield α-silyloxy-(*Z*)-alkenyl silanes. Interestingly, exposure of these products to additional silanes led to *anti*-silylated alkylcyclopropanes (Scheme 2.28).

Mechanistic studies were in agreement with a nucleophilic vinylic substitution (S_NV) pathway [125] for the formation of the alkene, which then formed the cyclopropane via an intramolecular S_N2' reaction induced by nucleophilic attack of the borohydride species. Furans are considered a chemical route for the transformation of biomass into chemicals as they are readily available from the condensative elimination of carbohydrates [126]; this study additionally highlights the broad synthetic utility of silylated products obtained from furans.

Oestreich and coworkers further enhanced the synthetic applicability of BCF/silane combinations through the demonstration that primary and secondary tosylates or triflates could be efficiently defunctionalized [127]. The use of 10 mol% BCF and a slight excess of Et_3SiH in CH_2Cl_2 at room temperature defunctionalized tosylates chemoselectively in the presence of bromides, silyl-protected alcohols (OTBDMS, $OSiEt_2Me$), and aryl ethers (Scheme 2.29.). This reaction also tolerated the presence of C=C double bonds rendering a straightforward deoxygenation, which could be used in natural product synthesis.

While studying the defunctionalization of bromides, tosylates, and triflates, Oestreich reported the intermediacy of phenonium ions, which occurred with aromatic 1,2- and 1,3-diols (Scheme 2.30). Aromatic systems trapped silyloxonium ions, leading to the formation of new C—C bonds. In contrast, aliphatic 1,2-diols lacked this mechanistic possibility, and instead the neighboring silyl ether group acts as the trapping nucleophile to generate an intermediate silyl epoxonium ion, which provided little regiocontrol on the reaction outcome.

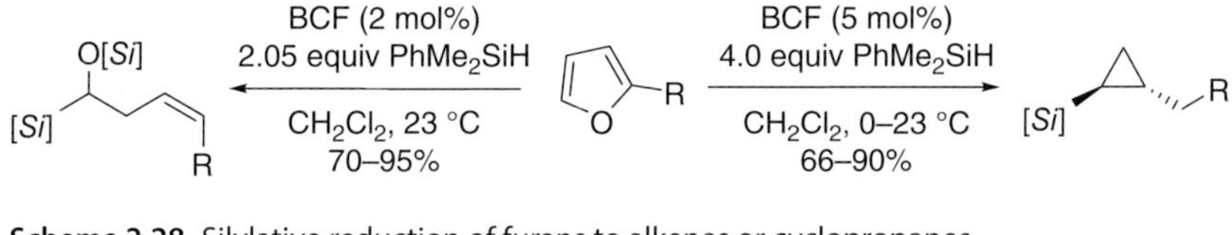

Scheme 2.28 Silylative reduction of furans to alkenes or cyclopropanes.

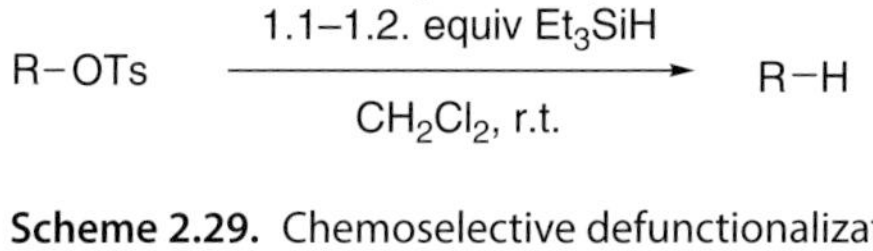

Scheme 2.29. Chemoselective defunctionalization of tosylates.

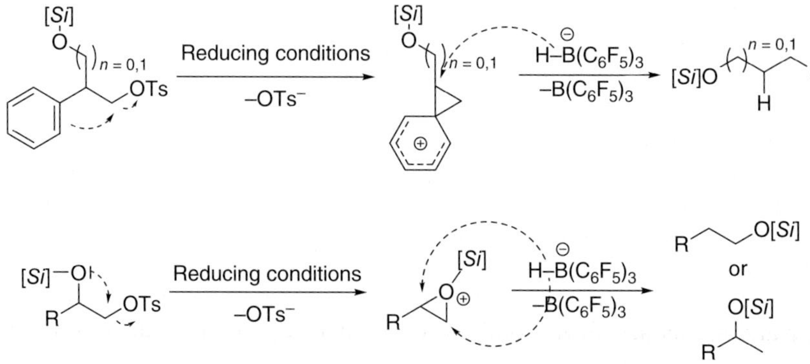

Scheme 2.30 Formation of intermediate phenoxonium and silylepoxonium ions in the defunctionalization of tosylates.

2.2.3.2.3 *Hydrodesulfuration of Thioethers (Silylative Desulfuration or Dehydrocarbonative Silylation)*

These reactions are basically the same as the deoxygenation reactions of ethers except that the sulfur atom is the nucleophile. Akiyama and coworkers reported the hydrodesulfuration (C—S bond cleavage) of sulfides using BCF and Et_3SiH [128]. This hydrodefunctionalization was efficient with secondary sulfides bearing vinylic, benzylic, and/or tertiary aliphatic groups (Scheme 2.31). No reaction was observed with primary benzylic, secondary aliphatic, or propargylic sulfides, but dithioacetals and dithioketals were successfully reduced to sulfides or hydrocarbons, respectively.

2.2.3.2.4 *Silylative Dehalogenation of Organohalides (Dehydrocarbonative Silylation)*

Stephan and a coworker reported an efficient BCF-catalyzed hydrodefluorination of C—F bonds using Et_3SiH [129]. Primary, secondary, and tertiary alkyl fluorides were quantitatively defunctionalized using 5 mol% BCF and 1.0 equiv of Et_3SiH (Scheme 2.32).

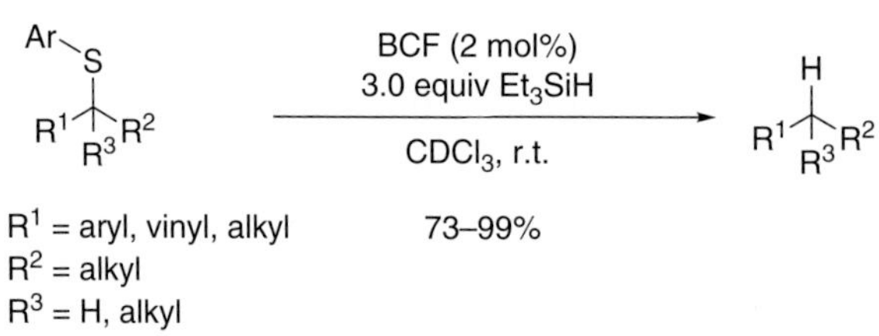

Scheme 2.31 Desulfuration of sulfides to hydrocarbons.

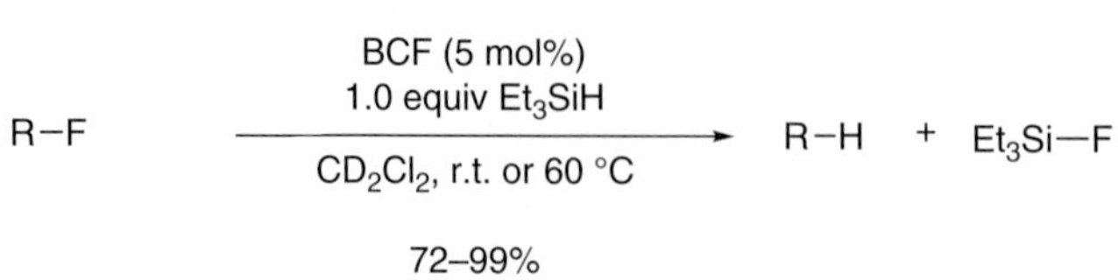

Scheme 2.32 Defluorination of organofluorines to hydrocarbons.

The presence of an ether functional group slowed the reaction presumably by acting as a competitive nucleophile for the η1-[B]–H–[Si] adduct or as a Lewis base for the BCF. Only monofluorinated carbons could be defunctionalized and CF_3 groups remained untouched under these reaction conditions. Different C—F bond activation modes are possible with Lewis acids [130], but in this case, even though not specifically proposed by Stephan, the organic fluoride may be the attacking nucleophile that generates the $[BCF–H]^-[R–F–SiR_3]^+$ reactive ion pair prior to a reductive recombination that reduces the C—F bond.

Oestreich and coworkers reported recently the silylative debromination of alkyl bromides [127]. While the corresponding alkyl chlorides are untouched, primary and secondary alkyl bromides were efficiently defunctionalized using BCF (10 mol%) and a slight excess of Et_3SiH (1.1–1.2 equiv) (Scheme 2.33). It stands to reason that silylbromonium ions are key intermediates.

2.2.3.3 Type III: The Reduction of Nonpolar π Bonds (A=A/A≡A)

Facilitated by the polarizability of the electron cloud, nonpolar π bonds (A=A/A≡A) may act as nucleophiles toward the electrophilic silicon atom of η1-[B]–H–[Si] (Scheme 2.34). The regioselectivity of these reductions is governed by the stability of the intermediary cations and by the steric bulk of the silane.

The recent advances in the catalysis of organosilicon compounds by transition-metal complexes significantly enhance the synthetic utility of methods that generate Si—C bonds [63]. Type III reductions are thus inherently valuable as they provide an additional avenue for the synthesis of organosilicon reagents.

2.2.3.3.1 Hydrosilylation of Alkenes and Alkynes

Gevorgyan and coworkers applied a variety of silanes in the BCF-catalyzed hydrosilylation of mono-, di-, and trisubstituted alkenes [131]. This *trans*-addition of silanes exhibited anti-Markovnikov regioselectivity, and the hydrosilylation products of aromatic silanes could be successfully converted to alcohols with stereoretention using the Tamao–Fleming oxidation (Scheme 2.35) [132].

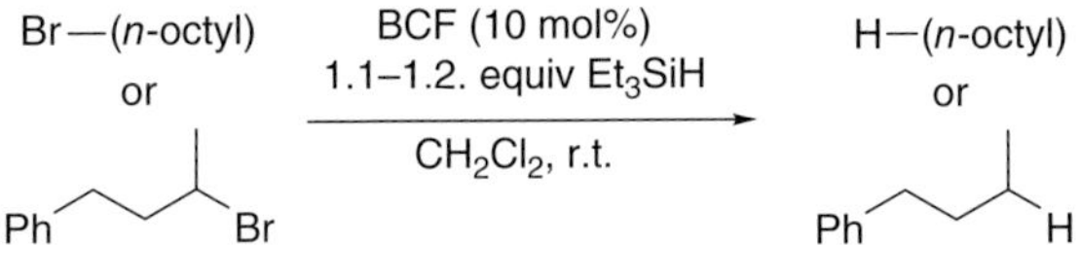

Scheme 2.33 Debromination of alkylbromides to alkanes.

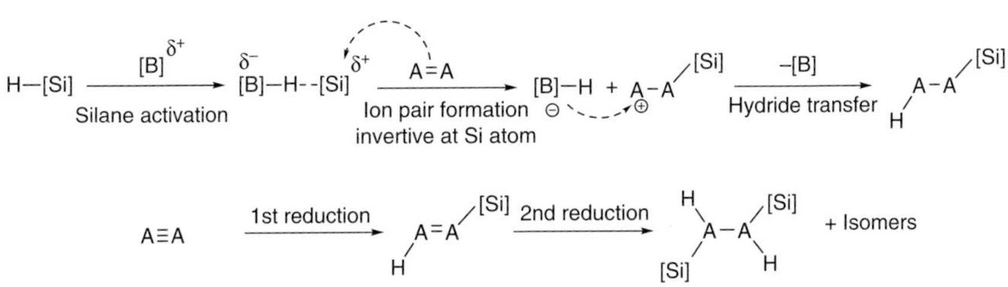

Scheme 2.34 The mechanistic steps and characteristics of type III reductions.

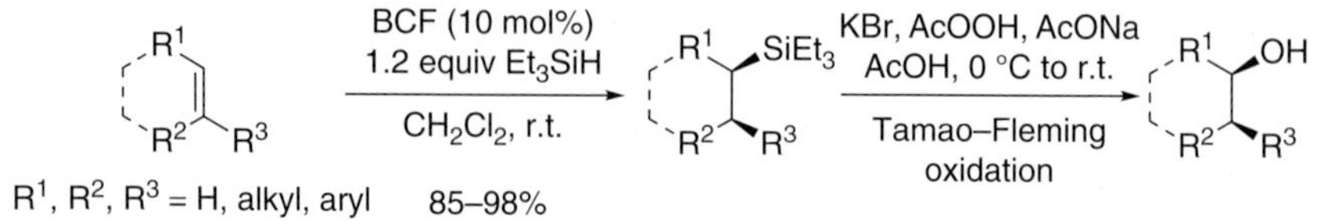

Scheme 2.35 Hydrosilylation of alkenes with subsequent Tamao–Fleming oxidation to alcohols.

Based on the Piers mechanism that has η^1-[B]–H–[Si] acting as a silylium source, Gevorgyan proposed that addition of silicon to the C=C double bond would generate a β-silicon-stabilized carbocation (via anti-Markovnikov silylation), which could subsequently be quenched by nucleophilic attack of the borohydride counterion from the less hindered side (resulting in *anti*-addition), all within an ion pair. The authors specifically stated that a free silylium cation should not be considered and deuterium labeling experiments ruled out the possibility of competitive hydride delivery by silanes.

While studying the reduction of enones, Piers observed that the stereospecific hydrosilylation of silyl enol ethers occurred if excess silane was used. Specifically, this substrate class was hydrosilylated to β-silyloxyalkylsilanes, with the stereo-defining step being the *anti*-addition of $[BCF–H]^-$ leading to *syn* products (Scheme 2.36).

There are also reports on the hydrosilylation of alkynes; however, the conversions were poor unless the alkyne was bonded to an EN atom or in conjugation with other π bonds. For example, Ingleson reported the *trans*-hydrosilylation of 1-phenylprop-1-yne with Ph_2SiH_2 (1.0 equiv) and 5 mol% BCF, yielding primarily *Z*-alkene (84 : 16 *Z*:*E*) (Scheme 2.37) [133]. Other phenylacetylene derivatives were also hydrosilylated in good spectral (*in situ*) yields, but the hydrosilylation of Ph—C≡C—Ph did not proceed under these conditions.

While studying the efficiency of his silane surrogates, Oestreich and coworkers reported that those conditions could indeed enable the hydrosilylation of Ph—C≡C—Ph [22]. Using the BCF analog $B(p\text{-H-}C_6F_4)_3$ as the catalyst and Et_3SiH as the reducing agent, 13% of the hydrosilylated product was observed after 48 hours at elevated temperatures (80 °C in benzene). Interestingly, the

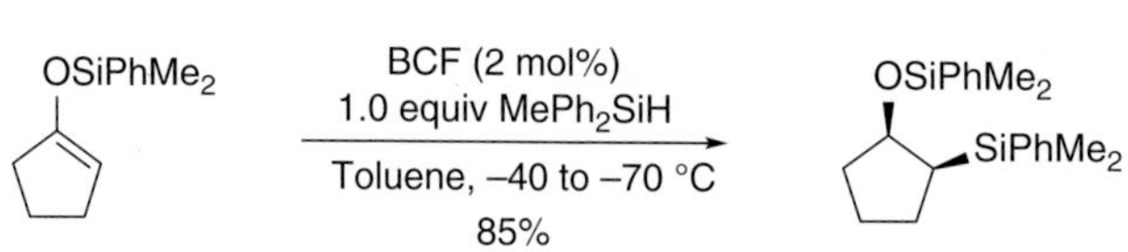

Scheme 2.36 Hydrosilylation of silyl enol ethers.

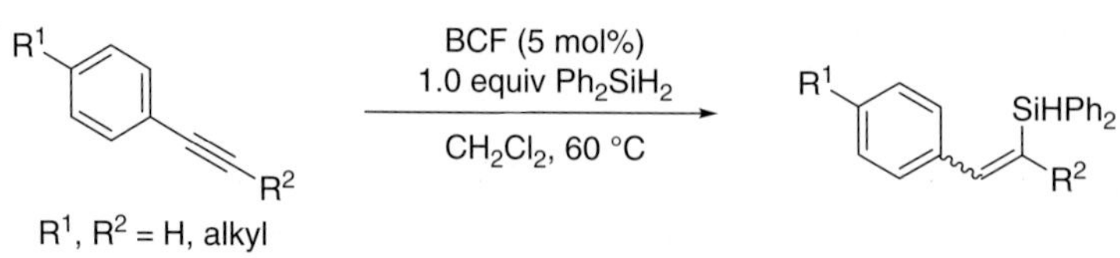

Scheme 2.37 Hydrosilylation of aromatic alkynes.

combination of the same catalyst with either a 1,4-cyclohexadiene silane surrogate (91% in CH_2Cl_2 at room temperature) or the BCF/Et_3SiH system (90% in benzene at 80 °C) was more successful in that particular reduction.

In contrast, Chang and coworkers demonstrated that the C≡C triple bonds of a variety of ynamides were regioselectively hydrosilylated by Et_2SiH_2 to *syn* silylated enamides in good yields using BCF catalysis (Scheme 2.38) [134]. Diphenylsilane and some monohydrosilanes also promote efficient conversions. The high regio- and stereoselectivity was explained by the initial formation of a silylated ketene iminium intermediate stabilized by the β-silicon effect, followed by hydride transfer from the borohydride species from the sterically less hindered side (*anti*-addition) rendering this reaction a net type III reduction.

The formation of silacycles when dienes, enones, enynes, ynones, or ene-imines were reacted with dihydrosilanes, bis-silanes, or disilanes under BCF catalytic conditions was first reported by Dussault and coworkers [135] and later expanded by Chang and coworkers [136]. These silacycles were then converted to 1,3-, 1,4-, 1,5-, or 1,6-diols or 1,4-aminoalcohols via the Tamao–Fleming oxidation (Scheme 2.39).

Depending on the substrate functionality, these reactions either featured a sequence of two type III or type I/type III hydrosilylative reductions. Especially interesting is the formation of seven-membered sila-cycles, which typically compete with intermolecular oligomerization processes.

In 2015, Chang and coworkers reported the application of BCF-catalyzed alkene hydrosilylation reactions for the synthesis of silicon-containing polymers [137]. Linear alternating polycarbosilanes could be generated by the copolymerization of dienes and aromatic disilanes or dihydrosilanes (Scheme 2.40). A related process was recently investigated computationally [138].

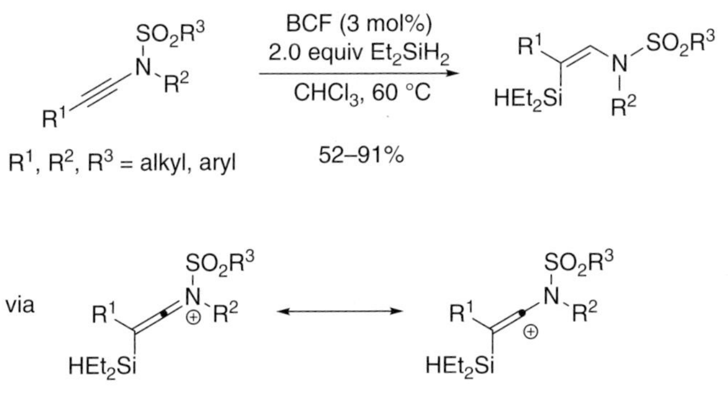

Scheme 2.38 Hydrosilylation of ynamides.

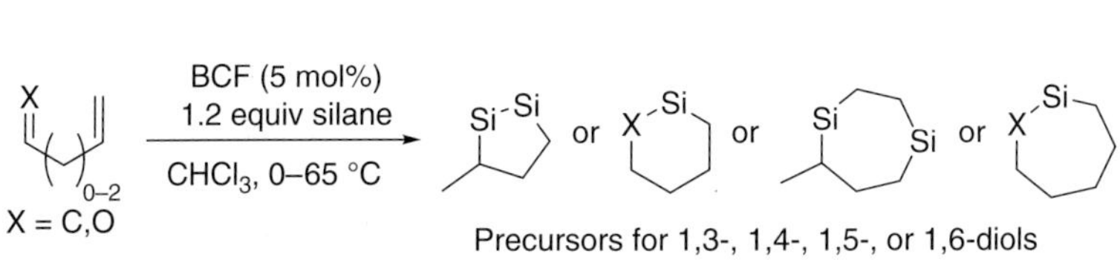

Scheme 2.39 Double hydrosilylation of polar and nonpolar π-bonds.

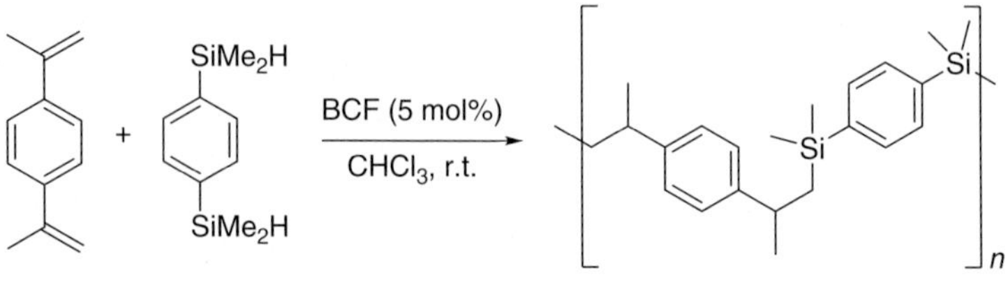

Scheme 2.40 Type III hydrosilylation in the synthesis of silicon polymers.

Another type III reduction was reported by Oestreich and coworkers in which silane surrogates [139] were used for the hydrosilylation of unactivated alkenes [140]. The alkene was activated for hydrosilylation by Pt and the resulting cyclohexa-2,5-diene-protected monohydrosilanes were converted to trihydrosilanes and benzene by BCF.

2.2.3.4 Type IV: The Reduction of Nonpolar σ Bonds (A—A)

Nonpolar σ bonds (A—A) can be polarized to react with the electrophilic silicon atom of the η^1-[B]–H–[Si] adduct (Scheme 2.41). The mechanistic steps for the reduction of polar and nonpolar σ bonds are the same if atom A has a lone pair or a neighboring π bond. However, a lack of this electronic support for atom A would require simultaneous nucleophilic attack and σ bond cleavage. And just as in the reduction of nonpolar π bonds, the regioselectivity of silylation versus hydrogenation is primarily governed by cation stability and steric aspects of the silane. Examples for the reduction of nonpolar σ bonds assisted by lone pairs or neighboring π bonds have been reported and reaction classes are discussed later.

2.2.3.4.1 *Sila-Friedel–Crafts-Type Reactions*

In these reactions, nonpolar aromatic C—H σ bonds are cleaved to yield silylated aromatic rings. The mechanism for these processes is analogous to electrophilic aromatic substations (S_EAr), in which the electrophilic silicon of the η^1-[B]–H–[Si] adduct acts as the electrophile for the nucleophilic π system of the aromatic ring. These reactions are the most prominent type IV reductions found in the literature.

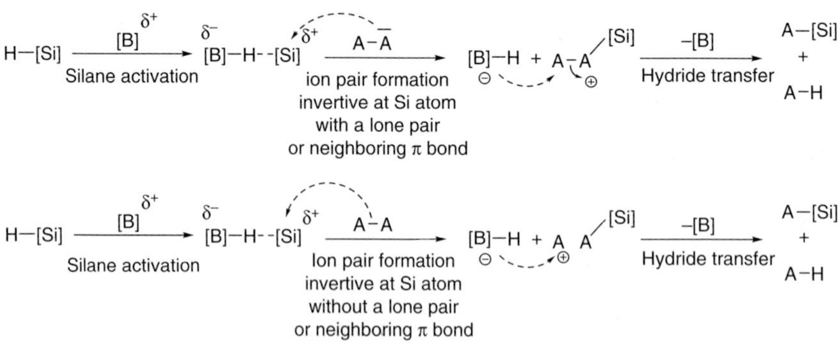

Scheme 2.41 The mechanistic steps and characteristics of type IV reductions.

As recently reviewed, the direct C—H silylation of aromatic rings using silanes and a borane catalyst was plagued by side reactions [141]. However, in 2014, Ingleson and a coworker reported the synthesis of siloles, specifically silafluorenes and silaindenes from *ortho*-silylated biphenyls and phenylacetylene derivatives using catalytic amounts of BCF and catalytic or stoichiometric quantities of 2,6-dichloropyridine, respectively (Scheme 2.42) [133]. The deliberate addition of catalytic quantities of the pyridine base facilitates deprotonation of silylated arenium ion intermediates, which reduces competitive side reactions [142].

Silafluorenes were formed via an intramolecular dehydrogenative sila-Friedel–Craft aromatic substitution in which the *in situ* formed silylium ion activated the neighboring aromatic ring. Silaindenes were formed via a two-step process. First, the aforementioned *trans*-hydrosilylation of phenylacetylene derivatives (type III reductions) to silylated *Z*-alkenes was realized, which then underwent the intramolecular dehydrogenative sila-Friedel–Crafts cyclization.

Hou and coworkers reported the aromatic dehydrogenative silylation of anilines yielding arylsilanes (Scheme 2.43) [143]. These sila-Friedel–Crafts reactions did not require the deliberate addition of a base, but the tertiary amine functionality found on the substrate likely played a similar role. BCF was used as a catalyst, and a variety of silanes (3.0 equiv) could be employed using chlorobenzene as the solvent. Even chlorosilanes were tolerated, but substrates bearing silyl ethers (OTIPS) and alkenes were reduced to hydrocarbons.

As an extension of their work, Zhang and coworkers reported recently the selective cross-metathesis of C—Si/Si—H bonds, leading to a variety of silylated aromatic compounds (Scheme 2.44) [144]. A variety of aromatic monohydrosilanes were successfully converted to Me_2Ar_2Si or to $Me_2ArR'Si$ driven by the formation of the volatile by-product Me_2SiH_2. *Meta*- or *para*-substitution of one aromatic ring with a tertiary amine was key for the success of most metathesis reactions, but indolyl and ferrocenyl substituents were also tolerated.

An interesting reactivity of *N*-methyl indoles with the BCF/silane system was recently reported by Zhang and coworkers [145]. The hydrosilylation of an aromatic C=C double bond of indoles was observed when Ph_2SiH_2 was activated by

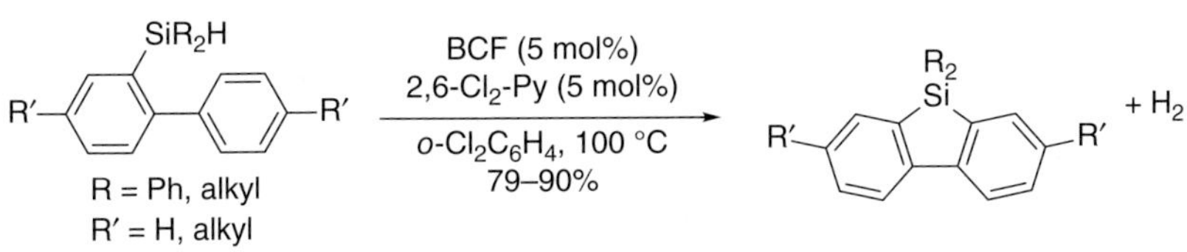

Scheme 2.42 Synthesis of silafluorenes and silaindenes.

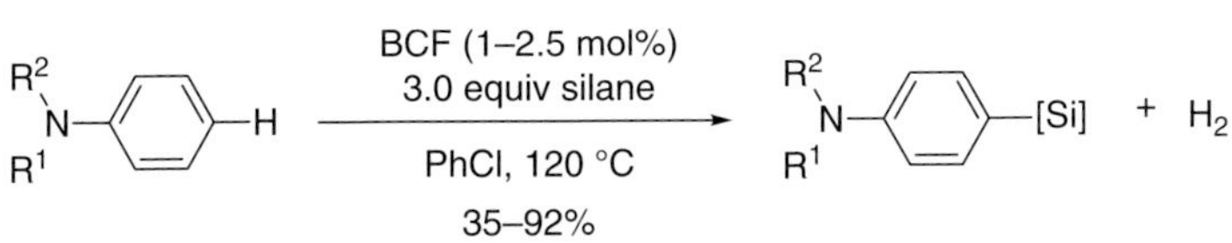

Scheme 2.43 Dehydrogenative silylation of anilines.

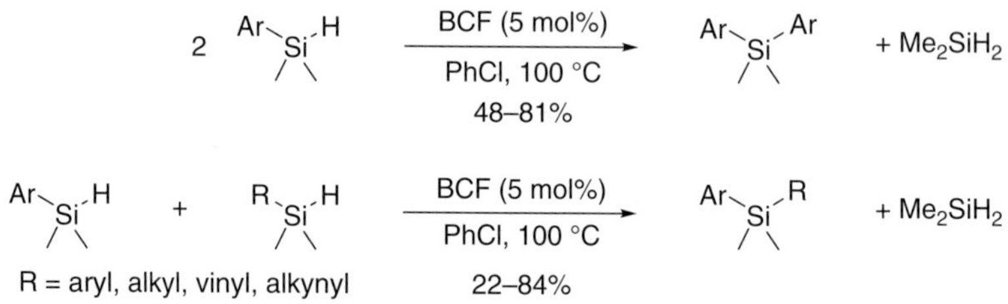

Scheme 2.44 Selective cross-metathesis of C—Si/Si—H bonds.

1 mol% BCF at room temperature in benzene-d6, but indolines (hydrogenated indoles) were obtained as by-products (Scheme 2.45).

This disproportionation reaction was explained by an interesting mechanism. After the electrophilic silicon of the η^1-[B]–H–[Si] adduct was attacked by the nucleophilic C=C double bond of indole, a C3-silylated indolinium ion that would in turn transfer a proton to another indole molecule (driven by rearomatization) was obtained, leading to a C3-silylated indole and a C3-protonated indolinium ion (Scheme 2.46). This completes a sila-Friedel–Crafts-type electrophilic substitution reaction (type IV). The iminium ion of the latter could further react with the borohydride species to form indoline via a type I reduction.

When the reaction mixture was heated to 120 °C, the dehydrogenation of the indoline by-product back to indole was noted, which allowed the design of convergent disproportionation processes using *N*-methyl indole derivatives, in which the formation and subsequent dehydrogenation of indoline was catalytic (Scheme 2.47). C3-silylated indoles were obtained in excellent yields and low BCF catalyst loading (0.01 mol%) could be used. An unexpected reduction of the

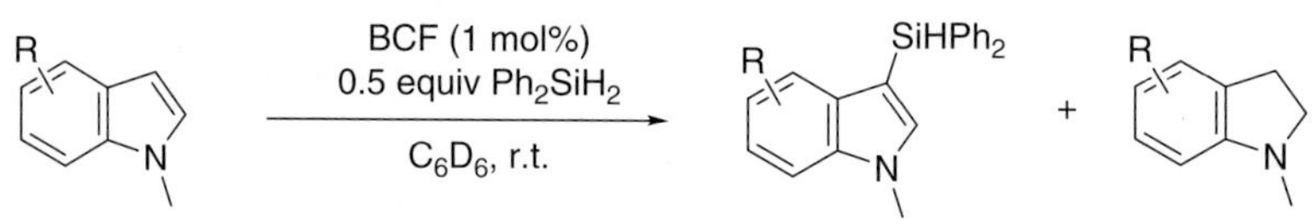

Scheme 2.45 Disproportionation of *N*-methyl indoles to indolines.

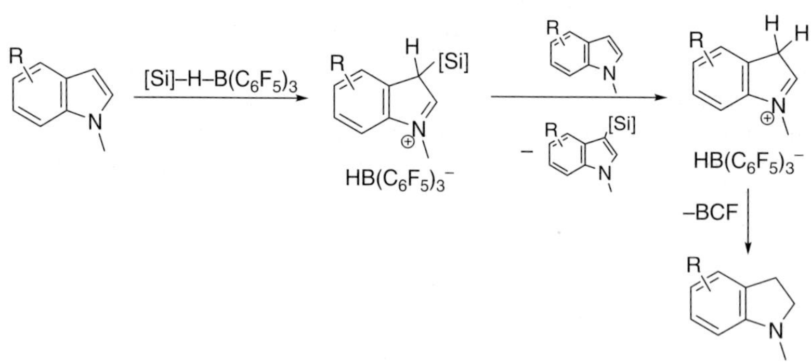

Scheme 2.46 Plausible mechanism for the formation of indolenes containing a type IV/type I reduction sequence.

Scheme 2.47 C3-silylation of indole derivatives.

indole double bond was also reported during studies of BCF/silane reductions with the natural compound mitragynine by Beng et al. [146].

Nucleophilic attack of nonpolar σ bonds to η^1-[Si]–H–[B] adducts without assisting lone pairs or neighboring π bonds has, to our knowledge, not been reported. Such reactions would include C—C single bond cleavage reactions or aliphatic C—H bond activations.

2.2.3.5 Combination of Reduction Types

BCF-catalyzed defunctionalization reactions with silanes have received considerable attention as they display potential applications in synthesis. Depending on their complexity, some examples involve multiple steps of the same or different reduction type, multiple polar π and σ bonds can be cleaved in one pot by sequential reactions with the key η^1-[B]–H–[Si] adduct. In this section, some of those examples are highlighted.

Shortly after Piers showcased the BCF-catalyzed activation of Ph_3SiH, Gevorgyan and Yamamoto and coworkers developed a protocol employing Et_3SiH [147, 148]. This method expanded the substrate scope to a series of carbonyl compounds, where aliphatic aldehydes, carboxylic acids, acid chlorides, and esters were reduced either to silyl ethers or even to alkanes depending on the structure of the substrate and the amount of silane used (Scheme 2.48).

These reductions cleave polar π and polar σ bonds and thus exhibit a mix of type I and type II reductions, the latter being dependent on the substrate. For example, a dehydrogenative silylation would be initially observed with carboxylic acids (type II) followed by cleavage of the π bond (type I). After two dehydrocarbonative silylations of silyl ethers (type II), the reduction sequence to the alkane is finalized.

Another pioneer in BCF-catalyzed reductions using silanes is Chandrasekhar. He reported the rapid defunctionalization of ketones using PMHS as the reducing agent (Scheme 2.49) [149].

A variety of aromatic, aliphatic, and conjugated ketones were reduced to the corresponding hydrocarbons in excellent yields at 5 mol% BCF catalyst loading using 3.0 equiv of PMHS. No reduction was observed with primary amides.

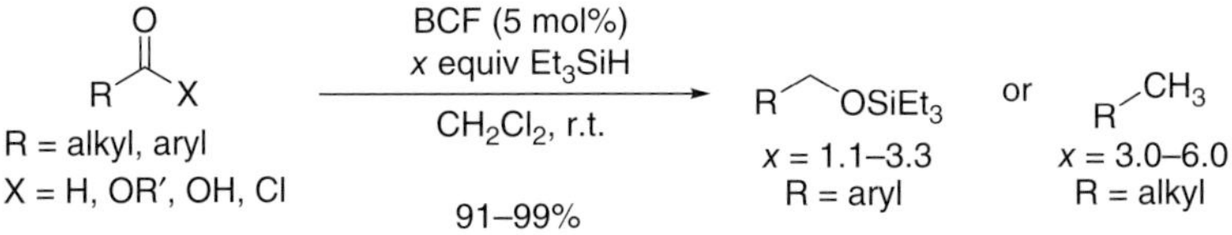

Scheme 2.48 Reduction of carbonyl compounds to alcohols and alkanes.

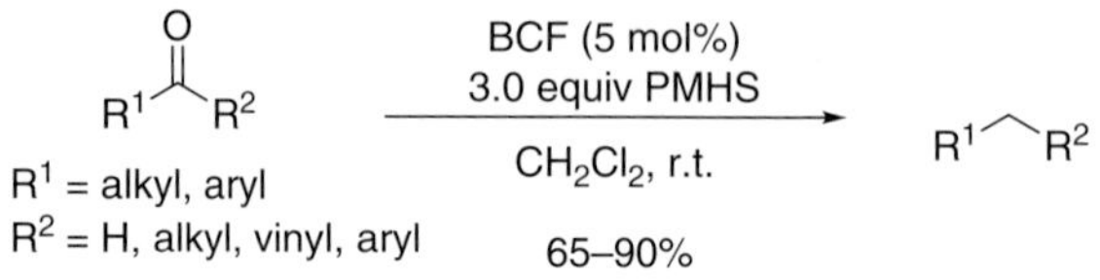

Scheme 2.49 Reductions of ketones and aldehydes to hydrocarbons using PMHS.

These reductions follow a type I (hydrosilylation of the carbonyl)/type II (dehydrocarbonative silylation of the silyl ether) sequence.

In 2015, Brookhart and coworkers demonstrated a selective reduction of carboxylic acids to disilyl acetals, which could be readily converted to the aldehyde with an acidic aqueous workup (Scheme 2.50) [150]. A wide range of aromatic and aliphatic acids were efficiently converted to their corresponding aldehydes following a type II (dehydrogenative silylation of the hydroxyl group)/type I (hydrosilylation of the carbonyl group) reduction sequence.

It should be noted that the identity of the reducing silane significantly affected the product distribution (disilyl acetals, silyl ethers, and over-reduced hydrocarbon). For example, when Et_3SiH was used in the reduction of hydrocinnamic acid, a selective reduction to disilyl acetals (>99%) was observed, whereas diphenylsilane gave primarily the silyl ether product (76%), and TMDS produced the hydrocarbon (>99%).

In 2014, Cantat and coworkers demonstrated that oxalic acid, which can be generated by the reductive C—C coupling of two CO_2 molecules, could be converted to ethane using PMHS or TMDS (Scheme 2.51) [151]. In addition, a chemoselective reduction to silylated 2-oxoacetic acid (glyoxolic acid, top left), glyoxal (bottom left), glycolic acid (not shown), and glycol aldehyde (top right) was demonstrated depending on solvent choice (CH_2Cl_2 or benzene) and silane

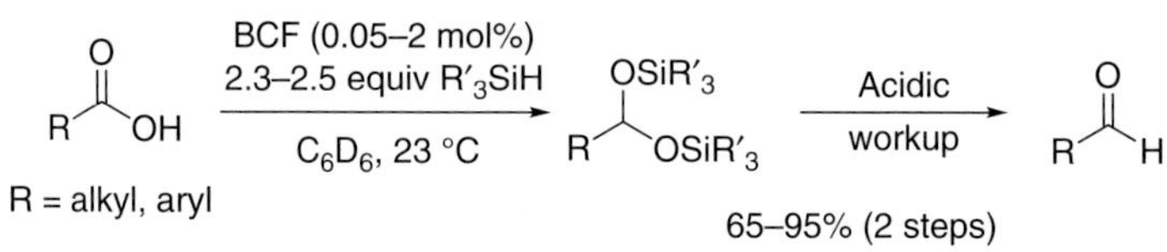

Scheme 2.50 Reduction of carboxylic acids to aldehydes.

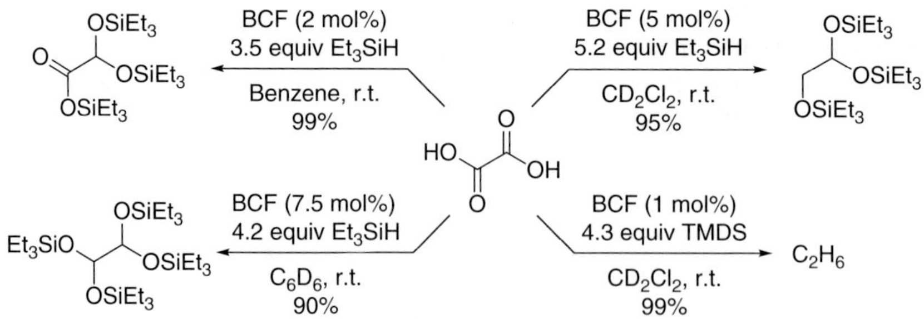

Scheme 2.51 Selective reductions of oxalic acid to reduction products at different oxidation stages.

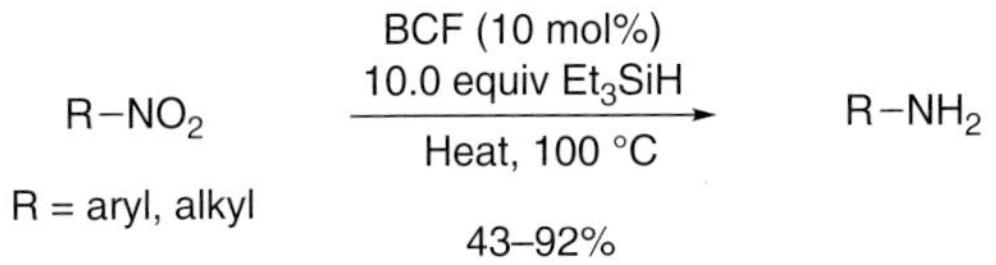

Scheme 2.52 Reduction of nitrated hydrocarbons to amines.

equivalents. These reductions follow several type II/type I reduction sequences depending on the product formed, and thus illustrates, yet again, the fine tunability of BCF-catalyzed reductive silylations through alterations in catalyst loading, solvent, and silane.

A BCF-catalyzed reduction of simple aromatic and aliphatic NO_2 groups with Et_3SiH was reported recently by Oestreich and a coworker [152]. Solvent-free conditions and high temperatures were key to realizing this reaction (Scheme 2.52). In some cases, competitive hydrodehalogenation reactions or the partial reduction of aromatic systems was observed. These reactions are comprised of a series of type I and type II reductions; however, the mechanism was not discussed in the paper and the order of π and σ bond reductions is unclear. The same set of reaction conditions could also be used for the reduction of sulfones and sulfoxides to sulfides (Scheme 2.53) [153].

Another type I/type II reduction sequence leading to the deoxygenation of sulfoxides and tertiary amine N-oxides was reported by Shi and coworkers (Scheme 2.54) [154]. Sulfides and amines were obtained using $PhSiH_3$ and BCF as the catalyst.

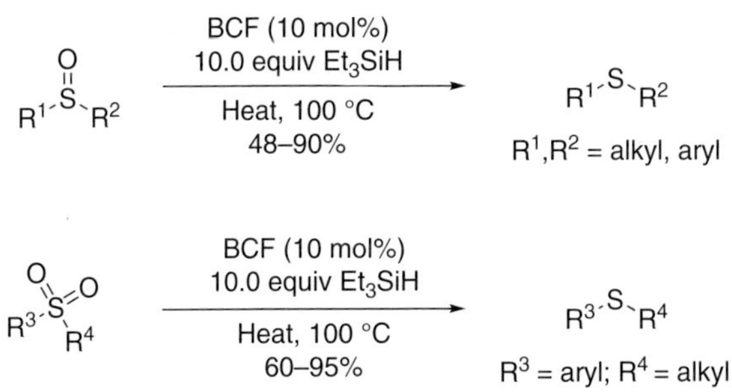

Scheme 2.53 Reduction of sulfones and sulfoxides to sulfides.

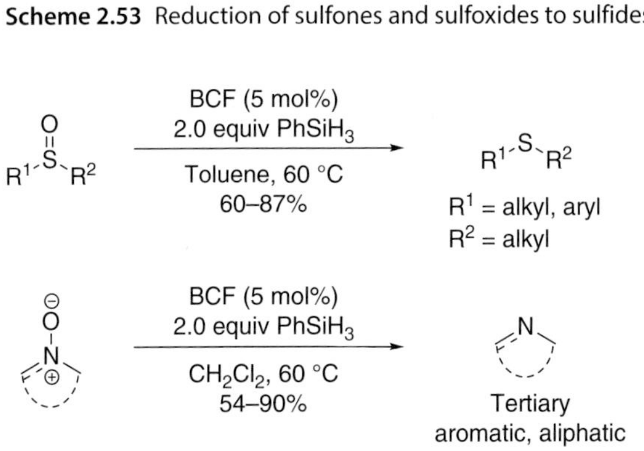

Scheme 2.54 Reduction of sulfoxides and N-oxides to sulfides and amines, respectively.

For the former, the authors proposed a type I/type II sequence analogous to the deoxygenation of carbonyl groups, with a type I hydrosilylation of the S=O π bond followed by the silylative hydrodesulfuration of the S—O—[Si] σ bond (type II). An acid–base reaction of the R_2SH^+ with the borohydride species liberating H_2 completes the catalytic cycle.

The reduction of tertiary amine N-oxides was proposed by Doris and coworkers to follow a different mechanism [155]. First, the type I hydrosilylation of the aromatic N-oxide ($C{=}N^+{-}O^-$) to break the C=N π bond by hydride transfer and form the CH—N—O—[Si] species, rearomatization of which leads to the tertiary amine and the silanol by-product (Scheme 2.55).

The reduction of organophosphorus compounds by Denis et al. showed that the reduction products were highly dependent on the silane used [156]. With tertiary silanes, only silylative C—O bond cleavage (reduction type II) of the corresponding phosphonates was observed (Scheme 2.56). More aggressive secondary or primary silanes complete the reduction to phosphines in fair yields (25–95%) at room temperature for both phosphonates and phosphinates.

An efficient deoxygenation of tertiary, secondary, and primary amides yielding amines was reported by Cantat and a coworker using 5–10 mol% BCF catalyst combined with PMHS (4.0 equiv) or TMDS (2.0 equiv) (Scheme 2.57) [157]. Key for successful reduction was heating toluene solutions for 18 hours at 100 °C. While this protocol worked well for tertiary and secondary amides, primary

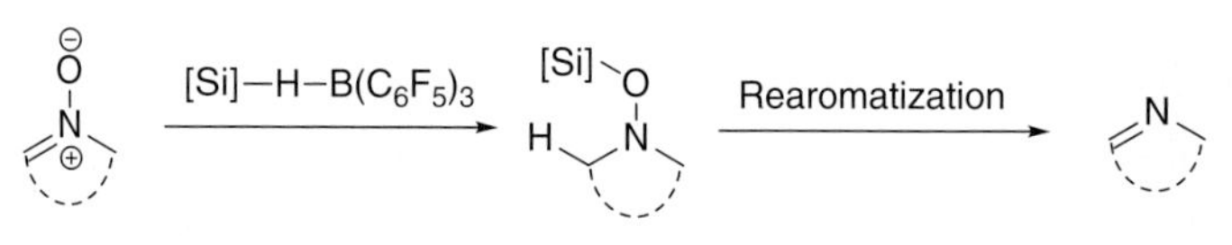

Scheme 2.55 Reduction of tertiary amine N-oxides.

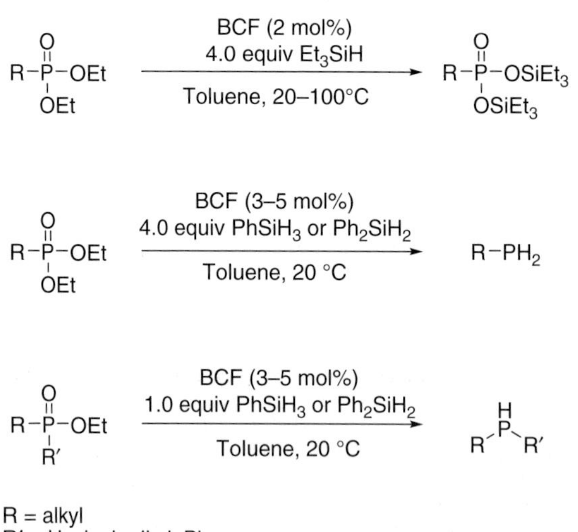

Scheme 2.56 Reduction of phosphonates and phosphinates to phosphines.

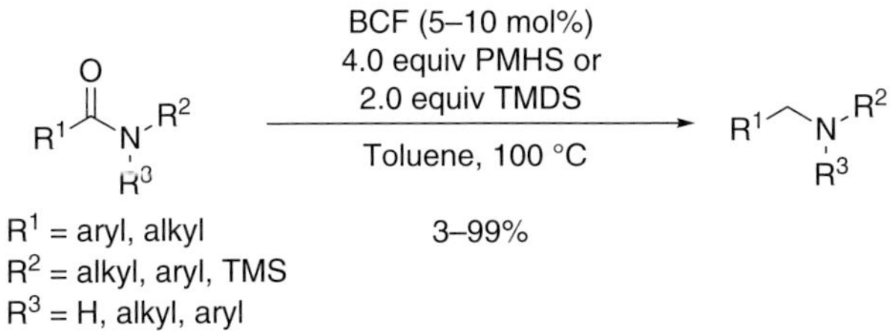

Scheme 2.57 Deoxygenation of amides to amines.

amides required a preliminary treatment with TMSCl before reduction to avoid the formation of product mixtures via reductive coupling reactions [158, 159]. A similar protocol for amide reduction was reported by Adronov and coworkers [160]; and by changing the catalytic borane species, a chemoselective amide reduction was achieved at room temperature [161].

Zhang and coworkers showed that cyclic imides were efficiently reduced to pyrrolidines by BCF/silane combinations [162]. The combination of $PhSiH_3$ (3.0 equiv) and 5 mol% BCF in dioxane at 110 °C converted aromatic, aliphatic, and polycyclic imides to the corresponding pyrrolidines, and followed a sequence of type I and type II reductions (Scheme 2.58). Alkenes, halogens, nitro groups, and heterocyclic compounds were untouched.

Okuda and coworkers showed that BPh_3 could also be used for the hydrosilylative conversion of CO_2 to silylformates in highly polar, aprotic solvents like CH_3CN or CH_3NO_2 using Ph_2SiH_2 as the reductant [163]. Interestingly, only one of the two Si—H bonds is cleaved (Scheme 2.59). The reduction of CO_2 by silanes is thermodynamically favored by the formation of strong Si—O bonds [164].

The reduction of CO_2 by BCF/silane systems was previously reported by Piers and coworkers, but their protocol required catalytic amounts of both the FLP $[TMPH]^+$(2,2,6,6-tetramethylpiperidinium)/$[HB(C_6F_5)_3]^-$ and BCF, using Et_3SiH as the reductant [165, 166]. CO_2 was successfully converted to methane by this catalytic system featuring a cascade of type I and type II reductions (Scheme 2.60).

Falivene and coworkers reported a tandem catalysis approach for the selective reduction of CO_2 to CH_4 using a combination of $Al(C_6F_5)_3$ (ACF) and BCF

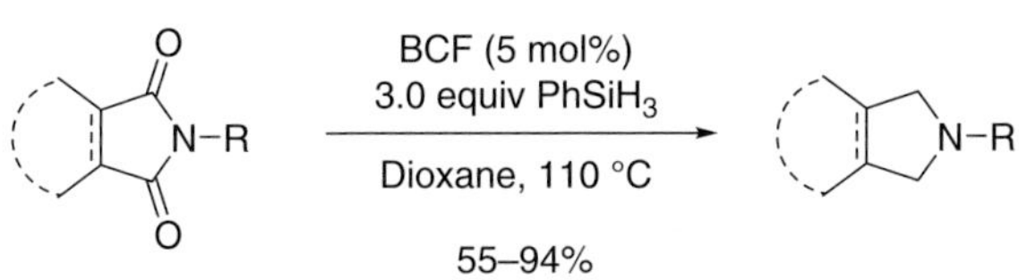

Scheme 2.58 Deoxygenation of imides to pyrrolidines.

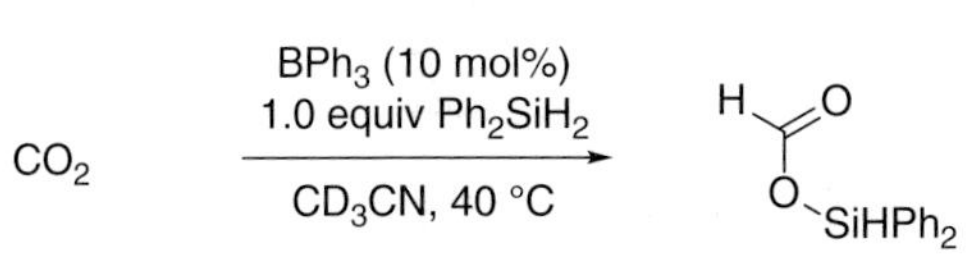

Scheme 2.59 Reduction of CO_2 to silylformates.

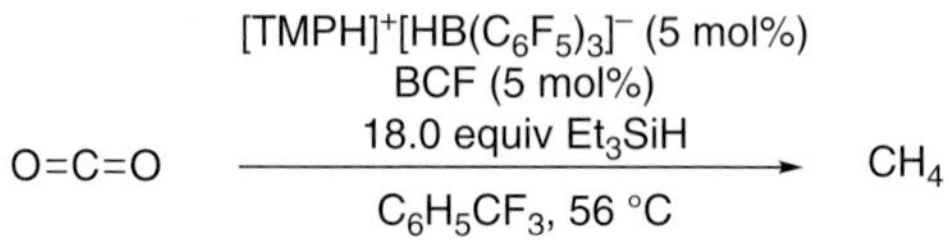

Scheme 2.60 Reduction of CO_2 to methane.

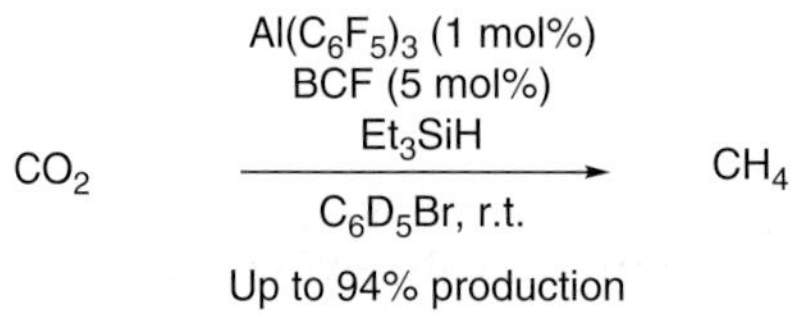

Scheme 2.61 Tandem catalytic reduction of CO_2 to methane.

catalysts with Et_3SiH as the reducing agent (Scheme 2.61) [167]. Detailed mechanistic studies pointed to an Al-catalyzed silylation of CO_2 to silyl formate, which was subsequently reduced in three B-catalyzed processes leading first to the disilyl ketal of formaldehyde, then a silyl methyl ether, and finally CH_4. A case was made that the catalysts operate synergistically and not individually in each reaction step. Other boron and other main-group element–catalyzed silylations of CO_2 were recently reported and reviewed [168, 169].

Cantat and a coworker demonstrated the silylative C—O bond cleavage of aryl alkyl ethers in lignin models [170] and lignin [171], leading to silylated aromatic alcohols. Diaryl ethers are untouched under these reaction conditions. Reviews on the reduction of C—O bonds in renewable feedstocks including CO_2 and biomass were recently published [108, 172]; computational investigations by Song and coworkers have also emerged [173].

Liu and coworkers demonstrated in 2015 that BCF-catalyzed hydrosilylation reactions could be used to methylate aromatic and aliphatic amines using CO_2 as the carbon source (Scheme 2.62) [174]; a transient formamide species was proposed. Replacement of the reaction solvent with polar dimethyl sulfoxide (DMSO) or dimethylformamide (DMF) by Chiang and coworkers realized an efficient catalyst-free version of this reaction [175]. Other methods were recently reviewed [176].

2.2.3.6 Mechanistic Variation of Reduction Types

2.2.3.6.1 Reductive Coupling Reactions

In 2005, Chandrashekar et al. reported the reductive etherification of carbonyl compounds using silyl ethers, BCF (1 mol%), and PMHS as the reducing agent

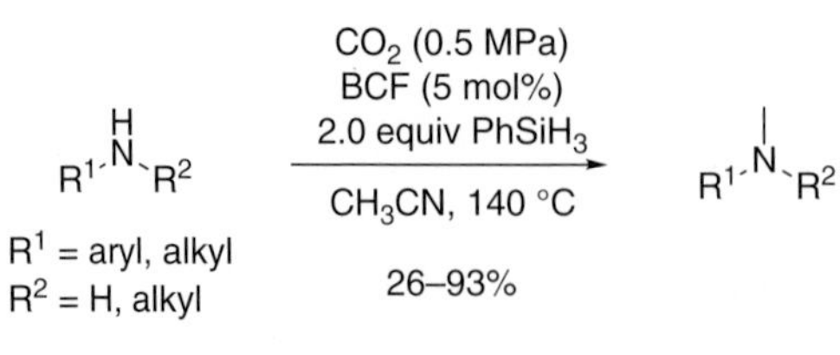

Scheme 2.62 Reductive methylation of amines with CO_2.

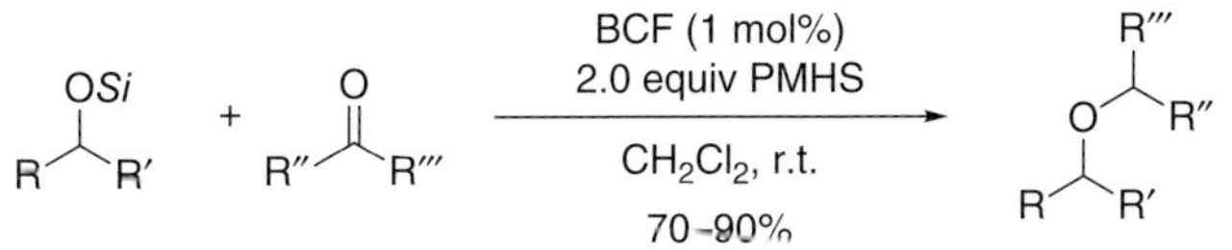

Scheme 2.63 Reductive etherification of carbonyl compounds.

under mild conditions (Scheme 2.63) [177]. A mechanism involving carbonyl activation by BCF was proposed.

In 2009, Chandrasekhar et al. reported the reductive alkylation of electron-rich arenes by aromatic or aliphatic aldehydes [178]. This coupling reaction yielded alkylated arenes by exposing the reactants to a BCF (5 mol%)/silane (PMHS, 3.0 equiv) combination in CH_2Cl_2 at room temperature (Scheme 2.64).

Even though the authors did not propose a reaction mechanism, two pathways are plausible. For example, nucleophilic trapping by the arene of the *in situ* formed silyloxonium ion (from the reaction of the η^1-[B]–H–[Si] adduct with the aldehyde) could occur. Subsequent dehydrogenative rearomatization of the Wheland complex by reaction with BCF–H^- would yield silylated benzyl alcohols, which can then be further deoxygenated via a type II reduction (dehydrocarbonative silylation) to alkyl-substituted arenes. A reversal of these steps would first reduce the aldehyde to provide silylated alcohols. These alkyl silyl ethers could be further silylated to provide disilyl oxonium cations, which could alkylate the electron-rich arene.

2.2.3.6.2 S_N2' Displacement of Silyloxonium Ions and Related Reactions

Chandrasekhar et al. reported the reductive dehydroxylation of Baylis–Hillman adducts leading to trisubstituted alkenes using catalytic BCF and PMHS as the reducing agent (Scheme 2.65) [179]. After a dehydrogenative silylation of the hydroxyl group, C—O bond cleavage by an S_N2' mechanism was observed (two type II reductions) to yield trisubstituted alkenes. This protocol benefited from a low BCF catalyst loading (0.5 mol%), short reaction times at room temperature, good chemoselectivity (nitro, ester, and TBS protecting groups are untouched),

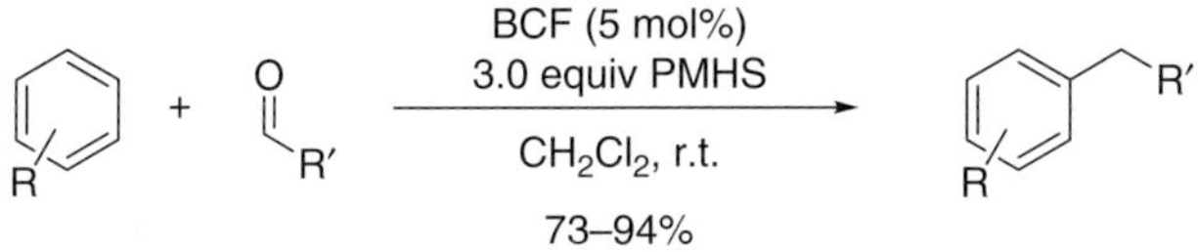

Scheme 2.64 Reductive coupling of aldehydes and arenes.

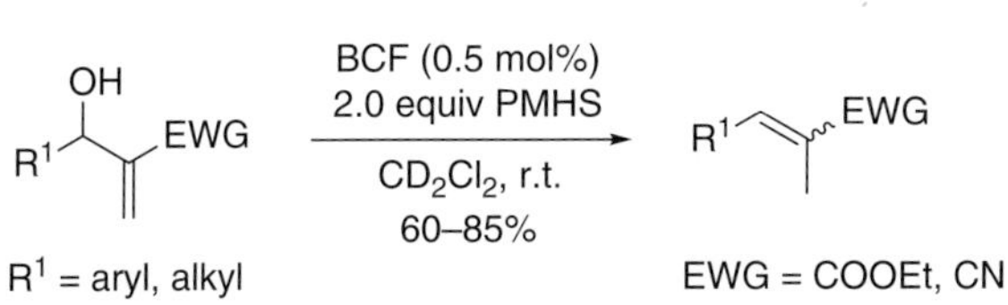

Scheme 2.65 S_N2' addition of the silane.

and yielded excellent *E*/*Z* selectively. *E* double bonds were obtained with ethyl esters as electron-withdrawing groups (EWGs), whereas nitriles exclusively yielded *Z* alkenes.

2.2.3.6.3 Formal Type III Reductions Utilizing Initial 1,4-Addition of the Silane

Piers and coworkers also demonstrated that BCF could catalyze the 1,4-reduction of α,β-unsaturated ketones, yielding silylated enol ethers when employing 1.0 equiv of $MePh_2SiH$ or other silanes as the reductant (Scheme 2.66) [180].

The 1,4-addition reduction of enones is likely initiated by reaction of the nucleophilic carbonyl oxygen with the electrophilic silicon of the η^1-[B]–H–[Si] adduct. The resonance-stabilized silyloxonium cation is preferentially trapped by the borohydride reducing agent, resulting in the formal 1,4-addition of the silanes. This process is dominant in many hydrosilylative C=C double bond reductions of heteroarenes and of α,β-unsaturated compounds, which on the surface appear to be type III reductions.

Chandrasekhar et al. reported the chemoselective reduction of C=C double bonds in conjugated systems (Scheme 2.67) [181]. Key for a successful protocol was the use of a 0.5 mol% loading of BCF and PMHS as the stoichiometric reductant. The highly chemoselective reduction of α,β-double bonds was realized for a variety of conjugated systems. Interestingly, complex product mixtures were obtained when the catalyst loading was increased to 1.0 mol%.

In 2009, Zhang and a coworker demonstrated the reduction of a variety of functional groups bearing C=X double bonds [182]. In this study, indoles, enamines, amides, secondary alcohols, ketones, α,β-unsaturated carboxylic acids, isocyanates, and enol ethers were completely reduced to either amines or alkanes

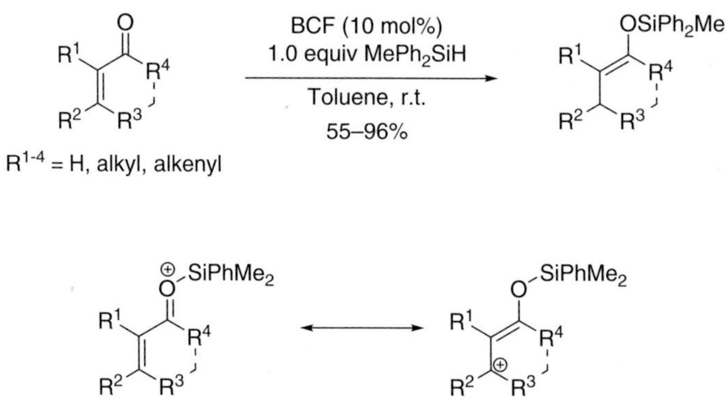

Scheme 2.66 Piers' enone reduction and hydrosilylation.

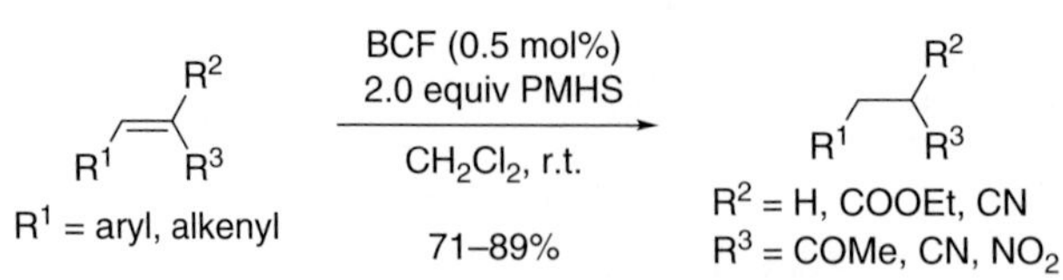

Scheme 2.67 Chemoselective reduction of conjugated C=C double bonds.

(with the exception of aromatic rings) using catalytic BCF and diphenylsilane Ph_2SiH_2 as the reductant (not shown).

Chang and a coworker reported the hydrosilylation of C=C double bonds in conjugation with esters and amides, yielding α-silylated compounds [183]. A variety of α,β-unsaturated esters, amides, lactones, and lactams were α-silylated using BCF as the catalyst and Me_2PhSiH as the reductant (Scheme 2.68). While the reduction of C=C double bonds in enones and enamides was reported earlier, only *O*-silylated products were obtained [180]. Mechanistic studies showed that esters formed monosilylated ketene acetal intermediates, which underwent a BCF-catalyzed migration of the *O*-silyl group to the α-carbon.

Chang and coworkers showed that pyridines were selectively reduced to various azacycles accompanied by the formation of a $C(sp^3)$—Si bond [184]. The use of BCF (5–10 mol%) and Et_2SiH_2 (4.0–9.0 equiv) in toluene at elevated temperatures (110 °C) provided an efficient protocol for piperidine synthesis (Scheme 2.69).

The outcome of this silylative reduction of pyridines was highly dependent on the substitution pattern of the pyridine substrates. While C-2 substituents led to complete reduction of the π-system, C-4 substitution yielded β,γ-unsaturated piperidines under the same reaction conditions. All products revealed silylation of C-5. The outcome of C-3-substituted pyridines could be tuned by the reducing

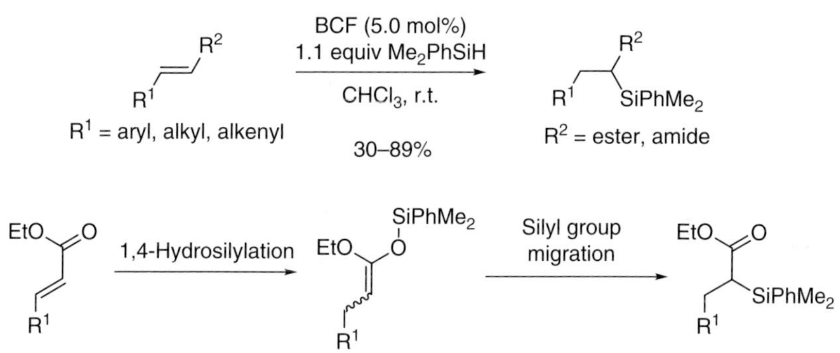

Scheme 2.68 Hydrosilylation of α,β-unsaturated carbonyl compounds to α-silylated products.

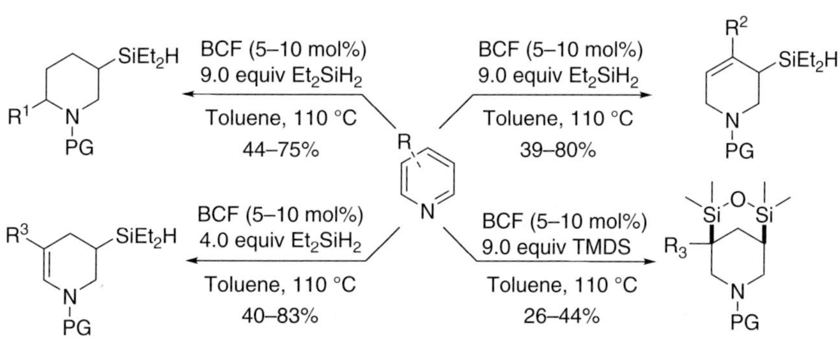

Scheme 2.69 Selective reduction of substituted pyridines.

silane. When Et_2SiH_2 was used, C-5 silylated α,β-unsaturated piperidines were obtained, but TMDS yielded fully reduced bicyclic siloxypiperidines featuring a disilyl ether bridge across C-3 and C-5 of the piperidine. It was also demonstrated that Tamao–Fleming oxidation could be used to gain hydroxylated piperidines (not shown).

The cascade reduction of pyridines to piperidines was further optimized by Wang and coworkers [185]. The addition of Ph_2NH (4.0 equiv) to the BCF (10 mol%)/Ph_2SiH_2 (2.0 equiv) system in toluene at elevated temperatures (110 °C) significantly enhanced the synthetic utility of BCF-catalyzed pyridine reductions.

The hydrosilylation of aromatic C=C double bonds of heteroarenes, such as quinolines, were also demonstrated by Park and coworkers [186]. This silylative dearomatization was shown to proceed with quinolines and isoquinoline derivatives in good overall yields. Key for successful conversion was the use of Et_2SiH_2 (4.0 equiv) in $CHCl_3$ at room or elevated temperatures (23–100 °C), and benefited from a low BCF catalyst loading (1 mol%). Other dihydrosilanes were also successfully employed.

In addition, Chang and coworkers reported in 2015 the chemoselective silylative reduction of α,β-unsaturated nitriles leading primarily to β-silylamines, but enamines could also be obtained when bulky silanes were used as the reducing agent (Scheme 2.70) [187]. A stepwise mechanism involving first the reduction of the nitrile and then the reduction of the double bond was proposed. In addition, because C–Si bonds are efficiently converted to alcohols, this methodology allows access to β-amino alcohols, which was demonstrated as well (not shown).

2.2.3.6.4 *Electrophilic Activation of C–C Multiple Bonds by BCF Coupled with Type I Reduction*

BCF can activate alkynes for nucleophilic cyclization [188]. For example, Melen and coworkers investigated stoichiometric and catalytic C–C and C–H bond formation with BCF via cationic intermediates [189]. In this study, the capability of BCF to act as a soft π acid, similarly to gold or platinum [190], was highlighted, which was also reported earlier [191]. Alkanoate esters were activated by BCF for

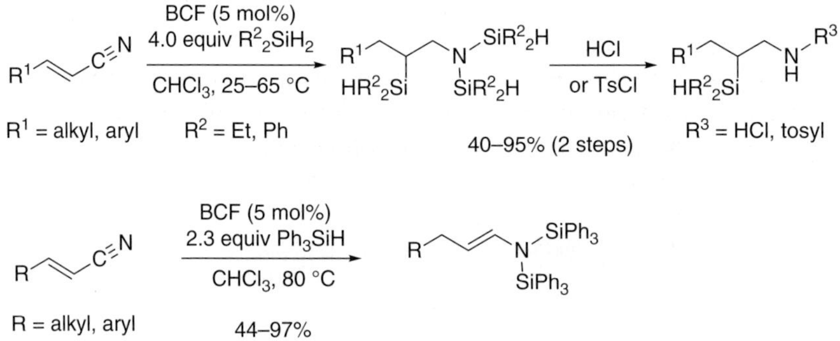

Scheme 2.70 Stepwise reduction of α,β-unsaturated nitriles.

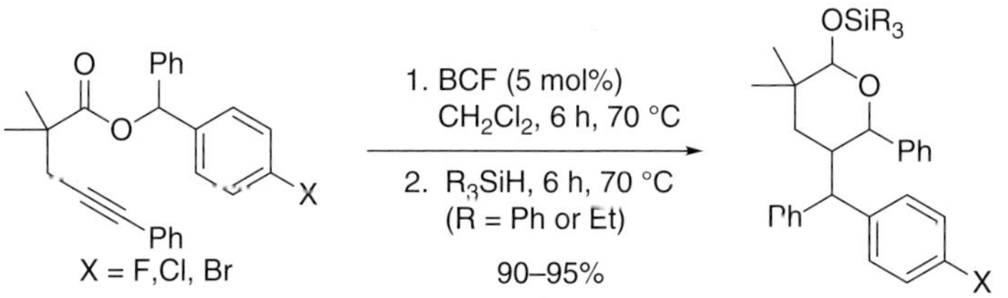

Scheme 2.71 Melen reduction of lactone/ester.

an intramolecular cyclization/carbocation rearrangement sequence followed by BCF-catalyzed hydrosilylation leading to the formation of silylated lactols in high yields (Scheme 2.71). This example shows that the catalytic electrophilic activation of C—C multiple bonds by BCF can be combined with BCF-catalyzed reductions using silanes (even in one pot).

2.2.3.6.5 *Intramolecular Silyoxy Migration*

Shimada and coworkers reported recently a BCF-catalyzed disilyl acetal rearrangement leading to disiloxanes (Scheme 2.72) [192]. The authors synthesized disilyl acetals from silyl acetates using Brookhart's $(Ir(coe)Cl)_2$ (coe = cylcooctene) catalyst [193]. Exposure of *in situ*–generated silyl acetals to dihydrosilanes under BCF catalytic conditions led to disiloxanes via a siloxyl/hydride rearrangement. By utilizing BCF's ability to cleave silyl ethers, this methodology was expanded to the synthesis of tri-, tetra-, and oligosiloxanes in one pot (not shown).

2.2.3.6.6 *The Role of BCF in Polymerizations*

Metallocenes are known as efficient catalysts for Ziegler–Natta-type polymerizations of alkenes [194]. Long before Piers' seminal studies of BCF catalyzed reductions, Marks and a coworker found that BCF was an excellent cocatalyst for metallocene-mediated alkene polymerizations [35]. Recently, the BCF/silane system has been shown to activate zirconocene dihalides for alkene polymerization [195] in the absence of aluminum activators or cocatalysts. The η^1-[Si]–H–[B] adduct converts zirconocene dihalide into zirconocene hydrides that are active polymerization initiators.

BCF-catalyzed reductions using silanes were also used to modify polyphenylsilanes [196], in the synthesis of catechol-functionalized polysiloxane building blocks [197], and applied in the curing of polysiloxanes [138].

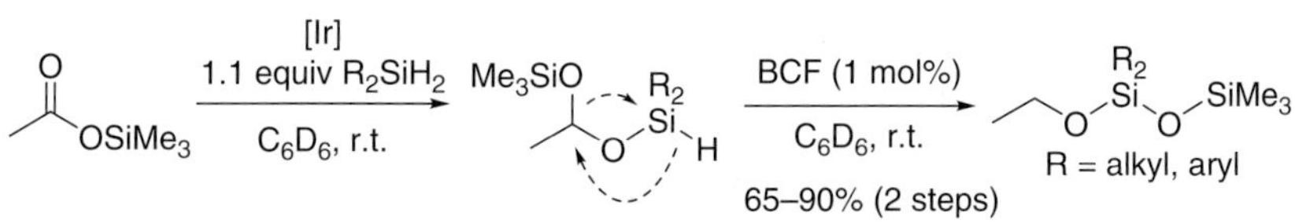

Scheme 2.72 Intramolecular silyloxy migration.

2.3 The Activation of Si—H Bonds by Aluminum Lewis Acids

Aluminum compounds rank among the earliest main-group Lewis acids applied in hydrosilylation catalysis. Since that time, the interest in the reductive hydrofunctionalization of organic compounds by Al Lewis acid catalysts has grown [198].

Simple Al salts, such as $AlCl_3$ and $EtAlCl_2$, were reported to activate silanes for the efficient hydrosilylation of alkenes and alkynes [199–203]. Various mechanisms were proposed for these reactions. Most studies observed selective *trans*-hydrosilylation of cyclic alkenes and internal alkynes; and terminal alkynes reacted regioselectively to the anti-Markovnikov product. Several mechanistic proposals consistent with the reaction outcome were reported. All mechanisms involve the formation of a silylium ion. One proposal suggested the activation of a Si—H bond by an electrophilic Al-alkyne π complex leading to a vinyl-aluminum species and a silylium ion, which stereospecifically transmetallate with aluminum to yield the *trans*-hydosilylated product. A second mechanism proposes the direct activation of the Si—H bond by the Al Lewis acid via an η^1-[Al]–H–[Si] adduct to yield an electrophilic silylium ion that could add to C=C multiple bonds. This mechanism has strong similarities to that discussed earlier for BCF-catalyzed reactions. Deuterium labeling studies supported the direct activation of silanes by Al [204]. Interestingly, when cationic β-diketiminato aluminum complexes were used as the catalyst, Nikonov and coworkers observed *cis*-hydrosilylations of alkynes but *trans*-addition of the silane to 1-methylcyclohexene [205]. These results show that different mechanistic pathways operate in Al-catalyzed hydrosilylations.

Simple Al salts were less efficient for the hydrosilylation of carbonyls. Instead, Laali and coworkers reported the reductive etherification of carbonyl compounds catalyzed by $Al(OTf)_3$ in the presence of silanes [89]. This reaction was accompanied by Friedel–Crafts alkylated by-products, suggesting the formation of silylcarboxonium ions that were insufficiently quenched by the Al–H species and reacted instead with the hydrosilylation product to ethers or with the aromatic solvent to provide alkylated arenes. It should be noted that this reactivity is also commonly observed with alkaline earth Lewis acids. In contrast, a more advanced bimetallic cationic B/Al complex reported by Bergman and a coworker was efficient for the hydrosilylation of aldehydes, ketones, lactones, and imines and was shown to directly activate the Si—H bond via an η^1-[Al]–H–[Si] adduct [206].

Tris(pentafluorophenyl)alane $Al(C_6F_5)_3$ (ACF) displays fascinating reactivity in these type of conversions and seems to operate more similar to boranes [207]. This alane congener of BCF is reported to be a significantly stronger Lewis acid than BCF, but it is also more difficult to handle due to its water sensitivity [208]. The coordination of Si—H bonds by alanes is known [209], and the formation of the more stable η^1-[Al]–H–[Si] complex was recently reported by Chen and a coworker, resembling the computed structure of the η^1-[B]–H–[Si] adduct between silanes and BCF [210]. As a result, ACF and BCF catalyze similar reaction types. But because of their difference in acidity, different outcomes emerge. First, the alane facilitated a ligand redistribution of two tertiary silanes, leading to a secondary and a quaternary silane (95% conversion after three hours). BCF was basically ineffective in catalyzing this reaction (3% conversion after 12 hours) [211]. Second, the alane was more active in the polymerization of methyl methacrylate than BCF. Third, the hydrosilylation of unactivated alkenes was more effective with alanes than BCF, rendering the superacidic alane catalyst more efficient for type III reductions, but BCF was a far better catalyst for type I reductions. This was explained by BCF's lower probability of deactivation by coordination to the carbonyl oxygen. Fourth, the alane complex was also more reactive than BCF for the defunctionalization of organic fluorides. Further applications of ACF for catalytic Si—H bond activations remain to be explored [64].

2.4 The Activation of Si—H Bonds by Group 14 Lewis Acids

2.4.1 Introduction

The most prominent group 14 Lewis acids are triaryl-stabilized carbocations and tri-coordinate silylium ions, which are isoelectronic to BCF and ACF. The central atom in all of these compounds has a trigonal planar coordination geometry and exhibit an empty p orbital which allows these compounds to function as Lewis acids. An important difference between group 13 and group 14 Lewis acids is the cationic character of the latter. Stabilization of their positive charge requires the presence of anions which affect their Lewis acidity and catalytic activity. This section discusses only the most prominent Lewis acids already mentioned. Other examples of Si and heavier group 14 Lewis acids were recently discussed in the literature [212].

2.4.2 Carbocations as Lewis Acids

The most famous group 14 Lewis acid is the triphenylmethyl ion (trityl cation), which was first recognized as a carbocation in 1902 by Baeyer and Villinger as part of their studies of triphenylmethane-derived dyes [213, 214]. The promising properties of the trityl cation as an efficient Lewis acid catalyst was recently discussed [215]; its ability to activate Si—H bonds was reported. For example, in the Bartlett–Condon–Schneider reaction, which is discussed in more detail, the trityl cation is used to abstract a hydride from silanes to generate electrophilic $[R_3Si]^+$ equivalents [216, 217]. This hydride abstraction reaction is considered irreversible due to the formation of the more stable R_3C—H bond when compared to the R_3Si—H bond [218]. But this reaction may be inhibited kinetically through steric barriers, as shown by Lambert and Zhao who reported no reaction between Mes_3Si–H and trityl cation [219]. The balance of thermodynamic preferences versus kinetic possibilities makes most trityl cation–mediated reactions with silanes stoichiometric in nature.

An illustrative example was reported by Kawashima and coworkers, who used the trityl cation to activate silanes for intramolecular sila-Friedel–Craft reactions of arenes [220, 221]. In this reaction, the trityl cation was only used to generate the more reactive silylium ion, which mediates the desired reaction. Because an acidic proton is the by-product, 2,6-lutidine was employed as a base to inhibit the reverse reaction. Other sterically crowded pyridine derivatives provided reaction conditions to mediate the dehydrogenative annulation (*cis* sila-arylation) of alkenes and alkynes across benzyl and naphthyl silanes [222, 223].

Interestingly, when ortho alkynic substituents were present in phenyl silanes, a base-free intramolecular hydrosilylation of alkynes was realized with catalytic amounts of trityl cation [224]. As before, the active species was an *in situ*–generated silylium ion, which upon addition to the alkyne generated a vinyl cation intermediate. This reactive carbocation abstracted the hydride of a second alkynic silane to propagate the chain reaction.

As illustrated by these examples, the trityl cation can only *initiate* silane activation but not *catalyze* it because of the thermodynamic irreversibility of hydride abstraction. It thus stands to reason that if catalytic amounts of trityl cation suffice, then it is the silylium ion that is likely the acting catalytic species. Examples where the trityl cation is used as an initiator to create catalytic silylium species are provided in the next section.

In an elegant approach, Ingleson designed aromatic heteroatom-stabilized carbocations that not only are sufficiently Lewis acidic to cleave Si—H bonds of silanes but also release the abstracted hydride when necessary. This "partial" activation of silanes reflects reactivity analogous to BCF by reversibly forming a central η^1-[catalyst]–H–[Si] adduct, which further reduces organic substrates. After discovering the promising Si–H/Si–D scrambling mediated by organic salts without the formation of reactive silylium species [225], Ingleson and coworkers turned to benzothiazolium salts, which established a more favorable equilibrium for catalytic silane activation. With this catalyst, the dehydrogenative silylation of alcohols as well as the reduction of aldehydes, ketones, and imines was accomplished [226].

2.4.3 Cationic Tri-coordinate Silylium Ions and Neutral Si(IV) Lewis Acids

Even though tri-coordinate silylium species are isoelectronic to the trityl cation, the inorganic character of the Si center is not to be underestimated when evaluating its reactivity. The higher electropositive character of Si when compared to carbon paired with the combination of a larger atom size and more diffuse third-row valence orbitals (which inhibit efficient delocalization by neighboring π systems) position the positive charge directly at the silicon center [218]. As a result, tri-coordinate silylium ions are highly reactive electrophiles and, depending on the countering anion, can even be used to create a super-Lewis acidic environment [227]. The diversity of structures and methods available for the controlled synthesis of silylium equivalents render this Lewis acid rather tunable for a variety of applications including for the activation of silanes [228, 229].

Silylium ions are typically generated by metathesis reactions where an electrophilic activator El^+ ($El^+ = Ag^+$, Na^+, $[R_3C]^+$, and others) is used to cleave a Si—Nu bond (Nu = halide, hydride). The aforementioned Bartlett–Condon–Schneider reaction employs trityl cations, such as $[Ph_3C]^+[B(C_6F_5)_4]^-$ or $[Ph_3C]^+[CHB_{11}H_5Br_6]^-$ (also called Reed's carborane-based trityl cation), to abstract a hydride from R_3Si–H and generate highly reactive $[R_3Si]^+$ equivalents. Even silanes with varying substituents, which typically suffer from an undesired redistribution of substituents, can be controlled to selectively produce specific silylium ions by choosing the right silane/counteranion combination [216]. It should be noted that most silylium ions generated this way coordinate either to the anion or solvent. This is because of the high reactivity of free silylium ions which are extremely difficult to obtain [230], explaining their suitability for challenging the catalytic transformations highlighted subsequently.

According to a seminal study by Ozerov, the silylium ion can catalyze the defluorination of organofluorides [231, 232]. This elegant approach takes advantage of stronger Si—F bond and the weaker Si—H bond when compared to its carbon congener. The active silylium Lewis acid catalyst was generated *in situ* from the reaction of Reed's trityl cation with Et_3SiH or Hex_3SiH. This fluorophilic species abstracted F^- from an alkyl fluoride liberating a strongly Lewis acidic carbocation, which cleaved the Si—H bond of another silane molecule to regenerate the silylium catalyst and yield the alkane product.

An alternative approach to utilize the Lewis acidity of silicon are the development of neutral Si(IV) Lewis acids. Pioneered by Tilley, it was shown that the neutral Si(IV) Lewis acid bis(perfluorocatecholato)silane can activate Et_3SiH for hydrosilylation reactions of aldehydes; this approach was tolerant of other functional groups [233]. And most recently, Greb reported the neutral silicon Lewis superacid bis(perchlorocatecholato)silane which catalytically defunctionalized 1-adamantyl fluoride with Et_3SiH or PHMS as the stoichiometric reductant [234].

2.5 The Activation of Si—H Bonds by Phosphorous-Based Lewis Acids

2.5.1 P(III) Lewis Acids

Phosphenium cations, which exhibit a Lewis acidic P(III) atom, have been known and their properties investigated since 1964 [235]. Vidović and coworkers demonstrated that phosphenium cations could activate Si—H bonds [236]; and Stephan and coworkers showed that bipyridine- and terpyridine-ligated phosphenium ions mediated catalytic hydrodefluorination reactions of primary, secondary, and tertiary organic fluorides with silanes [237]. It is not yet clear whether the bipyridine-ligated Lewis acid initiated the reaction by generating catalytically active silylium cation or if it operated as the catalyst; the available mechanistic data suggest the latter. It was proposed that this Lewis acid would first cleave the C—F bond of the substrate, yielding fluorinated catalyst and a carbocation. Silane could then transfer hydride to the latter and the resulting silylium species would regenerate the catalyst by fluoride abstraction from the organofluorine substrate.

2.5.2 P(V) Lewis Acids

Numerous P(V) Lewis acids have been synthesized. Recently, electrophilic phosphonium cations (EPCs) have received attention and been reported to effectively activate small molecules such as silanes [238]. For example, Oestreich and coworkers reported the reduction of the P=O bonds of phosphine oxides [239]. In this study, BCF and fluorinated EPCs were compared as catalytic activators for Si–H bonds of various silanes. Both catalysts were efficient for this reaction across a large substrate scope when $PhSiH_3$ was used as the reducing agent, but the EPCs were overall more reactive, tolerated lower catalyst loadings (2 mol%) than BCF (10 mol%), and proceeded at lower reaction temperatures. Chemoselective reduction of the P=O double bond was observed while halogens, ethers, ketones, and carboxylic acids remained untouched. The reduction of a P=S bond was also reported with both catalysts.

Compared to BCF, the Lewis acidity of EPC $[(C_6F_5)_3PF]^+[B(C_6F_5)_4]^-$ is significantly enhanced due to the positive charge on the P atom. Its low-lying antibonding σ^*-orbital allows for Si–H bond activation, which is believed to be mechanistically similar to the aforementioned Piers pathway for the activation of silanes by BCF [238]. As a result, EPCs are comparable to BCF and other borane catalysts in their function and activity, and numerous reactions involving silane activation (covered in Chapter 3), were also realized with the EPC $[(C_6F_5)_3PF]^+$ $[B(C_6F_5)_4]^-$ catalyst. These include hydrodefluorinations of fluoroalkanes [240, 241]; hydrosilylations of ketones, imines, and amides [242–244]; hydrosilylations of alkenes and alkynes [245]; dehydrocoupling of amines, acids, phenols, and thiols with silanes [246]; catalytic ketone hydrodeoxygenation using silanes [247]; and the reductive alkylation of arenes with aromatic or aliphatic CF_3 groups to CH_2-Ar fragments [248] or benzylfluorides [249].

The Lewis acidity of EPCs was further enhanced by installing EWGs on the P(V) center [44, 242, 250] or by neighboring two electrophilic centers in close proximity. The latter examples include a P(V) center on a pyridinium backbone [251], a B and P(V) center on a benzene backbone [252], and two P(V) centers on a ferrocene [253] or naphthalene backbone [254, 255]. These EPCs were efficient silane activators in a series of catalyzed reactions.

2.6 Summary and Conclusions

The controlled and highly tunable catalytic activation of Si–H bonds by BCF catalysts generates silylium ions of varying Lewis acidity, which has enabled the development of a variety of reductions, with a somewhat surprising functional group tolerance. These reductions have been utilized in targeted defunctionalization reactions useful in organic synthesis and in selective deoxygenations of biomass. The development of other main-group Lewis acids, specifically EPCs, will offer more tunability in the near future and promise to further enhance the application of controlled silane activations for achieving specific reaction outcomes.

Acknowledgments

D. W. appreciates the support of the Oak Ridge Associated Universities (ORAU) Travel Grants Program and Appalachian State's University Research Council (URC), Research Institute for Environment, Energy, & Economics (RIEEE), Office of Research, Office of Student Research, College of Arts and Sciences, and A.R. Smith Department of Chemistry and Fermentation Sciences. M.R. Gagné thanks the Department of Energy (Basic energy Science, DE-FG02–05ER15630) for their generous support.

References

1 Buff, H. and Wöhler, F. (1857). *Justus Liebig's Ann. Chem.* 102: 128.
2 Buff, H. and Wöhler, F. (1857). *Justus Liebig's Ann. Chem.* 103: 218–229.
3 Wöhler, F. (1858). *Justus Liebig's Ann. Chem.* 107: 112–119.
4 Mellor, J.W. (1947). *A Comprehensive Treatise on Inorganic and Theoretical Chemistry*, vol. VI, 219. Longmans, Green and Co.
5 Wöhler, F. (1863). *Justus Liebig's Ann. Chem.* 127: 257–274, 268.
6 Mendeleev, D.I. (1903). *The Principles of Chemistry*, St. Petersburg, 531.
7 Reynolds, J.E. (1909–1910). *Not. Proc. R. Inst.* 19: 642–650.
8 Reynolds, J.E. (1913). *Proc. R. Soc. A* 88: 37–48.
9 Moissan, H. (1904). *Traité de Minérale Chimie*, vol. 1, 34. Paris: Masson et Cie.
10 Mellor, J.W. (1947). *A Comprehensive Treatise on Inorganic and Theoretical Chemistry*, vol. VI, 215. Longmans, Green and Co.
11 Semenov, V.V. (2016). *Her. Russ. Acad. Sci.* 86: 466. https://doi.org/10.1134/S1019331616060071.
12 Koda, S. (1992). *Prog. Energy Combust. Sci.* 18: 513–528. https://doi.org/10.1016/0360-1285(92)90037-2.
13 Noorizadeh, S. and Shakerzadeh, E. (2009). *J. Mol. Struct.* 920: 110–113.
14 Walsh, R. (1981). *Acc. Chem. Res.* 14: 246–252.
15 Walsh, R. (1983). *J. Chem. Soc., Faraday Trans. 1* 79: 2233–2248.
16 Lambert, J.B., Wang, G.-T., Finzel, R.B., and Teramura, D.H. (1987). *J. Am. Chem. Soc.* 109: 7838–7845.
17 Lambert, J.B., Zhao, Y., Emblidge, R.W. et al. (1999). *Acc. Chem. Res.* 32: 183–190.
18 Simonneau, A. and Oestreich, M. (2013). *Angew. Chem. Int. Ed.* 52: 11905–11907.
19 Simonneau, A. and Oestreich, M. (2015). *Nat. Chem.* 7: 816–822.
20 Keess, S. and Oestreich, M. (2017). *Chem. Sci.* 8: 4688–4695.
21 Oestreich, M. (2016). *Angew. Chem. Int. Ed.* 55: 494–499.
22 Keess, S., Simmoneau, A., and Oestreich, M. (2015). *Organometallics* 34: 790–799.

23 Yuan, W., Orecchia, P., and Oestreich, M. (2017). *Chem. Commun.* 53: 10390–10393.

24 Doncaster, A.M. and Walsh, R. (1981). *Int. J. Chem. Kinet.* 13: 503–514.

25 Piers, W.E., Marwitz, A.J.V., and Mercier, L.G. (2011). *Inorg. Chem.* 50: 12252–12262.

26 Rupp, A.R., Johnstone, T.C., and Stephan, D.W. (2018). *Dalton Trans.* 47: 7029–7035.

27 Rubin, M., Gevorgyan, V., Chandrasekhar, S. et al. (2019). *e-EROS Encyclopaedia of Reagents for Organic Synthesis*, 1. Wiley.

28 Massey, A.G. and Park, A.J. (1964). *J. Organomet. Chem.* 2: 243–250.

29 Massey, A.G. and Park, A.J. (1966). *J. Organomet. Chem.* 5: 215–225.

30 Mitsui, H., Ilda, T., Ikeno, I. et al. (2000). European Patent Application EP 0995753 A2.

31 Sundararaman, A. and Jäkle, F. (2003). *J. Organomet. Chem.* 681: 134–142.

32 Laszlo, P. and Teston, M. (1990). *J. Am. Chem. Soc.* 112: 8750–8754.

33 Piers, W.E. and Chivers, T. (1997). *Chem. Soc. Rev.* 26: 345–354.

34 Fontaine, F.-G. and Stephan, D.W. (2017). *Philos. Trans. R. Soc. A* 375: 20170004. https://doi.org/10.1098/rsta.2017.0004.

35 Chen, E.Y.-X. and Marks, T.J. (2000). *Chem. Rev.* 100: 1391–1434.

36 Larson, J.R. and Melen, R.L. (2017). *Inorg. Chem.* 56: 8627–8643.

37 Welch, G.C., Cabrera, L., Chase, P.A. et al. (2007). *Dalton Trans.* 3407–3414.

38 Beckett, M.A., Brassington, D.S., Light, M.E., and Hursthouse, M.B. (2001). *J. Chem. Soc., Dalton Trans.* 1768–1772; and references therein.

39 Childs, R.F., Mulholland, D.L., and Nixon, A. (1982). *Can. J. Chem.* 60: 801–808.

40 Ullrich, M., Lough, A.J., and Stephan, D.W. (2010). *Organometallics* 29: 3647–3654.

41 Körte, L.A., Schwabedissen, J., Soffner, M. et al. (2017). *Angew. Chem. Int. Ed.* 56: 8578–8582; and references therein.

42 Bergquist, C., Bridgewater, B.M., Harlan, C.J. et al. (2000). *J. Am. Chem. Soc.* 122: 10581–10590.

43 Scott, D.J., Simmons, T.R., Lawrence, E.J. et al. (2015). *ACS Catal.* 5: 5540–5544.

44 Postle, S., Podgorny, V., and Stephan, D.W. (2016). *Dalton Trans.* 45: 14651–14657.

45 Mewald, M., Fröhlich, R., and Oestreich, M. (2011). *Chem. Eur. J.* 17: 9406–9414.

46 Mohr, J., Durmaz, M., Irran, E., and Oestreich, M. (2014). *Organometallics* 33: 1108–1111.

47 Paradies, J. (2017). *Top. Organomet. Chem.* https://doi.org/10.1007/3418_2016_173.

48 Mei, W., Feng, X., and Du, H. (2018). *Acc. Chem. Res.* 51: 191–201.

49 Parks, D.J. and Piers, W.E. (1996). *J. Am. Chem. Soc.* 118: 9440–9441.

50 Doyle, M.P., West, C.T., Donnelly, S.J., and McOsker, C.C. (1976). *J. Organomet. Chem.* 117: 129–140.

51 Fry, J.L., Orfanopoulos, M., Adlington, M.G. et al. (1978). *J. Org. Chem.* 43: 374–375.

52 Parks, D.J., Blackwell, J.M., and Piers, W.E. (2000). *J. Org. Chem.* 65: 3090–3098.

53 Rendler, S. and Oestreich, M. (2008). *Angew. Chem. Int. Ed.* 47: 5997–6000; and references therein.

54 Tsuchimoto, T. and Kawakami, Y. (2007). *Silicon Chem.* 3: 243–249.
55 Sakata, K. and Fujimoto, H. (2013). *J. Org. Chem.* 78: 12505–12512.
56 A new mechanistic tool to investigate Walden inversion on silicon was recently reported: Fallon, T. and Oestreich, M. (2015). *Angew. Chem. Int. Ed.* 54: 12488–12491.
57 Oestreich, M., Hermeke, J., and Mohr, J. (2015). *Chem. Soc. Rev.* 44: 2202–2220.
58 Houghton, A.Y., Hurmalainen, J., Mansikkamäki, A. et al. (2014). *Nat. Chem.* 6: 983–988.
59 Tolman, C.A. (1977). *Chem. Rev.* 77: 313–348.
60 Hester, D.M., Sun, J., Harper, A.W., and Yang, G.K. (1992). *J. Am. Chem. Soc.* 114: 5234–5240.
61 Zhang, S., Dobson, G.R., and Brown, T.L. (1991). *J. Am. Chem. Soc.* 113: 6908–6916.
62 Weicker, S.A. and Stephan, D.W. (2015). *Bull. Chem. Soc. Jpn.* 88: 1003–1016.
63 Komiyama, T., Minami, Y., and Hiyama, T. (2017). *ACS Catal.* 7: 631–651.
64 Lipke, M.C., Liberman-Martin, A.L., and Tilley, T.D. (2017). *Angew. Chem. Int. Ed.* 56: 2260–2294.
65 Gudz, A., Payne, P.R., and Gagné, M.R. (2017). *Organometallics* 36: 4047–4053.
66 Blackwell, J.M., Sonmor, E.R., Scoccitti, T., and Piers, W.E. (2000). *Org. Lett.* 2: 3921–3923.
67 Hog, D.T. and Oestreich, M. (2009). *Eur. J. Org. Chem.* 5047–5056.
68 Mewald, M. and Oestreich, M. (2012). *Chem. Eur. J.* 18: 14079–14084.
69 Hermeke, J., Mewald, M., and Oestreich, M. (2013). *J. Am. Chem. Soc.* 135: 17537–17546.
70 Chen, D., Leich, V., Pan, F., and Klankermayer, J. (2012). *Chem. Eur. J.* 18: 5184–5187.
71 Zhu, X. and Du, H. (2015). *Org. Biomol. Chem.* 13: 1013–1016.
72 Liu, Y. and Du, H. (2013). *J. Am. Chem. Soc.* 135: 6810–6813.
73 Fasano, V., Radcliffe, J.E., and Ingleson, M.J. (2016). *ACS Catal.* 6: 1793–1798.
74 Gyömöre, A., Bakos, M., Földes, T. et al. (2015). *ACS Catal.* 5: 5366–5372; and references therein.
75 Fasano, V. and Ingleson, M.J. (2018). *Synthesis* 50: 1783–1795.
76 Fasano, V. and Ingleson, M.J. (2017). *Chem. Eur. J.* 23: 2217–2224.
77 Pan, Y., Chen, C., Xu, X. et al. (2018). *Green Chem.* 20: 403–411.
78 Harrison, D.J., McDonald, R., and Rosenberg, L. (2005). *Organometallics* 24: 1398–1400.
79 Gandhamsetty, N., Jeong, J., Park, J. et al. (2015). *J. Org. Chem.* 80: 7281–7287.
80 Skjel, M.K., Houghton, A.Y., Kirby, A.E. et al. (2010). *Org. Lett.* 12: 376–379.
81 Asao, N., Ohishi, T., Sato, K., and Yamamoto, Y. (2002). *Tetrahedron* 58: 8195–8203.
82 Süsse, L., Hermeke, J., and Oestreich, M. (2016). *J. Am. Chem. Soc.* 138: 6940–6943.
83 Kang, B.C., Shin, S.H., Yun, J., and Ryu, D.H. (2017). *Org. Lett.* 19: 6316–6319.
84 Hirao, A., Itsuno, S., Nakahama, S., and Yamazaki, N. (1981). *J. Chem. Soc., Chem. Commun.* 315–317.
85 Corey, E.J., Bakshi, R.K., and Shibata, S. (1987). *J. Am. Chem. Soc.* 109: 5551–5553.

86 Wilkins, L.C. and Melen, R.L. (2016). *Coord. Chem. Rev.* 324: 123–139.
87 Peng, D., Zhang, M., and Huang, Z. (2015). *Chem. Eur. J.* 21: 14737–14741.
88 Hubbard, J.L. (1988). *Tetrahedron Lett.* 29: 3197–3200.
89 Bach, P., Albright, A., and Laali, K.K. (2009). *Eur. J. Org. Chem.* 1961–1966.
90 Benda, K., Regenhardt, W., Schaumann, E., and Adiwidjaja, G. (2009). *Eur. J. Org. Chem.* 1016–1021.
91 Blackwell, J.M., Foster, K.L., Beck, V.H., and Piers, W.E. (1999). *J. Org. Chem.* 64: 4887–4892.
92 Harrison, D.J., Edwards, D.R., McDonald, R., and Rosenberg, L. (2008). *Dalton Trans.* 3401–3411.
93 Lee, P.T.K., Skjel, M.K., and Rosenberg, L. (2013). *Organometallics* 32: 1575–1578.
94 Lee, P.T.K. and Rosenberg, L. (2016). *J. Organomet. Chem.* 809: 86–93.
95 Kuciński, K. and Hreczycho, G. (2017). *ChemCatChem* 9: 1868–1885.
96 Greb, L., Tamke, S., and Paradies, J. (2014). *Chem. Commun.* 50: 2318–2320.
97 Chulsky, K. and Dobrovetsky, R. (2017). *Angew. Chem. Int. Ed.* 56: 4744–4748.
98 Rubinstsztajn, S. and Cella, J.A. (2004). *Polym. Prepr.* 45: 635–636.
99 Brook, M.A. (2018). *Chem. Eur. J.* https://doi.org/10.1002/chem.201800123.
100 Chojnowski, J., Rubinsztajn, S., Cella, J.A. et al. (2005). *Organometallics* 24: 6077–6084.
101 Grande, J.B., Thompson, D.B., Gonzaga, F., and Brook, M.A. (2010). *Chem. Commun.* 46: 4988–4990.
102 Gevorgyan, V., Liu, J.-X., Rubin, M. et al. (1999). *Tetrahedron Lett.* 40: 8919–8922.
103 Gevorgyan, V., Rubin, M., Benson, S. et al. (2000). *J. Org. Chem.* 65: 6179–6186.
104 Roesler, R., Har, B.J.N., and Piers, W.E. (2002). *Organometallics* 21: 4300–4302.
105 Nimmagadda, R.D. and McRae, C. (2006). *Tetrahedron Lett.* 47: 5755–5758.
106 Nimmagadda, R.D. and McRae, C. (2006). *Tetrahedron Lett.* 47: 3505–3508.
107 Mack, D.J., Guo, B., and Njardarson, J.T. (2012). *Chem. Commun.* 48: 7844–7846.
108 Bender, T.A., Dabrowski, J.A., and Gagné, M.R. (2018). *Nat. Chem. Rev.* 2: 35–46.
109 Tarber, D.F. and Malcom, S.C. (2001). *J. Org. Chem.* 66: 944–953.
110 Tarber, D.F., Jiang, Q., Chen, B. et al. (2002). *J. Org. Chem.* 67: 4821–4827.
111 Vyvyan, J.R., Oaksmith, J.M., Parks, B.W., and Peterson, E.M. (2005). *Tetrahedron Lett.* 46: 2457–2460.
112 McGrath, N.A., Bartlett, E.S., Sittihan, S., and Njardarson, J.T. (2009). *Angew. Chem. Int. Ed.* 48: 8543–8546.
113 Adduci, L.L., McLaughlin, M.P., Bender, T.A. et al. (2014). *Angew. Chem. Int. Ed.* 53: 1646–1649.
114 Robert, T. and Oestreich, M. (2013). *Angew. Chem. Int. Ed.* 52: 5216–5218.
115 Seo, Y. and Gagné, M.R. (2018). *ACS Catal.* 8: 81–85.
116 Adduci, L.L., Bender, T.A., Dabrowski, J.A., and Gagné, M.R. (2015). *Nat. Chem.* 7: 576–581.
117 Adduci, L.L. (2014). Methods of carbohydrate functionalization and defunctionalization. PhD dissertation. University of North Carolina at Chapel Hill.

118 Zhang, J., Park, S., and Chang, S. (2017). *Angew. Chem. Int. Ed.* 56: 13757–13761.
119 Drosos, N. and Morandi, B. (2015). *Angew. Chem. Int. Ed.* 54: 8814–8818.
120 Drosos, N., Ozkal, E., and Morandi, B. (2016). *Synthesis* 27: 1760–1764.
121 Drosos, N., Cheng, G.-J., Ozkal, E. et al. (2017). *Angew. Chem. Int. Ed.* 56: 13377–13381.
122 Bender, T.A., Dabrowski, J.A., Zhong, H., and Gagné, M.R. (2016). *Org. Lett.* 18: 4120–4123.
123 Bender, T.A., Payne, P.R., and Gagné, M.R. (2017). *Nat. Chem.* https://doi.org/10.1038/nchem.2863.
124 Hazra, C.K., Gandhamsetty, N., Park, S., and Chang, S. (2016). *Nat. Commun.* 7: 13431.
125 Bernasconi, C.F. and Rappoport, Z. (2009). *Acc. Chem. Res.* 42: 993–1003.
126 Corma, A., Iborra, S., and Velty, A. (2007). *Chem. Rev.* 107: 2411–2502.
127 Chatterjee, I., Porwal, D., and Oestreich, M. (2017). *Angew. Chem. Int. Ed.* 56: 3389–3391.
128 Saito, K., Kondo, K., and Akiyama, T. (2015). *Org. Lett.* 17: 3366–3369.
129 Caputo, C.B. and Stephan, D.W. (2012). *Organometallics* 31: 27–30.
130 Stahl, T., Klare, H.F.T., and Oestreich, M. (2013). *ACS Catal.* 3: 1578–1587.
131 Rubin, M., Schwier, T., and Gevorgyan, V. (2002). *J. Org. Chem.* 67: 1936–1940.
132 Jones, G.R. and Landais, Y. (1996). *Tetrahedron* 52: 7599–7662.
133 Curless, L.D. and Ingleson, M.J. (2014). *Organometallics* 33: 7241–7246.
134 Kim, Y., Dateer, R.B., and Chang, S. (2017). *Org. Lett.* 19: 190–193.
135 Shchepin, R., Xu, C., and Dassault, P. (2010). *Org. Lett.* 12: 4772–4775.
136 Shin, K., Joung, S., Kim, Y., and Chang, S. (2017). *Adv. Synth. Catal.* 359: 3428–3436.
137 Kim, D.W., Joung, S., Kim, J.G., and Chang, S. (2015). *Angew. Chem. Int. Ed.* 54: 14805–14809.
138 Mathew, J., Eguchi, K., Nakajima, Y. et al. (2017). *Eur. J. Org. Chem.* 4922–4927.
139 Yuan, W., Smirnov, P., and Oestreich, M. (2018). *Chem* 4: 1443–1450.
140 Smirnov, P. and Oestreich, M. (2016). *Organometallics* 35: 2433–2434.
141 Bähr, S. and Oestreich, M. (2017). *Angew. Chem. Int. Ed.* 56: 52–59.
142 Curless, L.D., Clark, E.R., Dunsford, J.J., and Ingleson, M.J. (2014). *Chem. Commun.* 50: 5270–5272.
143 Ma, Y., Wang, B., Zhang, L., and Hou, Z. (2016). *J. Am. Chem. Soc.* 138: 3663–3666.
144 Ma, Y., Zhang, L., Luo, Y. et al. (2017). *J. Am. Chem. Soc.* 139: 12434–12437.
145 Han, Y., Zhang, S., He, J., and Zhang, Y. (2017). *J. Am. Chem. Soc.* 139: 7399–7407.
146 Beng, G.T., Meesala, R., Mordi, M.N., and Mansor, S.M. (2014). *Synth. Commun.* 44: 1291–1295.
147 Gevorgyan, V., Rubin, M., Liu, J.-X., and Yamamoto, Y. (2001). *J. Org. Chem.* 66: 1672–1675.
148 Bajracharya, G.B., Nogami, T., Jin, T. et al. (2004). *Synthesis* (2): 308–311.
149 Chandrasekhar, S., Reddy, C.R., and Babu, B.N. (2002). *J. Org. Chem.* 67: 9080–9082.

150 Bézier, D., Park, S., and Brookhart, M. (2013). *Org. Lett.* 15: 496–499.
151 Feghali, E., Jacquet, O., Thuéry, P., and Cantat, T. (2014). *Catal. Sci. Technol.* 4: 2230–2234.
152 Porwal, D. and Oestreich, M. (2016). *Eur. J. Org. Chem.* 3307–3309.
153 Porwal, D. and Oestreich, M. (2017). *Synthesis* 49: 4698–4702.
154 Ding, F., Jiang, Y., Gen, S. et al. (2017). *Eur. J. Org. Chem.* 3427–3430.
155 Donck, S., Gravel, E., Shaw, N. et al. (2015). *RSC Adv.* 5: 50865–50868.
156 Denis, J.-M., Forintos, H., Szelke, H., and Kelglevich, G. (2002). *Tetrahedron Lett.* 43: 5569–5571.
157 Blondiaux, E. and Cantat, T. (2014). *Chem. Commun.* 50: 9349–9352.
158 Das, S., Wendt, B., Moller, K. et al. (2012). *Angew. Chem. Int. Ed.* 51: 1662–1666.
159 Li, B., Sortais, J.-B., and Darcel, C. (2013). *Chem. Commun.* 49: 3691–3693.
160 Chadwick, R.C., Kardelis, V., Lim, P., and Adronov, A. (2014). *J. Org. Chem.* 79: 7728–7733.
161 Peruzzi, M.T., Mei, Q.Q., Lee, S.J., and Gagné, M.R. (2018). *Chem. Commun.* 54: 5855–5858.
162 Ding, G., Wu, X., Lu, B. et al. (2018). *Tetrahedron* 74: 1144–1150.
163 Mukherjee, D., Sauer, D.F., Zanardi, A., and Okuda, J. (2016). *Chem. Eur. J.* 22: 7730–7733.
164 Fernández-Alvarez, F.J., Aitani, A.M., and Oro, L.A. (2014). *Catal. Sci. Technol.* 4: 611–624.
165 Berkefeld, A., Piers, W.E., and Parvez, M. (2010). *J. Am. Chem. Soc.* 132: 10660–10661.
166 Sumerin, V., Schulz, F., Nieger, M. et al. (2008). *Angew. Chem. Int. Ed.* 47: 6001–6003.
167 Chen, J., Falivene, L., Caporaso, L. et al. (2016). *J. Am. Chem. Soc.* 138: 5321–5333.
168 Bontemps, S. (2016). *Coord. Chem. Rev.* 308: 117–130.
169 Liu, X.-F., Ma, R., Qiao, C. et al. (2016). *Chem. Eur. J.* 22: 16489–16493.
170 Feghali, E. and Cantat, T. (2014). *Chem. Commun.* 50: 862–865.
171 Feghali, E., Carrot, G., Thuéry, P. et al. (2015). *Energy Environ. Sci.* 8: 2734–2743.
172 Chauvier, C. and Cantat, T. (2017). *ACS Catal.* 7: 2107–2115.
173 Ma, G., Song, G., and Li, Z.H. (2017). *Phys. Chem. Chem. Phys.* 19: 28313–28322.
174 Yang, Z., Yu, B., Zhang, H. et al. (2015). *Green Chem.* 17: 4189–4193.
175 Niu, H., Lu, L., Shi, R. et al. (2017). *Chem. Commun.* 53: 1148–1151.
176 Fernández-Alvarez, F.J. and Oro, L.A. (2018). *ChemCatChem* 10: 4783–4796.
177 Chandrasekhar, S., Chandrasekhar, G., Babu, B.N. et al. (2004). *Tetrahedron Lett.* 45: 5497–5499.
178 Chandrasekhar, S., Khatun, S., Rajesh, G., and Reddy, C.R. (2009). *Tetrahedron Lett.* 50: 6693–6697.
179 Chandrasekhar, S., Chandrasekhar, G., Vijeender, K., and Reddy, M.S. (2006). *Tetrahedron Lett.* 47: 3475–3478.
180 Blackwell, J.M., Morrison, D.J., and Piers, W.E. (2002). *Tetrahedron* 58: 8247–8254.

181 Chandrasekhar, S., Chandrasekhar, G., Reddy, M.S., and Srihari, P. (2006). *Org. Biomol. Chem.* 4: 1650–1652.
182 Tan, M. and Zhang, Y. (2009). *Tetrahedron Lett.* 50: 4912–4915.
183 Kim, Y. and Chang, S. (2016). *Angew. Chem. Int. Ed.* 55: 218–222.
184 Gandhamsetty, N., Park, S., and Chang, S. (2015). *J. Am. Chem. Soc.* 137: 15176–15184.
185 Liu, Z.-Y., Wen, Z.-H., and Wang, X.-C. (2017). *Angew. Chem. Int. Ed.* 56: 5817–5820.
186 Gandhamsetty, N., Joung, S., Park, S.-W. et al. (2014). *J. Am. Chem. Soc.* 136: 16780–16783.
187 Gandhamsetty, N., Park, J., Jeong, J. et al. (2015). *Angew. Chem. Int. Ed.* 54: 6832–6836.
188 Issaian, A., Faizi, D.J., Bailey, J.O. et al. (2017). *J. Org. Chem.* 82: 8165–8178.
189 Soltani, Y., Wilkins, L.C., and Melen, R.L. (2017). *Angew. Chem. Int. Ed.* 56: 11995–11999.
190 Fürstner, A. and Davies, P.W. (2007). *Angew. Chem. Int. Ed.* 46: 3410–3449.
191 Wilkins, L.C., Lawson, J.R., Wieneke, P. et al. (2016). *Chem. Eur. J.* 22: 14618–14624.
192 Matsumoto, K., Sajna, K.V., Satoh, Y. et al. (2017). *Angew. Chem. Int. Ed.* 56: 3168–3171.
193 Cheng, C. and Brookhart, M.S. (2012). *Angew. Chem. Int. Ed.* 51: 9422–9424.
194 Hoff, R. and Mathers, R.T. (2010). *Handbook of Transition Metal Polymerization Catalysts* (Online ed.). Wiley.
195 Varga, V., Lamac, M., Horacek, M. et al. (2016). *Dalton Trans.* 45: 10146–10150.
196 Lee, P.T.K. and Rosenberg, L. (2017). *Dalton Trans.* 46: 8818–8826.
197 Heo, J., Kang, T., Jang, S.G. et al. (2012). *J. Am. Chem. Soc.* 134: 20139–20145.
198 Nikonov, G.I. (2017). *ACS Catal.* 7: 7257–7266.
199 Oertle, K. and Wetter, H. (1985). *Tetrahedron Lett.* 26: 5511–5514.
200 Yamamoto, K. and Takamae, M. (1990). *Synlett* 259–260.
201 Sudo, T., Asao, N., Gevorgyan, V., and Yamamoto, Y. (1999). *J. Org. Chem.* 64: 2494–2499.
202 Sudo, T., Asao, N., and Yamamoto, Y. (2000). *J. Org. Chem.* 65: 8919–8923.
203 Kato, N., Tamura, Y., Kashiwabara, T. et al. (2010). *Organometallics* 29: 5274–5282.
204 Song, Y.-S., Yoo, B.R., Lee, G.-H., and Jung, I.N. (1999). *Organometallics* 18: 3109–3115.
205 Jacobsson, K., Chu, T., and Nikonov, G.I. (2016). *ACS Catal.* 6: 7350–7356.
206 Koller, J. and Bergman, R.G. (2012). *Organometallics* 31: 2530–2533.
207 (a) Belgardt, T., Storre, J., Roesky, H.W. et al. (1995). *Inorg. Chem.* 34: 3821–3822. (b) Hair, G.S., Cowley, A.H., Jones, R.A. et al. (1999). *J. Am. Chem. Soc.* 121: 4922–4923.
208 Chen, E.Y.-X. (2012). *e-Encycl. Reag. Org. Synth.* https://doi.org/10.1002/047084289X.rn01382.
209 Groβekappenberg, H., Lühmann, N., Saak, W., and Müller, T. (2015). *Z. Anorg. Allg. Chem.* 641: 2543–2548.
210 Chen, J. and Chen, E.Y.-X. (2015). *Angew. Chem. Int. Ed.* 54: 6842–6846.

211 Feigl, A., Chiorescu, I., Deller, K. et al. (2013). *Chem. Eur. J.* 19: 12526–12536.

212 Fang, H., Wang, Z., and Fu, X. (2017). *Coord. Chem. Rev.* 344: 214–237.

213 Olah, G. (1970). *Science* 168: 1298–1311.

214 Baeyer, A. and Villinger, V. (1902). *Ber. Dtsch. Chem. Ges.* 35: 1189–1201.

215 Naidu, V.R., Ni, S., and Franzén, J. (2015). *ChemCatChem* 7: 1896–1905.

216 Corey, J.Y. (1975). *J. Am. Chem. Soc.* 97: 3237–3238.

217 Omann, L., Pudasaini, B., Irran, E. et al. (2018). *Chem. Sci.* 9: 5600–5607.

218 (a) Lambert, J.B., Kania, L., and Zhang, S. (1995). *Chem. Rev.* 95: 1191–1201. (b) Reed, C.A. (1998). *Acc. Chem. Res.* 31: 325–332.

219 Lambert, J.B. and Zhao, Y. (1997). *Angew. Chem. Int. Ed. Engl.* 36: 400–401.

220 Furukawa, S., Kobayashi, J., and Kawashima, T. (2009). *J. Am. Chem. Soc.* 131: 14192–14193.

221 Furukawa, S., Kobayashi, J., and Kawashima, T. (2010). *Dalton Trans.* 39: 9329–9336.

222 Arii, H., Yano, Y., Nakabayashi, K. et al. (2016). *J. Org. Chem.* 81: 6314–6319.

223 Arii, H., Kurihara, T., Mochida, K., and Kawashima, T. (2014). *Chem. Commun.* 50: 6649–6652.

224 Arii, H., Nakabayashi, K., Mochida, K., and Kawashima, T. (2016). *Molecules* 21: 999.

225 (a) Clark, E.R. and Ingleson, M.J. (2014). *Angew. Chem. Int. Ed.* 126: 11488–11491. (b) Lawrence, E.J., Clark, E.R., Curless, L.D. et al. (2016). *Chem. Sci.* 7: 2537–2543.

226 Fasano, V., Radcliffe, J.E., Curless, L.D., and Ingleson, M.J. (2017). *Chem. Eur. J.* 23: 187–193.

227 Schulz, A. and Villinger, A. (2012). *Angew. Chem. Int. Ed.* 51: 4526–4529.

228 Klare, H.F.T. (2017). *ACS Catal.* 7: 6999–7002.

229 Klare, H.F.T. and Oestreich, M. (2010). *Dalton Trans.* 39: 9176–9184.

230 Lambert, J.B., Zhao, Y., and Zhang, S.M. (2001). *J. Phys. Org. Chem.* 14: 370–379.

231 Douvries, C., Nagaraja, C.M., Chen, C.-H. et al. (2010). *J. Am. Chem. Soc.* 132: 4946–4953.

232 Douvries, C. and Ozerov, O.V. (2008). *Science* 321: 1188–1190; and references therein.

233 Liberman-Martin, A.A., Bergman, R.G., and Tilley, T.D. (2015). *J. Am. Chem. Soc.* 137: 5328–5331.

234 Maskey, R., Schädler, M., Legler, C., and Greb, L. (2018). *Angew. Chem. Int Ed.* 57: 1717–1720.

235 Bayne, J.M. and Stephan, D.W. (2016). *Chem. Soc. Rev.* 45: 765–774.

236 Dordevic, N., Ganguly, R., Petkovic, M., and Vidović, D. (2017). *Inorg. Chem.* 56: 14671–14681.

237 Chitnis, S.S., Krischer, F., and Stephan, D.W. (2018). *Chem. Eur. J.* 24: 6543–6546.

238 Stephan, D.W. (2017). *Angew. Chem. Int. Ed.* 56: 5984–5992.

239 Mehta, M., Garcia de la Arada, I., Perez, M. et al. (2016). *Organometallics* 35: 1030–1035.

240 Caputo, C.B., Hounjet, L.J., Dobrovetsky, R., and Stephan, D.W. (2013). *Science* 341: 1374–1377.

241 Caputo, C.B., Winkelhaus, D., Dobrovetsky, R. et al. (2015). *Dalton Trans.* 44: 12256–12264.

242 Holthausen, M.H., Mehta, M., and Stephan, D.W. (2014). *Angew. Chem. Int. Ed.* 53: 6538–6541.

243 Perez, M., Qu, Z.-W., Caputo, C.B. et al. (2015). *Chem. Eur. J.* 21: 6491–6500.

244 Augurusa, A., Mehta, M., Perez, M. et al. (2016). *Chem. Commun.* 52: 12195–12198.

245 Perez, M., Hounjet, L.J., Caputo, C.B. et al. (2013). *J. Am. Chem. Soc.* 135: 18308–18310.

246 Perez, M., Caputo, C.B., Dobrovetsky, R., and Stephan, D.W. (2014). *Proc. Natl. Acad. Sci. U.S.A.* 111: 10917–10921.

247 Mehta, M., Holthausen, M.H., Mallov, I. et al. (2015). *Angew. Chem. Int. Ed.* 54: 8250–8254.

248 Zhu, J., Perez, M., Caputo, C.B., and Stephan, D.W. (2016). *Angew. Chem. Int. Ed.* 55: 1417–1421.

249 Zhu, J., Perez, M., and Stephan, D.W. (2016). *Angew. Chem. Int. Ed.* 55: 8448–8451.

250 LaFortune, J.H.W., Johnstone, T.C., Perez, M. et al. (2016). *Dalton Trans.* 45: 18156–18162.

251 Bayne, J.M., Holthausen, M.H., and Stephan, D.W. (2016). *Dalton Trans.* 45: 5949–5957.

252 Moebius, J., vom Stein, T., and Stephan, D.W. (2016). *Chem. Commun.* 52: 6387–6390.

253 Mallov, I. and Stephan, D.W. (2016). *Dalton Trans.* 45: 5568–5574.

254 Holthausen, M.H., Bayne, J.M., Mallov, I. et al. (2015). *J. Am. Chem. Soc.* 137: 7298–7301.

255 Holthausen, M.H., Hiranandani, R.R., and Stephan, D.W. (2015). *Chem. Sci.* 6: 2016–2021.

3

Si—H Bond Activation by Transition-Metal Lewis Acids

Georgii I. Nikonov

Brock University, Chemistry Department, 1812 Sir Isaac Brock Way, St. Catharines, ON, L2S 3A1, Canada

The discovery of transition-metal-catalyzed hydrosilylation has brought significant and continuing interest in Si—H bond activation by transition metals [1]. The conventional mechanisms for the hydrosilylation of alkenes and ketones proposed by Chalk and Harrod [2] and Ojima et al. [3], respectively, are based on Si–H oxidative addition, substrate coordination, migratory insertion (either into the M—H or M—Si bond), and reductive elimination (forming Si—C or C—H bonds) steps. It was later realized that Lewis-acidic, and in particular cationic, metal complexes can operate via different mechanisms involving electrophilic activation of Si—H bonds [4]. Although the exact mechanistic details may differ, the unifying theme of these reactions is that the electrophilically activated silicon center (and not the metal) serves as the reaction site of the initial attack by the substrate [5]. Moreover, in many cases, this novel, Lewis acid metal-induced mode of silane activation provides unique chemo- and regioselectivity in the hydrosilylation of unsaturated substrates. The development of this field over the past 15 years has been very impressive and recently reviewed by Tilley and coworkers [6]. This short chapter highlights only a few selected results with emphasis on the mechanistic aspects of these reactions.

The first breakthrough happened three decades ago, when Luo and Crabtree discovered that the cationic, solvent-ligated complexes $[IrH_2S_2(PPh_3)_2]SbF_6$ (**1**, S = THF, CH_3OH, H_2O, Me_2CO) were very active catalysts (turnover frequency, TOF up to 50 000 h^{-1}) for the alcoholysis of silanes [7]. Based on detailed mechanistic and kinetic studies, the authors were able to propose a new mechanism involving the activation of silane through the formation of a cationic silane σ-complex **2** by coordination of the Si—H bond to the metal center without oxidative addition and subsequent nucleophilic attack by the alcohol on the η^2-$HSiR_3$ ligand (Scheme 3.1). Although good evidence for silane σ-complexes had been acquired by that time, all of them were neutral, whereas cationic analogs resisted isolation. Earlier attempts at preparing these species largely failed because of their high oxo-, nitro-, and halophilicity resulting in rapid abstraction of the silylium ion by solvents or counteranions (e.g. $[BF_4]^-$). Kinetic

Organosilicon Chemistry: Novel Approaches and Reactions, First Edition.
Edited by Tamejiro Hiyama and Martin Oestreich

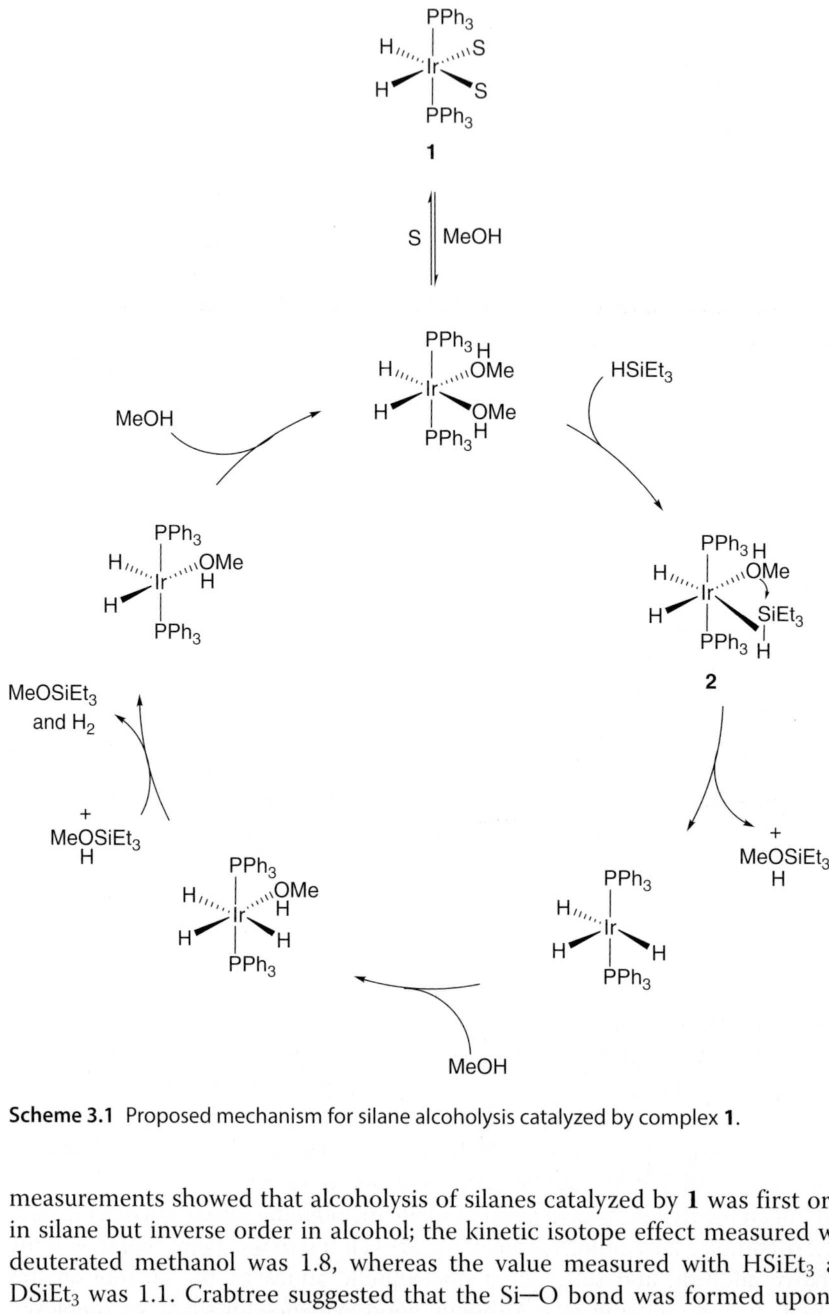

Scheme 3.1 Proposed mechanism for silane alcoholysis catalyzed by complex **1**.

measurements showed that alcoholysis of silanes catalyzed by **1** was first order in silane but inverse order in alcohol; the kinetic isotope effect measured with deuterated methanol was 1.8, whereas the value measured with $HSiEt_3$ and $DSiEt_3$ was 1.1. Crabtree suggested that the Si—O bond was formed upon an intramolecular attack of the coordinated methanol (by means of the second lone pair on oxygen) at the silane in intermediate **2** because the same reaction catalyzed by the related complex $[IrH(OH_2)(bq)L_2]^+$ (bq = 7,8-benzoquinolate), having only one coordination site, was 90 times slower relative to **1**.

The Brookhart group then showed that the iron complex $[Cp(CO)(Ph_3P)Fe(\eta^2\text{-}HSiEt_3)]^+$ (**3**), formed upon protonation of the silyl, methyl, or hydride precursors in the presence of silane, catalyzed the hydrolysis [8] and alcoholysis [9] of silanes. Significantly decreased activity in ethanol was observed in comparison with the Crabtree's system. This observation was attributed to rapid catalyst deactivation due to the formation of the alcohol adduct $[Cp(CO)(Ph_3P)Fe(OHEt)]^+$. In contrast, a less nucleophilic substrate, phenol, does not form such an adduct, and therefore phenolysis of silane proceeded at a good rate. Unlike the Crabtree catalyst **1**, all intermediates of the catalytic cycle, including the cationic silane complex **3** and triethylsilyl diethyloxonium ion, were detected by variable temperature NMR spectroscopy [10]. The use of the "innocent" borate counteranion $[BAr'_4]^-$ ($Ar' = 3,5\text{-}(F_3C)_2(C_6H_3)$) was crucial for stabilizing these highly reactive silylium species. It was suggested that the key catalytic event was the nucleophilic abstraction of the silylium ion by water or alcohol, which, unlike Crabtree's iridium catalyst, occurred intermolecularly because the fragment $[Cp(CO)(Ph_3P)Fe]^+$ provided only one reaction site. A density functional theory (DFT) study by Bühl and Mauschick supported and refined the mechanistic description of this reaction [11]. In particular, calculations confirmed the experimental observation that the rate-limiting step in the catalytic cycle was elimination of dihydrogen from the intermediate $[Cp(CO)(Ph_3P)Fe(\eta^2\text{-}H_2)]^+$ formed upon proton transfer from $[Et_3SiOHEt]^+$ to $Cp(CO)(Ph_3P)FeH$. Jagirdar and coworkers obtained experimental and computational evidence for the related electrophilic activation of the Si—H bond by the cationic complex $[Ru(P(OH)_3)(dppe)_2][OTf]_2$ (dppe = $Ph_2PCH_2CH_2PPh_2$), but its application as a catalyst was not reported [12].

The novel principle of silane activation by electrophilic metal complexes formulated in these earlier works was further extended to the hydrosilylation of ketones [13]. Dioumaev and Bullock reported that the cationic complexes $[Cp(CO)_2(IMes)M][B(C_6F_5)_4]$ (**4**, IMes = 1,3-bis(2,4,6-trimethylphenyl)imidazol-2-ylidene; M = Mo, W) catalyzed this process without any solvent, with the TOF reaching 2000 for the hydrosilylation of diethylketone with Me_2PhSiH mediated by the tungsten catalyst. Due to the significant decrease in polarity of the reaction media, >95% of the catalyst precipitated at the end of the reaction and thus could be recycled several times without a significant drop in activity. The authors proposed that the reaction proceeded by the ionic hydrosilylation mechanism outlined in Scheme 3.2. The cationic silyl hydride complex **5** was identified as the resting state of the catalyst. Abstraction of the silylium ion from **5** by the ketone generated an activated silylcarboxonium ion **7** and a neutral tungsten (II) hydride **6** which subsequently transferred the hydride to **7**, regenerating the catalyst.

Glaser and Tilley found that chloride abstraction from $Cp^*(^iPr_3P)Ru(H)(Cl)(SiH_2Ph)$ by $Li[B(C_6F_5)_4]\cdot 3Et_2O$ generated a cationic ruthenium complex that was able to catalyze the anti-Markovnikov *cis*-addition of $PhSiH_3$ to alkenes (hex-1-ene, ethene, cyclohexene, styrene, 1-methylcyclohex-1-ene) [14]. Based on the analogy with the stoichiometric reaction of the silylene complex $[Cp^*(^iPr_3P)Os(H)_2({=}SiHAr)][B(C_6F_5)_4]$ with hex-1-ene to give $[Cp^*(^iPr_3P)Os(H)_2({=}SiHexAr)][B(C_6F_5)_4]$, this ruthenium catalyst was originally formulated as the silylene complex **8** (Scheme 3.3a). The proposed mechanism of hydrosilylation involved insertion of olefin into the Si—H bond to give a secondary silylene species **9**, after

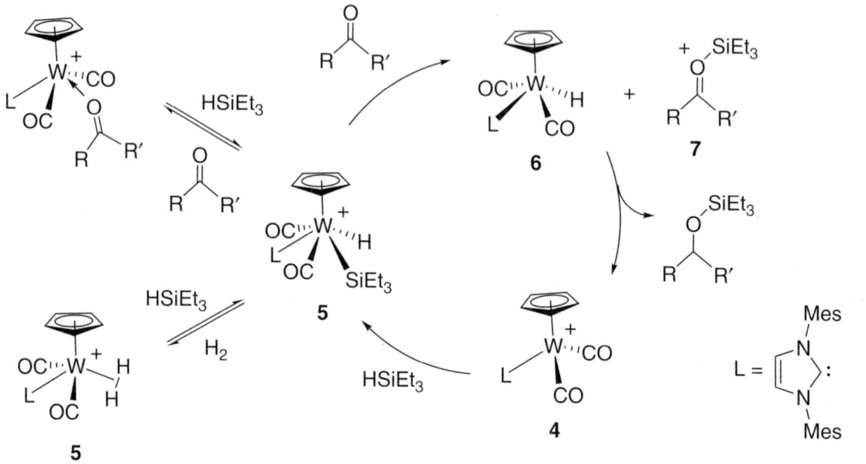

Scheme 3.2 Ionic hydrosilylation of ketone catalyzed by complex **4**.

which hydride migration and silane elimination produced the cationic complex **10**. The catalytic cycle is closed by double Si–H addition of a primary silane to regenerate the putative silylene complex. As such, this novel mechanism of hydrosilylation was analogous to hydroboration. Consistent with this mechanism was the observation that hydrosilylation was chemospecific, resulting exclusively in secondary silanes.

However, DFT calculations by Beddie and Hall [15] and Böhme [16] revealed that the silane addition product was not a silylene but an η^3-silane complex **11**, with two Si—H bonds coordinated to the metal center (Scheme 3.3b) [17, 18].

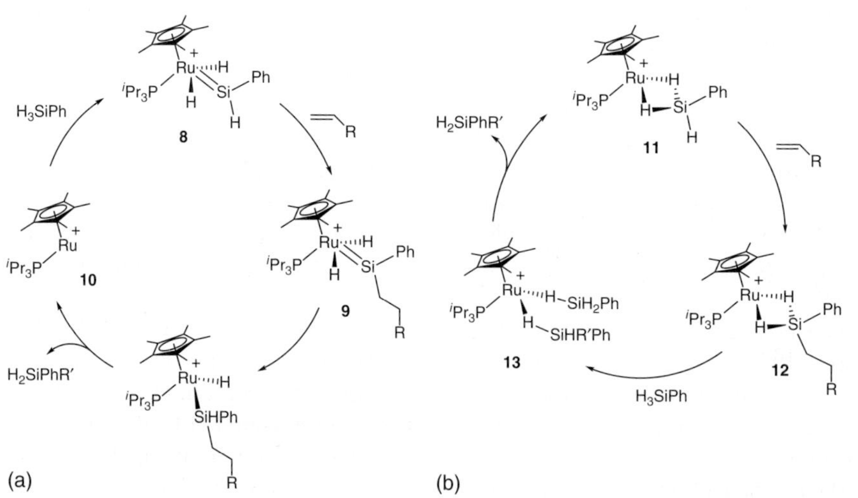

Scheme 3.3 Alkene hydrosilylation catalyzed by [Cp*(^{i}Pr$_3$P)Ru(H$_2$SiHAr)][B(C$_6$F$_5$)$_4$]. (a) Mechanism proposed by Glaser and Tilley. (b) Mechanism based on calculations by Beddie and Hall and by Böhme.

Calculations for the model system $[Cp(Me_3P)Ru(\eta^3\text{-}H_3SiPh)]^+$ further showed that bonding in the $Ru(\mu\text{-}H)_2Si$ fragment was rather flexible, such that opening the H—Ru—H bond angle by 25° required only 11–12 kJ/mol, whereas widening the H—Si—H angle by 13° needed 17–20 kJ/mol. As a result, coordination of ethylene to the silicon atom in the transition state resulted in nearly complete disruption of the bridging $Si—H^b(Ru)$ bonding, but once insertion into the terminal $Si—H^t$ bond was complete, the $\eta^3\text{-}H^b{}_2SiRAr$ fragment was restored (**12** in Scheme 3.3). Olefin insertion is the rate-determining step in the whole process. Of note is that alternative insertions of ethylene into the $Si—H^b$ bond or the $Ru—H^b$ bond to the bridging hydride are higher energy processes and require an extra 33 and 22 kJ/mol, respectively; therefore, the traditional Chalk–Harrod [2] and modified Chalk–Harrod [19] pathways of hydrosilylation are not realized. Unlike the original suggestion by Glaser and Tilley, the catalytic cycle is completed upon addition of another equivalent of silane to **12** to furnish the double η^1-silane complex **13** (Scheme 3.3), which then releases the product silane $H_2Si(C_2H_5)Ph$.

Related anti-Markovnikov hydrosilylation of alkenes was also observed with the cationic iridium silylene complex $[(PNP)(H)Ir({=}SiHMes)][B(C_6F_5)_4]$ (**14**) supported by Ozerov's PNP ligand [20, 21]. Lipke and Tilley further showed that the neutral ruthenium borate complexes $[PhB(CH_2PPh_2)_3]RuH(\eta^3\text{-}H_2SiRR')$ (**15**, RR′ = MePh and Ph_2) were efficient catalysts for the hydrosilylation of ketones with $MePhSiH_2$, Ph_2SiH_2, and $EtMe_2SiH$ at low catalyst loading (0.01–2.5 mol%) [22]. The latter reaction proceeded via analogous insertion of the C=O double bond into the Si—H bond of the electrophilically activated silane ligand. Ueno and coworkers then reported that cationic silylene complexes $[Cp(CO)_2Fe{=}SiR_2(NCCD_3)][B(Ar^F)_4]$ (**16**, Ar^F = 3,5-bis(trifluoromethyl)phenyl, R = *p*-tolyl or Me) catalyzed the hydrosilylation of aldehydes and ketones by the parent silyl complexes $Cp(OC)_2FeSiR_2H$ (**17**) in moderate to high yield [23]. Unlike the Tilley systems described, these reactions proceeded with secondary silylene complexes and likely occurred via the outer-sphere mechanism outlined in Scheme 3.4.

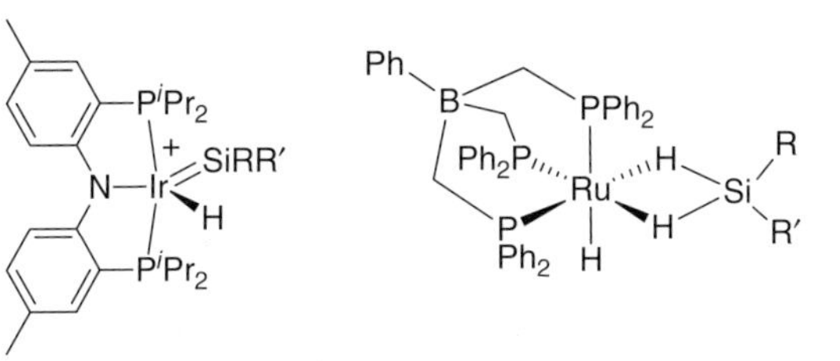

A different silylene-based mechanism was proposed by Gade et al. to explain the long-standing puzzle of catalytic hydrosilylation by rhodium complexes. That is, the classical Ojima mechanism, based on the sequence of Si–H oxidative addition to Rh(I), ketone insertion into the Rh—Si bond, and C–H reductive elimination, failed to explain the unusual rate enhancement and kinetic isotope effect observed when secondary silanes H_2SiR_2 were used instead of tertiary silanes.

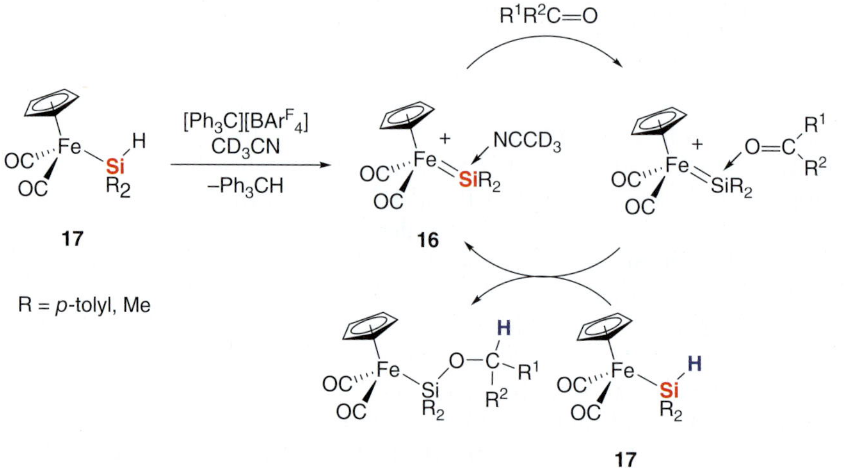

Scheme 3.4 Carbonyl hydrosilylation catalyzed by $[Cp(OC)_2Fe{=}SiR_2(NCCD_3)]^+$.

The special role of a secondary silane led Zheng and Chan to propose an alternative pathway based on carbonyl coordination to a secondary silyl ligand (Scheme 3.5, left) [24]. Gade and coworkers reported detailed mechanistic and DFT studies on the hydrosilylation of ketones catalyzed by the cationic rhodium complex **18** supported by a chiral oxazoline–N-heterocyclic carbene (NHC) ligand [25–27]. This catalyst also showed enhanced activity for secondary versus tertiary silanes and displayed an inversed isotope effect of 0.8, which was incompatible with the

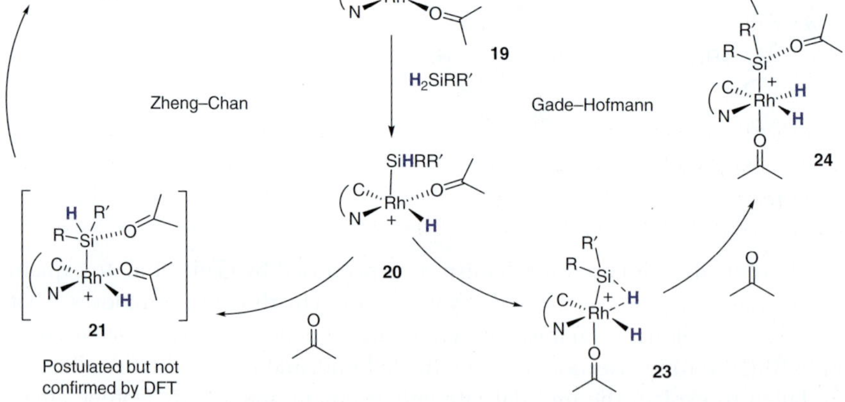

Scheme 3.5 Ketone hydrosilylation catalyzed by cationic oxazoline–carbene complex of Rh. Left: The Zhen–Chang mechanism. Right: The Gade–Hoffmann mechanism.

Ojima and Chang mechanisms. Under catalytic conditions, when excess ketone is present, the diene ligand of **18** is replaced by two molecules of ketone to give **19**. After oxidative addition of silane, the silyl ketone complex **20** is formed. Complexes **19** and **20** are pertinent to the Ojima (not shown in Scheme 3.5), Zheng–Chan (Scheme 3.5, left) and Gade–Hofmann mechanisms (Scheme 3.5, right) under discussion. The rate-determining step in the Ojima mechanism is ketone insertion into the Rh—Si bond, requiring a high barrier with $\Delta G^{\neq}$ = 132.7 kJ/mol. The key step in the Zhen–Chan scheme, coordination of ketone to the silyl ligand as in **21**, was not confirmed by DFT calculations. Instead, ketone inserts directly into the Si—H bond with a prohibitively high barrier of 195.7 kJ/mol to furnish the alkoxysilyl complex **22**. In contrast, the highest barrier in the Gade–Hofmann mechanism relative to the common intermediate **20** is only 77.6 kJ/mol. In this mechanism, **20** undergoes distortion of the silyl ligand to give **23**, which can be described either as an α-agostic silyl complex or a silylene with additional Si⋯H⋯Rh interactions. Consistent with both interpretations are the calculated Si–H and Rh–H distances to the bridging hydride of 1.68 and 1.87 Å. The attack of ketone at the silicon center of **23** results in complete cleavage of the Si—H bond and formation of the silylene adduct **24**. Both **23** and **24** were calculated to be very close in energy. Next, the hydride shifts to the carbonyl via a five-membered transition step requiring $\Delta G^{\neq}$ = 36.1 kJ/mol. In this catalytic cycle, silicon assumes the role of an intramolecular Lewis acid activator for the ketone substrate which is reduced by an outer-sphere mechanism. A mechanism similar to Gade–Hofmann may be operative in ketone hydrosilylation catalyzed by the secondary silylene complex [Cp*$(Me_3P)_2$Ir(=$SiPh_2$)(H)][B$(C_6F_5)_4$] [28]. The common theme of both the Tilley and Gade–Hofmann pathways is that a nonclassically coordinated silane or silyl ligand can be easily transformed into a silylene upon nucleophilic attack of the substrate.

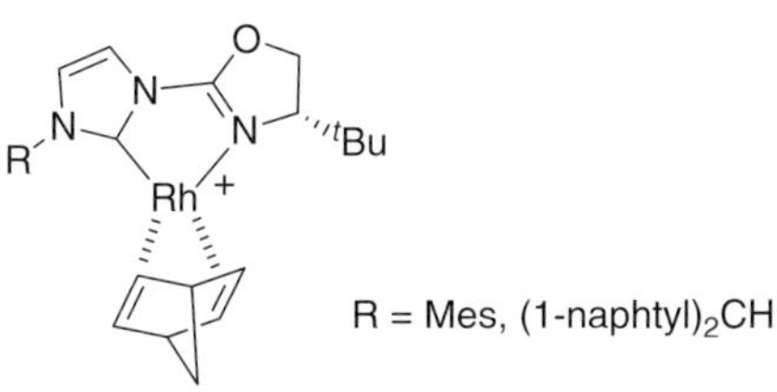

18

Abu-Omar and coworkers discovered that the oxorhenium(V) oxazoline complex **25** catalyzed the hydrosilylation of aromatic and aliphatic ketones at room temperature with low catalyst loading (0.1 mol%) with or without solvent [29]. In the latter case, akin to the Bullock system discussed earlier, the catalyst precipitates at the end of the reaction. The same catalyst **25** catalyzed the hydrolysis of silanes. The use of ^{18}O-labeled water in the stoichiometric experiment, **25**:Ph_2MeSiH:$H_2{}^{18}O$ = 1 : 1 : 1, resulted in only slightly reduced ^{18}O-enrichment in the silyl ether product (due to oxo exchange between $H_2{}^{18}O$ and **25**), confirming that Si—H addition across the Re=O bond is not relevant to this catalytic cycle. The latter fact is of significance because Toste and coworkers had previously showed that the related neutral complex Re(=O)$_2$I$(PPh_3)_2$ reacted with silanes to give $Re(=O)(OSiR_3)I(PPh_3)_2$ [30]. On the other hand, reaction of the neutral hydride $Re(O)(hoz)_2H$ (**26**, hoz = 2-(2¢-hydroxyphenyl)-2-oxazoline(−)) with the silyl oxonium ion $[PhHC{=}OSiEt_3]^+$ did result in hydride transfer and formation of silyl ether. However the hydricity of **26**, measured by the kinetics of its reaction with $[Ph_3C][B(C_6F_5)_4]$, was deemed to be insufficient. Thus, the rate constant for the hydride transfer from **26** was much lower than from $HSiEt_3$, 2.40 ± 0.05 versus $170 \pm 8\,M^{-1}\,s^{-1}$, respectively. Based on these observations, the ionic hydrosilylation mechanism proposed by Bullock was rejected. The mechanism in which complex **25** serves merely as an initiator to generate $[SiEt_3]^+$ was also ruled out because catalysis by the silylium ion has been shown to produce ethylbenzene in the reaction with acetophenone, which was not observed during catalysis with **25**. As an alternative mechanism, the authors suggested the sequence shown in Scheme 3.6a. However, DFT calculations by Wei did not confirm formation of the "σ-bond metathesis structure" **27**, whereas exploration of the ionic hydrosilylation route revealed reasonable barriers for all essential steps [31]. The reaction proceeded via formation of an η^1-silane complex **28**, which then rearranges into an isomer **28′** bearing the hydride trans to phenoxide ligand. The highest barrier is the transition state for silylium ion abstraction from **28′** by the carbonyl ($\Delta G^{\neq} = 17.4$ kcal/mol). Direct abstraction from **28** is a higher energy process because it would place the hydride and oxo ligands, two of the strongest trans influence ligands, opposite each other. The final step, hydride transfer from **26′** to the silyloxonium ion, is exergonic by 30.1 kcal/mol and is in fact relatively facile, requiring only 6.8 kcal/mol. In contrast, hydride transfer from $HSiMe_3$ is unfavorable by 3.0 kcal/mol. The ionic hydrosilylation mechanism is also likely applicable in other Re-catalyzed processes [32–34].

Conclusive proof of the intermediacy of η^1-silane complexes, such as **13** in Scheme 3.3, was garnered by Brookhart and coworkers [35]. These authors found that the cationic iridium complex $[(POCOP)Ir(H)(\kappa^1\text{-}O{=}CMe_2)][B(C_6F_5)_4]$ (**29**), where POCOP = $2,6\text{-}[^tBu_2PO]_2C_6H_3$), reacted with Et_3SiH in CD_2Cl_2 at 23 °C to give an isolable η^1-silane derivative $[(POCOP)Ir(H)(\eta^1\text{-}HSiEt_3)][B(C_6F_5)_4]$ (**30**) that was studied by X-ray diffraction, the first example of a structurally characterized cationic silane complex. Brookhart and a coworker then showed that **29** mediated a remarkable series of catalytic reductions involving silanes. Thus, **29** catalyzed the reduction of alkyl halides by Et_3SiH in chlorobenzene or in neat alkyl halide at very low catalyst loadings (0.5 mol% or less) operating at temperatures between 23 and 60 °C [36, 37]. Primary, secondary, and tertiary chlorides, bromides, and iodides as well as certain fluorides (e.g. 1-fluoropentane) undergo this reaction. Interestingly, the substrate reactivity order was RI < RCl < RBr, 2° RX < 1° RX, and $H_2CCl_2 < H_3CCl$, which was inconsistent with a radical mechanism. The unexpected sluggishness of the reduction of alkyl iodides was attributed to the strength of RI coordination in complexes **31**. In fact, the resting state of the catalyst in the reduction of alkyl iodides is the adduct **31**, whereas for other halides an equilibrium mixture of the halide complex **31** and the silane complex **30** forms, with the ratio dependent on the proportion of silane to halide and the nature of the halide. Kinetic studies of the reduction of CH_3I established that the reaction was zero order in the substrate (consistent with **31** being the resting state and the existence of the

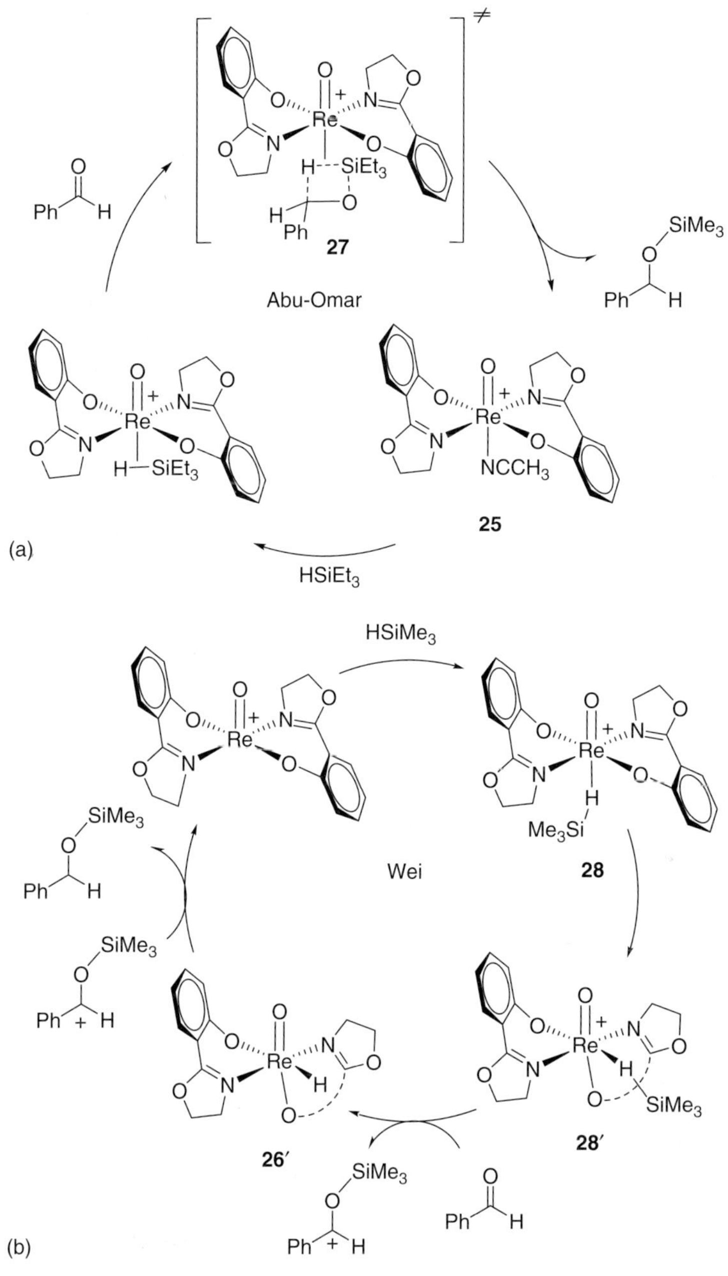

Scheme 3.6 Ketone hydrosilylation catalyzed by oxorhenium(V) oxazoline complex **25**. (a) Mechanism suggested by Abu-Omar et al. (b) Essential steps in DFT-calculated mechanism.

pre-equilibrium between **31** and **29**) and first order in silane. Based on these observations, an ionic hydrosilylation mechanism was proposed (Scheme 3.7). The reducing species for the silylhalonium ion was postulated to be the dihydride complex **32**, but subsequent studies suggested that it may be a different iridium hydride complex (*vide infra*).

Brookhart and coworkers further showed that the cationic iridium complex **29** is able to catalyze the hydrosilylation of a wide variety of nonactivated alkyl ethers ROR′, including primary, secondary, and tertiary alkyl ethers as well as aryl alkyl ethers, to give RH and Et_3SiOR' [38]. The reduction proceeded readily at room temperature with 1% catalyst load and with Et_3SiH as the reducing agent, whereas the second reduction of Et_3SiOR' to R′H and $(Et_3Si)_2O$ required longer reaction times and/or heating. Gagné and coworkers then demonstrated the successful application of this iridium catalyst in combination with Et_2SiH_2 toward the reduction of silyl-protected sugars to give a mixture of hexane isomers [39]. Mechanistic studies by Brookhart and coworkers identified the cationic η^1-silane complex **30** as the key intermediate and showed that this complex and Et_2O are in rapid equilibrium with the neutral dihydride $(POCOP)Ir(H)_2$ (**32**) and the oxonium ion $[Et_3SiOEt_2][B(C_6F_5)_4]$ (**33**) [30]. These two latter species were strongly favored for most substrates except for less basic ethers (e.g. anisole and $EtOSiEt_3$), in which case the cationic silane complex **30** was the resting state of the catalyst. An ionic hydrosilylation mechanism related to that shown in Scheme 3.7 was proposed to explain this chemistry. It was further suggested that hydride transfer from dihydride **32** to **33** was the rate-determining step. Small quantities of the silyl hydride $(POCOP)Ir(H)(SiEt_3)$, formed by the reaction of dihydride **32** with silane, were also identified by NMR. This silyl hydride complex was suggested to be less effective in hydride transfer than **32**.

The same iridium catalyst **29** has been successfully applied in the hydrosilylation of epoxides and cyclic ethers to furnish Et_3SiOR [40]. Because the latter species are much less prone to reduction (*vide supra*), highly chemoselective production of silyl-protected alcohols was accomplished. The reaction proceeds under mild conditions (chlorobenzene, 22 °C) at low catalyst loadings (0.25–0.50 mol%). Since the reaction likely occurs via the ionic hydrosilylation mechanism requiring

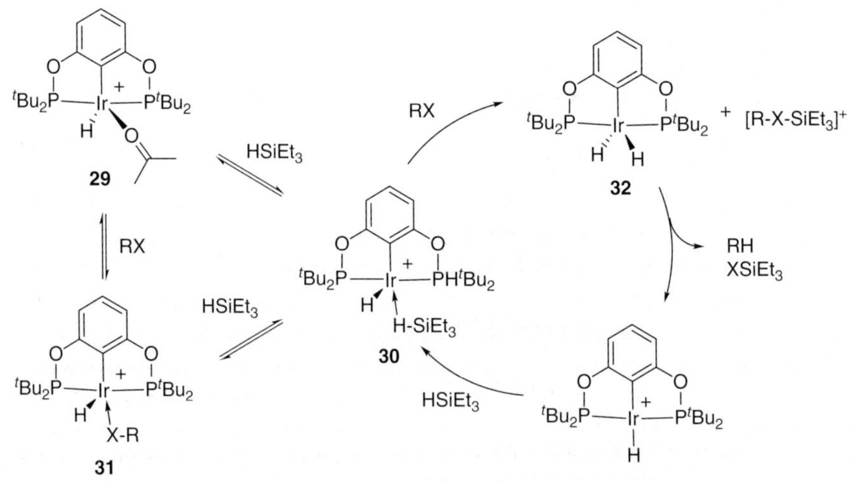

Scheme 3.7 Reduction of alkyl halides by Brookhart's complex **29**.

the intermediacy of a silylcarboxonium ion, for certain substrates structural rearrangements become possible. Thus, hydrosilylation of 2,3-dimethyl-2,3-epoxybutane (= 2,2,3,3-tetramethyloxirane) gave the silyl-protected secondary alcohol, $Et_3SiOCHMeCMe_3$, not the expected silyl-protected tertiary alcohol, $Et_3SiOCMe_2CHMe_2$, as a result of methyl-group migration to form a more stable oxonium-ion intermediate. Mechanistic studies further showed that the reaction proceeded via Lewis acid-induced isomerization of epoxides to ketones followed by hydrosilylation [32].

The latter process, the hydrosilylation of ketone and aldehydes by Et_3SiH mediated by the cationic complex **29**, was the subject of a separate study [41]. Quantitative reduction was accomplished at low catalyst loading (0.075–0.50 mol%) and mild conditions (20–30 minutes at 25 °C) even for very bulky ketones, such as diisopropyl ketone and methyl *tert*-butyl ketone. Moreover, **29** catalyzed the reduction of challenging substrates such as esters and amides. Thus, ethyl acetate was reduced to a mixture of Et_3SiOEt and Et_2O in a 2 : 1 ratio, the latter product likely resulting from the C—$OSiEt_3$ bond cleavage in the intermediate acetal $MeHC(OSiEt_3)(OEt)$. *N,N*-Diethyl acetamide was slowly hydrosilylated (42% conversion in 16 hours) with concurrent formation of Et_3N, suggesting C—$OSiEt_3$ bond cleavage in the intermediate hemiaminal $MeHC(OSiEt_3)(NEt_2)$. Mechanistic studies suggested that ketone complexes, $[(POCOP)Ir(H)(ketone)]^+$, related to **29** were the resting state of the catalyst existing in equilibrium with small amounts of the silane complex **30**. Consistent with this pre-equilibrium, initial-rate analysis revealed that the catalytic reaction rate was zero order in ketone and first order in silane, whereas competition experiments showed that more basic substrates, e.g. acetophenone versus 4′-(trifluoromethyl)acetophenone, were reduced faster. Altogether, these observations were in agreement with the ionic hydrosilylation mechanism shown in Scheme 3.8 (left).

DFT calculations by Wei et al. provided further support to the reaction sequence based on silylium ion abstraction from **30** to give an oxonium ion and iridium dihydride **32** followed by hydride transfer from the latter to the former to complete the reduction [42]. The rate-determining step was found to be silylium ion abstraction by the ketone with $\Delta G^{\neq}$ = 19.5 kcal/mol, whereas the hydride-transfer step was a fast process requiring only $\Delta G^{\neq}$ = 0.6 kcal/mol. The alternative Ojima mechanism required a high barrier with $\Delta G^{\neq}$ = 37.8 kcal/mol for carbonyl insertion into the Ir—H bond and thus was not kinetically competitive with the ionic hydrosilylation.

The Brookhart–Wei mechanism was challenged by Oestreich and coworkers who applied a silicon-stereogenic silane **34** as a mechanistic probe [43]. With this silane as the reductant, hydrosilylation of acetophenone with iridium catalyst **29** resulted in racemization, rationalized by an equilibrium between $[Et_3SiO{=}CMePh]^+$ and acetophenone prior to reduction. The reaction was found to depend crucially on the amount of silane used; it took 20 hours for 1 equiv of **34** but only minutes with 4 equiv of silane. Even more remarkably, reacting the dihydride **32** with silyloxonium species $[Et_3SiO{=}CMePh]$ $[B(C_6F_5)_4]$ resulted in hydride transfer to silicon along with formation of the ketone complex $(POCOP)Ir(H)(\kappa^1\text{-acetophenone})$ (**35**) and the silane complex **30** but only traces of the target hydrosilylation product $Et_3SiOCHMePh$. This

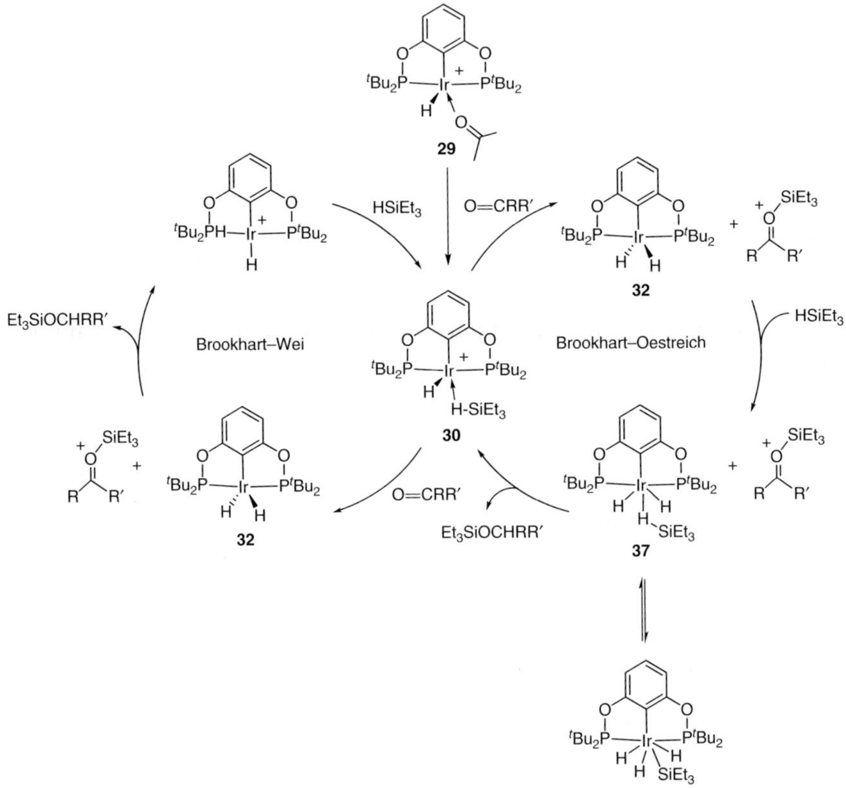

Scheme 3.8 Hydrosilylation of ketones catalyzed by Brookhart's complex **29**. Left: The Brookhart–Wei mechanism. Right: The Brookhart–Oestreich mechanism.

result suggests that the reverse reaction to give **30** requires a lower barrier than reduction to form the hydrosilylation product. The corresponding barriers in Wei's calculations for a model iridium complex (Me group at phosphorus instead of ^{i}Pr) were only 0.4 and 0.6 kcal/mol. Addition of silane to a mixture of **35** and **30** led to immediate production of silylated alcohol. Since the silane itself is not a sufficiently powerful reductant, it was concluded that its addition enhances the reducing ability of the iridium hydride. The active species was identified as the Ir(V) trihydride complex (POCOP)IrH_3SiR_3 (**36**) existing in equilibrium with the Ir(III) species (POCOP)$IrH_2(\eta^1\text{-}HSiR_3)$ (**37**, Scheme 3.8, right). The identity of these silyl species was assigned on the basis of ^{1}H NMR hydride chemical shifts, ^{1}H,^{31}P HSQC, and ^{1}H,^{29}Si HSQC, while the hapticity of silane complex **37** was established by DFT calculations. The reason why the Ir(V) hydride **36** is a better hydride donor than the Ir(III) hydride **32** was attributed to the higher hydricity of the former, gauged by the lower calculated ΔG for the hydride transfer to $[Ph_3C]^+$. It should be noted that Oestreich's refinement of the original Brookhart mechanism does not contradict Brookhart's experimental data but is at odds with the results of Wei's calculations predicting very facile hydride transfer from complex **34**.

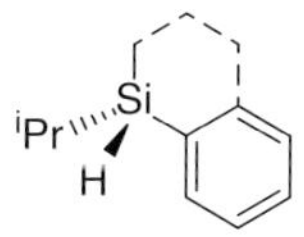

34

The intermediacy of Ir(V) silyl hydride **36** as a reductant in catalytic hydrosilylation was further demonstrated in the hydrosilylation of tertiary amides mediated by cation **29**. Brookhart and a coworker showed that 0.5 mol% of **29** catalyzed effectively the reduction of a variety of aromatic, aliphatic, heteroaromatic, and heterocyclic tertiary amides to the corresponding tertiary amines with 3 equiv Et_2SiH_2 in chlorobenzene at 60 °C [44]. These conditions tolerate the presence of olefin, alkyne, and nitrile moieties; however, keto and ester groups are readily reduced. $EtMe_2SiH$ can also be used, but longer reaction times are required, e.g. the reduction of PhC(=O)NMe_2 with Et_2SiH_2 took 1 hour, while reduction with $EtMe_2SiH$ required 48 hours. Monitoring the hydrosilylation of *N,N*-dimethylbenzamide by $EtMe_2SiH$ with ^{1}H NMR spectroscopy revealed that a $EtMe_2SiH$ analog of the cationic silane complex **30**, [(POCOP)Ir(H)(η^1-$HSiMe_2Et$)]$^+$, and the amide adduct (POCOP)Ir(H)(Me_2NC(=O)Ph), formed in a 69 : 31 ratio, were the resting state(s) of the catalyst. In contrast, the dominant species in the catalytic reduction of *N,N*-dimethylbenzamide with Et_2SiH_2 were the neutral Ir(III) complex (POCOP)Ir(H)($SiHEt_2$) (**38**) and the Ir(V) species (POCOP)Ir$(H)_3$($SiHEt_2$) (**39**), but these two complexes can only reduce the benzamide to the hemiaminal PhCH($OSiHEt_2$)NMe_2. Complete reduction to amine is only possible when an ammonium salt, e.g. [Et_3NH][$B(C_6F_5)_4$], is added. Complex **39** appears to be a better reducing agent because carrying out the hydrosilylation with an equilibrating mixture of **38** and **39** under hydrogen atmosphere shifting the equilibrium to **39**, allows for lower reaction temperature and shorter reaction times. A possible mechanism is given in Scheme 3.9.

The Brookhart group also reported that secondary amides can be effectively reduced by Et_2SiH_2 to imines and (with longer reaction times) to secondary amines using the iridium catalyst [$Ir(COE)_2Cl$]$_2$ at low catalyst load (0.5–1%) under mild conditions (room temperature in non-coordinating solvents or neat silane) with up to 1000 turnovers at room temperature and 99% conversion [45]. These reaction conditions tolerate such functional groups as aryl and alkyl halides, ethers, heterocycles, azo and amino groups, and olefins but are not compatible with nitro groups. Of particular note is that a *para*-ester-substituted *N*-butyl benzamide could be selectively reduced to an imine with complete retention of the typically more reactive ester group. Also interesting is that if a cyano group is present, the conversion of amide to imine is still fast, but further reduction is severely retarded. Mechanistic studies showed that the resting state of the catalyst is the silylene-bridged dimer **40** (Scheme 3.10), although it remains unclear if it is an Ir(V) tris(silyl) complex or an Ir(III) bis(silyl)silane complex. The proposed mechanism shown in Scheme 3.9 is a variant of ionic hydrosilylation. Abstraction of a silylium group from **40** generates an activated amino-siloxy cation which is then reduced by a hydride to furnish an O-silylated hemiaminal. Decomposition of the latter generates an imine, which can be then reduced in a second catalytic cycle.

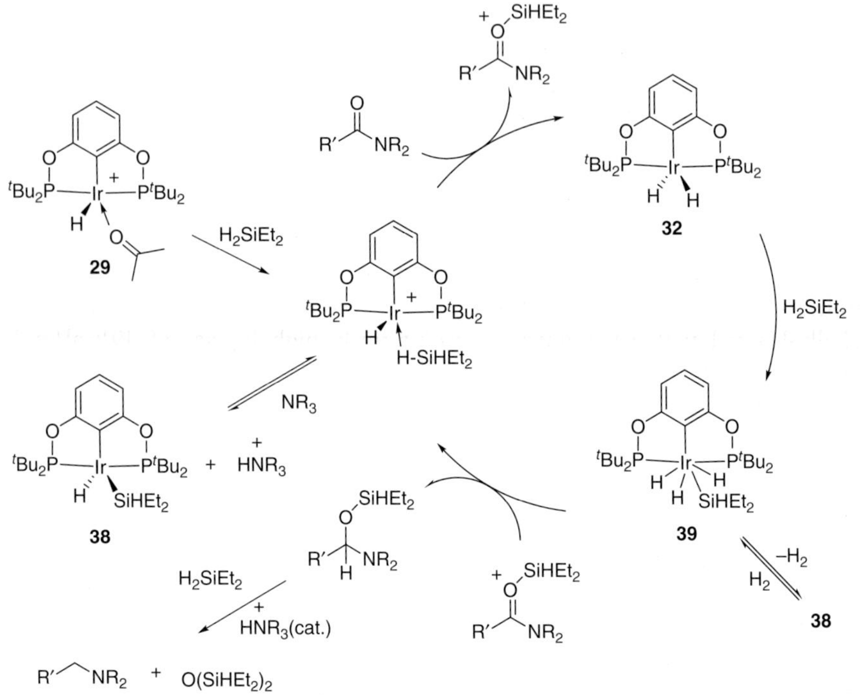

Scheme 3.9 Mechanism of tertiary amide hydrosilylation mediated by **29**.

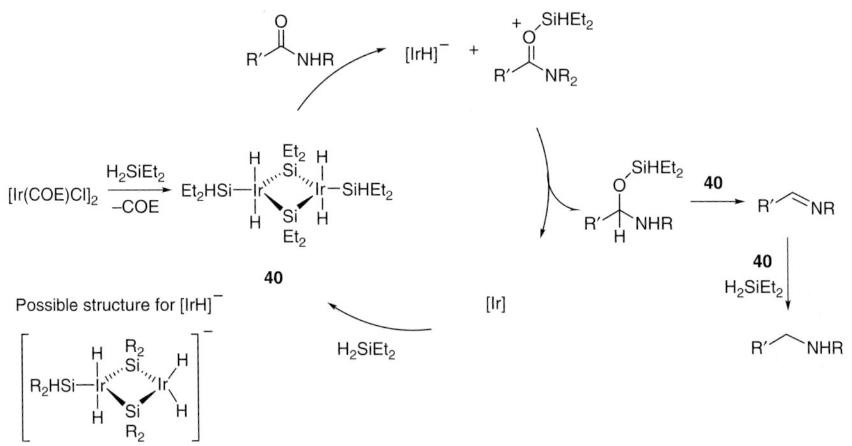

Scheme 3.10 Mechanism of tertiary amide hydrosilylation catalyzed by **40**.

Our group showed that 5 mol% of [Cp(^{i}Pr$_2$MeP)Fe(NCCH$_3$)$_2$][BF$_4$] catalyzed the hydrosilylation of benzaldehyde at room temperature in the course of three hours [46]. Wishing to enhance the catalyst efficiency, we applied the ruthenium analog [Cp(^{i}Pr$_3$P)Ru(NCCH$_3$)$_2$][B(C$_6$F$_5$)$_4$] (**41**) with the idea that it will provide more access to the coordination sphere of the metal. As hypothesized, the reaction time was shortened to 30 minutes and ketones were also reduced in the course of 18–24 hours, although the yields were low due to concomitant formation of silyl enol ethers [47]. Mechanistic studies showed that **41** did not react with ketones but readily reacted with hydrosilanes to give a series of surprisingly stable cationic silane σ-complexes $[Cp(^iPr_3P)Ru(NCCH_3)(\eta^2\text{-}HSiR_3]$ $[B(C_6F_5)_4]$ (**42**), only the third example of isolable compounds of this type. Even more importantly, application of **42** to catalysis showed that they too can mediate the hydrosilylation of carbonyls, proving the long-held belief that silane complexes can be intermediates in silane transformations. Our DFT studies on the model system $[Cp(Me_3P)Ru(NCMe)_2]^+$ further showed that the conventional Chalk–Harrod pathway requires a prohibitively high barrier of 49 kcal/mol, whereas the ionic hydrosilylation mechanism is much more feasible. The rate-determining step is the abstraction of silylium ion from the η^2-silane complex by the carbonyl oxygen atom (Scheme 3.11). Hydride transfer from the neutral hydride intermediate **43** to the silyl oxonium ion **45** proceeds via intermediate formation of the η^1-alkane complex **44** and requires only $\Delta G^{\neq} = 4.0$ kcal/mol. Wu and coworkers performed related studies for the real system $[Cp(^iPr_3P)Ru(NCMe)_2]^+$ and reached similar conclusions, despite the barrier for the rate-determining step being slightly higher, 22.6 versus 18.6 kcal/mol [48].

Oestreich and coworkers further refined this mechanism by arguing that the true reducing species in the mixture was the Ru(IV) silyl hydride $Cp(^iPr_3P)RuH_2SiR_3$ (**46**) [49]. Indeed, complexes akin to **46**, previously prepared by us [50, 51] and others [52, 53], were shown to reduce oxonium ions. However, given the fact that species **43** is very labile and has so far eluded isolation, it remains to be proved experimentally or computationally if the reaction goes by interception of **43** with the silane followed by hydride transfer from **46** to oxonium ion **45** or

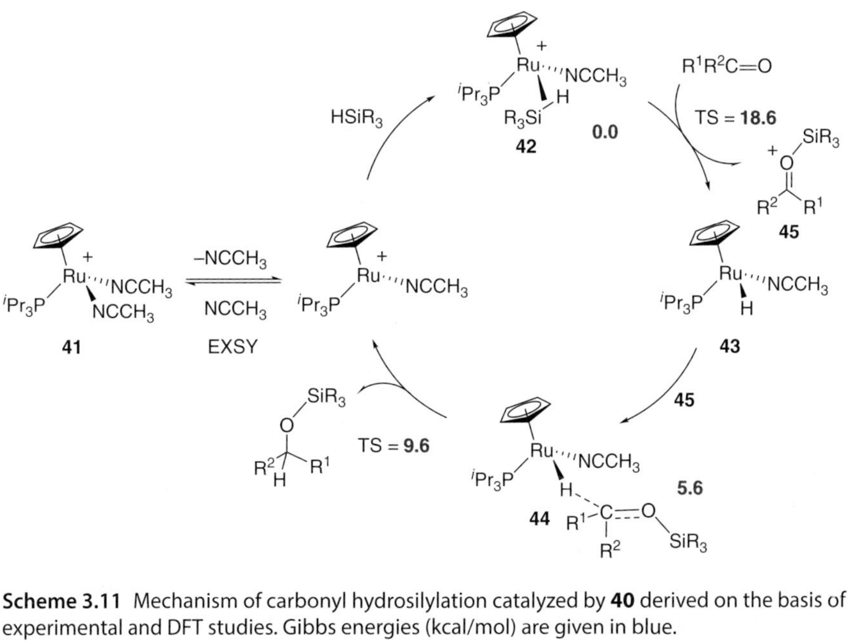

Scheme 3.11 Mechanism of carbonyl hydrosilylation catalyzed by **40** derived on the basis of experimental and DFT studies. Gibbs energies (kcal/mol) are given in blue.

by direct transfer from **43** to **45**. The argument for the first possibility is that silane is present in large excess, whereas the second pathway may be possible because both reactive species, **43** and **45**, are formed in a solvent cage.

An even more surprising discovery was made while monitoring the catalytic system of **41** and Me_2PhSiH by ^1H NMR spectroscopy. One day, Dmitry Gutsulyak came to my office, saying that he had observed an NMR splitting pattern for an ethyl group that could be only derived from the hydrosilylation of the acetonitrile ligand that dissociated from **41**. I did not believe him because nitrile hydrosilylation had been known to be a notoriously difficult process. However, Dmitry was persistent. He prepared $MeCH_2N(SiMe_2Ph)_2$ and showed that its spectrum was identical to the pattern observed in the catalytic mixture. To test this hypothesis, Dmitry attempted the catalytic hydrosilylation of a series of nitriles with 4 mol% of **41**, and the result was most surprising. Not only did the reaction proceed rapidly in $CDCl_3$ or CD_2Cl_2 at room temperature (e.g. four hours for PhCN) but the product was the imine $RHC{=}NSiMe_2Ph$, a substrate that is usually more reactive than the starting nitrile. Complete reduction of nitrile to amine could also be accomplished but required a much longer time (e.g. 66 hours for PhCN). Even more astounding, this catalytic protocol tolerated the presence of many other typically more reactive functionalities such as esters, ketones, and even aldehydes, suggesting that reduction of the cyano group could be performed chemoselectively without the need for costly protection/deprotection steps. The difference in the reactivity was such that reductions could be performed in acetone as the solvent! This unusual reactivity can be accounted for by the ionic hydrosilylation mechanism. The resting state of the catalyst is the silane complex **42** (Scheme 3.12). Nitrile abstracts the silylium ion better than do carbonyls to give the iminium ion **47**. Transfer of hydride from **43** or **46** generates the imine product and closes the catalytic cycle. The observed chemoselectivity is then explained by the higher nucleophilicity of nitriles, relative to carbonyls and imines, which results in rapid production of the iminium intermediate **47**.

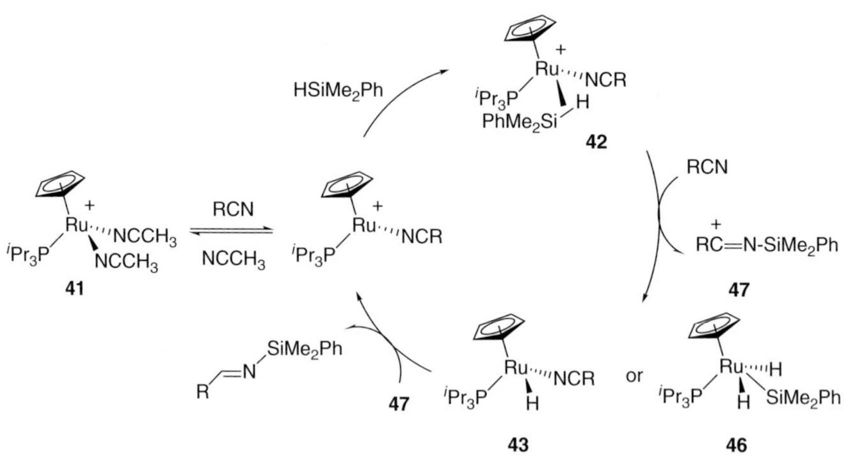

Scheme 3.12 Mechanism of nitrile hydrosilylation catalyzed by **41**.

Complex **41** also catalyzed the regio- and chemospecific hydrosilylation of pyridines to give *N*-silyl 1,4-dihydropyridines [54]. This result was significant because previously developed methods produced mixtures of 1,2- and 1,4-isomers that are difficult to separate [55]. Substitution in the 3 and 5 positions was tolerated, whereas substituents in the 2, 4, and 6 positions were not, despite quinoline being reduced in 50% yield. Although some tolerance to functional groups (e.g. chloride) was observed, it was much reduced in comparison to the case of nitriles, primarily due to the reactivity of dihydropyridines to carbonyls. For example, *N*-silyl dihydropyridine 1,4-$PhMe_2SiNC_5H_6$ (**48**) reacted with acetone to produce the hemiaminal 1,4-$PhMe_2SiO$-CMe_2-NC_5H_6 (**49**), which decomposed in hexanes to give the parent dihydropyridine 1,4-HNC_5H_6 and silylated enol. Mechanistic studies with Me_2PhSiD showed that the deuterium label is delivered exclusively to the 4-position, i.e. the reaction proceeded via direct production of 1,4-dihydropyridine and not through formation of the 1,2-isomer followed by rearrangement. Moreover, the hydrosilylation is reversible because addition of 3,5-lutidine to **48** in the presence of **41** resulted in an equilibrating mixture of two dihydropyridines (Scheme 3.13), whereas the transfer of silane to the nitrile PhCN to form PhHC=$NSiMe_2Ph$ was irreversible. Sun-Hwa Lee, who inherited this project from Dmitry, performed a kinetic study of the catalytic hydrosilylation of pyridine [56]. The reaction was first order in silane concentration but showed saturation behavior in the amount of pyridine. These kinetics suggest a pre-equilibrium step

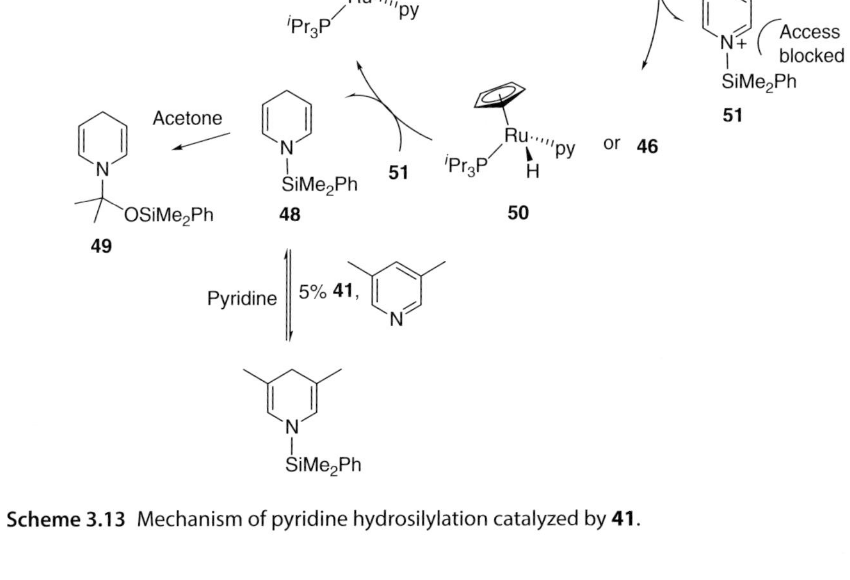

Scheme 3.13 Mechanism of pyridine hydrosilylation catalyzed by **41**.

involving the silane, which is in agreement with the mechanism shown in Scheme 3.12. Akin to the nitrile case discussed earlier, a cationic silane complex is formed upon addition of silane. Nucleophilic attack of pyridine, which is sensitive to substitution in the 2 and 6 positions, resulted in abstraction of the silylium ion and formation of the activated pyridinium salt. The observed 1,4-selectivity is likely due to the bulky silyl group blocking access to the 2 and 6 positions of the intermediate *N*-silyl pyridinium ion. The hydride-transfer step was further verified by the reaction of labile hydride **50** with the pyridinium salt $[(Et_3Si)(py)][B(C_6F_5)_4]$ which gave 1,4-$Et_3SiNC_5H_6$.

Since the ionic hydrosilylation mechanism requires accessibility to only one reaction site and the second site in **41** under catalytic conditions is likely occupied by pyridine, we reckoned that phenathroline-supported complexes can also function as catalysts for the same reaction. Indeed, Sun-Hwa showed that the complex $[Cp^*(phen)Ru(NCCH_3)]^+$ (**52**) catalyzed the hydrosilylation of a wider spectrum of N-heterocycles, including such challenging substrates as acridine, isoquinoline, and even phenanthroline (Scheme 3.14) [48]. Quinoline could also be reduced more efficiently with 5 mol% catalyst load with very high 1,4-regioselectivity but 2-substituted quinolines and phenanthrolines were inactive. Other applications of cationic complex **41** included catalytic chemoselective reduction of acid chlorides by silanes to aldehydes [57] and the related semi-catalytic conversion of secondary amides to imines [58]. In the latter case, the secondary amide was first converted stoichiometrically to an iminoyl chloride followed by catalytic reduction.

Highly Lewis-acidic metal complexes were also applied in the hydrosilylation of alkynes. Trost and Ball reported that cationic complexes $[Cp'Ru(NCMe)_3][PF_6]$ (Cp′ = Cp (**53**) and Cp* (**54**)) catalyzed the Markovnikov hydrosilylation of terminal alkynes and unusual *trans*-Si–H addition across alkynes [59, 60].

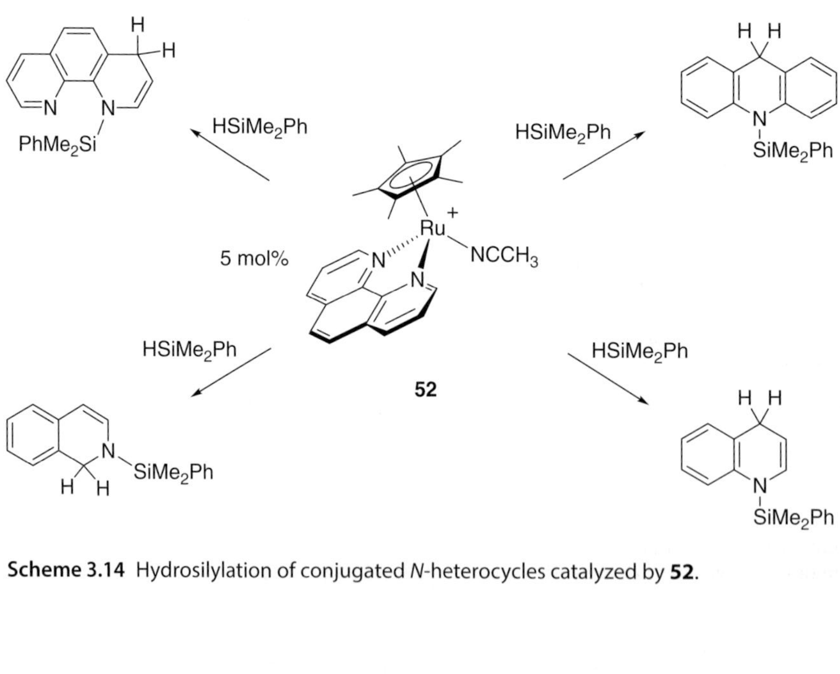

Scheme 3.14 Hydrosilylation of conjugated *N*-heterocycles catalyzed by **52**.

DFT calculations of the latter process revealed that although an intermediate cationic η[1]-silane complex is formed, the reaction proceeded via an inner-sphere mechanism involving an unusual metallocyclopropene intermediate that underwent a stereospecific counterclockwise rotation of the C_α—C_β bond followed by silyl migration to the carbene moiety [61]. Wu et al. further showed that the stereoselectivity of Si–H addition to silylalkynes depends on the ancillary ligand. While the Cp complex **53** favored *syn* addition, the use of the Cp* analog **54** led to *trans*-addition with the opposite regioselectivity, i.e. the silyl goes to the silyl end of alkyne [62]. All these reactions proceeded via inner-sphere pathways.

In contrast, the Oro group provided evidence for an outer-sphere ionic hydrosilylation of alkynes mediated by iridium and rhodium complexes **55** and **56**, respectively (Scheme 3.15) [63]. A series of terminal alkynes was converted in excellent yields and selectivities to the corresponding β-(*Z*)-vinylsilanes. Interestingly, the iridium complex **55** showed greater activity toward aromatic alkynes, whereas **56** was more active than its iridium analog toward aliphatic alkynes. The choice of solvent was also important as catalysis proceeded only in acetone and halted in other polar solvents (e.g. dichloromethane, methanol, or acetonitrile). A DFT study of the reaction mechanism revealed that the classical

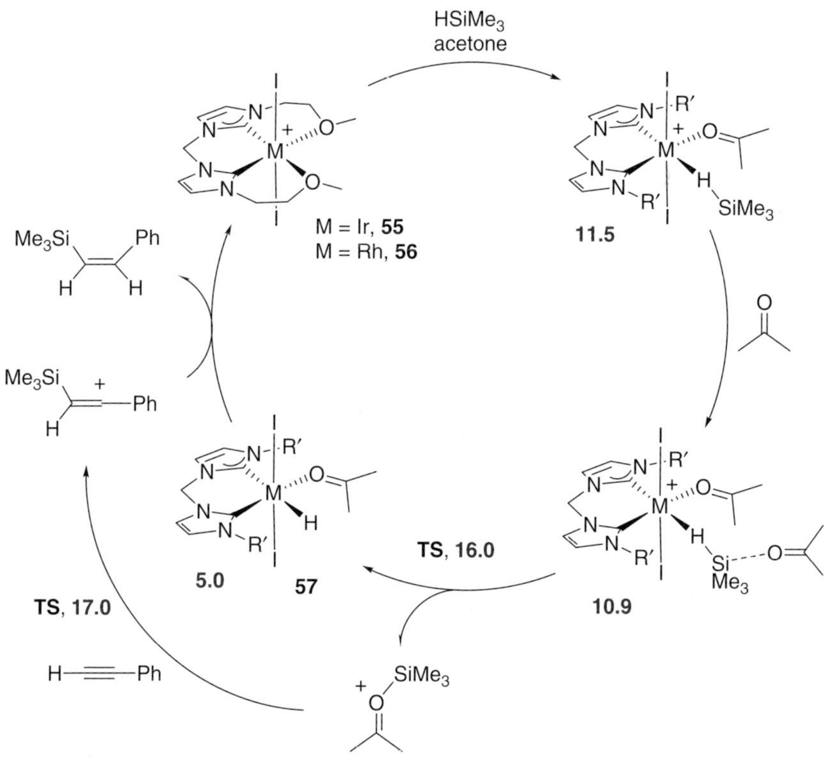

Scheme 3.15 Outer-sphere ionic hydrosilylation of alkynes catalyzed by complexes **55** and **56**. In blue are given free energies (in kcal/mol) calculated for **55**.

Ojima-type pathway was not applicable. Instead, the authors suggested an ionic hydrosilylation mechanism in which acetone serves as a silylium ion shuttle to the alkyne (Scheme 3.15). The final step, hydride transfer from complex **57** to the β-silyl carbocation intermediate, gives either the *Z*- or *E*-isomer of the alkene product. Although the *E*-isomer is more stable by 3.7 kcal/mol, the corresponding transition state lies 12.7 kcal/mol higher, thus explaining the observed stereoselectivity.

An alternative method for the electrophilic activation of Si—H bonds was proposed by the Oestreich, Ohki, and Tatsumi groups. These workers showed that the cationic ruthenium complex **58** ligated by a chelating arene-thiophenolate group reacted with silanes via a heterolytic cleavage of Si—H bonds across the Ru–S moiety to give a very active species that was able to catalyze the 3-silylation of indoles [64], Si–N coupling of silanes with pyrroles, indoles, carbazoles, anilines [65], enolyzable carbonyls [66], and imines [67] as well as the hydrodefluorination of CF_3 groups [68]. In particular, the Oestreich group further showed that **58** catalyzed the 1,4 regioselective reduction of pyridines and benzannulated pyridines with Me_2PhSiH, $EtMe_2SiH$, and $MePh_2SiH$ (Scheme 3.16) [69]. A wide spectrum of substrates including 3,5- and 4-substituted pyridines as well as quinolines, 2-substituted quinolines, isoquinolines, acridine, and phenanthridine can be reduced in excellent yields. Carboxyl and cyano groups were not compatible with this catalyst, but alkyl substitution in the 4-position was. The latter fact stands in contrast to our catalyst **41** which was inert toward 4-substituted pyridines. The ability of **58** to catalytically hydrosilylate 4-substituted pyridines allowed for distinction between a two-step radical reduction mechanism [70] or a one-step ionic reduction mechanism. That is, no other products were observed during the hydrosilylation of 4-cyclopropyl and 4-pentenyl derivatives even though they were susceptible to ring opening by the putative cyclopropylmethyl radical and spirocyclization of the hex-5-enyl radical, respectively. This observation revealed that neither of these radical events occurred in this catalytic system and therefore the ionic hydrosilylation mechanism was preferred. Oestreich and coworker further showed that complex **58** catalyzed the hydrosilylation of nitriles to either imines or amines, with the chemoselectivity dependent on the silane used, Et_3SiH or Me_2PhSiH, respectively [71].

Oestreich and coworkers conducted a detailed mechanistic investigation of Si–H activation by **58**. Stoichiometric reaction between $[(\eta^6(C),\kappa^1(S)\text{-SDmp})((4\text{-}C_6H_5F)_3P)Ru][BAr^F_4]$ (**58_F**, SDmp = 2,6-$Mes_2C_6H_3S$) and $HSiMePh_2$ allowed for isolation of the reactive hydrido silyl sulfonium intermediate **60** formed as a result of Si–H heterolytic cleavage across the Ru—S bond (Scheme 3.16) [72, 73]. An EXSY NMR experiment further showed that the Ru-bound hydride and the S-bound silyl group were in exchange with free silane, thus supporting the reversibility of Si—H bond addition. DFT calculations revealed that Si—H bond activation started with a barrierless formation of the η^1 or η^2 silane complexes **59**, depending on the silane ($MePh_2SiH$, Me_2PhSiH, $EtMe_2SiH$ or $^tBuMe_2SiH$) and phosphine (Me_3P versus Et_3P) used, followed by heterolysis of the Si—H bond on the Ru—S bond through a concerted four-membered transition state. This activation mode is thus reminiscent of a Noyori-type activation of

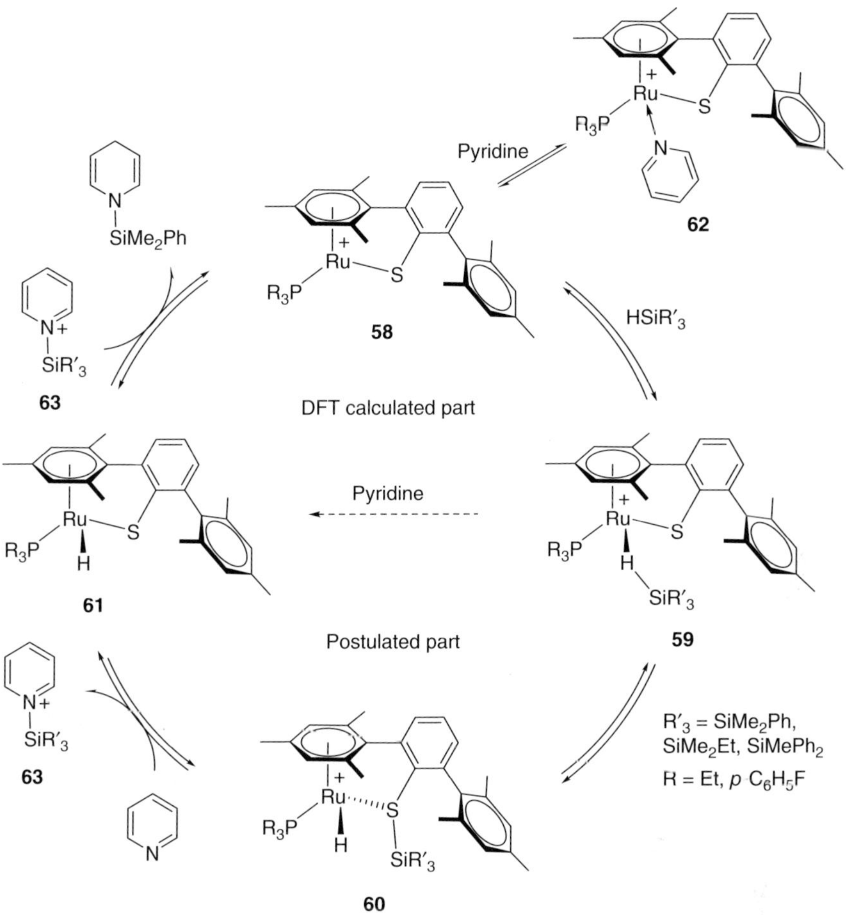

Scheme 3.16 DFT-supported and proposed mechanism of pyridine hydrosilylation catalyzed by **58**. The central dashed arrow shows the possibility of silylium ion abstraction from the silane σ-complex **59**.

dihydrogen or alcohols by bifunctional catalysts [74]. The inverse regioselectivity of the Si–H addition with the hydride going to sulfur and silyl binding to the metal was found to be disfavored by more than 30 kJ/mol (for R = Et and silane Me_2PhSiH). Cleavage of the Si—H bond required low barriers of 10–75 kJ/mol depending on the silane and led to a more thermodynamically favorable silyl sulfonium species **60**. Further mechanistic steps, silyl abstraction and hydride transfer, have not been elucidated and therefore it remains unanswered whether silyl abstraction takes place from the silane intermediate **59** (shown by the dashed arrow in the center of Scheme 3.16) or from **60**. One can expect that these divergent reaction pathways may also depend on the substrate (quinoline, indole, alcohol, or amine). In the case of NH and OH acidic substrates and N-alkyl indoles, the hydride-transfer step is replaced with proton abstraction but the mechanism for Si–H activation on the ruthenium center should remain the same. More recent mechanistic studies have showed that the resting state of the catalyst in the hydrosilylation of pyridine is the pyridine complex $[(\eta^6(C),\kappa^1(S)\text{-SDmp})((4\text{-}C_6H_5F)_3P)Ru(py)][BAr^F{}_4]$ (**62**) and, moreover, a stoichiometric reaction between the hydride **61** and the silylpyridinium ion **63**, shown as a step in Scheme 3.16 does not proceed in the absence of excess pyridine [75]. Stoichiometric studies and EXSY experiments show that all steps in the catalytic cycle are reversible and the overall process is driven by the thermodynamically favorable formation of *N*-silylated dihydropyridine.

The same mechanism can be applied to the catalytic hydrodefluorination of CF_3-substituted anilines by **58** (Scheme 3.17) [55]. Substrates containing the CF_3-group in the *para*-position reacted much faster than *ortho*-substituted derivatives (48 hours at 20 °C versus 72 hours at 100 °C), whereas the *meta* derivative did not react at all, correlating with the π-stabilizing ability of the anilino-group in the intermediate carbenium ion. Variation of the phosphine ligand showed that the preformed catalyst **58** with a more electron-donating phosphine (Et_3P) is a better catalyst compared to the Ph_3P and $(p\text{-}FC_6H_5)_3P$ derivatives, likely because of the much slower hydride transfer step for the latter complexes. Remarkably, the situation is opposite for the *in situ*-generated catalysts, which can be attributed to the assistance of the unreacted precursor $(\eta^6(C),\kappa^1(S)\text{-SDmp})(R_3P)RuCl$ by trapping the electron-deficient ruthenium cation **58** in the form of a dimer $[\{(\eta^6(C),\kappa^1(S)\text{-SDmp})(R_3P)Ru\}_2(\mu\text{-Cl})]^+$. The same effect can be achieved by adding a catalytic amount of NaOMe, which generates a significant amount of the hydride **61** that is able to stabilize **58** as the dimer $[\{(\eta^6(C),\kappa^1(S)\text{-SDmp})(R_3P)Ru\}_2(\mu\text{-H})]^+$ (**65**). Such a dimer was characterized by observing the bridging hydride as a peak in the ^1H NMR spectroscopy at −17.1 ppm (for the $(p\text{-}FC_6H_5)_3P$-substituted complex), which is shifted upfield relative to the monomeric hydride **61** (−8.6 ppm). The question, however, remains whether the incipient carbenium ion is free and undergoes direct intermolecular hydride transfer from **61** (or silane) or is coordinated to sulfur as shown in **64** and is reduced intramolecularly, as shown in Scheme 3.17.

Another highlight from the Oestreich group is the hydrosilylation of CO_2 catalyzed by **58** (Scheme 3.18) [76]. Using Et_3SiH as the reductant, chemoselective formation of bis(silyl)acetal $H_2C(OSiEt_3)_2$ was observed. However, with $EtMe_2SiH$ and depending on the temperature, the reduction halts at the formaldehyde oxidation stage in the form of $H_2C(OSiEtMe_2)_2$ (at 80 °C) or proceeds to the silylated methanol $H_3COSiEtMe_2$ (at 150 °C) but without over-reduction to methane. Although the activity is rather low, such that heating for several days is required, this result is significant as typically CO_2 reductions results either in formates or methane. For example, Brookhart et al. found that the cationic iridium silane complex **30** catalyzed the complete reduction of CO_2 into methane at very low catalyst load (0.0077%) and with a remarkable turnover number (TON) of ~8300 [77]. Conejero and coworkers reported that the NHC-supported cationic complex of Pt(II) (**66**) catalyzed the hydrosilylation of CO_2 to give the formate HC(=O)(OSiRR′H) (R = H, R′ = *n*-Bu; R = R′ = Et) [78]. DFT studies revealed that the reaction proceeded via intermediate formation of a cationic η^1-silane σ-complex. Silyl formate was also formed in the hydrosilylation of CO_2 catalyzed by iridium complex **67** (Scheme 3.19) [79].

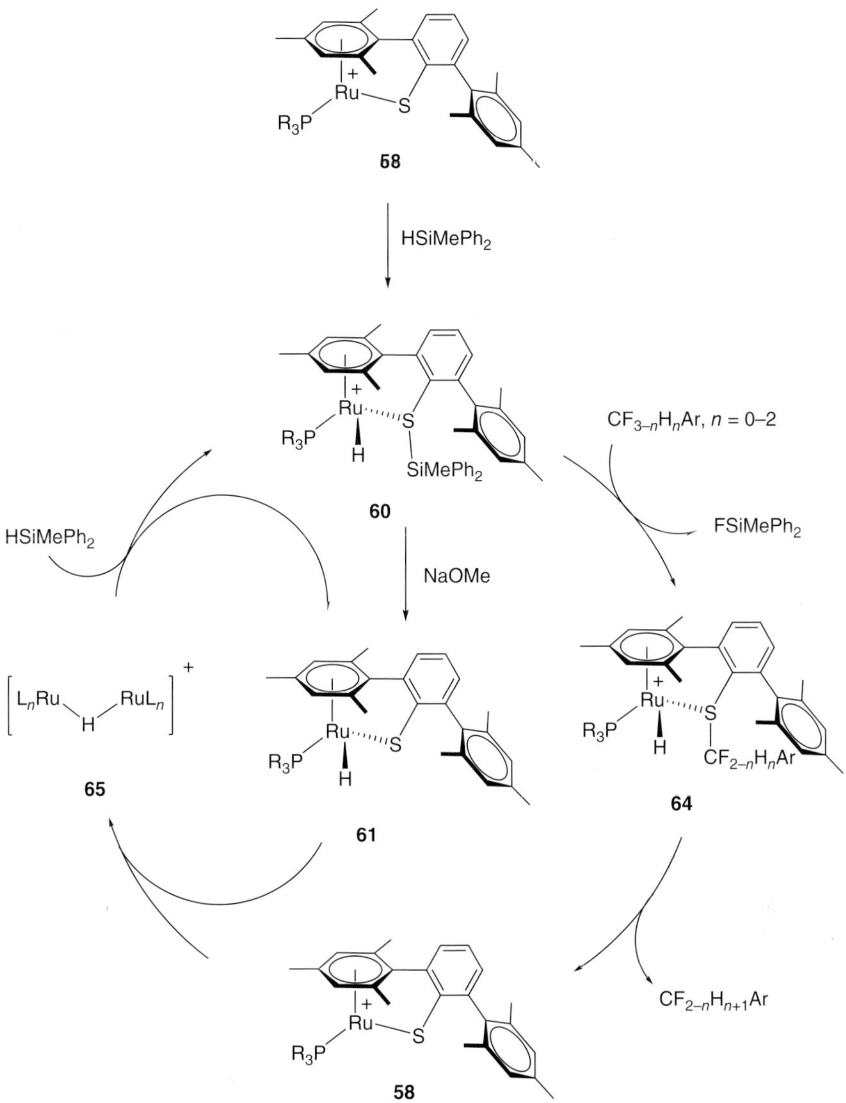

Scheme 3.17 Proposed mechanism for hydrodefluorination catalyzed by **58**.

Interestingly, DFT study by Oro and coworkers showed that catalysis proceeded via a combination of ionic hydrosilylation and Noyori-type bifunctional mechanisms. As such an oxygen atom from the triflate ligand abstracts the silylium ion from the intermediate silane complex **68** to give the hydride **69**. The latter can transfer the hydride and silylium moieties to CO_2 in a concerted, outer-sphere manner.

How do we know that a catalyst can operate by ionic hydrosilylation mechanism? Djukic and coworkers suggested to evaluate the propensity of a given metal–silyl complex to release a silylium ion $[SiR_3]^+$ by calculating the relative

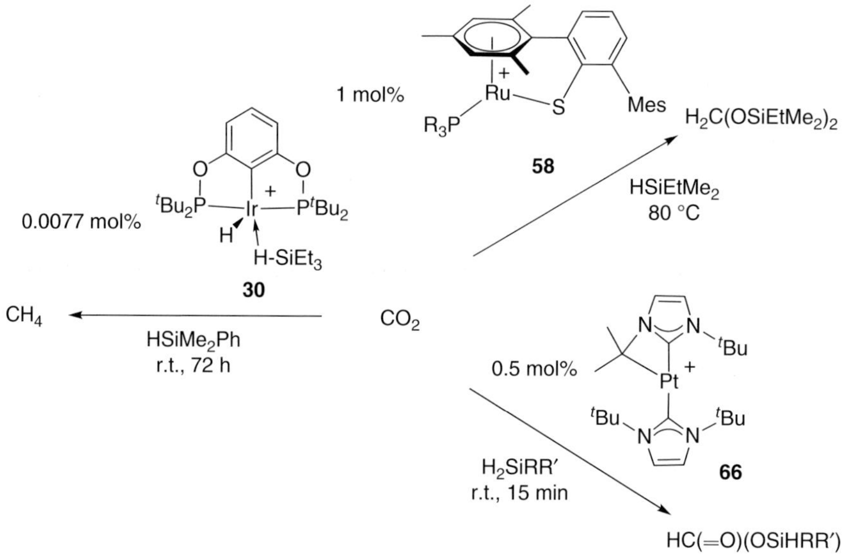

Scheme 3.18 Hydrosilylation of CO_2 catalyzed by cationic complexes.

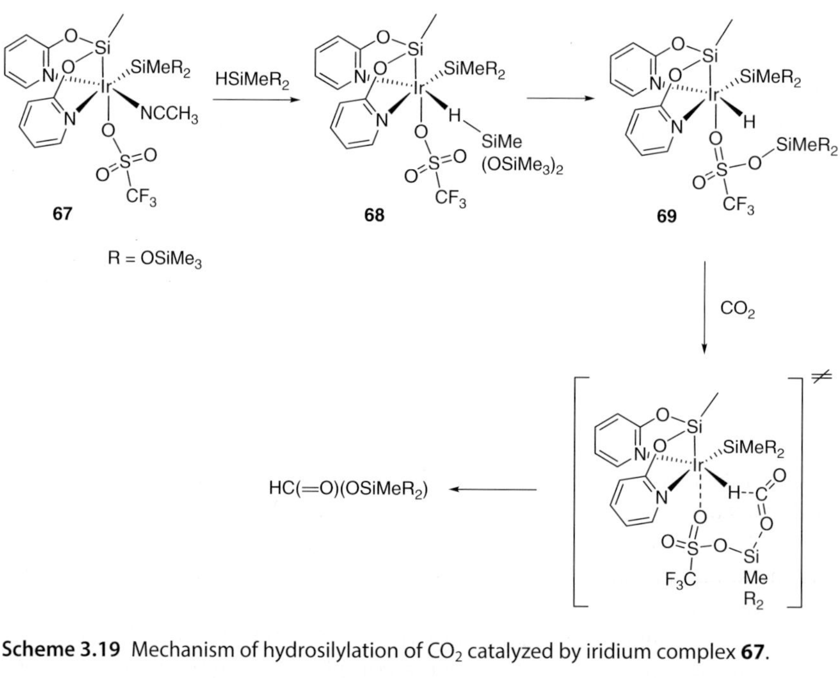

Scheme 3.19 Mechanism of hydrosilylation of CO_2 catalyzed by iridium complex **67**.

intrinsic silylicity descriptor *Π* (from the Greek word πυριτιο that means "elemental silicon") [80]. The latter can be defined as the ratio of the intrinsic silyl-to-triflate interaction energy of a silyltriflate, taken as a reference compound, relative to the silyl-to-metal interaction of the given complex, with intrinsic interaction energies being obtained from Energy Decomposition Analysis. For silicon compounds with $\Pi < 1$, the silicon center has a silicide character, i.e. is an X-type ligand. For silicon compounds with $\Pi > 1$, the silicon center has a silylium character and is a *Z*-type ligand. Analysis of a few dozens of known silyl and silane complexes showed good correlation between their catalytic activity and values of silylicity. Of the complexes discussed in this chapter, Brookhart's complex **30** has a very high Π (1.84) and can be described as an adduct Ir–H → [$SiEt_3$], consistent with the ease of abstraction of the silylium ion even by relatively weak bases, such as ethers. Our ruthenium complex **42** was found to have a lower silylicity with $\Pi = 1.43$. The QTAIM analysis revealed an interesting situation when a η^2-HSi silane complex and Ru–H → [$SiMe_2Ph$] alternatives could not be unequivocally distinguished.

In summary, Lewis-acidic metal complexes offer unique modes of activating Si—H bonds, opening up new opportunities for the catalytic transformations of unsaturated organic substrates with unusual chemo-, regio- and stereoselectivity. In particular, substrates (nitriles, amides, *N*-heterocycles) that until very recently were deemed unreducible by conventional hydrosilylation methods are now added to the catalytic repertoire of synthetic chemistry.

References

1 Corey, J.Y. (2016). *Chem. Rev.* 116: 11291–11435.
2 Chalk, A.J. and Harrod, J.F. (1965). *J. Am. Chem. Soc.* 87: 16–21.
3 Ojima, I., Nihonyanagi, M., Kogure, T. et al. (1975). *J. Organomet. Chem.* 94: 449–461.
4 Kubas, G.J. (2004). *Adv. Inorg. Chem.* 56: 127–177.
5 Robert, T. and Oestreich, M. (2013). *Angew. Chem. Int. Ed.* 52: 5216–5218.
6 Lipke, M.C., Liberman-Martin, A.L., and Tilley, T.D. (2017). *Angew. Chem. Int. Ed.* 56: 2260–2294.
7 Luo, X. and Crabtree, R.H. (1989). *J. Am. Chem. Soc.* 111: 2527–2535.
8 Scharrer, E., Chang, S., and Brookhart, M. (1995). *Organometallics* 14: 5686–5694.
9 Chang, S., Scharrer, E., and Brookhart, M. (1998). *J. Mol. Catal. A: Chem.* 130: 107–119.
10 Scharrer, E., Chang, S., and Brookhart, M. (1998). *Organometallics* 14: 5686–5694.
11 Bühl, M. and Mauschick, F.T. (2003). *Organometallics* 22: 1422–1431.
12 Nagaraja, C.M., Parameswaran, P., Jemmis, E.D., and Jagirdar, B.R. (2007). *J. Am. Chem. Soc.* 129: 5587–5596.
13 Dioumaev, V.K. and Bullock, M. (2003). *Nature* 424: 530–532.
14 Glaser, P.B. and Tilley, T.D. (2003). *J. Am. Chem. Soc.* 125: 13640–13641.
15 Beddie, C. and Hall, M.B. (2004). *J. Am. Chem. Soc.* 126: 13564–13565.
16 Böhme, U. (2006). *J. Organomet. Chem.* 691: 4400–4410.
17 Nikonov, G. (2001). *Angew. Chem. Int. Ed.* 40: 3353.
18 Nikonov, G. (2005). *Adv. Organomet. Chem.* 53: 217–309.
19 Seitz, F. and Wrighton, M.S. (1988). *Angew. Chem. Int. Ed. Engl.* 27: 289–291.

20 Calimano, E. and Tilley, T.D. (2008). *J. Am. Chem. Soc.* 130: 9226–9227.
21 Calimano, E. and Tilley, T.D. (2009). *J. Am. Chem. Soc.* 131: 11161–11173.
22 Lipke, M.C. and Tilley, T.D. (2014). *J. Am. Chem. Soc.* 136: 16387–16398.
23 Muraoka, T., Shimizu, Y., Kobayashi, H. et al. (2010). *Organometallics* 29: 5423–5426.
24 Zheng, G.Z. and Chan, T.H. (1995). *Organometallics* 14: 70–79.
25 César, V., Bellemin-Laponnaz, S., and Gade, L.H. (2004). *Angew. Chem. Int. Ed.* 43: 1014–1017.
26 César, V., Bellemin-Laponnaz, S., Wadepohl, H., and Gade, L.H. (2005). *Chem. Eur. J.* 11: 2862–2873.
27 Schneider, N., Finger, M., Haferkemper, C. et al. (2009). *Angew. Chem. Int. Ed.* 48: 1609–1613.
28 Klei, S.R., Tilley, T.D., and Bergman, R.G. (2002). *Organometallics* 21: 4648–4661.
29 Ison, E.A., Trivedi, E.R., Corbin, R.A., and Abu-Omar, M.M. (2005). *J. Am. Chem. Soc.* 127: 15374–15375.
30 Nolin, K.A., Krumper, J.R., Pluth, M.D. et al. (2007). *J. Am. Chem. Soc.* 129: 14684–14696.
31 Gu, P., Wang, W., Wang, Y., and Wei, H. (2013). *Organometalllics* 32: 47–51.
32 Du, G. and Abu-Omar, M.M. (2006). *Organometalllics* 25: 4920–4923.
33 Du, G., Fanwick, P.E., and Abu-Omar, M.M. (2007). *J. Am. Chem. Soc.* 129: 5180–5187.
34 Pérez, D.E., Smeltz, J.L., Sommer, R.D. et al. (2017). *Dalton Trans.* 46: 4609–4616.
35 Yang, J., White, P.S., Schauer, C.K., and Brookhart, M. (2008). *Angew. Chem. Int. Ed.* 47: 4141–4143.
36 Yang, J. and Brookhart, M. (2007). *J. Am. Chem. Soc.* 129: 12656–12657.
37 Yang, J. and Brookhart, M. (2009). *Adv. Synth. Catal.* 351: 175–187.
38 Yang, J., White, P.S., and Brookhart, M. (2008). *J. Am. Chem. Soc.* 130: 17509–17518.
39 McLaughlin, M.P., Adduci, L.L., Becker, J.J., and Gagné, M.R. (2013). *J. Am. Chem. Soc.* 135: 1225–1227.
40 Park, S. and Brookhart, M. (2011). *Chem. Commun.* 47: 3643–3645.
41 Park, S. and Brookhart, M. (2010). *Organometallics* 29: 6057–6064.
42 Wang, W., Gu, P., Wang, Y., and Wei, H. (2014). *Organometallics* 33: 847–857.
43 Metsänen, T.T., Hrobárik, P., Klare, H.F.T. et al. (2014). *J. Am. Chem. Soc.* 136: 6912–6915.
44 Park, S. and Brookhart, M. (2012). *J. Am. Chem. Soc.* 134: 640–653.
45 Cheng, C. and Brookhart, M. (2012). *J. Am. Chem. Soc.* 134: 11304–11307.
46 Gutsulyak, D.V., Kuzmina, L.G., Howard, J.A.K. et al. (2008). *J. Am. Chem. Soc.* 130: 3732–3733.
47 Gutsulyak, D.V., Vyboishchikov, S.F., and Nikonov, G.I. (2010). *J. Am. Chem. Soc.* 132: 5950–5951.
48 Yang, Y.-F., Chung, L.W., Zhang, X. et al. (2014). *J. Org. Chem.* 79: 8856–8864.
49 Fuchs, J., Klare, H.F.T., and Oestreich, M. (2017). *ACS Catal.* 7: 8338–8342.
50 Osipov, A.L., Gerdov, S.M., Kuzmina, L.G. et al. (2005). *Organometallics* 24: 587–602.

51 Gutsulyak, D.V., Osipov, A.L., Kuzmina, G. et al. (2008). *Dalton Trans.* 6843–6850.
52 Campion, B.K., Heyn, R.H., and Tilley, T.D. (1992). *Chem. Commun.* 1201–1203.
53 Rodriguez, V., Donnadieu, B., Sabo-Etienne, S., and Chaudret, B. (1998). *Organometallics* 17: 3809–3814.
54 Gutsulyak, D.V. and Nikonov, G.I. (2011). *Angew. Chem. Int. Ed.* 50: 1384–1387.
55 Osakada, K. (2011). *Angew. Chem. Int. Ed.* 50: 3845–3846.
56 Lee, S.-H., Gutsulyak, D.V., and Nikonov, G.I. (2013). *Organometallic* 32: 4457–4464.
57 Gutsulyak, D.V. and Nikonov, G.I. (2012). *Adv. Synth. Catal.* 354: 607–611.
58 Lee, S.-H. and Nikonov, G.I. (2014). *Dalton Trans.* 43: 8888–8893.
59 Trost, B.M. and Ball, Z.T. (2001). *J. Am. Chem. Soc.* 123: 12726–12727.
60 Trost, B.M. and Ball, Z.T. (2003). *J. Am. Chem. Soc.* 125: 30–31.
61 Chung, L.W., Wu, Y.D., Trost, B.M., and Ball, Z.T. (2003). *J. Am. Chem. Soc.* 125: 11578–11582.
62 Ding, S., Song, L.-J., Chung, L.W. et al. (2013). *J. Am. Chem. Soc.* 135: 13835–13842.
63 Iglesias, M., Sanz Miguel, P.J., Polo, V. et al. (2013). *Chem. Eur. J.* 19: 17559–17566.
64 Klare, H.F.T., Oestreich, M., Ito, J.-I. et al. (2011). *J. Am. Chem. Soc.* 133: 3312–3315.
65 Königs, C.D.F., Müller, M.F., Aiguabella, N. et al. (2013). *Chem. Commun.* 49: 1506–1508.
66 Königs, C.D.F., Klare, H.F.T., Ohki, Y. et al. (2012). *Org. Lett.* 14: 2842–2845.
67 Hermeke, J., Klare, H.F.T., and Oestreich, M. (2014). *Chem. Eur. J.* 20: 9250–9254.
68 Stahl, T., Klare, H.F.T., and Oestreich, M. (2013). *J. Am. Chem. Soc.* 135: 1248–1251.
69 Königs, C.D.F., Klare, H.F.T., and Oestreich, M. (2013). *Angew. Chem. Int. Ed.* 52: 10076–10079.
70 Shaw, A.P., Ryland, B.L., Franklin, M.J. et al. (2008). *J. Org. Chem.* 73: 9668–9674.
71 Wübbolt, S. and Oestreich, M. (2017). *Synlett* 28: 2411–2414.
72 Stahl, T., Hrobárik, P., Königs, C.D.F. et al. (2015). *Chem. Sci.* 6: 4324–4334.
73 Omann, L., Königs, C.D.F., and Oestreich, M. (2017). *Acc. Chem. Res.* 50: 1258–1269.
74 Grützmacher, H. (2008). *Angew. Chem. Int. Ed.* 47: 1814–1818.
75 Bähr, S. and Oestreich, M. (2018). *Chem. Eur. J.* 24: 5613–5622. https://doi.org/10.1002/chem.20175899
76 Metsänen, T.T. and Oestreich, M. (2015). *Organometallics* 34: 543–546.
77 Park, S., Bézier, D., and Brookhart, M. (2012). *J. Am. Chem. Soc.* 134: 11404–11407.
78 Ríos, P., Díez, J., López-Serrano, J. et al. (2016). *Chem. Eur. J.* 22: 16791–16795.
79 Lalrempuia, R., Iglesias, M., Polo, V. et al. (2012). *Angew. Chem. Int. Ed.* 51: 12824–12827.
80 Binh, D.H., Milovanović, M., Puertes-Mico, J. et al. (2017). *Chem. Eur. J.* 23: 17058–17069.

4

Metal–Ligand Cooperative Si—H Bond Activation

Francis Forster and Martin Oestreich

Institut für Chemie, Technische Universität Berlin, Strasse des 17. Juni 115, 10623 Berlin, Germany

4.1 Introduction

Biochemical processes continue to inspire synthetic chemists and are often the starting point for the development of new synthetic methodologies. Catalytic processes involving cooperative bond activation by metal–ligand cooperation has attracted particular attention [1]. Bond activation of small molecules can occur across the metal–ligand bond of a transition-metal frustrated Lewis pair [2]. Notably, cooperative metal–ligand interactions facilitate the heterolytic cleavage of Si—H bonds, with ancillary ligands serving as internal Lewis-basic sites that are directly involved in the Si—H bond activation process. These activations are usually described by a σ-bond metathesis (Scheme 4.1), and the nature of the metal–ligand combination highly determines the product distribution.

σ-Bond metathesis through transition state TS-**I** (Scheme 4.1a) leads to a metal silyl complex **II** with release of side product **III** (typically dihydrogen or methane). Conversely, the reaction through TS-**II** (Scheme 4.1b) generates a metal hydride **IV** accompanied by dissociation of silylated ligand **V** (typically a silyl (thio)ether or silyl amine).

The bond dissociation energy of their Si—H bond is significantly lower (~18 kcal/mol) than that of the H—H bond [3]. That ought to make Si—H bond heterolysis a more facile process. Moreover, the interaction with a *soft* sulfur atom ensures a higher residual Lewis acidity at the silicon atom compared to *harder* nitrogen or oxygen donors (Scheme 4.2) [4, 5]. A potential silicon cation transfer as well as turnover in catalytic processes is facilitated by this weak stabilization due to reversible coordination to the sulfur atom. This chapter discusses cooperative Si—H bond activation across M—L bonds (L = C, N, O, and S).

Organosilicon Chemistry: Novel Approaches and Reactions, First Edition.
Edited by Tamejiro Hiyama and Martin Oestreich

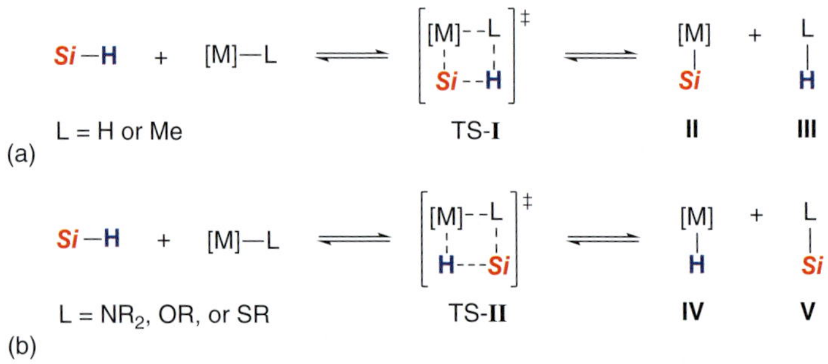

Scheme 4.1 Opposite bonding scenarios in the Si—H bond activation by transition-metal complexes depending on their ancillary ligands (*Si* = SiR_3 = triorganosilyl).

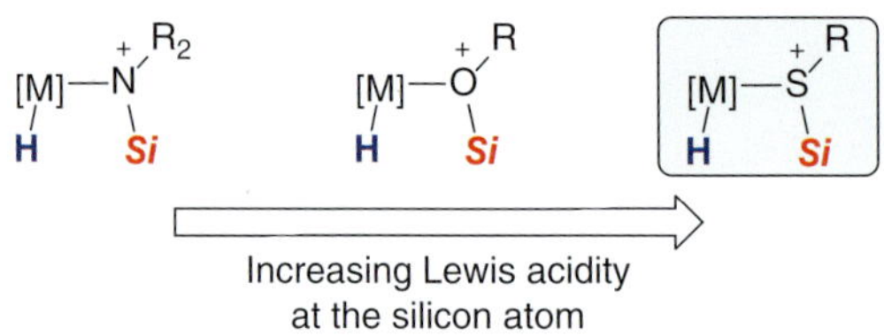

Scheme 4.2 Comparison of expected Lewis acidities of donor-stabilized silicon cations (*Si* = SiR_3).

4.2 Cooperative Si—H Bond Activation with Carbene Complexes Across M—C Double Bonds

Activation of hydrosilanes at M—C bonds was first demonstrated by Gessner and a coworker utilizing ruthenium–carbene complex **1** (Scheme 4.3) [6]. The Si—H bond of hydrosilanes was heterolytically cleaved, resulting in a ruthenium–silyl species and a newly formed C—H bond, as shown in [**1**·R_3SiH]. Piers and coworkers [7] and Iluc and a coworker [8] presented nickel complex [**2**·R_3SiH] and palladium complex [**3**·R_3SiH] that showed a similar mode of action. None of these complexes were tested in the catalytic transfer of the silyl group.

4.3 Cooperative Si—H Bond Activation at M—N Bonds

Several complexes for cooperative Si—H bond activation at M—N bonds were developed (Scheme 4.4). Tilley and a coworker first described tantalum(V) complexes [**4**·$PhSiH_3$] and [**5**·$PhSiH_3$] which were formed from the tantalum–imido complexes upon addition to the Ta—N double bond (not shown) [9]. Attack of the imido group at the hydrosilane resulted in a penta-coordinate silicon species, which subsequently transferred a hydride to the tantalum center. A similar mechanism is expected for the Si—H bond activation at Ti—N bonds, resulting in the formation of complex [**6**·$PhSiH_3$] developed by Andersen and coworkers [10a]. A related activation using a titanium hydrazide complex under formation of complex [**7**·$PhSiH_3$] was shown by Clot and coworkers [10b]. Fryzuk

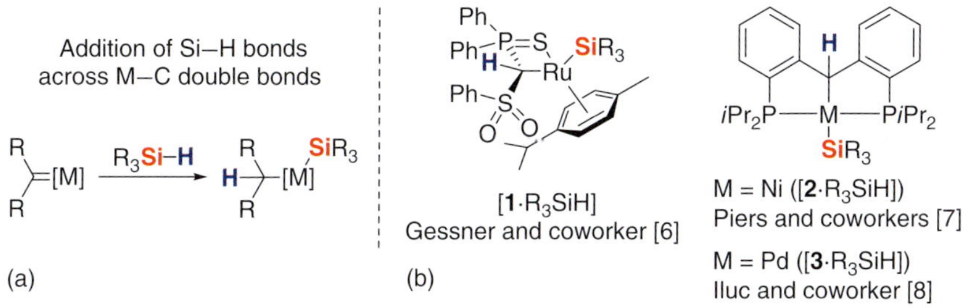

Scheme 4.3 Cooperative Si—H bond activation at metal–carbene complexes.

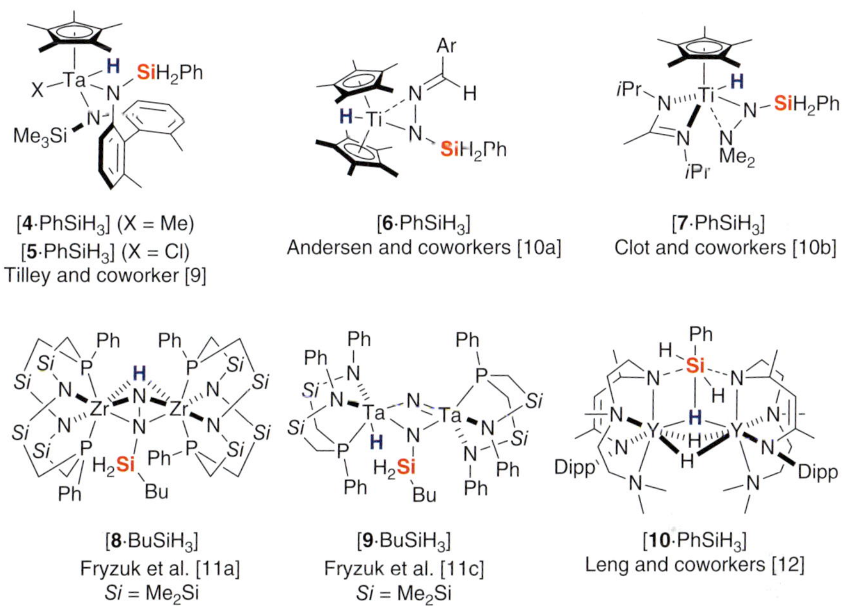

Scheme 4.4 Cooperative Si—H bond activation at M—N bonds.

et al. showed the Si—H bond activation with dinuclear zirconium(IV) complexes to form the addition product [**8**·BuSiH$_3$] [11a]. A similar example was later shown by the same group utilizing a dinuclear tantalum(V) complex, bridged by two hydrides (not shown), which releases dihydrogen upon the Si—H bond activation to form adduct [**9**·BuSiH$_3$] [11c]. Chen and coworkers then demonstrated the cooperative Si—H bond activation at Y—N bonds to form the dinuclear, hydride-bridged complex [**10**·PhSiH$_3$] [12]. However, all shown examples were not tested in catalysis.

4.4 Cooperative Si—H Bond Activation at M—O Bonds

Only a few examples of M—O bonds in metal–ligand cooperativity have been reported. Berke and coworkers demonstrated the heterolytic Si—H bond activation at a rhenium complex bearing nitrosyl ligands [13]. These ligands serve as Lewis bases and engage in cooperative Si—H bond activation to form a silylated nitrosyl ligand together with a rhenium–hydride as shown in complex $[\mathbf{11}\cdot Et_3SiH]^+[BAr^F{}_4]^-$ (Scheme 4.5a). Toste and coworkers then demonstrated cooperative Si—H bond activation at rhenium dioxo complex $[\mathbf{12}\cdot Me_2PhSiH]$ [14]. Complex **12** was shown to be an efficient catalyst for aldehyde and ketone hydrosilylation (Scheme 4.5b). Several differently substituted carbonyl compounds **13** were reduced to the silyl ethers **14** in the presence of catalyst **12**.

Rauchfuss and coworkers later reported the cooperative Si—H bond activation by alkoxoiron complex **15** (Scheme 4.6a) [15]. The tetradentate ligand in **15** participates in Fe—O bond cleavage, thereby splitting the Si—H bond of hydrosilanes to form adduct $[\mathbf{15}\cdot Ph_2SiH_2]$. Complex **15** was applied to aldehyde hydrosilylation of **16** and delivered silyl ether **17** in good yields together with disilyl ether **18** (Scheme 4.6b). Activation of monohydrosilanes was feasible, but subsequent catalysis failed.

4.5 Cooperative Si—H Bond Activation at M—S Bonds

4.5.1 Introduction

Metal–sulfur bonds [16], as in the [NiFe] hydrogenases, are able to split H_2 cooperatively (Scheme 4.7) [17]. The heterolysis of H_2 presumably occurs at the Ni—S(Cys) bond of the active site of [NiFe] hydrogenase, resulting in the formation of a nickel hydride and a protonated sulfur ligand. Inspired by this, chemists developed complexes with M—S bonds mimicking this structural motif for cooperative H—H and Si—H bond activation. These examples are discussed in the following chapters.

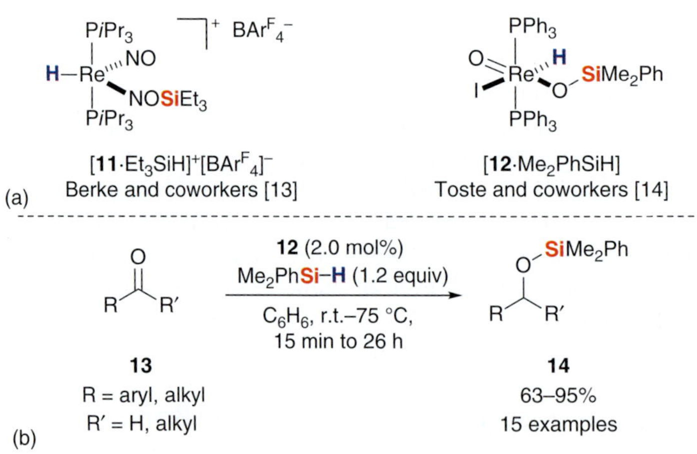

Scheme 4.5 Cooperative Si—H bond activation at a rhenium center and an ancillary nitrosyl ligand as well as at a Re–O bond (a) and carbonyl hydrosilylation with catalyst **12** (b).

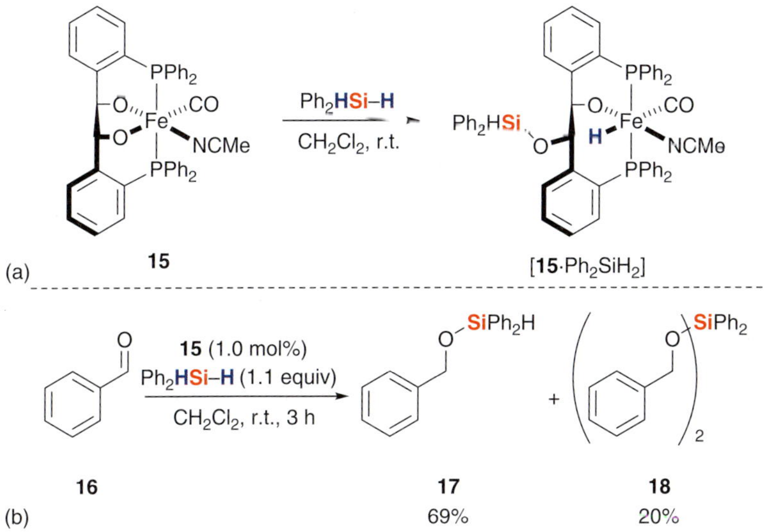

Scheme 4.6 Cooperative Si—H bond activation at Fe—O bonds (a) and hydrosilylation of benzaldehyde (**16**, b).

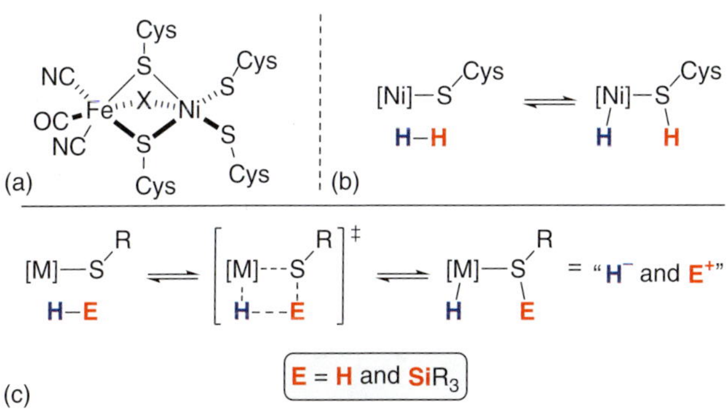

Scheme 4.7 Active site of [NiFe] hydrogenase (X = OH or O, Cys = cysteine) (a), proposed dihydrogen heterolysis (b). (c) Cooperative activation of E—H bonds (E = H and SiR_3) at M—S bonds.

4.5.2 Seminal Results in Cooperative Si—H Bond Activation Across M—S Bonds

The titanocene sulfido complex **19** developed by Andersen and coworkers was the first catalyst presented with a terminal M—S double bond which was applied to cooperative Si—H bond activation; the M—S bond facilitates the reversible splitting of both H—H (not shown) and Si—H bonds (Scheme 4.8) [18]. Deuterium labeling experiments showed that addition and elimination of the Si—H bond occur strictly in a pairwise manner.

Stradiotto and coworkers successfully applied this concept to catalytic hydrosilylation of C—O bonds using cationic rhodium(III) and iridium(III) complexes $[\mathbf{20}]^+[B(C_6F_5)_4]^-$ and $[\mathbf{21}]^+[B(C_6F_5)_4]^-$ with an indene-bridged bidentate P,S-ligand

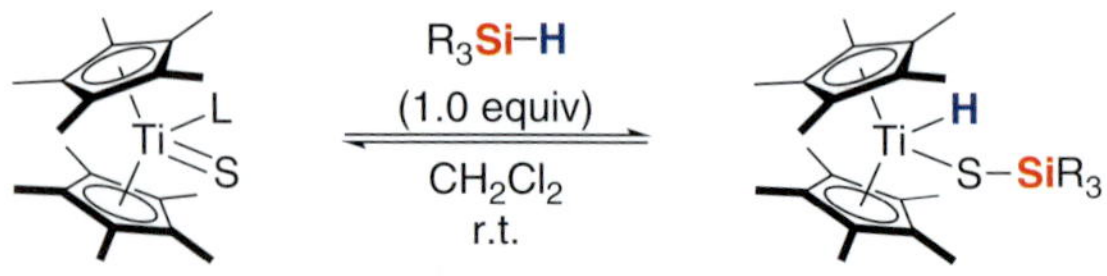

19 [**19**·R$_3$SiH]

Scheme 4.8 Heterolytic and reversible activation of hydrosilanes by titanocene sulfido complex **1** (L = pyridine).

(Scheme 4.9a) [19]. Activation of $PhSiH_3$ and Ph_2SiH_2 is possible with both complexes, but isolation of the rhodium congener [**20**]$^+$[B(C_6F_5)$_4$]$^-$ and its hydrosilane adduct [**20**·Ph_2SiH_2]$^+$[B(C_6F_5)$_4$]$^-$ was not feasible due to their high reactivities, despite NMR spectroscopic characterization being realized *in situ*. While iridium complex [**21**]$^+$[B(C_6F_5)$_4$]$^-$ was only poorly active in the hydrosilylation of ketones **22**, rhodium catalyst [**20**]$^+$[B(C_6F_5)$_4$]$^-$ showed high conversions in this reaction (Scheme 4.9b). By this, Stradiotto and coworkers demonstrated the first hydrosilylation catalyzed by a complex containing a polar M—S bond, capable of cooperative activation of Si—H bonds in hydrosilanes.

In the same year, Ohki, Tatsumi, and coworkers designed rhodium complex [**24**]$^+$[BAr$^F_4$]$^-$ (Ar^F = 3,5-(CF_3)$_2C_6H_3$) and iridium complex [**25**]$^+$[BAr$^F_4$]$^-$ containing Power's SDmp ligand [20] (Dmp = 2,6-dimesitylphenyl) enabling hydrogenase-like H_2 splitting at cryostatic temperatures (Scheme 4.10a) [21]. Although these complexes dissociate at room temperature and release dihydrogen at elevated temperature (regenerating the coordinatively unsaturated complexes), crystallographic proof for the dihydrogen adduct [**25**·H_2]$^+$[BAr$^F_4$]$^-$ was provided (Scheme 4.10b).

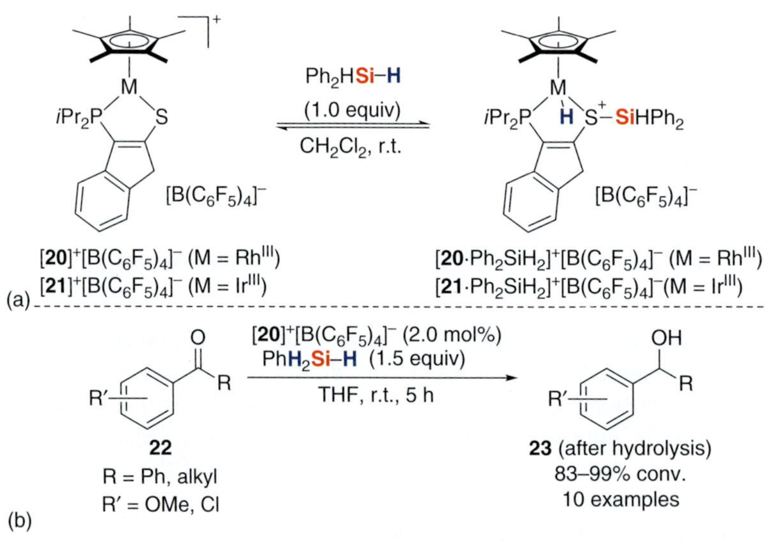

Scheme 4.9 (a) Cooperative Si—H bond activation at Rh—S and Ir—S bonds and (b) application in the catalytic hydrosilylation of ketones.

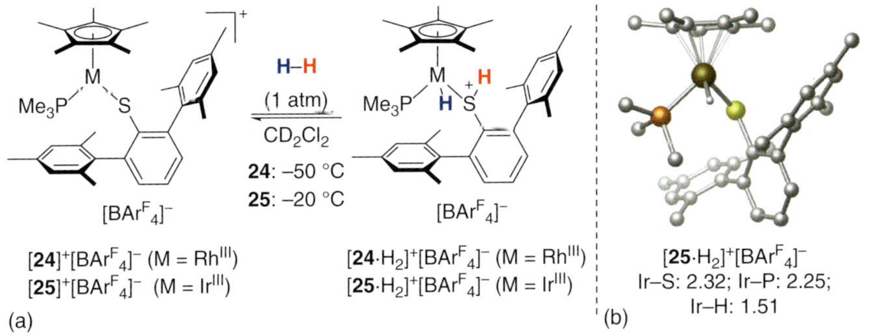

Scheme 4.10 (a) Cooperative dihydrogen activation at Rh—S and Ir—S bonds and (b) the molecular structure of dihydrogen addition product $[\mathbf{25}\cdot H_2]^+$ Counteranion in the crystal structure omitted for clarity. Bond lengths given in Å. Source: Reprinted with permission from Ref. [21] Copyright 2008 American Chemical Society.

Tethering one of the mesityl groups of the SDmp ligand to the metal center was crucial to prevent the observed dissociation of the SDmp ligand after the bond activation step. Ohki, Tatsumi, and coworkers accomplished this in ruthenium complexes $[\mathbf{26}]^+[BAr^F{}_4]^-$ (Scheme 4.11) and demonstrated the cooperative activation of H_2 by $[\mathbf{26}]^+[BAr^F{}_4]^-$ in the hydrogenation of acetophenone to 1-phenylethanol (not shown) [23]. These seminal findings in H—H bond activation encouraged them, in collaboration with the group of Oestreich, to employ complexes $[\mathbf{26}]^+[BAr^F{}_4]^-$ for cooperative Si—H [22, 24–27] bond activation (Scheme 4.11b) [28–30]. Complexes $[\mathbf{26}]^+[BAr^F{}_4]^-$ efficiently split Si—H bonds heterolytically into hydride and the corresponding silylium ion across their polar Ru—S bond. NMR spectroscopic analysis of the hydrosilane adducts $[\mathbf{26}\cdot R_3SiH]^+[BAr^F{}_4]^-$ showed characteristic hydride resonances of $\delta(^1H) \approx -8.0$ ppm and $^2J_{H,P}$ coupling constants of ~49 Hz. The corresponding sulfur-stabilized silylium ions had chemical shifts in the range of $\delta(^{29}Si) \approx 18$–42 ppm. X-ray diffraction secured the molecular structure of the

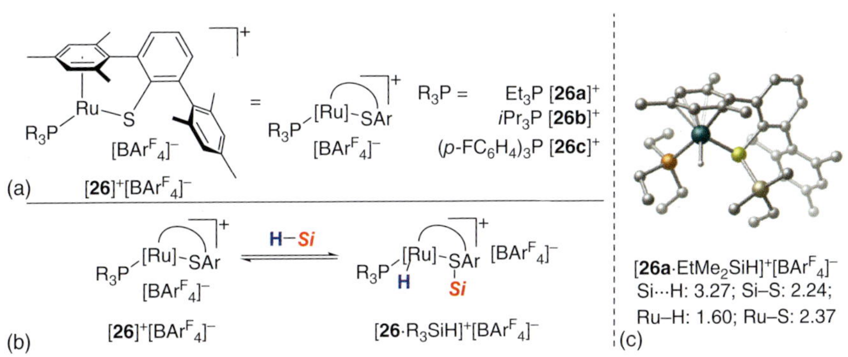

Scheme 4.11 (a) Tethered Ru—S complexes $[\mathbf{26}]^+[BAr^F{}_4]^-$ and (b) cooperative bond activation of Si—H bonds. (c) Molecular structure of $[\mathbf{26a}\cdot EtMe_2SiH]^+[BAr^F{}_4]^-$. Counteranion in the crystal structure omitted for clarity. Bond lengths given in Å. Source: Reproduced from Ref. [22b] with permission from the Royal Society of Chemistry.

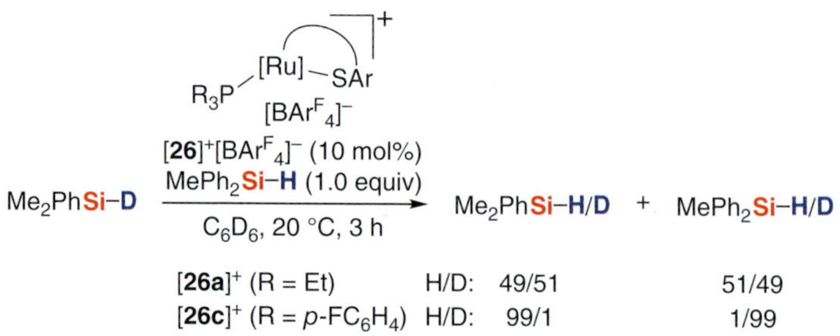

Scheme 4.12 ^{1}H/^{2}H scrambling experiments: the crucial role of the phosphine ligands.

hydrosilane adduct [**26a**·$EtMe_2SiH]^+[BAr^F{}_4]^-$ (Scheme 4.11c), revealing complete cleavage of the Si—H bond (Si···H distance: 3.27 Å).

Employing deuterium-labeled hydrosilanes in ^{1}H/^{2}H scrambling experiments further probed the Si—H bond activation step: in the presence of catalytic amounts of either complex $[\mathbf{26a}]^+$ or $[\mathbf{26c}]^+$, deuterium-labeled Me_2PhSiD was treated with non-deuterated $MePh_2SiH$ (Scheme 4.12) [22b]. Ruthenium(II) thiolate complex $[\mathbf{26a}]^+$ enabled complete ^{1}H/^{2}H exchange, whereas no scrambling was observed with catalyst $[\mathbf{26c}]^+$ even after three hours.

The ruthenium hydride in hydrosilane adduct [**26a**·$R_3SiH]^+[BAr^F{}_4]^-$ with Et_3P as ancillary ligand is expected to be a better hydride donor than in [**26c**·$R_3SiH]^+$ $[BAr^F{}_4]^-$ with $(p\text{-}FC_6H_4)_3P$ as ancillary ligand because of the stronger σ-donor ability of the trialkylphosphine compared to the electron-deficient *para*-fluorinated triarylphosphine. Conversely, [**26c**·$R_3SiH]^+[BAr^F{}_4]^-$ is likely a better R_3Si^+-transfer reagent compared to [**26a**·$R_3SiH]^+[BAr^F{}_4]^-$. Both the hydride donor strength and the R_3Si^+-transfer ability contribute to the scrambling. The ^{1}H/^{2}H exchange can either proceed through a σ-bond metathesis [31] solely involving the metal hydride (TS-**III**, Scheme 4.13a) [32] or through formation of hydronium ions ($[Si—H—Si]^+$) [33] by R_3Si^+-transfer to a free hydrosilane (Scheme 4.13b).

To distinguish between these two mechanisms, an enantioenriched silicon-stereogenic hydrosilane as stereochemical probe was reacted with catalyst $[\mathbf{26c}]^+[BAr^F{}_4]^-$. Complete retention of configuration at the silicon atom of the reisolated hydrosilane was observed after 24 hours (Scheme 4.14).[1] This result

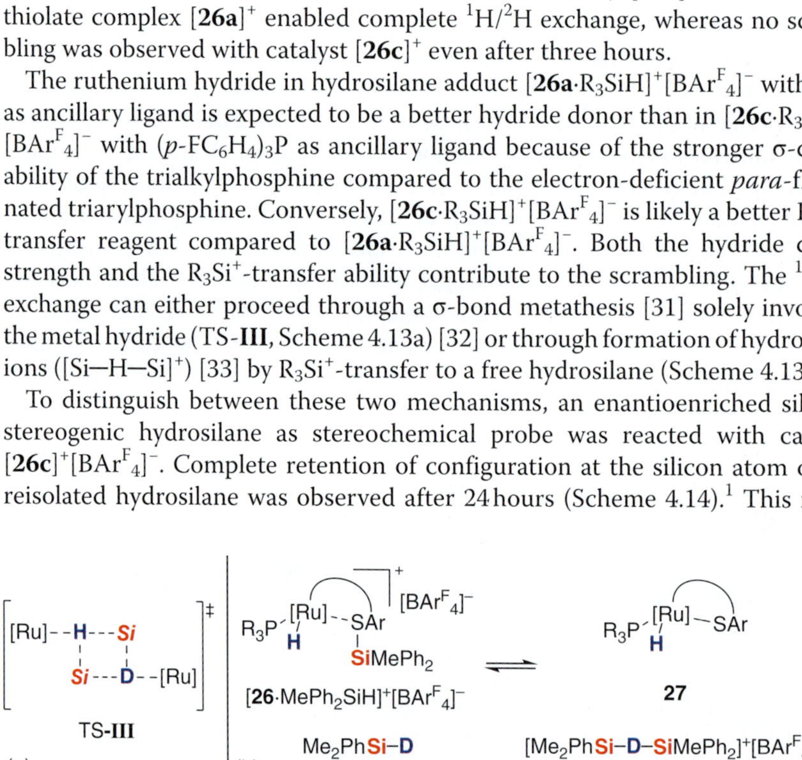

Scheme 4.13 Possible pathways for ^{1}H/^{2}H exchange.

[1] For a similar experiment using an enantioenriched silicon-stereogenic hydrosilane in the reaction with N-protected indoles, see Ref. [24a].

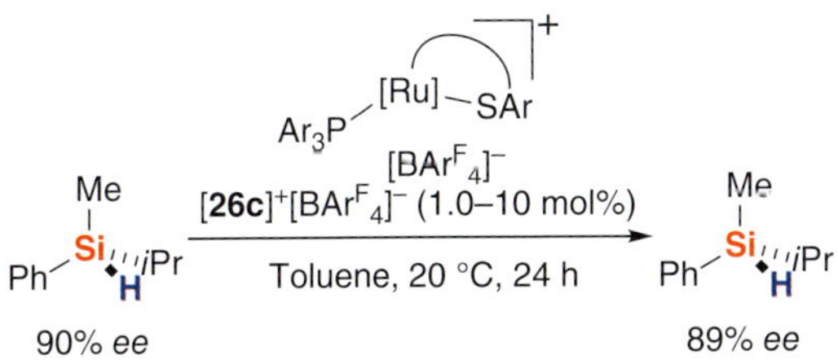

Scheme 4.14 Control experiment with a silicon-stereogenic hydrosilane: no support for the intermediacy of hydronium ions. $Ar_3P = (p\text{-}FC_6H_4)_3P$.

demonstrates that a pathway including hydronium ions is unlikely since that would result in racemization. The observed $^1H/^2H$ exchange is likely to be explained by the better hydride donor strength of $[\mathbf{26a}{\cdot}R_3SiH]^+[BAr^F{}_4]^-$.

The next chapters summarize the application of these sulfur-stabilized silylium ions in catalysis focusing on dehydrogenative coupling reactions.

4.5.3 Dehydrogenative C—H Silylation

A seminal achievement in catalysis utilizing ruthenium complexes $[\mathbf{26}]^+[BAr^F{}_4]^-$ involving sulfur-stabilized silicon electrophiles was accomplished by the dehydrogenative silylation of N-protected indoles **28** (**28** → **29**, Scheme 4.15). C3-silylated indoles **29** were obtained exclusively by cooperative Si—H bond activation in combination with electrophilic aromatic substitution (S_EAr). Substitution at C2 (not shown) as well as various alkyl, dimethylamino, and halogen substituents in different positions of the arene ring were tolerated.

C—H bond activation of indoles typically favors reaction at the C2 position but the abovementioned C3 silylation is, as expected for an S_EAr mechanism, the result of pure electronic control [34]. Blocking of the C3 position by a methyl group did not result in C2-silylated indoles (not shown). A pathway involving hydrosilylation followed by indoline-to-indole oxidation was excluded as no deuterium incorporation at C2 was observed by employing deuterated hydrosilanes (not shown). A plausible catalytic cycle commences with cooperative bond activation of the Si—H bond at the polar Ru—S bond (Scheme 4.16,

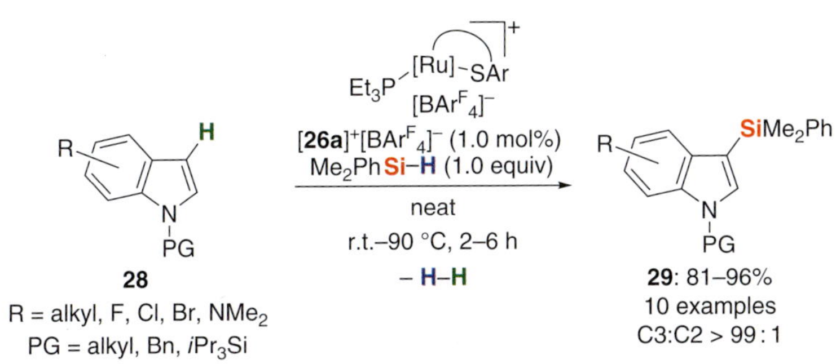

Scheme 4.15 Intermolecular electrophilic C—H silylation of N-protected indoles **28**.

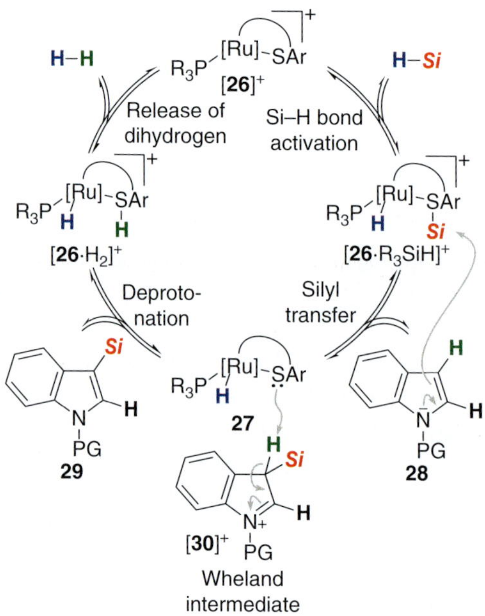

Scheme 4.16 Proposed mechanism for the intermolecular electrophilic C—H silylation of N-protected indoles **28** by $[\mathbf{26}]^+[BAr^F{}_4]^-$. Counteranions omitted for clarity.

$[\mathbf{26}]^+ \rightarrow [\mathbf{26}{\cdot}R_3SiH]^+$). Subsequent transfer of the silylium ion to indole **28** generates the Wheland intermediate $[\mathbf{30}]^+$ ($\mathbf{28} \rightarrow [\mathbf{30}]^+$) together with the neutral ruthenium hydride **27** ($[\mathbf{26}{\cdot}R_3SiH]^+ \rightarrow \mathbf{27}$). The intermediate $[\mathbf{30}]^+$ is likely deprotonated by the weakly basic sulfur atom in ruthenium hydride **27** and yields the C3-functionalized indoles **29** ($[\mathbf{30}]^+ \rightarrow \mathbf{29}$) as well as the dihydrogen adduct $[\mathbf{26}{\cdot}H_2]^+$ ($\mathbf{27} \rightarrow [\mathbf{26}{\cdot}H_2]^+$). This adduct liberates H_2 and then regenerates the active catalyst $[\mathbf{26}]^+$, thereby closing the catalytic cycle ($[\mathbf{26}{\cdot}H_2]^+ \rightarrow [\mathbf{26}]^+$).

Intramolecular C—H silylation of less nucleophilic benzenes was anticipated to proceed by the same mechanism which converts *ortho*-silylated biphenyls **31** into dibenzosiloles **32** (Scheme 4.17) [24c]. Applying rather forcing conditions, biphenyls **31** with a noteworthy number of functional groups successfully

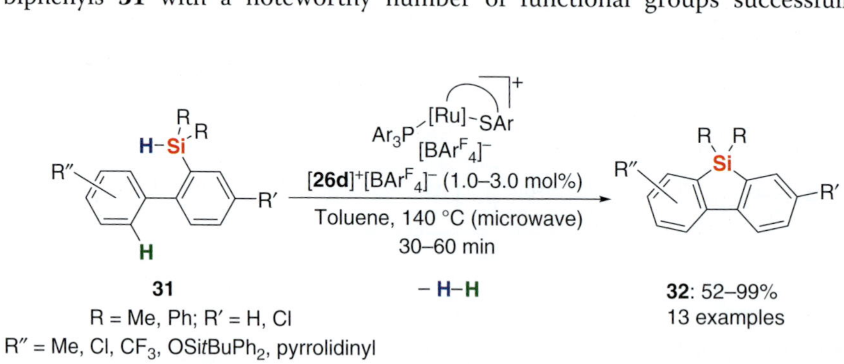

Scheme 4.17 Intramolecular electrophilic C—H silylation of arenes **31**. $Ar_3P = (p\text{-}FC_6H_4)_3P$.

underwent the ring closure and provided rapid access to dibenzosiloles **32** that are functionalized at both phenylene groups. A combination of inter- and intramolecular electrophilic C—H silylation was achieved in the catalytic synthesis of indole-fused benzosiloles by reacting 2-aryl-substituted indoles with dihydrosilanes (not shown) [24d].

4.5.4 Competing Dehydrogenative Coupling and Hydrosilylation

As shown in the C—H silylation of indoles (Scheme 4.16), the *in situ*–formed ruthenium hydride **27** is not sufficiently hydridic to transfer its hydride to the Wheland intermediate $[\mathbf{30}]^+$ but favors dehydrogenation. This observation led to the investigation of enolizable ketones **33** (X = O) under catalytic conditions with complexes $[\mathbf{26}]^+[BAr^F{}_4]^-$ and hydrosilanes (Scheme 4.18a, left) [25a]. Silyl enol ethers **34** were formed as the major products in the presence of catalyst $[\mathbf{26a}]^+[BAr^F{}_4]^-$ (**33** → **34**) instead of the *expected* silyl ethers **35**. A broad range of ketones with differently substituted aryl groups as well as purely aliphatic ketones participated in the catalysis. *Ortho*-substituted acetophenone derivatives favored the corresponding silyl enol ethers **34** with significantly higher selectivity, presumably due to increased steric hindrance. When ketones **33** reacted with deuterated hydrosilanes, deuterium incorporation in the α-position of the formed silyl ethers **35** was observed, likely due to subsequent hydrogenation of the initially formed silyl enol ether **34** by *in situ*–formed $[\mathbf{26a}{\cdot}DH]^+$ (**34** → **35**) [25d].

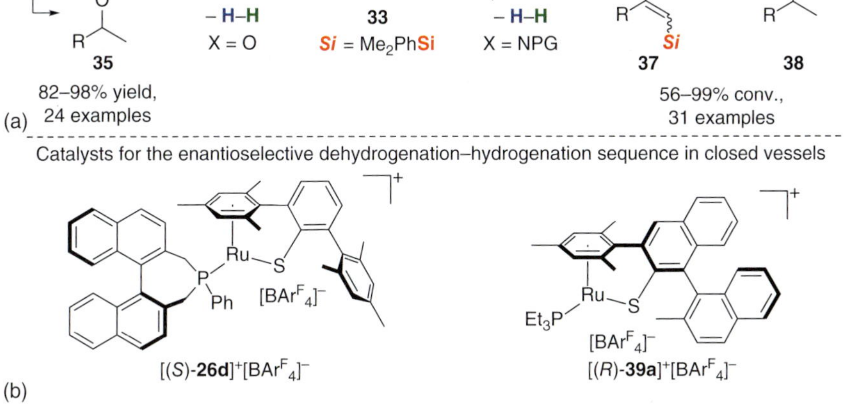

Scheme 4.18 (a) Dehydrogenative coupling of enolizable ketones (left) and imines (right) with hydrosilanes. Highlighted are the major products of catalysis in open vessels. (b) Chiral catalysts for enantioselective hydrogenations in closed vessels.

This methodology was later extended to enolizable ketimines **33** (X = NPG) (Scheme 4.18a, right) [25b]. To suppress reduction pathways, it was crucial to change to the bulkier catalyst system [**26b**]$^+$[BAr$^F_4$]$^-$ with *i*Pr$_3$P as ancillary ligand, which delivered N-silylated enamines **36** with high chemoselectivity (**33** → **36**). C-silylated N-silylated enamines **37** were formed only in trace amounts due to subsequent dehydrogenative silylation of **36** (**36** → **37**). The products of hydrogenation, amines **38**, were formed only in minor quantities (**36** → **38**). Net reduction of the C—X double bond by the aforementioned dehydrogenation–hydrogenation sequence was accomplished, when performing these reactions in closed vessels (**33** → **34** → **35** or **33** → **36** → **38**). More recently, enantioselective hydrogenations of ketones and ketimines were achieved with either [(*S*)-**26d**]$^+$[BAr$^F_4$]$^-$ coordinated with a chiral phosphine ligand [25d] or [(*R*)-**39a**]$^+$[BAr$^F_4$]$^-$ based on an axial chiral SDmp derivative (Scheme 4.18b) [25e]. Both systems delivered promising levels of enantioselection: ~55% *ee* with [(*S*)-**26d**]$^+$[BAr$^F_4$]$^-$ and ~40% *ee* [(*R*)-**39a**]$^+$[BAr$^F_4$]$^-$.

4.5.5 C–H Silylation by Hydrosilylation/Dehydrogenative Silylation/Retro-Hydrosilylation

The development of a 1,4-selective hydrosilylation of pyridines catalyzed by [**26**]$^+$[BAr$^F_4$]$^-$ (not shown) [26a] led to a domino process consisting of this hydrosilylation, the above-described dehydrogenative enamine silylation, and retro-hydrosilylation. This reaction sequence resulted in a formal *meta*-selective electrophilic aromatic substitution of pyridines **40** delivering C3-silylated pyridines **41** with reasonable functional-group tolerance (Scheme 4.19) [26b].

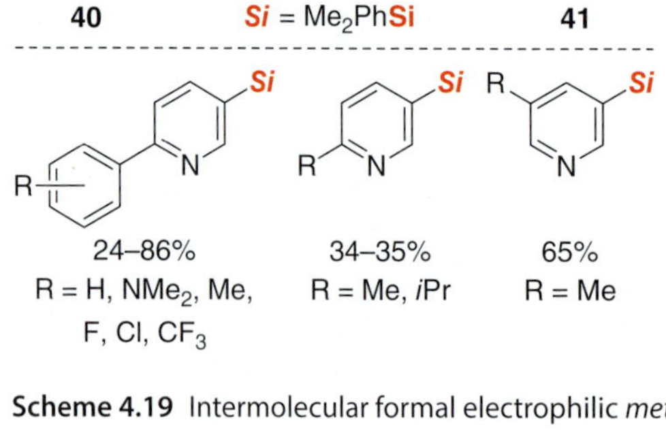

Scheme 4.19 Intermolecular formal electrophilic *meta*-C—H silylation of pyridines.

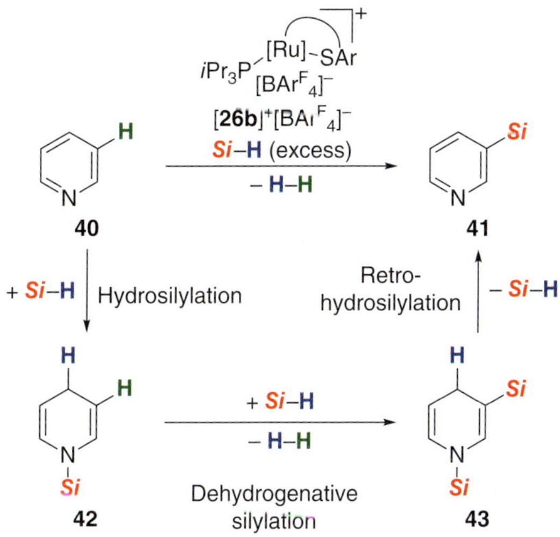

Scheme 4.20 Temporary dearomatization of pyridines, enabling their electrophilic *meta*-C—H silylation.

Careful ^{1}H NMR spectroscopic analysis of this three-step transformation provided convincing insight into the mechanism (Scheme 4.20). Pyridines **40** undergo the 1,4-hydrosilylation already at ambient temperature, resulting in 1,4-dihydropyridines **42**. At elevated temperatures, this newly formed N-silylated enamine unit is silylated in a dehydrogenative manner in its β-position to form *meta*-silylated 1,4-dihydropyridines **43** (cf. Scheme 4.18a, right). The *meta*-silylated 1,4-dihydropyridines **43** rearomatize at elevated temperature because of the reversibility of the 1,4-hydrosilylation [26c] to yield C3-silylated pyridines **41**.

4.6 Summary

This chapter gave an overview of complexes enabling cooperative Si—H bond activation with emphasis on application in catalysis. The cooperative catalysis at Ru—S bonds described herein is a powerful tool to form new carbon–main-group element bonds, and the low hydricity of the ruthenium(II) hydride with its adjacent basic sulfur ligand turned out to be crucial for efficient dehydrogenative couplings (Scheme 4.21). Besides the discussed examples, there are further applications of sulfur-stabilized silylium ions such as *N*-silylation of heterocycles and anilines [24b] and the hydrosilylation of CO_2 [25c] as well as in aldol reactions [25f]. Outside the scope of this book and, hence, this chapter are the generation of borenium [28], alumenium [29], and stannylium [30] ions by heterolytic E—H bond cleavage at Ru—S bonds (E = BR_2, AlR_2, or SnR_3). The high reactivity of these main-group cations was demonstrated in various catalytic reactions.

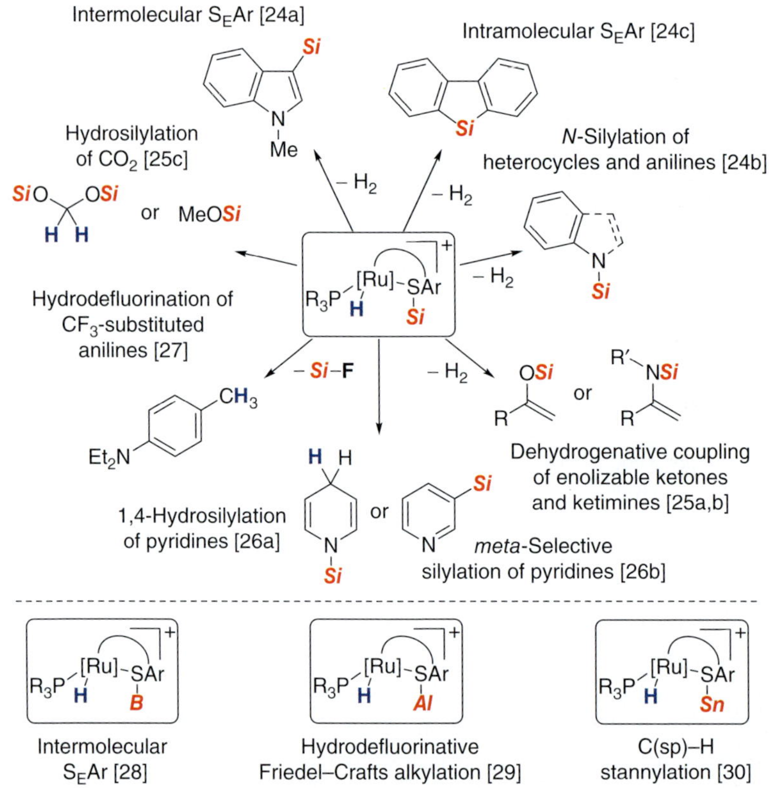

Scheme 4.21 Overview of ruthenium–thiolate complex-catalyzed reactions. (*Si* = SiR_3 or SiR_2, *B* = BR_2, *Al* = AlR_2, *Sn* = SnR_3).

References

1 For general reviews of metal–ligand cooperativity, see: (a) Khusnutdinova, J.R. and Milstein, D. (2015). *Angew. Chem. Int. Ed.* 54: 12236–12273. (b) Peters, R. (2015). *Cooperative Catalysis*. Weinheim: Wiley-VCH. (c) Grützmacher, H. (2008). *Angew. Chem. Int. Ed.* 47: 1814–1818.

2 For reviews of transition-metal frustrated Lewis pairs, see: (a) Berke, H., Jiang, Y., Yang, X. et al. (2013). *Top. Curr. Chem.* 334: 27–58. (b) Wass, D.F. and Chapman, A.M. (2013). *Top. Curr. Chem.* 334: 261–280. (c) Flynn, S.R. and Wass, D.F. (2013). *ACS Catal.* 3: 2574–2581.

3 (a) Kubas, G.J. (2005). *Catal. Lett.* 104: 79–101. (b) Walsh, R. (1981). *Acc. Chem. Res.* 14: 246–252.

4 Berlekamp, U.-H., Jutzi, P., Mix, A. et al. (1999). *Angew. Chem. Int. Ed.* 38: 2048–2050.

5 For silylsulfonium ions, see: (a) Prakash, G.K.S., Bae, C., Wang, Q. et al. (2000). *J. Org. Chem.* 65: 7646–7649. (b) Rohde, V.H.G., Pommerening, P., Klare, H.F.T., and Oestreich, M. (2014). *Organometallics* 33: 3618–3628.

6 Weisman, J. and Gessner, V.H. (2015). *Chem. Commun.* 51: 14909–14912.

7 LaPierre, E.A., Piers, W.E., Spasyuk, D.M., and Bi, D.W. (2016). *Chem. Commun.* 52: 1361–1364.

8 Comanescu, C.C. and Iluc, V.M. (2016). *Chem. Commun.* 52: 9048–9051.

9 Gountchev, T.I. and Tilley, T.D. (1997). *J. Am. Chem. Soc.* 119: 12831–12841.

10 (a) Kaplan, A.W., Polse, J.L., Ball, G.E. et al. (1998). *J. Am. Chem. Soc.* 120: 11649–11662. (b) Tiong, P.J., Nova, A., Clot, E., and Mountford, P. (2011). *Chem. Commun.* 47: 3147–3149.

11 (a) Fryzuk, M.D., Love, J.B., Rettig, S.J., and Young, V.G. (1997). *Science* 275: 1445–1447. (b) Basch, H., Musaev, D.G., Morokuma, K. et al. (1999). *J. Am. Chem. Soc.* 121: 523–528. (c) Fryzuk, M.D., MacKay, B.A., and Patrick, B.O. (2003). *J. Am. Chem. Soc.* 125: 3234–3235. (d) MacKay, B.A., Munha, R.F., and Fryzuk, M.D. (2006). *J. Am. Chem. Soc.* 128: 9472–9483.

12 Zhou, J., Chu, J., Zhang, Y. et al. (2013). *Angew. Chem. Int. Ed.* 52: 4243–4246.

13 Llamazares, A., Schmalle, H.W., and Berke, H. (2001). *Organometallics* 20: 5277–5288.

14 (a) Kennedy-Smith, J.J., A: Nolin, K., Gunterman, H.P., and Toste, F.D. (2003). *J. Am. Chem. Soc.* 125: 4056–4075. (b) Nolin, K.A., Krumper, J.R., Pluth, M.D. et al. (2007). *J. Am. Chem. Soc.* 129: 14684–14696. (c) Chung, L.W., Lee, H.G., Lin, Z., and Wu, Y.-D. (2006). *J. Org. Chem.* 71: 6000–6009.

15 Chu, W.-Y., Zhou, X., and Rauchfuss, T.B. (2015). *Organometallics* 34: 1619–1626.

16 For general reviews of transition metal–sulfur complexes, see: (a) Rauchfuss, T.B. (2004). *Inorg. Chem.* 43: 14–26. (b) Breysse, M., Furimsky, E., Kasztelan, S. et al. (2002). *Catal. Rev. Sci. Eng.* 44: 651–735. (c) Blower, P.J. and Dilworth, J.R. (1987). *Coord. Chem. Rev.* 76: 121–185. (d) Dubois, M.R. (1989). *Chem. Rev.* 89: 1–9.

17 For reviews of hydrogenases, see: (a) Lubitz, W., Ogata, H., Rüdiger, O., and Reijerse, E. (2014). *Chem. Rev.* 114: 4081–4148. (b) Gordon, J.C. and Kubas, G.J. (2010). *Organometallics* 29: 4682–4701. (c) Tard, C. and Pickett, C.J. (2009). *Chem. Rev.* 109: 2245–2274.

18 (a) Sweeney, Z.K., Polse, J.L., Andersen, R.A. et al. (1997). *J. Am. Chem. Soc.* 119: 4543–4544. (b) Sweeney, Z.K., Polse, J.L., Bergman, R.G., and Andersen, R.A. (1999). *Organometallics* 18: 5502–5510.

19 Hesp, K.D., McDonald, R., Ferguson, M.J., and Stradiotto, M. (2008). *J. Am. Chem. Soc.* 130: 16394–16406.

20 Ellison, J.J., Ruhlandt-Senge, K., and Power, P.P. (1994). *Angew. Chem. Int. Ed. Engl.* 33: 1178–1180.

21 Ohki, Y., Sakamoto, M., and Tatsumi, K. (2008). *J. Am. Chem. Soc.* 130: 11610–11611.

22 (a) For a recent account of cooperative catalysis at metal–sulfur bonds focusing on Si—H and B—H bond activation, see: Omann, L., Königs, C.D.F., Klare, H.F.T., and Oestreich, M. (2017). *Acc. Chem. Res.* 50: 1258–1269. (b) For a detailed analysis of the cooperative Si—H bond activation at the Ru—S bond, see: Stahl, T., Hrobárik, P., Königs, C.D.F. et al. (2015). *Chem. Sci.* 6: 4324–4334. (c) For a general account of cooperative catalysis at Ru—S bonds, see: Forster, F. and Oestreich, M. (2018). *CHIMIA Int. J. Chem.* 72: 584–588.

23 (a) Ohki, Y., Takikawa, Y., Sadohara, H. et al. (2008). *Chem. Asian J.* 3: 1625–1635. (b) For a detailed analysis of catalytic hydrogenations with complexes [**26**]$^+$[BAr$^F{}_4$]$^-$, see: Lefranc, A., Qu, Z.-W., Grimme, S., and Oestreich, M. (2016). *Chem. Eur. J.* 22: 10009–10016.

24 (a) For intermolecular silylation of heterocycles, see: Klare, H.F.T., Oestreich, M., Ito, J.-i. et al. (2011). *J. Am. Chem. Soc.* 133: 3312–3315. (b) For *N*-silylation of heterocycles and anilines, see: Königs, C.D.F., Müller, M.F., Aiguabella, N. et al. (2013). *Chem. Commun.* 49: 1506–1508. (c) For intramolecular silylation of biaryls, see: Omann, L. and Oestreich, M. (2015). *Angew. Chem. Int. Ed.* 54: 10276–10279. (d) For a double sila-Friedel–Crafts protocol, see: Omann, L. and Oestreich, M. (2017). *Organometallics* 36: 767–776.

25 (a) For the formation of silyl enol ethers, see: Königs, C.D.F., Klare, H.F.T., Ohki, Y. et al. (2012). *Org. Lett.* 14: 2842–2845. (b) For the formation of *N*-silyl enamines, see: Hermeke, J., Klare, H.F.T., and Oestreich, M. (2014). *Chem. Eur. J.* 20: 9250–9254. (c) For catalytic CO_2 reduction, see: Metsänen, T.T. and Oestreich, M. (2015). *Organometallics* 34: 543–546. (d) For an enantioselective version with chiral phosphine ligands, see: Bähr, S. and Oestreich, M. (2017). *Organometallics* 36: 935–943. (e) For an enantioselective version with a chiral thiolate ligand, see: Wübbolt, S., Maji, M.S., Irran, E., and Oestreich, M. (2017). *Chem. Eur. J.* 23: 6213–6219. (f) For aldol reactions, see: Pommerening, P., Stahl, T., and Oestreich, M. (2018). *Eur. J. Org. Chem.* 2290–2293.

26 (a) For 1,4-hydrosilylation of pyridines, see: Königs, C.D.F., Klare, H.F.T., and Oestreich, M. (2013). *Angew. Chem. Int. Ed.* 52: 10076–10079. (b) For formal *meta*-selective silylation of pyridines, see: Wübbolt, S. and Oestreich, M. (2015). *Angew. Chem. Int. Ed.* 54: 15876–15879. (c) For mechanistic studies, see: Bähr, S. and Oestreich, M. (2018). *Chem. Eur. J.* 24: 5613–5622.

27 Hydrodefluorination using sulfur-stabilized silylium ions: Stahl, T., Klare, H.F.T., and Oestreich, M. (2013). *J. Am. Chem. Soc.* 135: 1248–1251.

28 Cooperative activation of hydroboranes at polar Ru—S bonds: Stahl, T., Müther, K., Ohki, Y. et al. (2013). *J. Am. Chem. Soc.* 135: 10978–10981.

29 Cooperative activation of DIBAL—H at polar Ru—S bonds: Forster, F., Metsänen, T.T., Irran, E. et al. (2017). *J. Am. Chem. Soc.* 139: 16334–16342.

30 Cooperative activation of hydrostannanes at polar Ru—S bonds: Forster, F., López, V.M.R., and Oestreich, M. (2018). *J. Am. Chem. Soc.* 140: 1259–1262.

31 (a) Waterman, R. (2013). *Organometallics* 32: 7249–7263. (b) Perutz, R.N. and Sabo-Etienne, S. (2007). *Angew. Chem. Int. Ed.* 46: 2578–2592. (c) Lin, Z. (2007). *Coord. Chem. Rev.* 251: 2280–2291.

32 For a related racemization-free ^{1}H/^{2}H scrambling catalyzed by $B(C_6F_5)_3$, see: Rendler, S. and Oestreich, M. (2008). *Angew. Chem. Int. Ed.* 47: 5997–6000.

33 (a) Hoffmann, S.P., Kato, T., Tham, F.S., and Reed, C.A. (2006). *Chem. Commun.* 767–769. (b) Nava, M. and Reed, C.A. (2011). *Organometallics* 30: 4798–4800. (c) Connelly, S.J., Kaminsky, W., and Heinekey, D.M. (2013). *Organometallics* 32: 7478–7481.

34 For a review of sila-Friedel–Crafts reactions, see: Bähr, S. and Oestreich, M. (2017). *Angew. Chem. Int. Ed.* 56: 52–59.

5

Cationic Silicon-Based Lewis Acids in Catalysis

Polina Shaykhutdinova, Sebastian Keess, and Martin Oestreich

Institut für Chemie, Technische Universität Berlin, Strasse des 17. Juni 115, 10623 Berlin, Germany

5.1 Introduction

Cationic silicon-based Lewis acids had not found widespread application until the beginning of this century, although their pronounced electrophilicity indicated early their promising potential for the activation of Lewis-basic substrates [1]. It was, however, exactly this electron deficiency that impeded the handling of these highly reactive molecules in the condensed phase. Efforts of several research groups, particularly Reed's and Lambert's, directed toward the generation, isolation, and characterization of silicon cations led to a fundamental understanding of the properties of these Lewis acids [2]. This new knowledge and insight opened the door to synthetic applications of these highly reactive intermediates, and this chapter summarizes their use as catalysts in reactions as diverse as reduction, hydrodefluorination, and C—C bond formation [3].

5.2 Deoxygenation and Hydrosilylation of C=X Multiple Bonds

5.2.1 Deoxygenation of C=O Bonds

The treatment of carbonyl compounds with a hydrosilane in the presence of catalytic amounts of a silicon cation typically results in exhaustive reduction, i.e. deoxygenation. In an early example, the group of Kira reported the deoxygenation of C=O groups promoted by silicon cations generated from the corresponding hydrosilane and *in situ*–formed $Ph_3C^+[BAr^F_4]^-$ by Corey's silicon-to-carbon hydride transfer [4] (Scheme 5.1, top) [5]. The reaction of benzophenone (**1**) with Et_3SiH (**2**) in the presence of catalytic amounts of $Ph_3CCl/NaBAr^F_4$ furnished

Organosilicon Chemistry: Novel Approaches and Reactions, First Edition.
Edited by Tamejiro Hiyama and Martin Oestreich

alkane **3** as the product of complete reduction (**1** + **2** → **3**). Conversely, dialkyl-substituted **4** gave cyclododecene (**6**) through β-elimination of the oxonium-ion intermediate (**4** + **5** → **6**). These results were later confirmed by Piers and co-workers who obtained ethylbenzene (**9**) via reduction of acetophenone (**7**) with $i\mathrm{Pr_3SiH}$ (**8**) using preformed $\mathrm{Ph_3C^+[B(C_6F_5)_4]^-}$ as the initiator (**7** + **8** → **9**, Scheme 5.1, bottom) [6].

The catalytic cycle of the silicon cation–promoted deoxygenation is initiated by the formation of the donor-stabilized silicon cation $\mathbf{II}^+$ through trityl cation–mediated hydride abstraction from hydrosilane **I** ($\mathbf{I} \rightarrow \mathbf{II}^+$, Scheme 5.2) [5, 6]. Coordination of the Lewis-basic oxygen atom of carbonyl compound **III** to cation $\mathbf{II}^+$ generates silyloxocarbenium/silylcarboxonium ion $\mathbf{IV}^+$ ($\mathbf{II}^+ \rightarrow \mathbf{IV}^+$), which is subsequently reduced by hydrosilane **I** through the assumed four-membered transition state $\mathbf{V}^{\ddagger+}$, i.e. assisted by the carbonyl oxygen atom, to furnish oxonium ion $\mathbf{VI}^+$ ($\mathbf{IV}^+ \rightarrow \mathbf{V}^{\ddagger+} \rightarrow \mathbf{VI}^+$) [6]. Cation $\mathbf{VI}^+$ eventually dissociates to carbenium ion $\mathbf{VII}^+$ and disiloxane $\mathrm{R_3SiOSiR_3}$ ($\mathbf{VI}^+ \rightarrow \mathbf{VII}^+$), and hydride transfer from another molecule of hydrosilane **I** then liberates deoxygenated **III** (=**VIII**), thereby closing the catalytic cycle ($\mathbf{VII}^+ \rightarrow \mathbf{II}^+$).

Later, Khandelwal and Wehmschulte investigated the deoxygenative reduction of carbon dioxide promoted by silicon cations [7]. The developed method suffers from an unsatisfactory chemoselectivity, yielding mixtures of toluene and diphenylmethane as products of Friedel–Crafts alkylation of the benzene solvent, while the formation of methane was not observed (not shown).

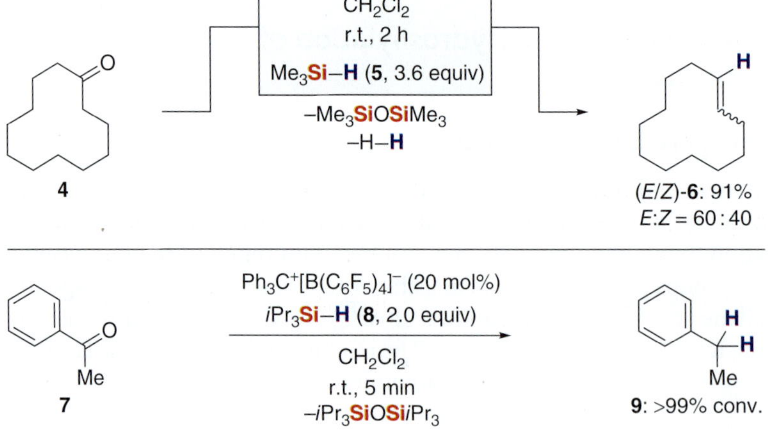

Scheme 5.1 Deoxygenation of carbonyl compounds promoted by an *in situ*–generated (top) or preformed (bottom) trityl cation.

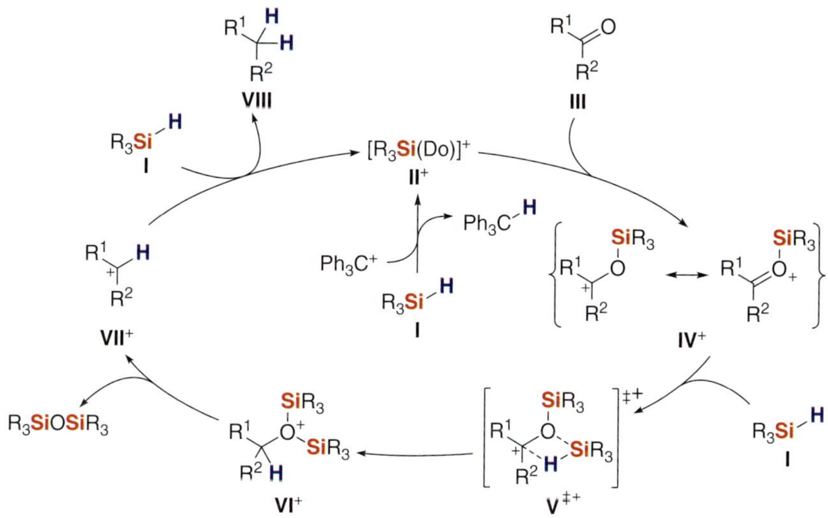

Scheme 5.2 Mechanism of the silicon cation–promoted deoxygenation of carbonyl compounds by the groups of Kira and Piers.

5.2.2 Hydrosilylation of C=O, C=N, C=C, and C≡C Bonds

The possibility of avoiding over-reduction in these reactions and reducing ketones chemoselectively to the silyl ethers was demonstrated by Müther and Oestreich (7 and **10–20** → **22–33**, Scheme 5.3) [8]. Key to the success was the application

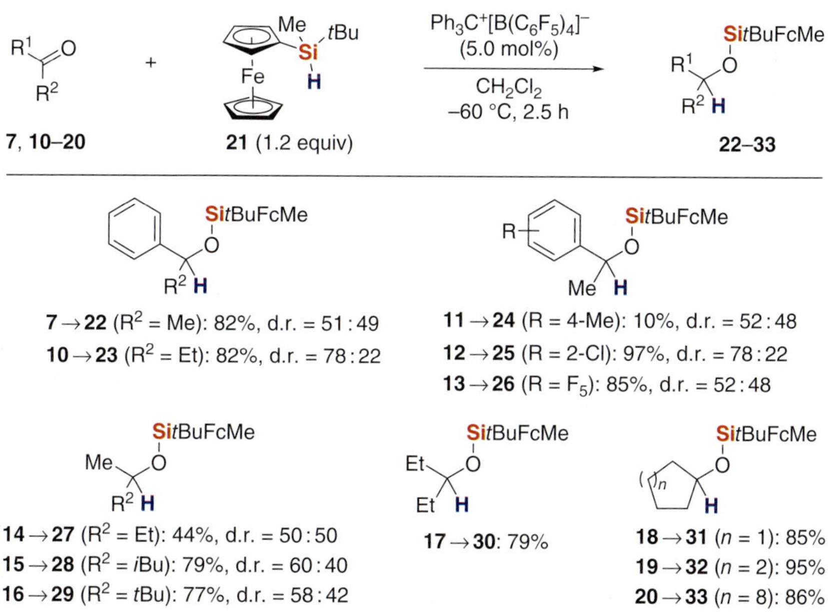

Scheme 5.3 Hydrosilylation of ketones mediated by silicon cations.

of ferrocenyl-substituted **21** that enabled the hydride transfer from hydrosilane **I** to oxocarbenium ion $\mathbf{IV}^+$ without the assistance of the oxygen atom of the carbonyl compound (cf. $\mathbf{V}^{\ddagger+}$, Scheme 5.2). The increased hydricity of **21** that facilitates this reduction step is likely based on the intramolecular stabilization of the silicon cation intermediate by the ferrocene backbone [9]. In addition, the augmented steric hindrance in the coordination sphere of the electrophilic silicon center likely prevents this silicon–oxygen interaction (cf. $\mathbf{V}^{\ddagger+}$, Scheme 5.2). With this strategy, a broad variety of alkyl aryl (**7** and **10–13**) as well as acyclic and cyclic dialkyl ketones (**14–20**) was reduced chemoselectively to the alcohol oxidation level (Scheme 5.3) [8]. Conversely, due to the diminished hydride affinities of the corresponding silylcarboxonium ion intermediates, diaryl-substituted ketones were not compatible with this hydrosilylation protocol (not shown).

The group of Oestreich also reported a method for the silicon cation–promoted reduction of imines (Scheme 5.4) [10] using the catalytic system previously applied to the hydrosilylation of ketones (cf. Scheme 5.3) [8]. Both Et_3SiH (**2**) and *t*BuFcMeSiH (**21**) were shown to be potent hydride donors to efficiently reduce the silyliminium-ion intermediates [10]. Different aryl- and alkyl-substituted *N*-tosylated aldimines **34–39** and **40–42** were compatible with this method, mostly furnishing amines **44–52** in high yields after hydrolysis. The authors also presented an example for the reduction of an alkyl aryl ketimine (**43** → **53**).

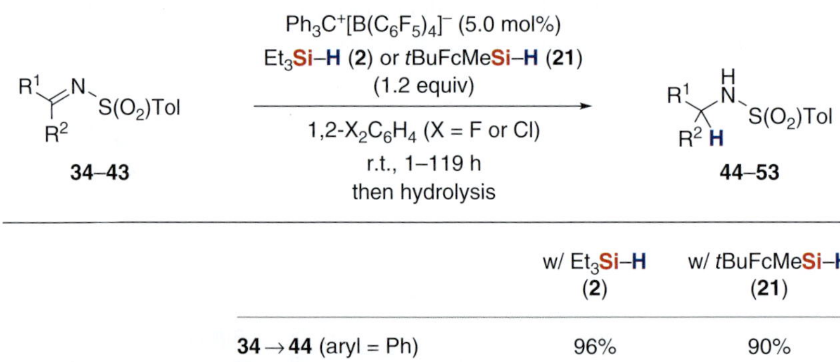

		w/ Et_3Si–H (**2**)	w/ *t*BuFcMeSi–H (**21**)
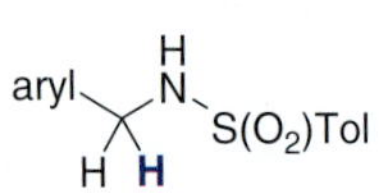	**34** → **44** (aryl = Ph)	96%	90%
	35 → **45** (aryl = 4-BrC_6H_4)	92%	61%
	36 → **46** (aryl = 2-tolyl)	78%	85%
	37 → **47** (aryl = 4-tolyl)	93%	59%
	38 → **48** (aryl = naphth-1-yl)	90%	93%
	39 → **49** (aryl = naphth-2-yl)	66%	63%
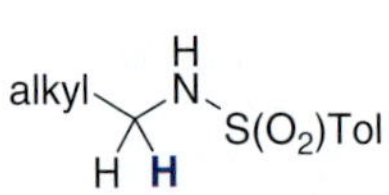	**40** → **50** (alkyl = cyclohexyl)	99%	98%
	41 → **51** (alkyl = pentan-3-yl)	96%	97%
	42 → **52** (alkyl = *t*Bu)	96%	95%
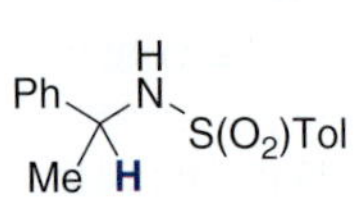	**43** → **53**	76%	19%

Scheme 5.4 Reduction of imines by silicon cation–promoted hydrosilylation.

Several years before the publication of these reports about C=X bond reduction, the group of Lambert had succeeded in the hydrosilylation of C=C bonds of 1,1-disubstituted alkenes using catalytic amounts of $Ph_3C^+[B(C_6F_5)_4]^-$ as initiator (Scheme 5.5) [11]. Additional stabilization of the β-silyl-stabilized [12] carbenium-ion intermediate by bisbenzylic (**54** → **57**) or benzylic/allylic (**55** → **58**) conjugation secured the formation of the hydrosilylation products **57** and **58**. Conversely, only decomposition of the bisallylic carbenium ion formed from trialkene **56** was observed (**56** → **59**). The group of Buriak later showed that various terminal alkenes as well as alkynes are compatible with the silicon cation-promoted hydrosilylation reaction (not shown) [13]. Subjection of hydride-terminated porous silicon to the reaction protocol enabled the synthesis of a broad range of functionalized surfaces. Müller and coworkers reported an intramolecular variant of the C=C bond hydrosilylation, and an intramolecular π-stabilized silanorbornyl cation was proposed as an intermediate (**60**–**62** → **63**–**65**, Scheme 5.6) [14].

All examples for the hydrosilylation of C=X bonds presented earlier underlie the same general mechanism (Scheme 5.7). Generation of the silicon electrophile **II**$^+$ by the Corey hydride transfer [4] (**I** → **II**$^+$) is followed by coordination of the donor-stabilized silicon cation **II**$^+$ to the σ-basic (for X = O and NR) or π-basic (for X = CR_2) substrate **IX** (**II**$^+$ + **IX** → **X**$^+$). Hydride reduction of the

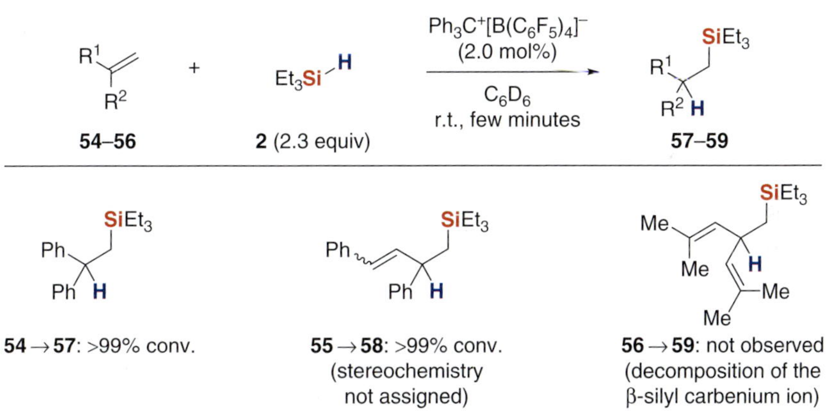

Scheme 5.5 Silicon cation-promoted intermolecular hydrosilylation of alkenes.

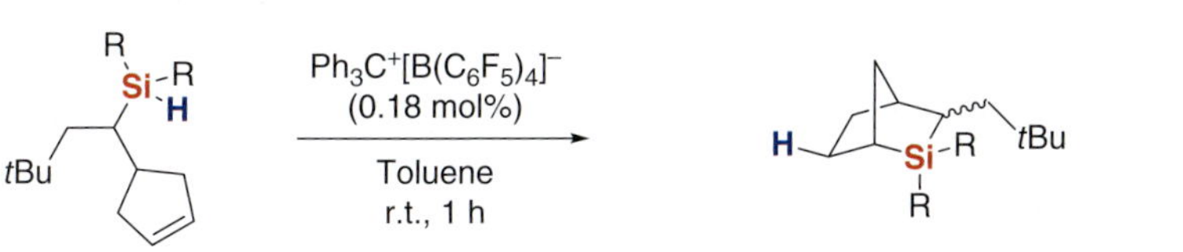

60 (R = H)
61 (R = Me)
62 (R = Ph)

63 (R = H): 88%, *exo*:*endo* = 72 : 28
64 (R = Me): 94%, *exo*:*endo* = 64 : 36
65 (R = Ph): 94%, *exo*:*endo* = 73 : 27

Scheme 5.6 Silicon cation-promoted intramolecular hydrosilylation of alkenes.

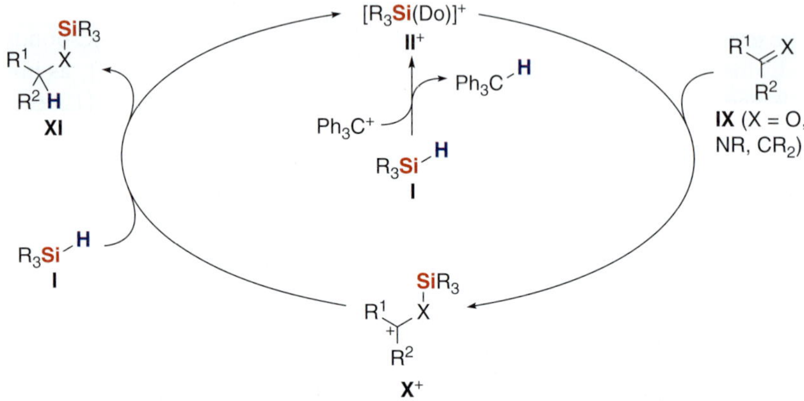

Scheme 5.7 Mechanism of the silicon cation–promoted hydrosilylation of C—X multiple bonds.

(heteroatom-stabilized) carbenium ion **X**$^+$ by hydrosilane **I** eventually releases the product **XI** and regenerates silicon cation **II**$^+$ (**X**$^+ \rightarrow$ **II**$^+$).

Mochida and coworkers developed a method that provides a synthetic access to silatetralins by a trityl cation–initiated reaction sequence between terminal alkyne **66** and benzyl-substituted hydrosilanes **67** and **68** (Scheme 5.8) [15]. Addition of the donor-stabilized silicon cation **XIII**$^+$ to alkyne **XIV** is not followed by rapid hydride transfer from another molecule of the hydrosilane **XII** to yield vinylsilane **XVI** (**XIII**$^+ \rightarrow$ **XV**$^+ \rightarrow$ **XVI**, gray route). Instead, an intramolecular Friedel–Crafts addition of the phenyl substituent to the vinyl cation in **XV**$^+$ furnishes Wheland intermediate **XVII**$^+$ (**XV**$^+ \rightarrow$ **XVII**$^+$) that eventually undergoes a proton shift to form the benzylic carbenium ion **XVIII**$^+$, which is additionally stabilized by the β-silicon effect [12] (**XVII**$^+ \rightarrow$ **XVIII**$^+$). Finally, hydride transfer from hydrosilane **XII** to carbenium ion **XVIII**$^+$ liberates silatetralin **XIX** and closes the catalytic cycle (**XVIII**$^+ \rightarrow$ **XIII**$^+$).

It was also shown that the use of stoichiometric amounts of a base such as 2,6-di-*tert*-butyl-4-methylpyridine shifts the chemoselectivity from proton transfer and subsequent hydride reduction (**XVII**$^+ \rightarrow$ **XVIII**$^+ \rightarrow$ **XIX**) to deprotonation of the Wheland intermediate **XVII**$^+$ (**XVII**$^+ \rightarrow$ **XX**) or the benzylic carbenium ion **XVIII**$^+$ (not shown) to afford the dehydrogenated product **XX** [15]. Stoichiometric amounts of the trityl cation were required for this transformation as a catalytic turnover by dihydrogen release from the hydrosilane **XII** and the protonated base [base·H]$^+$ to regenerate the silicon cation **XIII**$^+$ was not achieved [16]. However, this stoichiometric method enabled the synthesis of different 1,2-dihydro-2-silanaphthalene derivatives (not shown) [15].

Recently, Arii et al. exploited the concept of silicon cation–promoted hydrosilylation for the preparation of benzosiloles (Scheme 5.9) [17]. Intramolecular hydrosilylation of the C≡C bond of alkynyl-substituted arylsilanes **71–75** furnished benzosiloles **77–81** in moderate to high yields. A limitation of this method was found in the necessity of a silyl-substituted alkyne to stabilize the intermediate vinyl cation by the β-silicon effect (not shown) [12]. Accordingly, the *n*-butyl-substituted derivative **76** showed almost no reactivity (**76** → **82**).

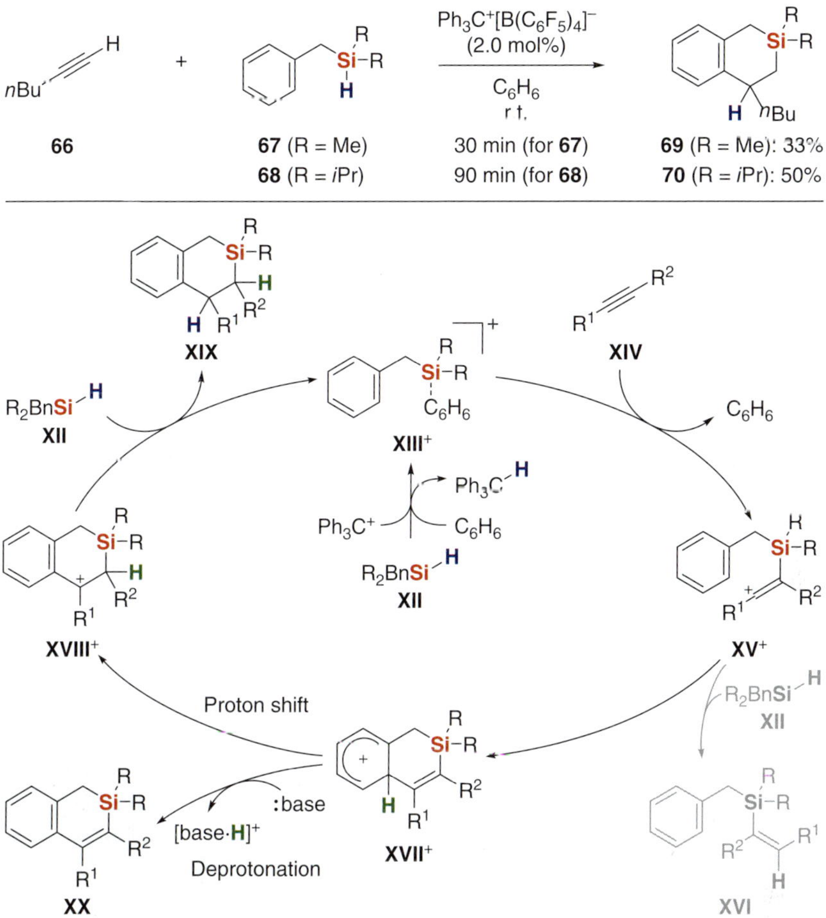

Scheme 5.8 Silicon cation–promoted synthesis of silatetralins by Mochida and coworkers.

5.3 C—F Bond Activation

The pronounced Lewis acidity and fluoride-ion affinity of silicon cations render these electrophiles ideal catalysts for the activation of C—F bonds [18]. Although carbon–fluorine bonds are among the strongest and most unreactive functional groups in organic chemistry, the normally even higher strength of silicon–fluorine bonds often provides a thermodynamic driving force for these transformations [19]. Consequently, cationic silicon Lewis acids have found widespread application in the arena of C—F bond activation in recent years [20].

5.3.1 Hydrodefluorination

The fundamental concept of silicon cation–promoted hydrodefluorination reactions is based on the capability of a donor-stabilized silicon cation **II**$^+$ to heterolytically cleave a C(sp^3)—F bond of a fluorinated hydrocarbon **XXI** and to

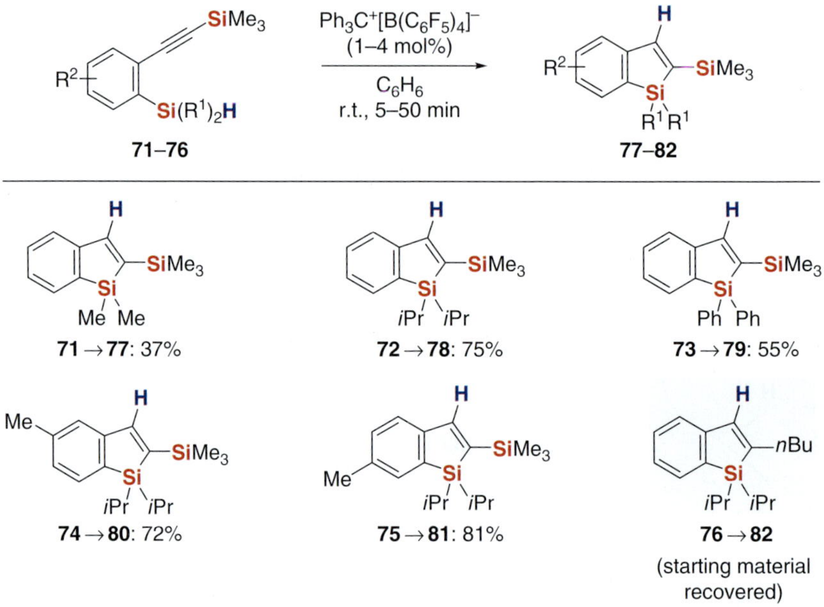

Scheme 5.9 Silicon cation–promoted synthesis of benzosiloles by Arii and coworkers.

form fluorosilane **XXII** along with the stabilized carbenium ion **XXIII**$^+$ (Scheme 5.10, **II**$^+$ → **XXIII**$^+$). Subsequent hydride transfer from a hydrosilane **I** to carbocation **XXIII**$^+$ then liberates the hydrocarbon **XXIV** and completes the catalytic cycle (**XXIII**$^+$ → **II**$^+$). The higher strength of a C(sp^3)—H bond relative to a Si—H bond contributes to the driving force of this formal Si—H/C—F metathesis.

Ozerov and coworkers developed an early example for a catalytic hydrodefluorination reaction of C(sp^3)—F bonds catalyzed by a silicon cation (Scheme 5.11) [21]. It was shown that the electron density at the silicon center of the silicon electrophile has a distinct effect on the success of the activation of the first out of three aliphatic C—F bonds in (trifluoromethyl)benzene (**83**). While

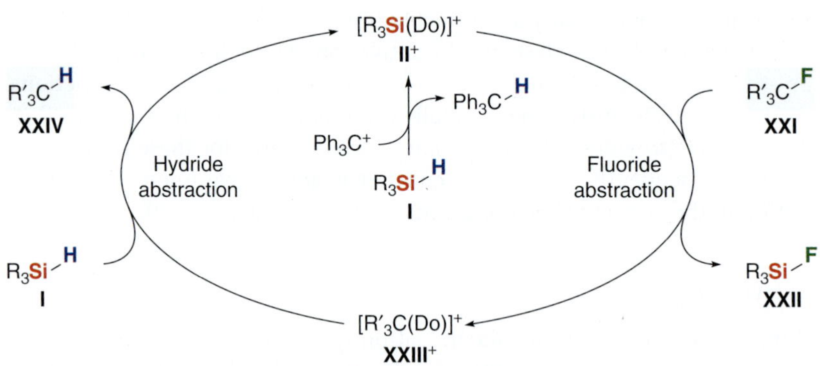

Scheme 5.10 Simplified mechanism of the silicon cation–promoted hydrodefluorination.

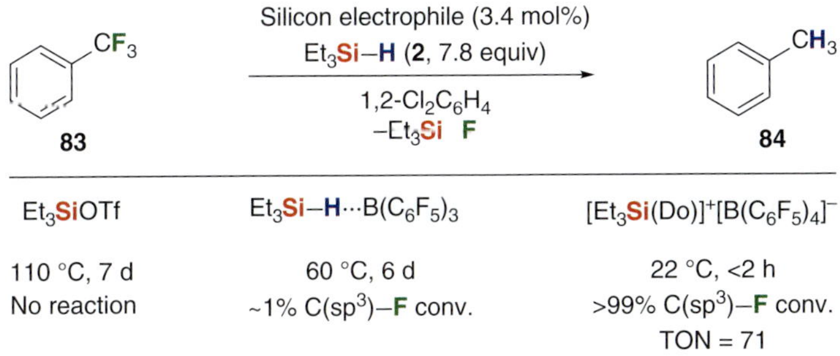

Scheme 5.11 Modifying the reactivity of the silicon electrophile in catalytic hydrodefluorination reactions.

neutral Et_3SiOTf (left) or $B(C_6F_5)_3$-activated Et_3SiH (middle) did not promote the hydrodefluorination of **83** even at elevated temperatures, cationic $[Et_3Si(Do)]^+$ furnished toluene (**84**) with quantitative conversion within less than two hours at room temperature (right). Various derivatives of (trifluoromethyl)benzene (**83**) were successfully transformed into their hydrocarbon analogs (**83** and **85–93** → **84** and **94–102**, Scheme 5.12) [21]. In contrast to transition-metal-catalyzed hydrodefluorination reactions that usually promote the activation of C(sp^2)—F bonds [20b, c], C(sp^2)—X bonds (X = F, Cl, and Br) were perfectly compatible with this reaction protocol (**86–92** → **95–101**). 1-Fluoropentane (**93**)

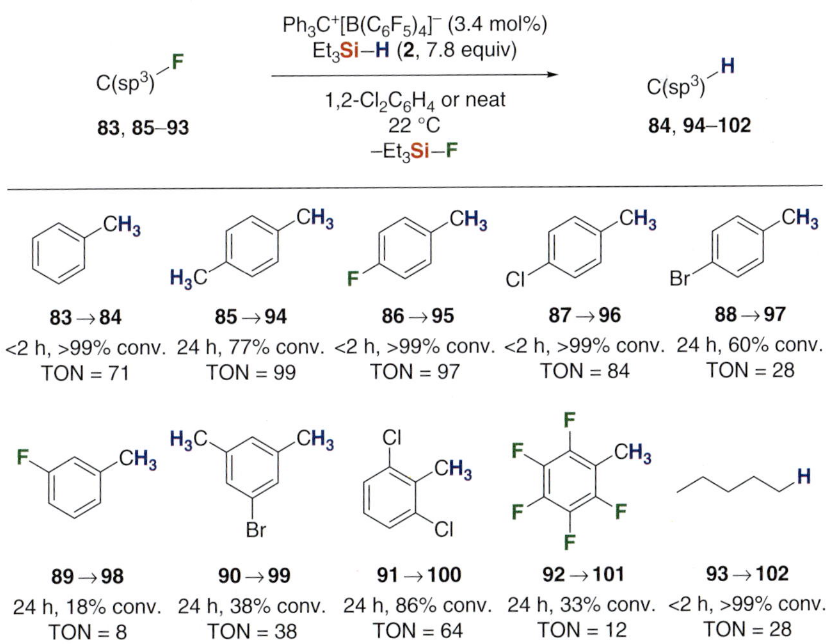

Scheme 5.12 Substrate scope of the silicon cation-promoted hydrodefluorination by Ozerov and coworkers.

also participated in the silicon cation–promoted hydrodefluorination (**93** → **102**), while perfluorinated hydrocarbons turned out to be unreactive (not shown).

The group of Müller pursued a related approach to C(sp^3)—F bond activation and developed a family of naphthalene-1,8-diyl-derived intramolecularly stabilized silicon cations **103^+–116^+**$[B(C_6F_5)_4]^-$ (Scheme 5.13) [22]. These Lewis acids were then tested in the hydrodefluorination of 1-fluorodecane (**117**) using stoichiometric amounts of Et_3SiH (**2**) as the hydride source (**117** → **118**). Hydronium (**103^+** and **104^+**) [22a, c], fluoronium (**105^+–107^+**) [22b], chalconium (**108^+–113^+**) [22d], and arenium (**114^+–116^+**) [22b] ions were shown to be potent catalysts for the activation of fluoroalkane **117**. It was demonstrated that both the nature of the tetryl atom and its substitution pattern exert no distinct effect on the performance of these Lewis acids as the turnover numbers (TONs) remained comparable for all these catalysts. Conversely, it was reported that the reactivity of the chalcogen-stabilized silicon cations **108^+–113^+** correlates with the Lewis basicity of the incorporated chalogen atom [22b]. The diminished TON when using $[CHB_{11}H_5Br_6]^-$ instead of $[B(C_6F_5)_4]^-$ as counteranion were attributed to the lower solubility of the *closo*-carborate in the solvent mixture 1,2-$Cl_2C_6H_4$/Et_3SiH.

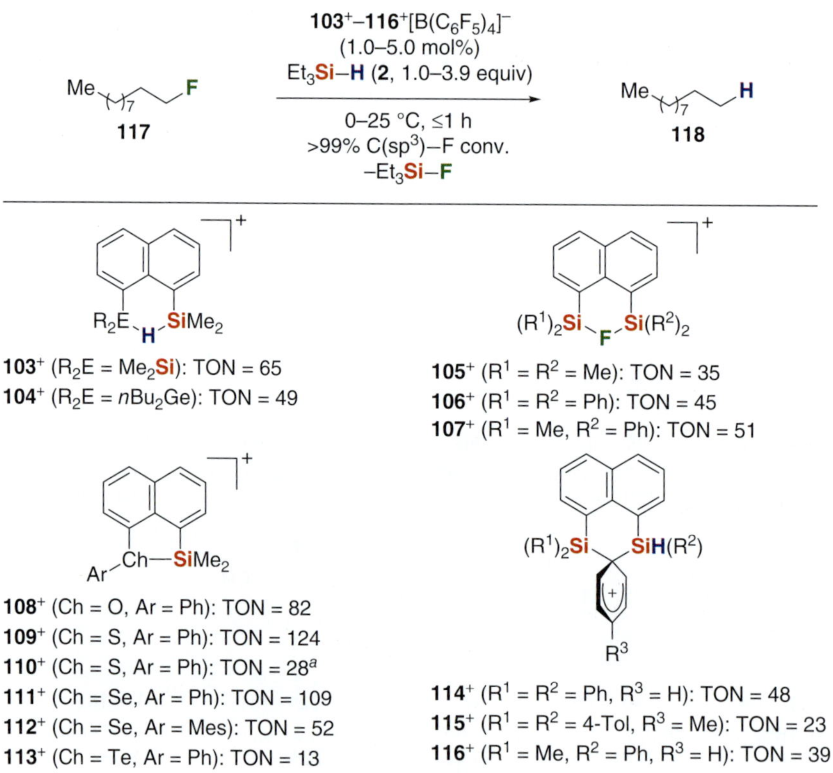

Scheme 5.13 Different intramolecularly stabilized silicon cations as catalysts for hydrodefluorination reactions. [a]$[CHB_{11}H_5Br_6]^-$ was used as anion.

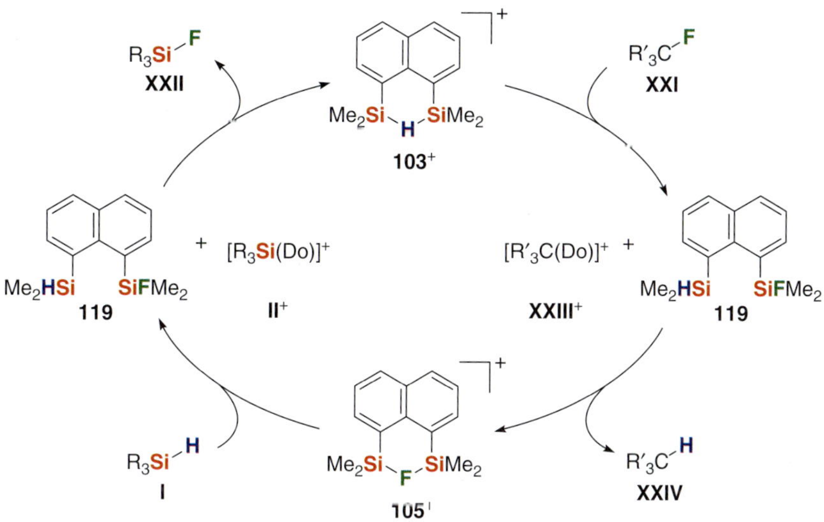

Scheme 5.14 Proposed mechanism for hydrodefluorination reactions catalyzed by intramolecularly stabilized silyl hydronium or fluoronium ions.

The group of Müller proposed different activation pathways for the hydrodefluorination catalyzed by either hydronium or fluoronium ions (Scheme 5.14) compared to chalconium ions (Scheme 5.15). The catalytic cycle for the activation with hydronium ion **103**$^+$ commences with fluoride abstraction from fluoroalkane **XXI** to furnish fluorosilane **119** (Scheme 5.14) [22a, c]. Hydride transfer from **119** to the intermediate carbenium ion **XXIII**$^+$ gives fluoronium ion **105**$^+$,

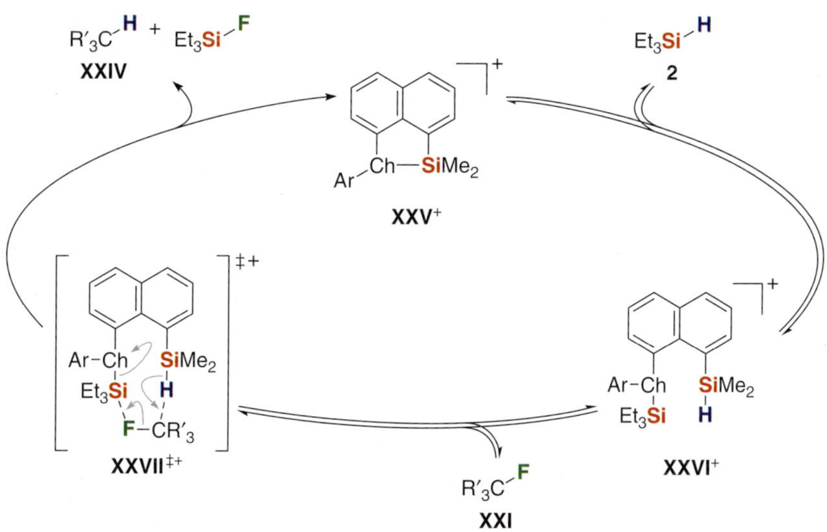

Scheme 5.15 Suggested mechanism for hydrodefluorination reactions catalyzed by chalogen-stabilized silicon cations.

which then again forms fluorosilane **119** by hydride abstraction from hydrosilane **I**. The intermediate donor-stabilized silicon cation $R_3Si(Do)^+$ (**II**$^+$) finally closes the catalytic cycle by fluoride abstraction from **119** to regenerate hydronium ion **103**$^+$. An analogous catalytic cycle is passed when starting with fluoronium ions **105**$^+$–**107**$^+$ (cf. Scheme 5.13) as the starting catalyst (Scheme 5.14, starting from **105**$^+$).

Conversely, based on mechanistic investigations on related hydrodefluorination reactions catalyzed by a cationic ruthenium–thiolate complex by Oestreich and coworkers (not shown) [23], a cooperative activation pathway was suggested for silyl chalconium ions (Scheme 5.15) [22d]. A reversible heterolytic cleavage of the Si—H bond of hydrosilane **2** by the stabilized silicon cation **XXV**$^+$ yields cation **XXVI**$^+$ (**XXV**$^+$ → **XXVI**$^+$). A concerted σ-bond metathesis of the activated hydrosilane with fluoroalkane **XXI** via the transition state **XXVII**$^{‡+}$ then closes the catalytic cycle, liberating hydrocarbon **XXIV** together with the fluorosilane (**XXVI**$^+$ → **XXVII**$^{‡+}$ → **XXV**$^+$). This proposed catalytic cycle is rationalized by the observed dependence of the reactivity on the Lewis basicity of the chalcogen atom (cf. Scheme 5.13) and the absence of rearrangement products.

The augmented fluorophilicity and Lewis acidity of cationic silicon Lewis acids have early unveiled the chemical stability of the counterion as pivotal in the generation, handling, and application of silicon cations [24]. Different research groups had observed slow degradation of the $[B(C_6F_5)_4]^-$ anion by fluoride abstraction [22a, 25], and that was detrimental to turnover in catalytic hydrodefluorination reactions (cf. Scheme 5.12 and 5.13) [21, 22a–c]. The group of Reed had solved this problem using halogenated carborate anions such as $[CHB_{11}I_{11}]^-$ in stoichiometric $C(sp^3)$–F activation reactions (not shown) [26]. Ozerov and coworkers took advantage of these results and investigated related carborates and a fully halogenated boranate as counteranions in the catalytic hydrodefluorination of perfluorotoluene (**92**). By this, these authors succeeded in increasing the turnover number by more than one order of magnitude (Scheme 5.16) [21, 27].

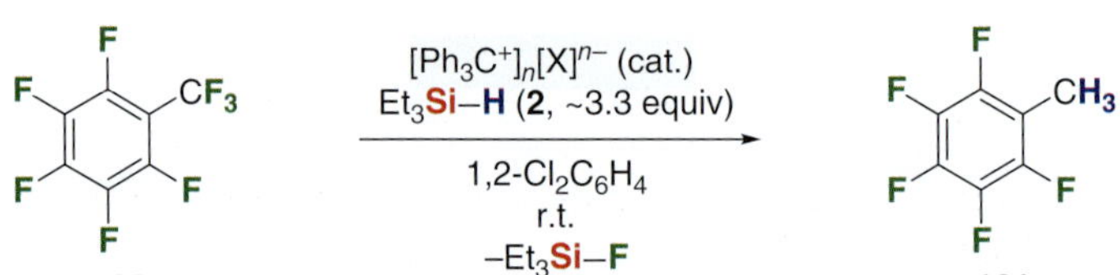

92 **101**

Ph_3C^+ $[B(C_6F_5)_4]^-$ (4.7 mol%)	Ph_3C^+ $[CHB_{11}H_5Br_6]^-$ (0.11 mol%)	Ph_3C^+ $[CHB_{11}H_5Cl_6]^-$ (0.11 mol%)	Ph_3C^+ $[CHB_{11}Cl_{11}]^-$ (0.11 mol%)	$[Ph_3C^+]_2$ $[B_{12}Cl_{12}]^{2-}$ (0.16 mol%)
Decomposition after TON = 19	TON = 220 after 1 h	TON = 880 after 1 h	TON = 370 after 1 h	TON = 110 after 24 h

Scheme 5.16 Effect of the stability of the counteranion in silicon cation-promoted hydrodefluorination reactions.

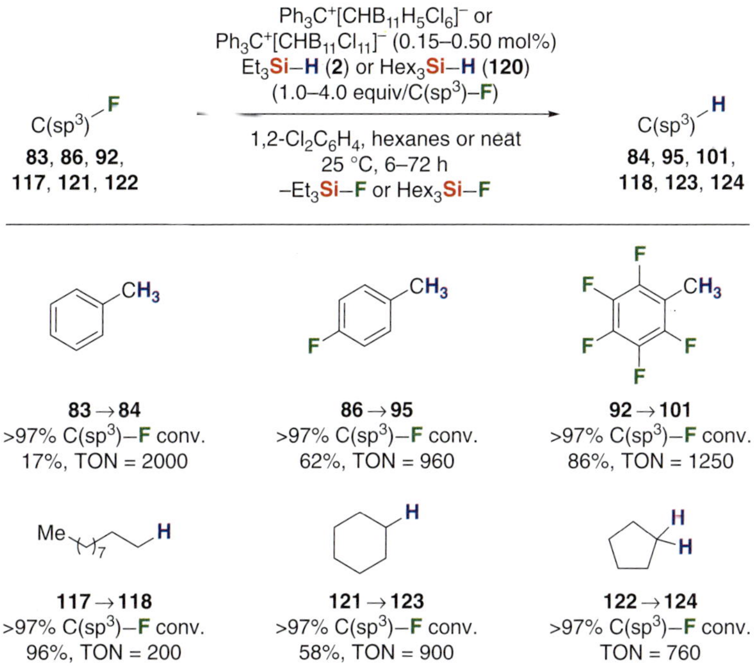

Scheme 5.17 Substrate scope of the hydrodefluorination of C(sp^3)—F bonds promoted by silicon cation carborate salts.

The same research group also investigated the scope of the hydrodefluorination reaction previously reported with $[B(C_6F_5)_3]^-$ as anion (cf. Scheme 5.12). The turnover numbers increased significantly for all substrates when using $[CHB_{11}H_5Cl_6]^-$ or $[CHB_{11}Cl_{11}]^-$ as counteranions (Scheme 5.17) [27a, b]. (Trifluoromethyl)benzene derivatives (**83**, **86**, and **92** → **84**, **95**, and **101**) as well as primary (**117** → **118**) and secondary (**121** and **122** → **123** and **124**) aliphatic fluoroalkanes were converted quantitatively into their respective hydrocarbon analogs. Conversely, tetrafluoromethane or other perfluorocarbons did not participate in this hydrodefluorination reaction (not shown).

Recently, Pan and Gabbaï reported a different method for the generation of silicon cations (Scheme 5.18) [28]. The stibonium salt $[Sb(C_6F_5)_4]^+[B(C_6F_5)_4]^-$ (**125**$^+[B(C_6F_5)_4]^-$) readily reacted with hydrosilane **2** to release acetonitrile-stabilized silicon cation **126**$^+$ (top), which can then participate in hydrodefluorination reactions as reported previously. Both alkyl fluoride **127** as well as benzyl trifluoride **83** were compatible with the reaction protocol and furnished *n*-octane (**128**) and toluene (**84**) with quantitative conversion, respectively (bottom). Control experiments showed that the stibonium cation **125**$^+$ alone cannot activate the C—F bonds in the substrates **127** or **83** and thus functions just as the initiator for the silicon cation-promoted hydrodefluorination.

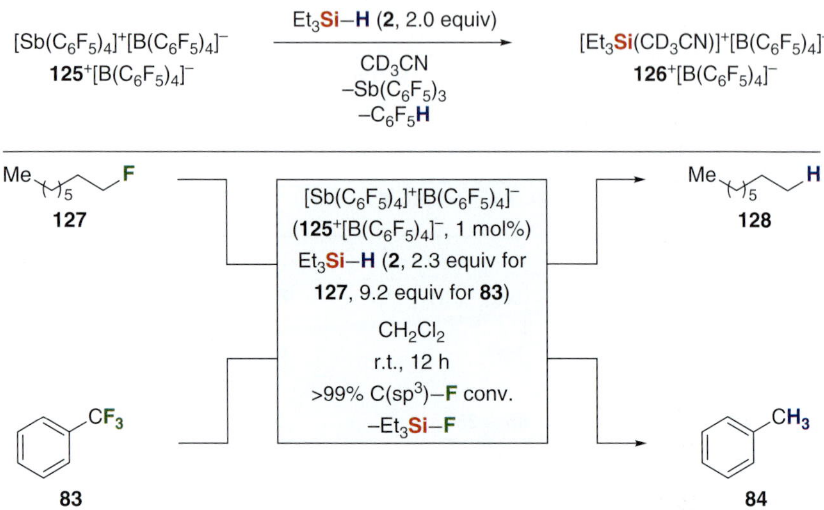

Scheme 5.18 Silicon cation–promoted hydrodefluorination initiated by a stibonium ion.

5.3.2 Defluorination Coupled with Electrophilic Aromatic Substitution (S_EAr)

The intermediacy of highly reactive carbenium ions in C—F bond activation reactions inevitably enables reaction pathways competing with hydride reduction, i.e. hydrodefluorination. Chemoselectivity issues are usually bypassed by evaluating the performance of hydrodefluorination reactions by the conversion of the substrate's C—F bond(s) or the respective turnover number, and not the yield of the formed hydrocarbons. Aside from rearrangements, Friedel–Crafts reactions with arene solvents or intramolecular attack by an aryl substituent are the most common in these transformations. This had already been noticed by the group of Ozerov in their early reports on hydrodefluorination reactions [21, 27a, b], and by Reed and coworkers as part of their studies of the stoichiometric generation of α-fluorinated carbeniumions [26]. Significant amounts of the Friedel–Crafts product **129** had already been formed when performing the reaction in deactivated 1,2-$Cl_2C_6H_4$ (**92** → **101** + **129**, Scheme 5.19a) [27a]. A chemoselective benzylation or alkylation of the more nucleophilic benzene was obtained when using it as the solvent (**86** → **130**, Scheme 5.19b, and **131** → **132**, Scheme 5.19c). In turn, carrying out the reaction in neat hydrosilane furnished a mixture of isomerized hydrodefluorination products (**133** → **134** + **135** + **136** + **123**, Scheme 5.19d).

The catalytic cycle of the silicon cation–promoted alkylation of arenes begins with the abstraction of a fluoride from **XXI** by fluorophile **II**$^+$ to generate carbenium ion **XXIII**$^+$ (**II**$^+$ → **XXIII**$^+$, Scheme 5.20). A different outcome compared to the hydrodefluorination pathway (gray route, **XXIII**$^+$ → **II**$^+$) is obtained as the nucleophilic attack by the arene outcompetes the hydride transfer from hydrosilane **I** to furnish Wheland complex **XXIX**$^+$ (**XXIII**$^+$ → **XXIX**$^+$). The donor-stabilized silicon cation **II**$^+$ is reformed via protonation of hydrosilane **I** by the

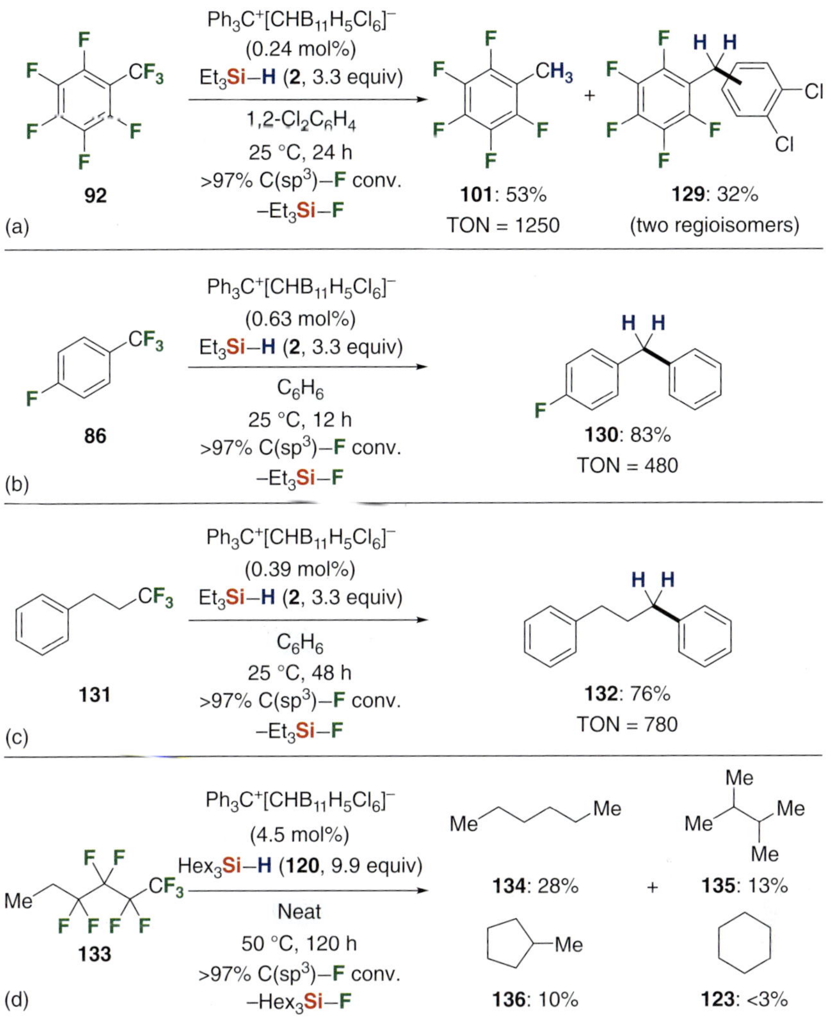

Scheme 5.19 Friedel–Crafts reactions with arene solvents (a–c) and isomerization reactions (d) as competing reaction channels in hydrodefluorination reactions.

strong Brønsted acid **XXIX**$^+$ [16, 27a], and dihydrogen as well as **XXVIII** are liberated (**XXIX**$^+ \rightarrow$ **II**$^+$).

Müller and coworkers successfully applied their intramolecular hydronium ion **103**$^+$, which was originally used in hydrodefluorination reactions (cf. Scheme 5.13) [22a], in the alkylation of arenes by designing a modified reaction protocol using excess amounts of benzene (Scheme 5.21) [29]. The chemoselectivity of these reactions is usually poor and side reactions, e.g. rearrangements or different degrees of alkylation (n = 1–3), could not be suppressed. Isomerization reactions of the intermediate primary carbenium ion are the main issue with primary fluoroalkanes (**117** → **137** + **138**), while secondary alkyl fluorides yield a mixture of products with various alkylation grades (**121** → **139**). Conversely, sterically

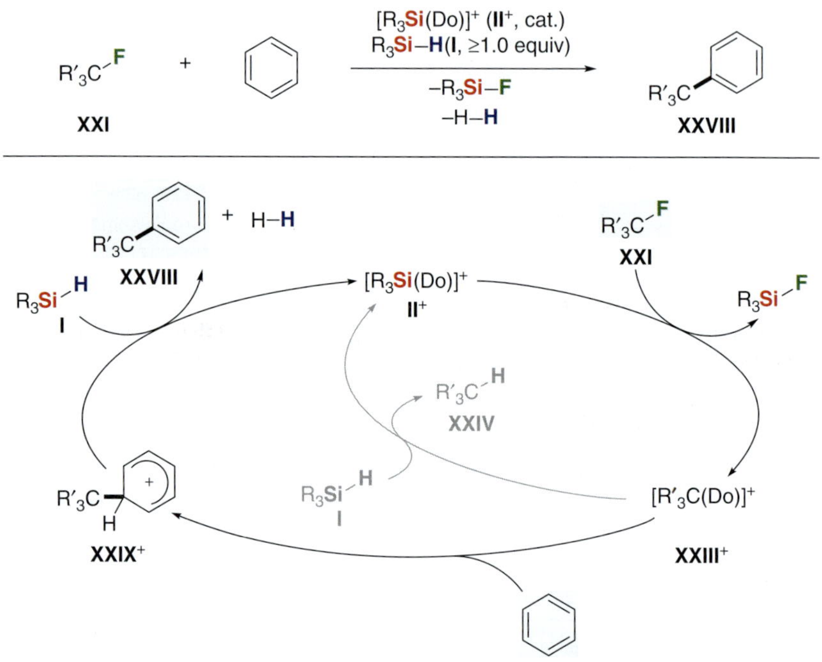

Scheme 5.20 Competing pathways of the silicon cation–promoted hydrodefluorination and S_EAr reactions.

hindered tertiary fluoroalkanes still favor hydrodefluorination over alkylation in benzene solvent (**140** → **141** + **142**). An attempt to suppress multiple alkylation by installing fluorine substituents on the aromatic substrate for blocking possible reaction sites and attenuating the reactivity toward monoselective S_EAr were not successful (not shown).

Reed, Siegel and coworkers found that silicon cations paired with the extremely weakly coordinating borate anion $[CHB_{11}Cl_{11}]^-$ are even capable of abstracting a fluoride from fluorobenzene (not shown) [30]. These stoichiometric experiments laid the foundation for catalytic intramolecular Friedel–Crafts arylations initiated by silicon cation–promoted $C(sp^2)$—F bond activation [31]. The catalytic cycle of these reactions is initiated by silicon cation–promoted fluoride abstraction to furnish high-energy phenyl cation **145**$^+$ (**143** + **144**$^+$ → **145**$^+$, Scheme 5.22), and subsequent intramolecular nucleophilic attack then gives Wheland intermediate **146**$^+$ (**145**$^+$ → **146**$^+$). The working group of Siegel developed a clever strategy to regenerate the catalytically active silicon cation without using stoichiometric amounts of a hydrosilane, i.e. releasing dihydrogen as byproduct (cf. Scheme 5.20). Instead, protonation of dimesityldimethylsilane (**147**) yields product **148** as well as β-silicon-stabilized Wheland intermediate **149**$^+$, that eventually collapses to liberate mesitylene (**150**) and silicon cation **151**$^+$ via a two-step protodesilylation sequence (**146**$^+$ → **149**$^+$ → **151**$^+$, Scheme 5.22).

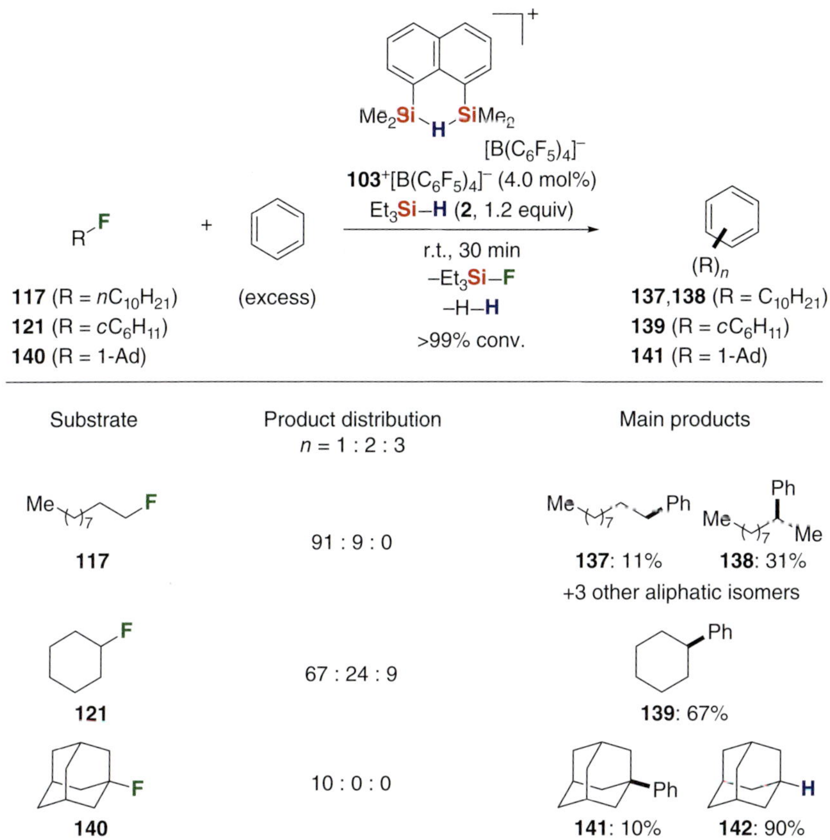

Scheme 5.21 S_EAr by intramolecular silyl hydronium ion–promoted C—F bond activation.

This approach enabled the synthesis of different polycyclic aromatic hydrocarbons via the formation of five-membered rings (**143** and **152–153** → **148** and **155–156**, Scheme 5.23) [31a]. Likewise, examples where six-membered rings were generated were reported, but stoichiometric amounts of the silicon cation were required in these reactions (not shown). However, the ring closure to form a strained four-membered ring was not successful (**154** → **157**).

The working group of Siegel and coworkers later showed that biaryl substrates with an aliphatic substituent in the ortho position also participate in a competitive electrophilic substitution of a C(sp^3)—H bond by the Wheland intermediate (Scheme 5.24) [31b]. These reactions usually furnished mixtures of the C(sp^3)–C(sp^2)- and C(sp^2)–C(sp^2)-coupled products when rings of the same size are formed (**158–160** → **163–165** + **168–170**). The chemoselectivity of the reaction was successfully steered toward the C(sp^3)–C(sp^2)-coupled polycycles in cases where the ring strain disfavored the C(sp^2)–C(sp^2) coupling (**161–162** → **166–167**).

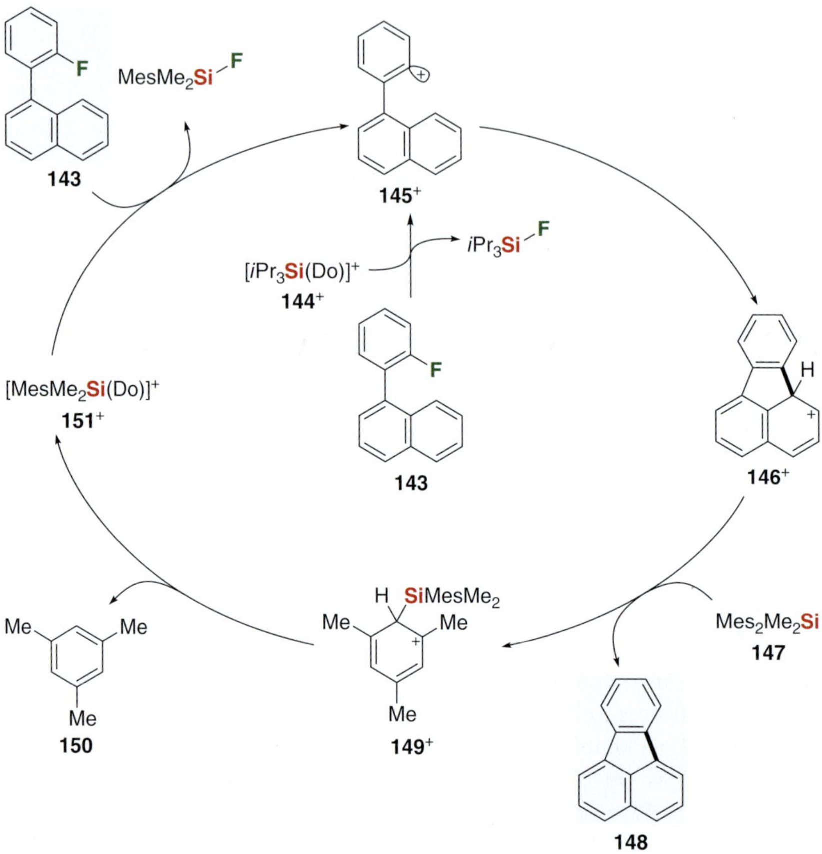

Scheme 5.22 Mechanism of the silicon cation–promoted C(sp^2)—F bond activation intercepted by an intramolecular Friedel–Crafts arylation.

Recently, Nelson and coworkers developed an impressive method for the arylation of C(sp^2)—H and C(sp^3)—H bonds by silicon cation–promoted activation of ortho-silylated fluoroarenes **173–184** (Schemes 5.25 and 5.26) [32]. Different halogen- (**173–177** → **185–188**, Scheme 5.25), aryl- (**178–181** → **189–192**), or alkyl-substituted (**182** → **193**) arenes were compatible with this reaction protocol and furnished the corresponding biphenyl derivatives in synthetically useful yields. In addition, double arylation to give terphenyl **194** was possible (**183** → **194**). The TBS-protected hydroxy functionality in **184** was equally tolerated and yielded the alcohol **195**, albeit in low yield (**184** → **195**).

The methodology was also applied to the Friedel–Crafts alkylation of ortho-silylated fluoroarenes (**181** and **196** → **139** and **197–201**, Scheme 5.26) [32]. Alkanes with chemically inequivalent, reactive C—H bonds such as *n*-pentane or *n*-hexane are burdened with poor chemoselectivity and yielded the arylation products **199** or **200** as a mixture of regioisomers. Intriguingly, methane (!) was

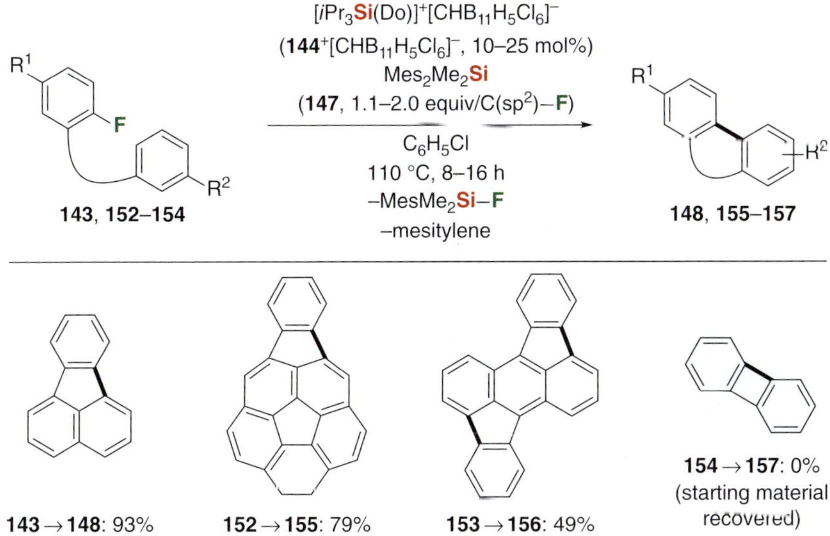

Scheme 5.23 Substrate scope of the silicon cation-promoted C(sp^2)—F bond activation intercepted by an intramolecular Friedel–Crafts arylation.

equally reactive under the reaction conditions and 1-methylnaphthalene (**201**) was obtained in moderate yield (**181** → **201**).

The proposed mechanism commences with the silicon cation-mediated abstraction of a fluoride from arene **196** to generate the β-silicon-stabilized phenyl cation **202**$^+$ (**196** → **202**$^+$, Scheme 5.27), which subsequently inserts into a C—H bond of alkane **XXX** to furnish Wheland complex **XXXI**$^+$ (**202**$^+$ → **XXXI**$^+$). Although the generation of arynes as reactive intermediates seems plausible, control experiments, i.e. formation of different regioisomers (cf. Scheme 5.25, **174** → **186** versus **176** → **188** as well as **178** → **189** versus **180** → **191**), indicate that aryne intermediates are not present in this reaction. A 1,2-hydride shift then forms β-silicon-stabilized Wheland intermediate **XXXII**$^+$ (**XXXI**$^+$ → **XXXII**$^+$) that liberates donor-stabilized silicon cation **203**$^+$ to reinitiate the catalytic cycle (**XXXII**$^+$ → **203**$^+$).

5.4 Friedel–Crafts C–H Silylation

Several catalytic methods for the C–H silylation with silicon electrophiles have been reported to date [33], but protocols that rely on cationic silicon Lewis acids remain rare. The group of Oestreich recently reported such a methodology where the Brønsted acid-promoted formation of a donor-stabilized silicon cation is applied to the catalytic Friedel–Crafts silylation of electron-rich (hetero) arenes (Scheme 5.28) [16]. The silylation of differently substituted N-protected indoles **204–210** occurred regioselectively in the C3 position and furnished the

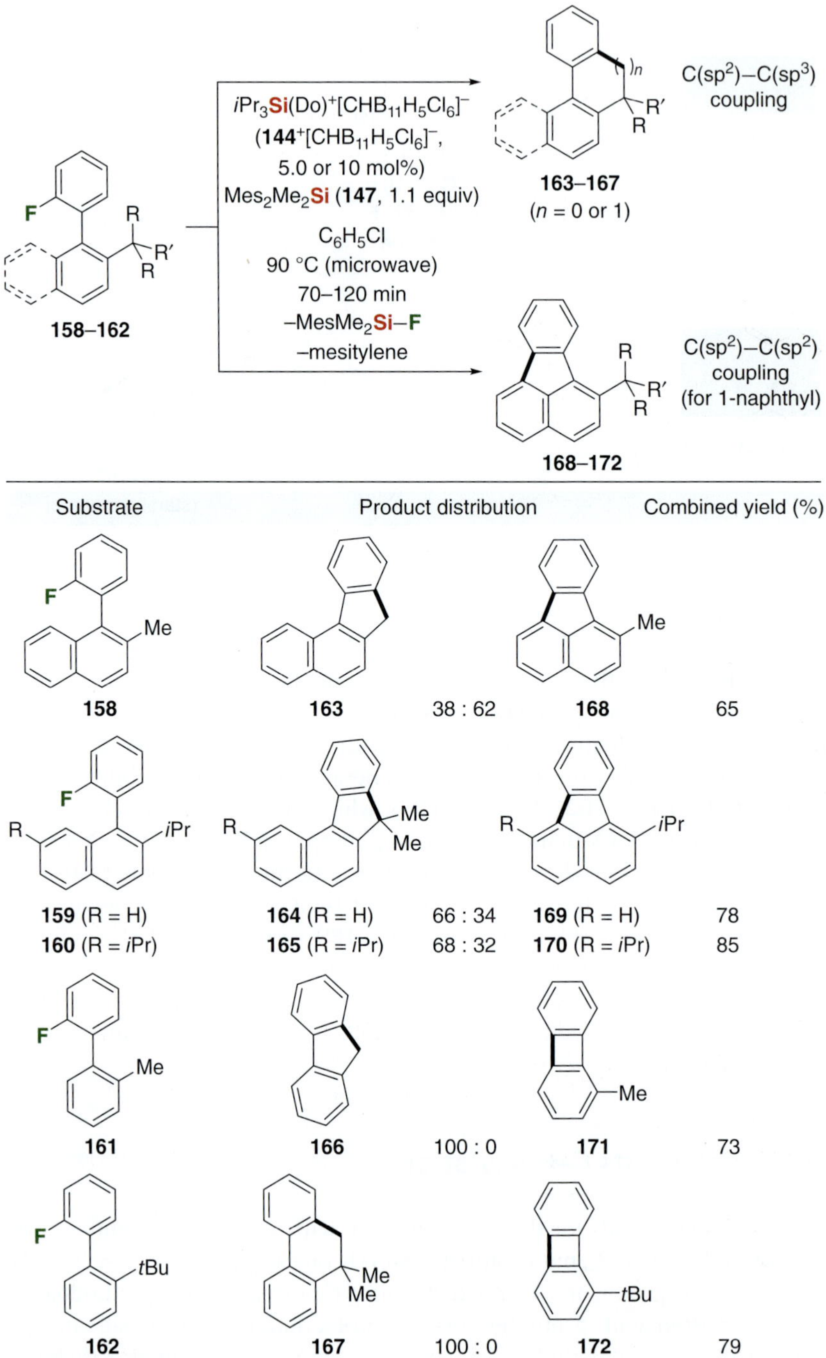

Scheme 5.24 Competing intramolecular C(sp^2) or C(sp^3) arylation via C(sp^2)—F bond activation.

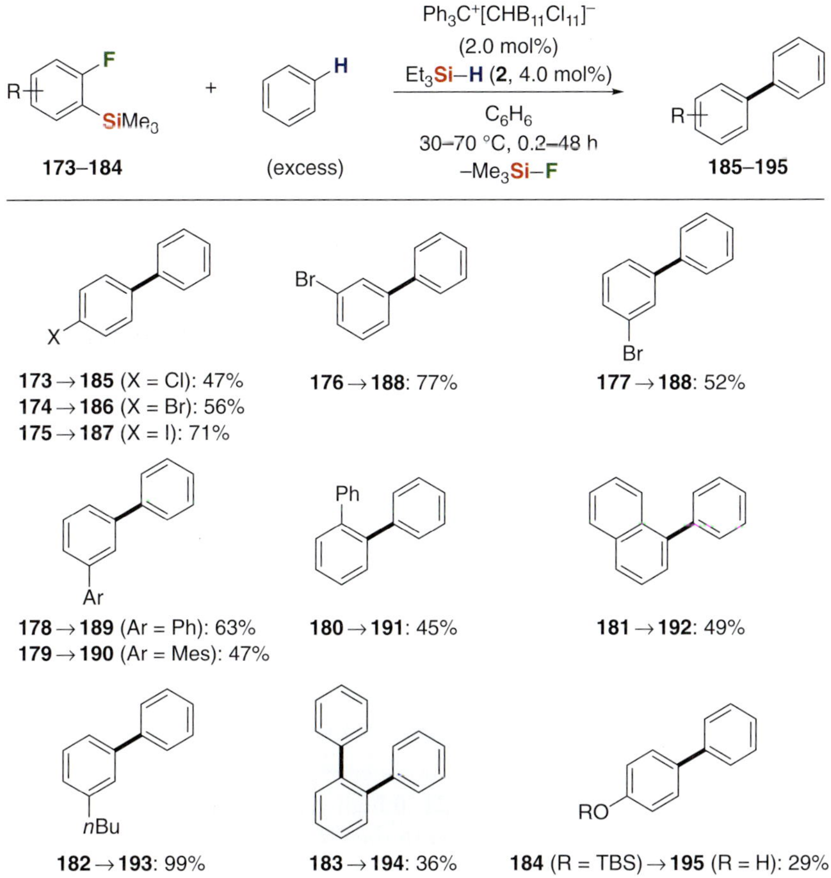

Scheme 5.25 Silicon cation-promoted arylation of $C(sp^2)$–H bonds by Nelson and coworkers.

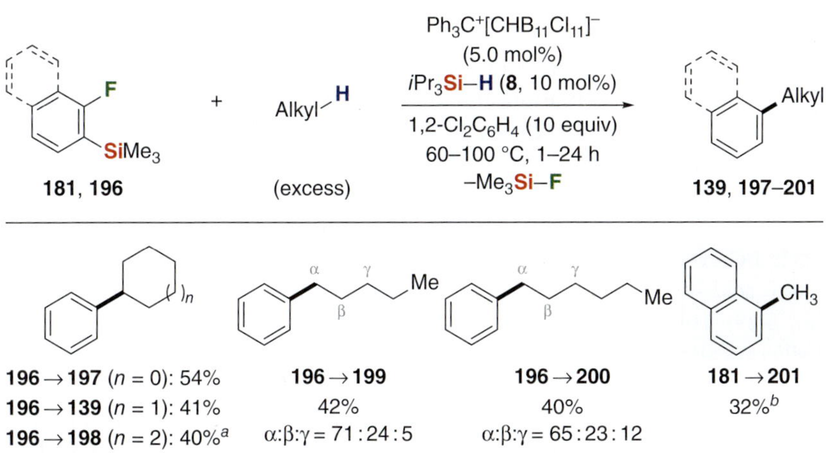

Scheme 5.26 Silicon cation-promoted arylation of $C(sp^3)$–H bonds by Nelson and coworkers. [a]Without 1,2-$Cl_2C_6H_4$. [b]Et_3SiH (**2**) used instead of *i*Pr_3SiH (**8**). C_6F_6 used as solvent.

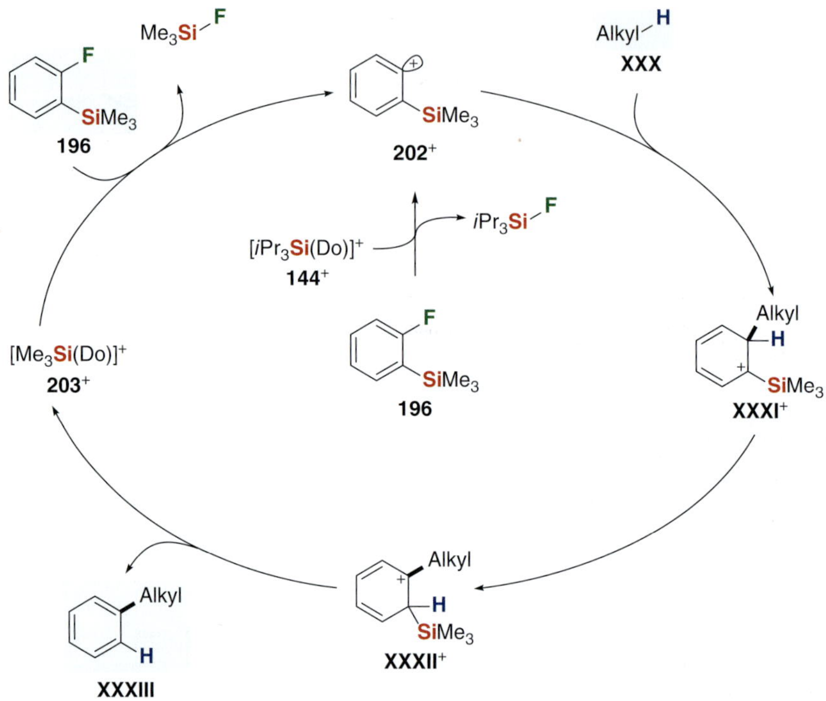

Scheme 5.27 Mechanism of the silicon cation–promoted arylation of aliphatic C—H bonds.

corresponding silylated indoles **215–221** in high yields (**204–210** → **215–221**). An attempt to direct the C–H silylation to the C2 position by installation of a methyl group at C3 shut down the reaction completely (not shown). This result indicates that an S_EAr mechanism is operative. The more challenging pyrrole **211** yielded a mixture of the two possible regioisomers (**211** → **222**). Other heterocycles such as benzofuran or benzothiophene were not compatible with this reaction protocol (not shown). Conversely, the aniline derivatives **223–225** as well as 1-methylindoline (**226**) and 1-methyl-1,2,3,4-tetrahydroquinoline (**227**) participated in the transformation, and the para-silylated products were obtained as single regioisomers (**223–227** → **228–232**). Blocking the para position with a methyl-substituent to steer the silylation to the ortho position did not yield any conversion of the substrate (not shown).

The proposed catalytic cycle of this C–H silylation of electron-rich (hetero) arenes is believed to start with the formation of siliconium ion **XXXIV**$^+$ by Brønsted acid–mediated protonation of hydrosilane **I** (**I** + **213**$^+$ → **XXXIV**$^+$, Scheme 5.29) [16]. Cation **XXXIV**$^+$ eventually collapses to form the donor-stabilized silicon cation **II**$^+$ and dihydrogen (**XXXIV**$^+$ → **II**$^+$). Indole **204** then adds nucleophilically to the silicon electrophile **II**$^+$ to furnish iminium ion **XXXV**$^+$ (**II**$^+$ → **XXXV**$^+$). Wheland complex **XXXV**$^+$ is a strong Brønsted acid, and protonation of another molecule of hydrosilane **I** regenerates silicon cation **XXXIV**$^+$ and releases the silylated indole **XXXVI** (**XXXV**$^+$ → **XXXVI**$^+$). A competing mechanistic pathway yields Wheland intermediate **233**$^+$ via protonation of

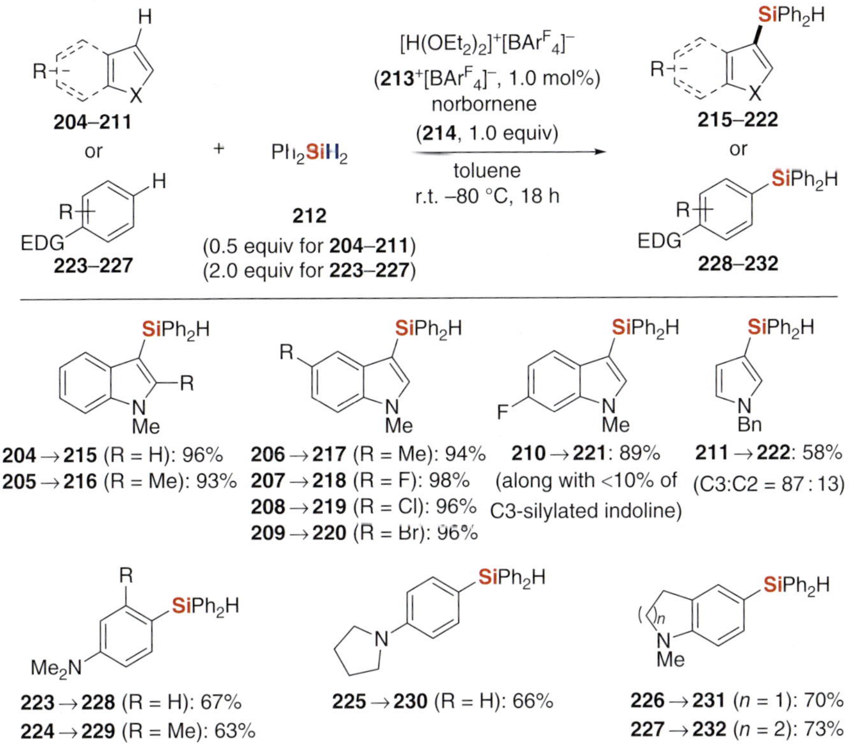

Scheme 5.28 Scope of the C–H silylation of electron-rich (hetero)arenes involving silicon cations.

another indole molecule **204** by **XXXV**$^+$ (**XXXV**$^+ \rightarrow$ **233**$^+$). The iminium ion in **233**$^+$ is then reduced by hydride transfer from hydrosilane **I** to furnish indoline **234** as byproduct while regenerating the silicon cation **II**$^+$ (**233**$^+ \rightarrow$ **II**$^+$).

5.5 Diels–Alder Reactions

Neutral silicon Lewis acids are routinely employed as catalysts for carbon–carbon bond formation [34]. However, despite their considerable electrophilicity, they are often not able to catalyze conventional but challenging transformations. In particular, efficient methods for Diels–Alder reactions of cyclohexa-1,3-diene (**235**) and common dienophiles are significantly underdeveloped compared to cycloadditions of the more reactive cyclopentadiene (**236**) [35]. In the search for more reactive silicon-based Lewis acids, the chemical shift of the silicon atom in the ^{29}Si NMR spectrum provides a qualitative estimate of their Lewis acidity. Especially, the nature of the counteranion can strongly influence the Lewis acid characteristic of the silicon electrophile. A significant counteranion dependence was demonstrated by Mathieu and Ghosez in the Diels–Alder reaction of cyclohexa-1,3-diene (**235**)

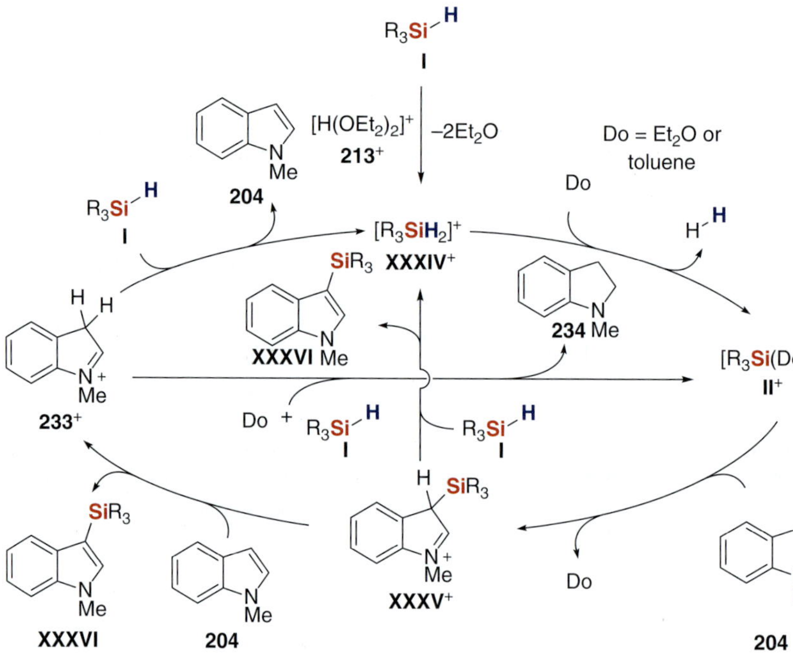

Scheme 5.29 Proposed mechanism of the silicon cation–promoted C–H silylation of electron-rich (hetero)arenes.

and methyl acrylate (**239**) promoted by the common silicon Lewis acids Me_3SiOTf (**237**) and Me_3SiNTf_2 (**238**) (Scheme 5.30) [36]. While Me_3SiOTf (**237**, ^{29}Si NMR: $\delta = 43.5$ ppm) did not catalyze the cycloaddition, Me_3SiNTf_2

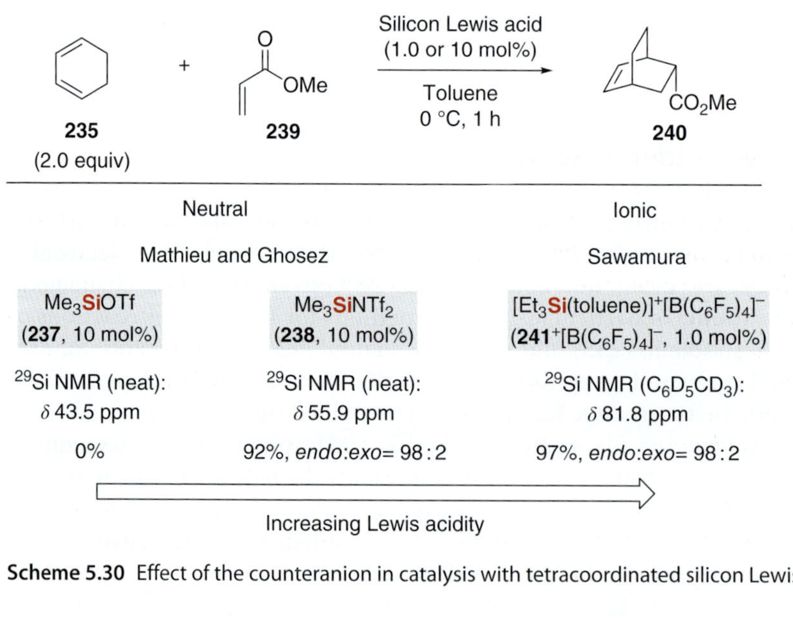

Scheme 5.30 Effect of the counteranion in catalysis with tetracoordinated silicon Lewis acids.

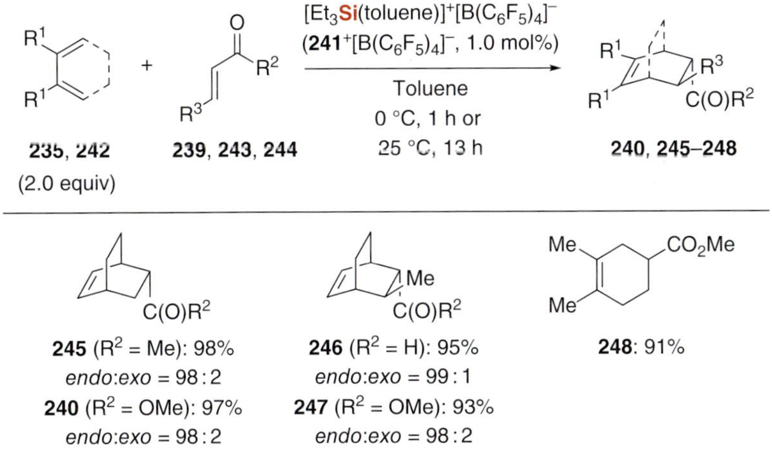

Scheme 5.31 Examples of $[Et_3Si(toluene)]^+[B(C_6F_5)_4]^-$-catalyzed Diels–Alder reactions by Sawamura and coworkers.

(**238**, ^{29}Si NMR: δ = 55.9 ppm) provided the cycloadduct **240** in very good yield and with excellent endo selectivity, although 10 mol% of the catalyst was required to achieve a near-quantitative yield (1.0 mol% furnished only 6% yield after one hour). Several years later, Sawamura and coworkers found even higher catalytic activity for the toluene-stabilized silicon cation $[Et_3Si(toluene)]^+$ (**241**$^+$, ^{29}Si NMR: δ = 81.8 ppm) with the less coordinating counteranion $[B(C_6F_5)_4]^-$ in the same Diels–Alder reaction [37]. The reduced catalyst loading of 1.0 mol% (compared to 10 mol% of Me_3SiNTf_2 (**238**)) illustrates the superiority of cationic **241**$^+$ over neutral **237** or **238** silicon Lewis acids. The potential of silicon cation **241**$^+$ for catalytic applications was showcased by several difficult Diels–Alder reactions, giving the cycloadducts in high yields and with excellent endo selectivities (**235** and **242** + **239**, **243**, and **244** → **240** and **245–248**, Scheme 5.31) [37].

A key challenge in the development of cationic silicon-based Lewis acids for catalysis lies in taming their pronounced electrophilicity without quenching their unique reactivity. As part of these examinations, the concept of intramolecular stabilization of the cationic silicon center was established. The group of Oestreich introduced the ferrocene-stabilized silicon cation **252**$^+$, which emerged as a superb catalyst for challenging Diels–Alder reactions of various cyclic and acyclic dienes and dienophiles (**235** and **242** + **239** and **249–251** → **240**, **248**, and **253–256**, Scheme 5.32) [38].

Modifying the ferrocene backbone of **252**$^+$ led to various planar-chiral silicon cations and was the starting point for the development of asymmetric variants of these difficult transformations. However, phenyl-substituted (S_p)-**258**$^+$ or annulated silicon cation (S_p)-**259**$^+$ did not induce any enantioselectivity in the model Diels–Alder reaction of cyclohexa-1,3-diene (**235**) and 3-acryloyloxazolidin-2-one (**257**) (**235** + **257** → **260**, Scheme 5.33) [39]. The installation of a chiral oxazoline moiety at the ferrocene unit resulted in the corresponding silicon cation (S_p,^{C}S,^{Si}RS)-**261**$^+$ (gray box), which was stabilized by the nitrogen

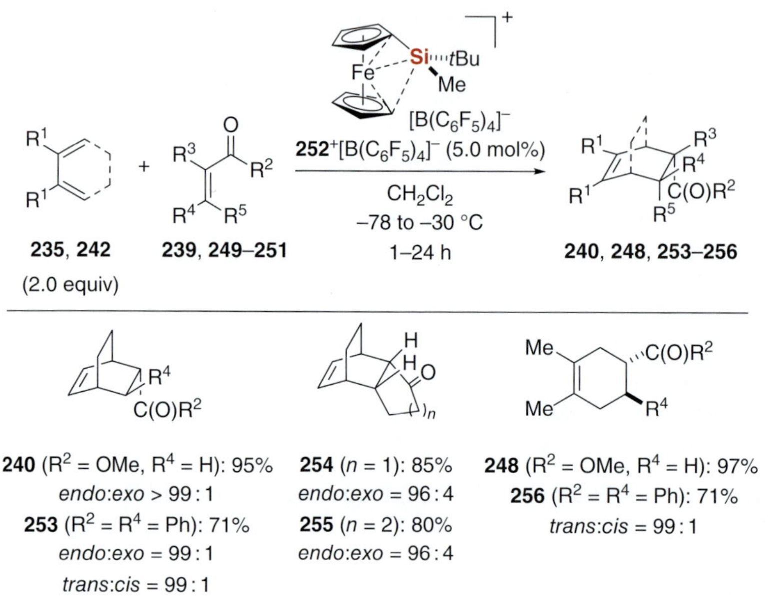

Scheme 5.32 Selected examples of Diels–Alder reactions catalyzed by ferrocene-stabilized silicon cation **252**$^+$.

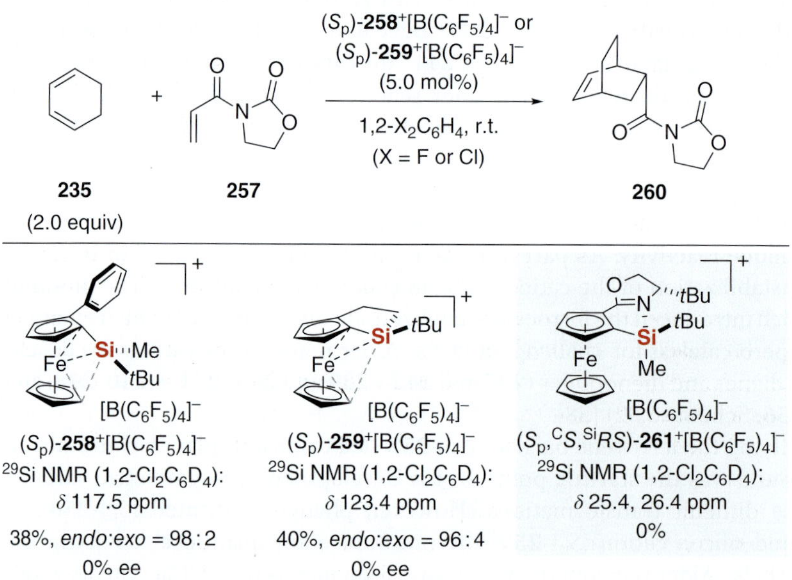

Scheme 5.33 Planar chiral, ferrocene-stabilized silicon cations (S_p)-**258**$^+$ and (S_p)-**259**$^+$, and oxazoline-stabilized $(S_p,{}^{C}S,{}^{Si}RS)$-**261**$^+$ as potential catalysts in asymmetric Diels–Alder reactions.

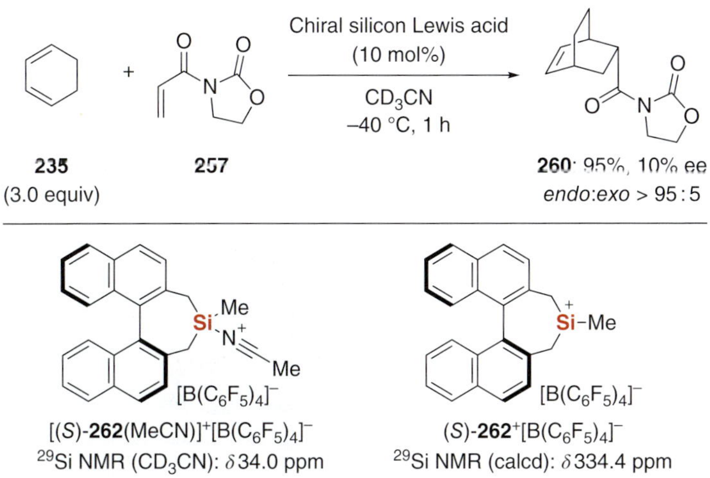

Scheme 5.34 Asymmetric Diels–Alder reaction with the silylnitrilium ion [(*S*)-**262**(MeCN)]$^+$.

heteroatom and not the ferrocenyl group. As a result of the high σ donor strength of the sp^2-hybridized nitrogen atom, no catalytic activity of the silicon compound was observed [40].

Prior to these studies on silicon cation–catalyzed asymmetric Diels–Alder reactions by Oestreich and coworkers, Helmchen and coworkers had already reported a silicon cation with an axially chiral binaphthyl backbone that enabled the asymmetric variant of the above Diels–Alder reaction (Scheme 5.34) [41]. Silylnitrilium ion [(*S*)-**262**(MeCN)]$^+$ (gray box), which is formed by intermolecular coordination of the acetonitrile solvent to the silicon cation, was shown by NMR spectroscopy to be the active catalyst rather than the free silylium ion (*S*)-**262**$^+$ [42], and promoted the reaction of 3-acryloyloxazolidin-2-one (**257**) with cyclohexa-1,3-diene (**235**) with an enantiomeric excess of 10% [41]. Although the achieved enantioselectivity is low, this example is an important pioneering piece of work for the application of cationic tetracoordinated silicon Lewis acids in catalysis [43].

As a lesson learned from the insufficient reactivity of the silicon cation (S_p,^{C}S,^{Si}RS)-**261**$^+$, new stabilizing units were sought. A family of sulfur-stabilized silicon cations was developed by the group of Oestreich and tested in the model Diels–Alder reaction of cyclohexa-1,3-diene (**235**) and chalcone (**249**) (Scheme 5.35). First, cyclic dithioacetals were examined as the electron-donating motifs. As a consequence of the silicon–sulfur interaction, three stereocenters on three different elements (silicon, sulfur, and carbon) were formed in the ferrocene-substituted silicon species **263**$^+$ (Scheme 5.35, top row, left); in addition to the existing planar chirality of the ferrocene unit, four diastereomers were formed in a single transformation [40]. Comparison of the ^{29}Si NMR shifts of the silicon cation **263**$^+$ with the neutral silicon compounds Me_3SiOTf (**237**) and Me_3SiNTf_2 (**238**), and cationic toluene-stabilized **241**$^+$ allows a qualitative estimation of their Lewis acidity. Thus, the stabilization of the electrophilic silicon atom by a soft sulfur donor leads to a more Lewis-acidic silicon species

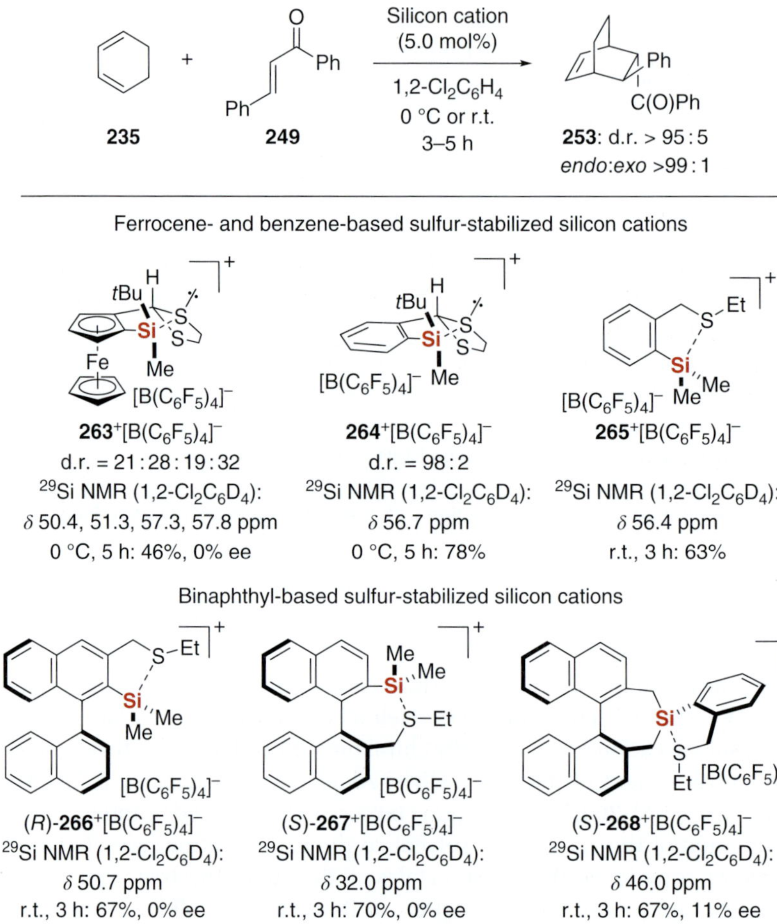

Scheme 5.35 Various sulfur-stabilized silicon cations as Lewis acid catalysts in the Diels–Alder reaction of cyclohexa-1,3-diene (**235**) and chalcone (**249**).

(^{29}Si NMR: δ 50.4, 51.3, 57.3, and 57.8 ppm for the four diastereomers) compared to other donors such as nitrogen or oxygen atoms. Further investigations into the nature of the silicon–sulfur interaction revealed that in the presence of a sulfur-containing substituent, a ferrocene backbone is no longer required. The benzene-based congener **264**$^+$ (Scheme 5.35, top row, middle) was synthesized and proved to be chemically stable [40]. The novel silicon cations **263**$^+$ and **264**$^+$ were able to catalyze the Diels–Alder reaction of cyclohexa-1,3-diene (**235**) and chalcone (**249**) to give the cycloadduct **253** in moderate to good yields and with excellent endo selectivity. However, no asymmetric induction was achieved with ferrocene-based **263**$^+$. Likewise, a simple alkyl thioether group was shown to sufficiently stabilize silicon cations, and the use of the cationic silicon species **265**$^+$ (Scheme 5.35, top row, right) in the model Diels–Alder reaction was successful: the reaction was carried out at room temperature and gave the desired product **253** in good yield and with excellent endo selectivity. The stabilizing motif was transferred to the chiral binaphthyl backbone with different substitution patterns. Silicon cations (*R*)-**266**$^+$ and (*S*)-**267**$^+$ (Scheme 5.35, bottom row, left and middle) were shown to catalyze that challenging cycloaddition, albeit without any enantioinduction. Silicon cation (*S*)-**268**$^+$, featuring the silicon atom as a part of a more rigid silepine ring, provided a low but promising enantioselectivity of 11% (Scheme 5.35, bottom row, right) [44].

The cationic silicon Lewis acid (*S*)-**268**$^+$ was used as a catalyst in a series of challenging diene/dienophile substrate combinations (Scheme 5.36) [44]. Reactions of cyclohexa-1,3-diene (**235**) with various cyclic and acyclic α,β-unsaturated carbonyl compounds **239**, **249–251**, **257**, and **269** afforded the cycloadducts **240**, **253–255**, **260**, and **272** in moderate to good yields and with very good endo selectivity but poor level of enantioselection. The best result was obtained with the oxazolidinone derivative **257**, improving the 10% ee previously achieved by Jørgensen and Helmchen using [(*S*)-**262**(MeCN)]$^+$ to 24% ee. The obtained yields and enantioinduction in the Diels–Alder reactions of the same selection of dienophiles **239**, **249–251**, **257**, and **269** with the more reactive cyclopentadiene (**236**) were largely similar to those seen with

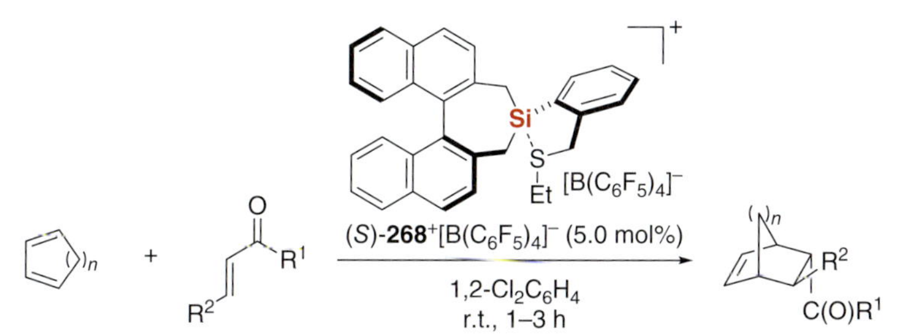

236 (*n* = 1) **235** (*n* = 2) **239**, **249–251**, **257**, **269** **240**, **253–255**, **260**, **270–276**

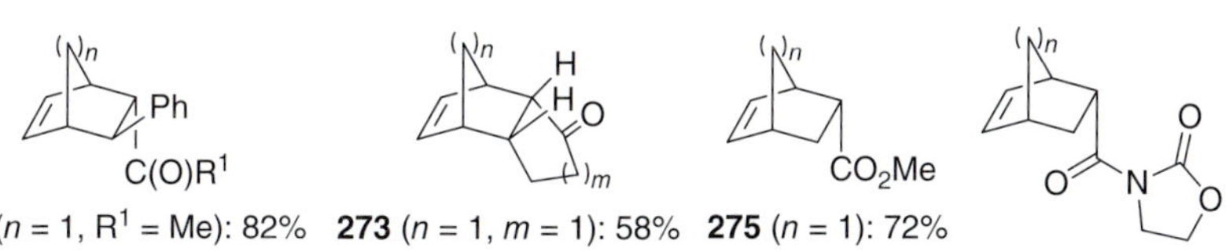

270 (*n* = 1, R^1 = Me): 82%
endo:*exo* > 95 : 5
7% ee
271 (*n* = 1, R^1 = Ph): 80%
endo:*exo* > 95 : 5
12% ee
272 (*n* = 2, R^1 = Me): 75%
endo:*exo* > 95 : 5
3% ee
253 (*n* = 2, R^1 = Ph): 67%
endo:*exo* > 95 : 5
11% ee

273 (*n* = 1, *m* = 1): 58%
endo:*exo* = 64 : 36
3% ee
274 (*n* = 1, *m* = 2): 59%
endo:*exo* = 77 : 23
0% ee
254 (*n* = 2, *m* = 1): 30%
endo:*exo* = 92 : 8
4% ee
255 (*n* = 2, *m* = 2): 32%
endo:*exo* > 95 : 5
1% ee

275 (*n* = 1): 72%
endo:*exo* = 87 : 13
10% ee
240 (*n* = 2): 83%
endo:*exo* = 95 : 5
14% ee

276 (*n* = 1): 78%
endo:*exo* > 95 : 5
20% ee
260 (*n* = 2): 52%
endo:*exo* = 95 : 5
24% ee

Scheme 5.36 Diels–Alder reactions of cyclopentadiene (**236**) and cyclohexa-1,3-diene (**235**) with various dienophiles catalyzed by sulfur-stabilized silicon cation (*S*)-**268**$^+$.

cyclohexa-1,3-diene (**235**). However, for the majority of examples the endo/exo selectivity dropped significantly.

The addition of another chiral unit to the sulfur tether in (*S*)-**268**$^+$ resulted in the diastereomeric bis(binaphthyl)-based silicon cations (*S*,*S*)-**277**$^+$ and (*S*,*R*)-**277**$^+$ that were again tested in the model Diels–Alder reaction of cyclohexa-1,3-diene (**235**) and chalcone (**249**), showing match/mismatch situations (34% ee with (*S*,*S*)-**277**$^+$ and 13% ee with (*S*,*R*)-**277**$^+$, Scheme 5.37, top row) [45]. The biphenyl-based congener (*RS*,*S*)-**278**$^+$ gave the cycloadduct **253** in good yield and with an enantioselectivity of 35% ee (bottom row, left), indicating that the binaphthyl silepine scaffold has no influence on the level of enantioinduction in these reactions [46]. However, a substantial decrease of enantioselectivity was obtained with the silaindane-based silicon cation (*S*)-**279**$^+$, and the cycloadduct **253** was isolated as a racemic mixture (bottom row, middle) [46]. Similarly, the use of the binaphthyl-based (*S*)-**280**$^+$ resulted in poor enantioselectivity in the Diels–Alder reaction of cyclohexa-1,3-diene

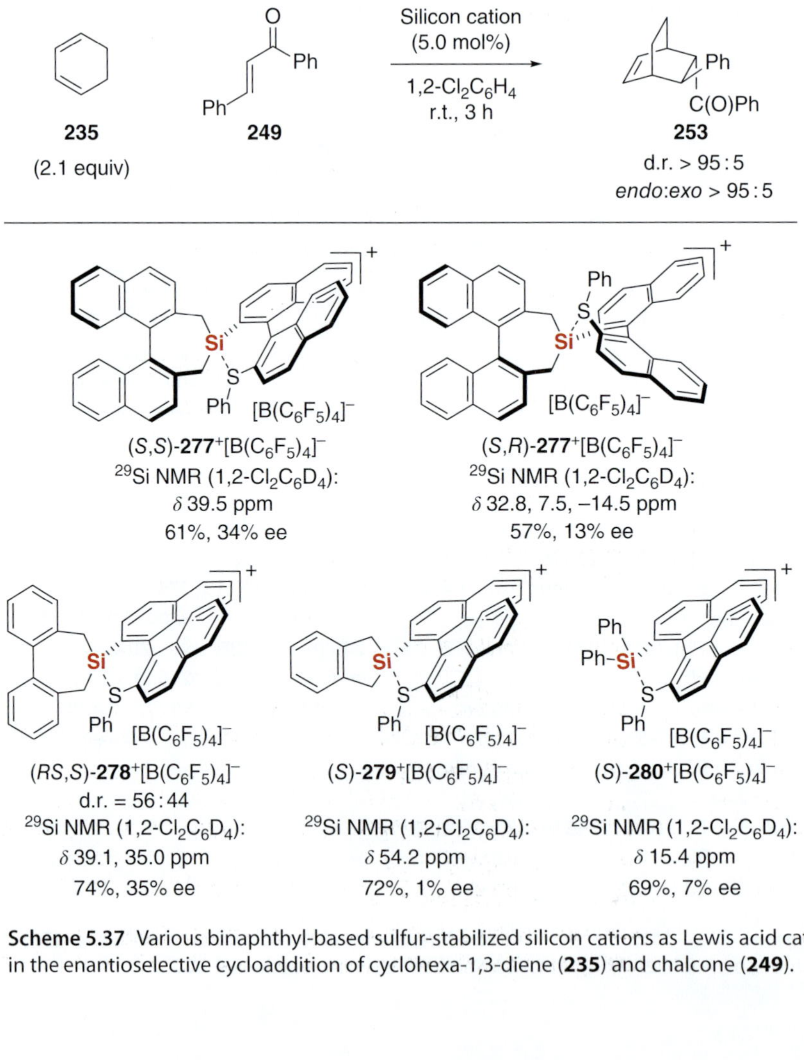

Scheme 5.37 Various binaphthyl-based sulfur-stabilized silicon cations as Lewis acid catalysts in the enantioselective cycloaddition of cyclohexa-1,3-diene (**235**) and chalcone (**249**).

(**235**) and chalcone (**249**), highlighting the importance of the cyclic backbone for achieving significant enantioinduction [46].

The ability of the Lewis acid (*S*,*S*)-**277**$^+$ to induce enantioinduction was then tested in Diels–Alder reactions of cyclohexa-1,3-diene (**235**) and various chalcone derivatives **249**, **269** and **278–285** (Scheme 5.38) [45]. An aryl group attached to the carbonyl carbon atom of the dienophile was shown to be essential for enantioinduction (**249** → **253** versus **269** → **272**). Chalcones with para substitution at the phenyl group gave moderate but higher enantioselectivities than that for parent chalcone (**249**) (**278–280** → **286–288**). The opposite effect on enantioselectivity was obtained by attaching an ortho-substituent

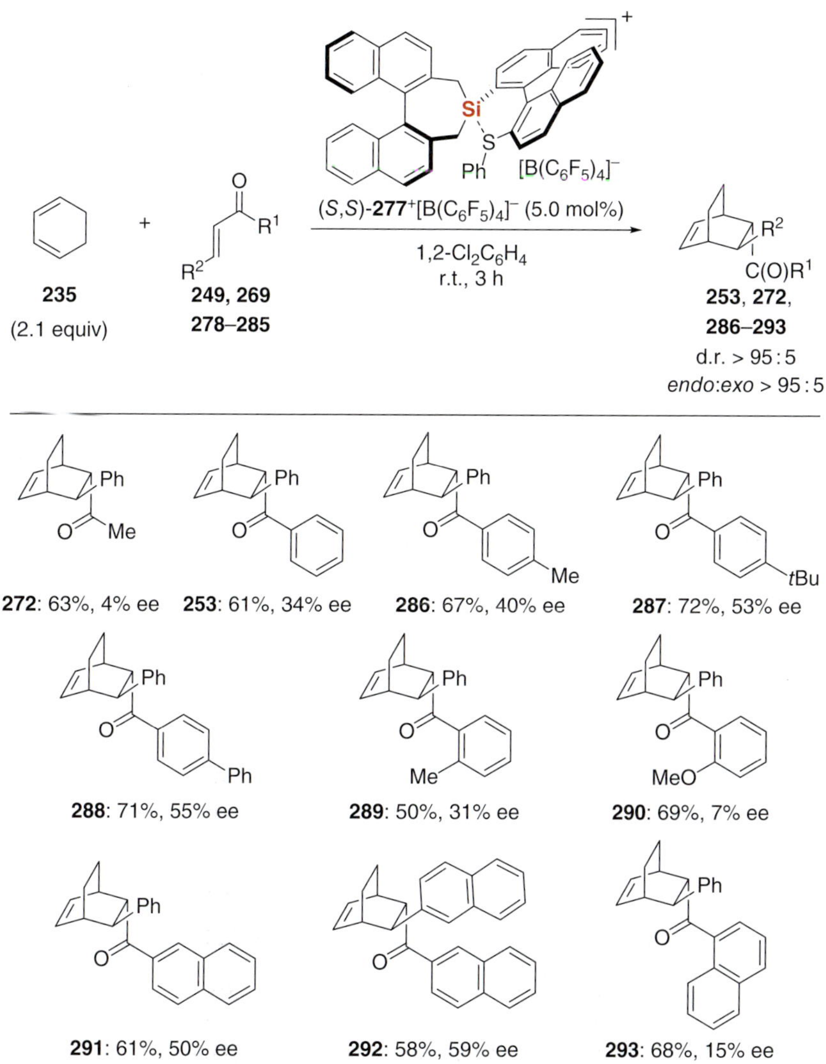

Scheme 5.38 Enantioselective Diels–Alder reactions of cyclohexa-1,3-diene (**235**) and chalcones catalyzed by the sulfur-stabilized silicon cation (*S*,*S*)-**277**$^+$.

(**281**–**282** → **289**–**290**). An increase of the enantiomeric excess was also promoted with β-naphthyl groups, as in the dienophiles **283** and **284** (**283** → **291** and **284** → **292**). Conversely, chalcone **285** with an α-naphthyl substituent led to a diminished enantioinduction (**285** → **293**).

Another part of this systematic investigation by Oestreich and coworkers focused on the effect of substituents on the stabilizing aryl thioether unit on the enantioinduction [46]. The bis(binaphthyl)-based silicon cations **294**$^+$ and **295**$^+$ and their analogous biphenyl-based congeners **296**$^+$ and **297**$^+$ with sterically demanding thioether groups were prepared and applied in the benchmark Diels–Alder reaction (Scheme 5.39). Depending on the substitution pattern at

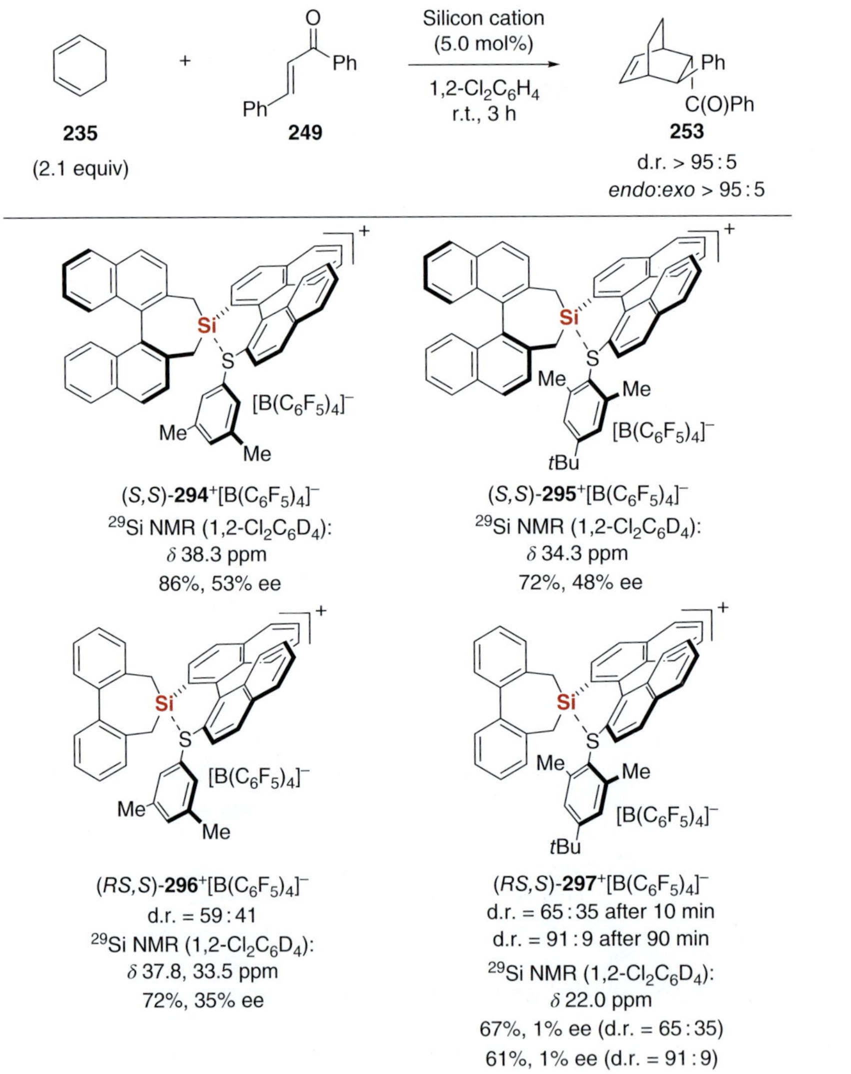

Scheme 5.39 Biphenyl- and binaphthyl-based sulfur-stabilized silicon cations with sterically demanding thioether groups and their performance as Lewis acid catalysts in the Diels–Alder reaction of cyclohexa-1,3-diene (**235**) and chalcone (**249**).

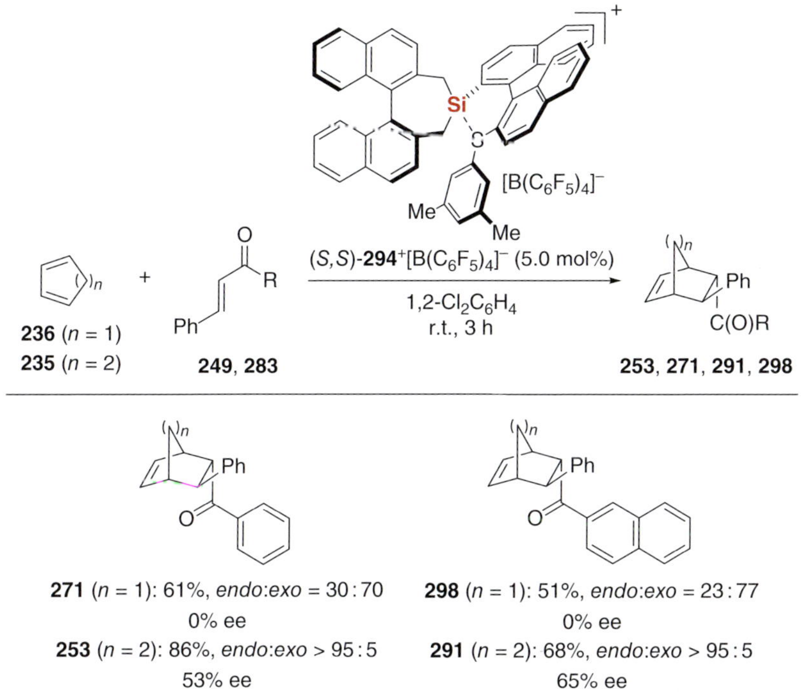

Scheme 5.40 Enantioselective Diels–Alder reactions of cyclopentadiene (**235**) or cyclohexa-1,3-diene (**235**) with different chalcone derivatives catalyzed by (S,S)-**294**$^+$.

the phenyl thioether group, the chemical stability, NMR spectroscopic behavior, and performance as chiral silicon Lewis acids varied.

The use of the Lewis acid (S,S)-**294**$^+$ in the Diels–Alder reactions of chalcones **249** and **283** with cyclohexa-1,3-diene (**235**) gave the cycloadducts **253** and **291** in moderate yields and with excellent endo selectivity (Scheme 5.40). Interestingly, partial reversal of the diastereoselectivity of the cycloadducts **271** and **298**, and absence of asymmetric induction were observed in the reaction with cyclopentadiene (**236**) [46].

Highly enantioselective Diels–Alder reactions of cyclopentadiene (**236**) with cinnamates **299**–**313** catalyzed by the silylcarboxonium ion **316**$^+$ were reported by the group of List (**236** + **299**–**313** → **317**–**331**, Scheme 5.41) [47]. The *in situ*-generated chiral ion pair from (S)-**314** and the silyl ketene acetal **315** promoted the cycloaddition to furnish various norbornene derivatives with high enantiomeric excesses and excellent diastereoselectivities.

5.6 Mukaiyama Aldol and Related Reactions

The Mukaiyama aldol and Michael reactions are further valuable methods for transition-metal-free C—C bond formation [48]. As part of their investigations on applications of silicon cations as highly reactive Lewis acids in catalysis (cf. Scheme 5.31), Sawamura and coworkers developed a methodology for

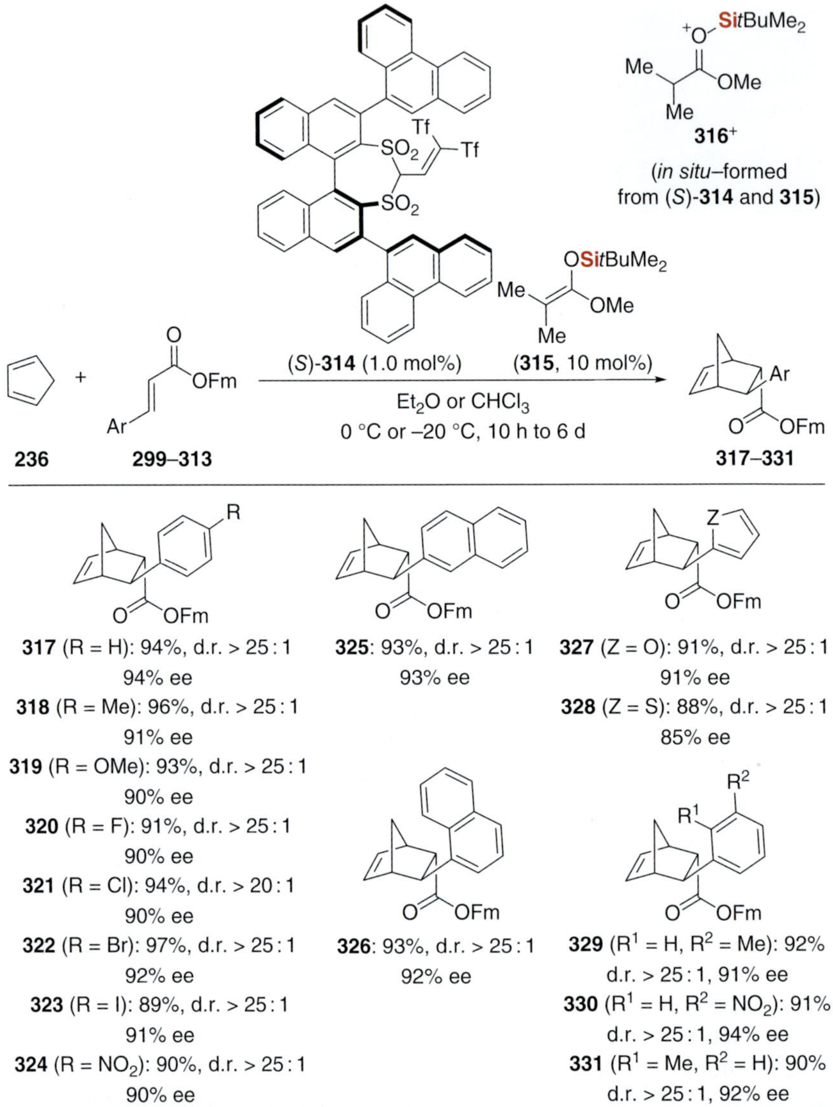

Scheme 5.41 Silylcarboxonium ion–promoted, asymmetric counteranion-directed catalysis of the Diels–Alder reactions of cyclopentadiene (**236**) and cinnamates.

the $[Et_3Si(toluene)]^+$-promoted Mukaiyama aldol reaction (**7**, **17**, **19**, and **332** + **333–336** → **337–341**, Scheme 5.42) [37]. Acyclic (**17** → **337**) or cyclic (**19** → **338**) dialkyl-substituted ketones or aromatic ketones (**7** → **339** and **341**) and aldehydes (**332** → **340**) participated and furnished β-hydroxy ketones in

Scheme 5.43 Silyloxonium ion–promoted asymmetric Mukaiyama–Michael reaction by List and coworkers. [a]With *p*-xylene as solvent. [b]With toluene/1,4-dioxane (3 : 1) as solvent at rt. [c]With methylcyclohexane as solvent at −40 °C.

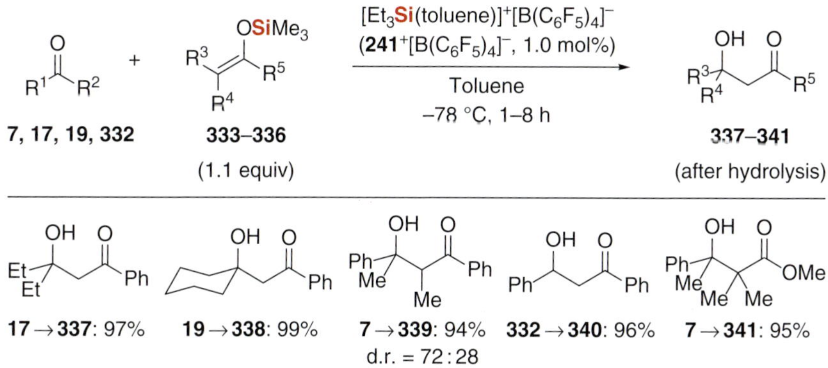

Scheme 5.42 Silylium ion-promoted Mukaiyama aldol reaction by Sawamura and coworkers.

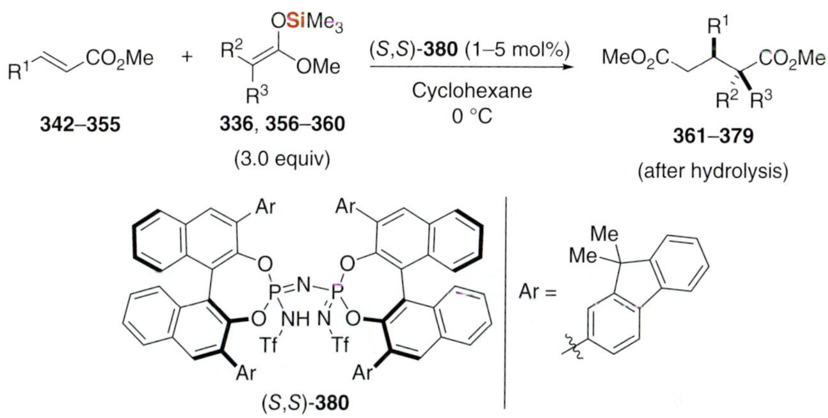

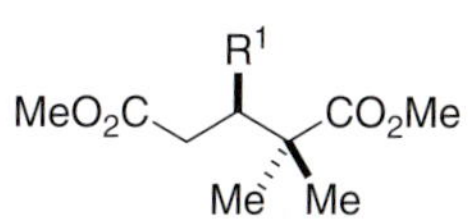

361–374

R^1	Yield (%)	ee (%)
Ph (**342** → **361**):	97	94
4-MeC_6H_4 (**343** → **362**):	95	94
3-MeC_6H_4 (**344** → **363**):	95	96
2-MeC_6H_4 (**345** → **364**):	92	93
4-FC_6H_4 (**346** → **365**):	99	95
4-ClC_6H_4 (**347** → **366**):	97	95
4-BrC_6H_4 (**348** → **367**):	98	96
Naphth-1-yl (**349** → **368**):	92	89
4-$MeOC_6H_4$ (**350** → **369**):	95	91
4-NCC_6H_4 (**351** → **370**):[a]	89	95
4-$O_2NC_6H_4$ (**352** → **371**):[b]	76	90
Fur-2-yl (**353** → **372**):	94	89
PhC_2H_4 (**354** → **373**):[c]	92	94
Hept-1-yl (**355** → **374**):[c]	94	90

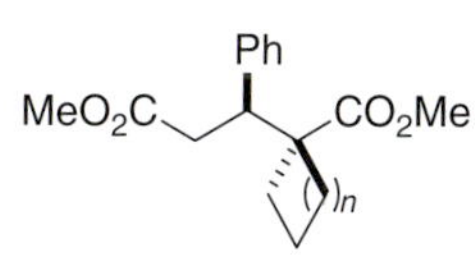

375–377

n = 1 (**342** → **375**): 99%, 93% ee
n = 2 (**342** → **376**): 97%, 92% ee
n = 3 (**342** → **377**): 95%, 96% ee

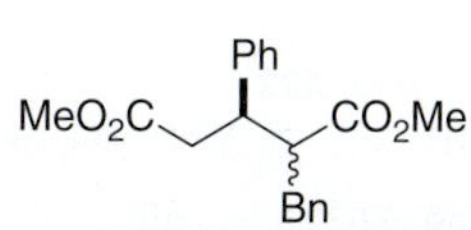

(*Z*)-**359** → *trans*-**378**: 96%, 97% ee, d.r. = 62 : 38
(*E*)-**359** → *cis*-**378**: 98%, 98% ee, d.r. = 92 : 8

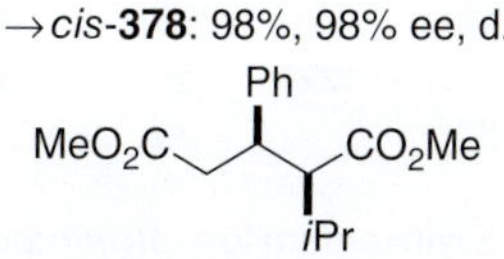

(*E*)-**360** → *cis*-**379**: 98%, 99% ee, d.r. = 96 : 4

excellent yields. Likewise, a masked α-disubstituted carboxylic ester reacted smoothly (**336** → **341**).

The group of List recently reported an asymmetric variant of the Mukaiyama–Michael reaction using an *in situ*–formed silylcarboxonium ion as the catalyst (cf. Scheme 5.41, gray box) and the principle of asymmetric counterion-directed catalysis [49] for the induction of chirality (Scheme 5.43) [50]. A wide variety of

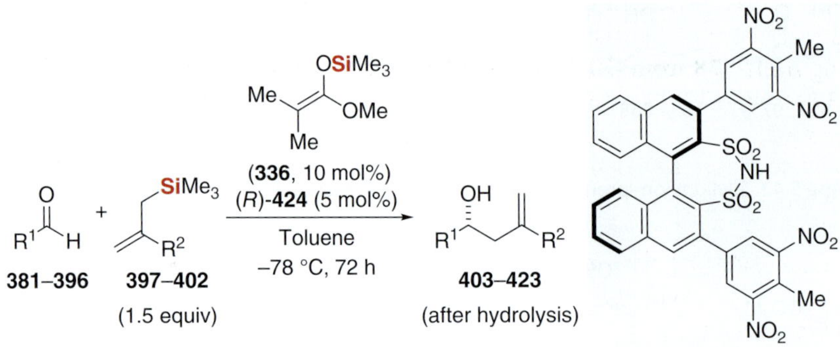

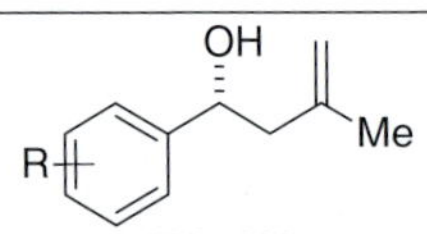

403–413

R	Yield (%)	ee (%)
H (**381** → **403**):	82	84
4-Me (**382** → **404**):	84	86
3-Me (**383** → **405**):	82	92
2-Me (**384** → **406**):	59	63
3,5-Me_2 (**385** → **407**):	96	96
3,5-Et_2 (**386** → **408**):	98	90
3-MeO (**387** → **409**):	92	95
3-MeS (**388** → **410**):	96	91
4-F-3,5-Me_2 (**389** → **411**):	>99	96
4-Br (**390** → **412**):	97	75
3-Vinyl (**391** → **413**):	>99	94

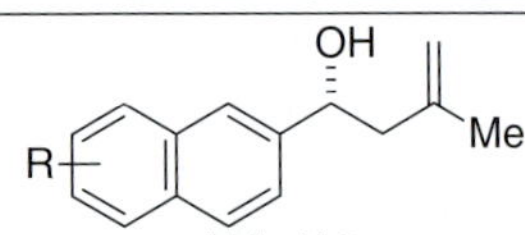

414–416

R	Yield (%)	ee (%)
H (**392** → **414**):	90	91
6-Br (**393** → **415**):	82	91
6-MeO (**394** → **416**):	80	83

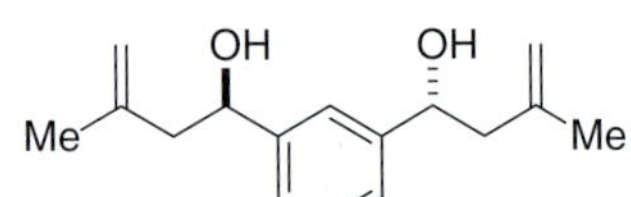

395 → **417**: 93%
98% ee, d.r. = 85.5 : 14.5[a]

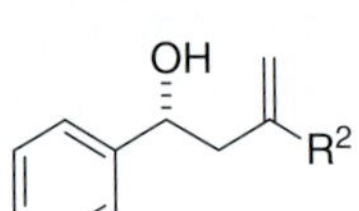

419–423

R^2	Yield (%)	ee (%)
Hept-1-yl (**398** → **419**):	93	90
$PhCH_2CH_2$ (**399** → **420**):	91	83
Ph (**400** → **421**):	83	86
4-MeC_6H_4 (**401** → **422**):	95	87
H (**402** → **423**):	90	25

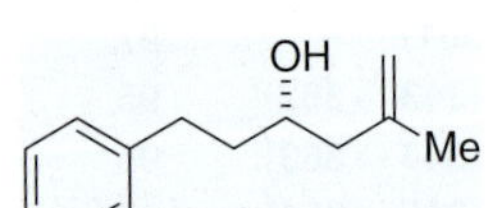

396 → **418**: 20%, 2% ee[b]

Scheme 5.44 Silyloxonium ion–promoted asymmetric Hosomi–Sakurai reaction by List and coworkers. [a]With 3 equiv of **397**. [b]With CH_2Cl_2 as solvent at 5 °C and without **336**.

α,β-unsaturated carboxylic esters **342–355** were compatible with the developed reaction protocol and gave access to chiral 1,3-dicarboxylic esters **361–379** in both high yields and enantioselectivities (**342–355** + **336** and **356–360** → **361–379**). However, a β-isopropyl-substituted cinnamate showed only low reactivity and a diminished enantiomeric excess, while the cognate *tert*-butyl-substituted derivative was completely unreactive (not shown). Differently substituted silyl ketene acetals **356–358** were also subjected to the reaction conditions and gave equally excellent yields and enantioselectivities (**342** + **356–358** → **375–377**). In addition, List and coworkers showed that this transformation was stereospecific, giving *trans*-**378** from (*Z*)-**359** in moderate diastereoselectivity, and *cis*-**378** or *cis*-**379** from (*E*)-**359** or (*E*)-**360**, respectively, in very good diastereomeric ratios (gray box).

Prior to these reports on asymmetric Diels–Alder and Mukaiyama–Michael reactions by List and coworkers (cf. Schemes 5.41 and 5.43), the same group developed a related strategy for an asymmetric Hosomi–Sakurai reaction (Scheme 5.44) [51]. The *in situ* generation of a silyl carboxonium ion (cf. **316**$^+$, Scheme 5.41) enabled the nucleophilic attack of allylic silanes at differently substituted aldehydes to furnish the respective homoallylic alcohols in synthetically useful enantiomeric excesses (**381–396** + **397–402** → **403–423**, Scheme 5.44). The overall substrate scope for aryl-substituted aldehydes was broad (**381–395** + **397** → **403–417**), and benzothiophen-5-yl or naphth-1-yl was also tolerated as substituents (not shown). Conversely, alkyl-substituted aldehydes did not provide any enantioselectivity (**396** + **397** → **418**, gray box). The β-substituent in the allylic silane was required to obtain high enantioselectivities (**381** + **398–401** → **419–422**), while unsubstituted **402** yielded the respective homoallylic alcohol **423** only in low enantiomeric excess (**381** + **402** → **423**).

References

1 For overviews of the history of silicon cations, see: (a) Müller, T. (ed. D. Scheschkewitz) (2014). *Structure and Bonding*, vol. 155, 107–162. Berlin: Springer-Verlag. (b) Kochina, T.A., Vrazhnov, D.V., Ignatyev, I.S. et al. (2011). *J. Organomet. Chem.* 696: 1331–1340. (c) Lee, V.Y. and Sekiguchi, A. (2007). *Acc. Chem. Res.* 40: 410–419. (d) Kochina, T.A., Vrazhnov, D.V., Sinotova, E.N., and Voronkov, M.G. (2006). *Russ. Chem. Rev.* 75: 95–110. (e) Müller, T. (2005). *Adv. Organomet. Chem.* 53: 155–215. (f) Lambert, J.B., Zhao, Y., and Zhang, S.M. (2001). *J. Phys. Org. Chem.* 14: 370–379. (g) Reed, C.A. (1998). *Acc. Chem. Res.* 31: 325–332. (h) Lambert, J.B., Kania, L., and Zhang, S. (1995). *Chem. Rev.* 95: 1191–1201. (i) Corriu, R.J.P. and Henner, M. (1974). *J. Organomet. Chem.* 74: 1–28.

2 Großekappenberg, H., Reißmann, M., Schmidtmann, M., and Müller, T. (2015). *Organometallics* 34: 4952–4958.

3 (a) Klare, H.F.T. (2017). *ACS Catal.* 7: 6999–7002. (b) Lee, V.Y. and Sekiguchi, A. (ed. V.Y. Lee) (2017). *Organosilicon Compounds*, vol. 1, 197–230. Oxford: Academic Press. (c) Schulz, A. and Villinger, A. (2012). *Angew. Chem. Int. Ed.* 51: 4526–4528. (d) Klare, H.F.T. and Oestreich, M. (2010). *Dalton Trans.* 39: 9176–9184.

4 Corey, J.Y. (1975). *J. Am. Chem. Soc.* 97: 3237–3238.
5 Kira, M., Hino, T., and Sakurai, H. (1992). *Chem. Lett.* 555–558.
6 Parks, D.J., Blackwell, J.M., and Piers, W.E. (2000). *J. Org. Chem.* 65: 3090–3098.
7 (a) Khandelwal, M. and Wehmschulte, R.J. (2012). *Angew. Chem. Int. Ed.* 51: 7323–7326. For a stoichiometric reduction, see: (b) Schäfer, A., Saak, W., Haase, D., and Müller, T. (2012). *Angew. Chem. Int. Ed.* 51: 2981–2984.
8 Müther, K. and Oestreich, M. (2011). *Chem. Commun.* 47: 334–336.
9 Müther, K., Fröhlich, R., Mück-Lichtenfeld, C. et al. (2011). *J. Am. Chem. Soc.* 133: 12442–12444.
10 Müther, K., Mohr, J., and Oestreich, M. (2013). *Organometallics* 32: 6643–6646.
11 (a) Lambert, J.B. and Zhao, Y. (1996). *J. Am. Chem. Soc.* 118: 7867–7868. (b) Lambert, J.B., Zhao, Y., and Wu, H. (1999). *J. Org. Chem.* 64: 2729–2736. (c) Lambert, J.B., Liu, C., and Kouliev, T. (2002). *J. Phys. Org. Chem.* 15: 667–671.
12 Lambert, J.B. (1990). *Tetrahedron* 46: 2677–2689.
13 Schmeltzer, J.M., Porter, L.A. Jr., Stewart, M.P., and Buriak, J.M. (2002). *Langmuir* 18: 2971–2974.
14 Steinberger, H.-U., Bauch, C., Müller, T., and Auner, N. (2003). *Can. J. Chem.* 81: 1223–1227.
15 Arii, H., Kurihara, T., Mochida, K., and Kawashima, T. (2014). *Chem. Commun.* 50: 6649–6652.
16 Chen, Q.-A., Klare, H.F.T., and Oestreich, M. (2016). *J. Am. Chem. Soc.* 138: 7868–7871.
17 Arii, H., Nakabayashi, K., Mochida, K., and Kawashima, T. (2016). *Molecules* 21: 999/1–999/7.
18 Gusev, D.G. and Ozerov, O.V. (2011). *Chem. Eur. J.* 17: 634–640.
19 (a) Brook, M.A. (2000). *Silicon in Organic, Organometallic and Polymer Chemistry*. New York: Wiley. (b) Walsh, R. (1981). *Acc. Chem. Res.* 14: 246–252. (c) Becerra, R. and Walsh, R. (eds. Z. Rappoport and Y. Apeloig) (1998). *The Chemistry of Organic Silicon Compounds*, vol. 2, 153–180. Chichester: Wiley.
20 (a) Shen, Q., Huang, Y.-G., Liu, C. et al. (2015). *J. Fluorine Chem.* 179: 14–22. (b) Ahrens, T., Kohlmann, J., Ahrens, M., and Braun, T. (2015). *Chem. Rev.* 115: 931–972. (c) Kuehnel, M.F., Lentz, D., and Braun, T. (2013). *Angew. Chem. Int. Ed.* 52: 3328–3348. (d) Stahl, T., Klare, H.F.T., and Oestreich, M. (2013). *ACS Catal.* 3: 1578–1587. (e) Meier, G. and Braun, T. (2009). *Angew. Chem. Int. Ed.* 48: 1546–1548. (f) Amii, H. and Uneyama, K. (2009). *Chem. Rev.* 109: 2119–2183.
21 Scott, V.J., Celenligil-Cetin, R., and Ozerov, O.V. (2005). *J. Am. Chem. Soc.* 127: 2852–2853.
22 (a) Panisch, R., Bolte, M., and Müller, T. (2006). *J. Am. Chem. Soc.* 128: 9676–9682. (b) Lühmann, N., Hirao, H., Shaik, S., and Müller, T. (2011). *Organometallics* 30: 4087–4096. (c) Kordts, N., Borner, C., Panisch, R. et al. (2014). *Organometallics* 33: 1492–1498. (d) Kordts, N., Künzler, S., Rathjen, S. et al. (2017). *Chem. Eur. J.* 23: 10068–10079.
23 (a) Stahl, T., Klare, H.F.T., and Oestreich, M. (2013). *J. Am. Chem. Soc.* 135: 1248–1251. (b) Stahl, T., Hrobárik, P., Königs, C.D.F. et al. (2015). *Chem. Sci.* 6: 4324–4334.

24 (a) Engesser, T.A., Lichtenthaler, M.R., Schleep, M., and Krossing, I. (2016). *Chem. Soc. Rev.* 45: 789–899. (b) Krossing, I. and Raabe, I. (2004). *Angew. Chem. Int. Ed* 43: 2066–2090. (c) Strauss, S.H. (1993). *Chem. Rev.* 93: 927–942.

25 Farooq Ibad, M., Langer, P., Schulz, A., and Villinger, A. (2011). *J. Am. Chem. Soc.* 133: 21016–21027.

26 Douvris, C., Stoyanov, E.S., Tham, F.S., and Reed, C.A. (2007). *Chem. Commun.* 1145–1147.

27 (a) Douvris, C. and Ozerov, O.V. (2008). *Science* 321: 1188–1190. (b) Douvris, C., Nagaraja, C.M., Chen, C.-H. et al. (2010). *J. Am. Chem. Soc.* 132: 4946–4953. (c) Gu, W. and Ozerov, O.V. (2011). *Inorg. Chem.* 50: 2726–2728.

28 Pan, B. and Gabbaï, F.P. (2015). *J. Am. Chem. Soc.* 136: 9564–9567.

29 Lühmann, N., Panisch, R., and Müller, T. (2010). *Appl. Organomet. Chem.* 24: 533–537.

30 Duttwyler, S., Douvris, C., Fackler, N.L.P. et al. (2010). *Angew. Chem. Int. Ed.* 49: 7519–7522.

31 (a) Allemann, O., Duttwyler, S., Romanato, P. et al. (2011). *Science* 332: 574–577. (b) Allemann, O., Baldridge, K.K., and Siegel, J.S. (2015). *Org. Chem. Front.* 2: 1018–1021.

32 Shao, B., Bagdasarian, A.L., Popov, S., and Nelson, H.M. (2017). *Science* 355: 1403–1407.

33 Bähr, S. and Oestreich, M. (2017). *Angew. Chem. Int. Ed.* 56: 52–59.

34 Dilmann, A.D. and Ioffe, S.L. (2003). *Chem. Rev.* 103: 733–772.

35 Remarkable approaches to facilitate challenging Diels–Alder reactions: (a) Furuta, K., Shimizu, S., Miwa, Y., and Yamamoto, H. (1989). *J. Org. Chem.* 54: 1481–1483. (b) Ahrendt, K.A., Borths, C.J., and MacMillan, D.W.C. (2000). *J. Am. Chem. Soc.* 122: 4243–4244. (c) Corey, E.J., Shibata, T., and Lee, T.W. (2002). *J. Am. Chem. Soc.* 124: 3808–3809. (d) Liu, W., You, F., Mocella, C.J., and Harman, W.D. (2006). *J. Am. Chem. Soc.* 128: 1426–1427.

36 (a) Mathieu, B. and Ghosez, L. (1997). *Tetrahedron Lett.* 38: 5497–5500. (b) Mathieu, B. and Ghosez, L. (2002). *Tetrahedron* 58: 8219–8226.

37 Hara, K., Akiyama, R., and Sawamura, M. (2005). *Org. Lett.* 7: 5621–5623.

38 (a) Klare, H.F.T., Bergander, K., and Oestreich, M. (2009). *Angew. Chem. Int. Ed.* 48: 9077–9079. (b) Schmidt, R.K., Müther, K., Mück-Lichtenfeld, C. et al. (2012). *J. Am. Chem. Soc.* 134: 4421–4428. (c) Nödling, A.R., Müther, K., Rohde, V.H.G. et al. (2014). *Organometallics* 33: 302–308.

39 Schmidt, R.K., Klare, H.F.T., Fröhlich, R., and Oestreich, M. (2016). *Chem. Eur. J.* 22: 5376–5383.

40 Rohde, V.H.G., Pommerening, P., Klare, H.F.T., and Oestreich, M. (2014). *Organometallics* 33: 3618–3628.

41 Johannsen, M., Jørgensen, K.A., and Helmchen, G. (1998). *J. Am. Chem. Soc.* 120: 7637–7638.

42 Olah, G.A., Rasul, G., and Prakash, G.K.S. (1999). *J. Am. Chem. Soc.* 121: 9615–9617.

43 Sakaguchi, Y., Iwade, Y., Sekikawa, T., Minami, T., and Hatanaka, Y. (2013). *Chem. Commun.* 49: 11173–11175.

44 Rohde, V.H.G.Müller, M.F. and Oestreich, M. (2015). *Organometallics* 34: 3358–3373.
45 Shaykhutdinova, P. and Oestreich, M. (2016). *Organometallics* 35: 2768–2771.
46 Shaykhutdinova, P., Kemper, S., and Oestreich, M. (2018). *Eur. J. Org. Chem.* 2896–2901.
47 Gatzenmeier, T., van Gemmeren, M., Xie, Y. *et al.* (2016). *Science* 351: 949–952.
48 (a) Matsuo, J.-I. and Murakami, M. (2013). *Angew. Chem. Int. Ed.* 52: 9109–9118 (b) Carreira, E.M. and Kvaerno, L. (2009). *Classics in Stereoselective Synthesis.* Weinheim: Wiley-VCH.
49 Mahlau, M. and List, B. 2013. *Angew. Chem. Int. Ed.* 52: 518–533.
50 Gatzenmeier, T., Kaib, P.S.J., Lingnau, J.B. *et al.* (2018). *Angew. Chem. Int. Ed.* 57: 2464–2468.
51 Mahlau, M., García-García, P., and List, B. (2012). *Chem. Eur. J.* **18**: 16283–16287.

6

Transition-Metal-Catalyzed C—H Bond Silylation

Yoshiya Fukumoto and Naoto Chatani

Osaka University, Department of Applied Chemistry, Faculty of Engineering, 2-1 Yamadoaka, Suita, Osaka 565-0871, Japan

6.1 C(sp)—H Bond Silylation

The fact that the oxidative addition of C(sp)—H bonds to transition-metal complexes proceeds much more easily than that of C(sp^2)—H and C(sp^3)—H bonds has resulted in a decreased interest in transition-metal-catalyzed C(sp)—H bond silylation with silylating reagents, such as hydrosilanes and disilanes, leading to the production of alkynylsilanes. This is, in fact, a difficult task, since the reaction of terminal alkynes with hydrosilanes under conditions of transition-metal catalysis gives, in almost all cases, cis-addition products as the result of the addition of a hydrogen atom and a silyl group across the C≡C bonds, a process that is generally referred to as the hydrosilylation of alkynes [1]. In fact, the possible formation of a C–H silylation product was first indicated when a small amount of by-product in the hydrosilylation of 1-hexyne with triethylsilane, catalyzed by $RhCl(PPh_3)_3$ was noted [2]. Therefore, inhibiting the addition of the silylating reagent to a C≡C bond in terminal alkynes was a key to the success of direct, catalytic C(sp)–H silylation.

To the best of our knowledge, the first selective C(sp)–H silylation of terminal alkynes with hydrosilanes was reported by Voronkov et al. in 1980, using a combination of $H_2PtCl_6 \cdot 6H_2O$ and LiI in a large excess (100 equiv) as the catalyst system (Scheme 6.1) [3]. The product distribution of the C–H silylation and hydrosilylation products was dramatically switched, depending on whether LiI was present or absent in the reaction. It was also revealed later that other iodide sources such as I_2 and Et_3SiI led to further improvement in the proportion of alkynylsilane product [4]. The proposed reaction mechanism involves the following steps: the reaction of a hydrosilane with HI, which is initially generated in the reaction system, leading to the formation of an iodosilane along with the evolution of H_2, and the reaction of the iodosilane with a terminal alkyne to afford the alkynylsilane product and HI, the latter of which was employed in the first step.

Organosilicon Chemistry: Novel Approaches and Reactions, First Edition.
Edited by Tamejiro Hiyama and Martin Oestreich

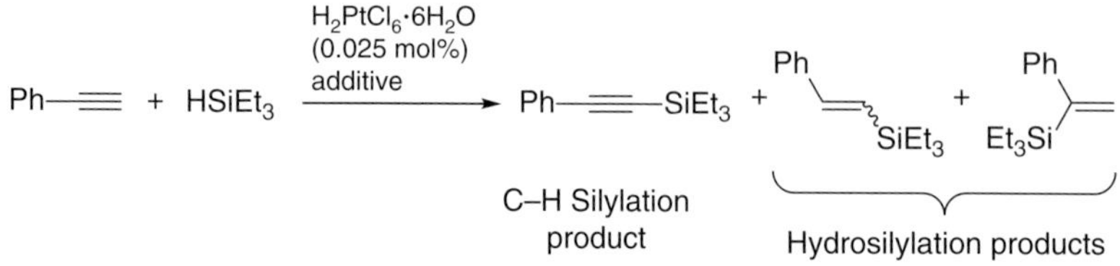

(6.1)

Additive	Product ratio		
None	0	72.2	27.8
LiI	72.7	15.1	12.2
I_2	82.9	9.2	7.9
Et_3SiI	93.4	5.8	Trace

Scheme 6.1 The H_2PtCl_6·$6H_2O/I^-$ or I_2-catalyzed C(sp)–H silylation of phenylacetylene.

Harrod and a coworker demonstrated the reaction of aryl-substituted acetylenes with trihydrosilanes or dihydrosilanes in the presence of a catalytic amount of CuCl and tertiary amines (Eq. (6.2)) [5]. This catalytic system was exploited for use in the copolymerization of dialkynylbenzenes and phenylsilane. Takaki et al. reported that the azametallacyclopropane complexes, Yb(η^2-Ph_2CNR)(hmpa)$_n$, also showed catalytic activity in the dehydrogenative silylation of alkyl-substituted alkynes (Eq. (6.3)), the use of which produced oligomers of diphenylsilane and 1,ω-diynes [6]. No hydrosilylation product was detected in either of the reactions. These reactions appear to involve the formation of an alkynylmetal species (**I$_2$**), formed by the reaction of a pre-catalyst with a terminal alkyne, as a key intermediate in the catalytic cycle (Scheme 6.2). The resulting species then reacts with a hydrosilane to afford the dehydrogenative silylation product and a hydridometal species (**II$_2$**), the latter of which reacts with another terminal alkyne molecule to regenerate **I$_2$** along with the evolution of H_2.

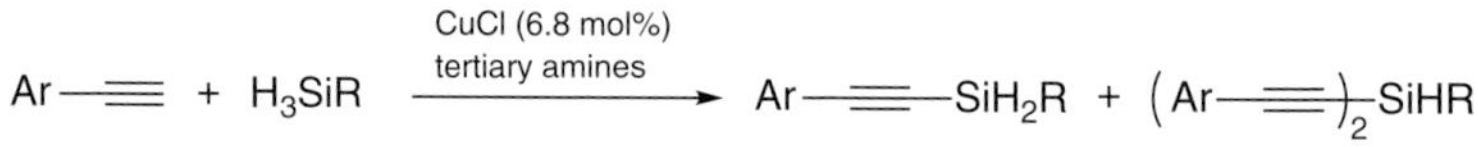

(6.2)

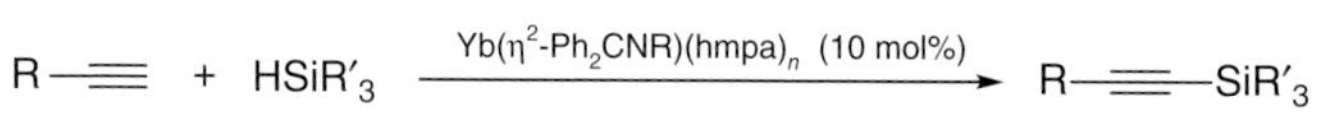

(6.3)

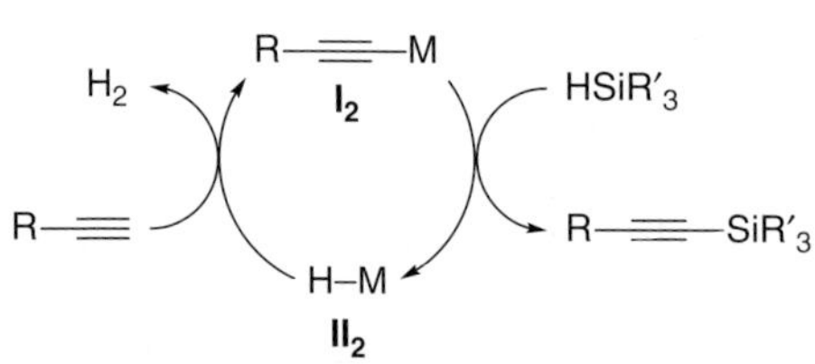

Scheme 6.2 Proposed reaction mechanism for the dehydrogenative silylation of terminal alkynes catalyzed by CuCl/tertiary amines or Yb(η^2-Ph_2CNR)(hmpa)$_n$.

Fuchikami and a coworker reported the $Ir_4(CO)_{12}/PPh_3$-catalyzed C(sp)–H silylation (Eq. (6.4)) [7]. The reaction was carried out using terminal alkyne in excess based on the fact that the hydrogen atoms on the terminal alkyne carbon and on the silicon atom were not released from the reaction mixture in the form of

H_2, but, rather, as a hydrogenation product of the terminal alkyne. The use of alkenes, such as diethyl fumarate and diethyl maleate, as hydrogen acceptors, was found to be effective for the reaction using an equimolar amount of terminal alkynes with hydrosilanes. Although the report contained no information regarding the reaction mechanism, Oro and coworkers proposed a mechanism for the related $[Ir(OMe)(diolefin)]_2/PAr_3$-catalyzed reaction, which involves the formation of a silyliridium species followed by the oxidative addition of a terminal alkyne to the iridium center and subsequent reductive elimination to afford the product [8].

$$R{-}{\equiv} + HSiR'_3 \xrightarrow[\text{Alkenes}]{Ir_4(CO)_{12}\ (1\ mol\%),\ PPh_3} R{-}{\equiv}{-}SiR'_3 \quad (6.4)$$

Mizuno and coworkers found that gold supported on manganese oxide-based octahedral molecular sieves (OMS-2) was an effective heterogeneous catalyst for the dehydrogenative coupling of terminal alkynes and hydrosilanes (Eq. (6.5)) [9]. The reaction was performed under 1 atm of O_2, and the two hydrogen atoms produced in the reaction were removed from the reaction system via the formation of H_2O. The proposed reaction mechanism involves the initial formation of an electrophilic silicon species $\mathbf{II_3}$ leading to the cis-addition of Au-SiR′$_3$ to the C≡C bond to form $\mathbf{III_3}$, its isomerization to a trans-adduct ($\mathbf{IV_3}$), and β-hydride elimination to form a (π-alkyne)AuH_2 complex $\mathbf{V_3}$. This type of mechanism was originally proposed by Crabtree and a coworker [10]. Finally, the reaction of $\mathbf{V_3}$ with O_2 is proposed to afford the product and H_2O, with $\mathbf{I_3}$ being regenerated (Scheme 6.3).

$$R{-}{\equiv} + HSiR'_3 \xrightarrow[O_2]{Au/OMS\text{-}2\ (1.9\ mol\%)} R{-}{\equiv}{-}SiR'_3 \quad (6.5)$$

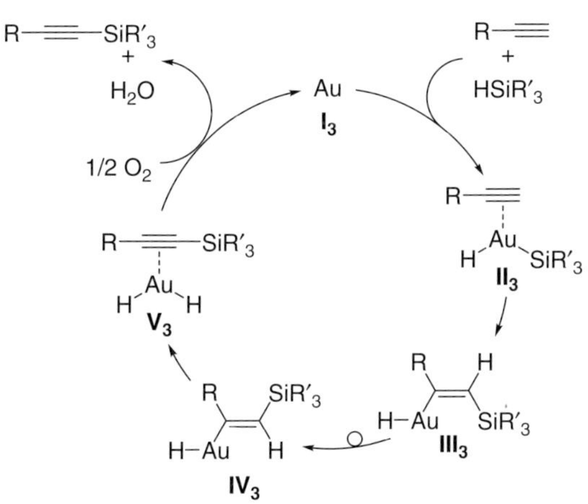

Scheme 6.3 Proposed reaction mechanism for the dehydrogenative silylation of terminal alkynes catalyzed by heterogeneous Au/OMS-2.

Marciniec et al. demonstrated that trimethylvinylsilane functioned as a silylating reagent in the $RuHCl(CO)(PCy_3)_2$-catalyzed silylation of terminal alkyne C(sp)—H bonds (Eq. (6.6)) [11]. The reaction mechanism, shown in Scheme 6.4, has been proved to involve the addition of a Ru–H species ($\mathbf{I_4}$) to the C=C bond of a vinylsilane and the subsequent release of ethylene by β-silyl elimination to form a Ru–Si species (**III$_4$**) [12]. The resulting **III$_4$** then reacts according to Crabtree's mechanism, as is also involved in Scheme 6.3, to produce the desired product along with the regeneration of **I$_4$**. These authors also reported the C–H silylation of terminal alkynes with iodotrimethylsilane catalyzed by $[IrCl(CO)_2]_2$/NEt^iPr_2 [13].

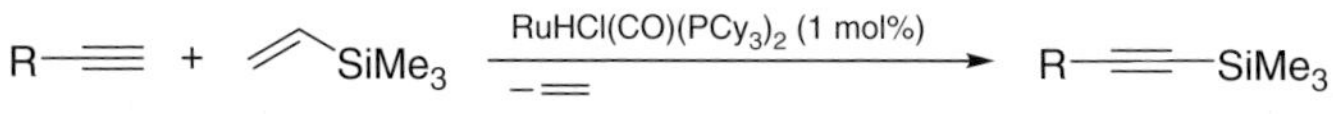

(6.6)

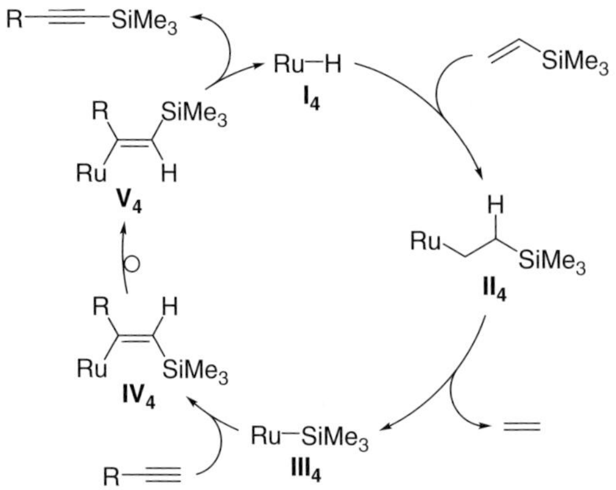

Scheme 6.4 Mechanism for the exchange reaction of the alkyne terminal proton for the silyl group in the vinylsilane, catalyzed by $RuHCl(CO)(PCy_3)_2$.

6.2 C(sp^2)—H Bond Silylation

The first report on catalytic C(sp^2)—H bond silylation was, as far as we know, described in 1982 by Curtis and coworkers, which was found during a study of the catalytic redistributive oligomerization of disiloxane [14]. When a solution of pentamethyldisiloxane and Vaska's complex, $IrCl(CO)(PPh_2)_2$, in benzene, was allowed to react, the Si–H phenylation product, pentamethylphenyldisiloxane, was formed in low yield, along with some redistribution products (Eq. (6.7)). Tanaka and coworkers then reported the direct silylation of arenes, which were used as solvents, with triethylsilane or hexamethyldisilane in the presence of $RhCl(CO)(PMe_3)_2$ under UV irradiation (Eq. (6.8)) [15]. However, the maximum turnover numbers (TONs) of the reactions were 3.76 for $HSiEt_3$, and 3.23 for $Me_3SiSiMe_3$, respectively, and biphenyl was also formed as a by-product in both reactions.

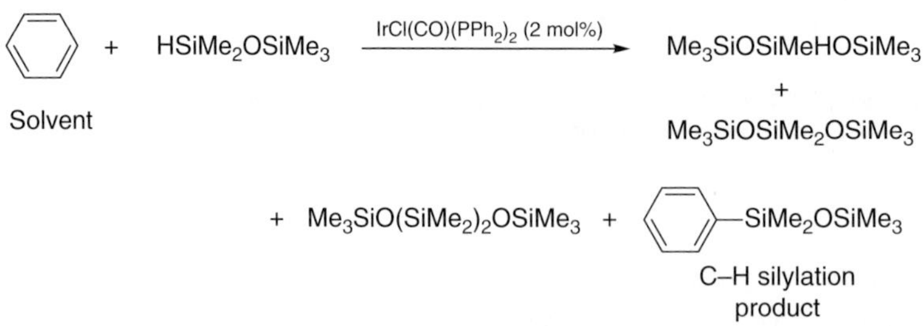

(6.7)

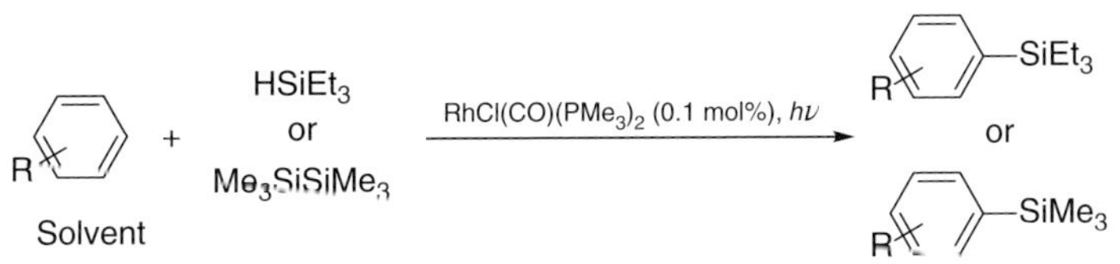

(6.8)

Since the discovery by Murai et al. in 1993 of the $RuH_2(CO)(PPh_3)_2$-catalyzed regioselective C–H alkylation of aromatic ketones at the *ortho* position with alkenes [16], chelation-assistance by directing groups have been recognized as one of the most powerful strategies for the catalytic regioselective functionalization of C—H bonds. Tanaka and coworkers first reported the chelation-assisted C–H silylation of *N*-alkyl aromatic imines with disilanes catalyzed by a combination of $Pt_2(dba)_3$ and $P(OCH_2)_3CEt$ (Eq. (6.9)) [17]. The proposed mechanism starts from the oxidative addition of a disilane to the platinum(0) species $\mathbf{I_5}$ to form the platinum(II) species $\mathbf{II_5}$ (Scheme 6.5). The coordination of the imine nitrogen in the substrate to the platinum(II) center in $\mathbf{II_5}$ then occurs to form $\mathbf{III_5}$, which is the important step, in that it causes the subsequent oxidative addition of the *ortho* C—H bond to the platinum(II) center to give $\mathbf{IV_5}$. The successive reductive elimination from $\mathbf{IV_5}$ results in the formation of a hydrosilane derivative and a monosilylated product. The mechanism also applies to the monosilylated product to afford a disilylated derivative.

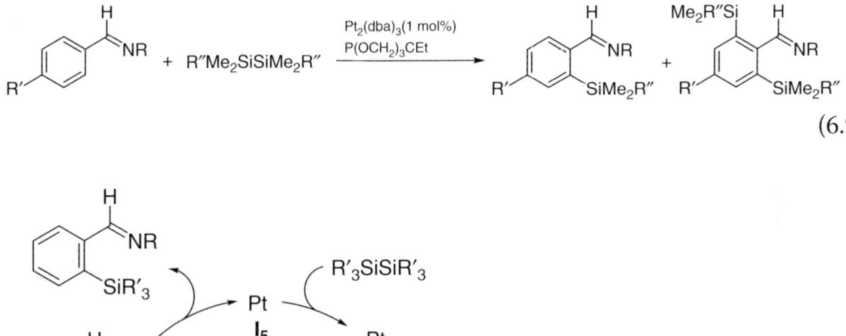

(6.9)

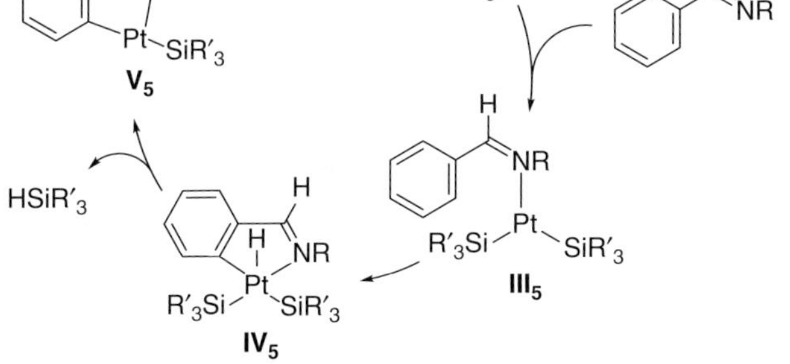

Scheme 6.5 Mechanism for the $Pt_2(dba)_3/P(OCH_2)_3CEt$-catalyzed $C(sp^2)$–H silylation of aromatic imines with disilanes.

The chelation-assisted regioselective silylation of aryl C—H bonds with disilanes was extended by Chatani et al., who reported a $[RhCl(cod)]_2$-catalyzed, pyridine-directed *ortho*-silylation with hexamethyldisilane (Eq. (6.10)) [18]. While a mixture of mono- and disilylated products was obtained in the reaction of the unsubstituted substrate, 2-phenylpyridine, the introduction of substituents such as a methyl group at the 3-position on the pyridine ring or, a methyl or other functional groups at the *meta* position on the phenyl ring, resulted in the selective production of the monosilylated product. The reaction mechanism appears to also involve the coordination of the pyridine nitrogen in the substrate to the rhodium center. The inhibition of the second silylation of 3-methyl-2-phenylpyridine can be attributed to steric congestion between the 3-methyl group and the silyl group that had already been introduced in the formation of the metallacycle intermediate $\mathbf{II_6}$ (Scheme 6.6).

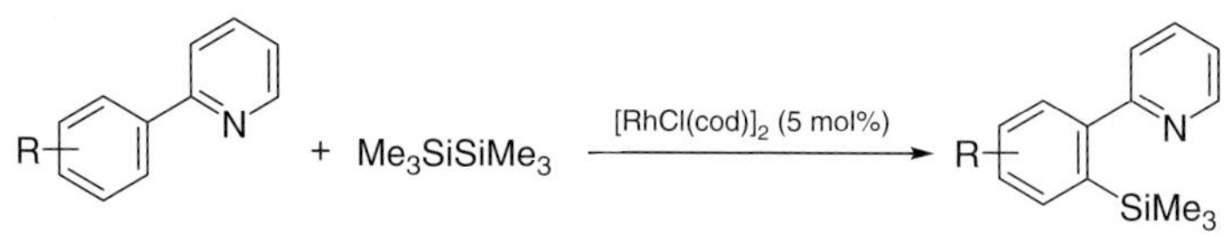

(6.10)

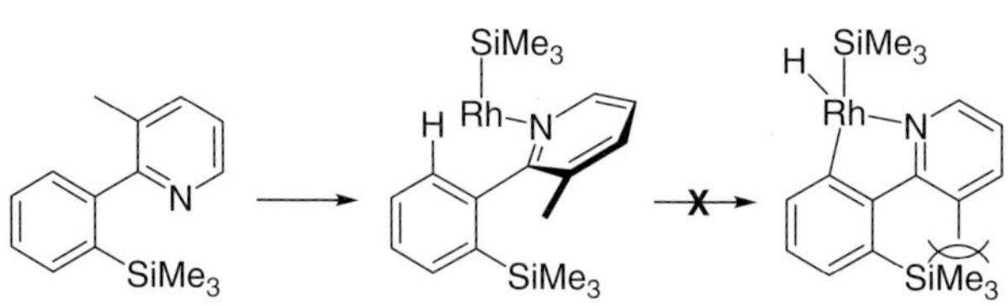

$\mathbf{I_6}$ $\mathbf{II_6}$

Scheme 6.6 Steric repulsion by a 3-methyl-2-pyridinyl group in the second silylation.

Kanai and coworkers reported the $Pd(OAc)_2$-catalyzed *ortho* C–H silylation of benzamide derivatives bearing the 8-quinolyl group on the amide nitrogen (Eq. (6.11)) [19]. The proposed reaction mechanism, which involves the formation of a *N,N′*-bidentate palladium species, $\mathbf{I_7}$, as a key intermediate, by the reaction of a substrate with $Pd(OAc)_2$, is depicted in Scheme 6.7. Chelation by the two nitrogen atoms to the palladium center enhances the strength of their binding [20], compared with monodentate directing groups, thereby facilitating the subsequent C—H bond cleavage by the palladium center to form $\mathbf{II_7}$. The σ-bond metathesis between Pd—N^{amide} bond in $\mathbf{II_7}$ and the Si—Si bond in a disilane, and *N*-to-*O* silicon migration affords $\mathbf{III_7}$. The reductive elimination from $\mathbf{III_7}$ yields a Pd(0) species and the initial, O-silylated product, which undergoes hydrolysis to afford the product. Oxidation of the Pd(0) species by Ag_2CO_3 in the presence of AcOH regenerates the $Pd(OAc)_2$. A related work in which an oxalyl amide group was used in place of the 8-quinolyl amide group as a directing group was reported by Wen and coworkers (Eq. (6.12)) [21]. Elias and coworkers reported the use of a 2-(methylthio)phenylamide group as the directing group in the palladium-catalyzed C–H silylation of the cyclopentadienyl group in (η^5-Cp)Co(C_4Ph_4) with $Me_3SiSiMe_3$ [22].

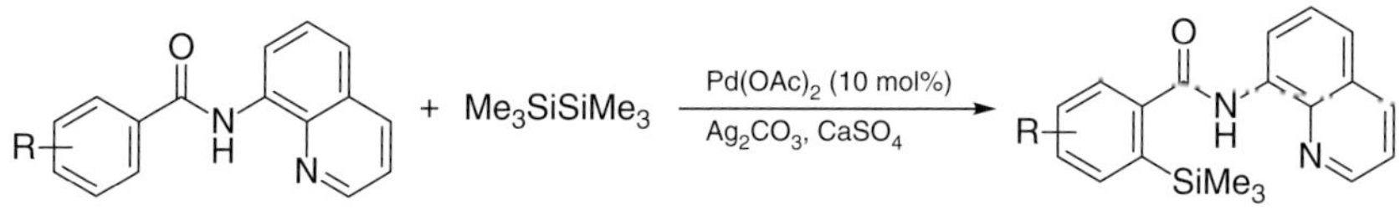

(6.11)

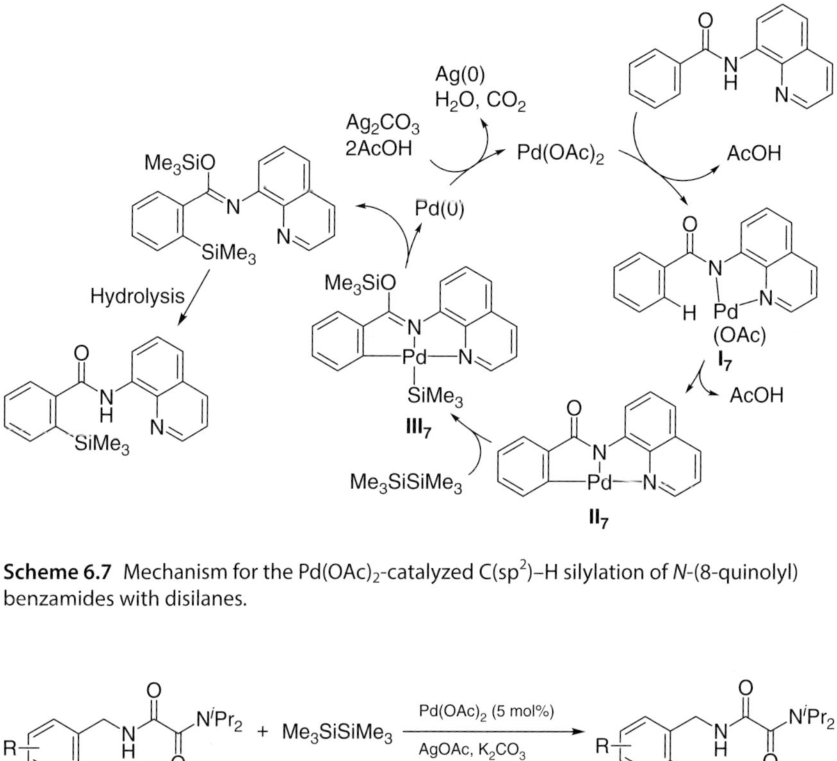

Scheme 6.7 Mechanism for the Pd(OAc)$_2$-catalyzed C(sp^2)–H silylation of *N*-(8-quinolyl) benzamides with disilanes.

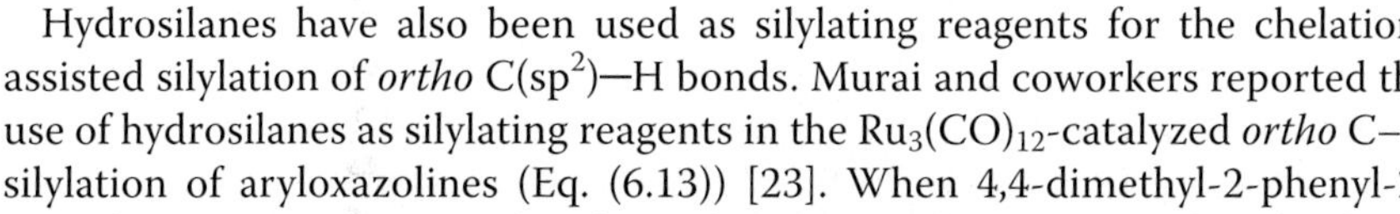

(6.12)

Hydrosilanes have also been used as silylating reagents for the chelation-assisted silylation of *ortho* C(sp^2)—H bonds. Murai and coworkers reported the use of hydrosilanes as silylating reagents in the $Ru_3(CO)_{12}$-catalyzed *ortho* C–H silylation of aryloxazolines (Eq. (6.13)) [23]. When 4,4-dimethyl-2-phenyl-2-oxazoline was simply reacted with triethylsilane in the presence of $Ru_3(CO)_{12}$ as the catalyst, the expected *ortho*-silylation product was obtained in only 7% yield. On the other hand, the addition of *tert*-butylethylene as a hydrogen acceptor improved the product yield to 93%. Scheme 6.8 shows the proposed reaction mechanism that appeared in their subsequent paper [24]. The successive oxidative addition of the *ortho* C—H bond in the aryloxazoline and a hydrosilane to the Ru(0) species $\mathbf{I_8}$ forms a Ru(IV) species $\mathbf{III_8}$. The insertion of an olefin into the Ru—H bond followed by the reductive elimination of the alkane resulted in the formation of the species $\mathbf{VI_8}$, and the final reductive elimination from $\mathbf{VI_8}$ affords the silylation product, with $\mathbf{I_8}$ being regenerated. The rapid H/D scrambling between the deuterium atom at the *ortho* position in the deuterated oxazoline and the hydrogen atom on the silicon atom provides support for the proposed formation of the intermediate $\mathbf{III_8}$. This result also indicates that the oxidative addition of the aryl C—H bond is not the rate-determining step and that a rapid equilibrium occurs prior to C—Si bond formation. The authors also revealed that various types of nitrogen-containing directing groups, which were shown in Scheme 6.9, can be used to enhance the regioselective silylation of aromatic C—H bonds [23–25]. The reaction of 2-phenylpyridines was also reported in this study, in advance of the use of a disilane reported by Chatani et al. [18]. Li and coworkers reported that the combination of $[IrCl(cod)]_2$/NEt^iPr_2 also catalyzed the C(sp^2)–H silylation of arenes and heteroarenes bearing various types of nitrogen-containing heterocyclic functional groups and an *N,N*-dimethylcarboxamide group as directing groups [26].

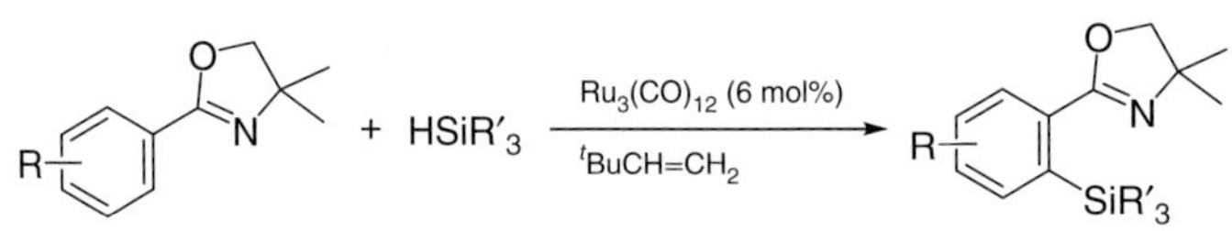

(6.13)

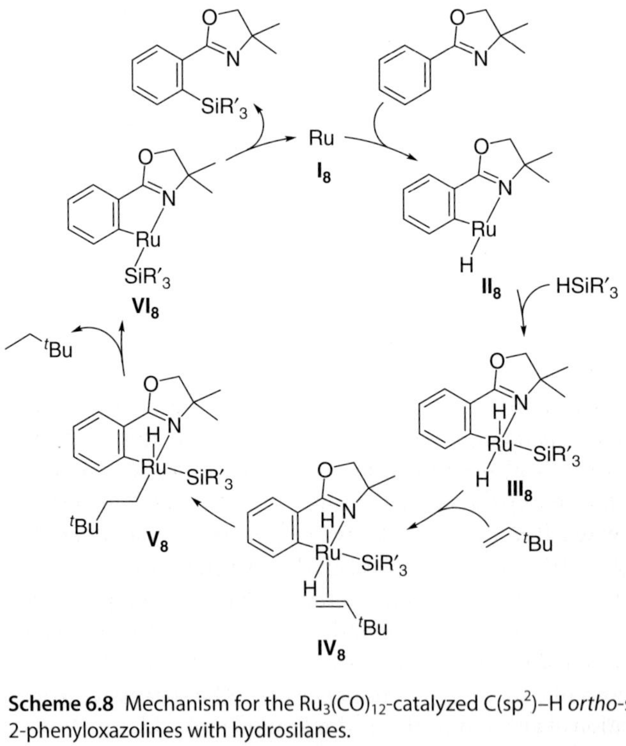

Scheme 6.8 Mechanism for the $Ru_3(CO)_{12}$-catalyzed C(sp^2)–H *ortho*-silylation of 2-phenyloxazolines with hydrosilanes.

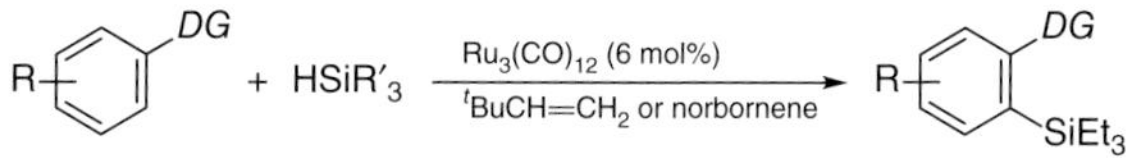

(6.14)

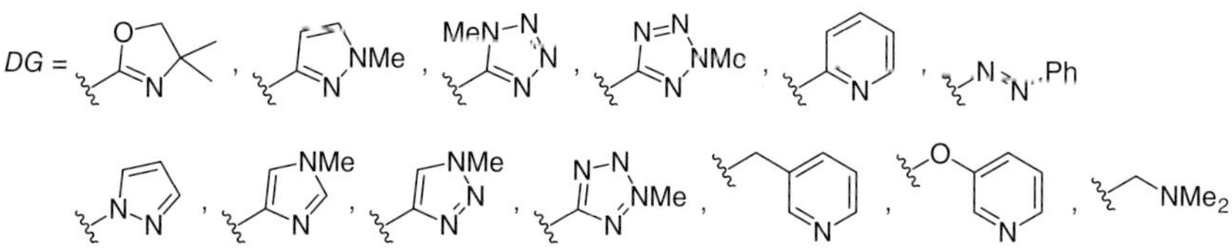

Scheme 6.9 Available directing groups for the $Ru_3(CO)_{12}$-catalyzed *ortho*-silylation of aryl C(sp^2)—H bonds with hydrosilanes.

Suginome and coworkers developed the use of a directing-group-assisted methodology for the reaction of diaminophenylborane derivatives (Scheme 6.10) [27]. These starting materials can easily be prepared by the reaction of boronic acids with 2-pyrazol-5-ylaniline or anthranilamide. After the completion of the C–H silylation, the resulting silyl group could be transformed into other functional groups such as alcohol, halides, and aryl groups, and the diaminophenylboryl groups remained intact under the reaction conditions. On the other hand, diamine moieties were removed by hydrolysis to give the corresponding boronic acids, which could be substituted by aryl group by a Suzuki–Miyaura coupling reaction.

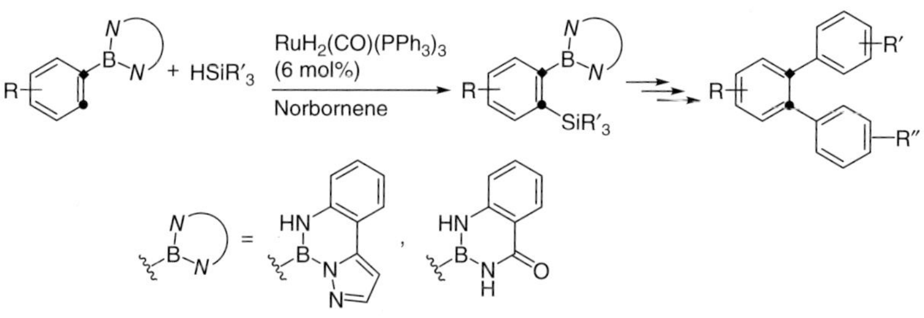

(6.15)

Scheme 6.10 C(sp^2)–H silylation of diaminophenylboranes catalyzed by $RuH_2(CO)(PPh_3)_2$.

Mashima and coworkers reported that iridium complexes bearing *N*-aryl carbene ligands (Ir-NHCs) catalyzed the C–H silylation of 2-phenylpyridines with triethylsilane (Eq. (6.16)) [28]. *N*-Phenylpyrazole and aromatic imines also reacted under the same reaction conditions to afford the corresponding *ortho*-silylated products. After an intensive mechanistic study of the catalytic reaction, it was proposed that the reaction involved the formation of a bis(cyclometalated)iridium complex **II**$_{11}$, which is formed by the reactions of an Ir-NHC with 2-phenylpyridine and subsequently with a hydrosilane, as a key catalytic species (Scheme 6.11). **II**$_{11}$ is in equilibrium with **III**$_{11}$ in the presence of the hydrosilane. Norbornene traps the two hydrogen atoms on the iridium center in **III**$_{11}$ and that at the *ortho* position of the *N*-aryl group in the carbene ligand, via the formation of **IV**$_{11}$, to form **V**$_{11}$. The new C—Si bond formation occurs through the reaction of **V**$_{11}$ with another hydrosilane molecule via σ-complex-assisted metathesis (σ-CAM) [29], together with the formation of an Ir—H bond to form **VII**$_{11}$. The product is released by ligand exchange for another molecule of 2-phenylpyridine, and **VIII**$_{11}$ is formed.

Finally, the hydrosilane undergoes reductive elimination followed by the oxidative addition of *ortho* C—H bond in the coordinated substrate, resulting in the regeneration of $\mathbf{II}_{11}$. A reaction mechanism involving C—Si bond formation via σ-CAM was also proposed by Iglesias and coworkers in their [Ir(IPr)H_2Py_3]BF_4-catalyzed C–H silylation of 2-phenylpyridines with hydrosilanes [30].

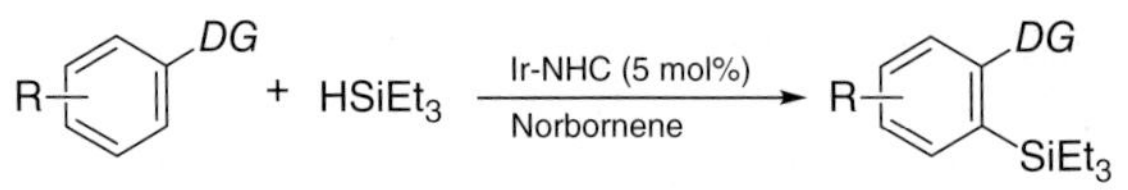

(16)

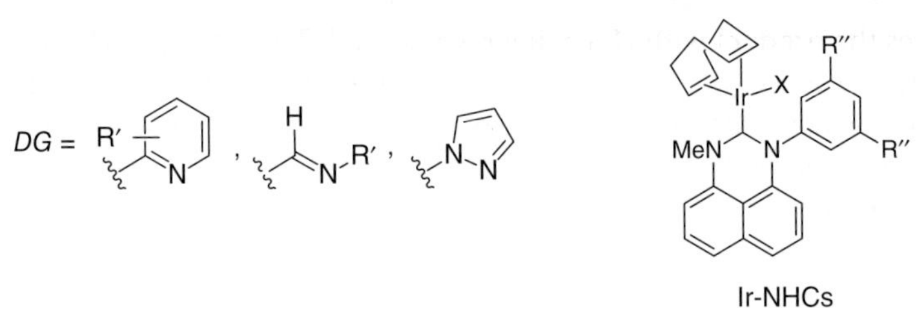

(6.16)

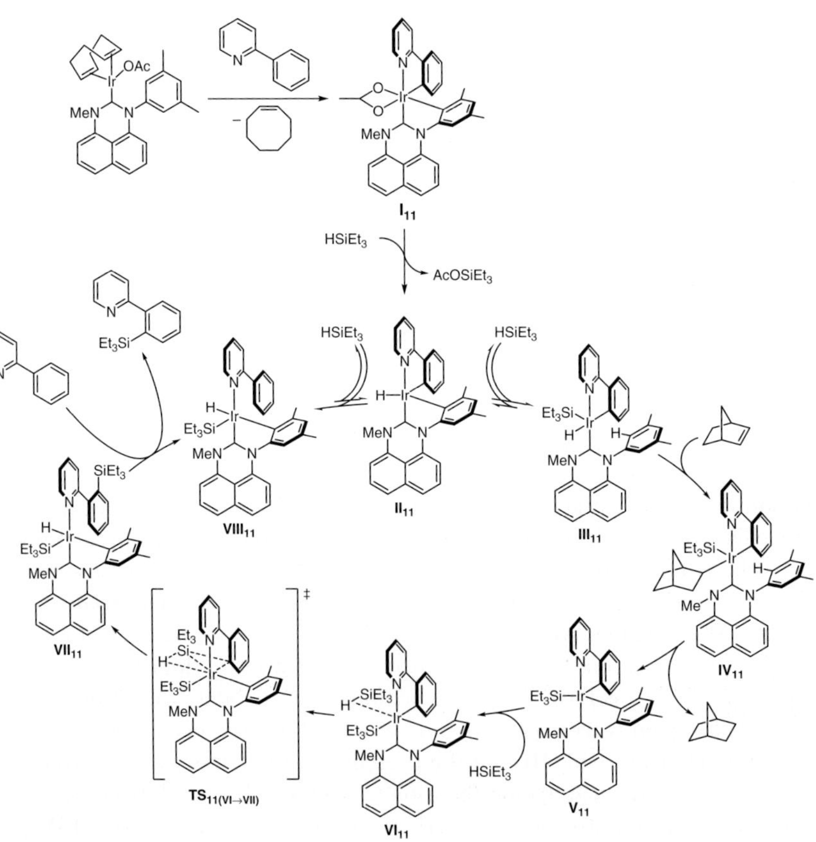

Scheme 6.11 Mechanism for the Ir-NHC-catalyzed C(sp^2)–H *ortho*-silylation of 2-phenylpyridines with hydrosilanes. All structural formulas to which the double dagger symbol is attached are essential transition state structures to study the C–H functionalization chemistry more deeply.

Kanai and coworkers found that 2-phenylpyridines were silylated with fluorohydrosilanes in the presence of Ir(acac)(cod) as the catalyst (Eq. (6.17)) [31]. Although the reaction proceeded in the absence of a hydrogen acceptor, the addition of an acceptor resulted in an improved product yield. The authors proposed that the reaction mechanism, in the absence of the hydrogen acceptor, starts from the interaction between the Lewis-acidic fluorinated silicon atom and the Lewis-basic pyridine nitrogen to form $\mathbf{I}_{12}$, followed by the oxidative addition of H—Si bond in $\mathbf{I}_{12}$ to $\mathbf{II}_{12}$ to afford a hydridoiridium species $\mathbf{III}_{12}$ (Scheme 6.12). The oxidative addition of the *ortho* C—H bond to the iridium center occurs to form an iridacycle $\mathbf{IV}_{12}$ or $\mathbf{V}_{12}$, and the subsequent reductive elimination gives the product with $\mathbf{II}_{12}$ being regenerated. The release of H_2 precedes the formation of $\mathbf{IV}_{12}$ or follows that of $\mathbf{V}_{12}$. Other nitrogen-containing functional groups, shown below Eq. (6.17), were also employed as directing groups for the reaction.

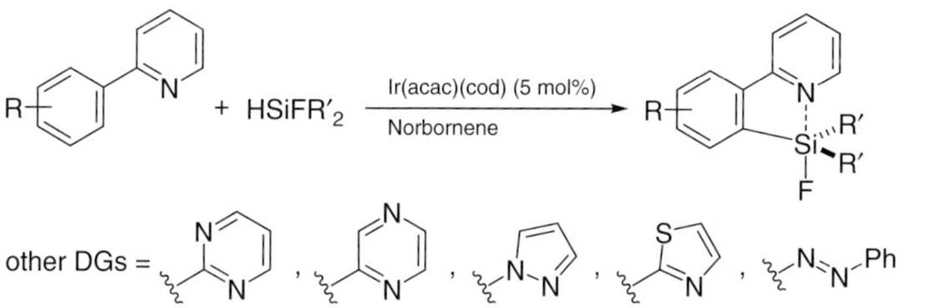

(6.17)

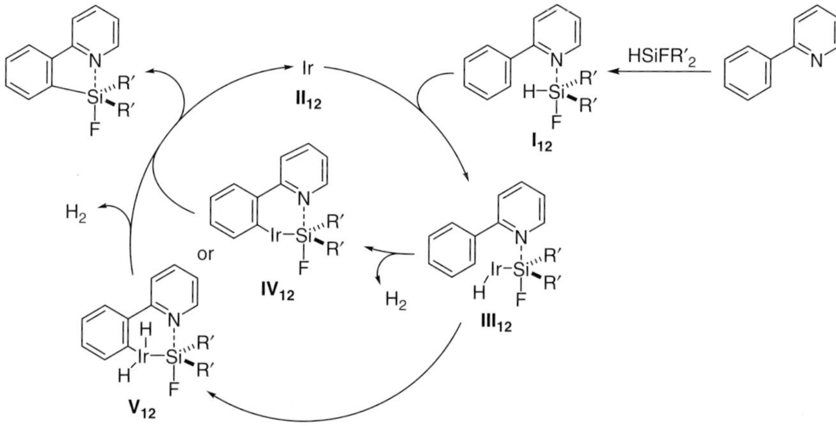

Scheme 6.12 Mechanism for the Ir(acac)(cod)-catalyzed C(sp^2)–H *ortho*-silylation of 2-phenylpyridines with $HSiFR'_2$.

Murata and coworkers reported that the ruthenium-catalyzed *ortho*-selective C–H silylation of aryloxazolines proceeded even in the absence of a hydrogen acceptor at a higher temperature, i.e. at 200 °C (Eq. (6.18)) [32]. Among hydrosilanes examined, $HSiMe_2(OSiMe_3)$ gave the best results. *N-tert*-Butylimino and 3-methylpyridyl groups were also available as directing groups for the reaction. A theoretical study suggested that the reductive elimination of H_2 took place via the σ-CAM of the *ortho* C—H bond with the Ru—H bond in $\mathbf{II}_{13}$, and the C—Si bond formation step from $\mathbf{IV}_{13}$ also proceeded similarly (Scheme 6.13).

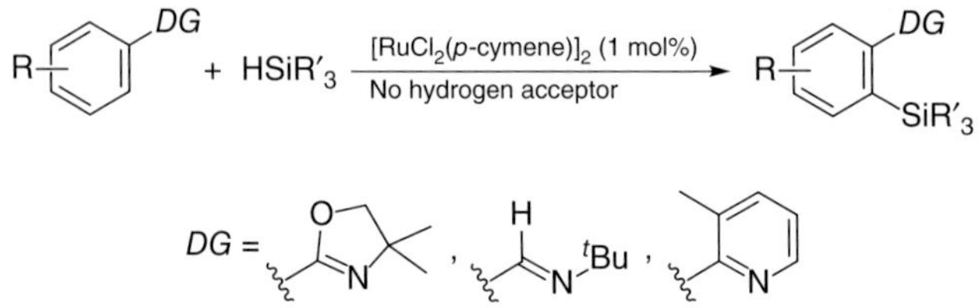

(6.18)

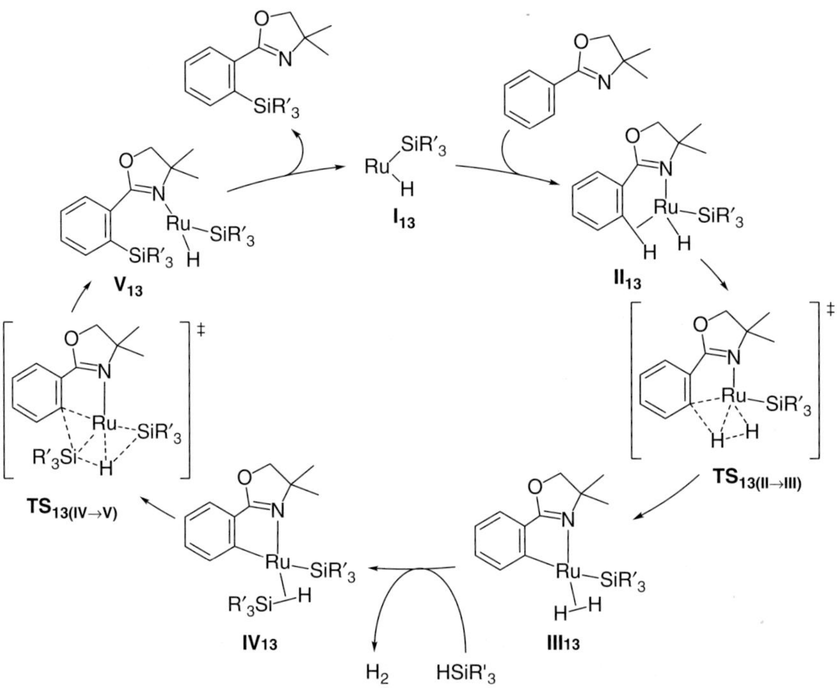

Scheme 6.13 Mechanism for the $[RuCl_2$ (*p*-cymene)$]_2$-catalyzed acceptorless $C(sp^2)$–H *ortho*-silylation of 2-phenyloxazolines with hydrosilanes. All structural formulas to which the double dagger symbol is attached are essential transition state structures to study the C–H functionalization chemistry more deeply.

C—H bond silylations promoted by directing groups other than nitrogen-containing functionalities are discussed subsequently. Hou and coworkers reported the dehydrogenative silylation of anisole derivatives at the *ortho* position catalyzed by a half-sandwich scandium complex (Eq. (6.19)) [33]. The addition of a hydrogen acceptor was not required for the reaction. A proposed mechanism, based on a study of a series of stoichiometric reactions, is shown in Scheme 6.14.

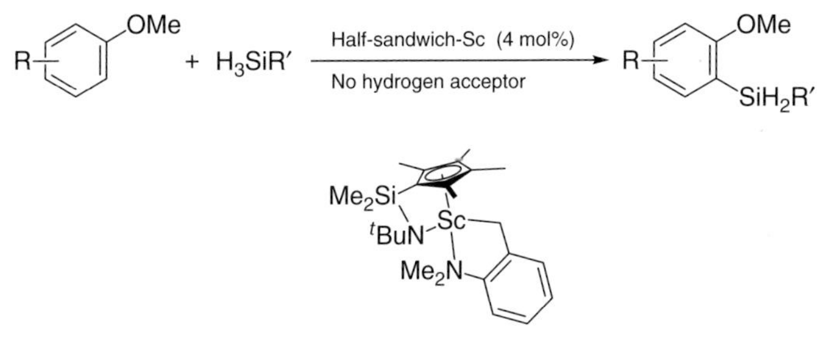

Half-sandwich-Sc

(6.19)

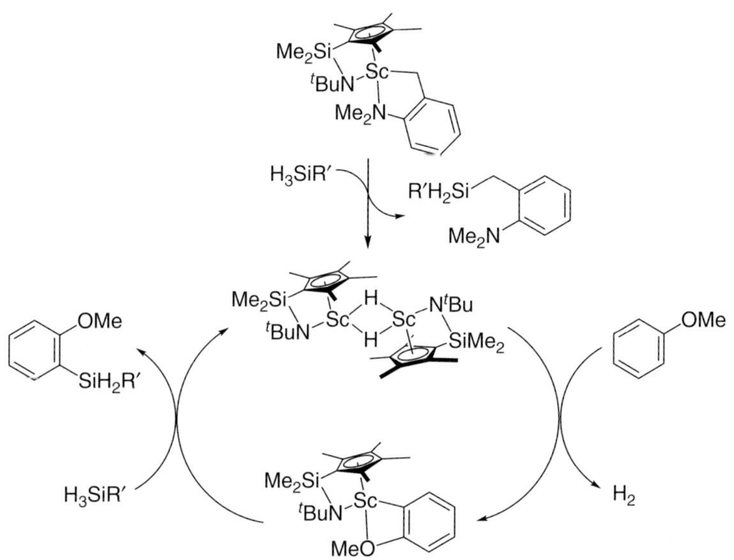

Scheme 6.14 Mechanism for the half-sandwich-Sc-catalyzed acceptorless C(sp^2)–H *ortho*-silylation of anisoles with hydrosilanes.

Tobita and coworkers reported the *ortho* C–H silylation of internal arylalkynes with hydrosilanes catalyzed by ruthenium complexes bearing a xanthene-based *Si,O,Si*-chelate ligand, xantsil, to afford *ortho*-silylated (*E*)-styrene derivatives (Eq. (6.20)) [34]. The alkyne group served as both a directing group and a hydrogen acceptor in this reaction.

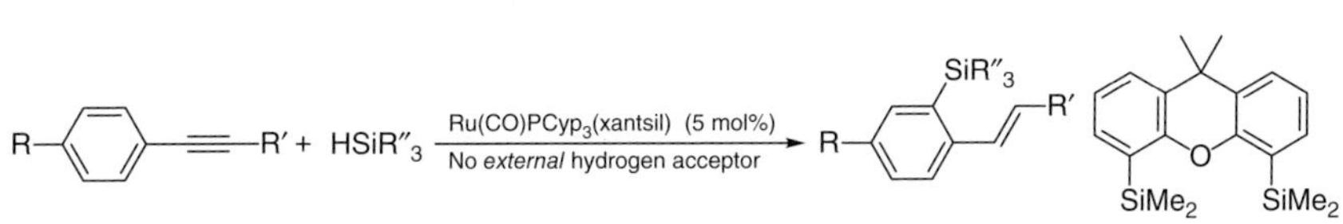

Xantsil

(6.20)

Carbonyl-group-directed *ortho* C–H silylations with vinylsilanes under $Ru_3(CO)_{12}$ catalysis was reported by Murai and coworkers (Eq. (6.21)) [35]. Although the reaction was limited to five-membered heteroarenes such as furans, pyrroles, and thiophenes, it was observed to take place at a position β to the directing carbonyl group, e.g. an amide, an ester, and a ketone. The C2—H bond in the 3-acylheteroarenes was regioselectively silylated. The cleavage of the C—H bond by the ruthenium center in $\mathbf{I_{15}}$ followed by the insertion of a C=C bond in the vinylsilane into the Ru—H bond gives β-(silyl)ethylruthenium species $\mathbf{III_{15}}$. After β-silyl elimination from $\mathbf{III_{15}}$ to form $\mathbf{IV_{15}}$ with the release of ethylene, followed by reductive elimination affords the silylation product with the regeneration of $\mathbf{I_{15}}$ (Scheme 6.15).

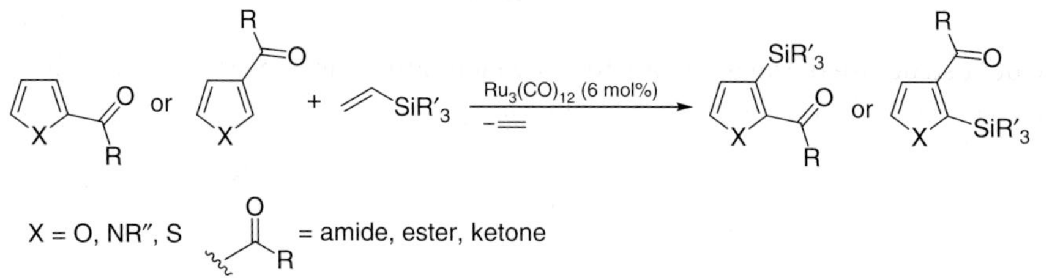

(6.21)

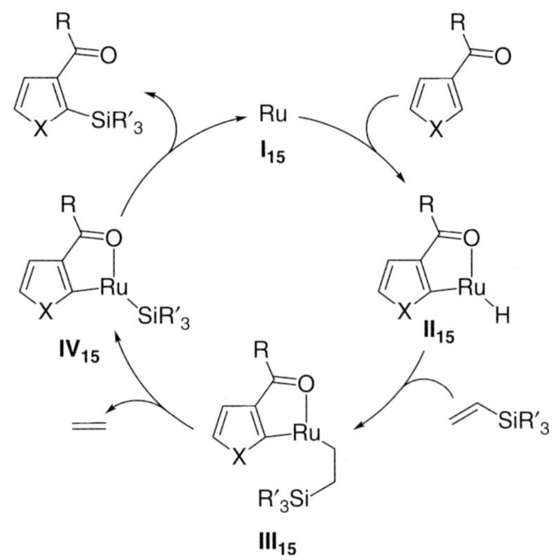

Scheme 6.15 Mechanism for the $Ru_3(CO)_{12}$-catalyzed *ortho* C(sp^2)–H silylation of 2-heteroarenes with vinylsilanes.

Recent advances in the chelation-assisted C–H silylation involves remote *meta*- and *para*-selective reactions with disilanes. Maiti and coworkers reported the use of a nitrile-based directing group in the *meta*-selective C–H silylation, in the presence of a $Pd(OAc)_2$ catalyst combined with *N*-acetylglycine (Eq. (6.22)) [36]. The catalytic system was applied to the reaction with $Me_3GeGeMe_3$ to produce the corresponding *meta*-germyl products.

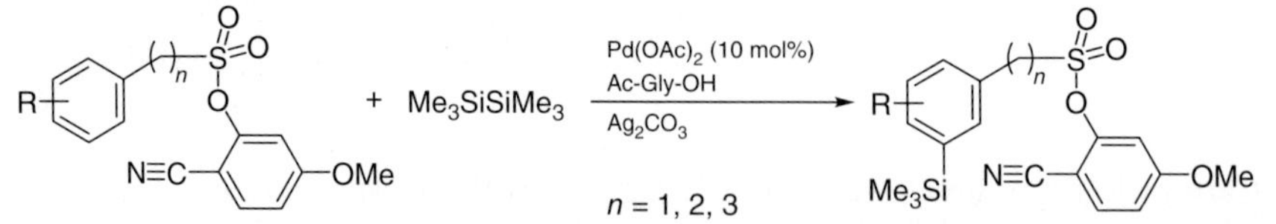

(6.22)

Maiti and coworkers also designed a new nitrile-based template tethered by a siloxy group for use in the $Pd(OPiv)_2$-catalyzed *para*-selective direct silylation of C—H bonds with disilanes (Eq. (6.23) in Scheme 6.16) [37]. *N*-Cbz-Leucine was found to be the ligand of choice for the success of the reaction. DFT calculations showed that a 16-membered metalation transition state to afford the *para*-silylation product was more stable than the 15-membered one for the formation of the *meta*-silylation product, due to decreased ring strain and distortion. NOE experiments and DFT calculations suggested that the hydrogen-bonding interaction between the oxygen atoms in the template and hexafluoroisopropyl alcohol (HFIP), used as the solvent, occurred, which would be expected to stabilize the bent structure of the template, thereby facilitating the approach of the palladium center to the *para* C—H bond for cleavage.

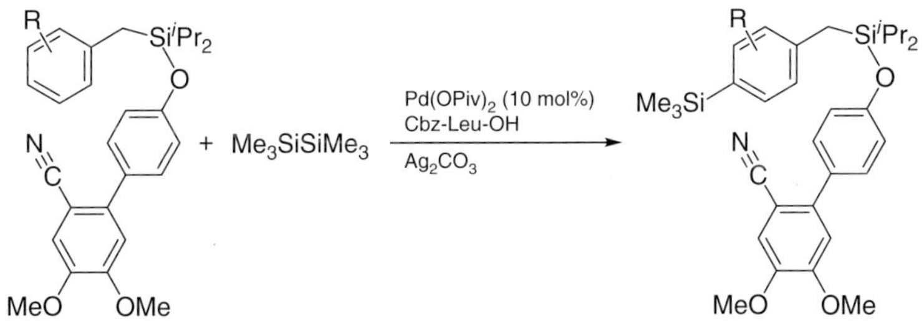

(6.23)

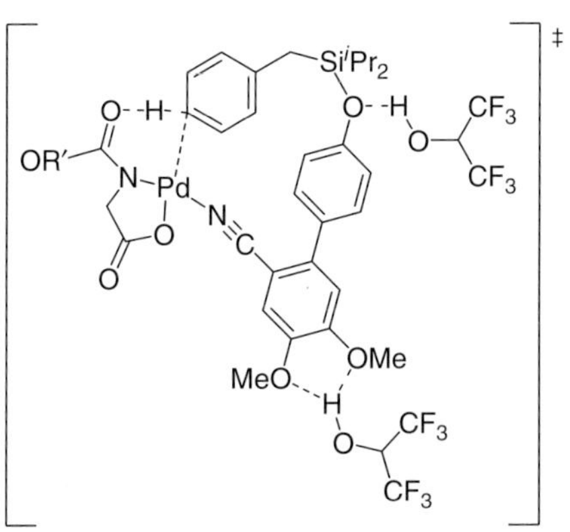

Scheme 6.16 The $Pd(OPiv)_2$/Cbz-Leu–OH-catalyzed *para*-C(sp^2)–H silylation with hexamethyldisilane. All structural formulas to which the double dagger symbol is attached are essential transition state structures to study the C–H functionalization chemistry more deeply.

The direct silylation of vinyl C—H bonds was also attained by Zhang and coworkers in the $Pd(OAc)_2$-catalyzed reaction of *N*-8-quinolylacrylamide derivatives with hexamethyldisilane, leading to the stereoselective production of *Z*-vinylsilanes (Eq. (6.24)) [38]. The reaction appears to proceed via a mechanism similar to that proposed for the palladium-catalyzed reaction of *N*-8-quinolylbenzamides, depicted in Scheme 6.7, except for the oxidation of the Pd(0) species to $Pd(OAc)_2$ with 1,4-benzoquinone in place of Ag_2CO_3, and the formation of the final product by alcoholysis with the resulting formation of 1,4-hydroquinone, and not hydrolysis.

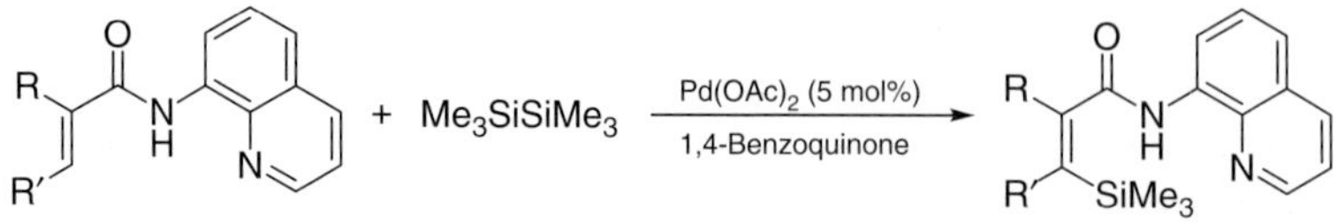

(6.24)

In parallel with the evolution of the chelation-assisted *ortho*-selective C(sp^2)–H silylations of arenes, considerable efforts have also been made in the development of non-chelation-assisted intermolecular silylations. In the early stages of this study, the use of a large excess of the arene and elevated reaction temperatures were required to produce good yields of the desired products, probably due to low efficiency of the catalytic system. The silylation of monosubstituted benzenes with silylating reagents resulted in the production of a mixture of *meta*- and *para*-regioisomers, which, with a few exceptions, afforded *para*-substituted products. On the other hand, a number of reactions of 1,2- and 1,3-disubstituted arenes have been found to proceed via a *meta*-selective process. These regioselectivities can be attributed to the steric effects of the substituents on the arenes.

Ishikawa et al. reported that, when certain solvents were used, the $Ni(PEt_3)_4$-catalyzed reaction of a cyclic disilane afforded σ-bond metathesis compounds of arene C—H and Si—Si bonds (Eq. (6.25) in Scheme 6.17) [39]. The reaction mechanism appears to involve the generation of a metallacycle $\mathbf{I}_{17}$, which is in equilibrium with an η^2-complex $\mathbf{II}_{17}$. A related work using *o*-di(hydrosilyl)benzenes instead of the benzodisilacyclobutene catalyzed by $Pt_2(dba)_3$ was subsequently reported by Tanaka and coworkers (Eq. (6.26) in Scheme 6.17) [40]. Deuterium labeling experiments using bis(deuteriodimethylsilyl)benzene showed that no deuterated compound was formed. This result indicates that the reaction also appears to proceed via the formation of $\mathbf{I}_{17}$ (and/or $\mathbf{II}_{17}$).

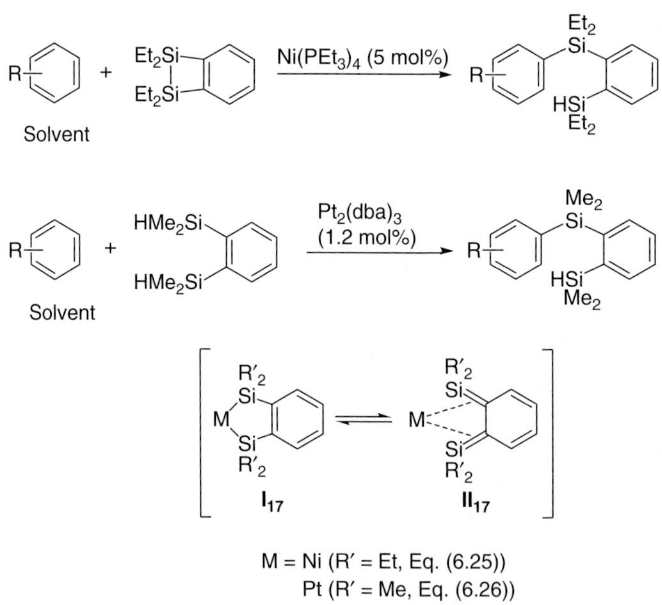

(6.25)

(6.26)

Scheme 6.17 C–H silylation of arenes with cyclic disilane catalyzed by $Ni(PEt_3)_4$ and *o*-di(hydrosilyl)benzenes catalyzed by $Pt_2(dba)_3$.

Berry and coworkers reported that a simple alkyl-substituted hydrosilane, triethylsilane, could be employed in the C–H silylation of solvent arenes with the aid of some rhodium and ruthenium complexes and *tert*-butylethylene as a hydrogen acceptor (Eq. (6.27) in Scheme 6.18) [41]. However, the reaction was accompanied by the dimerization of triethylsilane to afford [1-(diethylsilyl)ethyl]triethylsilane. Hartwig and a coworker demonstrated that $Tp^{Me_2}PtMe_2H$ (Tp^{Me_2} = tris(3,5-dimethylpyrazol-1-yl)borate) was an effective catalyst for the acceptorless C–H bond silylation of arenes with a variety of trialkylsilanes [42]. Murata et al. reported that a hydrotrisiloxane, $HSiMe(OSiMe_3)_2$, could be used in the related $PtCl_2/Tp^{Me_2}K$-catalyzed C–H bond silylation of arenes [43]. Ishiyama et al. reported that the $[Ir(OMe)(cod)]_2$/dtbpy system (dtbpy = 4,4′-di-*tert*-butyl-2,2′-bipyridyl), which could be successfully used for $C(sp^2)$–H bond borylation [44], was also applicable to silylation with a disilane, $(^tBuF_2Si)_2$ [45]. Their subsequent study, dealing with modifying the reaction conditions using $(^sBuF_2Si)_2$ and dipphen (dipphen = 2,9-diisopropylphenanthroline), instead of $(^tBuF_2Si)_2$ and dtbpy, respectively, resulted in a decrease in the amount of the added substrate arenes needed to 10 equiv, relative to the disilane [46]. The authors also reported the $[Ir(OMe)(cod)]_2$/dmphen-catalyzed C–H silylation of arenes with a trialkoxy-substituted hydrosilane, 1-hydrosilatrane (dmphen = 2,9-dimethylphenanthroline) [47]. These results are summarized in Scheme 6.18.

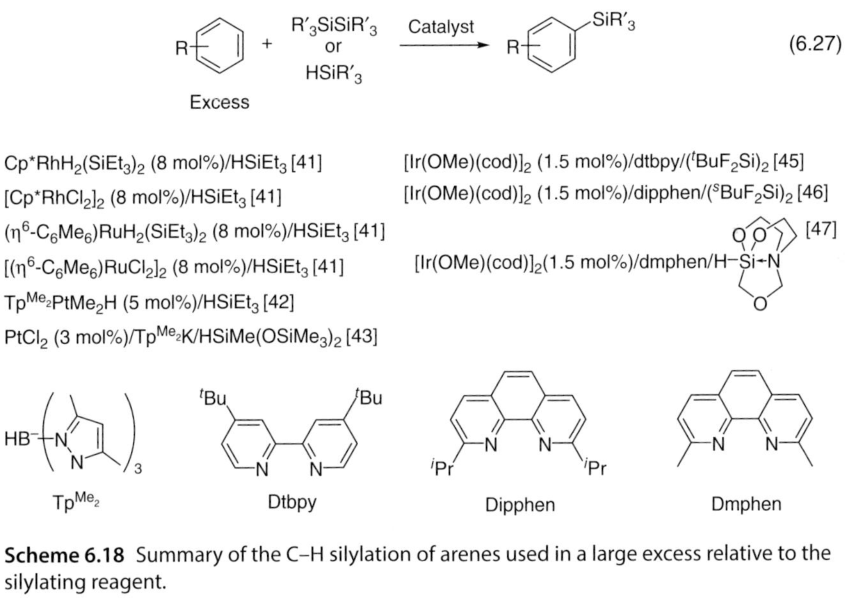

Scheme 6.18 Summary of the C–H silylation of arenes used in a large excess relative to the silylating reagent.

As also mentioned earlier, these non-chelation-assisted C–H silylations required a large excess of arene substrates relative to the silylating reagents. This stoichiometry is a limitation for synthetic applications because the arene is the more valuable reaction component. Hartwig and a coworker reported the $[Rh(OH)(coe)_2]_2$/bisphosphine-catalyzed C–H silylation of arenes with $HSiMe(OSiMe_3)_2$ under the reaction conditions, in which the arene is the limiting reagent (Scheme 6.19) [48]. The regioselectivity appears to be determined primarily by the steric effects of the substituents. In almost all cases, the reaction of 1,2-disubstituted benzenes afforded the silylation products at the 4-position as the sole products. *o*-Cymine was silylated at the *meta* position to the less-hindered methyl group with an 82 : 18 4-/5-selectivity. The electronic nature of the substituents also affected the selectivity of the reaction. C–H silylation of 2-(trifuluoromethyl)anisole occurred at the electron-rich 4-position with a 98 : 2 4-/5-selectivity. Mechanistic studies on the rhodium-catalyzed C–H silylation resulted in the proposed reaction mechanism, as shown in Scheme 6.20. A hydridorhodium species $\mathbf{I_{20}}$ is initially formed by the reaction of a rhodium precursor with a bisphosphine ligand and a hydrosilane. Next, the C=C bond in cyclohexene, which is used as a hydrogen acceptor, inserts into the Rh—H bond in $\mathbf{I_{20}}$, resulting in the formation of a cyclohexylrhodium complex, $\mathbf{III_{20}}$. The oxidative addition of a hydrosilane to $\mathbf{III_{20}}$ forms $\mathbf{IV_{20}}$ and the subsequent reductive elimination of cyclohexane from $\mathbf{IV_{20}}$ affords a silylrhodium complex, $\mathbf{V_{20}}$, the latter step of which appears to be the rate-determining step. The oxidative addition of the arene C—H bond followed by reductive elimination yields the C–H silylation product with $\mathbf{I_{20}}$ being regenerated. Based on the results of deuterium labeling experiments using deuteriosilane, the product-forming reductive elimination would be predicted to be faster at the more electron-rich position on the arene, which explains the 4-selectivity in the C—H bond silylation of 2-(trifluoromethyl) anisole.

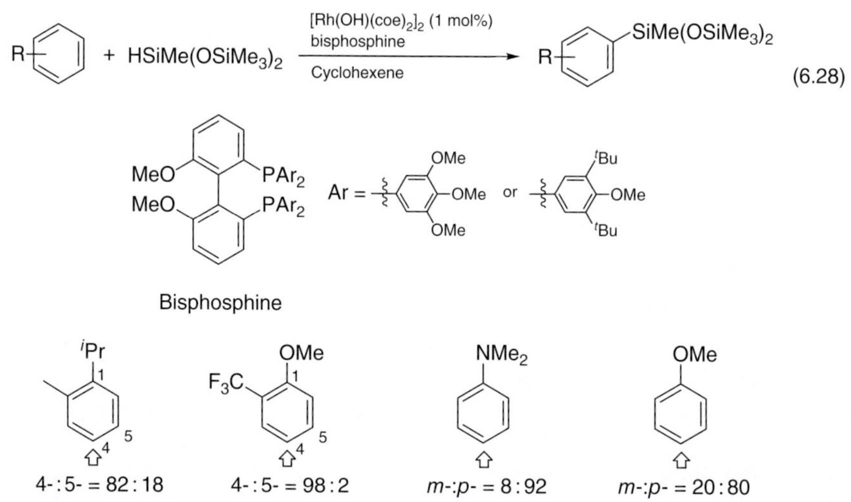

Scheme 6.19 $[Rh(OH)(cod)]_2$/bisphosphine-catalyzed intermolecular C–H silylation of arenes with $HSiMe(OSiMe_3)_2$.

Hartwig and a coworker later reported the $[Ir(OMe)(cod)]_2$/tmphen-catalyzed C–H silylation of arenes with a $HSiMe(OSiMe_3)_2$ (tmphen = 2,4,7-trimethylphenanthroline) [49]. The regioselectivity of reactions with unsymmetrical

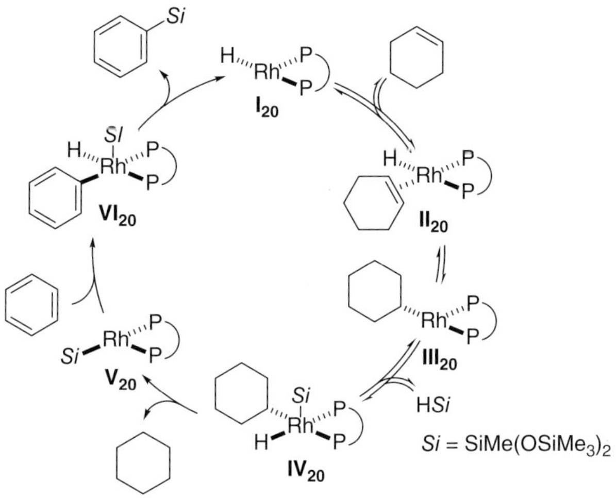

Scheme 6.20 Mechanism for the [Rh(OH)(cod)]$_2$/bisphosphine-catalyzed C(sp^2)–H silylation of arenes with HSiMe(OSiMe$_3$)$_2$.

1,2-disubstituted arenes was lower than that of the [Rh(OH)(coe)$_2$]$_2$/biphosphine-catalyzed reactions. For example, an 80 : 20 4-/5-selectivity was observed when 2-(trifuluoromethyl)anisole was present in the iridium-catalyzed reaction. On the other hand, the ability of the system to tolerate various types of functional groups was much higher. The [Ir(OMe)(cod)]$_2$/tmphen system also catalyzed the reaction of azurenes with HSiEt$_3$, leading to the resioselective production of 2-silylazurens [50]. The [Rh(nbd)$_2$]BF$_4$/(*S*,*S*)-^{i}Pr-bpe catalyst system, reported by Lee et al., was also found to be effective for the dehydrogenative silylation of arenes with HSiEt$_3$ ((*S*,*S*)-^{i}Pr-bpe = 1,2-bis[(2*S*,5*S*)-2,5-diisopropylphospholan-1-yl]ethane) [51].

Heteroarenes were also found to be accessible as substrates for the direct C–H silylation with disilanes or hydrosilanes under iridium catalysis. Ishiyama et al. reported the [Ir(OMe)(cod)]$_2$/tbphen-catalyzed C–H silylation of five-memberd heteroarenes with (^{t}BuF$_2$Si)$_2$ (tbphen = 2-*tert*-butyl-1,10-phenanthroline), which required the use of a large excess of substrate heteroarenes relative to the disilane, in terms of the efficiency of the reaction [52]. While thiophene and furan derivatives were exclusively silylated at the 2-position (Eq. (6.29a)), the reaction of *N*-triisopropylsilyl-substituted pyrrole and indole gave 3-silylated products (Eq. (6.29b)). The regioselective silylation of five-membered heterocycles, including *N*-non-protected and *N*-methyl imidazole derivatives, with an excess of HSiEt$_3$ at the 2-position was also reported by Falck and a coworker when a combination of [Ir(OMe)(cod)]$_2$ and dtbpy was used as the catalyst [53]. The only exception involved the reaction of *N*-tosylindole, which gave the 3-silylated product. The abovementioned catalytic systems, Hartwig's [Ir(OMe)(cod)]$_2$/tmphen [49] and Lee and Choi's [Rh(nbd)$_2$]BF$_4$/(*S*,*S*)-^{i}Pr-bpe [51], were used to catalyze the reaction of five- and six-membered heteroarenes with

$HSiMe(OSiMe_3)_2$ and $HSiEt_3$, respectively. A switch in regioselectivity, caused by the substituent on the nitrogen atom in the pyrrole and indole derivatives, was also observed in these cases. These results can also be explained by the steric repulsion by the bulky substituent on the nitrogen atom in the five-membered nitrogen heterocycles. These results are summarized in Scheme 6.21.

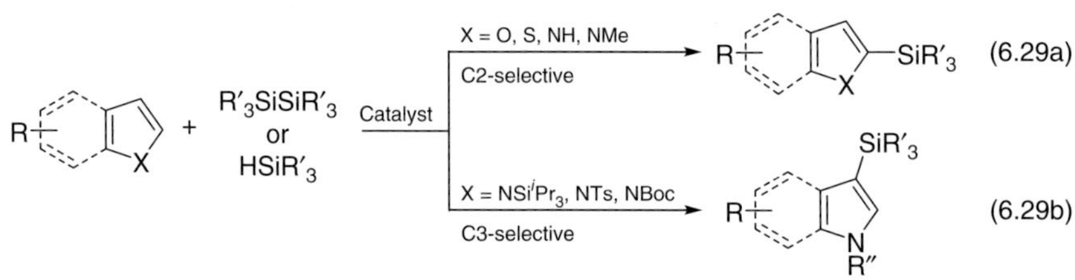

$[Ir(OMe)(cod)]_2$ (1.5 mol%)/tbphen/$(^tBuF_2Si)_2$ (excess heteroarene) [52]

$[Ir(OMe)(cod)]_2$ (5 mol%)/dtbpy/$HSiEt_3$ [53]

$[Ir(OMe)(cod)]_2$ (1.5 mol%)/tmphen/$HSiMe(OSiMe_3)_2$ [49]

$[Rh(nbd)_2]BF_4$ (2 mol%)/(*S*,*S*)-^{i}Pr-bpe/$HSiEt_3$ [51]

(POCOP^{i}Pr)RuH(nbd) (0.05–0.5 mol%)/$HSiMe(OSiMe_3)_2$ or $HSiEt_3$ (X = O, S) [54]

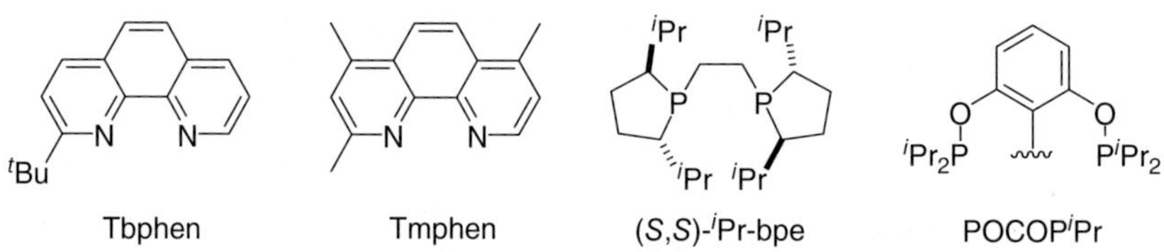

Scheme 6.21 Summary of the C2- or C3-selective C–H silylation of heteroarenes.

Contrary to the abovementioned examples of the C2-selective silylation of *N*-methylindoles, Oestreich and coworkers reported that some ruthenium thiolate complexes, shown in Scheme 6.22, catalyzed the regioselective silylation of *N*-methylindoles at the C3-position (Eq. (6.30), $BAr^F{}_4 = B[3,5\text{-}(CF_3)_2C_6H_3]_4$) [55]. The proposed reaction mechanism involves the formation of an electrophilic silicon species [56] as a key intermediate, which is formed by the heterolytic H—Si bond cleavage by the ruthenium thiolate complex (Scheme 6.23). The silyl group on the sulfur atom in $\mathbf{II_{23}}$ is transferred onto the nucleophilic C3 carbon in the indole to afford a hydridoruthenium complex $\mathbf{III_{23}}$ and an iminium species $\mathbf{IV_{23}}$. Proton abstraction at the C3 position in $\mathbf{IV_{23}}$ by the sulfur atom in $\mathbf{III_{23}}$ then yields the product with concomitant formation of the complex $\mathbf{V_{23}}$. Finally, the release of $\mathbf{H_2}$ resulted in the regeneration of the complex $\mathbf{I_{23}}$. The involvement of the cationic silicon species was supported by an experiment with a chiral silane, in which a racemic product was obtained. Sunada et al. also reported the C3-selective silylation of *N*-methylindoles catalyzed by a bis-η^2-(H–Si)-chelate iron complex [57].

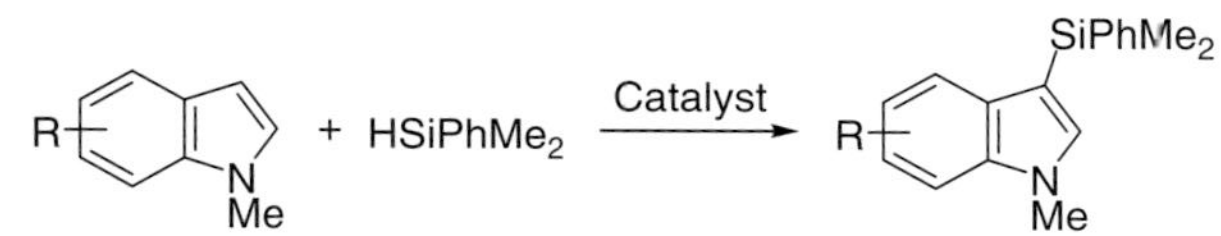

(6.30)

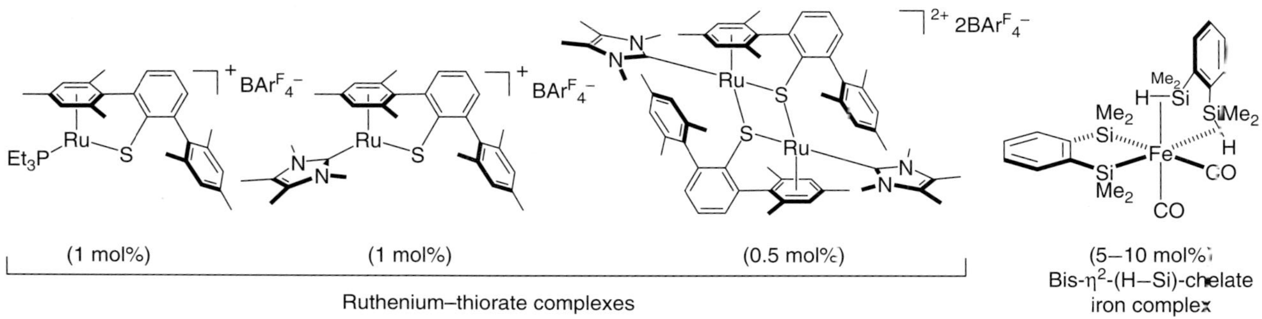

Scheme 6.22 Catalytic C3-silylation of *N*-methylindoles with $HSiPhMe_2$.

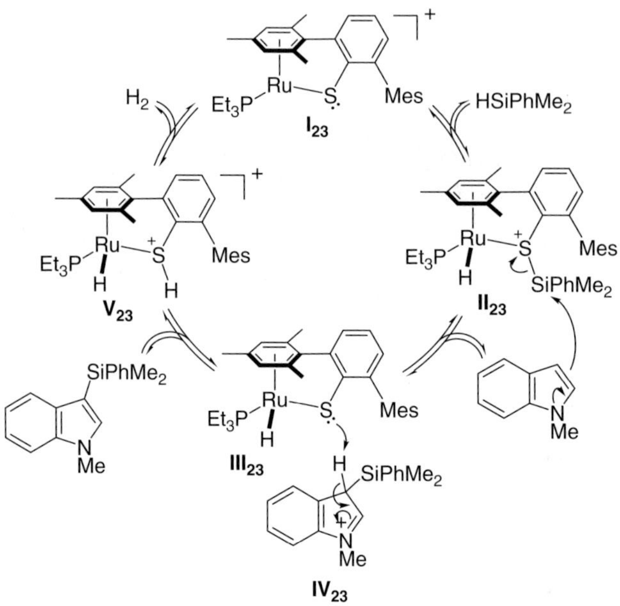

Scheme 6.23 Mechanism for the ruthenium thiolate complex–catalyzed regioselective silylation of *N*-methylindoles at the C3-position with $HSiPhMe_2$.

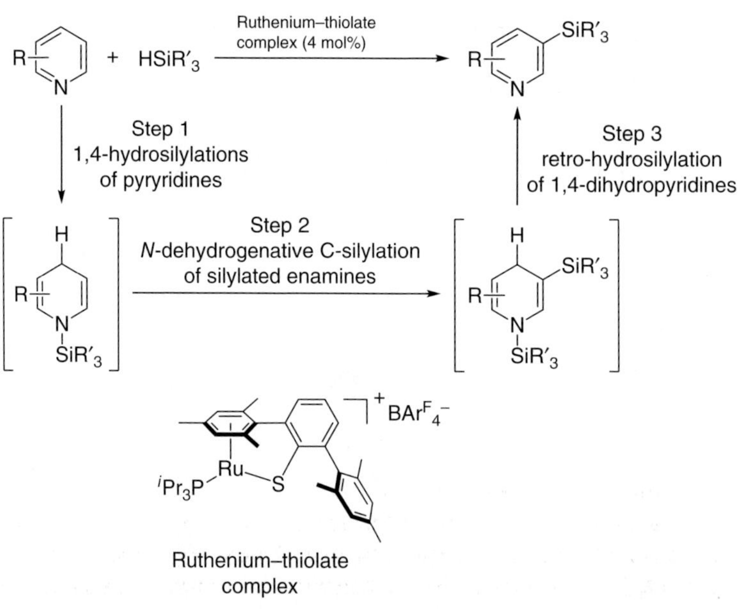

(6.31)

Scheme 6.24 Three-step C–H silylation of pyridines at the 3-position catalyzed by the ruthenium–thiolate complex.

Oestreich and a coworker reported the regioselective C–H silylation of pyridines at the 3-position (Scheme 6.24). The reaction proceeded via a three-step sequence involving the 1,4-hydrosilylation of pyridines (step 1), dehydrogenative C-silylation of *N*-silylated enamines (step 2), and retro-hydrosilylation of 1,3-bis(trimethylsilyl)-1,4-dihydropyridine (step 3) catalyzed by the ruthenium–thiolate complex in all steps [58].

Johnson and a coworker reported the Ni(η^2-$H_2C{=}CHSiMe_3)_2(^i Pr_2Im)$-catalyzed C–H silylation of polyfluoroarenes with trimethylvinylsilane (Eq. (6.32), $^i Pr_2Im$ = 1,3-di(isopropyl)imidazol-2-ylidene) [59]. It was proposed that the key step in the reaction mechanism involved the cleavage of the arene C—H bond and the formation of a new bond between the resulting hydrogen atom and the internal sp^2-carbon in the vinylsilane takes place simultaneously via the transition state $\mathbf{TS_{25(II\rightarrow III)}}$ (Scheme 6.25).

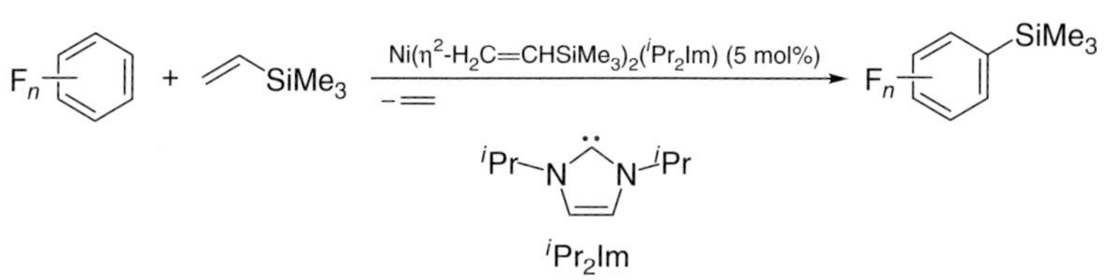

(6.32)

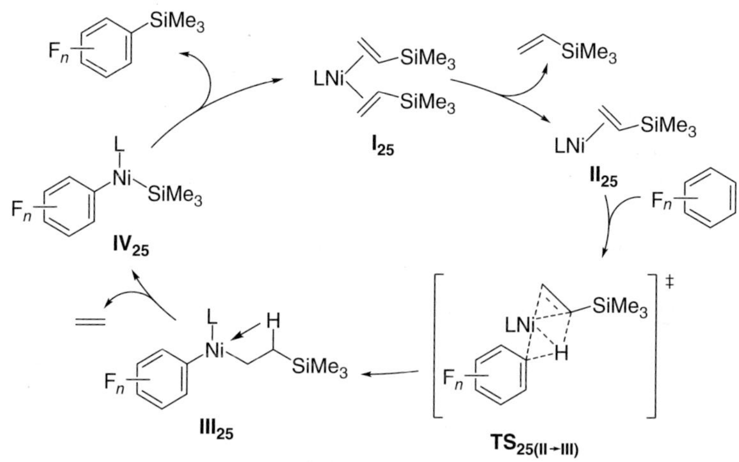

Scheme 6.25 Mechanism for the Ni(η^2-$H_2C{=}CHSiMe_3)_2(^i Pr_2Im)$-catalyzed C–H silylation of polyfluoroarenes with trimethylvinylsilane. All structural formulas to which the double dagger symbol is attached are essential transition state structures to study the C–H functionalization chemistry more deeply.

The hydrosilyl group itself was also found to function as the directing group for the intramolecular C(sp^2)—H bond cleavage at the *ortho* position, providing access to a silametallacycle intermediate [60]. In addition, if the tether in the length of the resulting intermediate is appropriate for the reductive elimination of the silicon and carbon atoms from the intermediate, the silacyclic compound would be produced (Eq. (6.33)). Since the first report involving a series of reactions as catalytic steps, i.e. the $Tp^{Me_2}PtMe_2H$-catalylzed cyclization of ω-silylalkylbenzenes, was described by Hartwig and a coworker [42], a number of related intramolecular C–H silylations have been developed. Substrates tethered by alkoxy [61], siloxy [62], aryl [63], and arylamino [64] groups were amenable for the reaction, the results of which are summarized in Scheme 6.26. The cyclization of *ortho*-biarylsilanes has been particularly studied, because it provides a profitable method for the construction of complex π-conjugated molecules with an incorporated silicon atom such as tetrasila[8]circulenes [63d] and sila[*n*]helicenes [63e].

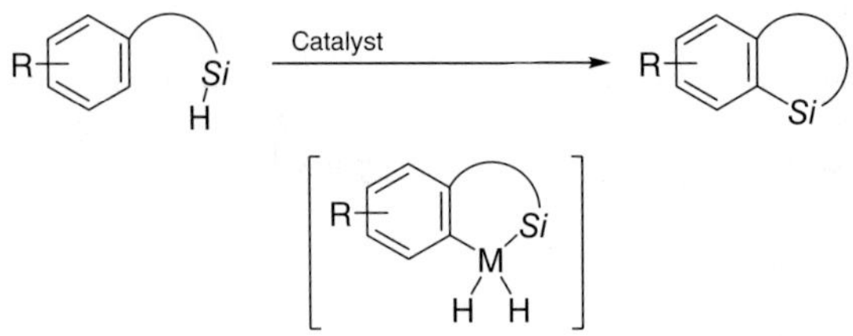

(6.33)

Takai et al. applied the rhodium/(*R*)-BINAP-catalyzed enantioselective intramolecular C–H silylation of bis(biaryl)dihydrosilanes to produce axially chiral spiro-9-silabifluorenes (Eq. (6.34) in Scheme 6.27) [65]. The enantioselective synthesis of planar-chiral benzosilolometallocenes by the rhodium-catalyzed C–H silylation of 2-(hydrosilylaryl)ferrocenes was independently reported by three groups (Eq. (6.35)) [66].

The intermolecular hydrosilylation or dehydrogenative silylation of functional groups on the arene ring with dihydrosilanes to generate silane-tethered arenes containing one Si—H bond, followed by the one-pot intramolecular *ortho* C–H silylation, was employed for the construction of diverse silacyclic scaffolds. Functional groups including aldehydes [61, 67], ketones [61, 67], esters [67–69], alcohols [61], amines [70], and alkenes [71] were amenable for this strategy. A summary of products produced by this two-step sequence is shown in Scheme 6.28.

Ryberg and coworkers extended the two-step strategy to enantioselective reactions starting from diaryl ketones or diarylmethanol derivatives, which involved the desymmetrization of diarylmethoxyhydrosilanes via rhodium-catalyzed asymmetric C–H silylations. Some of the phosphine ligands shown in Scheme 6.29 were found to be effective for the reaction [72]. Hartwig and coworkers later achieved the iridium-catalyzed desymmetrization of symmetric diarylmethoxyhydrosilanes and the kinetic resolution of unsymmetrical substrates in the presence of a chiral pyridinyloxazoline ligand (Eq. (6.38)) [73].

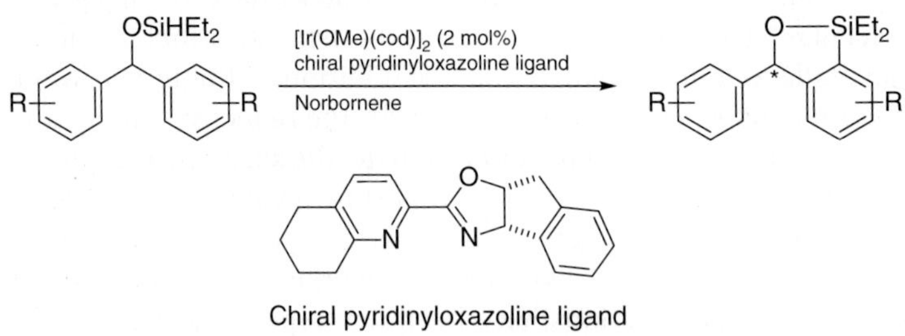

(6.38)

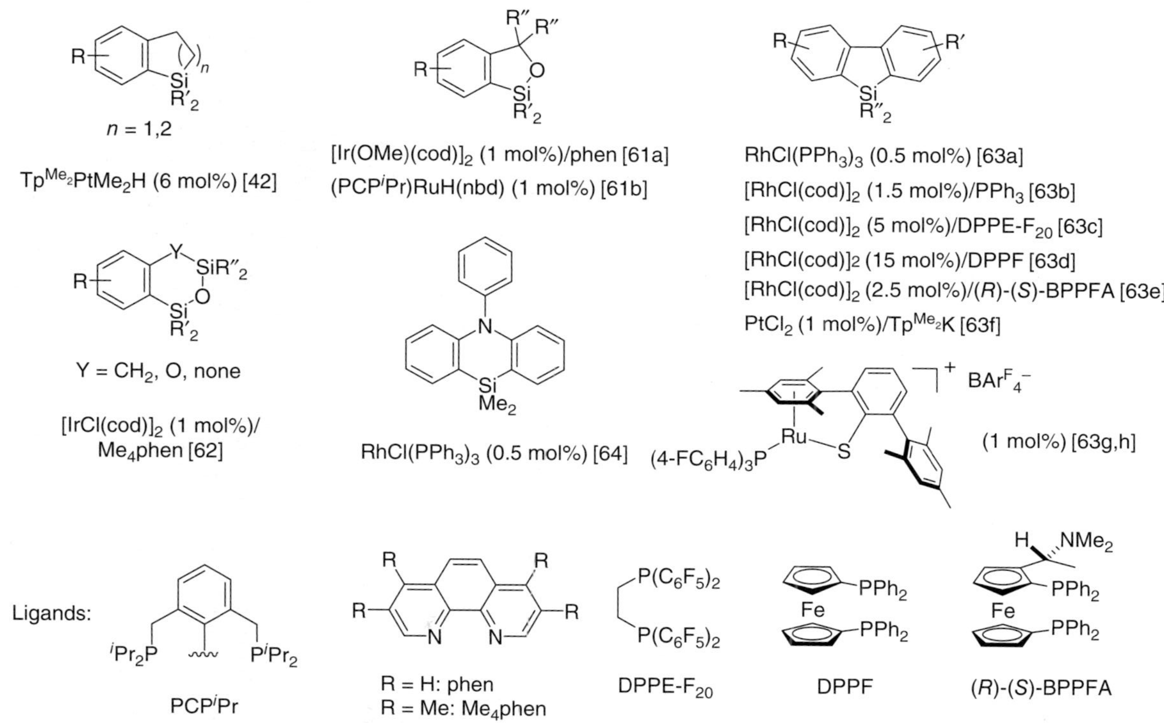

Scheme 6.26 Summary of products produced in catalytic intramolecular $C(sp^2)$–H silylations shown in Eq. (6.33).

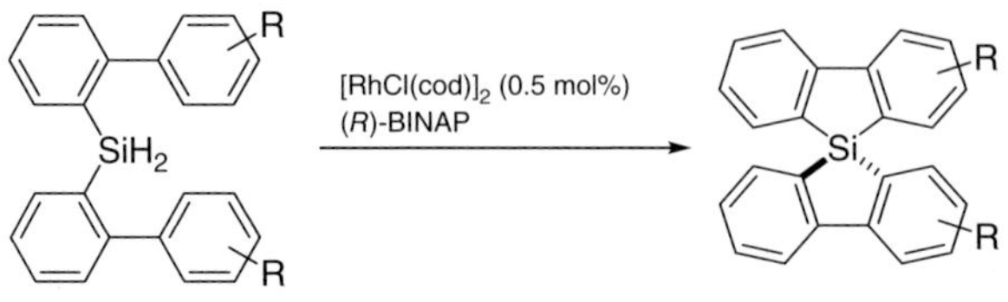

Up to 95% ee

(6.34)

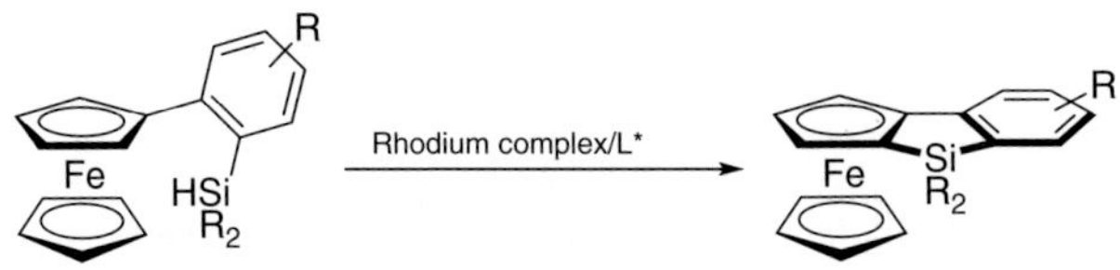

(6.35)

$[RhCl(coe)_2]_2$ (10 mol%)/chiral diene	Up to 86% ee [66a]
$[RhCl(cod)]_2$ (2.5 mol%)/ (R)-DTBM-SEGPHOS	Up to 93% ee [66b]
$[RhCl(cod)]_2$ (2.5 mol%)/ (R)-TMS-SEGPHOS	Up to 97% ee [66c]

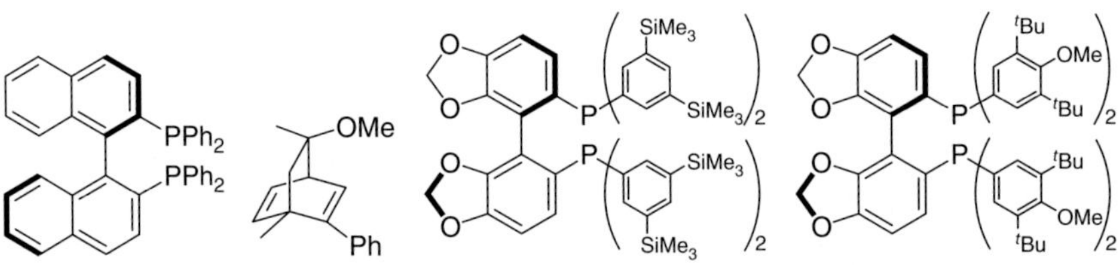

(R)-BINAP Chiral diene (R)-TMS-SEGPHOS (R)-DTBM-SEGPHOS

Scheme 6.27 Enantioselective catalytic intramolecular C(sp^2)–H silylations.

The rhodium-catalyzed intramolecular C(sp^2)–H silylation by biarylsilacyclobutanes leading to the production of dibenzosiloles was reported by He and coworkers (Eq. (6.39)) [74]. The driving force for this reaction is the high ring strain in the silacyclobutane, which can undergo facile oxidative addition to the rhodium center to form the five-membered silametallacycle $\mathbf{II_{30}}$ (Scheme 6.30). β-Hydride elimination followed by the elimination of HCl gives an alkene-coordinated rhodium complex $\mathbf{IV_{30}}$. Cleavage of the C—H bond occurs to afford a six-membered silametallacycle $\mathbf{V_{30}}$, and reductive elimination gives the allylsilole $\mathbf{VI_{30}}$ as an initial product. The insertion of the C=C bond into the Rh—H bond and subsequent protoderhodation produces the final product.

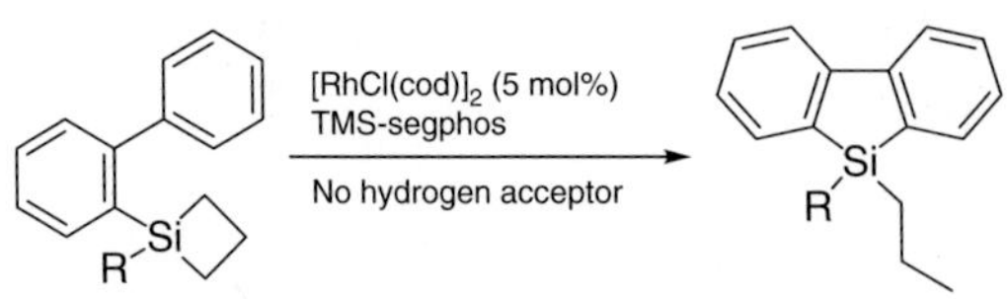

(6.39)

Subsequently, the authors reported the enantioselective synthesis of arylpropylsilole derivatives by the rhodium/(*R*)-TMS-segphos-catalyzed reaction of biarylhydrosilacyclobutanes with arenes or heteroarenes (Scheme 6.31) [75].

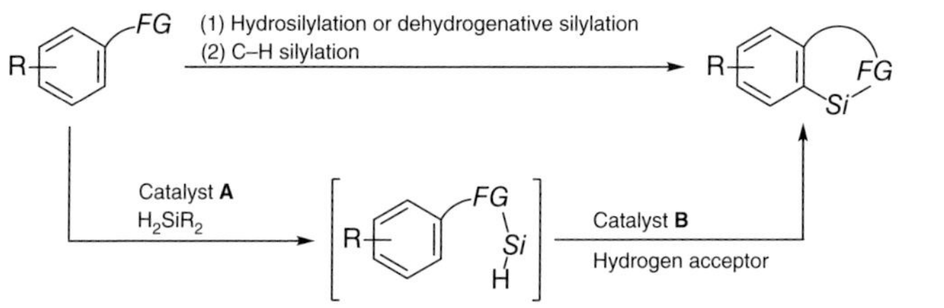

(6.36)

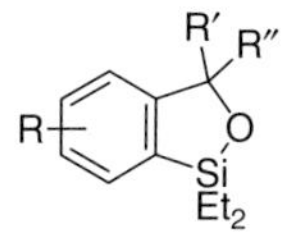

FG = ketones (R″ = H), alcohols [61]

(1) $[Ir(OMe)(cod)]_2$ (0.05 mol%)/H_2SiEt_2
(2) $[Ir(OMe)(cod)]_2$ (1 mol%)/phen

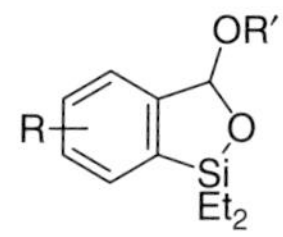

FG = methoxycarbonyl [68]

(1) $[IrCl(coe)_2]_2$ (0.1 mol%)/H_2SiEt_2
(2) $[RhCl(nbd)]_2$ (0.4 mol%)/P(*p*-OMeC_6H_4)$_3$

FG = aldehydes (R′, R″ = H), ketones (R″ = H) [67]

(1) $[RhCl(nbd)]_2$ (0.2–0.4 mol%)/P(*p*-OMeC_6H_4)$_3$
(2) No additional complex

FG = methoxycarbonyl (R′, R″ = H) [67]

(1) $[IrCl(coe)_2]_2$ (0.1 mol%)/$H_2Si^iPr_2$
(2) $[RhCl(nbe)_2]_2$ (0.1 mol%)/P(*p*-OMeC_6H_4)$_3$

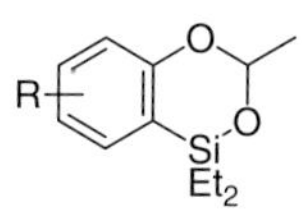

FG = acetoxy [69]

(1) $[IrCl(coe)_2]_2$ (0.4 mol%)/H_2SiEt_2
(2) $[RhCl(nbd)]_2$ (0.1 mol%)/P(*p*-OMeC_6H_4)$_3$

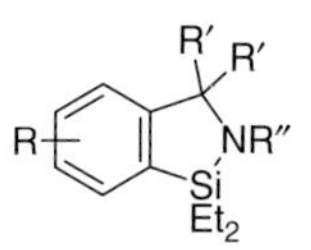

FG = amines [70]

(1) $[Ir(OMe)(cod)]_2$ (0.5 mol%)/H_2SiEt_2
(2) $[Ir(OMe)(cod)]_2$ (1.5 mol%)/Me_4phen

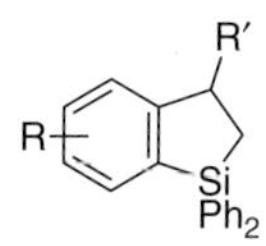

FG = alkenes [71]

(1) $NiBr_2(PPh_3)_2$ (2 mol%)/H_2SiPh_2
(2) $[Ir(OMe)(cod)]_2$ (2 mol%)/dtbpy

Scheme 6.28 Summary of products produced in catalytic two-step C(sp²)–H silylations.

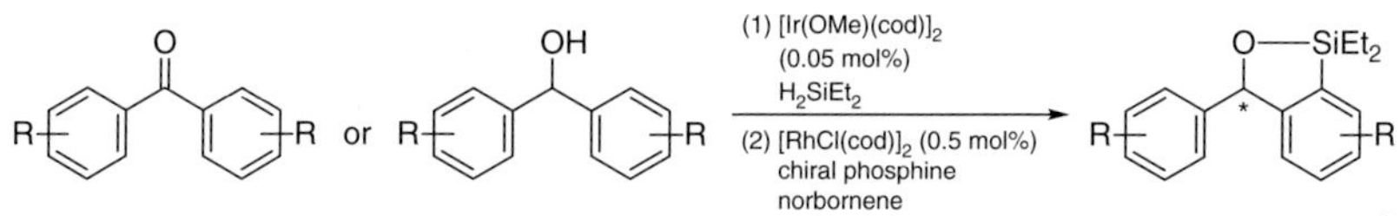

(6.37)

Chiral phosphine ligands

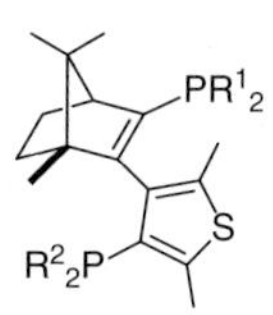

catASium

R^1 = Ph, R^2 = Ph
R^1 = 3,5-$Me_2C_6H_3$, R_2 = Ph

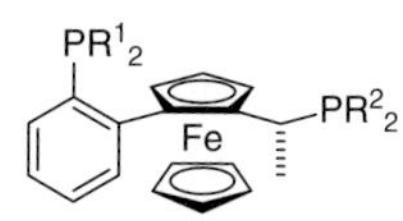

Walphos

R^1 = Ph, R^2 = Cy
R^1 = Ph, R^2 = 2-norbornyl
{ R^1 = 4-OMe-3,5-$Me_2C_6H_3$,
R^2 = 3,5-$(CF_3)_2C_6H_3$

Scheme 6.29 Enantioselective two-step desymmetrization of diarylketones or diarylmethylalcohols.

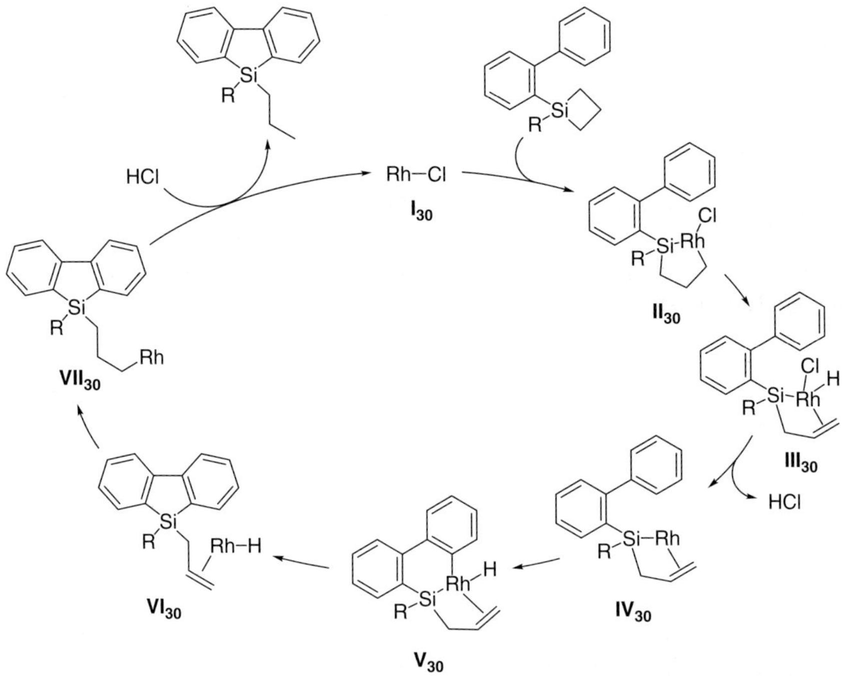

Scheme 6.30 Mechanism for the $[RhCl(cod)]_2$/TMS-segphos-catalyzed intramolecular ring expansion of silacyclobutanes leading to dibenzosiloles.

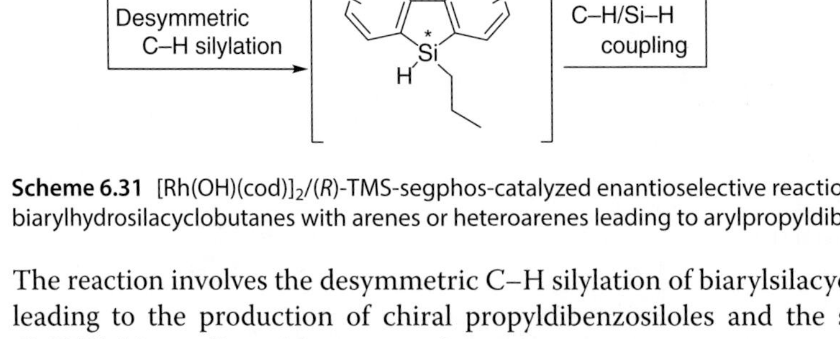

(6.40)

Scheme 6.31 $[Rh(OH)(cod)]_2$/(*R*)-TMS-segphos-catalyzed enantioselective reaction of biarylhydrosilacyclobutanes with arenes or heteroarenes leading to arylpropyldibenzosiloles.

The reaction involves the desymmetric C–H silylation of biarylsilacyclobutanes, leading to the production of chiral propyldibenzosiloles and the subsequent C–H/Si–H coupling with arenes or heteroarenes.

6.3 C(sp^3)—H Bond Silylation

While, as mentioned in the previous section, considerable progress has been made regarding the silylation of C(sp^2)—H bonds, using hydrosilanes, disilanes, or silacyclobutanes as silylating reagents, C(sp^3)—H bond silylation is a less-developed area. The formation of benzylic C(sp^3)–H silylation products were observed as by-products in early studies of C(sp^2)–H silylation, including Tanaka's rhodium/UV-catalyzed reaction with hydrosilanes [15] and platinum-catalyzed reaction with *o*-di(hydrosilyl)benzenes [40], and Ishikawa's nickel-catalyzed reaction with benzodisilacyclobutenes [39], when toluene and xylenes were used as the substrates. Ishikawa et al. reported a nickel-catalyzed C(sp^2)–H silylation in which mesitylene was site-selectively reacted at the benzylic position to afford 3,5-dimethylbenzylsilane, but in low yield (Eq. (6.41)).

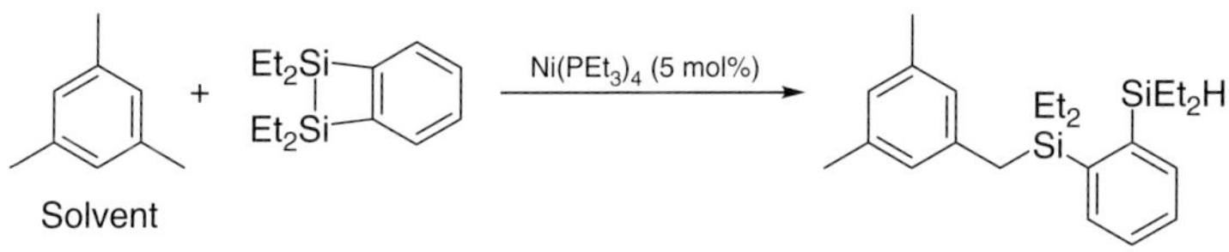

(6.41)

Berry and coworkers reported that the oligomerization of trimethylsilane proceeded in the presence of a ruthenium catalyst, $RuH_3(SiMe_3)(PMe_3)_3$ [76]. They subsequently reported that some rhodium and ruthenium complexes, such as $Cp^*RhH_2(SiEt_3)_2$, $[Cp^*RhCl_2]_2$, (η^6-*p*-cymene)$RuH_2(SiEt_3)_2$, and [(η^6-*p*-cymene)$RuCl_2]_2$, catalyzed the dehydrogenative dimerization of triethylsilane [77]. The formation of a η^2-silene hydride complex was proposed as the key catalytic species in the catalytic cycle, followed by the insertion of a Si=C bond into the M—Si bond and reductive elimination to yield the product (Scheme 6.32).

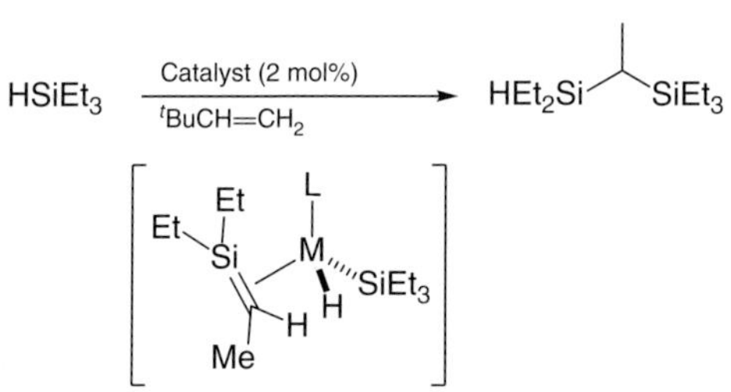

(6.42)

Scheme 6.32 Catalytic dimerization of $HSiEt_3$ via the formation of a η^2-silene hydride complex.

Tilly and coworker reported the intermolecular C–H silylation of methane with hydrosilanes catalyzed by Cp^*_2ScMe [78]. Although the methylated silanes were produced with no detectable formation of other side products by the catalytic system, the system is not altogether suitable for practical use, owing to the slow methane conversion. The reaction was proposed to involve Si—Me bond formation via σ-bond metathesis between the Sc—Me and Si—H bonds in the reaction mechanism (Scheme 6.33).

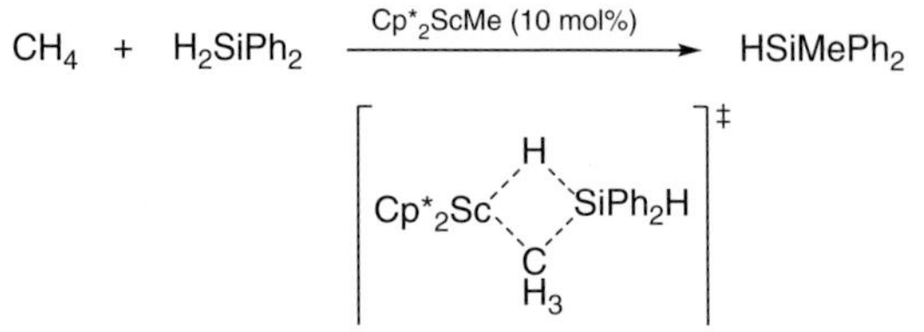

(6.43)

Scheme 6.33 Cp^*_2ScMe-catalyzed C–H silylation of methane with H_2SiPh_2.

Fukumoto and coworkers reported the $Ir_4(CO)_{12}$-catalyzed regioselective silylation of C(sp^3)—H bonds in 4-alkylpyridines at the benzylic position with hydrosilanes (Eq. (6.44)) [79]. The low product yields in the reaction of 2-substituted 4-methylpyridine derivatives were improved markedly by the addition of other pyridine derivatives, such as 3,5-dimethylpyridine, as additives. The coordination of a hydrosilane to an iridium carbonyl complex $\mathbf{I_{34}}$ to form the electrophilic silicon species, either an η^1-silane iridium complex $\mathbf{II_{34}}$ or a σ-silane iridium complex $\mathbf{II_{34}}'$, appears to initially occur in the reaction (Scheme 6.34). The silyl group in $\mathbf{II_{34}}$ migrates to the nitrogen atom in the pyridine substrate to yield a silylpyridinium iridate species. The abstract of a proton at the benzylic position of the *N*-silylpyridinium species by a hydridoiridate species then takes place to afford a dihydridoiridium complex $\mathbf{IV_{34}}$ and the silylenamine $\mathbf{V_{34}}$, the latter of which reacts with either the electrophilic silicon species $\mathbf{II_{34}}$ or the silylpyridinium species $\mathbf{III_{34}}$ to furnish $\mathbf{VI_{34}}$, and the subsequent elimination of the silyl group from the nitrogen in $\mathbf{VI_{34}}$ produces the product. $\mathbf{IV_{34}}$ reacts with norbornene to give norbornane, along with the regeneration of $\mathbf{I_{34}}$. It is thought that the added 3,5-dimethylpyridine functions as a transporter of the silyl group to the 2,4-dimethylpyridine after the formation of the *N*-silyl-3,5-dimethylpyridinium species. The authors revealed that the reaction system could be applied to the C(sp^3)–H silylation of 2-alkylpyridines, when higher reaction temperature was used [80], and the cationic (POCOP^{t}Bu)IrHCl/NaBAr$^F{}_4$ catalytic system was effective for the regioselective C(sp^3)–H silylation of 2-alkyl-1,3-azoles with hydrosilanes [81].

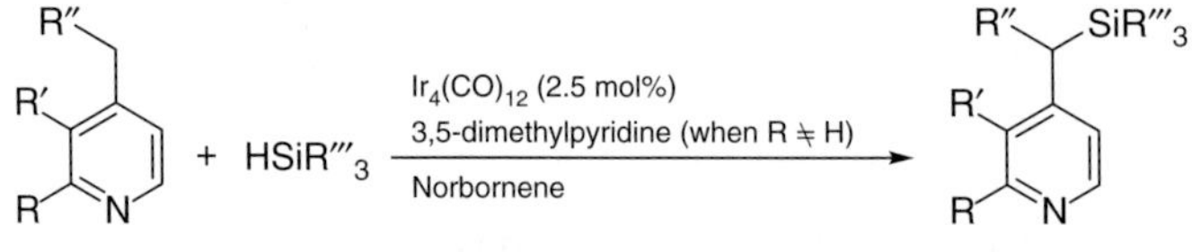

(6.44)

Chelation assistance by directing groups in the transition-metal-catalyzed C—H bond functionalization is also an efficient strategy for the regioselective silylation of C(sp^3)—H bonds. A number of directing groups that have been found to participate in the C(sp^2)–H silylation can also be used in C(sp^3)–H silylations. Kakiuchi et al. demonstrated that a variety of nitrogen-containing functional groups, which are shown in Scheme 6.35, worked as directing groups in the $Ru_3(CO)_{12}$-catalyzed benzylic C(sp^3)–H silylation with hydrosilanes [82]. However, the reaction appears to have only limited applicability and

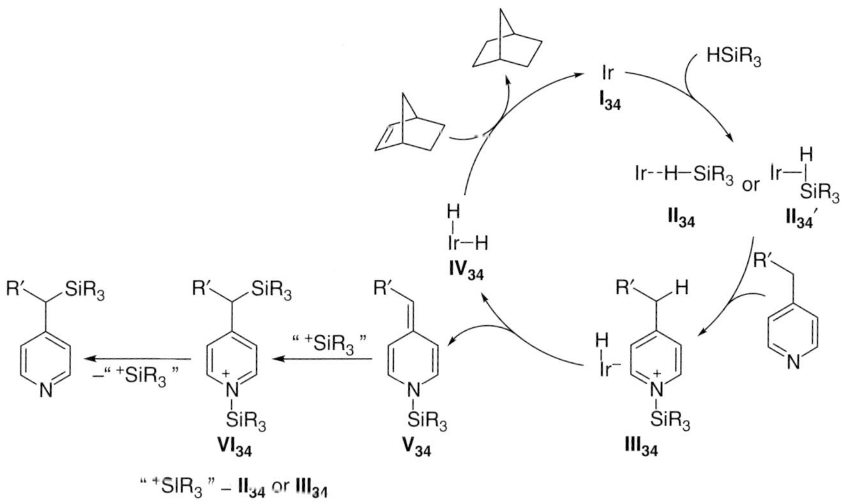

Scheme 6.34 Mechanism for the $Ir_4(CO)_{12}$-catalyzed C(sp^3)–H silylation of 4-alkylpyridines at the benzylic-position.

is restricted to methyl C—H bonds. An aryl C—H bond at the *ortho* position was found to react preferentially prior to the benzylic C—H bond in the case of 2-(2-tolyl)pyridine. Sato et al. reported that $[IrCl(cod)]_2$ was also effective for use in the regioselective silylation of benzylic methyl C—H bonds, which did not require the addition of a hydrogen acceptor [83]. Silylation of the methyl group in 8-methylquinoline was also achieved using the same reaction systems (Eq. (6.46)) [82, 83].

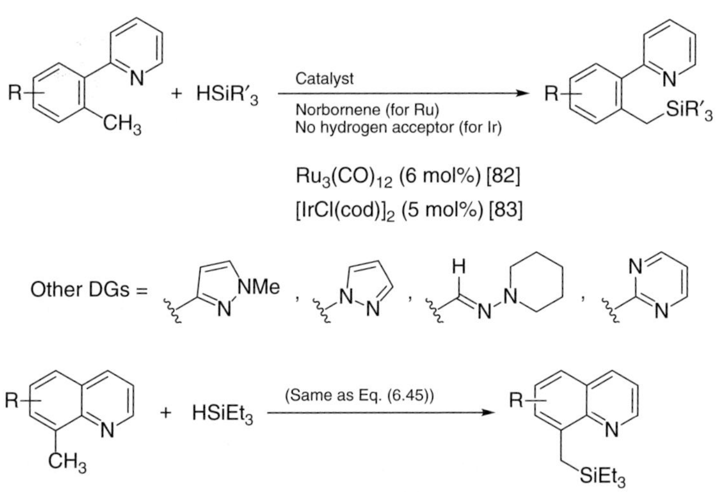

(6.45)

(6.46)

Scheme 6.35 Chelation-assisted benzylic C(sp^3)–H silylations.

Sato et al. reported that the regioselective C(sp^3)–H silylation of 2-ethylpyridine at the β-position in the ethyl group, i.e. the methyl C—H bond (Eq. (6.47)) [84]. The silylation of the C—H bond in the methyl group in 2-(*N,N*-dimethylamino)pyridine also proceeded under the same reaction conditions. The same paper reported that $[RhCl(cod)]_2$ functioned as a more active catalyst for the C–H silylation of 2-(*N,N*-dimethylamino)pyridine and afforded a mixture of mono- and disilylated products. The rhodium complex also catalyzed the reaction of 2-methoxypyridines to produce 2-(silylmethoxy)pyridine. These results are summarized in Scheme 6.36, including the results for other reactions in which $Ru_3(CO)_{12}$ [85], $[Cp^*RuCl]_4$ [86], and $[IrCl(cod)]_2/NEt^iPr_2$ [26] were used as catalysts. 2-Oxazolyl and 1-pyrazolyl groups were also available as directing groups in the $[Cp^*RuCl]_4$-catalyzed reaction [86]. All of the reactions described represent the dehydrogenative silylation of a methyl group, and the only exception involves that of 2-(*N*-benzylamino)-3-methylpyridine at the benzylic position leading to the desired product, but in low yield [84].

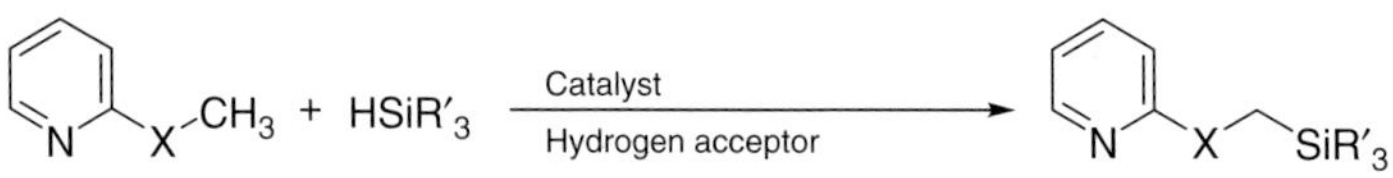

(6.47)

$[Ir(OMe)(cod)]_2$ (5 mol%) [84] (*no hydrogen acceptor*)	X = *C*, *N*
$[RhCl(cod)]_2$ (5 mol%) [84]	X = *N*, *O*
$Ru_3(CO)_{12}$ (0.67 mol%) [85]	X = *C*
$[Cp^*RuCl]_4$ (1.25 mol%) [86]	X = *C*, *N*
$[IrCl(cod)]_2$ (2 mol%)/$^iPr_2EtN$ [26]	X = *C*, *N*

Scheme 6.36 Chelation-assisted C(sp^3)–H silylation of 2-substituted pyridines at the β-position.

Suginome and coworkers reported the ruthenium-catalyzed C–H silylation of the methyl group in diaminomethylboranes shown in Scheme 6.37 [87]. In the case of ethyl-substituted borane, both α-CH_2 and β-CH_3 bonds were silylated with no regioselectivity being detected.

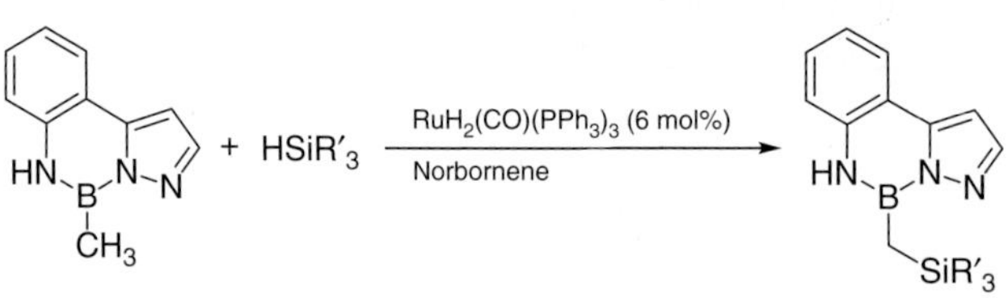

(6.48)

Scheme 6.37 $RuH_2(CO)(PPh_3)_2$-catalyzed C(sp^3)–H silylation of diaminomethylboranes.

$Pd(OAc)_2$-catalyzed 8-aminoquinoline-directed regioselective C–H silylation with hexamethyldisilane could be applied to a C(sp^3)—H bond at the β-position in alkylcarboxamide derivatives (Scheme 6.38). In earlier work by Kanai and

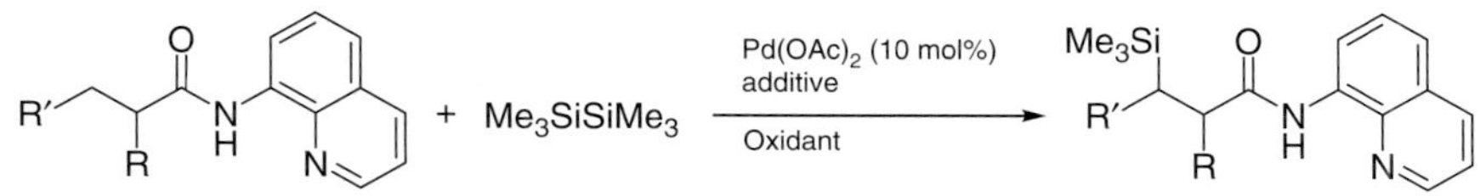

(6.49)

Additive/oxidant

N-Boc-Val-OH/Ag_2CO_3 [88]

None/Ag_2CO_3 and 2,6-dimethoxy-1,4-benzoquinone [88]

(*S*)-BINA-PO_2H/$AgCO_3$ [88]

None/1,4-benzoquinone [89]

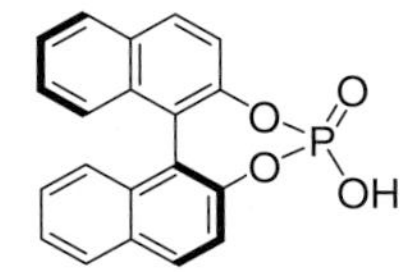

(*S*)-BINA-PO_2H

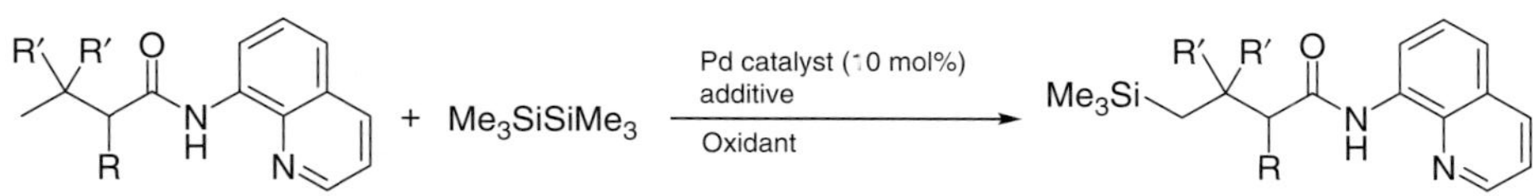

(6.50)

Pd catalyst/additive/oxidant

$Pd(OAc)_2$/none/1,4-benzoquinone [89]

$Pd(OPiv)_2$/2-chloroquinoline/Ag_2CO_3 [90]

Scheme 6.38 Palladium-catalyzed regioselective C(sp^3)–H silylations of *N*-(8-quinolyl)carboxamides with hexamethyldisilane at the β- and γ-positions.

coworkers, the silylation of methylene groups, such as cyclohexyl, cyclopentyl, and β-phenethyl groups, were successful, but product yields were moderate to poor [19]. Shi's group [88] and Zhang's group [89] independently reported that methyl groups could be efficiently silylated when the reaction conditions were appropriately modified (Eq. (6.49)). However, in these cases, the yields of products in which the methylene groups in simple alkyl carboxamides were silylated at the β-position were improved but the improvement was not substantial. Shi and coworkers reported that phenylalanine derivatives bearing substituents on the phenyl group were diastereoselectively silylated at the β-position [88]. Interestingly, Zhang and coworkers found that the γ-methyl group-selective silylation proceeded when L-valine and L-isoleucine derivatives was used as substrates (Eq. (6.50)) [89]. The same type of γ-C–H silylation of β-substituted and β,β-disubstituted butyric acid derivatives catalyzed by a combination of $Pd(OPiv)_2$ and 2-chloroquinoline was also reported by Sunoj and coworkers [90].

As is the case of $C(sp^2)$—H silylation, the intramolecular silylation of $C(sp^3)$—H bonds has also been developed. These reactions can be classified as the single reaction of isolated silane substrates and two-step reactions via the in situ generation of silane substrates. The products produced in the former reaction are summarized in Scheme 6.39. Hartwig and coworkers showed one example of the Pt-catalyzed C–H silylation of tributylsilane to yield dibutylsilacyclopentane [42]. Hartwig's group then reported the $[Ir(OMe)(cod)]_2/Me_4phen$-catalyzed intramolecular silylation of methyl or methylene C—H bonds located γ to the oxygen atom in the alkoxy group of alkoxysilanes to produce five-membered oxasilacycles [91]. The reaction of the methylene C—H bond required more severe reaction conditions, such as the use of a higher reaction temperature of 120 °C and a much heavier catalyst loading of 4 mol% [91b], compared with that of methyl C—H bonds at 80–100 °C and 0.5–1 mol% catalyst loading [91a]. Gevorgyan and coworkers reported the $[Ir(OMe)(cod)]_2$-catalyzed cyclization of *tert*-butyl(pyridin-2-yl)methylalkylsilane leading to the production of silacyclopentane derivatives [92]. The fact that the reaction of the benzyl analog of the substrate did not give the cyclization product at all indicates that the (pyridin-2-yl) group is essential for the reaction to proceed, probably due to the stabilization of the key intermediate by chelation with silicon and pyridine nitrogen atoms to the iridium center. Takai and coworkers reported that rhodium complexes also functioned as catalysts for the $C(sp^3)$–H silylation [63a, 93], and among them, the combination of $[RhCl(cod)]_2$ and DTBM-segphos were the best catalyst system for the dehydrogenative silylation of *o*-alkylphenylsilanes leading to the formation of dihydrobenzosiloles [93b]. Huang and coworkers extended this type of reaction to other substrates bearing methoxy, dimethylamino, trimethylsilyl, and trimethylgermyl groups by using a pincer-ruthenium complex as the catalyst [94]. Hartwig and coworkers reported that six-membered silacycles could be constructed by the rhodium/xantophos-catalyzed C–H silylation at a primary C—H bond δ to an oxygen atom [95]. Their computational studies indicated that the barrier for the reductive elimination from the seven-membered metallacycle to afford the desired product is lower than that for the six-membered metallacycle formed by cleavage of the methylene C—H bond located γ to the oxygen atom.

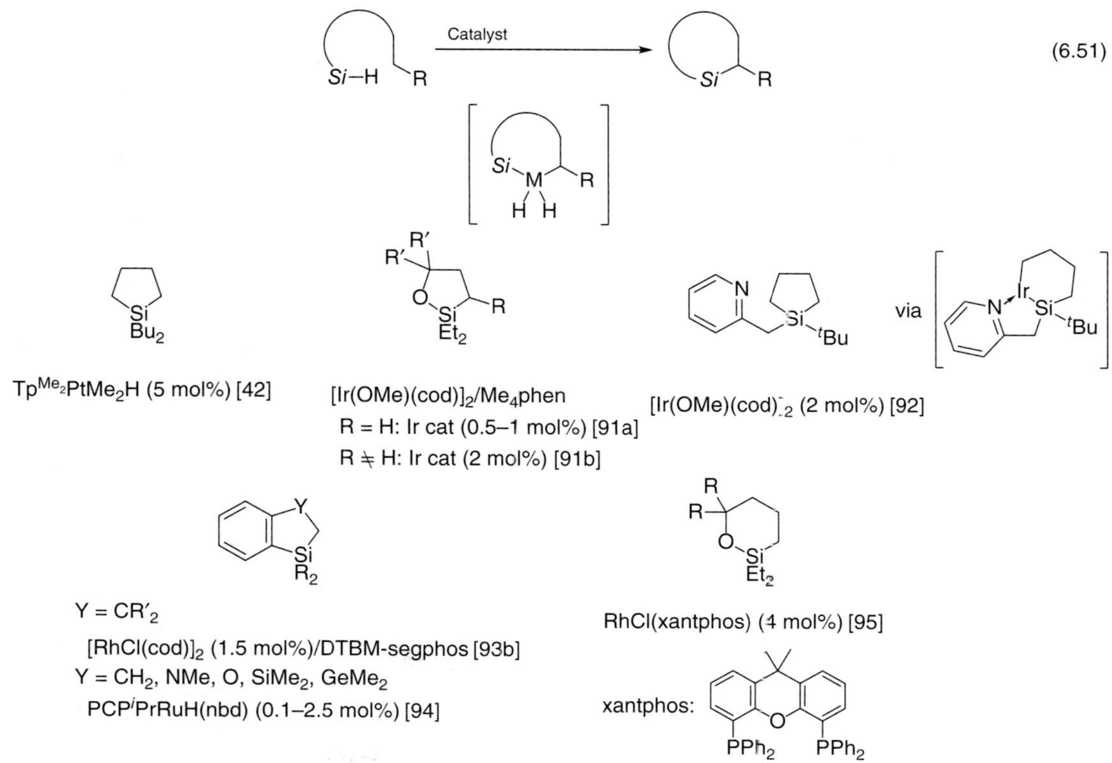

(6.51)

Scheme 6.39 Summary of products produced in catalytic intramolecular C(sp^3)–H silylations.

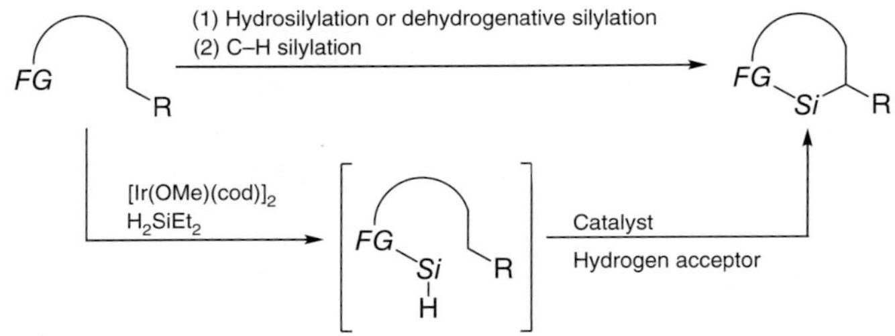

(6.52)

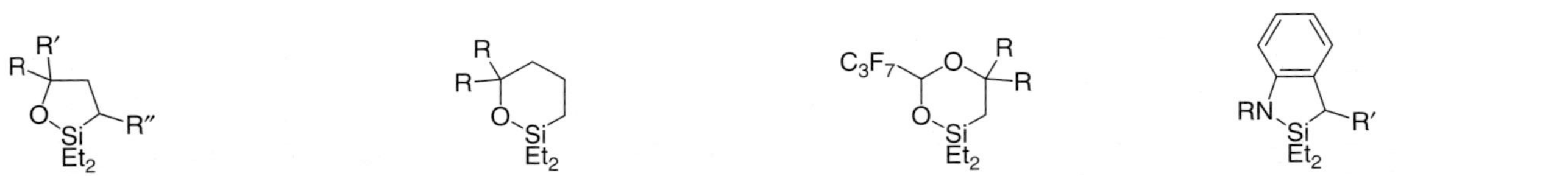

FG = ketones (R = H) or alcohols
$R'' = H$ [91a]
(1) Ir catalyst (0.05 mol%)
(2) $[Ir(OMe)(cod)]_2$ (0.5 mol%)/ Me_4phen

$R'' \neq H$ [91b]
(1) Ir catalyst (0.05 mol%)
(2) $[Ir(OMe)(cod)]_2$ (2 mol%)/ Me_4phen

FG = alcohols [95]
(1) Ir catalyst (0.2 mol%)
(2) RhCl(xantphos) (4 mol%)

FG = perfluoropropanoyl [96]
(1) Ir catalyst (1 mol%)
(2) $[Ir(OMe)(cod)]_2$ (2 mol%)/ Me_4phen

FG = amines [70]
(1) Ir catalyst (1 mol%)
(2) $[Ir(OMe)(cod)]_2$ (1.5 mol%) Me_4phen

Scheme 6.40 Summary of products produced in catalytic two-step C(sp^3)–H silylations.

Reactions leading to the formation of oxasilacyclic compounds from alkyl silyl ethers, as shown in Scheme 6.39, were extended to the two-step procedure using alcohols or ketones as starting materials (Scheme 6.40) [91]. The primary C–H silylation, which occurs β to the oxygen atom in alkyl esters of perfluorobutyric acid, was achieved by the catalytic hydrosilylation of a C=O bond with H_2SiEt_2 and subsequent catalytic regioselective silylation of the C—H bond [96]. The reason for the choice of a perfluoropropyl group as the directing group is due to its contribution to the stabilization of an acetal, formed by the initial hydrosilylation, by the electron-withdrawing nature of the perfluoropropyl group. An amino group on the aryl ring could also be applied to the strategy, in which the C—H bond at the benzylic position was silylated, leading to the production of dihydrobenzoazasilole derivatives [70].

Two types of enantioselective intramolecular C(sp^3)–H silylations have been reported by Hartwig's group (Scheme 6.41). Desymmetrization of an 1,1-dimethylalkyl groups at the *ortho* position in arylhydrosilanes to yield chiral dihydrobenzsiloles was catalyzed by the combination of $[Ir(OMe)(cod)]_2$ and the chiral pyridinyloxazoline ligand (Eq. (6.53)) [97], the latter of which was also used for enantioselective C(sp^2)–H silylation [73]. The enantioselective silylation of cyclopropyl C—H bonds in the cyclopropylmethyl silyl ether catalyzed by $[RhCl(cod)]_2$/(*S*)-DTBM-segphos to produce the chiral 3-oxabicyclo[3.1.0]hexanes was attained (Eq. (6.54)) [98].

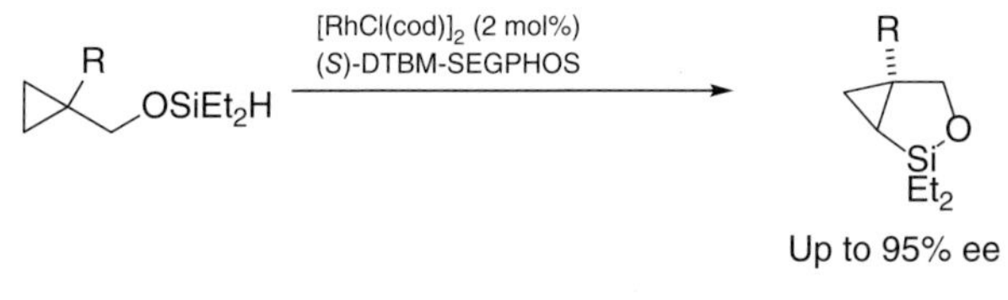

(6.53)

(6.54)

Scheme 6.41 Enantioselective catalytic intramolecular C(sp^3)–H silylations.

References

1 Books on the transition metal-catalyzed hydrosilylation of alkynes, see: (a)Marciniec, B. (ed.) (1992). *Comprehensive Handbook on Hydrosilylation.* Oxford: Pergamon.(b) Marsiniec, B. (ed.) (2009). *Hydrosilylation: A Comprehensive Review on Recent Advances.* Berlin: Springer-Verlag.

2 Dickers, H.M., Haszeldine, R.N., Mather, A.P., and Parish, R.V. (1978). *J. Organomet. Chem.* 161: 91–95.

3 Voronkov, M.G., Druzhdzh, P.V., and Pukhnarevich, V.B. (1980). *Zh. Obshch. Khim.* 50: 1662.
4 Voronkov, M.G., Ushakova, N.I., Tsykhanskaya, I.I., and Pukhnarevich, V.B. (1984). *J. Organomet. Chem.* 264: 39–48.
5 Liu, H.Q. and Harrod, J.F. (1990). *Can. J. Chem.* 68: 1100–1105.
6 Takaki, K., Kurioka, M., Kamata, T. et al. (1998). *J. Org. Chem.* 63: 9265–9269.
7 Shimizu, R. and Fuchikami, T. (2000). *Tetrahedron Lett.* 41: 907–910.
8 Fernandez, M.J., Oro, L.A., and Manzano, B.R. (1988). *J. Mol. Catal.* 45: 7–15.
9 Yamaguchi, K., Wang, Y., Oishi, T. et al. (2013). *Angew. Chem. Int. Ed.* 52: 5627–5630.
10 Jun, C.-H. and Crabtree, R.H. (1993). *J. Organomet. Chem.* 447: 177–187.
11 Marciniec, B., Dudziec, B., and Kownacki, I. (2006). *Angew. Chem. Int. Ed.* 45: 8180–8184.
12 Marciniec, B. (2007). *Acc. Chem. Res.* 40: 943–952.
13 (a) Kownacki, I., Marciniec, B., Dudziec, B., and Kubicki, M. (2011). *Organometallics* 30: 2539–2545. (b) Kownacki, I., Orwat, B., Marciniec, B., and Kownacka, A. (2014). *Tetrahedron Lett.* 55: 548–550.
14 Gustavson, W.A., Epstein, P.S., and Curtis, M.D. (1982). *Organometallics* 1: 884–885.
15 Sakakura, T., Tokunaga, Y., Sodeyama, T., and Tanaka, M. (1987). *Chem. Lett.* 16: 2375–2378.
16 Murai, S., Kakiuchi, F., Sekine, S. et al. (1993). *Nature* 366: 529–531.
17 (a) Williams, N.A., Uchimaru, Y., and Tanaka, M. (1995). *J. Chem. Soc., Chem. Commun.* 1129–1130. (b) Williams, N.A., Uchimaru, Y., and Tanaka, M. (2003). *Dalton Trans.* 236–243.
18 Tobisu, M., Ano, Y., and Chatani, N. (2008). *Chem. Asian J.* 3: 1585–1591.
19 Kanyiva, K.S., Kuninobu, Y., and Kanai, M. (2014). *Org. Lett.* 16: 1968–1971.
20 For reviews on bidentate chelation-assisted C—H functionalizations, see: (a) Daugulis, O., Roane, J., and Tran, L.D. (2015). *Acc. Chem. Res.* 48: 1053–1064. (b) Chatani, N. (2018). *Bull. Chem. Soc. Jpn.* 91: 211–222. and references cited therein.
21 Chen, C., Guan, M., Zhang, J. et al. (2015). *Org. Lett.* 17: 3646–3649.
22 Deb, M., Singh, J., Mallik, S. et al. (2017). *New J. Chem.* 41: 14528–14538.
23 Kakiuchi, F., Igi, K., Matsumoto, M. et al. (2001). *Chem. Lett.* 30: 422–423.
24 Kakiuchi, F., Matsumoto, M., Tsuchiya, K. et al. (2003). *J. Organomet. Chem.* 686: 134–144.
25 Kakiuchi, F., Igi, K., Matsumoto, M. et al. (2002). *Chem. Lett.* 31: 396–397.
26 Wang, H., Wang, G., and Li, P. (2017). *Org. Chem. Front.* 4: 1943–1946.
27 (a) Ihara, H. and Suginome, M. (2009). *J. Am. Chem. Soc.* 131: 7502–7503. (b) Ihara, H., Koyanagi, M., and Suginome, M. (2011). *Org. Lett.* 13: 2662–2665. (c) Koyanagi, M., Eichenauer, N., Ihara, H. et al. (2013). *Chem. Lett.* 42: 541–543.
28 Choi, G., Tsurugi, H., and Mashima, K. (2013). *J. Am. Chem. Soc.* 135: 13149–13161.
29 A review on the σ-CAM, see: Perutz, R.N. and Sabo-Etienne, S. (2007). *Angew. Chem. Int. Ed.* 46: 2578–2592.
30 Rubio-Pérez, L., Iglesias, M., Munárriz, J. et al. (2017). *Chem. Sci.* 8: 4811–4822.
31 Wakaki, T., Kanai, M., and Kuninobu, Y. (2015). *Org. Lett.* 17: 1758–1761.

32 Sakurai, T., Matsuoka, Y., Hanataka, T. et al. (2012). *Chem. Lett.* 41: 374–376.
33 Oyamada, J., Nishiura, M., and Hou, Z. (2011). *Angew. Chem. Int. Ed.* 50: 10720–10723.
34 (a) Komuro, T., Kitano, T., Yamahira, N. et al. (2016). *Organometallics* 35: 1209–1217. (b) Kitano, T., Komuro, T., Ono, R., and Tobita, H. (2017). *Organometallics* 36: 2710–2713.
35 Kakiuchi, F., Matsumoto, M., Sonoda, M. et al. (2000). *Chem. Lett.* 29: 750–751.
36 Modak, A., Patra, T., Chowdhury, R. et al. (2017). *Organometallics* 36: 2418–2423.
37 Maji, A., Guin, S., Feng, S. et al. (2017). *Angew. Chem. Int. Ed.* 56: 14903–14907.
38 Pan, J.-L., Chen, C., Ma, Z.-G. et al. (2017). *Org. Lett.* 19: 5216–5219.
39 (a) Ishikawa, M., Sakamoto, H., Okazaki, S., and Naka, A. (1992). *J. Organomet. Chem.* 439: 19–21. (b) Ishikawa, M., Okazaki, S., Naka, A., and Sakamoto, H. (1992). *Organometallics* 11: 4135–4139.
40 Uchimaru, Y., El Sayed, A.M.M., and Tanaka, M. (1993). *Organometallics* 12: 2065–2069.
41 Ezbiansky, K., Djurovich, P.I., LaForest, M. et al. (1998). *Organometallics* 17: 1455–1447.
42 Tsukada, N. and Hartwig, J.F. (2005). *J. Am. Chem. Soc.* 127: 5022–5023.
43 Murata, M., Fukuyama, N., Wada, J.-I. et al. (2007). *Chem. Lett.* 36: 910–911.
44 Ishiyama, T., Takagi, J., Ishida, K. et al. (2002). *J. Am. Chem. Soc.* 124: 390–391.
45 Ishiyama, T., Sato, K., Nishio, Y., and Miyaura, N. (2003). *Angew. Chem. Int. Ed.* 42: 5346–5348.
46 Saiki, T., Nishio, Y., Ishiyama, T., and Miyaura, N. (2006). *Organometallics* 25: 6068–6073.
47 Ishiyama, T., Saiki, T., Kishida, E. et al. (2013). *Org. Biomol. Chem.* 11: 8162–8165.
48 (a) Cheng, C. and Hartwig, J.F. (2014). *Science* 343: 853–857. (b) Cheng, C. and Hartwig, J.F. (2014). *J. Am. Chem. Soc.* 136: 12064–12072.
49 Cheng, C. and Hartwig, J.F. (2015). *J. Am. Chem. Soc.* 137: 592–595.
50 Murai, M., Takami, K., Takeshima, H., and Takai, K. (2015). *Org. Lett.* 17: 1798–1801.
51 Lee, K.-S., Katsoulis, D., and Choi, J. (2016). *ACS Catal.* 6: 1493–1496.
52 Ishiyama, T., Sato, K., Nishio, Y. et al. (2005). *Chem. Commun.* 5065–5067.
53 Lu, B. and Falck, J.R. (2008). *Angew. Chem. Int. Ed.* 47: 7508–7510.
54 Fang, H., Guo, L., Zhang, Y. et al. (2016). *Org. Lett.* 18: 5624–5627.
55 (a) Klare, H.F.T., Oestreich, M., Ito, J.-I. et al. (2011). *J. Am. Chem. Soc.* 133: 3312–3315. (b) Bähr, S., Simonneau, A., Irran, E., and Oestreich, M. (2016). *Organometallics* 35: 925–928.
56 For reviews on the catalytic reactions with hydrosilanes involving electrophilic silicon species, see: (a) Robert, T. and Oestreich, M. (2013). *Angew. Chem. Int. Ed.* 52: 5216–5218. (b) Bähr, S. and Oestreich, M. (2017). *Angew. Chem. Int. Ed.* 56: 52–59. (c) Lipke, M.C., Liberman-Martin, A.L., and Tilley, T.D. (2017). *Angew. Chem. Int. Ed.* 56: 2260–2294.
57 Sunada, Y., Soejima, H., and Nagashima, H. (2014). *Organometallics* 33: 5936–5939.
58 Wübbolt, S. and Oestreich, M. (2015). *Angew. Chem. Int. Ed.* 54: 15876–15879.

59 Elsby, M.R. and Johnson, S.A. (2017). *J. Am. Chem. Soc.* 139: 9401–9407.

60 (a) Boebel, T.A. and Hartwig, J.F. (2008). *J. Am. Chem. Soc.* 130: 7534–7535. (b) Robbins, D.W., Boebel, T.A., and Hartwig, J.F. (2010). *J. Am. Chem. Soc.* 132: 4068–4069. (c) Cho, S.H. and Hartwig, J.F. (2013). *J. Am. Chem. Soc.* 135: 8157–8160. (d) Cho, S.H. and Hartwig, J.F. (2014). *Chem. Sci.* 5: 694–698. (e) Larsen, M.A., Cho, S.H., and Hartwig, J.F. (2016). *J. Am. Chem. Soc.* 138: 762–765.

61 (a) Simmons, E.M. and Hartwig, J.F. (2010). *J. Am. Chem. Soc.* 132: 17092–17905. (b) Fang, H., He, Q., Liu, G., and Huang, Z. (2017). *Synlett* 28: 2468–2472.

62 Lin, Y., Jiang, K.-Z., Cao, J. et al. (2017). *Adv. Synth. Catal.* 359: 2247–2252.

63 (a) Ureshino, T., Yoshida, T., Kuninobu, Y., and Takai, K. (2010). *J. Am. Chem. Soc.* 132: 14324–14326. (b) Murai, M., Okada, R., Asako, S., and Takai, K. (2017). *Chem. Eur. J.* 23: 10861–10870. (c) Mitsudo, K., Tanaka, S., Isobuchi, R. et al. (2017). *Org. Lett.* 19: 2564–2567. (d) Serizawa, Y., Akahori, S., Kato, S. et al. (2017). *Chem. Eur. J.* 23: 6948–6952. (e) Murai, M., Okada, R., Nishiyama, A., and Takai, K. (2016). *Org. Lett.* 18: 4380–4383. (f) Murata, M., Takizawa, M., Sasaki, H. et al. (2016). *Chem. Lett.* 45: 857–859. (g) Omann, L. and Oestreich, M. (2015). *Angew. Chem. Int. Ed.* 54: 10276–10279. (h) Omann, L. and Oestreich, M. (2017). *Organometallics* 36: 767–776.

64 Li, H., Wang, Y., Yuan, K. et al. (2014). *Chem. Commun.* 50: 15760–15763.

65 (a) Kuninobu, Y., Yamauchi, K., Tamura, N. et al. (2013). *Angew. Chem. Int. Ed.* 52: 1520–1522. (b) Murai, M., Takeuchi, Y., Yamauchi, K. et al. (2016). *Chem. Eur. J.* 22: 6048–6058.

66 (a) Shibata, T., Shizuno, T., and Sasaki, T. (2015). *Chem. Commun.* 51: 7802–7804. (b) Murai, M., Matsumoto, K., Takeuchi, Y., and Takai, K. (2015). *Org. Lett.* 17: 3102–3105. (c) Zhang, Q.-W., An, K., Liu, L.-C. et al. (2015). *Angew. Chem. Int. Ed.* 54: 6918–6921.

67 Hua, Y., Jung, S., Roh, J., and Jeon, J. (2015). *J. Org. Chem.* 80: 4661–4671.

68 Hua, Y., Asgari, P., Dakarapu, U.S., and Jeon, J. (2015). *Chem. Commun.* 51: 3778–3781.

69 (a) Hua, Y., Asgari, P., Avullala, T., and Jeon, J. (2016). *J. Am. Chem. Soc.* 138: 7982–7991. (b) Asgari, P., Dakarapu, U.S., Nguyen, H.H., and Jeon, J. (2017). *Tetrahedron* 73: 4052–4061.

70 Li, Q., Driess, M., and Hartwig, J.F. (2014). *Angew. Chem. Int. Ed.* 53: 8471–8474.

71 (a) Kuznetsov, A. and Gevorgyan, V. (2012). *Org. Lett.* 14: 914–917. (b) Kuznetsov, A., Onishi, Y., Inamoto, Y., and Gevorgyan, V. (2013). *Org. Lett.* 15: 2498–2501.

72 (a) Lee, T., Wilson, T.W., Berg, R. et al. (2015). *J. Am. Chem. Soc.* 137: 6742–6745. For mechanistic studies, see: (b) Lee, T. and Hartwig, J.F. (2017). *J. Am. Chem. Soc.* 139: 4879–4886.

73 Su, B., Zhou, T.-G., Li, X.-W. et al. (2017). *Angew. Chem. Int. Ed.* 56: 1092–1096.

74 Zhang, Q.-W., An, K., Liu, L.-C. et al. (2016). *Angew. Chem. Int. Ed.* 55: 6319–6323.

75 Zhang, Q.-W., An, K., Liu, L.-C. et al. (2017). *Angew. Chem. Int. Ed.* 56: 1125–1129.

76 Procopio, L.J. and Berry, D.H. (1991). *J. Am. Chem. Soc.* 113: 4039–4040.

77 Djurovich, P.I., Dolich, A.R., and Berry, D.H. (1994). *J. Chem. Soc., Chem. Commun.* 1897–1898.

78 (a) Sadow, A.D. and Tilley, T.D. (2003). *Angew. Chem. Int. Ed.* 42: 803–805. (b) Sadow, A.D. and Tilley, T.D. (2005). *J. Am. Chem. Soc.* 127: 643–656.

79 Fukumoto, Y., Hirano, M., and Chatani, N. (2017). *ACS Catal.* 7: 3152–3156.

80 Fukumoto, Y., Hirano, M., Matsubara, N., and Chatani, N. (2017). *J. Org. Chem.* 82: 13649–13655.

81 Hirano, M., Fukumoto, Y., Matsubara, N., and Chatani, N. (2018). *Chem. Lett.* 47: 385–388.

82 Kakiuchi, F., Tsuchiya, K., Matsumoto, M. et al. (2004). *J. Am. Chem. Soc.* 126: 12792–12793.

83 Mita, T., Michigami, K., and Sato, Y. (2012). *Org. Lett.* 14: 3462–3465.

84 Mita, T., Michigami, K., and Sato, Y. (2013). *Chem. Asian J.* 8: 2970–2973.

85 Li, W., Huang, X., and You, J. (2016). *Org. Lett.* 18: 666–668.

86 Kon, K., Suzuki, H., Takada, K. et al. (2016). *ChemCatChem* 8: 2202–2205.

87 Ihara, H., Ueda, A., and Suginome, M. (2011). *Chem. Lett.* 40: 916–918.

88 Liu, Y.-J., Liu, Y.-H., Zhang, Z.-Z. et al. (2016). *Angew. Chem. Int. Ed.* 55: 13859–13862.

89 Pan, J.-L., Li, Q.-Z., Zhang, T.-Y. et al. (2016). *Chem. Commun.* 52: 13151–13154.

90 Deb, A., Singh, S., Seth, K. et al. (2017). *ACS Catal.* 7: 8171–8175.

91 (a) Simmons, E.M. and Hartwig, J.F. (2012). *Nature* 483: 70–73. (b) Li, B., Driess, M., and Hartwig, J.F. (2014). *J. Am. Chem. Soc.* 136: 6586–6589.

92 Ghavtadze, N., Melkonyan, F.S., Gulevich, A.V. et al. (2014). *Nat. Chem.* 6: 122–125.

93 (a) Kuninobu, Y., Nakahara, T., Takeshima, H., and Takai, K. (2013). *Org. Lett.* 15: 426–428. (b) Murai, M., Takeshima, H., Morita, H. et al. (2015). *J. Org. Chem.* 80: 5407–5414.

94 Fang, H., Hou, W., Liu, G., and Huang, Z. (2017). *J. Am. Chem. Soc.* 139: 11601–11609.

95 Karmel, C., Li, B., and Hartwig, J.F. (2018). *J. Am. Chem. Soc.* 140: 1460–1470.

96 Bunescu, A., Butcher, T.W., and Hartwig, J.F. (2018). *J. Am. Chem. Soc.* 140: 1502–1507.

97 Su, B. and Hartwig, J.F. (2017). *J. Am. Chem. Soc.* 139: 12137–12140.

98 Lee, T. and Hartwig, J.F. (2016). *Angew. Chem. Int. Ed.* 55: 8723–8727.

7

Transition-Metal-Free Catalytic C—H Bond Silylation

David P. Schuman[1], Wen-Bo Liu[2], Nasri Nesnas[3], and Brian M. Stoltz[1]

[1] *California Institute of Technology, Division of Chemistry and Chemical Engineering, 1200 E. California Blvd., Pasadena, CA, 91125, USA*
[2] *Engineering Research Center of Organosilicon Compounds & Materials, Ministry of Education, College of Chemistry and Molecular Sciences, Wuhan University, 299 Bayi Rd., Wuhan, Hubei, 430072, China*
[3] *Florida Institute of Technology, Department of Biomedical and Chemical Engineering and Sciences, 150 W. University Blvd., Melbourne, FL, 32901, USA*

7.1 Introduction

Although transition-metal-free catalytic C–H silylation has been known for some time (Section 7.2.1) it has only recently been heavily investigated. We begin by describing an early example of catalytic, transition-metal-free C–H silylation before transitioning to the current state of this field.

7.2 Lewis Acid

7.2.1 BCl_3 Catalyst

In the late 1950s, one of the earliest relevant examples of such reactivity involving the BCl_3-catalyzed C–H silylation of benzene was reported by Dow Chemists (Scheme 7.1) [1–4].

While this reaction demonstrates a direct C–H silylation, the very forcing reaction conditions significantly limit the scope of this chemistry. Furthermore, the authors showed that in a matter of hours, the reaction reaches an equilibrium of starting material, product, and various silane-containing by-products. Further investigations indicated that this reaction might proceed through a Friedel–Crafts-type pathway (Scheme 7.2a) or through compound **10** via an initial borylation, followed by a phenyl group transfer (Scheme 7.2b).

Later publications by the same researchers included only limited mechanistic investigations, but these studies indicated neither nucleophilic aromatic substitution nor radical pathways were the operative mechanism in this unusual reactivity [3].

Organosilicon Chemistry: Novel Approaches and Reactions, First Edition.
Edited by Tamejiro Hiyama and Martin Oestreich

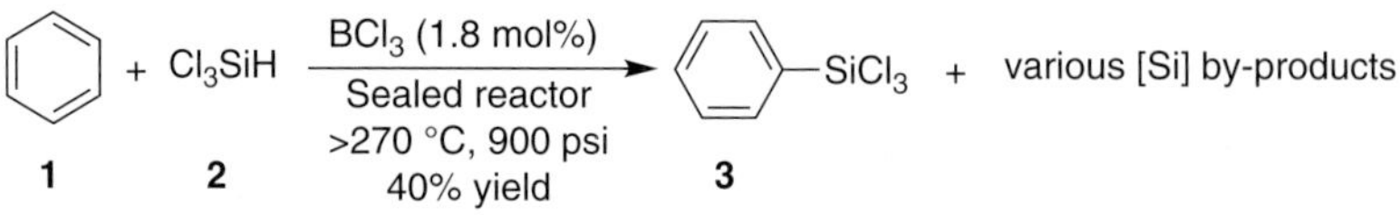

Scheme 7.1 Early example of C–H silylation catalyzed by Lewis acid.

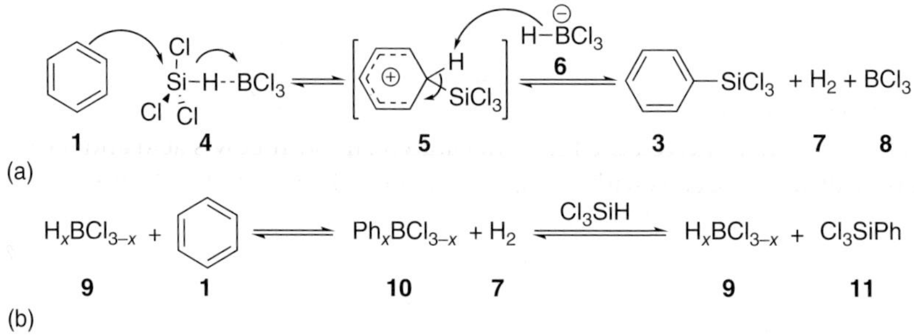

Scheme 7.2 Possible reaction pathways for BCl_3-catalyzed C–H silylation. (a) Direct silylation. (b) Successive borylation/arylation.

This report of Lewis acid–catalyzed C–H silylation highlights a number of key challenges in the development of such a method. The catalyst can often activate both the [Si]—H bond in the starting material and the [Si]—C bond in the resulting product. Furthermore, the forcing conditions (i.e. high temperature and pressure) exacerbate the issue of selective bond activation and limit functional group compatibility. Thus, future studies would need to address these issues for higher selectivity and yield.

7.2.2 $B(C_6F_5)_3$, a "Frustrated" Lewis Acid Catalyst

Bulky Lewis acids have been shown to activate a variety of Si—H and H—H bonds via the formation of frustrated Lewis pairs (FLP) [5]. A recent report from Ingleson detailed the reaction of heterocycles with hydrosilanes activated by $B(C_6F_5)_3$ (Schemes 7.3, 7.17) [6]. The $B(Ar)_3$–hydrosilane FLP **13** can go on to

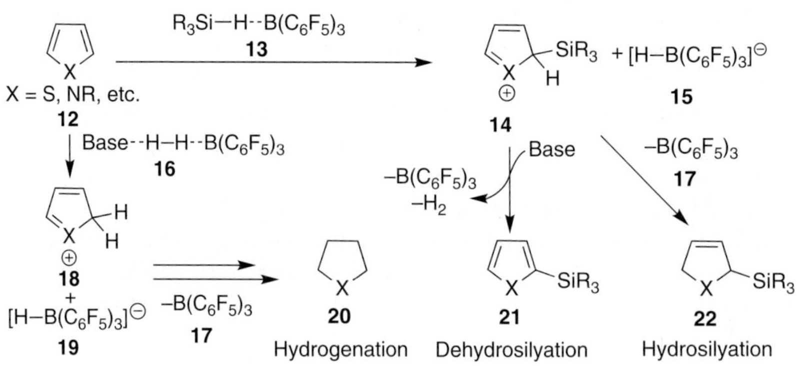

Scheme 7.3 Competitive (de)hydrosilylation and hydrogenation using Lewis acid catalysis.

react with substrate **12** with the possible products of hydrogenation (**20**), dehydrosilylation (**21**), and hydrosilylation (**22**).

A significant challenge in this reaction is ensuring a rapid deprotonation of the arenium cation **14** to avoid hydrosilylation (**14**–**22**). Hydrogen gas must be evolved to render the reaction catalytic, but the electrophilic reaction intermediates (**14**, **18**) and Lewis acid catalyst (**17**) necessitate the use of a suitably bulky, yet weakly nucleophilic base. The resulting buildup of hydrogen may lead to competitive hydrogen activation via **16** and substrate hydrogenation to afford **20**. Toward this end, the authors found 2,6-dichloropyridine (Cl_2—py) to be a suitable base for the silylation of 2-methylthiophene.

While the authors rendered the reaction catalytic in both Lewis acid and base, the reaction always proceeds with the formation of hydrogenated and/or hydrosilylated by-products **25** (Table 7.1, entries 2–5). The buildup of such byproducts was shown to be detrimental to the reaction as the inclusion of tetrahydrothiophene inhibits product formation (entry 6), likely due to competitive coordination of tetrahydrothiophene to the Lewis acid catalyst. Efforts to limit the hydrogenation pathway by conducting the reaction under static vacuum results in decreased overall yield (entry 1 versus entry 7). The authors further demonstrated the reaction scope using an N-protected indole substrate **26** (Table 7.2), although the same issue of competitive reduction (hydrogenation only, in the case of indole substrates) limited overall utility of this method.

The report by Ingleson demonstrated an initial investigation into Lewis acid–catalyzed C–H silylation and documents many of the challenges present in such a reaction manifold.

Table 7.1 Selected examples from reaction optimization experiments.

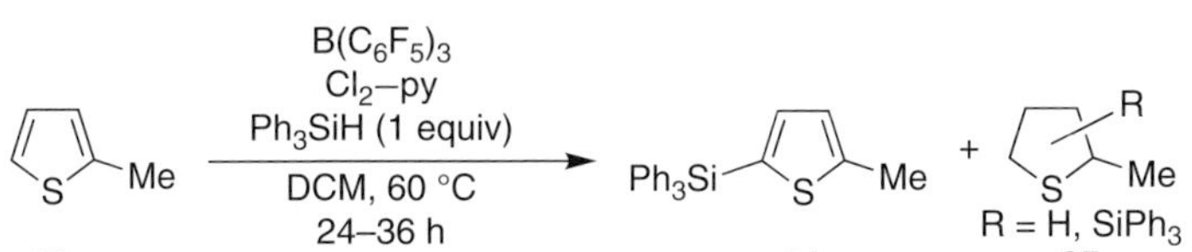

Entry	$B(C_6F_5)_3$ (mol%)	Cl_2—py (mol%)	**24** (% yield)	**25** (% yield)
1	100	100	51	33
2	20	20	56	34
3	5	5	42	18
4	5	100	51	27
5[a)]	5	5	46	32
6[b)]	100	100	0	0
7[c)]	100	100	33	Unreported

a) 1.5 equiv silane used.
b) 1 equiv tetrahydrothiophene added.
c) Conducted under vacuum.

Table 7.2 FLP-catalyzed electrophilic C–H silylation of heteroaromatics.

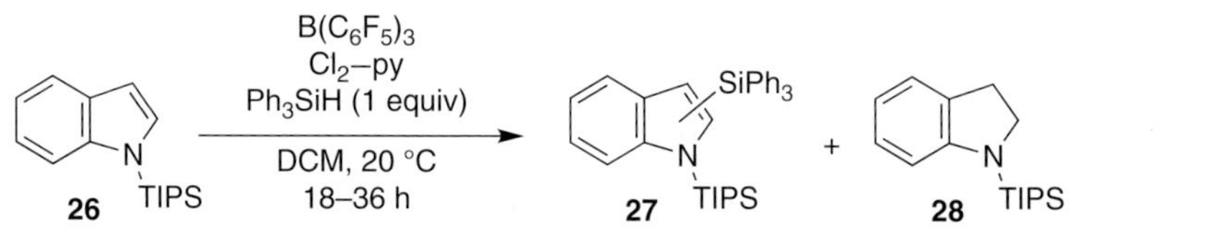

Entry	$B(C_6F_5)_3$ (mol%)	Cl_2–py (mol%)	**27** (% yield)	**28** (% yield)
1	100	100	59	19
2	100	0	30	21

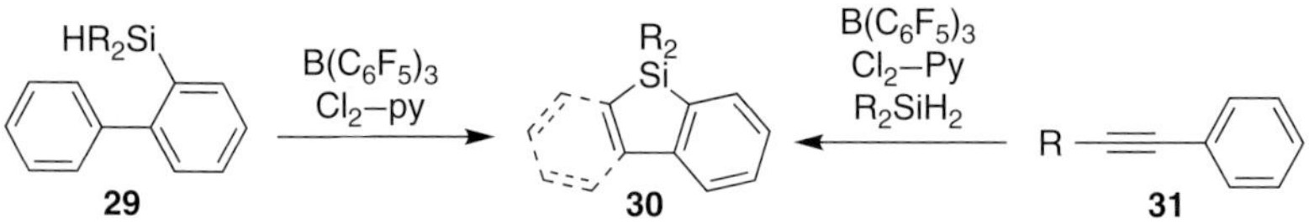

Scheme 7.4 $B(C_6F_5)_3$-catalyzed intra/intermolecular silylation.

A following report, again from the laboratory of Ingleson, demonstrated that a biphenyl-silyl scaffold undergoes intramolecular C–H silylation with a Lewis acid/base catalyst system (Scheme 7.4, left) [7]. They were also able to effect a tandem hydrosilylation, dehydrosilylation procedure to form cyclized product **30** from aryl alkyne starting material **31** (Scheme 7.4, right).

The authors proposed a mechanistic cycle that begins with Lewis acid activation of the pendant silane and nucleophilic attack by the aryl ring to afford cationic intermediate **32**. This intermediate is deprotonated by the base to regenerate aromaticity in product **30**, while the protonated base can go on to react with Lewis acid complex **15** to generate hydrogen gas and regenerate both base and Lewis acid catalyst **17** (Scheme 7.5).

Based on the observations in their previous studies, [6] the authors hypothesized that the use of biphenyl-silyl scaffold **29** would limit undesired reduction by biasing the system toward rearomatization of the silolane product (**32**–**30**), as shown in their proposed mechanism Scheme 7.5). The authors were able to realize this strategy, render the reaction catalytic, and apply their optimized conditions toward the synthesis of a variety of substituted siloles (**34**) (Table 7.3).

Unfortunately, this methodology did not translate to the cyclization of heterocyclic substrates **35** (Scheme 7.6). Instead of forming the desired silole product **37**, only the initial C–H silylation product **36** was detected. Through further experimentation, the authors believed the catalyst was entirely consumed in an off-cycle hydrogen FLP process and therefore could not affect the ring-closing cyclization.

Interestingly, the authors were able to design a tandem one-pot hydrosilylation/dehydrosilylation procedure in order to convert internal alkynes **38** to

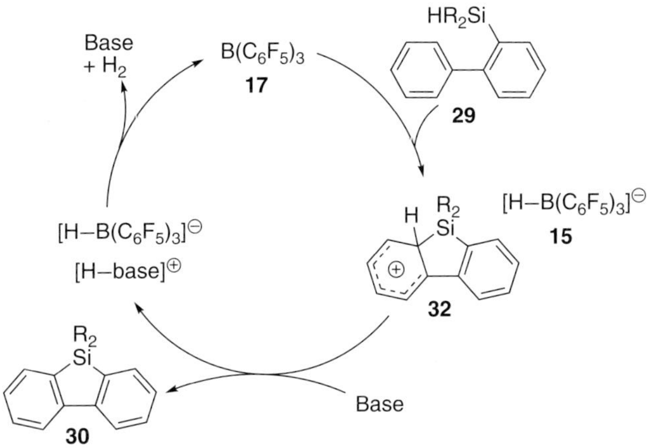

Scheme 7.5 Proposed intramolecular silylation mechanism.

Table 7.3 Optimization and substrate scope of intramolecular silole synthesis.

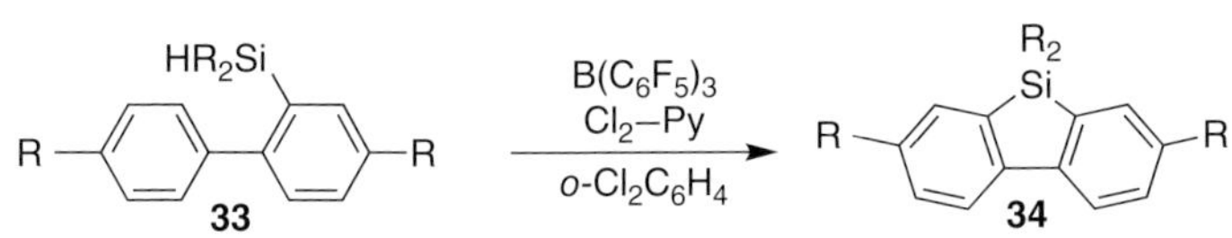

Entry	R	R′	$B(C_6F_5)_3$ (mol%)	Cl_2–py (mol%)	*t* (h)	*T* (°C)	Yield (%)
1	Ph	H	100	100	18	60	99
2	Ph	H	10	10	24	100	99
3	Ph	H	5	5	96	100	99
4	Ph	H	10	0	168	100	50
5	Ph	*t*-Bu	5	5	5	100	99
6	*i*-Bu	H	5	5	96	100	96

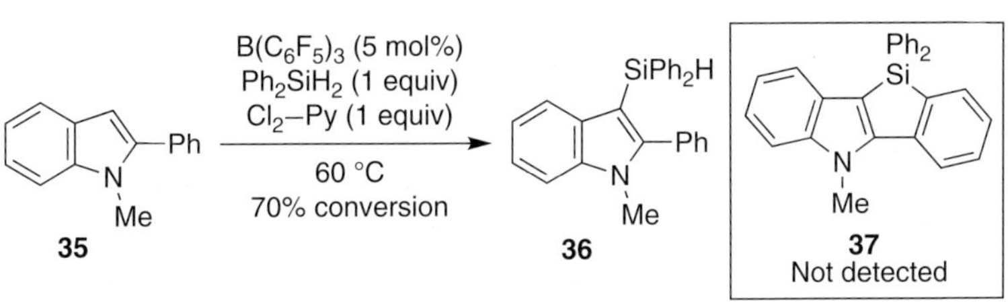

Scheme 7.6 Attempts toward heterocycle cyclization.

silaindenes **40** (Table 7.4). This reaction proceeds through an initial alkyne hydrosilylation, which may occur in either a syn or trans process leading to *trans*- and/or *cis*-**39**. Only *cis*-**39** can proceed to form the silole product **40**.

Table 7.4 One-pot hydrosilylation/dehydrosilylation.

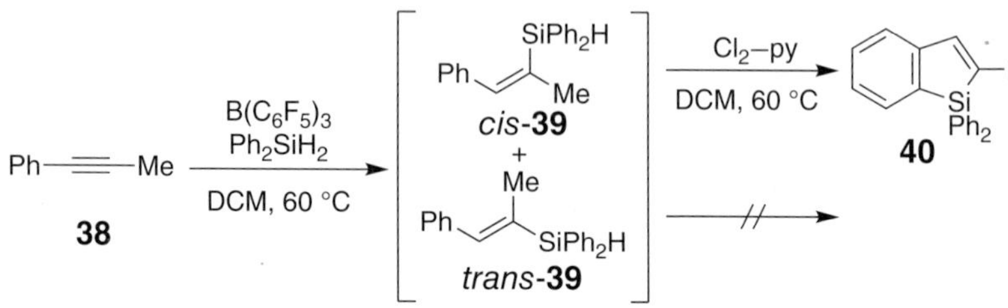

Entry	$B(C_6F_5)_3$ (mol%)	*t* (h)	*cis* (%)	*trans* (%)	Cl_2—py (mol%)	*t* (h)	Yield (%)
1	100	5	84	16	100	72	84
2	5	5	85	15	5	48	23
3	5	4	84	16	100	72	70
4	5	—	—	—	100	72	50

While this chemistry provided an intriguing application of the $B(C_6F_5)_3/Cl_2$—py system, competing reduction pathways and catalyst inhibition remained a major limitation.

Given these limitations, it was surprising when the Hou group recently reported an aromatic C–H silylation using $B(C_6F_5)_3$ alone, with no added base and a relatively broad scope (Tables 7.5 and 7.6) [8, 9]. This reaction tolerates a variety of hydrosilanes **42**, including chlorosilanes (entries 11 and 12), furnishing the dehydrosilylated product in moderate to good yields (Table 7.5).

Turning to the matter of substrate scope, a variety of anilines and *N*-aryl substrates are able to undergo C–H silylation exclusively at the para position relative to nitrogen (Table 7.6).

Table 7.5 Scope of silanes in $B(C_6F_5)_3$catalyzed system

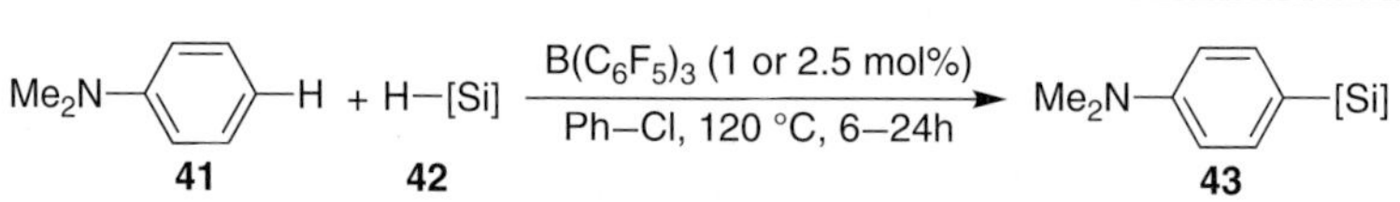

Entry	Silane	Yield (%)	Entry	Silane	Yield (%)
1	Ph_2SiH_2	84	7	*n*-Bu_3SiH	61
2	$PhSiH_3$	80	8	$PhMe_2SiH$	85
3	$PhMeSiH_2$	82	9	Ph_2MeSiH	80
4	Et_2SiH_2	65	10	Ph_3SiH	67
5	$EtMe_2SiH$	68	11	Ph_2SiClH	51[a)]
6	Et_3SiH	60	12	$PhSiClH_2$	70[a)]

a) Yield over two steps.

Table 7.6 Substrate scope of C–H silylation.

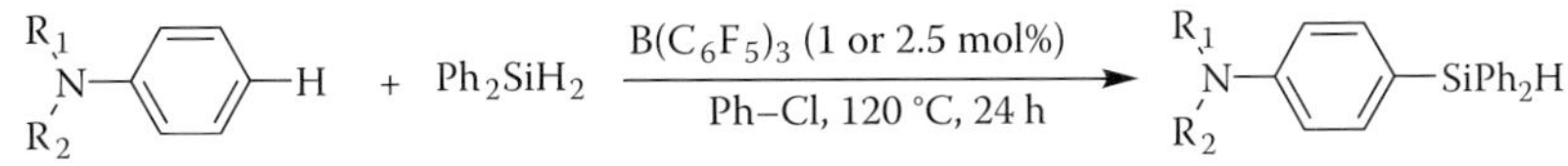

44 45 46

Entry	Product	Yield (%)
1	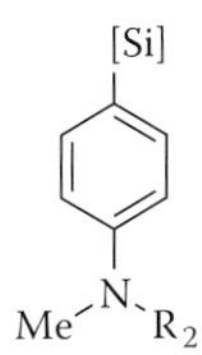 R_2 = Et	82%
2	*i*-Pr	81%
3	*n*-C_6H_{13}	80%
4	TMS	78%
5	allyl	75%
6	Ph	41%
7	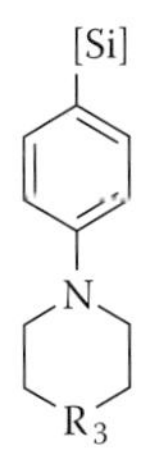 R_3 = CH_2	90%
8	O	73%
9	NMe	73%
10		81%
11	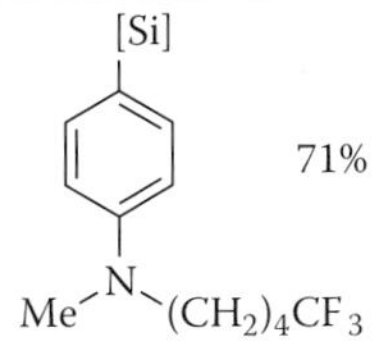	71%
12	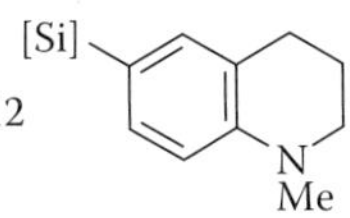	72%
13	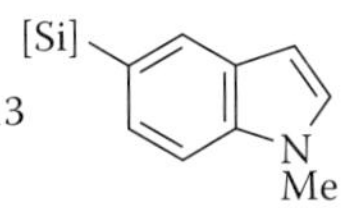	34%
14	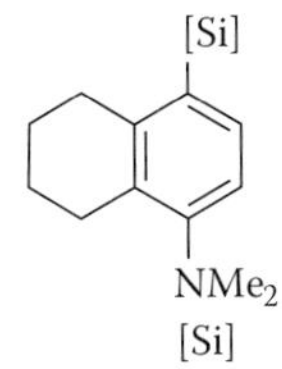	90%
15	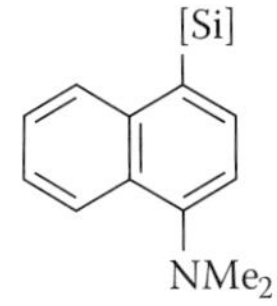	45%
16	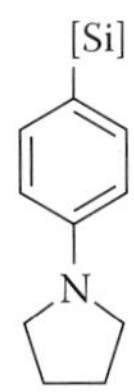	86%

The authors noted the interesting C-5 silylation of *N*-methylindole (Table 7.6, entry 13), which is difficult to access using other silylation reactions, but did not provide the rationale as to why C-5 silylation was favored over C-3 silylation.

While a thorough mechanistic study was not undertaken, the authors proposed a mechanism based on the current literature of $B(C_6F_5)_3$-catalyzed silylation (Scheme 7.7). The hydrosilane forms the activated FLP **13**, which is attacked by the electron-rich arene **41**. The resulting Wheland intermediate (**48**) is deprotonated by $[H–B(C_6F_5)_3]^{\ominus}$(**15**) to form the silylated product **49**, hydrogen gas, and regenerate the catalyst **17**. This specific reaction does not require an exogenous base or result in reduction products (i.e. hydrosilylation and hydrogenation) that plague similar systems discussed here. One can postulate that the electron-rich nature of the substrate may play a role or that the tertiary amine substrate itself may act as a base during the reaction, in a stepwise deprotonation/hydrogen release mechanism (see Figure 1).

A following report by Zhang and coworkers describes reactivity similar to that of Ingleson and Hou [10]. Zhang and coworkers do not initially aim to reduce the hydrogenation by-product often observed in these types of silylation reactions, but instead uses this to help drive the reaction to completion in a type of disproportionation reaction (**52** and **53**, Table 7.7). While the silylated product yields are limited to 50% of the starting material, a variety of indole substrates (**51**) were shown to undergo silylation in good relative yields. The reactions are

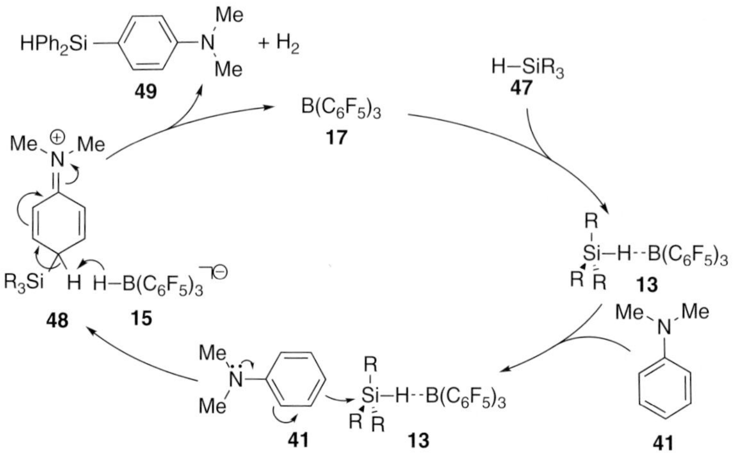

Scheme 7.7 Proposed mechanism of $B(C_6F_5)_3$-catalyzed C–H silylation in aniline-type.

Table 7.7 Substrate scope of disproportionation C–H silylation.

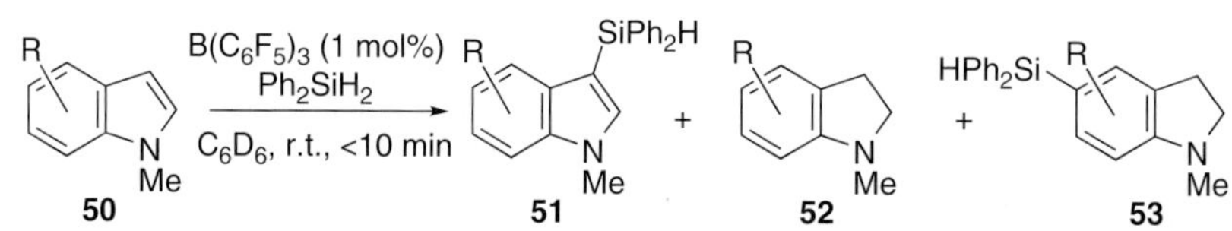

Entry	R	Equiv Ph_2SiH_2	Yield **51** (%)	Yield **52** (%)	Yield **53** (%)
1	H	0.5	43	45	2
2	5-Me	0.5	49	48	0
3	6-Br	0.5	48	47	1
4	6-F	0.5	41	42	12
5	5-Ph	0.5	45	47	0
6	H	2	41	41	18
7	5-Me	2	49	49	0
8	6-Br	2	44	46	9
9	6-F	2	20	26	52
10	5-Ph	2	49	49	0

complete in short periods of time and, by modifying the reaction conditions, can be used to form double C–H silylation products (**54**, Table 7.8). The authors show NMR studies and a proposed mechanism that is in line with the previous FLP mechanisms discussed earlier.

Zhang and coworkers were able to render this reaction a convergent disproportionation, where the indoline formed in the reaction (**52**) is reoxidized to indole (**51**) and can reenter the catalytic cycle (Table 7.9). While this did

Table 7.8 Substrate scope of double-silylation selective disproportionation C–H silylation.

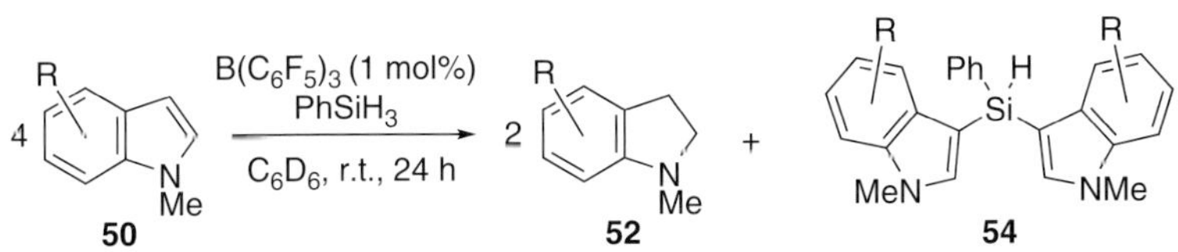

Entry	R	Yield **52** (%)	Yield **54** (%)
1	H	45	44
2	5-Me	43	45
3	6-Br	46	44
4	6-F	43	37
5	5-Ph	49	43

Table 7.9 Substrate scope of convergent disproportionation C–H silylation.

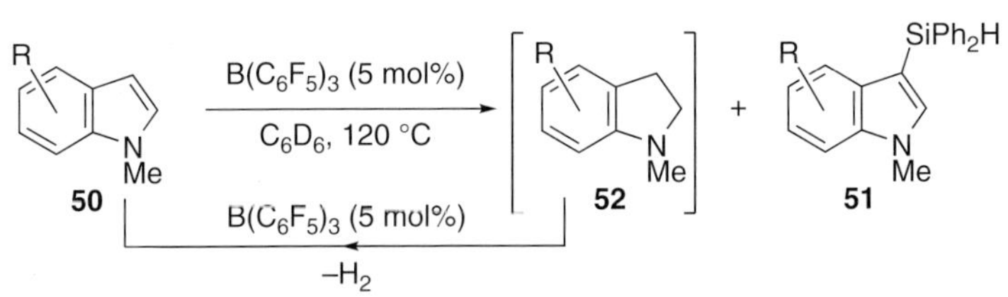

Entry	R	Silane[a)]	Time (h)	Yield **51** (%)
1	H	Ph_2MeSiH	2.5	86
2	H	Ph_3SiH	9	94
3	5-Me	Ph_2SiH_2	48	69
4	5-Me	Ph_3SiH	2.5	93
5	5-Me	Ph_2MeSiH	2.5	95
6	5-F	Ph_3SiH	2.5	92
7	5-Cl	Ph_3SiH	2.5	91
8	5-Br	Ph_3SiH	2.5	83
9	5-Ph	Ph_3SiH	2.5	99

a) 2 equiv silane used.

require significant heating compared to the disproportionation reaction, the reaction yields significantly increased. This method is specific to indole substrates, but the authors were able to solve many of the typical issues associated with using an FLP silylation catalyst.

7.2.3 Lewis Acid Conclusions

While Lewis acid catalysis provides a rapid and simple method for the activation of [Si]—H bonds, challenges exist in this mode of reactivity. Reduction via hydrogen activation or hydrosilylation can compete with product formation and potentially lead to catalyst poisoning. Nevertheless, this reaction manifold remains an interesting and evolving area in transition-metal-free C–H silylation chemistry.

7.3 Brønsted Acid

Hydrosilanes may also be activated by protonation and subsequent loss of hydrogen to generate an activated silyl cation intermediate **56** (Scheme 7.8), which can then be trapped by a nucleophilic substrate in a manner similar to Lewis acid catalysis (Section 7.2). Inspired by stoichiometric examples of such reactions, the Oestreich group sought to develop a catalytic C–H silylation variant [11]. They were successful in utilizing a Brønsted acid for the catalytic C–H silylation of aromatic heterocycles.

During reaction optimization, they observed degradation of the silane using TfOH and settled on the use of Brookhart's acid, $\{[H(OEt_2)_2]^{\oplus}[BAr^F_4]^{\ominus}\}$ (**58**), for further optimization studies (Table 7.10).

The control reaction demonstrates that the $[BAr^F_4]^{\ominus}$ counterion displayed no catalytic activity alone (Table 7.10, entry 2). Interestingly, the inclusion of norbornene (nbe) and lowering the catalyst loading both led to increased yield (entries 6–8). In this reaction, norbornene may act as a proton scavenger, and together with lower catalyst loadings, help balance proton removal from the reaction intermediate with hydrogen release. Both of these processes could serve to limit the overall proton concentration in solution, thus inhibiting protodesilylation of the product.

The reaction scope in both silane and substrate was investigated (Table 7.11). Mono and diphenylsilanes (entries 1, 2, and 6) afford good yields of the corresponding silylated products, but triphenyl (entry 3), tris-C(sp^3) alkyl (entry 4), and dialkoxysilanes (entry 5) all performed poorly. Aryl halides, nitrogen-containing heterocycles, and anilines undergo silylation smoothly; however, oxygen- and sulfur-containing heterocycles or anisoles did not react under these conditions. Furthermore, the reaction does not tolerate substitution at the site of reactivity (i.e. C-3 of indole or para position of aniline).

While only a single transition-metal-free catalytic example has been reported, the use of Brønsted acids as a C–H silylation catalyst provides an interesting and orthogonal route compared to other methods. Unfortunately, this class of

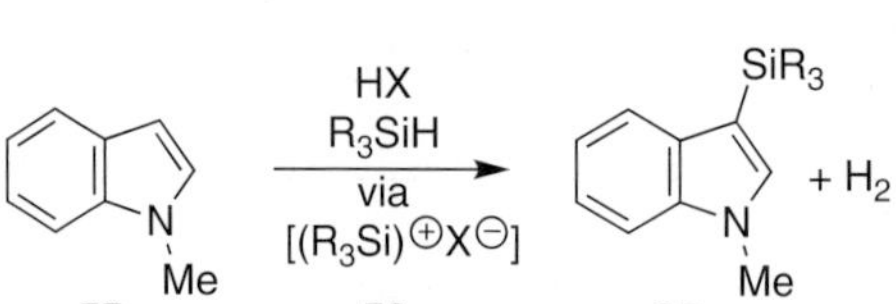

55 56 57

Scheme 7.8 Overview for the use of acid as dehydrosilylation catalyst.

Table 7.10 Optimization of acid-catalyzed indole silylation.

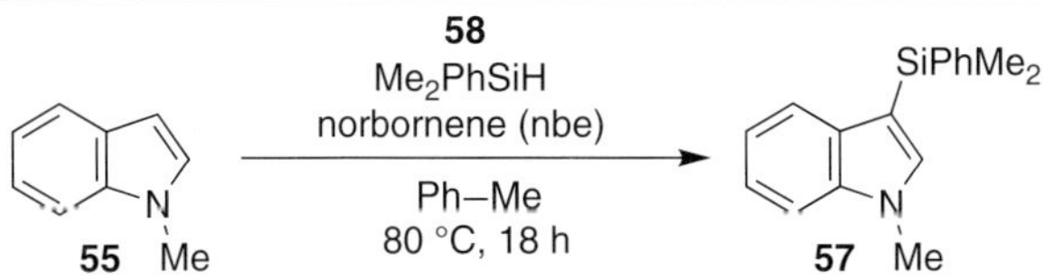

$[H(OEt_2)_2]^{\oplus}[BAr^F{}_4]^{\ominus}$
58,
Brookhart's acid

Entry[a]	**58** (mol%)	nbe (equiv)	Yield (%)
1	—	—	NR
2[b]	1	—	NR
3	1	—	70
4	2	—	77
5	4	—	53
6	2	0.5	85
7	2	1	95
8	1	1	97

a) 2 equiv substrate, 1 equiv silane.
b) $NaBAr^F{}_4$ used as catalyst

Table 7.11 Selected examples of silane and substrate scope.

Entry	Product		Yield (%)	Entry	Product	Yield (%)
1	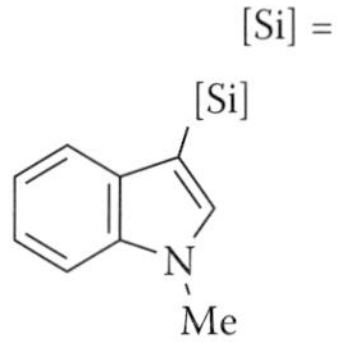	[Si] = Me_2PhSi	96	12	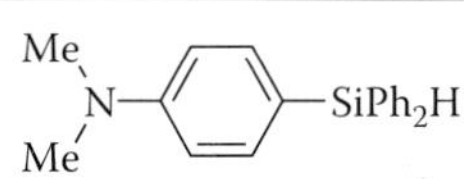	67
2		$MePh_2Si$	93			
3		Ph_3Si	8			
4		Et_3Si	7	13	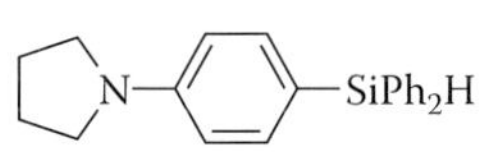	66
5		$(EtO)_2MeSi$	0			
6		Ph_2SiH	96			
7	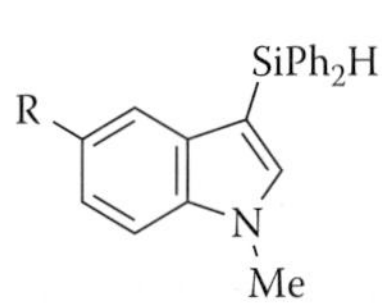	R = Me	94			
8		F	98	14	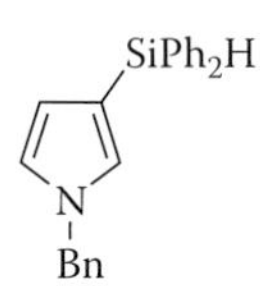	58
9		Cl	96			
10		Br	96			
11[b]	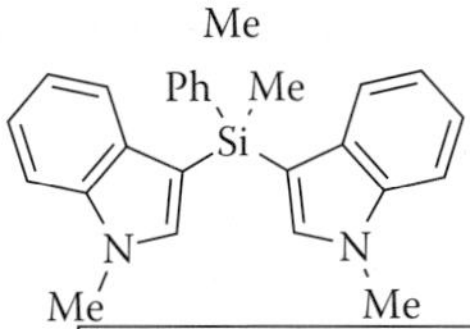		73	15	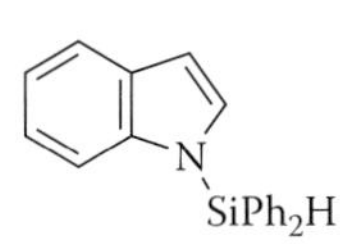	75

Unreactive substrates

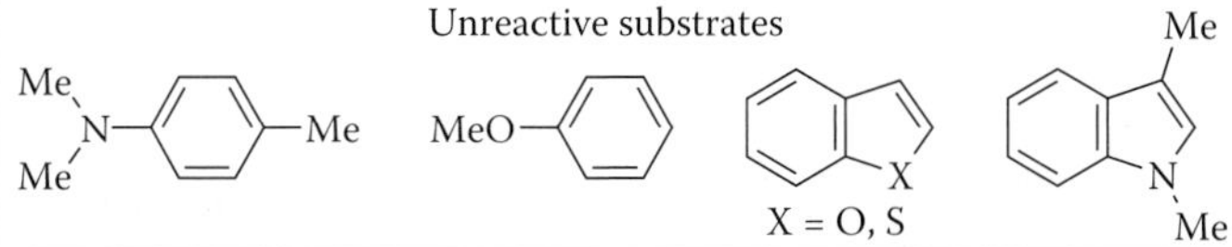

a) 2 equiv substrate, 1 mol% **58**, 1 equiv [Si]–H, 1 equiv nbe, PhMe, rt to 80 °C, 18 h.
b) 3 equiv *N*-methylindole **(55)** used.

catalyst faces limited substrate scope and functional-group compatibility, but may continue to develop as a C–H silylation methodology.

7.4 Brønsted Base

7.4.1 Early Example of Catalytic C-H Silylation by Brønsted Base

Taking the term "catalytic" in a broader sense allows us to present an interesting publication from Fornarini involving gas-phase, mechanistic investigations of silyl cations and aromatic nucleophiles in the presence of a catalytic base (Scheme 7.9) [12]. The author proposed that Wheland intermediate **61** can be accessed by the reaction of arenes with silyl cations generated by irradiation. Depending on the base identity or concentration used, the base could engage the activated silyl group of the Wheland intermediate **61**, resulting in a formal desilylation event and reforming starting material **59**. Alternatively, the base could deprotonate the Wheland intermediate to form the silylated product **60**. While this report is not of particular synthetic relevance, it provides an interesting mechanistic and initial study for C–H silylation by Brønsted bases.

7.4.2 Fluoride/Base Catalysis

An intriguing base-catalyzed C–H silylation method was proposed by Kondo and a coworker, whereby suitable choice of an activating reagent and leaving group on silicon could result in the generation of a base upon addition of a nucleophile to silicon (Scheme 7.10a) [13, 14]. The authors achieved the desired reactivity using a fluoride base as the initiator and the Ruppert–Prakash reagent (**65**) as the tethered base and silylating agent (Scheme 7.10b).

A variety of fluoride salts were found to afford the C–H silylation product (Table 7.12, entries 1–4) and the reaction performs well using 1,3-dimethyl-2-imidazolidinone (DMI) or 1,2-dimethoxyethane (DME) as a solvent (entries 8 and 11 versus 10). Further investigation showed the reaction was rapid and resulted in high product yields after only a few hours (entries 9 versus 8 and 6 versus 3).

With these optimized conditions in hand, the authors demonstrated that a number of heterocyclic and substituted aromatic compounds undergo silylation in good to excellent yields (Table 7.13). Aryl halides were well tolerated (entries 4, 5, 11, and 12) as well as both electron donating (entry 2) and withdrawing groups (entries 6–8, 10, and 13).

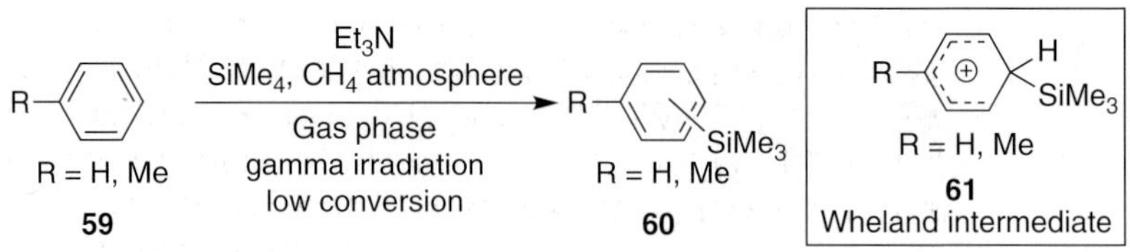

Scheme 7.9 Early example of gas-phase direct silylation by catalytic Brønsted base.

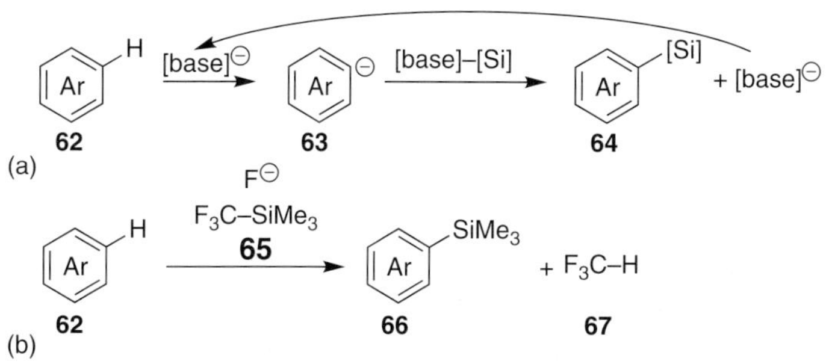

Scheme 7.10 a) Inspiration for base-catalyzed C–H silylation b) Final reaction overview.

Table 7.12 Optimization of fluoride-catalyzed C–H silylation.

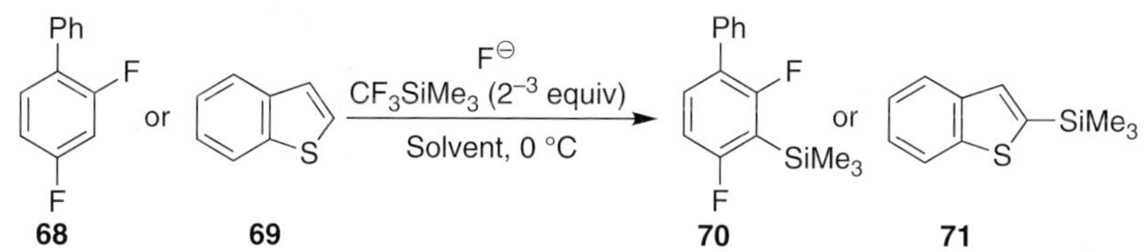

Entry	Substrate	Fluoride Source (mol%)	Solvent	Time (h)	Yield (%)
1	**69**	KF (50)	DMI	24	51
2	**69**	RbF (50)	DMI	24	86
3	**69**	CsF (50)	DMI	24	83
4	**69**	$(Me_4N)F$ (50)	DMI	24	Trace
5	**69**	—	DMI	24	0
6	**69**	CsF (50)	DMI	2	97
7	**68**	CsF (50)	DMI	2	100
8	**68**	CsF (20)	DMI	2	100
9	**68**	CsF (10)	DMI	2	50
10	**68**	CsF (20)	DMF or Ph–Me	2	0
11	**68**	CsF (20)	DME	2	86
12	**68**	CsF (20)	DME	1	90

The authors proposed a mechanism based on their mechanistic studies of the present system, combined with their previous studies of similar in situ generation of reactive bases (Scheme 7.11) [13, 14]. The reaction begins by initial generation of a pentacoordinate silicon **72** (X = F) from the [$F^{\ominus}$] catalyst, which produces a [$CF_3^{\ominus}$]-equivalent **72** to deprotonate the arene **62**. The resulting anion **63** proceeds to react with the Ruppert–Prakash reagent (**65**), resulting in the silylated product **66** and $CF_3^{\ominus}$ (**73**). It is possible that the $CF_3^{\ominus}$ anion (**73**) directly deprotonates the arene **62** (Scheme 7.11, dashed line), thereby completing the catalytic reaction, or the $CF_3^{\ominus}$ anion (**73**) may form a pentacoordinate silicate **72**

Table 7.13 Substrate scope of fluoride-catalyzed C–H silylation.

Entry	Product	Yield (%)	Entry	Product	Yield (%)
1	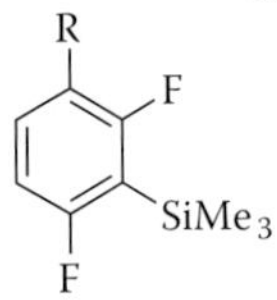 R = H	85	9	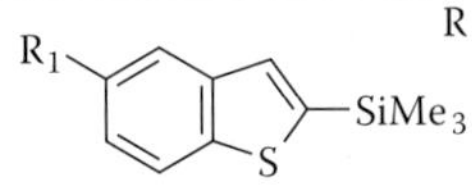 R_1 = Me	81
2	OMe	71	10	CN	60
3	Me	57	11	Br	91
4	Br	100	12	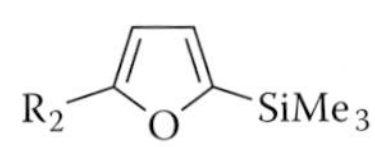 R_2 = I	46
5	I	71	13	CN	84
6	CN	75	14	CO_2Et	0
7	NO_2	76			
8	$C(O)NMe_2$	71			

See Refs. [11, 12] for conditions

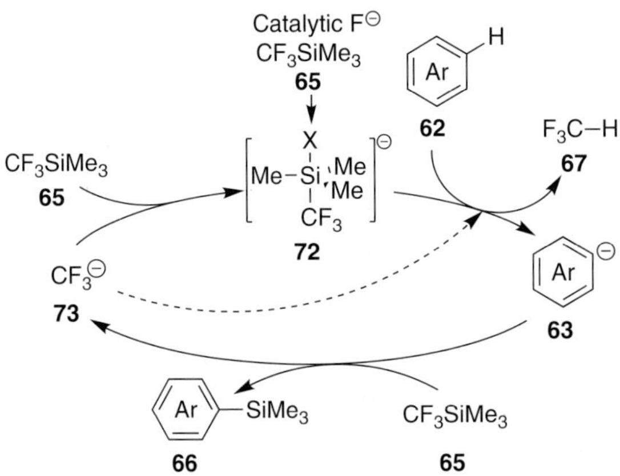

Scheme 7.11 Proposed mechanism for Brønsted base-catalyzed reaction.

($X = CF_3$) and complete the cycle as drawn. Although the authors were able to detect CF_3H during the catalytic reaction, further mechanistic study is required to determine these final steps of the reaction mechanism.

7.4.3 Brønsted Base-Catalyzed C–H Silylation of Alkynes

Direct dehydrogenative, C—Si bond formation may be achieved by deprotonation of an alkyne with a Brønsted base, followed by electrophilic trapping of the alkynyl carbanion with a silane. This process may be rendered catalytic if the pentacoordinate silicate intermediate acts as a base to deprotonate another equivalent of the starting alkyne. In 1969, such an example was reported by Calas and Bourgeois, wherein NaH was used as a catalyst for the dehydrogenative C–H silylation of 1-hexyne (**74**) and Et_3SiH; however, this was the only example disclosed (Scheme 7.12) [15].

Further independent studies by the research groups of Itoh and Ono found that heterogenous Brønsted bases (i.e. MgO, KNH_2/Al_2O_3, and KF/Al_2O_3), alkali

Scheme 7.12 Dehydrosilylation of terminal alkynes using Brønsted base.

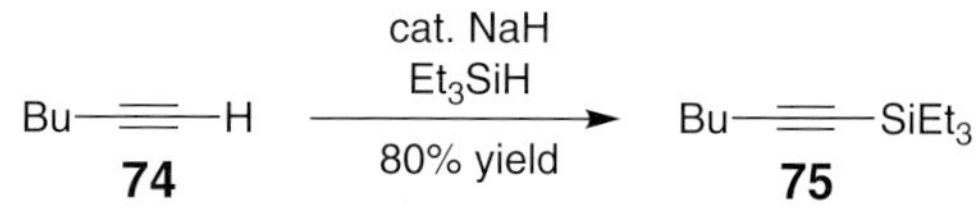

Table 7.14 Base screen for dehydrogenative cross coupling of alkynes and hydrosilanes.

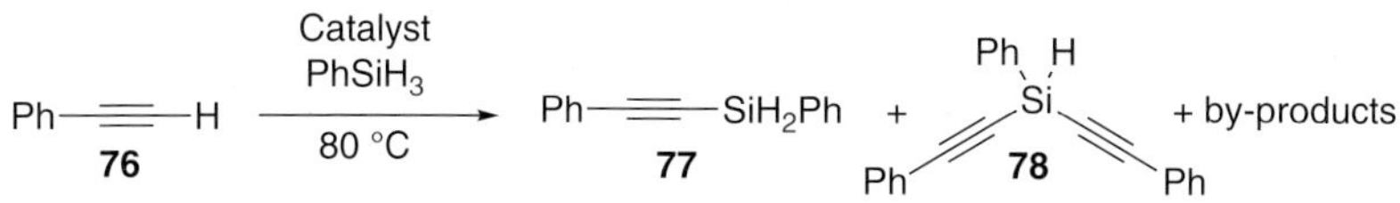

Entry	Catalyst	mol%	Solvent	Yield **77** (%)	Yield **78** (%)
1[a)]	MgO	>400	Benzene	43	24
2[b)]	$LiAlH_4$	5	Diglyme (120 °C)	42	9
3[c)]	LiOEt	0.5	Diglyme	34	6
4[c)]	NaOEt	0.6	Diglyme/THF (10/1)	46	18
5[c)]	KOEt	0.7	Diglyme/Et_2O (10/1)	6	<1
6[c)]	$Ba(OMe)_2$	0.5	Diglyme	49	25
7[c)]	$Ba(OC_8H_{17})_2$	0.6	Toluene (40 °C)	73	6
8[c)]	*n*-BuLi	0.6	Diglyme	40	7
9[c)]	PhLi	0.7	Diglyme	45	10
10[c)]	$Ba(CCPh)_2$	0.4	Diglyme	54	30
11[c)]	$LiN(SiMe_3)_2$	0.6	Diglyme	43	13

a) See Ref. [16].
b) See Ref. [17].
c) See Ref. [18].

metal hydrides, metal alkoxides, alkylmetals, and metal amides could all effect C–H silylation (Table 7.14)[16–20][1].

Preliminary mechanistic investigations conducted by Itoh supported an ionic reaction pathway (Scheme 7.13). Deprotonation of the terminal alkyne by strong

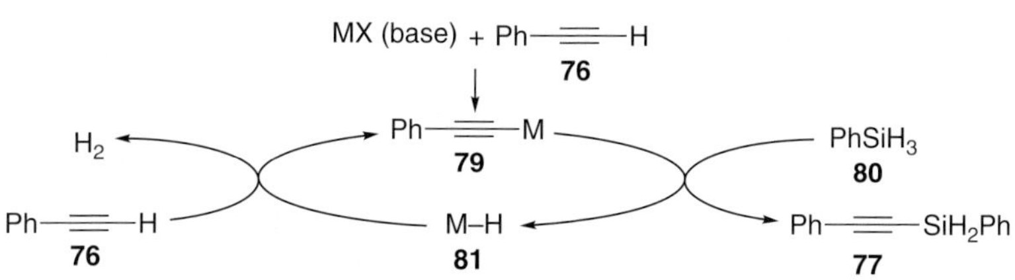

Scheme 7.13 Proposed mechanism of base-catalyzed direct C–H silylation of alkynes

1 Note-A single example of C(sp^2) silylation using benzene was also reported with the same catalyst system described in Ref. [19].

Table 7.15 Reaction optimization and silane scope.

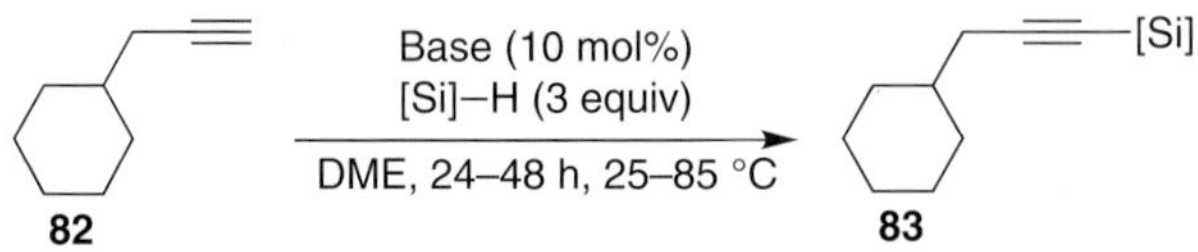

Entry	Base	Silane	Yield (%)	Entry	Base	Silane	Yield (%)
1	KO*t*-Bu	Et_3SiH	89	8	LiOH	$PhMe_2SiH$	0
2	NaO*t*-Bu	Et_3SiH	46	9	NaOH	*t*-Bu_2SiH_2	91
3	LiO*t*-Bu	Et_3SiH	<1	10	KOH	*i*-Bu_3SiH	69
4	Pyridine	Et_3SiH	0	11	NaOH	$(EtO)_3SiH$	68
5	KOH	Et_3SiH	95	12	KOH	*i*-Pr_2(*o*-py)SiH	78
6	KOH	$PhMe_2SiH$	89	13	NaOH	Me_2(*o*-py)SiH	78
7	NaOH	$PhMe_2SiH$	93				

base generates a metal acetylide, which can then react with the hydrosilane to form the desired product and a metal hydride, thereby completing the catalytic cycle. (Note that in this reaction deprotonation may occur from a pentacoordinate silicate similar to that in Scheme 7.11, rather than by regeneration of a metal hydride.)

These methods typically result in low yields of the desired products along with various by-products. Moreover, the scope of both the alkynes and silanes are very limited.

This reaction was recently revisited by Stoltz and Grubbs, who developed a facile and practical dehydrosilylation of terminal alkynes using hydroxide catalysts (Table 7.15) [21]. This work was inspired by their previously established KO*t*-Bu-catalyzed C–H silylation of aromatic heterocycles (see subsequent text for discussion of KO*t*-Bu-catalyzed C–H silylation, 22). A wide variety of hydrosilanes were found to be amenable to C–H silylation including aryl, alkyl, bulky triisobutyl, alkoxy, and pyridyl silanes (entries 9–13).

Interestingly, the authors found varied success for either KOH or NaOH between substrates, but there did not appear to be a direct correlation between choice of substrate or silane and catalyst performance. Therefore, a series of substrates were subjected to catalytic reactions using NaOH, KOH, and KO*t*-Bu to compare the yield of dehydrosilylation product (Table 7.16, entries 1–18). Upon exploration of the substrate scope, the reaction was found to be compatible with a variety of synthetically useful functional groups including heterocycles (entries 1–6, 25), secondary and tertiary amines (entries 27 and 22), and a variety of aryl functional groups (entries 7–12, 19–24, 28). By changing reaction conditions, mono- and bis-silylated diynes (entries 29, 30) and symmetrical or unsymmetrical sila-diynes (entries 30, 31) could be synthesized in useful yields. Overall, the reaction has been shown to be compatible with a wide variety of both hydrosilanes and alkyne substrates.

Table 7.16 Comparison of alkyne dehydrosilylaton catalysts and substrate scope.

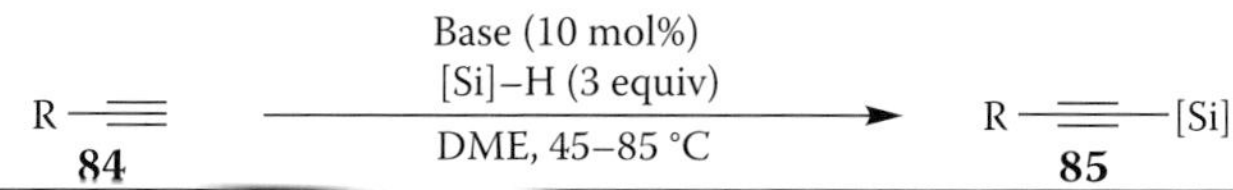

Entry	Product	Base	Yield (%)
1	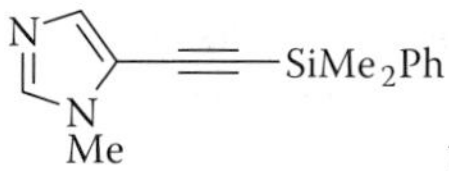	NaOH	82
2		KOH	19
3		KO*t*-Bu	<5
4	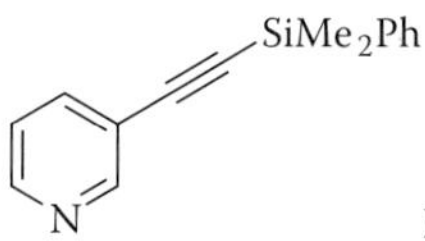	NaOH	77
5		KOH	7
6		KO*t*-Bu	0
7	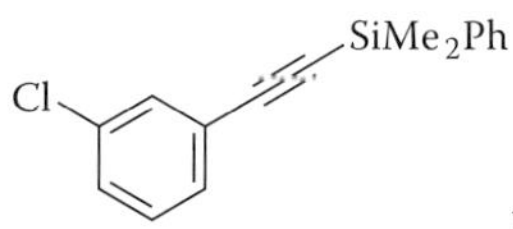	NaOH	90
8		KOH	38
9		KO*t*-Bu	<5
10	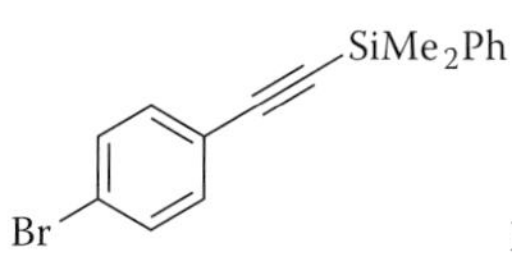	NaOH	52
11		KOH	<5
12		KO*t*-Bu	0
13	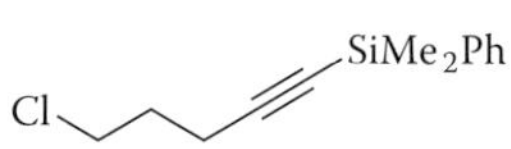	NaOH	79
14		KOH	0
15		KO*t*-Bu	0
16	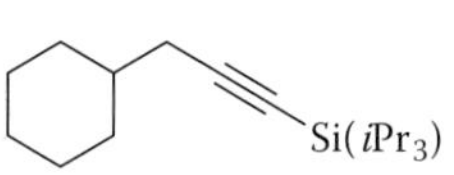	NaOH	0
17		KOH	69
18		KO*t*-Bu	0

Entry	Product[a)]		Yield (%)
19[a)]	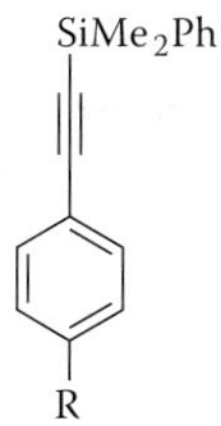	R = H	89
20[a)]		F	88
21[a)]		Br	52
22[a)]		NMe$_2$	99
23[a)]		Me	92
24[a)]		MeO	91
25[a)]	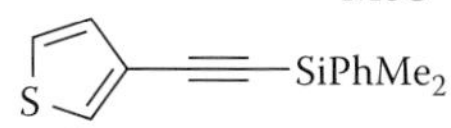		93
26[a)]	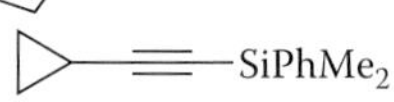		70
27[a)]	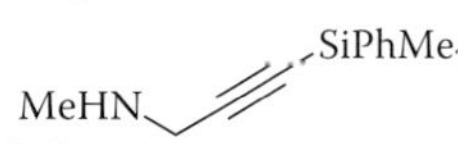		80
28[a)]	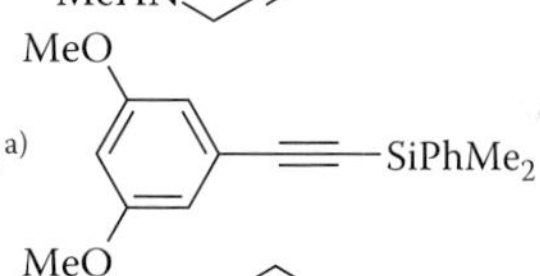		95
29[b)]	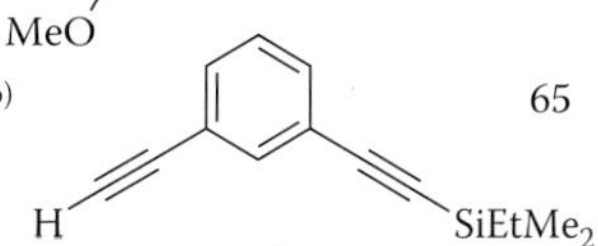		65
30[c)]	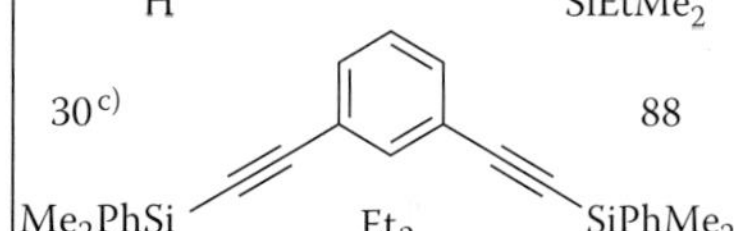		88
31[d)]	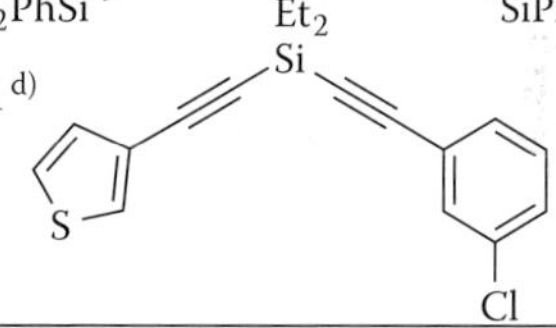		58

a) NaOH used.
b) 20 mol% NaOH, 0.33 equiv Me$_2$PhSiH, yield based on [Si]–H.
c) 20 mol% NaOH, 3 equiv Me$_2$PhSiH.
d) 10 mol% NaOH, 3 equiv Et$_2$SiH$_2$ then filter/conc. 10 mol% NaOH, 1.5 equiv *m*-Cl aryl alkyne.

7.5 Radical Dehydrosilylation

7.5.1 "Electron" as a C–H Silylation Catalyst

If we again take a liberal interpretation of the term catalyst, an interesting report from the laboratory of Studer details the use of an electron as a

Table 7.17 Optimization of intramolecular C–H silylation.

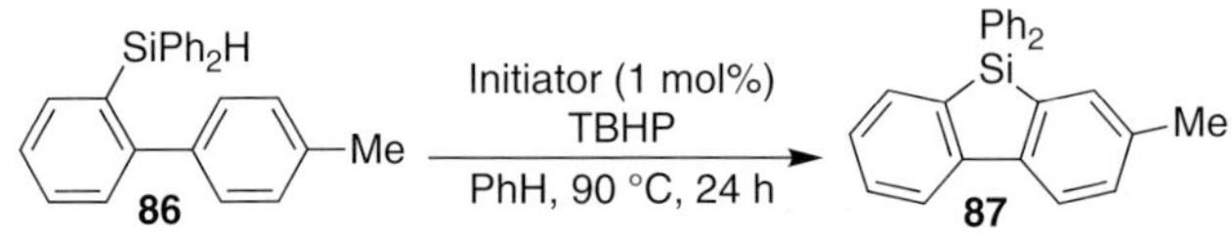

Entry	TBHP (equiv)	Initiator	Yield (%)	Entry	TBHP (equiv)	Initiator	Yield (%)
1	2.2	FeCP$_2$	34	5	3.3	TBAI	65
2	2.2	FeCl$_3$	37	6	4.4	TBAI	62
3	2.2	CuI	66	7	5.5	TBAI	58
4	2.2	TBAI	55	8[b)]	3.3	TBAI	46

a) TBHP, *t*-BuOOH; TBAI, (Bu)$_4$NI.
b) 0.4 mol% initiator.

catalyst in an intramolecular C–H silylation reaction of biarylsilanes **86** (Table 7.17) [23, 24].

Their method uses a substoichiometric radical initiator and excess peroxide as an oxidant. They were able to optimize the reaction to use low loadings of the radical initiator (entry 8), but unfortunately had to use a large excess of the peroxide oxidant (Table 7.17) and observed lower yields than related reactions using $B(C_6F_5)_3$ catalysts (see Section 7.2.2).

The authors went on to demonstrate the substrate scope of their intramolecular C–H silylation reaction (Table 7.18). The reaction was shown to tolerate some substitution on both aryl rings of the biphenyl, including a variety of halide substitution (entries 4, 5, 8, 10) and even pyridine (entry 9). Both electron-rich (entries 7 and 11) and electron-poor substrates (entries 4 and 6) resulted in moderate formation of the cyclized product.

Based on their observations and reaction conditions, they proposed a single-electron-transfer (SET)-type mechanism, whereby an electron acts as a catalyst in their reaction cycle (Scheme 7.14). While this reaction does require superstoichiometric peroxide oxidant, it is still an interesting reaction to discuss as an introduction to catalytic, radical C–H silylation. It must also be mentioned that this reaction does show significant similarities to the classical Minisci reaction, the addition of alkyl radicals to protonated heterocycles [25].

The group of Li reported a similar reaction at higher temperature without the use of catalytic TBAI (Scheme 7.15, top) [26]. A series of silafluorenes **96** were synthesized by radical, intramolecular cyclization of biphenyl-2-hydrosilane substrates. A cascade process was also established to synthesize silaindenes by first radical addition to alkynes followed by radical cyclization (Scheme 7.15, bottom). While these strategies proved to be quite limited in scope, the reactions serve as an interesting study in radical C–H silylation methods.

Table 7.18 Selected substrate scope of intramolecular C–H silylation.

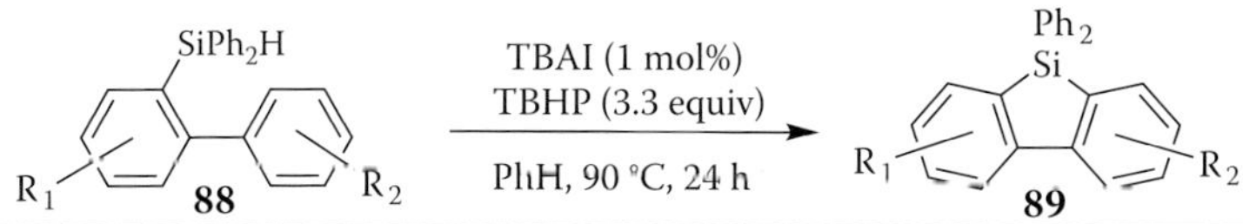

Entry	Product		Yield (%)	Entry	Product	Yield (%)
1	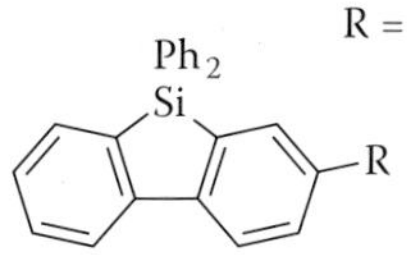	R = Me	55			
2		H	66	9	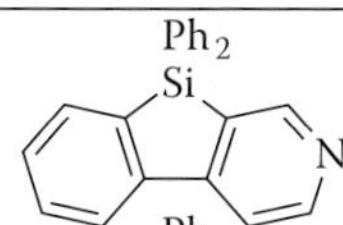	39
3		*t*-Bu	71			
4		F	67			
5		Cl	77			
6		CF_3	75	10	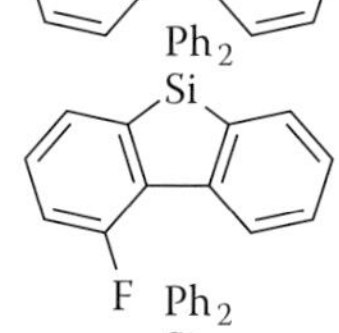	55
7		OMe	68			
8	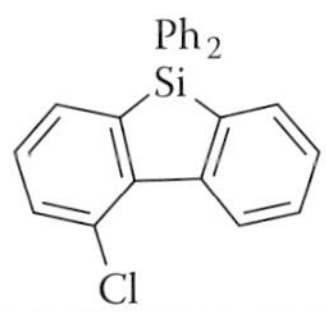		75	11	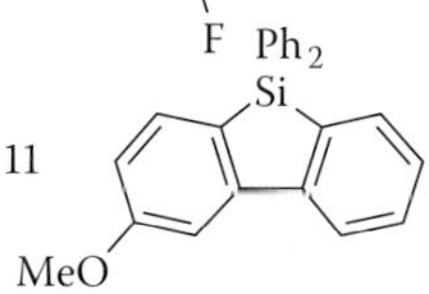	70

a) TBHP, *t*-BuOOH; TBAI, $(Bu)_4NI$.

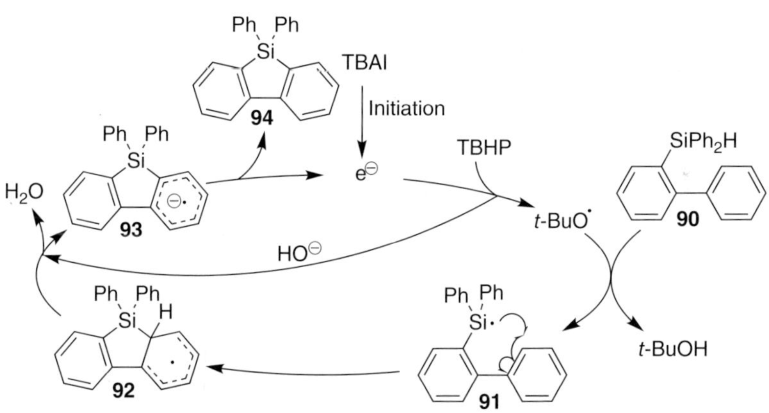

Scheme 7.14 Proposed mechanism for intramolecular C–H silylation.

7.5.2 Discovery of Unusual KO*t*-Bu-Catalyzed C–H Silylation

The laboratory of Grubbs discovered the interesting direct C–H silylation side products **102–106** (Table 7.19) while working on a system for the reductive cleavage of aryl ethers **100** [27].

Using substoichimetric base (entry 4), C–H silylated dibenzofurans **102**, **103**, and **104** form in reasonable yield. Inspired by this initial result, they sought to further develop this chemistry.

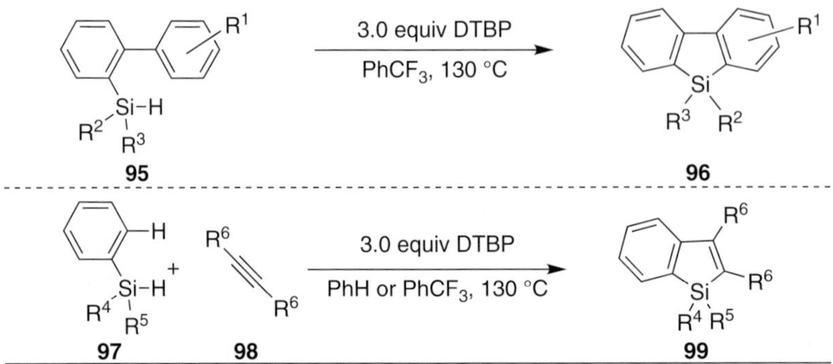

DTBP = t-BuOOt-Bu. See Ref. [26] for details

Scheme 7.15 Uncatalyzed radical C–H silylation/annulation.

Table 7.19 Optimization of reductive aryl ether cleavage.

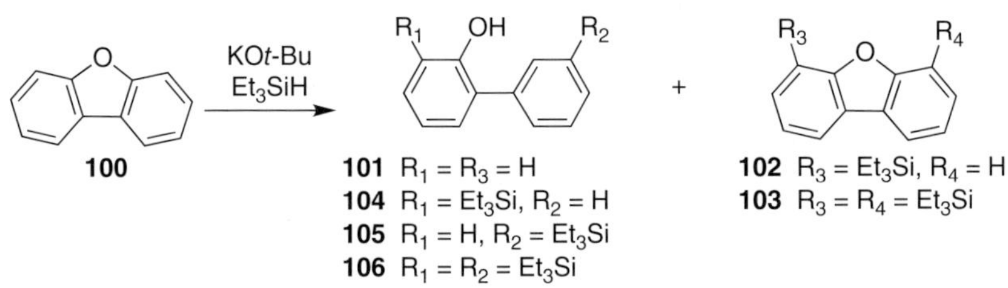

Entry	Et_3SiH (equiv)	KO*t*-Bu (equiv)	Solvent	*T* (°C)	Yield (%)					
					101	**102**	**103**	**104**	**105**	**106**
1	5	2	PhMe	100	38	16	10	21	2	7
2	5	5	PhMe	100	63	10	2	20	-	2
3	3	3	Mesitylene	165	85	3	-	5	2	-
4	5	0.5	PhMe	100	12	48	20	9	-	1
5	5	2	Mesitylene / 1,4-cylohexadiene	165	95	-	-	-	-	-

7.5.2.1 KO*t*-Bu-Catalyzed C–H Silylation Methodology

In a collaborative effort between the Grubbs and Stoltz groups, a detailed reaction optimization was conducted and a wide array of aromatic heterocycles **107** (Scheme 7.16) were found to undergo dehydrogenative C–H silylation to afford the silylated product **108** [22].

During the reaction optimization, the authors quickly found that catalytic amounts of KO*t*-Bu were able to furnish high yields of the silylated products (Table 7.20, entries 2 versus 4). Other *t*-butoxy bases with smaller counterions

Scheme 7.16 Overview of KO*t*-Bu catalyzed C–H silylation.

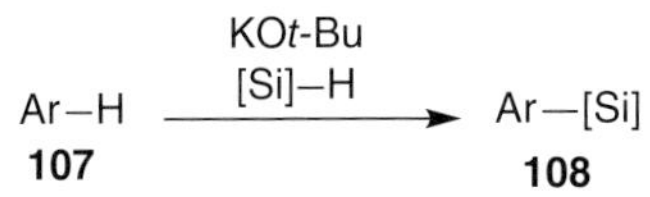

Table 7.20 Selected examples from optimization of C–H silylation.

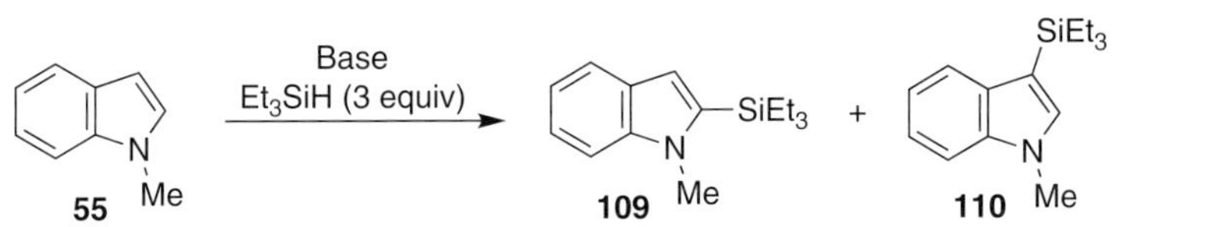

Entry	Base	mol%	T (°C)	t (h)	Solvent	Yield **109** (%)	Ratio **109 : 110**
1	Li/NaO*t*-Bu	100	25	16	THF	0	—
2	KO*t*-Bu	100	25	16	THF	67	>20 : 1
3	KHMDS	100	25	16	THF	44	>20 : 1
4	KO*t*-Bu	20	45	60	THF	98	4 : 1
5	KO*t*-Bu	20	45	60	MTBE	89	>20 : 1
6	KO*t*-Bu	20	45	60	DME	95	3.4 : 1
7	KO*t*-Bu	20	45	48	neat	88	>20 : 1

(i.e. $Li^{\oplus}$ or $Na^{\oplus}$) were not competent catalysts (entry 1) for this reaction, but other bases with a potassium cation do afford a product, albeit with decreased yield (entries 3 versus 2). While the reaction works in a number of ethereal solvents (entries 4–6), they were pleased to find that the reaction also worked well in the absence of solvent (entry 7).

With these optimized conditions in hand, a large substrate scope was shown to undergo facile C–H silylation, with only a small subset shown here (Table 7.21). A variety of protected indoles (entries 1–8, 11–13), azaindoles (entries 14–17), thiophene and furan (entries 9 and 10), and other heterocycles undergo silylation in good to excellent yields. Interestingly, 5,6-fused heterocycles undergo silylation preferentially at the C-2 position (α for thiophene and furan) rather than the typically more nucleophilic C-3 position.

To further demonstrate the synthetic utility of this methodology, the silylation reaction was shown to effect tandem inter/intramolecular silylation to from cyclic silanes **111** (Scheme 7.17) and late-stage silylation of pharmaceutically active compounds **112** and **113**.

In addition to the synthetic applications previously discussed, this methodology was shown to be applicable for the ultra-deep desulfurization of fuels [28] and was further developed to the alkyne C–H methods discussed in Section 7.4.3 [21] and a catalytic O–H silylation method [29].

Table 7.21 Representative substrate scope of the KO*t*-Bu-catalyzed C–H silylation reaction.

Entry	Product		Yield (%)
1	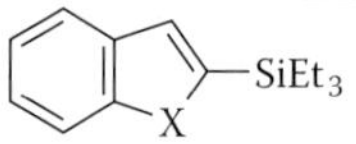	X = S	93
2[b)]		N-SEM	67
3		N-Me	78
4		N-Bn	82
5[c)]		N-MOM	55
6	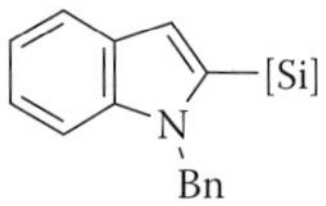	[Si] = $SiPhMe_2$	75
7		$SiEt_2H$	76
8		$SiEtMe_2$	71
9	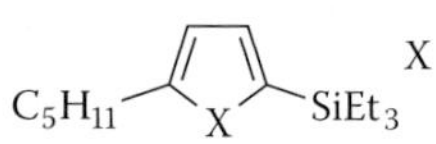	X = S	92
10		O	95
11	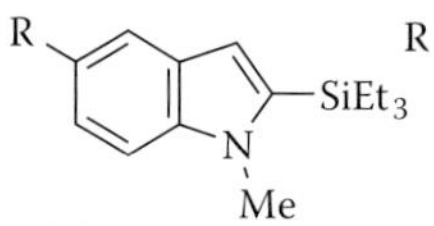	R = OMe	64
12		OBn	68
13		CH_2OMe	48

Entry	Product	Yield (%)
14[d)]	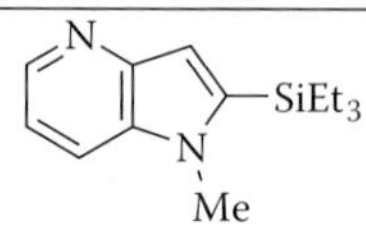	33
15	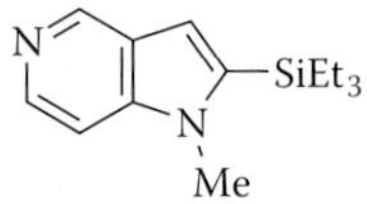	31
16	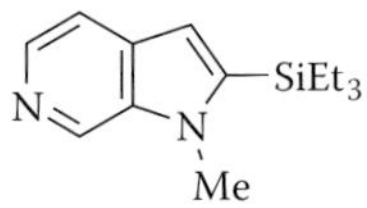	50
17	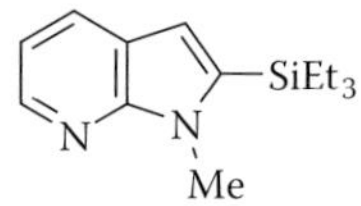	71

a) 20 mol% KO*t*-Bu, 3 equiv base, THF or neat, 25–60 °C, C2:C3 ratio > 20 : 1.
b) 14 : 1 C2:C3.
c) 10 : 1 C2:C3.
d) 6 : 1 C2:C3.

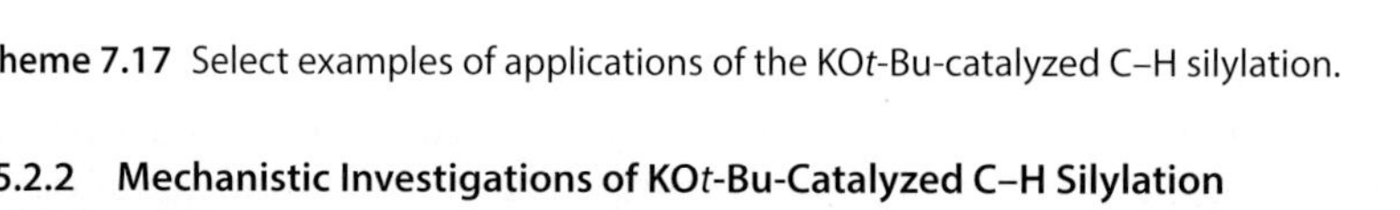

111 43% (from 1-phenylpyrrole) **112** 62% (from antihististamine thenalidine) **113** 57% (from antiplatelet ticlopidine)

See Ref. [22] for reaction conditions

Scheme 7.17 Select examples of applications of the KO*t*-Bu-catalyzed C–H silylation.

7.5.2.2 Mechanistic Investigations of KO*t*-Bu-Catalyzed C–H Silylation and Related Chemistry

During the course of the initial disclosure for this novel reaction, a brief mechanistic investigation was conducted and the initial results indicated that the reaction might occur by a radical mechanism [22]. To date, three reports exploring this reaction mechanism have been published. One indicates that a radical-type reaction mechanism may be operative under these reaction conditions (Scheme 7.18) [30]. A key aspect of this mechanism is the generation of a silyl radical **115**, which then adds to the heterocycle forming a stabilized radical heterocycle **114**. A hydrogen atom removal event then regenerates aromaticity and generates another silyl radical, thereby closing the catalytic cycle.

A variety of techniques were used to probe the active reaction mechanism, a few of which are highlighted here. The authors demonstrated a very large KIE

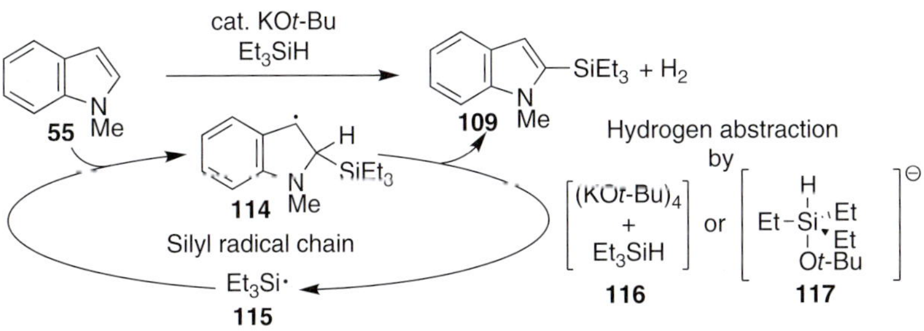

Scheme 7.18 Overview of radical mechanism for C–H silylation reaction.

effect at the C-2 position of indole H/D–**55** (Scheme 7.19), showed the silylation event is reversible by reacting **109** with $EtMe_2Si$–H to afford a mixture containing **109**, **118**, and **55**, and that cyclopropyl radical trap substrates will open only at the presumed site of radical formation **119** versus **122–124**.

An extensive computational investigation was conducted to determine the energetic feasibility of such a reaction pathway and while many aspects between experimental and computational results are in agreement, the radical initiation event is inconclusive (see Ref. [30] for full discussion).

During the course of the mechanistic investigation, it became apparent that a second reaction pathway might be possible. This second mechanism focused on

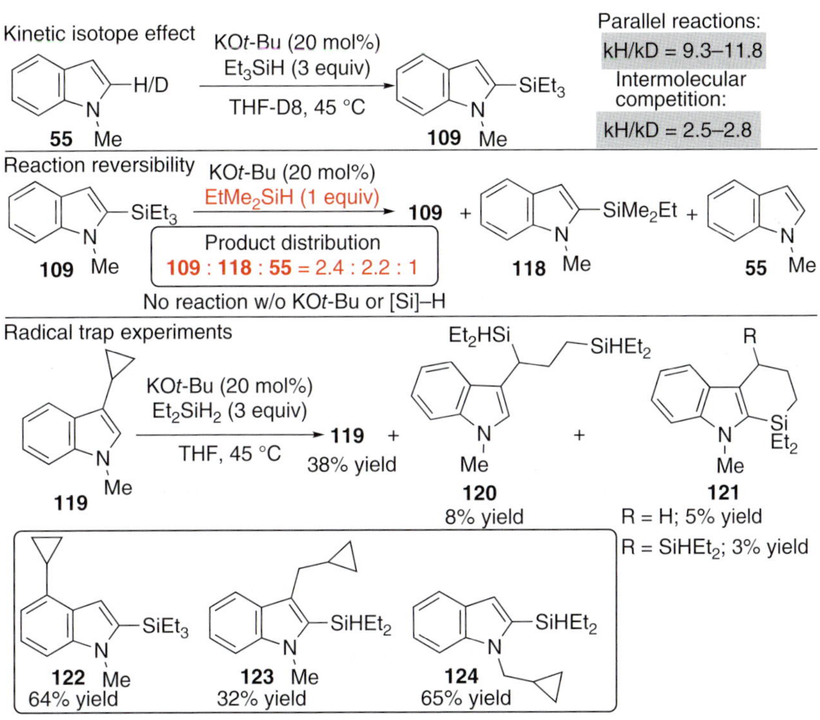

Scheme 7.19 Overview of mechanistic investigations.

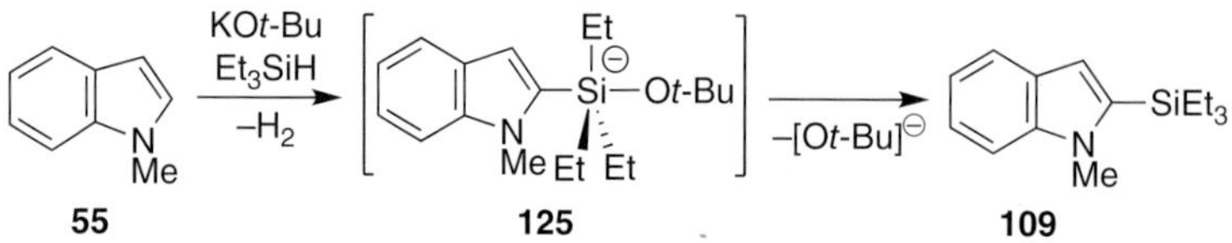

Scheme 7.20 Overview for ionic reaction mechanism.

the same KO*t*-Bu-catalyzed C–H silylation reaction but proposes an ionic/neutral mechanism (Scheme 7.20) [31].

The Zare group was able to use desorption electrospray ionization-mass spectrometry (DESI-MS) as a method to identify reaction intermediates. They were able to identify both the deprotonated indole substrate **126**, **127**, and **D-126** (Scheme 7.21) and the silylation intermediate formed by nucleophilic addition of indole to coordinated silane after loss of hydrogen (**125**). Furthermore, subjecting a mixture of $Li^{\oplus}$, $Na^{\oplus}$, and $K^{\oplus}$ solutions to DESI-MS with the indole substrate, the authors were able to detect the bis- and tris-coordinated π-arene complexes (Scheme 7.21c). These complexes may help to weaken the corresponding C—H bond and facilitate deprotonation.

Given these results, and a multitude of computational and other studies that are left out here due to space constraints, the authors propose a mechanism whereby the arene is coordinated by potassium cation **128** (Scheme 7.22), which facilities deprotonation by pentacoordinate silicate **117** to afford coordinated anion **129** or free anion **127**. The deprotonated arene **127** then attacks the silyl

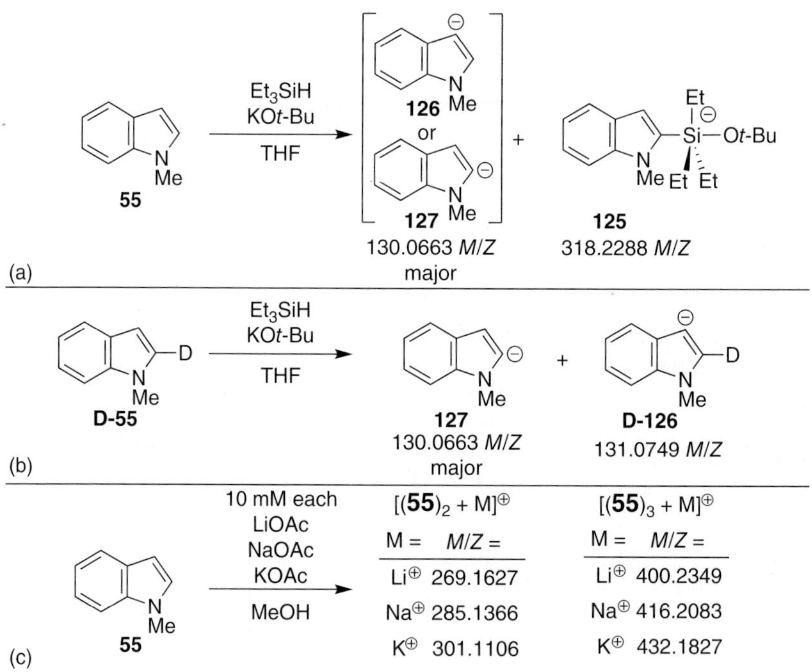

Scheme 7.21 a) Reaction intermediates observed by DESI-MS b) Regioselectivity of substrate deprotonation c) Competitive π-arene complexation of substrate.

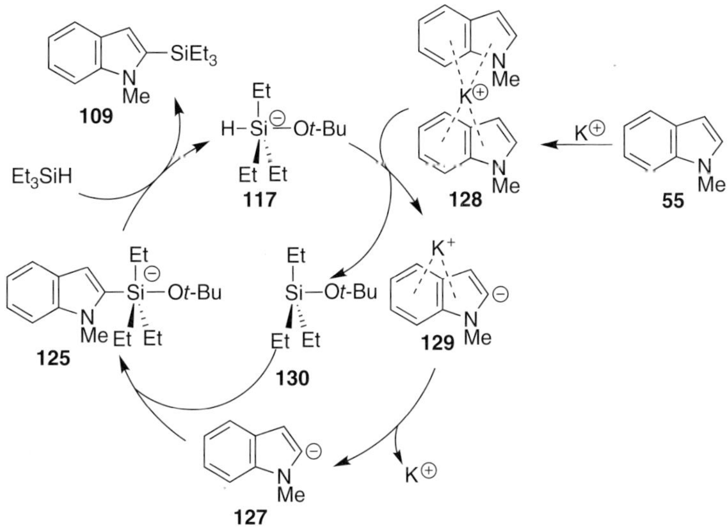

Scheme 7.22 Proposed ionic mechanism.

ether **130**, generating the silylated anion **125**, which, after release of *t*-butoxide and subsequent regeneration of the pentacoordinate silicon **117**, forms the silylated product **109**.

The neutral mechanism (not shown here, see Ref. [31]) involves a similar transformation but no discrete ions are formed. While neither the radical nor ionic/neutral mechanism can disprove the other, both have shown significant mechanistic evidence and may be operative under different reaction conditions.

A third mechanistic study has recently been reported by the groups of Murphy and Tuttle, which uses modified KO*t*-Bu-catalyzed silylation reaction conditions to effect electron transfer and hydride transfer reactions (Scheme 7.23) [32].

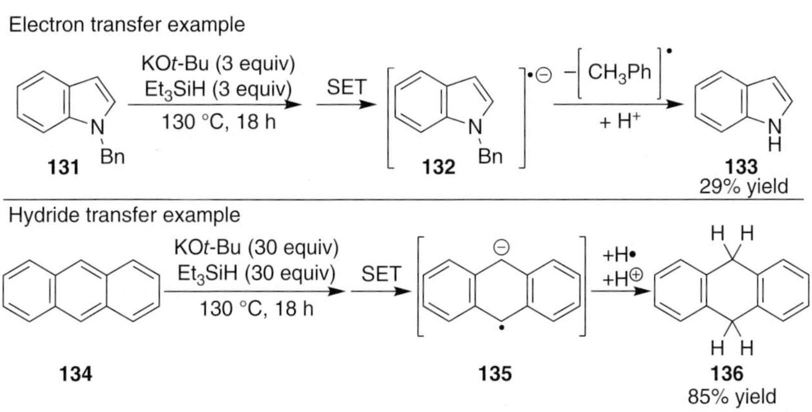

Scheme 7.23 Overview of electron transfer and hydride transfer in Stoltz–Grubbs silylation system proposed by Murphy and Tuttle et al.

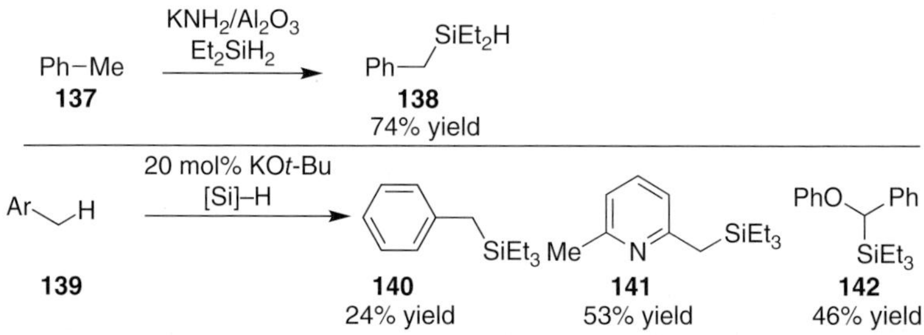

Scheme 7.24 Examples of C(sp^3)–H silylation.

While this is an interesting report related to the previously mentioned KO*t*-Bu-catalyzed C–H silylation methodology, it is unclear if either of these mechanisms (i.e. electron transfer or hydride transfer) are active under catalytic conditions or whether this is an opportunity to access new types of reactivity.

7.6 C(sp^3)–H Silylation

To our knowledge, only two examples of catalytic, transition-metal-free C(sp^3)–H silylation reactions have been reported in the literature. Both catalyst systems have already been presented here (see Sections 7.4.3 and 7.5.1.2, respectively) and both systems are limited to the activation of benzylic C—H bonds (Scheme 7.24) [19, 22].

7.7 Conclusion

This chapter has focused on detailing the major developments in transition-metal-free, C–H silylation reactions. One broad category of such catalysts involves the generation of an electrophilic silicon which is trapped by the nucleophilic substrate. FLPs, especially using $B(C_6F_5)_3$, have seen widespread use in this mode of C–H silylation, as they both activate silicon and may act as a base or hydride transfer reagent. While these catalysts have some common issues, they have proved to be useful in C–H silylation. Development of new Lewis acid catalysts may help suppress H_2 activation, Lewis pair formation with substrate or by-product, and expand the scope to less nucleophilic substrates.

The sole report by Oestreich detailing the use of Brønsted acid in analogous reactivity is exciting as this may provide an orthogonal method of silane activation and expand reactivity to less nucleophilic substrates. It goes without saying that many complex molecules are incompatible with a strong acid, and selective protonation of the silicon will be key to the further development of such acid catalyst systems.

Brønsted base silylation catalysts allow access to classes of substrates commonly made through a metalation and trapping approach but without the need for pyrophoric or otherwise undesired reagents. The key to this reactivity is the in situ generation of a transient, significantly stronger base than the catalyst itself. Especially if the Si—H bond of hydrosilanes can be used as a hydride base, this mode of catalysis is likely to see significant future use due to the familiar reactivity (i.e. versus a [Si]–X, base system) but with increased safety and decreased waste.

The final mode of catalysis explored is radical-type C—H bond silylation. In recent years, this mode of catalysis has been a major focus. This reactivity allows the activation of a wide variety of aromatic and even some aliphatic C—H bonds using simple reaction protocols. Catalytic generation of such a radical intermediate does pose challenges with functional-group compatibility (i.e. carbonyl, nitro, cyano, some olefins, etc.), which has limited the reaction scope. Future catalyst development may help avoid unwanted reactivity by modulating the reduction potential of the radical intermediate or decreasing the radical lifetime and concentration (i.e. by precomplexation of substrate and silane before radical generation).

A particular focus was devoted to discussing the different nature of the reaction catalysts and mechanisms, when known. Understanding the mechanism and inherent limitations will serve to guide the design of future C–H silylation catalysts. Given the interest this field has experienced in the recent years, we are excited to see what future developments bring.

References

1 Barry, A.J., Gilkey, J.W., and Hook, D.E. (1959). *Ind. Eng. Chem. Res.* 51: 131–138.

2 Barry, A.J., Gilkey, J.W., and Hook, D.E. (1959). *Metal-Organic Compounds: Advances in Chemistry*, vol. 23, 246–264. Washington, DC: American Chemical Society.

3 Wright, A. (1978). *J. Organomet. Chem.* 145: 307–314.

4 For a review on related Catalytic Friedel–Crafts reactivity see: Bähr, S. and Oestreich, M. (2017). *Angew. Chem. Int. Ed.* 56: 52–59.

5 (a) Piers, W.E., Marwitz, A.J.V., and Mercier, L.G. (2011). *Inorg. Chem.* 50: 12252–12262. (b) Stephan, D.W. (2015). *J. Am. Chem. Soc.* 137: 10018–10032.

6 Curless, L.D., Clark, E.R., Dunsford, J.J. et al. (2014). *Chem. Commun.* 50: 5270–5272.

7 Curless, L.D. and Ingleson, M.J. (2014). *Organometallics* 33: 7241–7246.

8 Ma, Y., Wang, B., Zhang, L. et al. (2016). *J. Am. Chem. Soc.* 138: 3663–3666.

9 Similar reactivity was described by Prof. Oestreich, but their optimized system involved a transition-metal-based catalyst: Yin, Q., Klare, H.F.T., and Oestreich, M. (2016). *Angew. Chem. Int. Ed.* 55: 3204–3207.

10 Han, Y., Zhang, S., He, J., and Zhang, Y. (2017). *J. Am. Chem. Soc.* 139: 7399–7407.

11 For stoichiometric examples see: Chen, Q.-A., Klare, H.F.T., and Oestreich, M. (2016). *J. Am. Chem. Soc.* 138: 7868–7871.

12 Fornarini, S. (1988). *J. Org. Chem.* 53: 1314–1316.

13 Sasaki, M. and Kondo, Y. (2015). *Org. Lett.* 17: 848–851.

14 Nozawa-Kumada, K., Osawa, S., Sasaki, M. et al. (2017). *J. Org. Chem.* 82: 9487–9496.

15 Calas, R. and Bourgeois, P. (1969). *C. R. Acad. Sci., Paris, Ser. C* 268: 72–74.

16 Itoh, M., Mitsuzuka, M., Utsumi, T. et al. (1994). *J. Organomet. Chem.* 476: C30–C31.

17 Ishikawa, J.-I., Inoue, K., and Itoh, M. (1998). *J. Organomet. Chem.* 552: 303–311.

18 Ishikawa, J.-I. and Itoh, M. (1999). *J. Catal.* 185: 454–461.

19 Baba, T., Kato, A., Yuasa, H. et al. (1998). *Catal. Today* 44: 271–276.

20 Itoh, M., Kobayashi, M., and Ishikawa, J. (1997). *Organometallics* 16: 3068–3070.

21 Toutov, A.A., Betz, K.N., Schuman, D.P. et al. (2017). *J. Am. Chem. Soc.* 139: 1668–1674.

22 Toutov, A.A., Liu, W.-B., Betz, K.N. et al. (2015). *Nature* 518: 80–84.

23 Leifert, D. and Studer, A. (2015). *Org. Lett.* 17: 386–389.

24 For a stoichimetric example of related reactivity see: Du, W., Kaskar, B., Blumbergs, P. et al. (2003). *Bioorg. Med. Chem.* 11: 451–458.

25 For a review on radical addition to heterocycles, including Minisci Reactions, see the following and references cited within; Tauber, J., Imbri, D., and Opatz, T. (2014). *Molecules* 19: 16190–16222.

26 Xu, L., Zhang, S., and Li, P. (2015). *Org. Chem. Front.* 2: 459–463.

27 Fedorov, A., Toutov, A.A., Swisher, N.A. et al. (2013). *Chem. Sci.* 4: 1640–1645.

28 Toutov, A.A., Salata, M., Fedorov, A. et al. (2017). *Nat. Energy* 2: 17008.

29 Toutov, A.A., Betz, K.N., Haibach, M.C. et al. (2016). *Org. Lett.* 18: 5776–5779.

30 Liu, W.-B., Schuman, D.P., Yang, Y.-F. et al. (2017). *J. Am. Chem. Soc.* 139: 6867–6879.

31 Banerjee, S., Yang, Y.-F., Jenkins, I.D. et al. (2017). *J. Am. Chem. Soc.* 139: 6880–6887.

32 Smith, A.J., Young, A., Rohrbach, S. et al. (2017). *Angew. Chem. Int. Ed.* 56: 13747–13751.

8

Silyl-Heck, Silyl-Negishi, and Related Reactions

Sarah B. Krause and Donald A. Watson

University of Delaware, Department of Chemistry and Biochemistry, 202 Lammot DuPont Lab, Newark, DE 19716, USA

8.1 Introduction

Organosilanes have wide applications in synthetic chemistry, materials science, medicine, agrochemicals, and other industries. The formation of allyl-, vinyl-, and alkylsilanes from silicon-halide reagents is attractive as these starting materials are widely abundant and inexpensive. Over the past several years, palladium- and nickel-catalyzed cross-coupling of silicon electrophiles with organic nucleophiles have emerged as a new approach to the preparation of certain classes of organosilanes. The development of these C—Si bond-forming reactions reflects the now commonplace use of cross-coupling technologies using carbon-based electrophiles.

Early studies on the transition-metal activation of silicon–halogen bonds, particularly from the Tanaka group, revealed that various transition metals can undergo oxidative addition into the silicon–halogen bond. More recently, this activation has been applied in cross-coupling reactions for the synthesis of organosilanes. Seminal work by Tanaka and Murai, and more recently by Watson, explored cross-coupling of silicon electrophiles with alkenes or alkynes to form unsaturated organosilanes. Further work by Watson gave access to alkylsilanes via palladium-catalyzed silyl-Negishi and silyl-Kumada reactions. This chapter reviews the development and current state of these transformations, inclusive of work published through the end of 2017.

8.1.1 Activation of Silicon–Halogen Bonds

The cross-coupling of silicon electrophiles relies on the activation of silicon–halogen bonds as the first step in the mechanism of these reactions. Several early examples demonstrated the oxidative addition of silicon–halogen bonds with transition-metal catalysts.

Organosilicon Chemistry: Novel Approaches and Reactions, First Edition.
Edited by Tamejiro Hiyama and Martin Oestreich

8.1.1.1 Oxidative Addition to Platinum Complexes

Tanaka first described the stoichiometric oxidative addition of the Si—X bond of trimethylsilyl bromide (Me_3SiBr) and trimethylsilyl iodide (Me_3SiI) to zero-valent platinum complexes. These complexes were isolated and characterized by single-crystal X-ray diffraction (Scheme 8.1a) [1]. The resulting bisphosphine complexes are square planar with a *trans*-relationship between the silicon atom and the halide. They also found that oxidative addition of trimethylsilyl chloride (Me_3SiCl) does not occur, even at higher temperatures. This can be explained by the stronger Si—Cl bond (98 kcal/mol) compared to Si—Br (76 kcal/mol) and Si—I (54 kcal/mol) [2]. In subsequent studies, when di- and trichlorosilanes were analyzed, it was found that reactivity toward oxidative addition increased with increasing amounts of chlorides on the silicon atom, observing oxidative addition of polychlorosilane (Scheme 8.1b) [3].

The oxidative addition of Me_3SiI and Me_3SiBr was also demonstrated with a bipyridine-ligated platinum(II) complex **1** to form the platinum(IV)–silyl complex **2** (Scheme 8.2) [4]. This platinum(IV)–silyl complex was characterized by single-crystal X-ray diffraction, showing a *trans*-relationship between the silicon atom and the iodine atom.

8.1.1.2 Oxidative Addition to Palladium Complexes

Until very recently, no oxidative-addition complexes of silicon–halogen bonds to palladium had been reported. In 2009, Cloke and coworkers reported the first example of a complex resulting from oxidative addition of a silicon–halogen bond to a palladium center (Scheme 8.3) [5]. The addition of Me_3SiI to palladium(0) complex **3** supported by an N-heterocyclic carbene (NHC) resulted in the formation of the oxidative-addition complex **4** after 40 days, which was characterized by single-crystal X-ray diffraction. In this case, the complex is formed as a μ-iodo dimer, bearing two square planar metal centers.

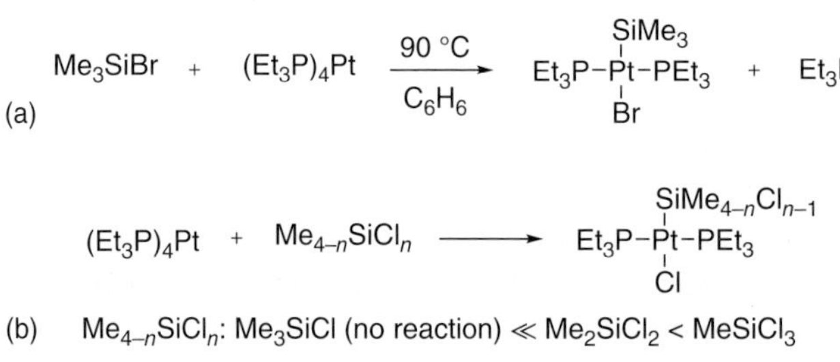

Scheme 8.1 Oxidative addition of Si—Br (a) and Si—Cl (b) bonds to platinum complexes.

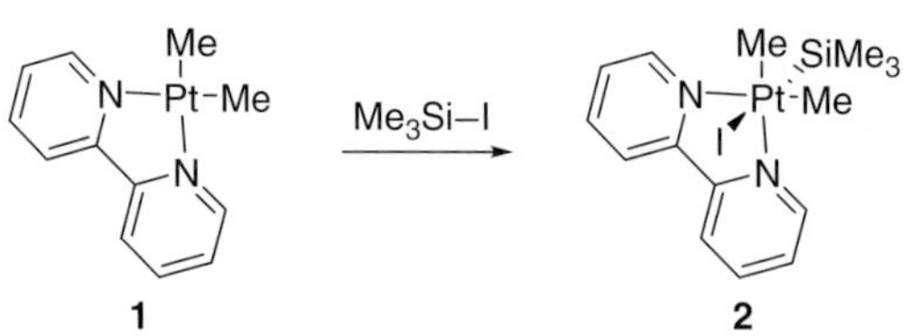

1

2

Scheme 8.2 Oxidative addition of a Si—I bond to bipyridine-ligated platinum complex.

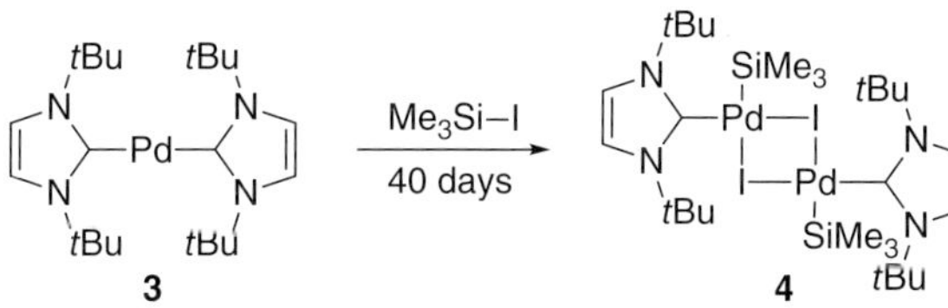

Scheme 8.3 Oxidative addition of a Si—I bond to an NHC–palladium complex.

The Watson and Murai groups have also recently reported oxidative-addition complexes of silicon–halogen bonds to palladium. This is discussed subsequently.

8.1.1.3 Oxidative Addition to Iridium and Rhodium Complexes

Only a few reactions of group 9 metals with silyl halides have been described. Milstein and coworkers reported the oxidative addition of Si—Cl bonds to an iridium complex [6]. The oxidative addition of methyltrichlorosilane ($MeSiCl_3$) to complex **5** led to the formation of the octahedral silanide complex **6**, which was characterized by single-crystal X-ray diffraction. In contrast, the reaction with Me_3SiCl led to the iridium-hydride complex **7** (Scheme 8.4). It was proposed that Me_3SiCl oxidatively adds to the iridium complex leading to intermediate **8**, which is then followed by β-hydride elimination to form the observed iridium-hydride complex and dimethyl silene.

In an interesting example, Tanaka reported the halide exchange of both bis(triphenylphosphine)rhodium(I) carbonyl chloride ($(CO)(Ph_3P)_2RhCl$) and bis(triphenylphosphine)iridium(I) carbonyl chloride ($(CO)(Ph_3P)_2IrCl$) with Me_3SiI to form complexes **9** (Scheme 8.5). The mechanism of this transformation may involve oxidation addition to the metal center to form the metal(IV) intermediate **10**; however, a four-centered σ-bond-metathesis mechanism may also account for this reactivity.

Recently, however, the oxidative addition of silicon–halogen bonds to rhodium complexes was conclusively described by Ozerov and coworkers (Scheme 8.6) [7]. Oxidative addition of Me_3SiI to the rhodium(I) complex **11** proceeds to completion (line a). Similar to the earlier trend observed with platinum complexes by Tanaka and coworkers, the rate of oxidative addition increases with decreased bond strength. In the case of silyl halides with a stronger silicon–halogen bond,

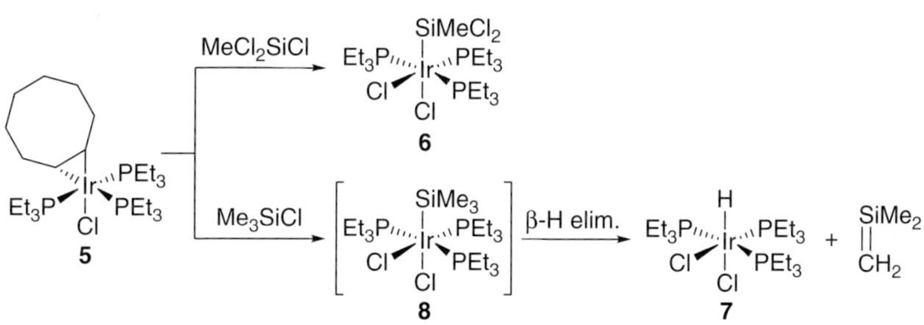

Scheme 8.4 Oxidative addition of Si—Cl bonds to iridium.

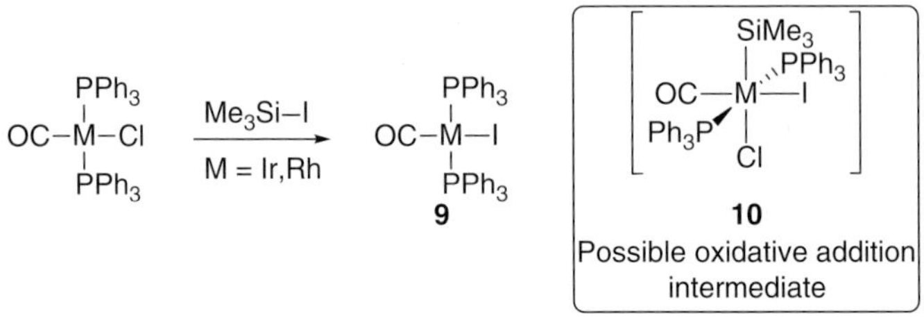

Scheme 8.5 Oxidative addition of Si—I bonds to iridium and rhodium, respectively.

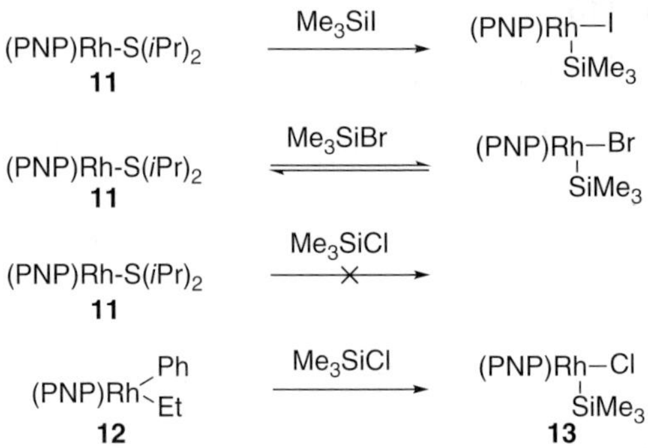

Scheme 8.6 Trends of rhodium oxidative addition.

such as Me_3SiBr, the oxidative addition is equal in energy to the binding of the isopropyl sulfide ligand (line b), and the oxidative addition of Me_3SiCl using complex **11** did not occur (line c). However, using complex **12**, the irreversible reductive elimination of ethylbenzene drives oxidative addition of the Si—Cl bond, forming complex **13** (line d).

8.2 Silyl-Heck Reactions

Allyl- and vinylsilanes are important intermediates in synthetic chemistry. Among many applications, Hiyama cross-coupling [8], Hosomi–Sakurai allylation [9], and Fleming–Tamao oxidation [10] are particularly important.

Numerous methods exist for the preparation of both allyl- and vinylsilanes; however, these methods have limitations that make preparation cumbersome. For example, several methods require manipulation of prefunctionalized substrates. Several methods also require the use of harsh nucleophiles, which limits the functional-group tolerance [11]. Hydrosilylation of alkynes is a mild alternative. However, this requires access to an alkyne starting material, which has a limited commercial availability. Finally, several methods have been developed which introduce the silicon center indirectly using C—C bond-forming reactions, such as alkene metathesis and C—C bond-forming cross-coupling reactions. However, these methods require an unsaturated organosilane starting material, which must be prepared by other methods.

A method to prepare allyl- and vinylsilanes from simple alkene starting materials is highly attractive. Simple alkenes are inexpensive, widely available, and stable reagents. However, few methods exist to directly attach silyl groups to alkenes to form allyl- or vinylsilanes. The silyl-Heck reaction allows for the direct conversion of simple, terminal alkenes, i.e. α-olefins and styrenes, to allyl- and vinylsilanes using silicon electrophiles. The reaction proceeds through a mechanism similar to the Heck arylation reaction (Scheme 8.7) [12]. Oxidative addition of a silicon electrophile to a low-valent, late transition-metal complex (Pd or Ni) produces the oxidative-addition complex **14**. Migratory insertion of an alkene leads to the palladium(II) intermediate **15**, which undergoes β-hydride elimination to give the allyl- or vinylsilane product (**16**).

Although the potential of using the silyl-Heck reaction for the preparation of unsaturated organosilanes was recognized over 25 years ago, it was not until recently that practical protocols for this reaction have been developed.

8.2.1 Early Silyl-Heck Studies

Early studies from the Tanaka group laid the foundation for silyl-Heck reactions. Tanaka and coworkers was the first to recognize the importance of the oxidative-addition complexes of Me_3SiI and Me_3SiBr and related intermediates in Heck-like reactions. Initial attempts focused on the stoichiometric reaction with the platinum bromide complex **17** and styrene. Although traces of the desired product **18** were observed, the main product was Me_3SiBr, indicating that the oxidative addition to the Si—Br bond was reversible (Scheme 8.8) [3].

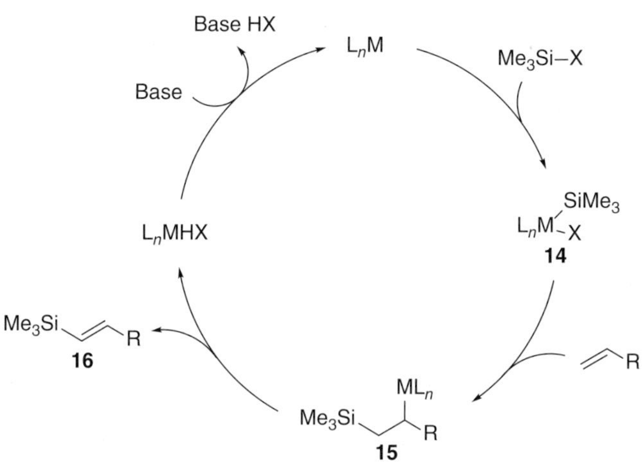

Scheme 8.7 The silyl-Heck reaction mechanism.

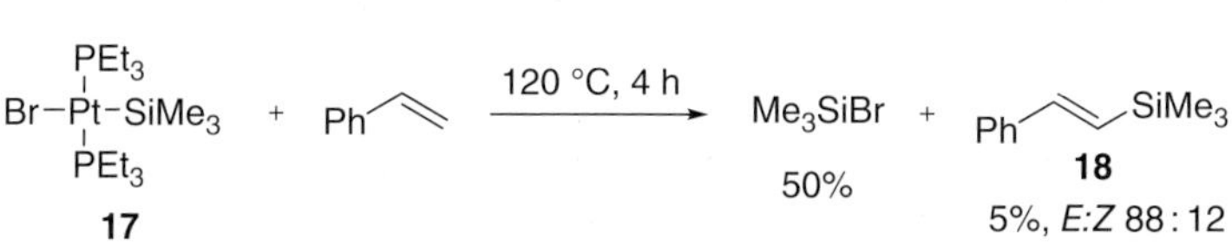

Scheme 8.8 Reaction of Tanaka's oxidative-addition complex.

In a subsequent study, Heck-like reactivity was observed with Me_3SiI using a palladium catalyst [13]. After prolonged heating at 120 °C in triethylamine (Et_3N), bis(triethylphosphine)palladium(II) dichloride ($(Et_3P)_2PdCl_2$) was found to catalyze the reaction of Me_3SiI with excess styrene to form vinylsilane **19** in modest yield (54% based on Me_3SiI) as a single regioisomer with >99 : 1 *E*:*Z* selectivity (Scheme 8.9). This report represented the seminal example of a silyl-Heck reaction. However, it also highlighted the challenges with making the reaction synthetically practical: low yield (particularly with respect to alkene), inverted stoichiometry (olefin is used in large excess), and limited olefin scope (only three simple styrene derivatives were studied).

8.2.2 Multicomponent Coupling

In a related process, Murai and coworkers developed a three-component reaction that proceeds through a similar mechanism involving oxidative addition of a silicon–halogen bond to palladium(0) (Scheme 8.10) [14]. It is presumed that oxidative addition of Me_3SiI to the palladium(0) complex followed by migratory insertion of the terminal alkyne leads to the vinylpalladium species **20**, which then undergoes transmetalation with an acetylenic tin reagent. Reductive elimination produces the vinylsilane product **21** and regenerates the palladium(0) catalyst.

Using this reaction, a variety of trisubstituted vinylsilanes were formed in good yields with good regio- and stereoselectivity (Scheme 8.11). Both aromatic and

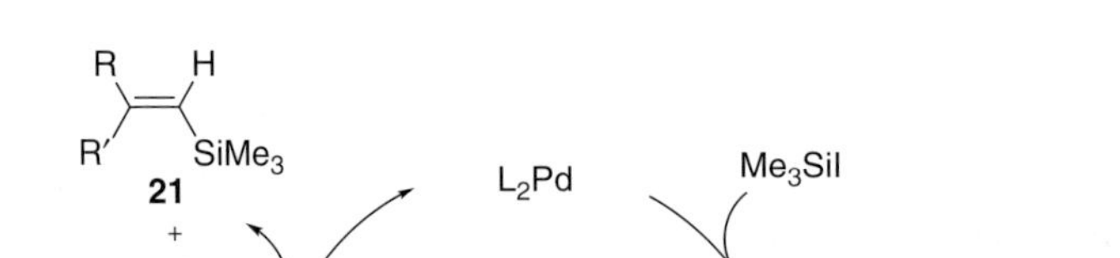

Scheme 8.9 The first example of a silyl-Heck reaction.

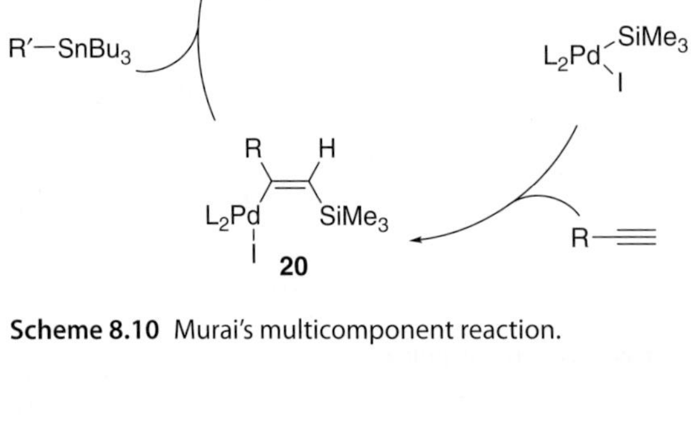

Scheme 8.10 Murai's multicomponent reaction.

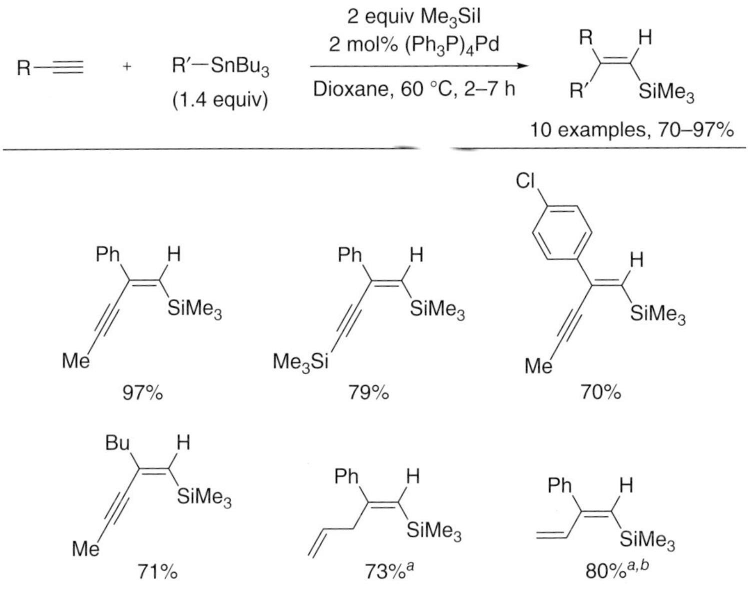

[a] 8 mol% $(Ph_3P)_4Pd$.
[b] 2 equiv $R'SnBu_3$.

Scheme 8.11 Scope of Murai's multicomponent reaction.

aliphatic terminal alkynes were tolerated as well as various organostannanes including alkynyl-, allyl-, and vinylstannanes.

Further investigations into this reaction found that zinc transmetalating agents, such as diethylzinc (Et_2Zn), can be utilized in the reaction to form trisubstituted alkenes in good yields and selectivities, thereby significantly expanding the scope of substitution on both the alkyne and nucleophile in the reaction (Scheme 8.12) [15].

8.2.3 Improved Silyl-Heck Reaction Conditions

Our group recognized the utility of Tanaka's results as an entry into the synthesis of unsaturated organosilanes. We hypothesized that a general protocol for converting terminal alkenes into allyl- or vinylsilanes would be possible with the development of high-yielding and mild reaction conditions by continued investigation of the catalytic conditions [16].

Identification of an effective ligand was key in the development of a general and practical transformation. Using (1,5-cyclooctadiene)bis(trimethylsilylmethyl) palladium(II) ((COD)Pd($CH_2TMS)_2$) as the palladium pre-catalyst, the reaction of 4-*tert*-butylstyrene and Me_3SiI was examined as a model system for systematically determining the effects of different phosphine ligands (Scheme 8.13). To maximize the potential for developing general reaction conditions, our initial studies were conducted at a lower temperature (80 °C), in solvent (toluene), and used the alkene as the limiting reagent. Triethylphosphine (Et_3P) as used by

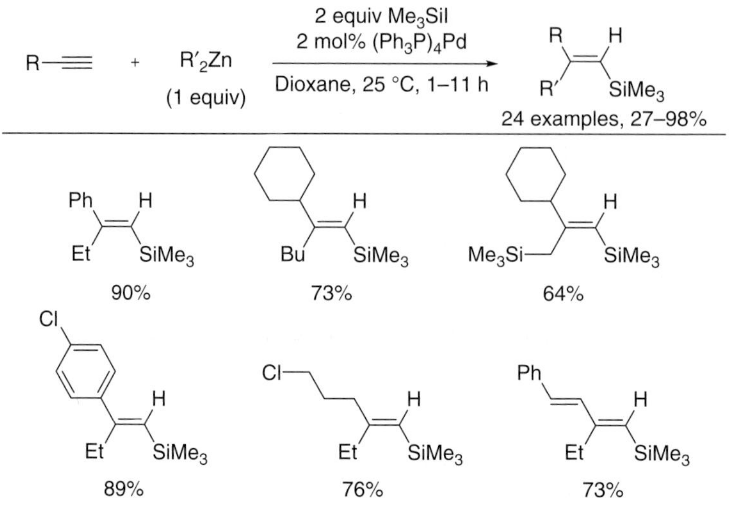

Scheme 8.12 Scope of the multicomponent cross-coupling reaction with zinc nucleophiles.

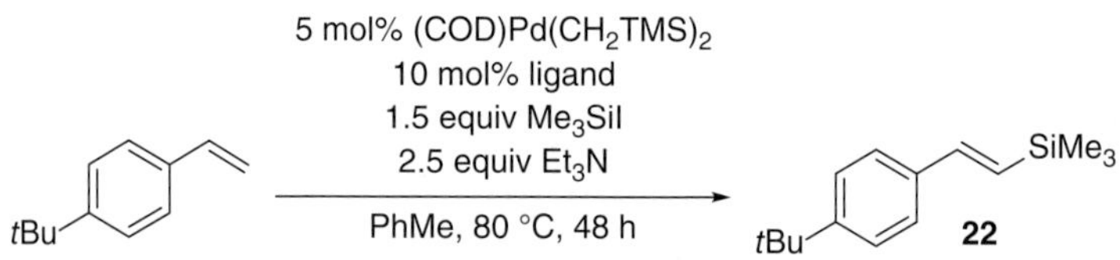

Entry	Ligand	Yield (%)	Entry	Ligand	Yield (%)
1	Et_3P	6	5	Cy_3P	4
2	Ph_3P	4	6	Cy_2PhP	27
3	dppe	0	7	$CyPh_2P$	65
4	$t\mathrm{Bu}_3P$	1	8	$t\mathrm{BuPh}_2P$	80

Scheme 8.13 Identification of a ligand for the silyl-Heck reaction.

Tanaka and triphenylphosphine (Ph_3P) as used by Murai and coworkers in related three-component reactions did not give more than trace yields of the desired product **22** under the reaction conditions (entries 1 and 2). Bidentate ligands, such as 1,2-bis(diphenylphosphino)ethane (dppe), were also ineffective in catalyzing the reaction (entry 3). Larger and more electron-rich trialkylphosphine ligands, which have been shown, aid in the oxidative-addition and electron-exchange processes through the formation of low-valent palladium complexes, also did not provide significant yield of the desired product (entries 4 and 5). This is presumably due to steric demands that would be encountered in the oxidative addition products of Me_3SiI using these large ligands. In contrast, phosphine ligands containing both alkyl and phenyl groups provided significantly improved results (entries 6–8). For example, Cy_2PhP provided 27% of the desired product (entry 6). Ultimately, optimization studies showed that *tert*-butyldiphenylphosphine ($tBuPh_2P$) was the optimal ligand for the reaction (entry 8). We attribute the success of the mixed aryl/alkyl ligands to their being sufficiently large and electron-rich enough to support low-valent palladium(0), while still providing sufficient space around the palladium(II) center to accommodate the large $SiMe_3$ group. After optimization of reaction conditions (most notably by lowering the temperature and concentration), a wide range of vinylsilanes were formed from styrene derivatives in good yields (Scheme 8.14). After further minor optimizations, the functional-group tolerance of this reaction was explored. The reaction is tolerant of a variety of functional groups, including halides, ethers, ketones, esters, and heterocycles. In all cases, as was observed by Tanaka, the reaction is selective for terminal silylation producing exclusively the *E*-vinylsilane.

Utilization of chlorosilanes in the silyl-Heck reaction is desirable due to their greater availability and functional-group compatibility and commercial availability; however, activation of the Si—Cl bond has proved to be difficult due to its high bond strength (113 kcal/mol). While attempts to directly use chlorosilanes in the silyl-Heck reaction did not succeed, we were able to show that Me_3SiCl can be utilized if lithium iodide (LiI) is added to the reaction. It is assumed that *in situ* halide exchange occurs, leading to the formation of Me_3SiI in the reaction solution [17]. Under the conditions shown in Scheme 8.15, using Me_3SiCl, 4-*tert*-butyl styrene was silylated in 94% yield. Moreover, the reaction of Me_3SiBr is sluggish, and Me_3SiOTf does not react using our palladium-catalyzed conditions.

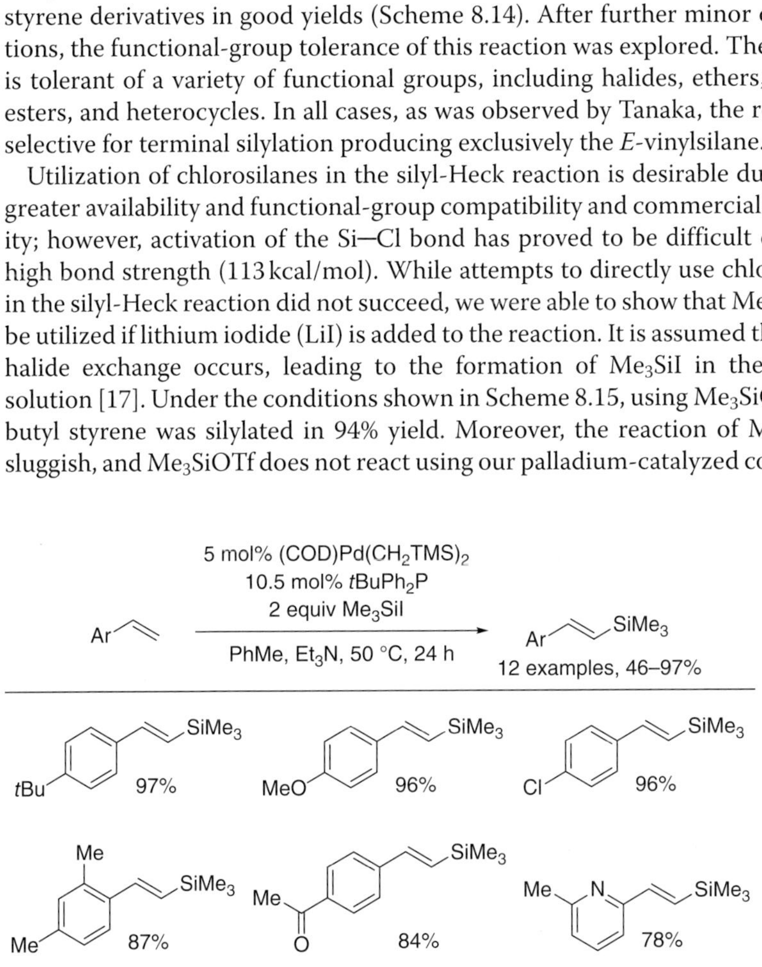

Scheme 8.14 Scope of the silyl-Heck reaction with styrene substrates.

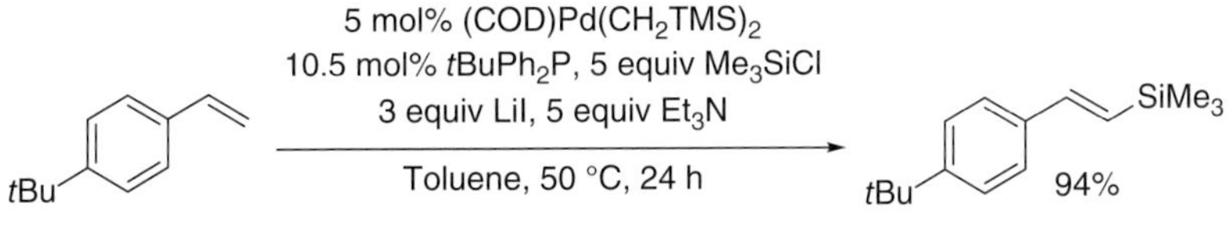

Scheme 8.15 The silyl-Heck reaction with Me_3SiCl.

However, with added LiI, both reagents were shown to participate in the silyl-Heck reaction leading to yields comparable to those from the direct reaction with Me_3SiI.

The use of other, non-styrenyl, alkene substrates in silyl-Heck reactions was also investigated. In particular, terminal alkenes bearing allylic hydrogen atoms (α-olefins) were of interest, as either allyl- or vinylsilanes could emerge as products of the silyl-Heck mechanism. Using the conditions developed for use with styrenes, we found that α-olefins prefer to form allylsilanes. Initially, low yields were observed due to isomerization of the starting material to unreactive internal alkenes. With slightly modified conditions (changing the reaction solvent and lowering the temperature), however, isomerization of the starting material could be reduced to give synthetically useful yields of a variety of allylsilanes with good to excellent selectivity for the *E*-allylsilane (Scheme 8.16). In nearly all cases, the selectivity for allylsilanes over vinylsilanes was close to 95 : 5.

Alkene isomerization likely originates from the Pd–H species. Attempts to sequester those metal hydrides (e.g. by modulating the base) proved unsuccessful. Therefore, to minimize isomerization under reaction conditions, we reasoned that the speed of the productive reaction needed to be increased so that it could outcompete the isomerization reaction. In cross-coupling reactions involving carbon electrophiles, numerous studies have shown that more electron-rich ligands favor oxidative addition and larger ligands lead to more reactive catalysts by preventing over-coordination of the metal center. As such, we wished to investigate larger and more electron-rich ligands in the silyl-Heck reaction as a means to overcome the background alkene isomerization. We hypothesized that oxidative addition was likely a challenging step in the silyl-Heck reaction and that larger, more electron-rich ligands would thereby improve the reaction outcomes. Through rational ligand design, a series of diaryl-*tert*-butyl phosphine ligands were prepared and analyzed by their effectiveness in the silyl-Heck reaction of dec-1-ene and Me_3SiI (Scheme 8.17) [18]. Consistent with our hypothesis, the electron density of the ligands had a dramatic effect on

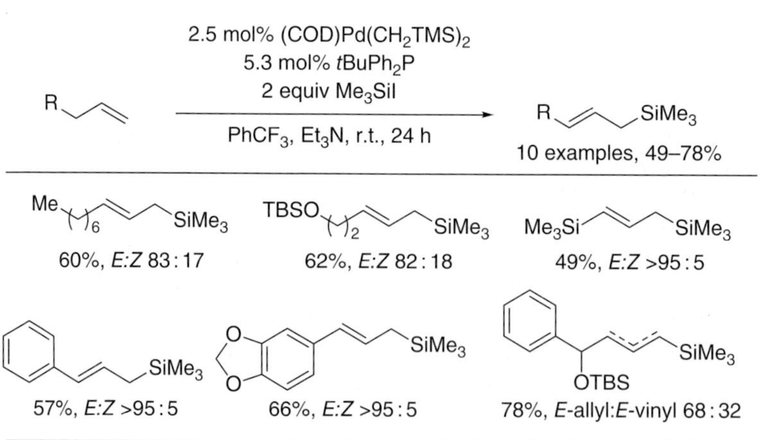

Scheme 8.16 Scope of the silyl-Heck reaction to form allylsilanes.

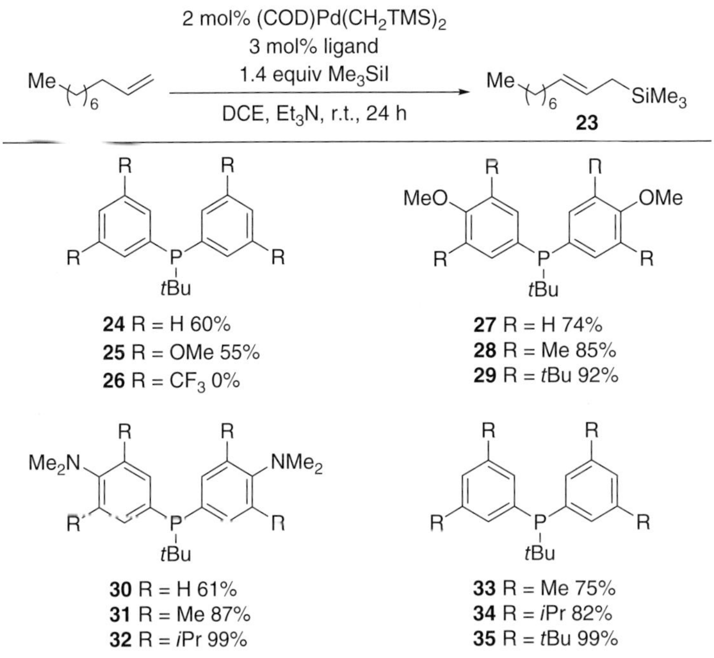

Scheme 8.17 Optimization of the silyl-Heck reaction through rational ligand design.

the yield of the desired product **23** formed. For example, the *p*-anisyl-substituted ligand **27** increased the yield of the desired product in the reaction, whereas ligands **25** and **26** with electron-withdrawing groups in *meta*-positions decreased the yield of the desired product. Further increasing the electron-donor ability of the ligand with alkyl groups in the *meta*-position as in **28** and **29** continued to improve the reaction. Ligands with a dimethylamino group in the *para*-position were also studied. Ligand **30** lacking *meta*-substitution gave similar results to ligand **24**, and variant **31** with methyl substitution in the *meta*-position gave similar results to the methoxy variant **28**. However, with *meta*-isopropyl substitution as in **32**, nearly quantitative yields of the desired product **23** were observed. While this was a very improved result, examination of the X-ray structure of **32** suggested that the large isopropyl groups force the dimethylamino group into a conformation where the nonbonding electrons on the nitrogen atom are perpendicular to the π-system of the aromatic ring. As such, we questioned how much electronic benefit was being provided from the dimethylamino group. To study this, a series of ligands with only alkyl substitution at the *meta*-positions were prepared. Ligands **33** and **34** bearing *meta*-methyl and -isopropyl groups showed significant improvement over $t\text{BuPh}_2\text{P}$, and ligand **35** bearing *meta-tert*-butyl groups gave nearly quantitative yield of the desired product. This ligand was identified as the optimal for the reaction because it is easily prepared in a single step on a multi-gram scale, and is stable to storage on the bench under air for extended periods of time.

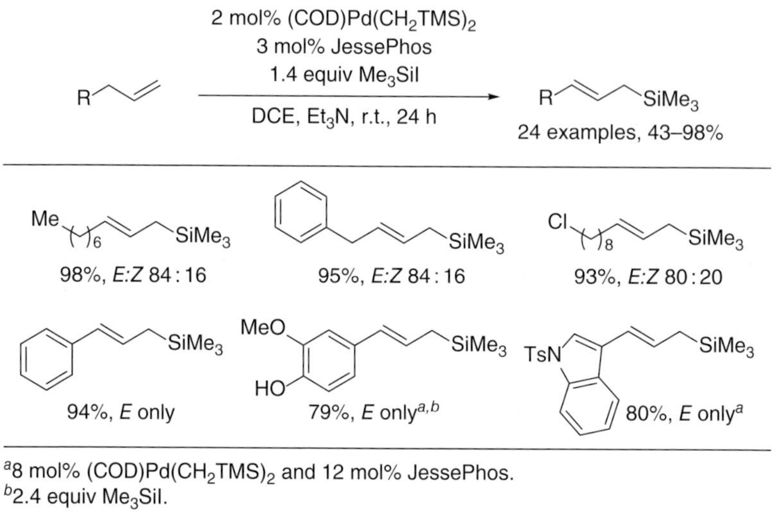

Scheme 8.18 Scope of the silyl-Heck reaction with the second-generation ligand.

Using ligand **35** (subsequently named **JessePhos**) and dichloroethane (DCE) as the reaction solvent, the amount of Me_3SiI and the reaction temperature could both be lowered. Under these conditions, there was almost no isomerization of the terminal alkene starting material. A variety of allylsilanes were prepared from α-olefins in good yields and with similar high selectivity for the *E*-allylsilane as was observed with previous conditions (Scheme 8.18).

8.2.4 Mechanistic Considerations

To better understand the silyl-Heck reaction, numerous mechanistic investigations have been undertaken. The formation of allylsilanes appears to be kinetic in origin. To investigate this, vinylsilanes were added to an allylsilane-forming silyl-Heck reaction to determine whether the allylsilane products of the silyl-Heck reaction were resulting from isomerization of vinylsilanes. Vinylsilane **36** was added to the silyl-Heck reaction of hex-1-ene and Me_3SiI (Scheme 8.19).

Scheme 8.19 Kinetic stability of vinylsilanes under silyl-Heck reaction conditions.

Allylsilane **37** was formed in greater than 95% yield. Importantly, vinylsilane **36** remained unchanged during the reaction, and allylsilane **38**, which would be formed from the isomerization of vinylsilane **36**, was not observed. This suggests that the observed allyl/vinyl ratios in the products of the silyl-Heck reaction are kinetic in nature and implicate β-hydride elimination from the alkylpalladium species as the selectivity-determining step.

To better understand the selectivity for allyl- versus vinylsilanes, we investigated the ratio of products formed in the silyl-Heck reaction of *tert*-butyl-substituted alkene **39** (Scheme 8.20). The reaction produces a mixture of allyl- and vinylsilane products in a 72 : 28 ratio. In principle, either steric or electronic effects could determine the selectivity during β-hydride elimination. Considering the putative allylpalladium intermediate **42** that forms en route to **40**, β-hydride elimination could occur toward either the *tert*-butyl group or the $SiMe_3$ group (Scheme 8.21). In a typical silyl-Heck reaction lacking a fully substituted carbon center, steric repulsion should favor allylsilane formation. However, in the case of **42**, we predicted that steric selectivity would favor the formation of vinylsilane **41** due to the greater steric influence of the *tert*-butyl group compared to the $SiMe_3$ group (*A*-value = 4.9 versus 2.5 kcal/mol). In contrast, because of the highly electropositive nature of silicon, the hydrogen atoms α to the silicon center bear decreased hydricity compared to those on an alkyl chain (the α-silicon effect). As such, electronic control would favor β-hydride elimination toward the *tert*-butyl group. The experimentally observed allyl/vinyl ratio of the product shows that product formation favors allylsilane **40**. Although the observed ratio of allylsilane/vinylsilane is lower than is typically observed (c. 95 : 5), the reaction still favors the allyl isomer and strongly suggests a significant electronic contribution to the selectivity of the silyl-Heck reaction.

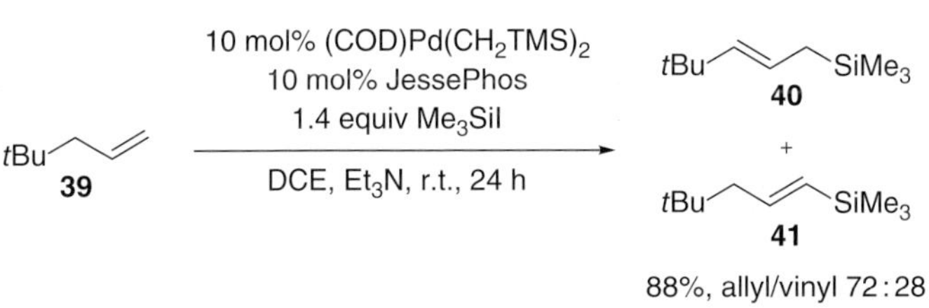

Scheme 8.20 Investigation of allyl/vinyl ratio in the silyl-Heck reaction.

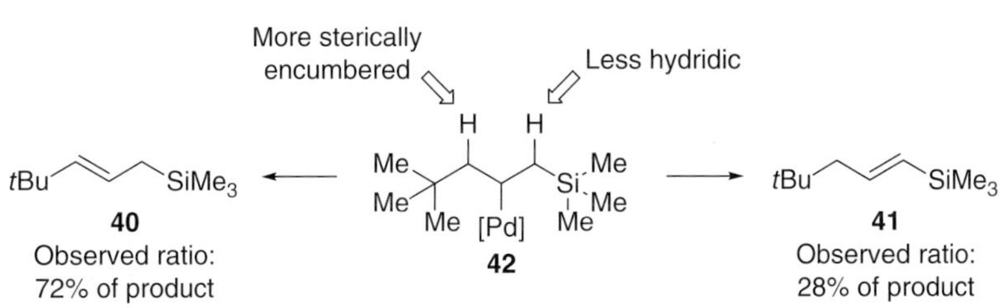

Scheme 8.21 Steric versus electronic control in β-hydride elimination.

We also wished to better understand the nature of the active catalyst. Our optimization studies showed that the silyl-Heck reaction is particularly sensitive to the ligand/metal (L/M) ratio used. Optimal reaction conditions use 1 : 1 to 1.5 : 1 L/M ratios, and significant decreases in the yields of reactions were observed with an increase of the L/M ratio to 2 : 1. These results suggest that monoligated palladium(0) species are involved in the oxidative addition in the reaction. The existence of these low-valent species was confirmed with the isolation of the oxidative-addition complex **43** (Scheme 8.22). This distorted T-shaped monodentate phosphine complex, with an open coordination site opposite from the silicon ligand, was fully characterized by NMR and single-crystal X-ray crystallography. This was the first characterized example of a monomeric palladium complex resulting from the oxidative addition of a silicon–halogen bond. This structure, along with the previously mentioned importance of the L/M ratio, strongly suggests that the active catalyst in the silyl-Heck reaction is a monophosphine–palladium complex. Importantly, this structure also reveals why this class of (alkyl)(diaryl)phosphines are so effective in the silyl-Heck reaction; the aromatic groups are able to rotate into a nearly coplanar conformation, creating space to accommodate the large silyl ligand.

8.2.5 Pre-catalyst Investigations

After an efficient ligand was developed for the silyl-Heck reaction, our attention turned toward improving the palladium pre-catalysts. The optimized conditions for the formation of both vinyl- and allylsilanes required (COD)Pd(CH_2TMS)$_2$ as a palladium pre-catalyst. Although this is easily prepared on a gram scale and is air stable, it is slightly thermally sensitive and decomposes after standing at room temperature. To increase the utility of the silyl-Heck reaction, alternative pre-catalyst systems were investigated. First, the conditions for the formation of vinylsilane **22** from 4-*tert*-butylstyrene were analyzed (Scheme 8.23) [19]. Optimized silyl-Heck conditions for the reaction showed that (COD)Pd(CH_2TMS)$_2$ gave a quantitative yield (entry 1) of the desired product; however, changing the palladium source to tris(dibenzylideneacetone)dipalladium(0)

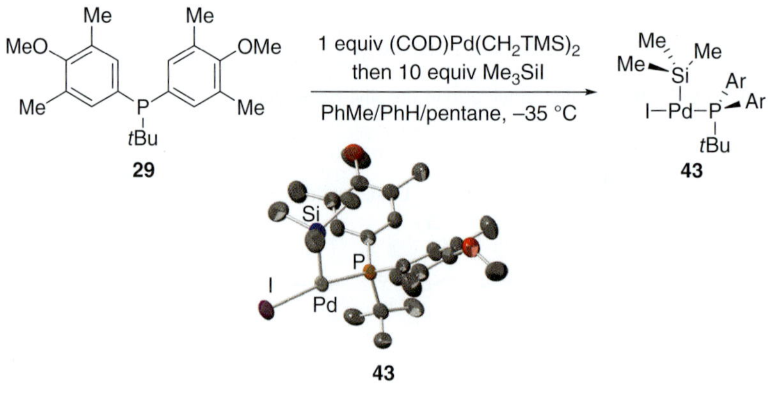

Scheme 8.22 Formation of oxidative addition complex **43**.

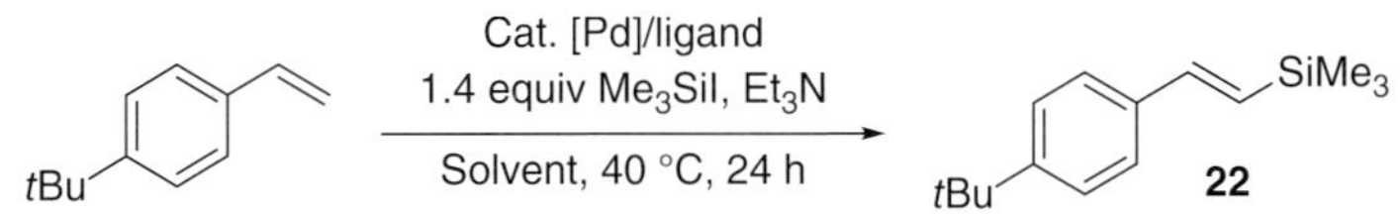

Entry	Solvent	Ligand	[Pd]	Yield (%)
1	PhMe	10.5 mol% *t*BuPh$_2$P	5 mol% (COD)Pd(CH$_2$TMS)$_2$	98
2[a]	PhMe	10.5 mol% *t*BuPh$_2$P	2.5 mol% Pd$_2$(dba)$_3$	12
3	DCE	5 mol% *t*BuPh$_2$P	2.5 mol% Pd$_2$(dba)$_3$	75
4	DCE	7.5 mol% JessePhos	2.5 mol% Pd$_2$(dba)$_3$	91
5	DCE	5 mol% JessePhos	2.5 mol% Pd$_2$(dba)$_3$	99

[a]2.0 equiv Me$_3$SiI, 50 °C

Scheme 8.23 Effect of palladium pre-catalyst and ligand on yield.

(Pd$_2$(dba)$_3$) dramatically decreased the yield of the product formed during the reaction (entry 2). Somewhat unexpectedly, a significant increase in yield was observed with a change in the reaction solvent to DCE (entry 3); however, the yields remained much lower than with the original catalyst system employing (COD)Pd(CH$_2$TMS)$_2$. In contrast, the use of **JessePhos** significantly increased the yield of the desired product (entry 4). After further optimization of reaction conditions, vinylsilane **22** was obtained in quantitative yield (entry 5).

Using these reaction conditions, a variety of vinylsilanes could be obtained via the silyl-Heck reaction using a widely abundant, commercially available, and bench-stable palladium pre-catalyst (Scheme 8.24). We viewed this as a significant practical advance in silyl-Heck chemistry. A wide range of functional groups were well tolerated using these conditions. In all cases, the yields of products were comparable to those obtained with the first-generation conditions and only the *E*-vinylsilane was observed. Further, due to the greater reactivity of the Pd$_2$(dba)$_3$/JessePhos system, silyl iodides containing groups larger than methyl

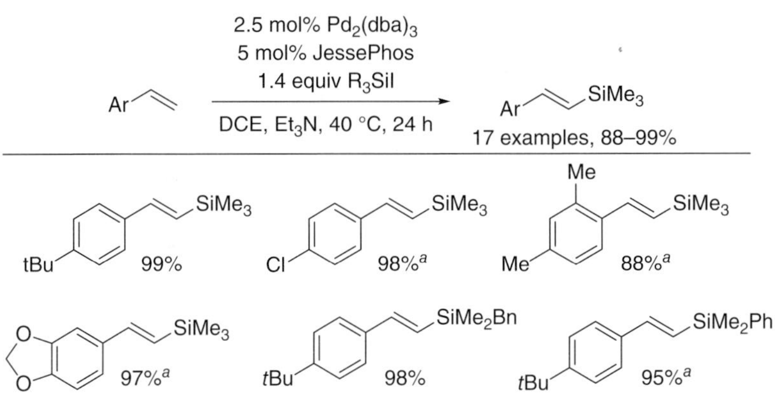

[a]2.0 equiv R$_3$SiI.

Scheme 8.24 Scope of the second-generation silyl-Heck reaction to form vinylsilanes.

could also be activated under these conditions, leading to vinylsilanes with potentially different reactivities in subsequent transformations. Although this ligand does not completely solve steric limitations (trace yield observed with triethylsilyl iodide), vinylsilanes containing benzyl, phenyl, and furyl groups were prepared in high yields. These types of allyl groups have shown benefit in downstream chemistry compared to $SiMe_3$.

Although this pre-catalyst system works well for the formation of vinylsilanes, it is significantly less effective for the formation of allylsilanes than the conditions employing $(COD)Pd(CH_2TMS)_2$ as a palladium pre-catalyst. Using dec-1-ene as a substrate, a significant amount of isomerized starting alkene was observed, leading to a decreased yield of product **23** (Scheme 8.25).

One integral aspect of these new conditions was the superiority of DCE as the reaction solvent, which is unusual in Pd(0)/Pd(II) chemistry. We wished to further understand these solvent dependencies. This investigation began by studying the pre-catalyst components employed for the formation of allylsilanes [20]. The combination of $(COD)Pd(CH_2TMS)_2$ and JessePhos in an inert solvent, such as toluene, led to the formation of the Pd(0) complex $(JessePhos)_2Pd$ (Scheme 8.26). This is in accord with similar studies from the Buchwald group using diaryl phosphine ligands [21]. In contrast, the same reaction preformed in DCE led to the formation of the Pd(II) complex $(JessePhos)_2PdCl_2$. Both of these complexes were isolated and characterized by single-crystal X-ray diffraction.

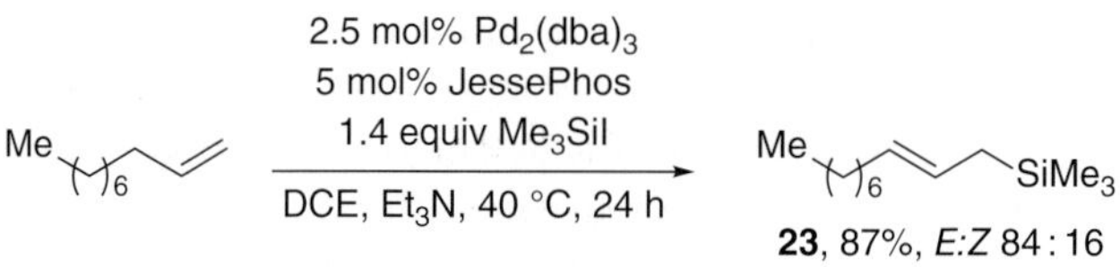

Scheme 8.25 Reaction of olefins using $Pd_2(dba)_3$.

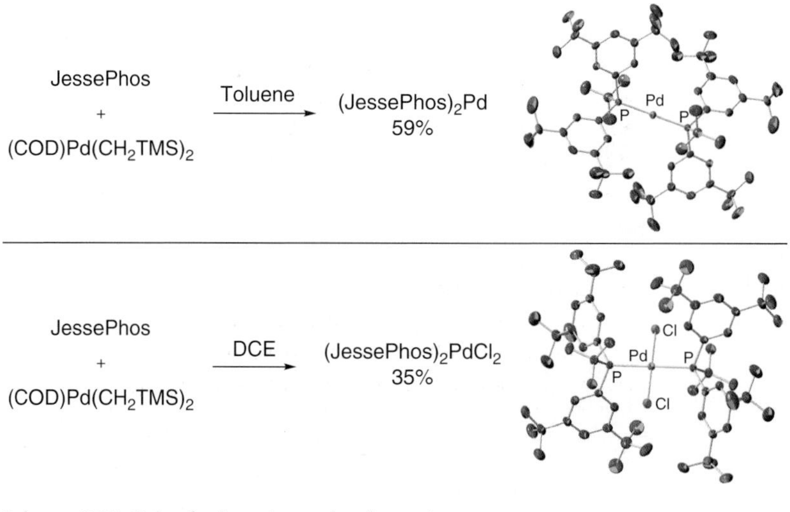

Scheme 8.26 Role of solvent in catalyst formation.

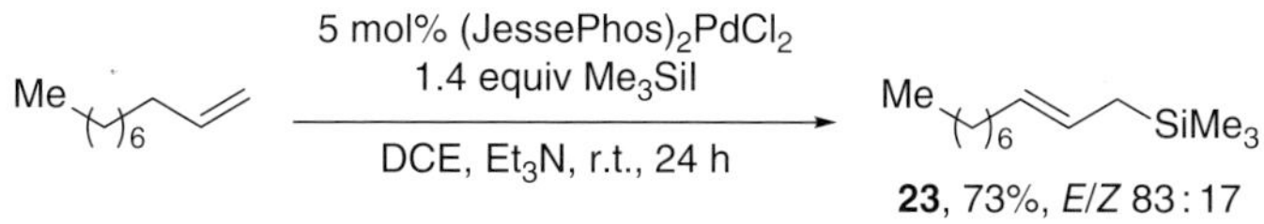

Scheme 8.27 The silyl-Heck reaction with $(JessePhos)_2PdCl_2$.

$(JessePhos)_2PdCl_2$ complex was then investigated as a single-component pre-catalyst for silyl-Heck reactions. With dece-1-ne as a substrate, a 73% yield of allylsilane **23** was obtained, with the mass balance being mainly the unreacted starting material (Scheme 8.27). The decrease in efficiency of this single-component pre-catalyst from previous conditions can be attributed to the 2 : 1 L/M ratio of the complex, which has been shown to be less than ideal for silyl-Heck reactions. The catalytic relevance of $(JessePhos)_2PdCl_2$ was then studied. A combination of $(JessePhos)_2PdCl_2$ and Me_3SiI in DCE led to the formation of a new complex, $[(JessePhos)PdI_2]_2$, which was characterized by single-crystal X-ray diffraction (Scheme 8.28). Presumably driven by the Si—Cl bond strength, $(JessePhos)_2PdCl_2$ undergoes halide metathesis with Me_3SiI to provide the Pd(II) I_2 complex, as well as Me_3SiCl and free JessePhos, which was observed by ^{31}P NMR. Due to the size of the iodo ligand, the complex sheds an extra phosphine ligand and is stabilized by the bridging halide centers. $[(JessePhos)PdI_2]_2$ was then investigated for its use as a single-component pre-catalyst in silyl-Heck reactions. With a 1 : 1 L/M ratio, $[(JessePhos)PdI_2]_2$ was found to be a competent single-component pre-catalyst for the silyl-Heck reaction, giving a 98% yield of allylsilane **23** (Scheme 8.29). The increased efficiency of this pre-catalyst is likely due to the 1 : 1 L/M ratio versus the 2 : 1 L/M ratio of $(JessePhos)_2PdCl_2$.

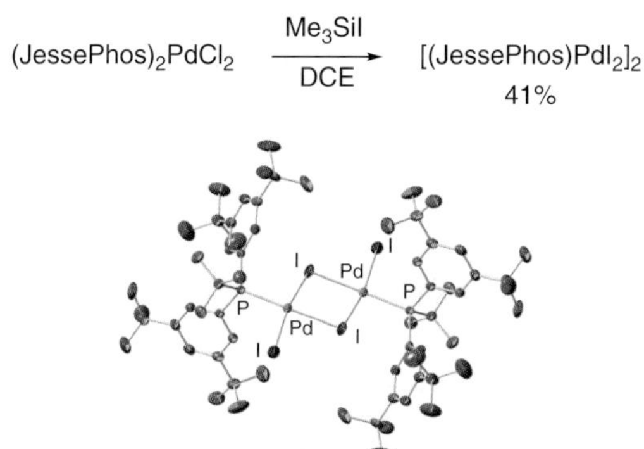

Scheme 8.28 The formation of $[(JessePhos)PdI_2]_2$.

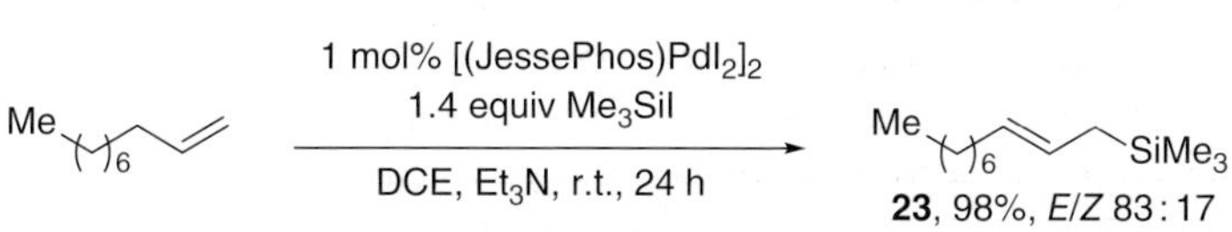

Scheme 8.29 The silyl-Heck reaction with $[(JessePhos)PdI_2]_2$.

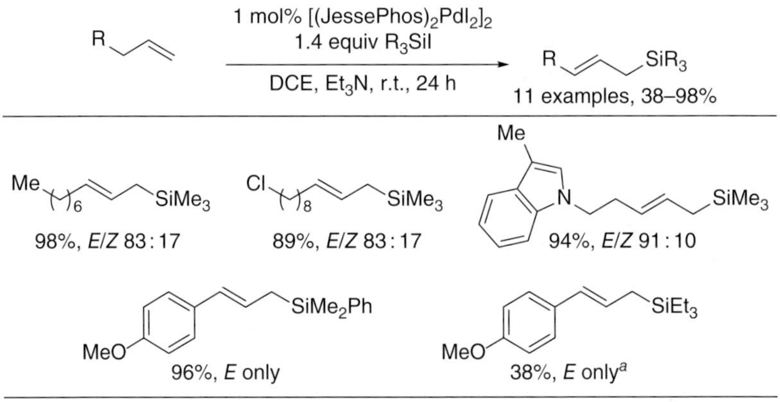

[a]72 h.

Scheme 8.30 Scope of allylsilane formation with a single-component pre-catalyst.

With palladium loadings of 1–2%, no isomerization of the alkene starting material was observed. Further, this pre-catalyst is easy to prepare in one step from PdI_2. It is air, moisture, and temperature stable, which further increases the utility of the silyl-Heck reaction for the formation of allylsilanes. Using this pre-catalyst, a variety of allylsilanes were prepared in comparable yields to the (COD) $Pd(CH_2TMS)_2$/JessePhos conditions (Scheme 8.30). *E/Z* ratios for the products of the reaction are similar to those previously reported and the functional-group tolerance of the reaction remained unchanged. Furthermore, previously unreactive larger silyl groups, including the triethylsilyl group (Et_3Si), were able to be incorporated into the reaction to form more functionalized allylsilanes for the first time.

These conditions provide a unified set of conditions for the formation of both allyl- and vinylsilanes, which further increases the utility of the silyl-Heck reaction. Using $[(JessePhos)PdI_2]_2$ vinylsilanes were also prepared in comparable yields to those previously reported, giving only the *E*-vinylsilane.

8.2.6 The Formation of Silyl Ethers and Disiloxanes via the Silyl-Heck Reaction

Silyl ethers and disiloxanes have shown superior reactivity as substrates for Hiyama–Denmark cross-coupling reactions and Tamao–Fleming oxidations. Our group envisioned using the silyl-Heck reaction as a mild and functional-group-tolerant method to convert simple alkenes to vinyl-substituted silyl ethers and disiloxanes. As discussed earlier, Si—OTf bonds were able to be activated and to participate in the silyl-Heck reaction with the use of iodide additives, and so we began by analyzing dimethylsilyl ditriflate ($Me_2Si(OTf)_2$) for the conversion of 4-*tert*-butyl styrene to vinyl-substituted silyl ether **44** (Scheme 8.31) [22]. Without the addition of an iodide additive, the reaction gave only a trace yield of product **44** (entry 1); but with stoichiometric lithium iodide (LiI) added, a 78% yield of **44** was formed (entry 2). By increasing the amount of $Me_2Si(OTf)_2$ to

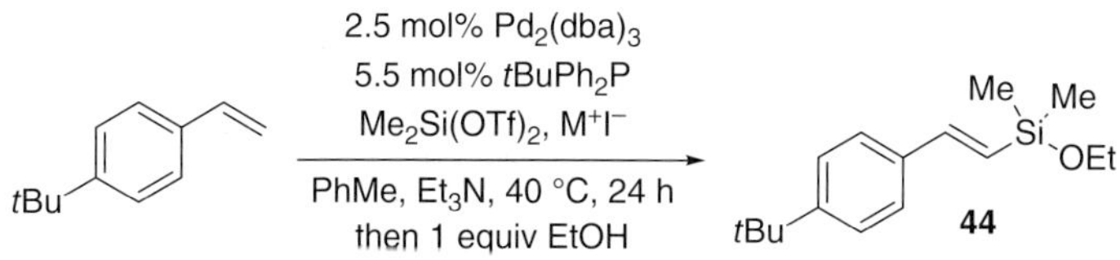

Entry	$Me_2Si(OTf)_2$ (equiv)	M^+I^- (mol%)	Yield (NMR)
1	1	LiI (0)	6
2	1	LiI (100)	78
3	1.1	LiI (5)	98
4[a]	1.1	NaI (5)	99

[a]1.25 mol% $Pd_2(dba)_3$, 2.75 mol% tBuPh_2P, 35 °C.

Scheme 8.31 Optimization of iodide additives for the formation of vinyl-substituted alkoxysilanes/silyl ethers.

1.1 equiv, an almost quantitative yield of the desired product was obtained (entry 3). Unlike previous silyl-Heck reactions, the cheaper and more readily handled sodium iodide (NaI) was equally effective as LiI and allowed for a decrease in pre-catalyst loading and temperature. Using these conditions, multiple vinyl-substituted silyl ethers were able to be prepared *via* the silyl-Heck reaction, by quenching with various alcohols, including secondary and tertiary alcohols (Scheme 8.32). Good yields of products were obtained and only the *trans*-isomer was observed in all cases.

Disiloxanes were also formed using this reaction by quenching with water instead of an alcohol (Scheme 8.33). While increased substitution on the alkene was not tolerated using the reported catalyst for these reactions, increased steric hindrance on the arene did not adversely affect the reaction. This method is uniquely advantageous, as only a single regiomeric and geometric isomer is

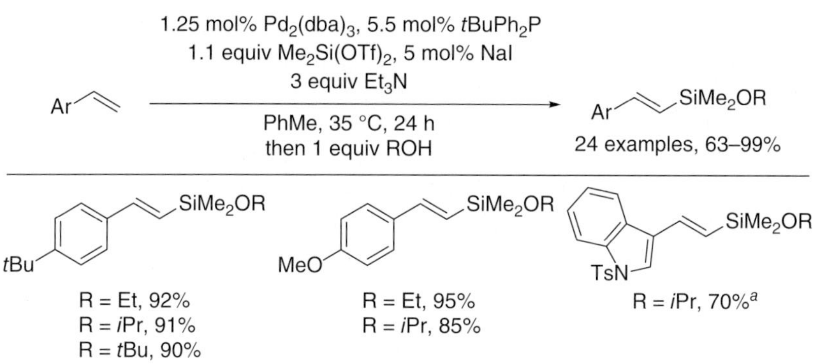

[a]48 h, 50 °C.

Scheme 8.32 Scope of the silyl-Heck reaction to form vinyl-substituted alkoxysilanes/silyl ethers.

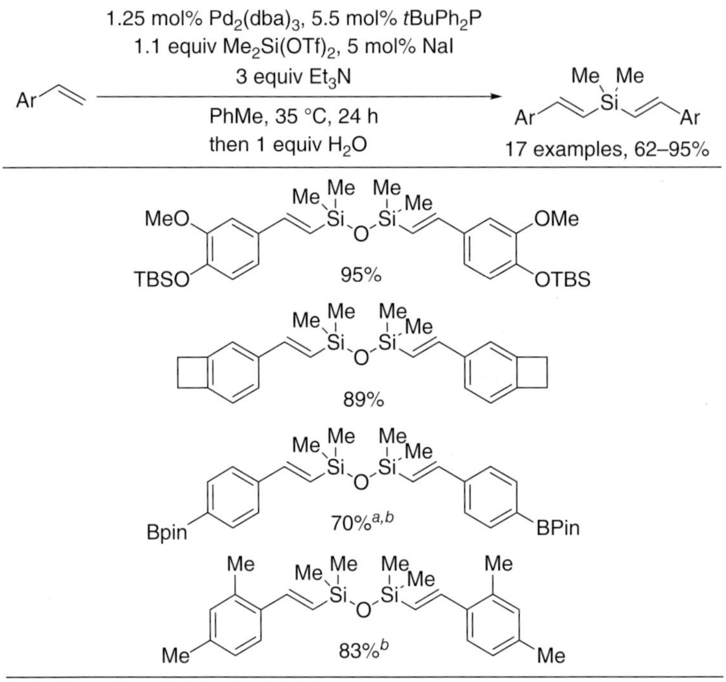

[a]2.5 mol% $Pd_2(dba)_3$, 11 mol% tBuPh_2P, 48 h.
[b]50 °C.

Scheme 8.33 Scope of the silyl-Heck reaction to form disiloxanes.

obtained. For example, direct Heck arylation of electron-rich styrenes leads to a mixture of regioisomeric products, with particularly poor selectivity when using triflate electrophiles [12, 23]. For example, Heck arylation of *p*-methoxystyrene with phenyltriflate (PhOTf) gives a 57 : 43 mixture of α and β products **45** and **46** in 81% yield (Scheme 8.34a). In contrast, Hiyama–Denmark coupling of alkoxy-substituted vinylsilane **47** proceeded in a 95% yield to give exclusively the stilbene product (Scheme 8.34b). Since **47** can be produced in high yield as a single isomer from the silyl-Heck reaction, this two-step protocol is superior in overall yield and avoids the difficult separation of regioisomeric products.

In addition, because of the intermediacy of the vinylsilyl triflate, this strategy provides a divergent approach to differentially substituted vinylsilanes. For example, quenching the reaction with allyl- or benzylmagnesium chloride leads to the corresponding dimethyl(vinyl)silanes **48** and **49** in high yields (Scheme 8.35).

8.2.7 The Nickel-Catalyzed Silyl-Heck Reaction

Silyl triflates are much more abundant than are silyl iodides; however, they fail to participate in the silyl-Heck reaction without iodide additives. We speculated that the inability of palladium to insert into the Si—OTf bond was due to the strong silicon–oxygen bond. Nickel catalysts have shown to be efficient at the activation of strong carbon–oxygen bonds, so nickel-based catalysts were

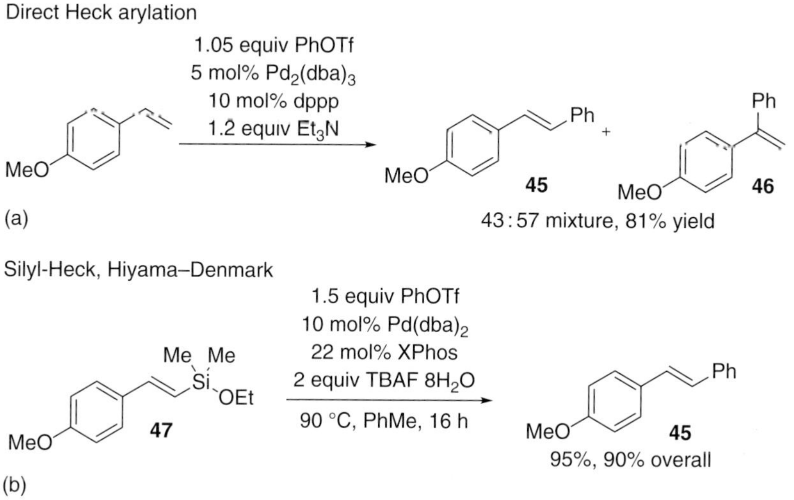

Scheme 8.34 (a) Silyl-Heck/Hiyama–Denmark (b) direct Heck.

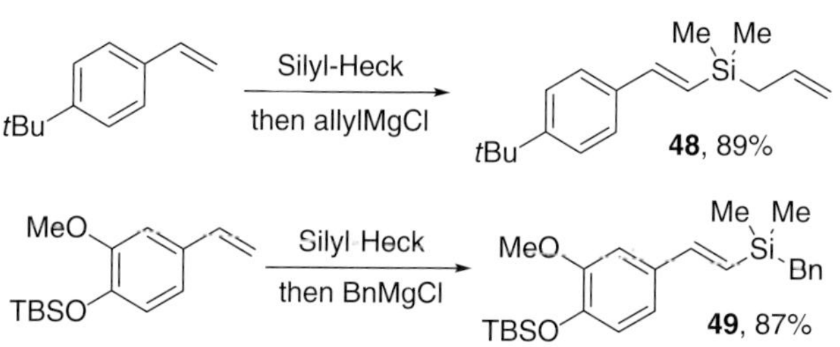

Scheme 8.35 Preparation of diverse vinylsilanes.

investigated for use in activating silyl triflates in silyl-Heck reactions [24]. With 4-*tert*-butylstyrene as a substrate, bis(1,5-cyclooctadiene)nickel(0) (Ni(COD)$_2$) was investigated as a pre-catalyst with a variety of phosphine ligands (Scheme 8.36) to form vinylsilane **22**. Interestingly, mixed alkyl/aryl ligands, as used in palladium-catalyzed silyl-Heck reactions, were ineffective in the transformation (entries 1 and 2). However, moderately bulky trialkyl phosphine ligands, which were ineffective with palladium catalysis, provided highly active catalysts (entries 3 and 4). Very large ligands, such as tri-*tert*-butylphosphine (*t*Bu$_3$P), however, were ineffective (entry 5). Ultimately, *tert*-butyldicyclohexylphosphine (*t*BuCy$_2$P) was found to be the optimal ligand, and by decreasing the L/M ratio, a high yield of vinylsilane **22** was obtained (entries 6 and 7).

These conditions were then used for the formation of vinylsilanes from styrene derivatives with various substitution on the arene (Scheme 8.37). Although substrates containing fluoride substitution were well tolerated, aryl groups with heavier halides were not compatible with these conditions. When a slightly smaller trialkylphosphine ligand (Cy$_3$P) was used, silyl triflates larger than Me$_3$SiOTf also participated in this reaction (Scheme 8.38). Multiple alkyl(dimethyl)silyl triflates, as well as methyldiphenylsilyl (**50**) and triethylsilyl

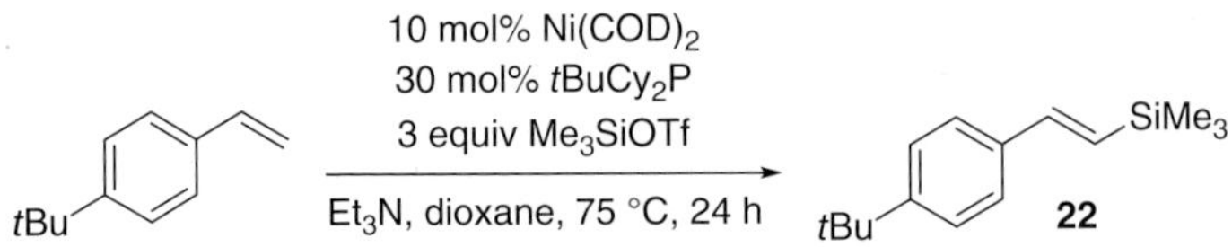

Entry	Ligand	Yield (%)
1	tBuPh_2P	12
2	Cy_2PhP	11
3	Cy_3P	57
4	tBu$_2$CyP	55
5	tBu$_3$P	7
6	tBuCy$_2$P	71
7[a]	tBuCy$_2$P	90

[a] 15 mol% tBuPCy$_2$.

Scheme 8.36 Identification of a ligand for the nickel-catalyzed silyl-Heck reaction.

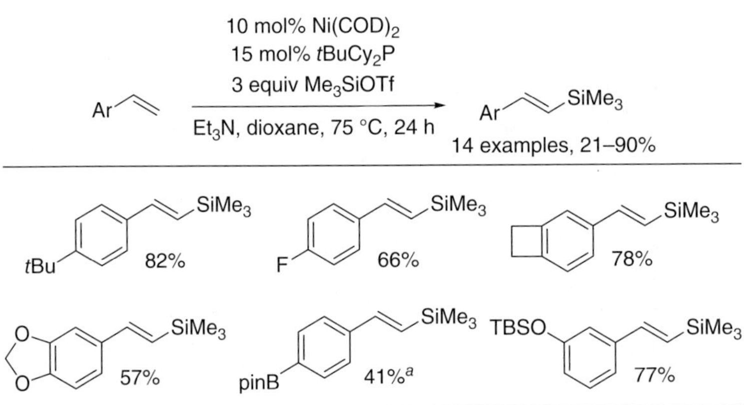

[a] Cy_3P used in place of tBuCy$_2$P.

Scheme 8.37 Scope of the nickel-catalyzed silyl-Heck reaction.

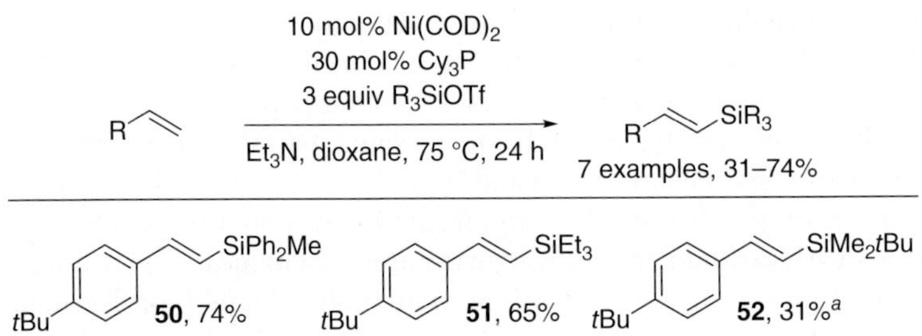

[a] nBu$_3$P in place of Cy_3P, 105 °C.

Scheme 8.38 Scope of larger silyl triflates.

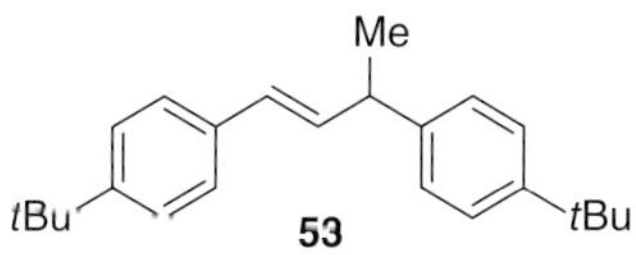
53

Figure 8.1 By-product of nickel-catalyzed silyl-Heck reactions.

triflate (**51**) were effective substrates. Even tBuMe$_2$SiOTf was shown to react, although the reaction requires an increase in temperature and a slightly smaller ligand (**52**). In the case of reactions using larger silyl triflates, the major byproduct is alkene **53**, which presumably forms via a metal hydride-mediated Heck-type pathway (Figure 8.1). Minor amounts of similar dimers are also observed as by-products in reactions using Me$_3$SiOTf; however, their formation is less significant. This suggests that the dimerization pathway becomes more competitive with increasing steric bulk of the silyl triflate and is possibly due to the difficulty of oxidative addition.

8.3 Silyl-Negishi Reactions

Alkylsilanes have a multitude of applications in basic science, medicine, and other industries [25]. However, limited methods for their preparation have been developed. In particular, the direct alkylation of silicon electrophiles with alkyl nucleophiles has remained a challenge due to competitive Si—H bond formation. This is particularly true for sterically encumbered systems, such as branched (secondary) alkyl nucleophiles [11, 26]. Although elegant approaches for the synthesis of branched alkyl silanes via the cross-coupling of silicon nucleophiles and carbon electrophiles have recently been described [27, 28], the coupling of sterically hindered alkyl nucleophiles with silicon electrophiles was not known.

A palladium-catalyzed cross-coupling reaction was envisioned between silyl electrophiles and alkyl nucleophiles (Scheme 8.39) [29]. The catalytic cycle, analogous to both the silyl-Heck and Negishi reactions, begins with oxidative addition of a silyl iodide to Pd(0) to generate intermediate **54**. Transmetalation with an alkylorganometallic gives intermediate **55** and **56**, which undergo reductive elimination to give the desired alkylsilane product **57** and isomerized product **58** and regenerate the catalyst.

The initial investigation was focused on the reaction of alkylzinc iodides, due to their greater stability and functional-group tolerance versus other alkyl nucleophiles, and silyl iodides. As a model system, isopropylzinc iodide and dimethylphenylsilyliodide(Me$_2$PhSiI)werereactedinthepresenceofapalladiumpre-catalyst, and the formation of the desired product **59** and isomerized product **60** were monitored (Scheme 8.40). Et$_3$N was identified as an additive that suppressed the formation of the isomerized product **60**. With Ph$_3$P as a ligand, the desired product was obtained in 77% yield with good selectivity (entry 1). Increasing the steric bulk with tri-*ortho*-tolylphosphine ((o-tol)3P) led to a decrease in yield (entry 2). However, the isomeric ligand tri-*para*-tolylphosphine ((p-tol)3P) led to both an increase in yield and selectivity (entry 3). In contrast, the electron-deficient

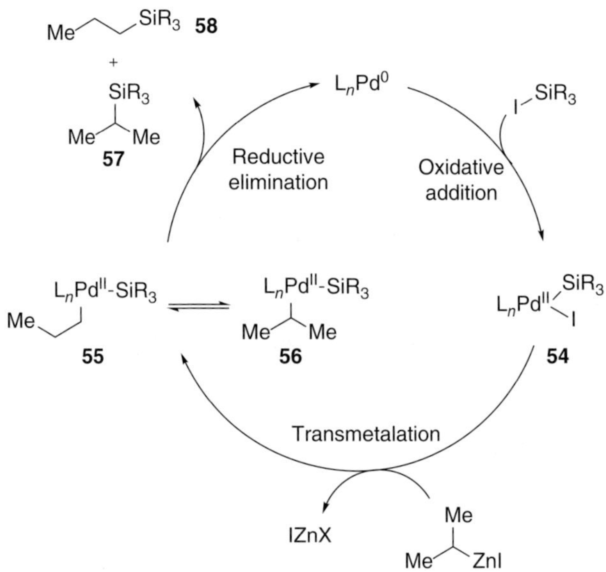

Scheme 8.39 Mechanism of the silyl-Negishi reaction.

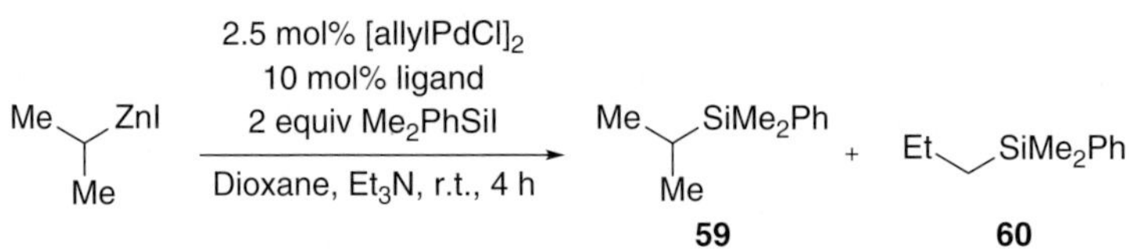

Entry	Ligand	**59 + 60**	**59 : 60**
1	Ph_3P	77	99 : 1
2	$(o\text{-tol})_3P$	48	98 : 2
3	$(p\text{-tol})_3P$	91	>99 : 1
4	$(p\text{-}CF_3C_6H_4)_3P$	42	87 : 13
5	$(p\text{-}MeOC_6H_4)_3P$	88	>99 : 1
6	Ph_2tBuP	83	93 : 7
7	JessePhos	94	98 : 2
8	DrewPhos	99	>99 : 1

Scheme 8.40 Identification of a ligand of the silyl-Negishi reaction.

ligand $(p\text{-}CF_3C_6H_4)_3P$ greatly reduced the yield and increased isomerization (entry 4). The electron-rich ligand $(p\text{-}MeOC_6H_4)_3P$ gave comparable yield and selectivity to $(p\text{-tol})_3P$ (entry 5). However, this ligand gave inconsistent results due to the instability of the ligand under the Lewis-acidic reaction conditions. In an attempt to avoid ligand decomposition, we investigated the electron-rich Ar_2tBuP scaffold that was successful in the silyl-Heck reactions. $tBuPh_2P$ produced only moderate yield and gave a considerable amount of isomerization (entry 6). However, with the larger and more electron-rich ligand, JessePhos, an

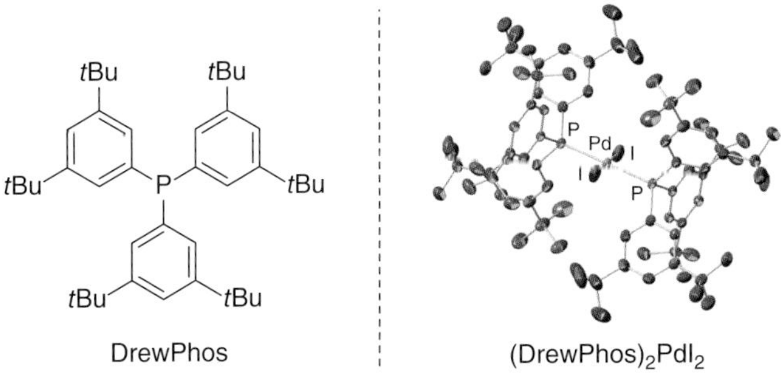

Figure 8.2 DrewPhos ligand and single-component pre-catalyst.

increase in yield and a decrease in isomerization was observed (entry 7). Employing a triaryl ligand based on the 3,5-di-*tert*-butylarene (DrewPhos, Figure 8.2) consistently produced near-quantitative yield of the desired product and no detectable isomerization (entry 8). The success of this ligand is thought be due to its electron richness as well as its steric bulk, which is known to promote reductive elimination. To further improve the reaction, a single-component pre-catalyst, $(DrewPhos)_2PdI_2$, was prepared (Figure 8.2). This pre-catalyst was highly efficient and allows for palladium loadings as low as 1 mol%.

Using these conditions, alkylsilanes were formed with a variety of substitution from both the zinc and silane coupling partners. Both primary and secondary alkylzinc iodides and alkylzinc bromides gave excellent yields of alkylsilane products (Scheme 8.41). In the case of primary alkylzinc iodides, a significant

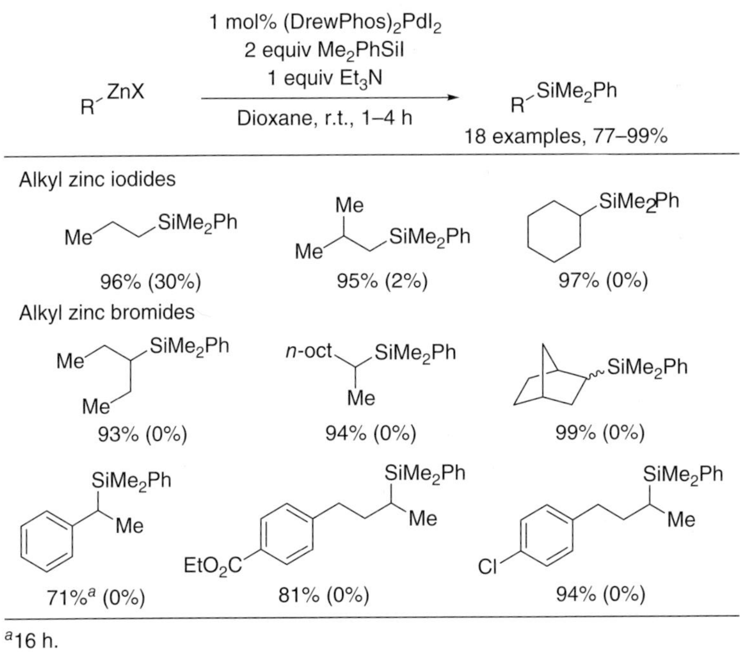

Scheme 8.41 Scope of alkylzinc halides in the silyl-Negishi reaction.

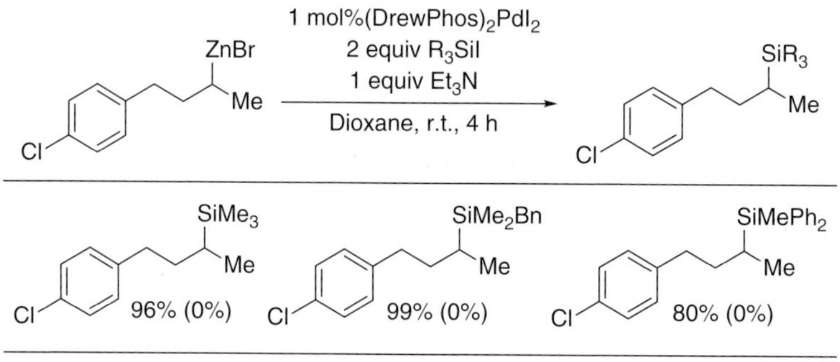

Scheme 8.42 Scope of silyl iodides in the silyl-Negishi reaction.

background reaction was observed (background reactivity shown in parentheses). However, with increased steric hindrance or in the case of the less reactive alkylzinc bromides, very little to none of the background reaction was observed. In all cases, the catalyzed reaction still proceeded in good yield after four hours. The reactions require the use of 1,4-dioxane as solvent; however, a method for solvent exchange from the commercially available tetrahydrofuran (THF) solutions of alkylzinc halides was developed, which greatly expands the amount of readily available nucleophiles for this transformation. Various substitution patterns on the silicon coupling partner, including phenyl, benzyl, and ethyl groups, were also tolerated in the reaction, giving good yields of alkylsilane products (Scheme 8.42). The different substitution patterns on the silicon atom in these products allow for altered reactivity and downstream transformations.

The steric limitations of this method were investigated through the reaction of the sterically demanding β,β-dimethyl- and β,β,β-trimethyl-substituted zinc iodides **61** and **62** with Me_2PhSiI (Scheme 8.43). **61** reacted to form the desired product **63** in good yield with no observable isomerization. However, with even a further sterically demanding **62**, a decrease in yield of the desired product **65** was observed with a significant amount of undesired isomerization product **66**, showing the limitations to this reaction. Presumably, this occurs through β-hydride elimination and reinsertion to the sterically favorable terminal carbon

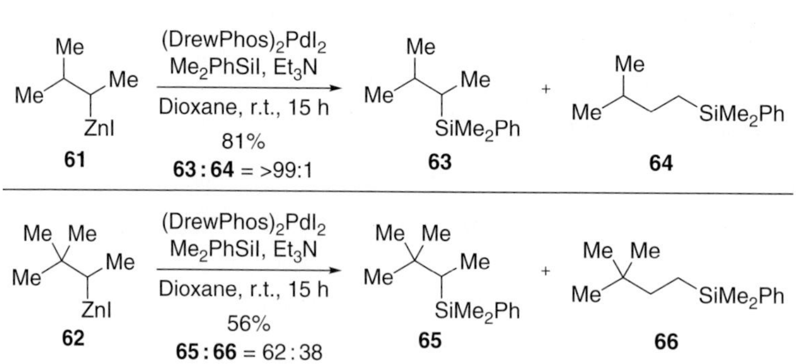

Scheme 8.43 Examination of steric limitations of the silyl-Negishi reaction.

atom. This shows that even with strained, secondary neopentylic nucleophiles, β-hydride elimination is still not the predominant pathway.

8.4 Silyl-Kumada–Corriu Reactions

Although the silyl-Negishi reaction allows for the formation of alkylsilanes with good yields and selectivities, it is limited to the use of silyl-iodide electrophiles. Silyl chlorides are significantly more stable to air and moisture and are more functional-group tolerant than are silyl iodides. They are also much more widely available, as they are the product of the Muller–Rochow "direct" process [30]. We envisioned that different classes of alkyl nucleophiles may be successful in cross-coupling with silyl-chloride electrophiles [31].

The reaction of alkyl Grignard reagents with dimethylphenylsilyl chloride ($Me_2PhSiCl$) was investigated (Scheme 8.44). Using isopropyl magnesium chloride as the alkyl nucleophile and THF as the reaction solvent, modest reactivity was observed. However, the reaction resulted predominantly in the undesired linear alkylsilane product (entry 1). Reactivity was increased using *i*PrMgCl·LiCl and *i*PrMgBr, although the linear product **60** continued to be favored (entries 2 and 3). In contrast, changing the reaction solvent to diethyl ether (Et_2O), only the desired branched product **59** was observed (entry 4). Reactivity was further increased with *i*PrMgBr and *i*PrMgI nucleophiles to give a quantitative yield, selectively, of the desired product (entries 5 and 6).

A wide range of Grignard reagents were tolerated as alkyl nucleophiles in the reaction (Scheme 8.45). Although primary Grignard reagents showed modest background reactivity, the catalyzed reactions produced the desired product in near-quantitative yields. Acyclic and cyclic secondary alkyl Grignard reagents **67** and **68** worked well as substrates for the reaction, as did the more sterically hindered **69** and aryl substrates, e.g. **70**. Various substitutions on the silicon atom were also tolerated (Scheme 8.46). A variety of alkyl- and aryldimethylsilyl chlorides were well tolerated, including alkyl chlorides, showing the selectivity of

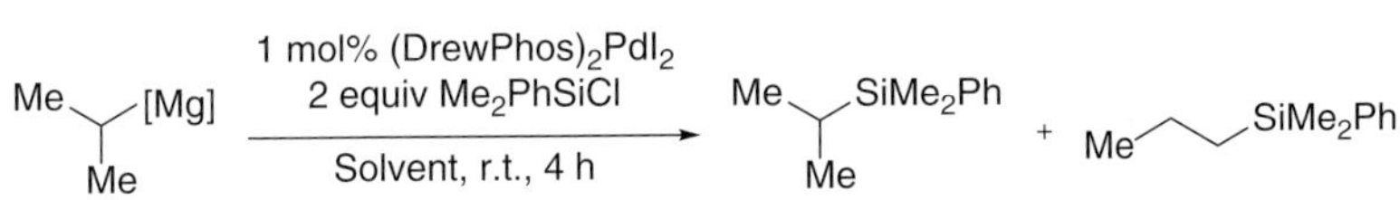

59 **60**

Entry	[Mg]	Solvent	**59 + 60**	**59 : 60**
1	MgCl	THF	45	38 : 62
2	MgCl·LiCl	THF	92	31 : 69
3	MgBr	THF	96	34 : 66
4	MgCl	Et_2O	24	>99 : 1
5	MgBr	Et_2O	99	>99 : 1
6	MgI	Et_2O	99	>99 : 1

Scheme 8.44 Optimization of the silyl-Kumada–Corriu reaction.

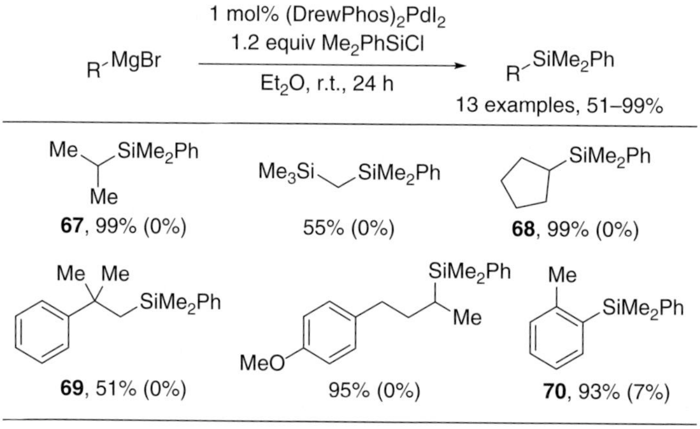

Scheme 8.45 Scope of nucleophiles in the silyl-Kumada–Corriu reaction.

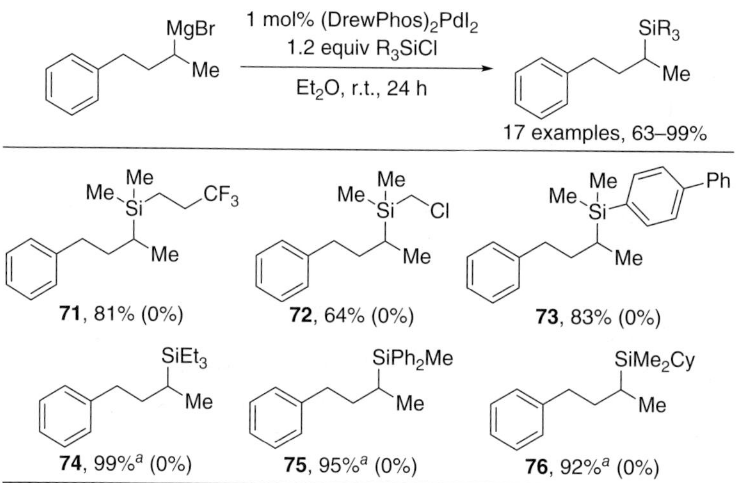

[a]2 equiv R_3SiCl, Bu_2O, 50 °C.

Scheme 8.46 Scope of silyl chlorides in the silyl-Kumada–Corriu reaction.

Si—Cl bond activation over that of C—Cl bond activation (**71**, **72**, **73**). With an increase in temperature and dibutyl ether (Bu_2O) as a reaction solvent, larger silyl chlorides were also good substrates for the reaction (**74**, **75**, **76**).

8.5 Summary and Conclusions

Over the past several years, silicon electrophiles have emerged as competent coupling partners in palladium- and nickel-catalyzed cross-coupling reactions. Standing on the shoulders of pioneering work from Tanaka and Murai, recent developments have demonstrated that these reagents can provide mild and effective entries into a broad range of both unsaturated and saturated organosilanes, including allyl-, vinyl-, and alkylsilanes. Although this recent progress has demonstrated the viability of cross-coupling reactions analogous to those well-established using carbon electrophiles, i.e. the venerable Heck, Negishi, and Kumada–Corriu reactions, much work in the area remains. For example, many of the mechanistic details related to the silyl cross-coupling reactions have yet to be adequately studied. In addition, many additional transformations that relate to the early silyl cross-coupling reactions can be envisioned. Work in this area promises to continue to expand and develop in the coming years.

References

1 Yamashita, H., Hayashi, T., Kobayashi, T. et al. (1988). *J. Am. Chem. Soc.* 110: 4417.
2 Walsh, R. (1981). *Acc. Chem. Res.* 14: 246.
3 Yamashita, H., Tanaka, M., and Goto, M. (1997). *Organometallics* 16: 4696.
4 (a) Levy, C.J., Puddephatt, R.J., and Vittal, J.J. (1994). *Organometallics* 13: 1559. (b) Levy, C.J., Vittal, J.J., and Puddephatt, R.J. (1996). *Organometallics* 15: 2108.
5 Esposito, O., Roberts, D.E., Cloke, F.G.N. et al. (2009). *Eur. J. Inorg. Chem.* 2009: 1844.
6 Zlota, A.A., Frolow, F., and Milstein, D. (1989). *J. Chem. Soc., Chem. Commun.* 1826.
7 Gatard, S., Chen, C.-H., Foxman, B.M., and Ozerov, O.V. (2008). *Organometallics* 27: 6257.
8 Hatanaka, Y. and Hiyama, T. (1988). *J. Org. Chem.* 53: 918.
9 Hosomi, A. and Sakurai, H. (1976). *Tetrahedron Lett.* 17: 1295.
10 Jones, G.R. and Landais, Y. (1996). *Tetrahedron* 52: 7599.
11 Murakami, K., Yorimitsu, H., and Oshima, K. (2009). *J. Org. Chem.* 74: 1415.
12 Oestreich, M. (ed.) (2008). *The Mizoroki-Heck Reaction*, 568. Chichester: Wiley.
13 Yamashita, H., Kobayashi, T.-A., Hayashi, T., and Tanaka, M. (1991). *Chem. Lett.* 20: 761.
14 Chatani, N., Amishiro, N., and Murai, S. (1991). *J. Am. Chem. Soc.* 113: 7778.
15 Chatani, N., Amishiro, N., Morii, T. et al. (1995). *J. Org. Chem.* 60: 1834.
16 McAtee, J.R., Martin, S.E.S., Ahneman, D.T. et al. (2012). *Angew. Chem. Int. Ed.* 51: 3663.
17 Olah, G.A., Narang, S.C., Gupta, B.G.B., and Malhotra, R. (1979). *J. Org. Chem.* 44: 1247.
18 McAtee, J.R., Yap, G.P.A., and Watson, D.A. (2014). *J. Am. Chem. Soc.* 136: 10166.
19 McAtee, J.R., Krause, S.B., and Watson, D.A. (2015). *Adv. Synth. Catal.* 357: 2317.
20 Krause, S.B., McAtee, J.R., Yap, G.P.A., and Watson, D.A. (2017). *Org. Lett.* 19: 5641.
21 Lee, H.G., Milner, P.J., Colvin, M.T. et al. (2014). *Inorg. Chim. Acta* 422: 188.
22 Martin, S.E.S. and Watson, D.A. (2013). *J. Am. Chem. Soc.* 135: 13330.
23 Fristrup, P., Le Quement, S., Tanner, D., and Norrby, P.-O. (2004). *Organometallics* 23: 6160.
24 McAtee, J.R., Martin, S.E.S., Cinderella, A.P. et al. (2014). *Tetrahedron* 70: 4250.

25 (a) Bassindale, A. and Gaspar, P.P. (eds.) (1991). *Frontiers of Organosilicon Chemistry*. Royal Society of Chemistry. (b) Larson, G.L. (1996). *Advances in Silicon Chemistry*. JAI Press. (c) Auner, N. and Weis, J. (2003). *Organosilicon Chemistry V*. Wiley-VCH.

26 (a) Lacout-Loustalet, M.B., Dupin, J.P., Metras, F., and Valade, J. (1971). *J. Organomet. Chem.* 31: 187. (b) Lennon, P.J., Mack, D.P., and Thompson, Q.E. (1989). *Organometallics* 8: 1121. (c) Murakami, K., Hirano, K., Yorimitsu, H., and Oshima, K. (2008). *Angew. Chem. Int. Ed.* 47: 5833. (d) Morita, E., Murakami, K., Iwasaki, M. et al. (2009). *Bull. Chem. Soc. Jpn.* 82: 1012.

27 Chu, C.K., Liang, Y., and Fu, G.C. (2016). *J. Am. Chem. Soc.* 138: 6404.

28 (a) Xue, W., Qu, Z.-W., Grimme, S., and Oestreich, M. (2016). *J. Am. Chem. Soc.* 138: 14222. (b) Scharfbier, J. and Oestreich, M. (2016). *Synlett* 27: 1274.

29 Cinderella, A.P., Vulovic, B., and Watson, D.A. (2017). *J. Am. Chem. Soc.* 139: 7741.

30 Kalchauer, W. and Pachaly, B. (2008). *Handbook of Heterogeneous Catalysis*. Weinheim: Wiley-VCH.

31 Vulovic, B., Cinderella, A.P., and Watson, D.A. (2017). *ACS Catal.* 8113.

9

Transition-Metal-Catalyzed Cross-coupling of Organosilicon Compounds

Tamejiro Hiyama[1], Yasunori Minami[1], and Atsunori Mori[2]

[1] *Chuo University, Research and Development Initiative, 1-13-27 Kasuga, Bunkyo-ku, Tokyo 112-8551, Japan*
[2] *Kobe University, Department of Chemical Science and Engineering, 1-1 Rokkodai, Nada, Kobe 657-8501, Japan*

9.1 Introduction

In modern organic synthesis, the cross-coupling reaction by transition-metal catalysis has become a powerful tool to straightforwardly and selectively form a variety of carbon–carbon bonds consisting of the framework in organic molecules. Since the discovery of the cross-coupling reaction of Grignard reagents (RMgX) with organic halides (R′X) using nickel/phosphine catalysts, as reported by Kumada and Tamao in 1972, a wide range of organometallic reagents have been shown to be applicable. In addition to such reactive metallic reagents as lithium (Murahashi) and aluminum (Nozaki-Oshima, Negishi), as well as zinc (Negishi, Normant), zirconium (Negishi), iron (Kochi), and copper (Normant), group 3 and 4 elements have been shown to undergo similar coupling reactions. Some are named after the principal chemists: Suzuki–Miyaura (boron), Migita–Kosugi–Stille (tin), and Hiyama (silicon). These main-group reagents are generally less nucleophilic than the representative alkaline and alkaline earth organometallics but show reactive enough nucleophilic behavior by activation with a nucleophilic additive. It is also advantageous for such main-group reagents to exhibit remarkable tolerance toward various functional groups. The observed chemoselectivity is the salient feature of such organometallic reagents, and thus, the cross-coupling reaction using these main-group reagents is now employed widely in the synthesis of pharmaceuticals, agrochemicals, and advanced organic materials with complex carbon skeletons [1–3].

9.1.1 Historical Background of the Cross-coupling with Organosilicon Reagents

The cross-coupling of organosilicon reagents was referred to for the first time by Kumada and Tamao in 1982. They used hexacoordinate silicate **1** bearing a styryl

Organosilicon Chemistry: Novel Approaches and Reactions, First Edition.
Edited by Tamejiro Hiyama and Martin Oestreich

group and five fluorines and showed the coupling reaction with phenyl iodide proceeded in the presence of a palladium catalyst to afford stilbene **2** [4] (Eq. (9.1)). Pentacoordinate silylcatecholate **3** was later shown by Hosomi et al. to undergo the cross-coupling under milder conditions, thereby giving styrene derivative **4** [5] (Eq. (9.2)). They used **3** as the coupling reagent inspired by the mechanism suggested by Hatanaka and Hiyama and a coworker [6].

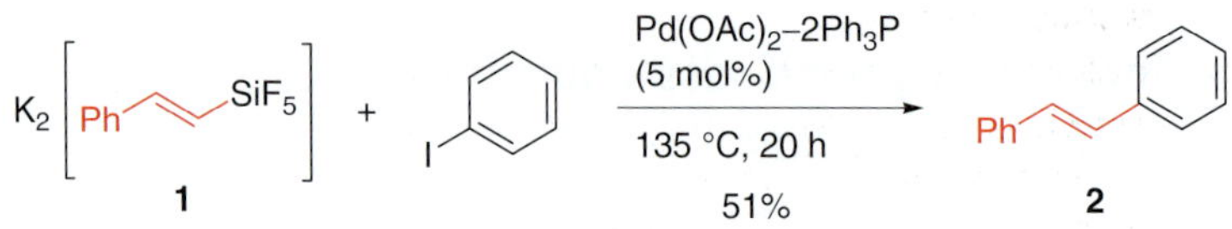

(9.1)

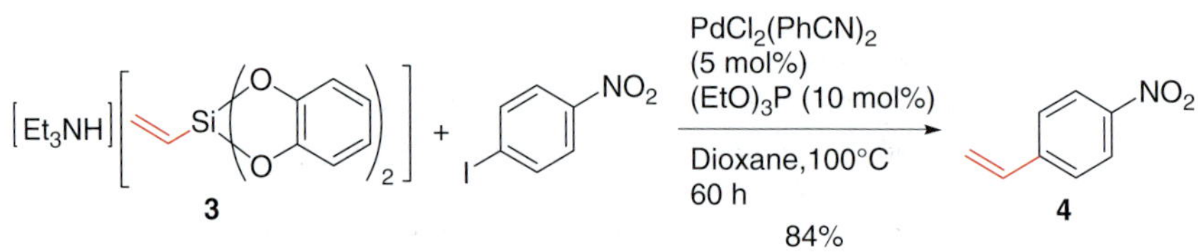

(9.2)

Neutral tetracoordinate organosilane was shown to undergo cross-coupling by Hatanaka and Hiyama in 1988. Vinyltrimethylsilane **5** reacted with aryl iodide catalyzed by palladium in the presence of $(Et_2N)_3S^+(Me_3SiF_2)^-$ (TASF, tris(dimethylamino)sulfonium difluorotrimethylsilicate) as a fluoride ion source to give the corresponding vinylated product **6** in excellent yields [6]. The intermediacy of pentacoordinate silicate is suggested, which is derived from the nucleophilic attack of a fluoride ion to **5**.

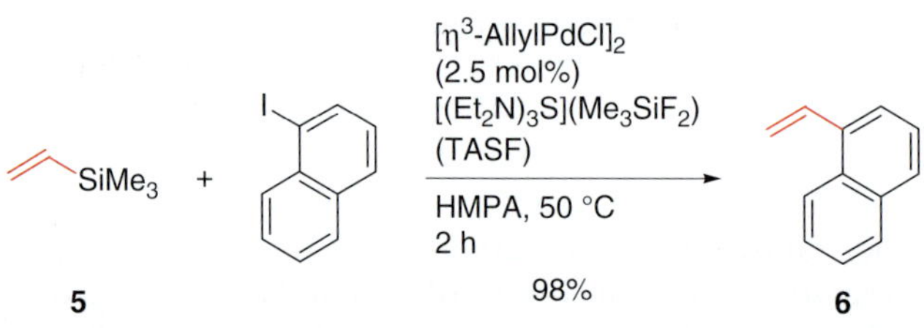

(9.3)

The scope of the cross-coupling of organosilicon reagents was successively studied by Hiyama and Hatanaka [1, 7]: Alkenyl$SiR_nF_{(3-n)}$ **7**, bearing fluorine atoms on the silicon atom, was shown to undergo cross-coupling with aryl and alkenyl halides, where one or two fluorine atoms on the silicon atom are required for the smooth coupling reaction, whereas trialkyl- and trifluorosilanes failed the reaction (Eq. (9.4)). The reaction of aryl(oligofluoro)silanes with allylic carbonates also proceeds, where difluorosilanes are the most effective, to give coupled product **10** [8] (Eq. (9.5)). Alkyltrifluorosilanes **11** were shown to couple with several aryl bromides and iodides using tetra-*n*-butylammonium fluoride (TBAF) in excess [9, 10] (Eq. (9.6)). On the other hand, cross-coupling of alkynylsilane **13** is shown to be potent enough to give enyne **14** [6] (Eq. (9.7)).

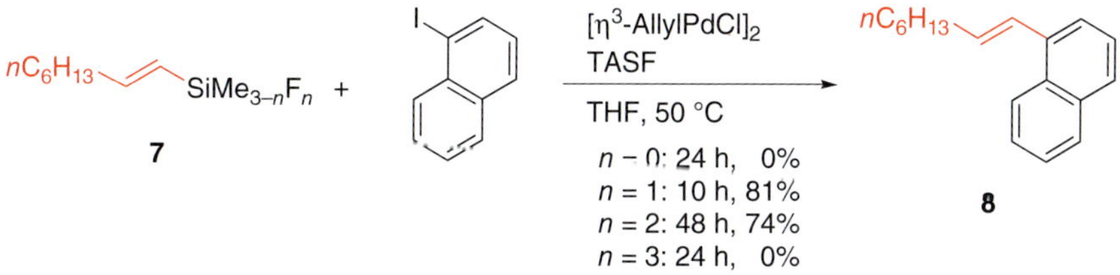

(9.4)

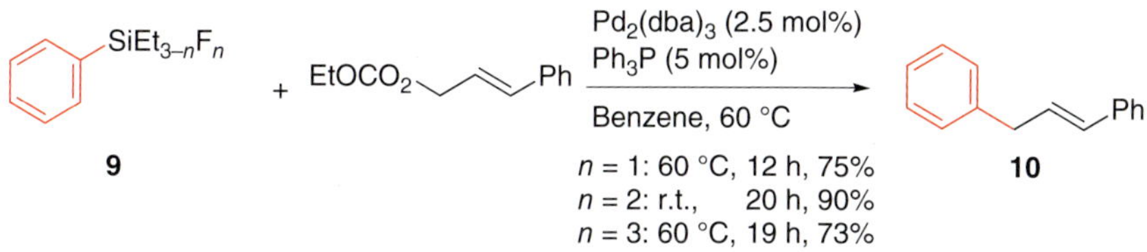

(9.5)

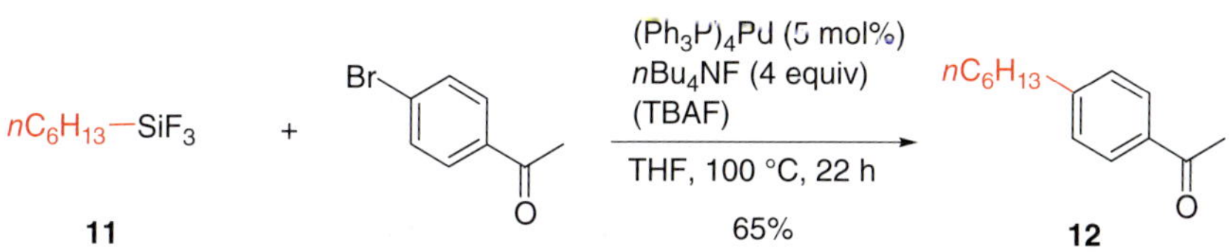

(9.6)

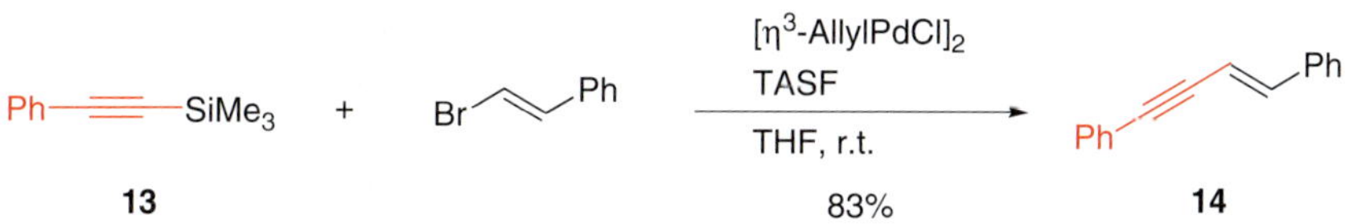

(9.7)

The key to successful cross-coupling of organosilicon reagents **15** is attributed to the *in situ*-formed pentacoordinate silicates **16** by a nucleophilic attack of a fluoride ion to silicon. The resulting **16** exhibits an organometallic character to allow facile transmetalation with a palladium catalyst to afford organopalladium **17**, whose reductive elimination gives rise to the coupled products, regenerating palladium(0) (Eq. (9.8)). For the cross-coupling of vinyltrimethylsilane with vinyl iodide, density functional theory (DFT) calculation shows that the transmetalation step is proposed to proceed via the reaction of vinyl–Pd–F complex **18**, formed through substitution of an iodide ligand to a fluoride ligand, with ligated vinylsilane to form fluorosilane-ligated divinylpalladium complex **19** or attack of a fluoride ion to vinylsilane-ligated vinyl–Pd–I complex **20** to form divinyl complex **21** [11] (Eqs. (9.9) and (9.10)). Kinetic experiments suggest that the reaction of aryl–Pd–F complex **22** with aryltrimethoxysilane is a crucial step in a rate-determining transmetalation step [12] (Eq. (9.11)). Moreover, reductive elimination has been shown to be accelerated by a fluoride ion.

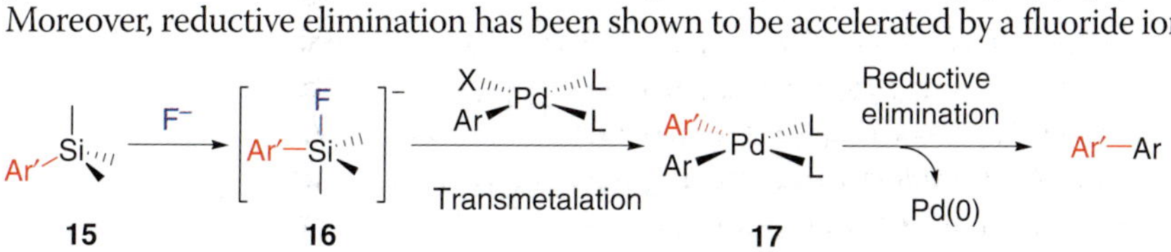

(9.8)

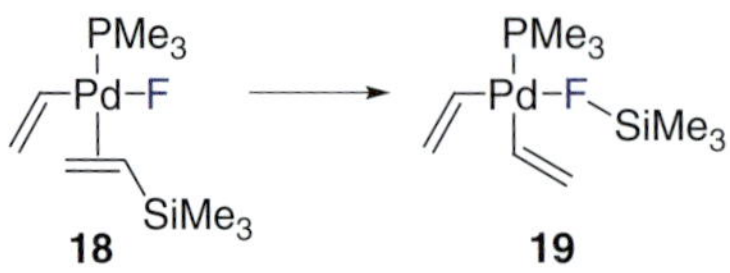

(9.9)

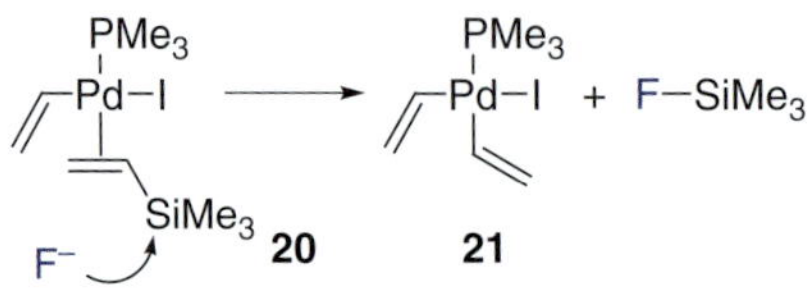

(9.10)

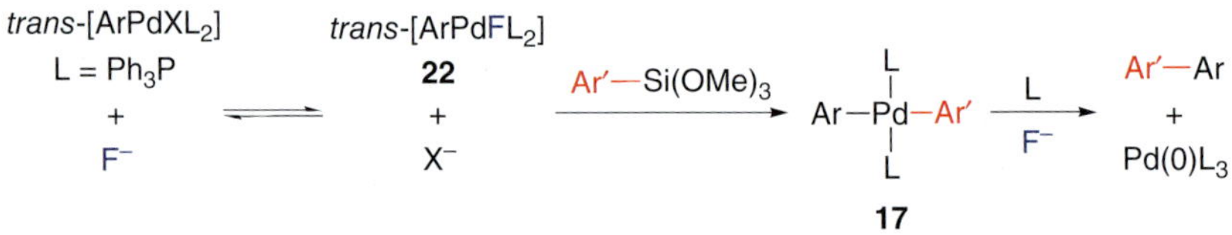

(9.11)

Palladium-catalyzed allylic substitution of organosilicon reagent **23** is also shown to take place, affording **10** [13] (Eq. (9.12)). When allylic carbonates **24** or diene monoxides (vinyl epoxides) **25** are employed as substrates, the coupling reaction proceeds without a fluoride ion because an alkoxide ion generated by the reaction of palladium(0) with **24** or **25** works as a potent nucleophilic activator in a manner similar to the case in Eq. (9.5) [8].

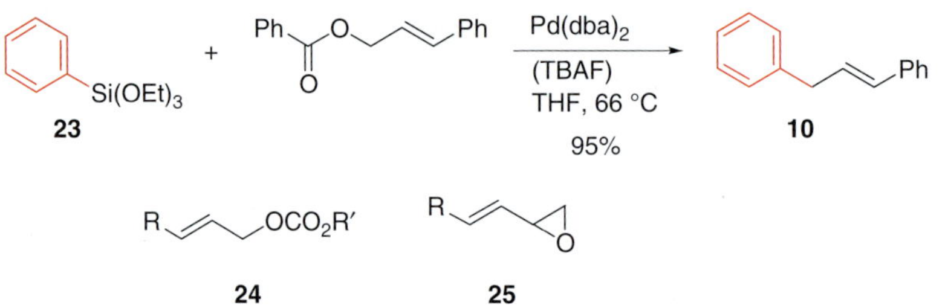

(9.12)

DeShong and a coworker studied systematically the cross-coupling of aryl(trimethoxy)silane **23′** with aryl halides for synthesis of biaryls [14]. He showed that the catalyst system consisting of $Pd(OAc)_2$ and $2Ph_3P$ was shown to be effective for smooth biaryl formation [15] (Eq. (9.13)). Aryl- and alkenyl(trialkoxy)silanes [$RSi(OR')_3$] are readily available by nucleophilic substitution of the corresponding organometallic reagents (RM) with $Si(OR')_4$ [16, 17] and hydrosilylation of alkynes with $HSi(OR')_3$, respectively [18]. $RSi(OR')_3$ is a reagent of choice particularly for biaryl synthesis.

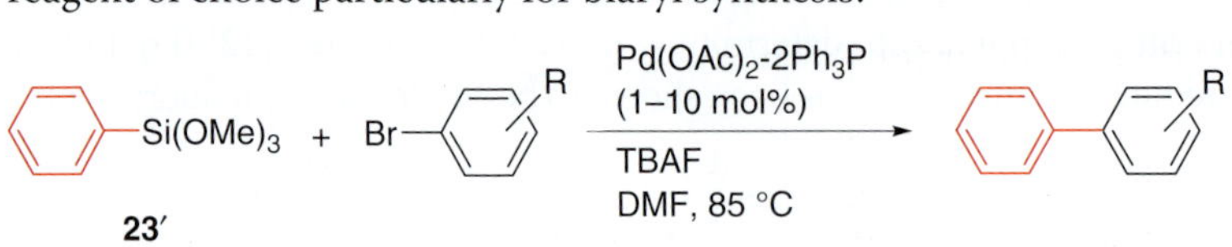

R = 4-COMe: 83% (3 mol%); 4-CH_3: 77% (3 mol%);
2,6-Me_2: 85% (10 mol%)

(9.13)

In addition to combinations of fluorosilanes or alkoxysilanes with a fluoride ion activator, chlorosilanes **26**, which are also ubiquitous and inexpensive organosilanes, undergo cross-coupling with several organic halides, where NaOH was shown to be an alternatively effective activator, to give biaryl **27** [19] (Eq. (9.14)).

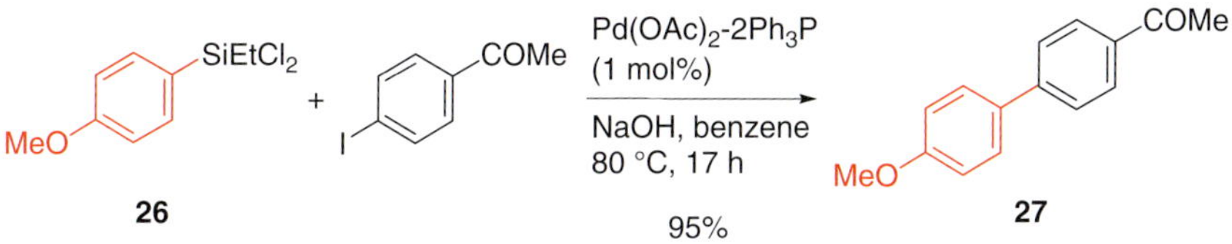

(9.14)

The tris(trimethylsilyl)silyl (TTMSS) group, called the super silyl group, is a bulky silyl substituent and has a unique electronic character versatile for various organic reactions. Of course, alkenyl[tris(trimethylsilyl)]silanes are applicable to the cross-coupling with iodoarenes. For example, reaction of iodobenzene with β-(*Z*)-tris(trimethylsilyl)silyl styrene (**28**) in the presence of a $(Ph_3P)_4Pd$ catalyst, TBAF, NaOH and hydrogen peroxide took place to give stilbene **29** ($E/Z = 3/97$) with retention of configuration [20, 21] (Eq. (9.15)). It is assumed that the Si—Si bonds in the TTMSS group are oxidized by H_2O_2 to provide a siloxane intermediate which is an effective cross-coupling reagent, as described earlier.

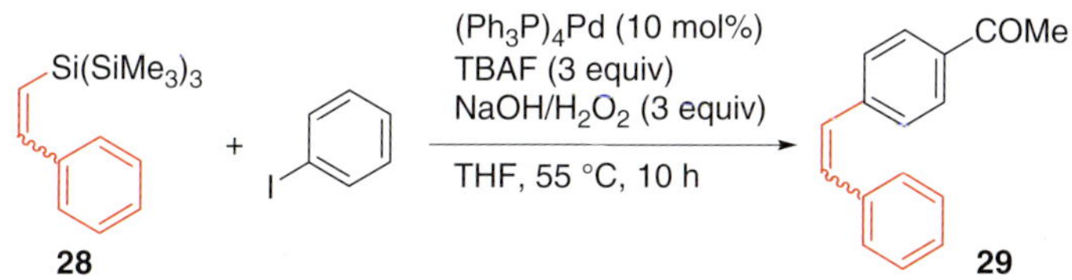

From (*Z*)-TTMS-silane: 90% (*E*/*Z* = 3/97)
From (*E*)-TTMS-silane: 83% (*E* only) (9.15)

In this chapter, we briefly review the recent progress in the silicon-based cross-coupling chemistry through transition-metal catalysis with the emphasis on metallic species, involving ligand design, structures of organosilicon nucleophile, leaving group of organic electrophiles, and synthetic applications to the concise synthesis of biologically active molecules.

9.2 Improvements in the Cross-coupling Reaction of Organosilicon Compounds

9.2.1 Ligand Design for the Palladium Catalyst

In the studies on the cross-coupling of $RSi(OR')_3$ with aryl chlorides, DeShong initially used the Buchwald-type bulky and electron-rich phosphines and N-heterocyclic carbenes (NHCs) to attain good to excellent yields of R–Ar [2, 15]. Later, they demonstrated ligands **30**–**35** and complexes **36**–**40** to assist the palladium-catalyzed coupling with aryl chlorides [22–32]. These ligands are characterized by electron donation to palladium and thus accelerate the coupling reaction with aryl chlorides, particularly with unactivated substrates. Najera showed that the use of palladium complex **41** with an oxime-derived bidentate ligand exhibits remarkably high turnover numbers and allows to recycle the catalyst for the reaction of $ArSi(OR)_3$ with aryl bromides and chlorides [33–35] (Figure 9.1).

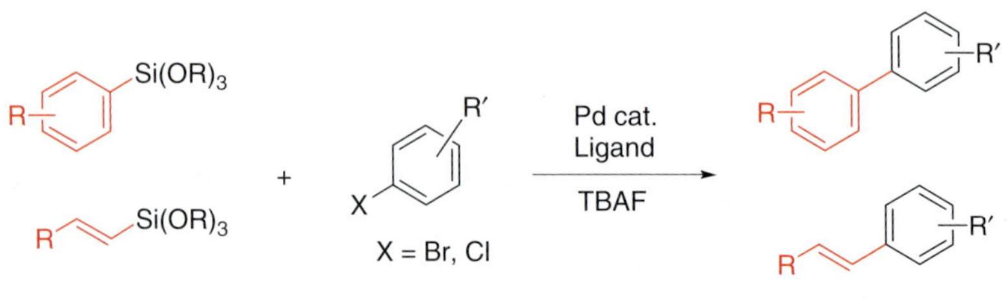

9.2.2 Variation of Palladium Catalysts and Additive Systems

Nanoparticles of palladium prepared in aqueous or organic solvents catalyze the cross-coupling reaction of $ArSi(OR)_3$ with organic halides [36–41]. Pd/C is also applicable to the reaction [42–44]. Palladium supported on polystyrene has been reported to similarly catalyze the cross-coupling [45]. The heterogeneous reactions proceed under conditions similar to those of homogeneous catalysis for the representative Hiyama coupling, although somewhat harsh conditions are required. A silica-supported magnetic ferrite nanoparticle ($SiO_2@Fe_3O_4$), in which an organic-group-bearing phosphine moiety is supported on the silica surface, can be employed as a ligand of palladium catalysts to accelerate the silicon-based coupling reaction. The catalyst was readily removed by applying an external magnet after the reaction was completed and reused up to 10 times without significant loss of efficiency, i.e. chemical yield [46].

In addition to the palladium catalysts discussed, other transition-metal complexes assist the cross-coupling. Especially, the combined use of palladium and copper is known to be particularly effective [47, 48]. For example, mesityltriethoxysilane (**42**) cross-couples with *p*-iodo-4-trifluorobenzene in the presence of a $(Ph_3As)_2PdCl_2$ catalyst, CsF, and CuF_2 in a stoichiometric amount and gives coupled product **43** in high yield, whereas the reaction in the absence of the copper source does not afford **43** [47] (Eq. (9.16)). Of note, $(Ph_3As)PdCl_2(IDM)$ (IDM = 1,3-dimethylimidazol-2-ylidene) is the most effective palladium catalyst for this coupling reaction. Transmetalation from arylsilanes to copper and then to palladium is proposed to effect the cross-coupling.

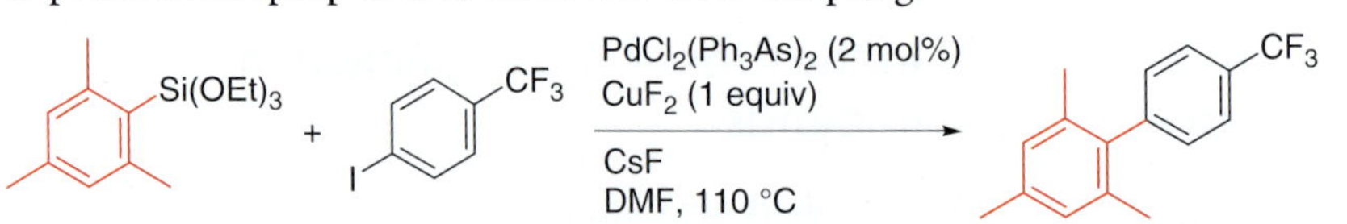

42 **43**

88%
<5% (absence of CuF_2)

>98% (use of $[PdCl_2(IDM)(Ph_3As)]$)

(9.16)

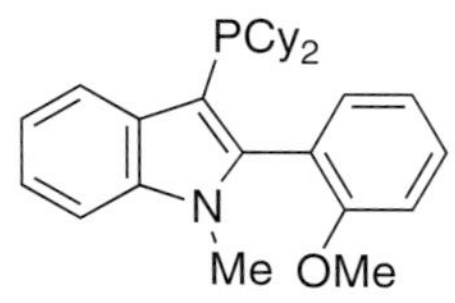

30[22]

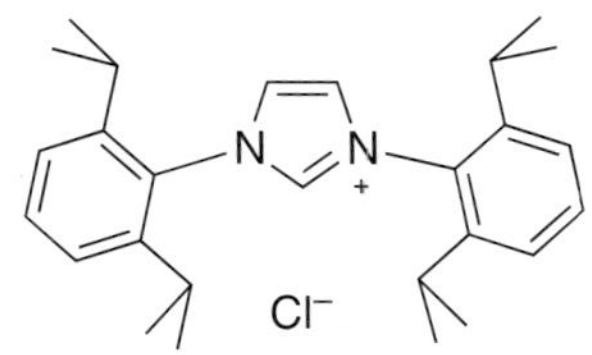

31, IPr[24]

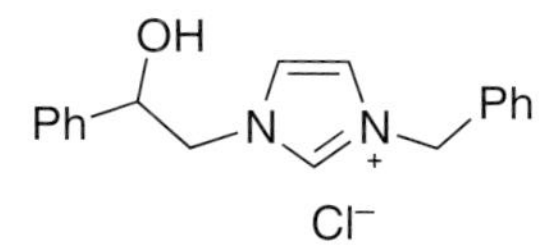

32[26]

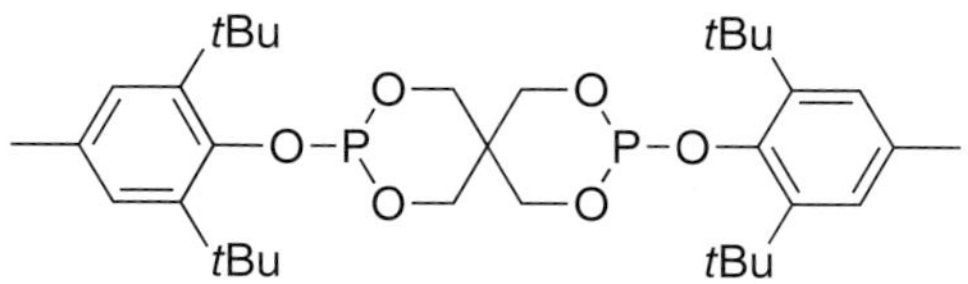

33[29]

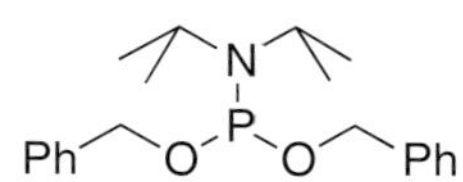

34[30]

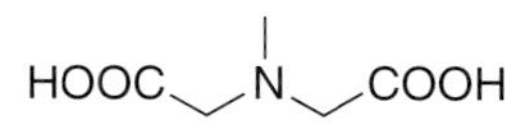

35[31]

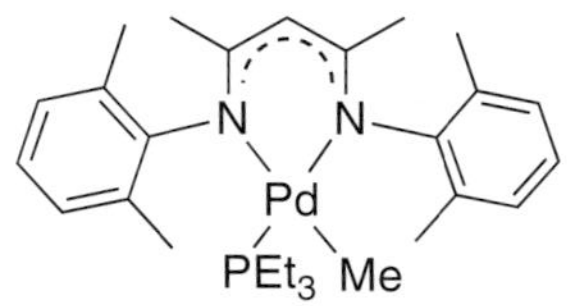

36[27]

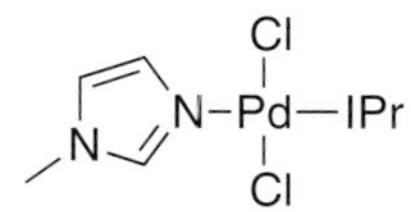

37[25]

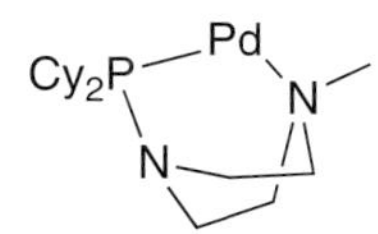

38[23]

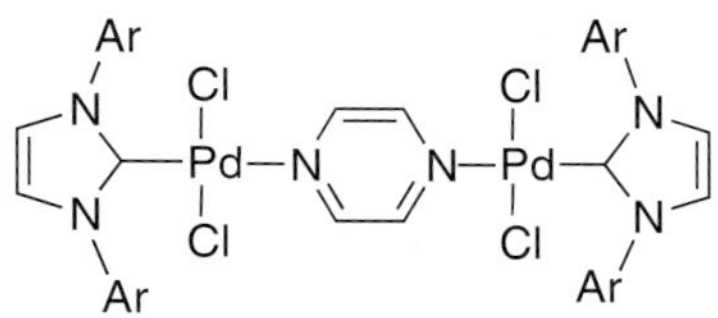

39 (Ar = Mes, Dipp)[27]

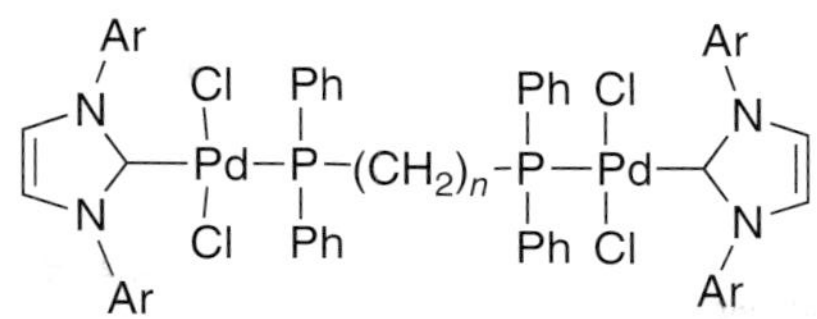

40 (n = 2,4,6; Ar = Mes, Dipp)[29]

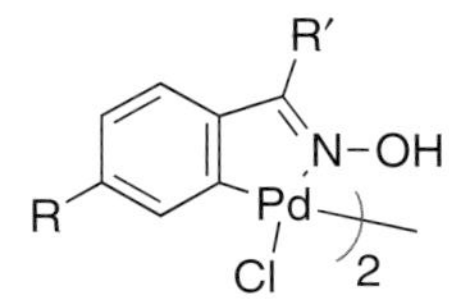

41 (R = OH, R′ = Me;
R = Cl, R′ = p-Cl-C_6H_4)[33–35]

Figure 9.1 Ligands and palladium catalysts which accelerate the coupling reaction of organosilanes with aryl bromides and chlorides.

9.2.3 Alternative Electrophiles and Metal Catalysts

Arenediazonium salts are excellent electrophiles, because neutral molecular nitrogen is an excellent leaving group. Thus, arenediazonium salts readily cross-couple with $PhSi(OEt)_3$ to give biphenyl **44** upon palladium catalysis in polar protic solvents [49–51] (Eq. (9.17)). As trifluoromethanesulfonate (triflate, TfO^-) is a similarly excellent leaving group, aryl and alkenyl triflates also participate in the silicon-based cross-coupling. Tosylates ($OSO_2C_6H_4$-*p*-Me: OTs) **45a** and mesylates (OSO_2Me: OMs) **45b** also are potent organic electrophiles, particularly when a Buchwald-type bulky phosphine ligand is employed for the palladium catalyst [52, 53] (Eq. (9.18)). Aryl sulfinates also are potent to cross-couple with arylsilanes. When organosulfinates are employed for the reaction with phenyl(triethoxy)silane using TBAF and $PdCl_2$ as a catalyst, the coupling reaction proceeds smoothly, liberating SO_2 and giving the coupled product **47** [54] (Eq. (9.19)). Acid halides react with arytrifluorosilanes smoothly in the presence of $Pd(dba)_2$/tri-*tert*-butylphosphine catalyst and cesium fluoride, thereby giving unsymmetrical benzophenone derivatives [55] (Eq. (9.20)). This protocol is applicable to not only typical acid chlorides but also to acid fluorides as stable electrophiles toward a nucleophile such as H_2O. This reaction is triggered by the oxidative addition of the acyl—X bond to the palladium(0) complex followed by the transmetalation of pentacoordinate silicate. Thiol esters (ArC(O)SPh) are applicable to the reaction with organosilanes to form diarylketones [56]. Alternatively, a cross-coupling of aryl trimethylsilyl ketones with aryl bromides results in ketone formation [57].

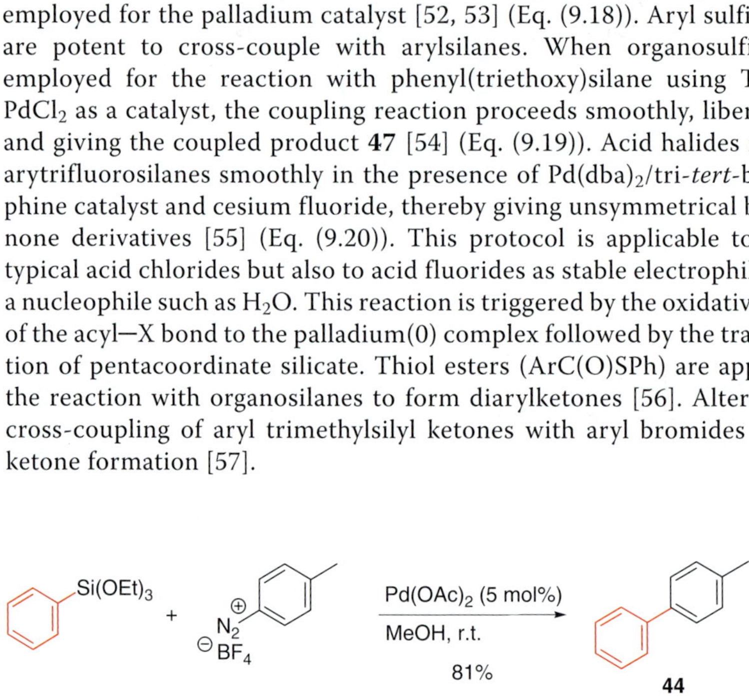

(9.17)

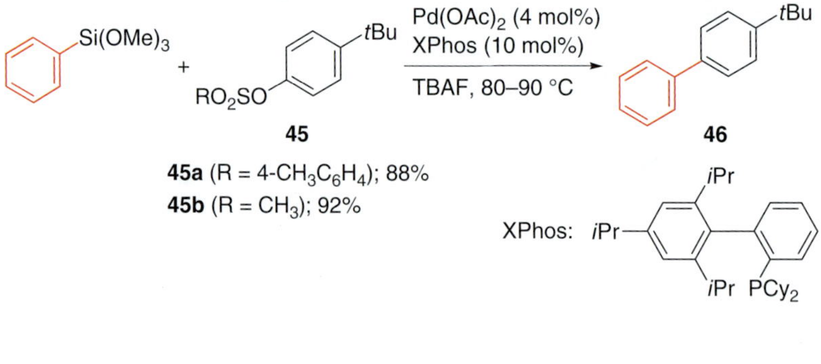

(9.18)

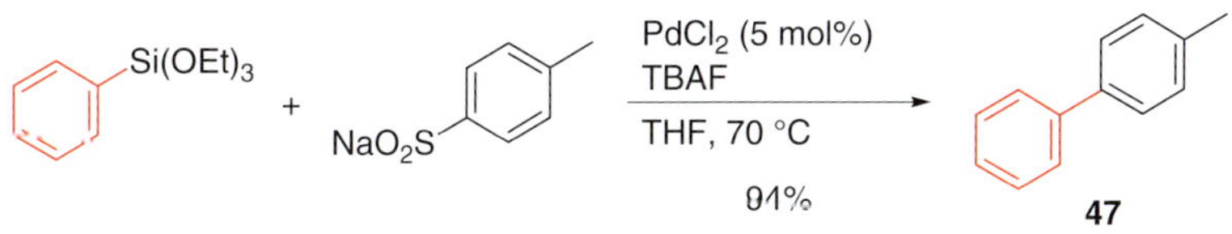

(9.19)

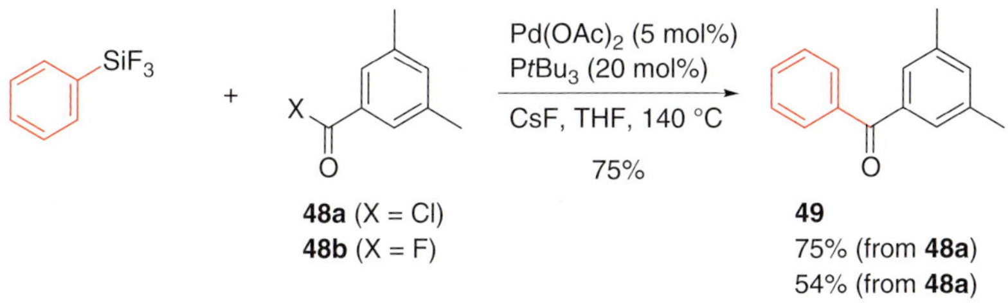

(9.20)

Appropriate ligand design for palladium catalysts expands the possibility of the cross-coupling reaction of organosilanes with alkyl halides, as demonstrated by Fu and coworkers [58, 59], who considered that β-hydride elimination of the intermediate alkylpalladium halide (**50**), generated by oxidative addition of the alkyl halide to the palladium(0) catalyst, should definitely be responsible for the side reaction. They found that bulky phosphine ligands were effective in overcoming this problem. In fact, the reaction of *n*-dodecyl bromide (**51**) with trimethoxy(phenyl)silane in the presence of TBAF catalyzed by 4 mol% of $PdBr_2$/(tBu)$_2$MeP (10 mol%) gives the coupled product dodecylbenzene (**52**) (Eq. (9.21)). Unactivated and activated secondary alkyl bromides are also shown to undergo the cross-coupling with aryltrifluorosilanes [59, 60]. For example, cyclohexyl bromide (**53**) reacts with trifluoro(phenyl)silane when $NiCl_2$·glyme (10 mol%) and norephedrine (**54**, 15 mol%) are employed in the presence of CsF, and the coupling reaction proceeds at 60 °C in *N,N*-dimethylacetamide (DMA) as a solvent to give cyclohexylbenzene (**55**) [59] (Eq. (9.22)). Asymmetric cross-coupling of aryl- or alkenyltrimethoxysilanes **56** with α-bromo esters was accomplished using $NiCl_2$·glyme/chiral (*S,S*)-diamine ligand **57** catalyst and $Bu_4N^+(F_2SiPh_3)^-$ (tetrabutylammonium difluorotriphenylsilicate, TBAT), thereby giving α-aryl carboxylic acid derivatives [61] (Eq. (9.23)). The resulting 2-arylcarboxylate esters were reduced to the corresponding alcohols without racemization.

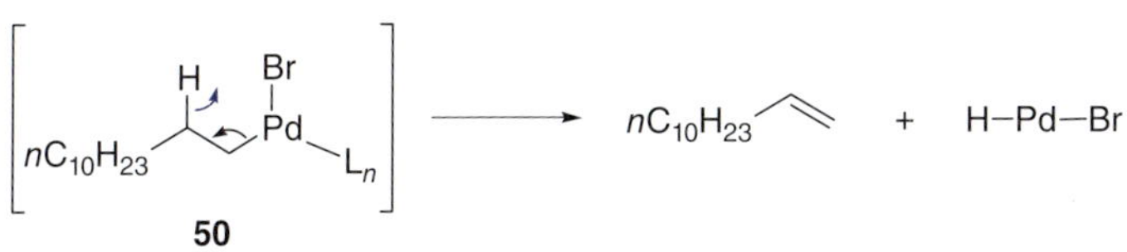

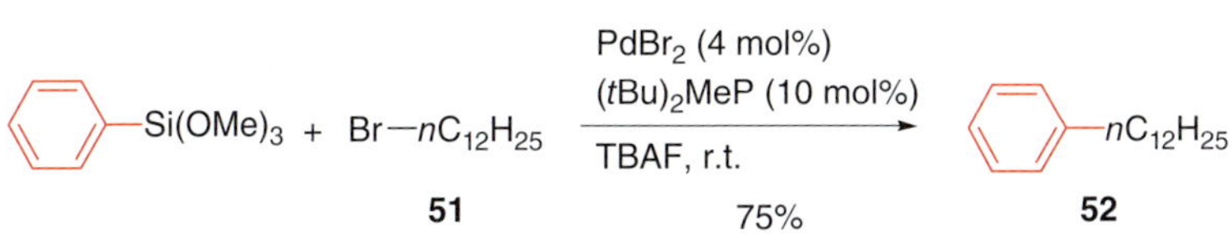

(9.21)

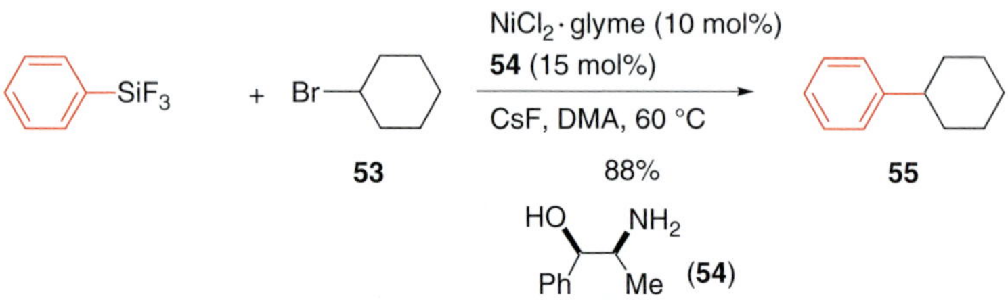

(9.22)

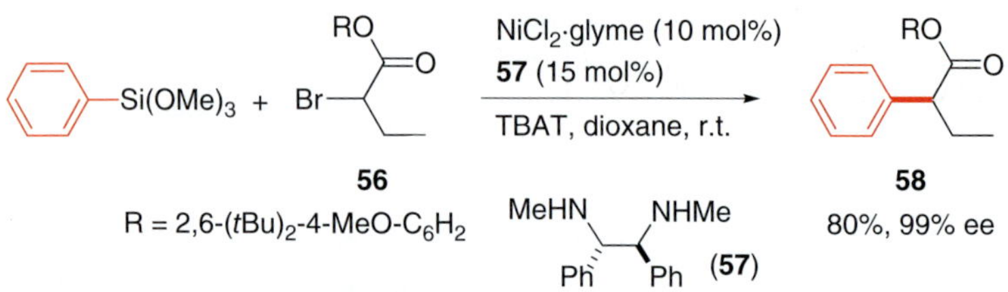

(9.23)

In the presence of not only group 10 transition-metal complexes but also other metal catalysts, the coupling of organosilicon reagents proceeds smoothly. The use of CuI and (2-NMe_2-C_6H_4)Ph_2P made the reaction of aryl(triethoxy)silanes with iodoarenes feasible, particularly in the presence of CsF [62] (Eq. (9.24)). When iodopyridines or iodothiophenes are used, the coupling proceeds in the absence of phosphine ligand [63]. Triethoxy(vinyl)silane **59** cross-couples with 1-bromo-2-phenylethyne to give 1,3-enyne **60** [64] (Eq. (9.25)).

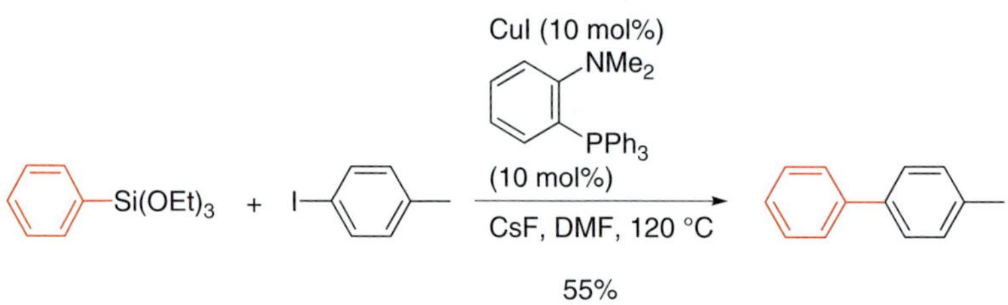

(9.24)

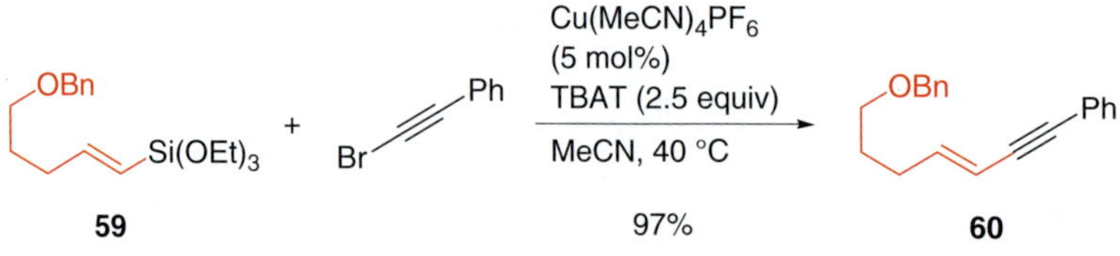

(9.25)

Copper also smoothly catalyzes the coupling of organo(hydro)silanes with organic halides bearing an sp^3 carbon–halogen bond [65, 66]. Indeed, the reaction of aryldimethylsilane **61** with allyl, benzyl, and alkyl was attributed to the intermediate pentacoordinate silicate **63** generated by the reaction of ethylene glycol, base, and a copper salt, leading to the formation of organocopper ArCu which reacts with organic halides R—Br (Eq. (9.26)).

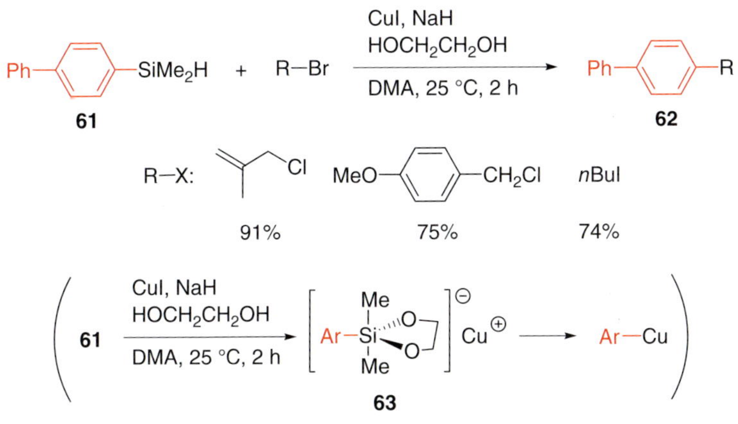

(9.26)

Cleavage of a C—C bond in the secondary alcohol of 2-arylpyridine derivative **64** followed by cross-coupling takes place when a rhodium catalyst $[Cp^*Rh(CH_3CN)_3][SbF_6]_2$ is employed. The reaction of **64** with $PhSi(OMe)_3$ in the presence of 5 mol% of the catalyst and AgF as an additive undergoes substitution of **64** at the carbon atom adjacent to the 2-pyridyl group by stirring at 90 °C for 16 hours to produce phenylated product **65** [67] (Eq. (9.27)). The coupling reaction of $PhSi(OMe)_3$ and 2-nitrobenzoic acid (**66**) also proceeds through decarboxylative C—C bond cleavage with a palladium catalyst under oxidative conditions to give biphenyl **67** [68] (Eq. (9.28)).

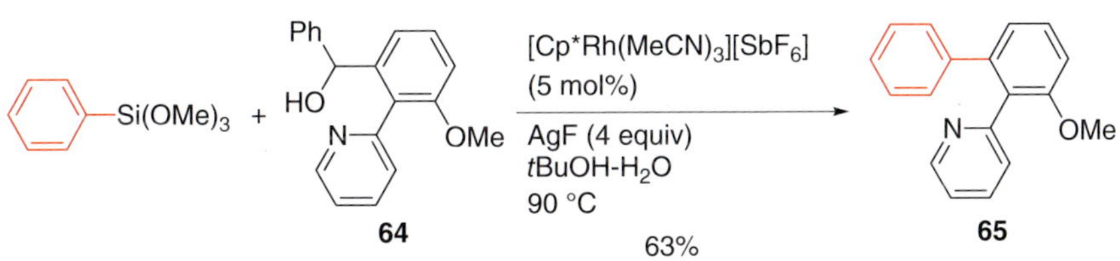

(9.27)

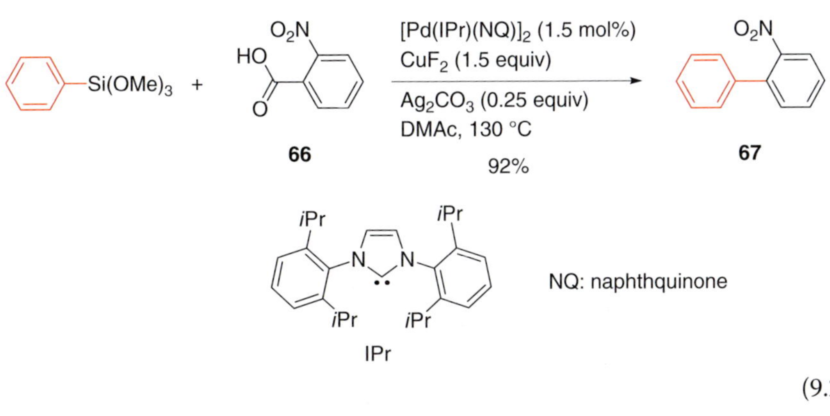

(9.28)

Transition-metal catalysis combined with photoredox catalysis allows mild alkyl transfer from a silicon reagent to the coupling partner through single-electron transfer (SET). Alkyl(bis-catecholato)silicates **68**, which release alkyl radicals upon photoirradiation, are good reagents for this purpose. Goddard et al. reported that **68**, having a potassium counterion complexed by 18-crown-6 (18-C-6), underwent alkylation of bromo(hetero)arenes using a nickel/dtbpy and **Ir1** dual-catalytic system under blue LED irradiation [69–71] (Eq. (9.29)). Various alkyl groups can be connected to arene substrates. Organic photoredox catalysts such as **4CzIPN**, instead of **Ir1**, are effective for this alkylation coupling reaction [72, 73]. Bromoalkanes are applicable as an electrophile for the cross-coupling of **68**·K[18-crown-6] with $NiCl_2$(cod)/dtbpy/**Ir1** as catalysts to provide various alkyl–alkyl coupling products, although homocoupled products from bromoalkanes are accompanied [74] (Eq. (9.30)).

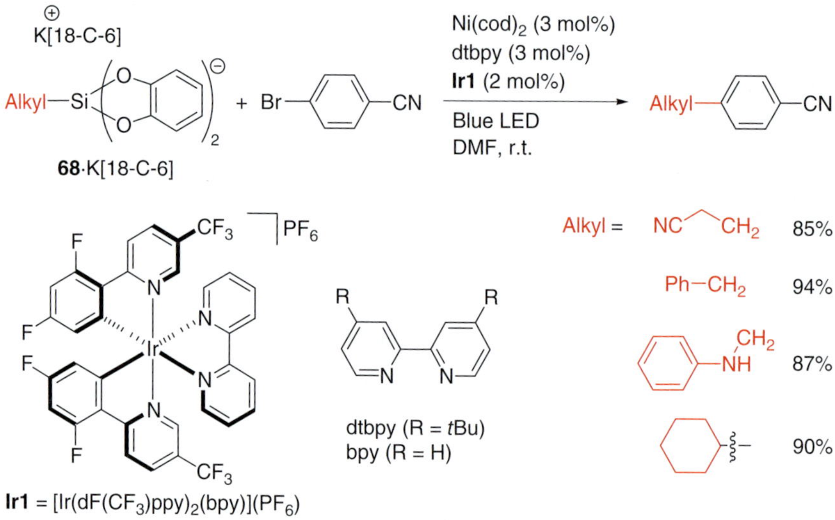

(9.29)

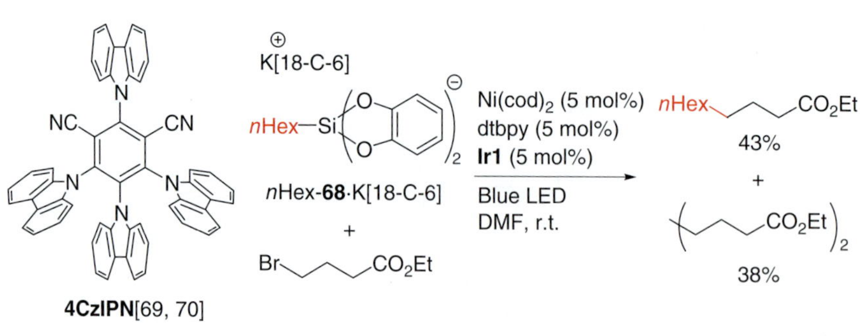

(9.30)

Subsequently, Molander and coworkers showed cross-coupling of secondary ammonium salt **68** with haloarenes in the presence of $NiCl_2$(dme)/dtbpy and $[Ru(bpy)_3](PF_6)_2$ under irradiation of LED light to afford the corresponding alkylated products [75] (Eq. (9.31)). Similarly, a variety of alkylsilicates **68** (alkyl = *n*-alkyl, *sec*-alkyl, and benzyl) undergo the reaction not only with haloarenes but also with haloalkenes [76], aryl triflates, tosylates, and mesylates [77], and brominated 2,1-borazaronaphthalenes **69** and **70** [78]. When bromofluoroiodobenzene is used as an aryl electrophile, the iodine is selectively replaced by an alkyl group in preference to bromine or fluorine, giving alkylation product **71** [79] (Eq. (9.32)). This reaction system is applicable to thioetherification of bromoarenes under the combination of an isobutylsilicate with an alkanethiol [80].

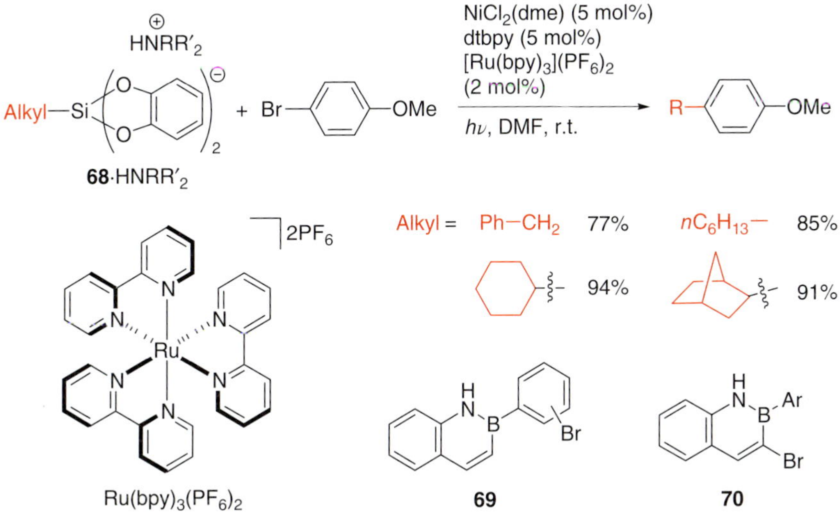

(9.31)

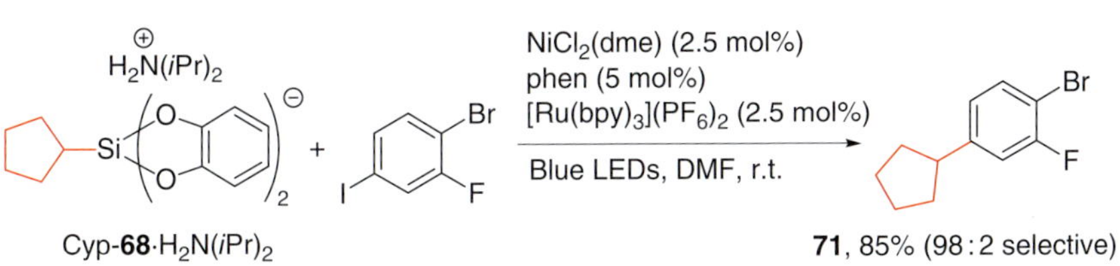

(9.32)

Taking an iridium(III) complex as a representative photochemical catalyst, the photochemical process is discussed as follows. At first, iridium(III) is excited by photoirradiation to give iridium (III)*, which oxidizes the alkylsilicate by SET to generate an alkyl radical and iridium(II). The alkyl radical couples with nickel(0) to form an alkyl–Ni(I) complex. Subsequent oxidative addition of haloarenes furnishes an alkyl(aryl)Ni(III)–X complex. Reductive elimination produces the coupled product and Ni(I)–Br. The nickel(I) complex

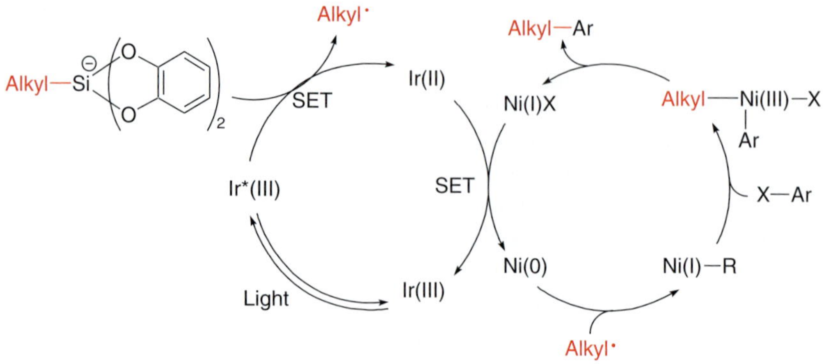

Figure 9.2 Mechanism of photochemically driven iridium(III)/nickel(0)-catalyzed coupling of alkylsilicate and ArX.

is reduced by iridium(II) through SET to give iridium(III) and nickel(0), respectively, (Figure 9.2).

9.2.4 Cross-coupling Reaction of Functionalized Organosilicon Reagents

Ketene trimethylsilyl acetal **72** successfully cross-couples, as reported in 1991, with phenyl triflate at the α-position of the carbonyl group using a palladium catalyst [81, 82]. When 0.5–2 mol% palladium catalyst, formed by the reaction of (η^3-C_4H_7)PdOAc and a bidentate phosphine 1,1′-bis(diphenylphosphino) ferrocene (DPPF), was employed for the reaction of (*E*)-$CH_3CH{=}C(OMe)OSiMe_3$ (**72**) and PhOTf (Tf = SO_2CF_3), methyl α-arylpropanoate **73** was produced [81] (Eq. (9.33)). Enol silyl ethers **74** derived from α-fluoropropiophenone [83], amides **75** [84], lactones **76** [84], and **77** [85] were shown to undergo α-arylation reaction with aryl bromides or triflates.

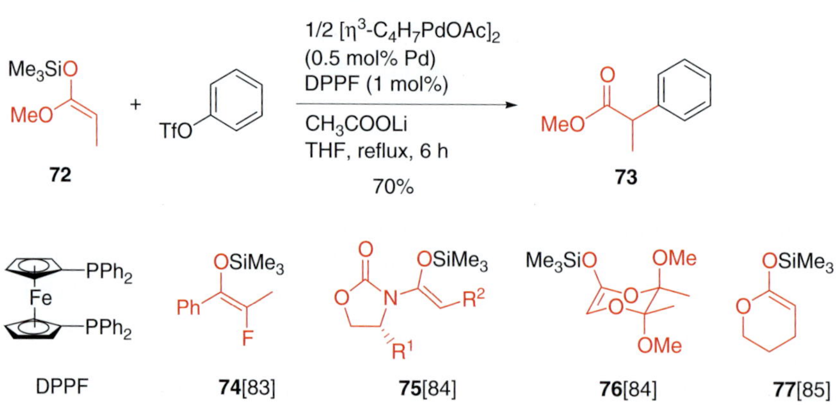

(9.33)

An enantioselective version of the arylation reaction was successfully carried out with a palladium catalyst bearing a chiral ligand [86–88]. The reaction of ketene silyl acetal **78** with 6-methoxy-2-naphthyl triflate in the presence of a palladium catalyst bearing chiral monophosphine ligand **79** effects enantioselective arylation and, after hydrolysis, gives (*S*)-naproxen (**81**), a nonsteroidal *anti*-inflammatory agent, in a highly enantioselective manner (92% ee) [87] (Eq. (9.34)). Chiral ligand **80** based on ferrocene was employed for enantioselective arylation with ketene silyl acetals [88].

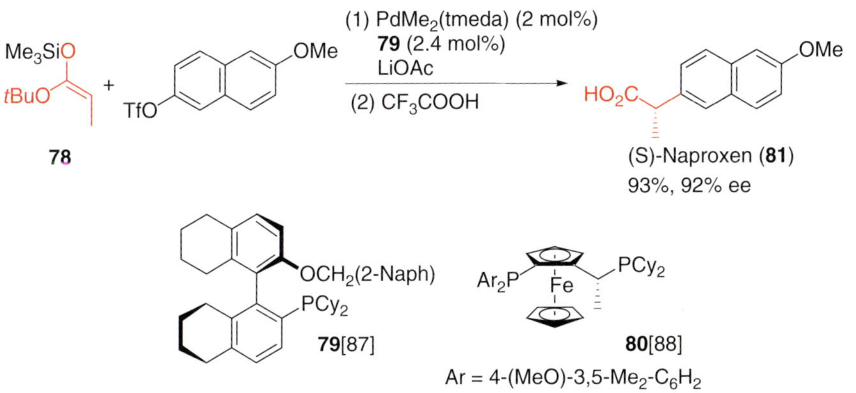

(9.34)

Arylation at the β-position of ester carbonyl was achieved with ketene silyl acetal **82**. The reaction of **82** with aryl bromide catalyzed by a palladium and bulky 1-(2-dimethylaminophenyl)-2-dicyclohexylphosphinoimidazole (**83**) ligand in the presence of ZnF_2 induced arylation of the β-C—H bond of ester to give **84** in 58% yield [89] (Eq. (9.35)).

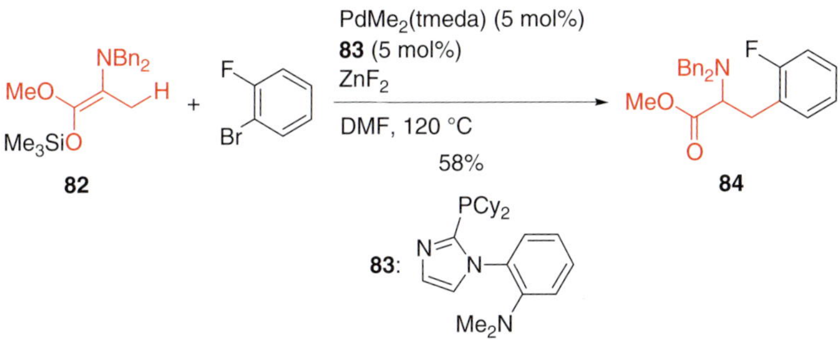

(9.35)

9.2.5 Cross-coupling Reaction of Organosilanes Through Directed C—H Bond Activation

Inspired by the recent progress in the transition-metal-mediated C–H functionalization of organic molecules using a wide range of nucleophiles and electrophiles

[90–94], the introduction of an organic group with organosilicon reagents has been studied worldwide to date. Organosilanes $ArSi(OR)_3$ are employed as a nucleophile in the transition-metal-mediated functionalization of several types of C—H bonds. When an acetamide group is employed as a directing group, the reaction at *ortho*-C—H bond of *N*-aryl acetamide **85** successfully proceeds with $PhSi(OMe)_3$, as reported in 2007 by Shi and coworkers [95]. Using 5 mol% $Pd(OAc)_2$ in the presence of $Cu(OTf)_2$/AgF (2 equiv), they isolated the coupled product **86** (Eq. (9.36)). Furthermore, C—C bond formation at *ortho*-C—H bonds of cyclic enamide **87** [96] and benzoyl amide **88** [97], C2—H bonds of azole **89** [98, 99] and indole **90** [100], *ortho*-C—H bonds of 2-phenylpyridine (**91**) [101] and enaminone **92** [102], pentafluorobenzene **93** [103], and indoline **94** [104] is catalyzed by a palladium or nickel. The reaction of α-*N*-phthalimido(Phth)amide **95** with $PhSi(OEt)_3$ takes place at β-C(sp^3)—H bond in the amide group to give coupled product **97** with the aid of a palladium catalyst and quinoline ligand **96** [105] (Eq. (9.37)).

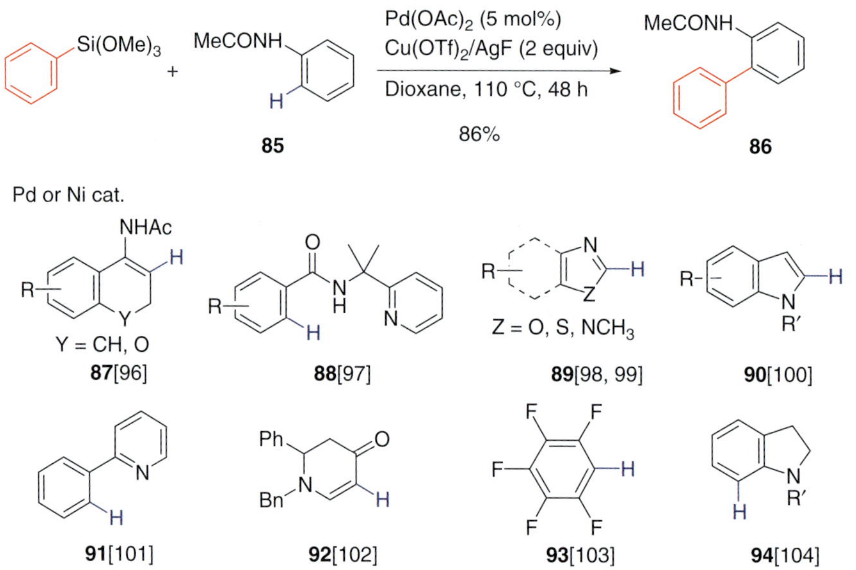

(9.36)

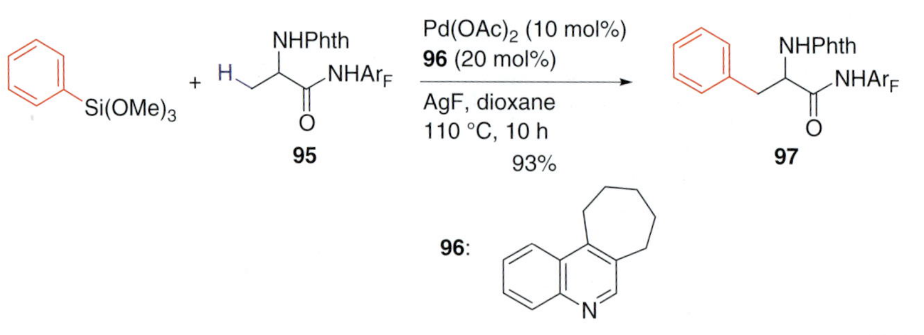

(9.37)

Some other transition-metal complexes work as a catalyst for the C–H functionalization reactions with organosilanes. Ruthenium catalyst $[RuCl_2(p\text{-cymene})]_2$ (5 mol%)/$AgSbF_6$ (20 mol%) in the presence of CuF_2 (3 equiv) as an additive is effective for the reaction of benzamide **98** [106, 107] and imine **99** [108] at the C—H bonds from the directing group to afford phenylated product **100** and **101**, respectively, (Eq. (9.38)). Rhodium $[Cp^*RhCl_2]_2$ (2 mol%)/$AgSbF_6$ (8 mol%) promotes the C–H arylation reaction of *N*-(2-pyrimidyl)indole **102** with $ArSi(OMe)_3$ [109], $PhCH_2Si(OEt)_3$ [110], and allyl-$Si(OMe)_3$ [110] efficiently in the presence of AgF-$Cu(OAc)_2$ (2 equiv). The reaction of **102** with $PhSi(OMe)_3$ is shown in Eq. (9.39) [109].

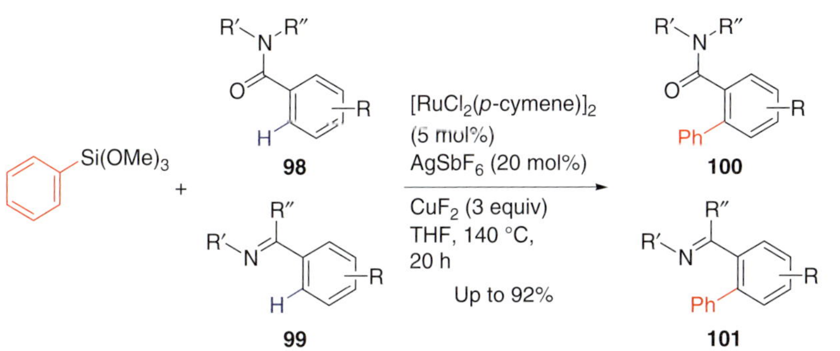

(9.38)

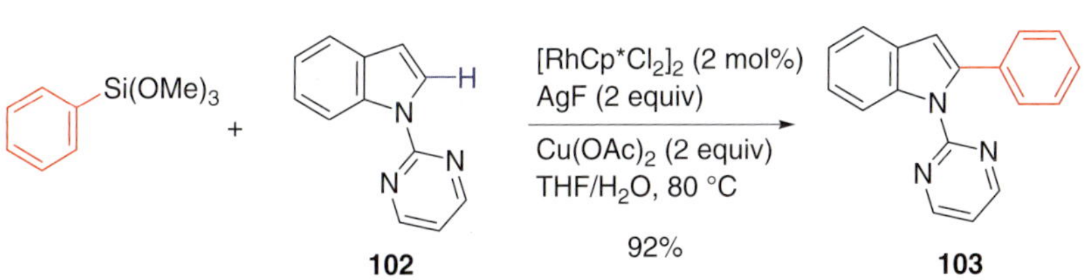

(9.39)

The coupling reaction at a heteroatom—H bond with an organosilicon reagent is also effected by transition-metal catalysis. An N—H bond of primary and secondary amines reacts with aryl- and vinyl(trimethoxy)silanes under mild conditions. The reaction of benzimidazole (**104**) with organosilanes in the presence of stoichiometric amounts of $Cu(OAc)_2$ and TBAF proceeds at room temperature under aerobic conditions to give vinylated (**105**) and arylated (**106**) products [111] (Eq. (9.40)). Aromatic and aliphatic amines (**107**, **108**, **109**), amides (**110**), tosyl amide (**111**), and 2,2,2-trifluoroethylamine (**112**) have been shown to react under similar conditions [111, 112].

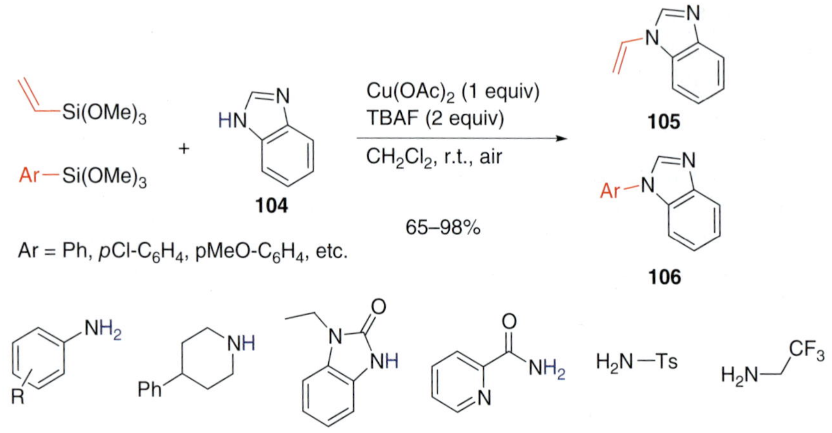

107[111, 112] **108**[111] **109**[111] **110**[111] **111**[112] **112**[112]

(9.40)

Sulfoximine (**113**) reacts with $PhSi(OMe)_3$ to give *N*-phenylsulfoximine (**114**) using CuI and TBAF (10 mol% each) under O_2 at room temperature in dichloromethane [113, 114] (Eq. (9.41)).

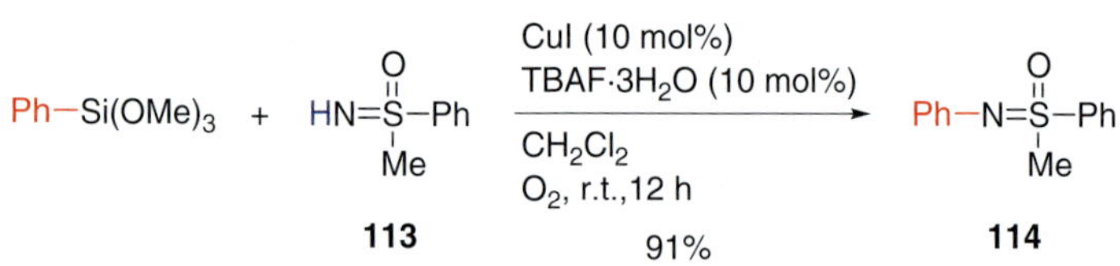

(9.41)

The formation of a C—P bond also proceeds by palladium catalysis in the presence of Ag_2CO_3/KF using diethyl phosphite (**115**) under the oxidative conditions, affording coupled product **116** [115] (Eq. (9.42)).

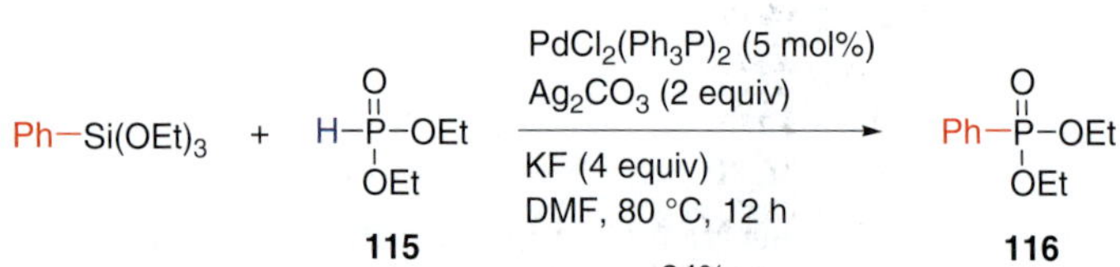

(9.42)

9.2.6 Tandem Reaction Involving Silicon-Based Cross-coupling

Combined use of the coupling reaction of organosilicon reagents with well-established reactions expands the synthetic utility of the coupling reaction. Alkenylsilanes are prepared stereoselectively by a variety of reactions such as hydrosilylation, hydrometalation of silylalkynes, or silylation of alkenylmetals with an appropriate electrophilic silane reagent.

An example of the collaborative use of hydrosilylation and cross-coupling is shown in Eqs. (9.43) and (9.44). Denmark employed cyclic alkenylsilyl ether **118**, obtained by intramolecular hydrosilylation of alkyne **117**, for the cross-coupling reaction with aryl 1-idodonaphthalene by a palladium catalyst and obtained arylated alkenyl alcohol **119** after removal of the silyl protecting group [116]. Ring-closing metathesis of siloxy-tethered diene **120** with Schrock catalyst ([$(CF_3)_2MeCO]_2Mo(=CHCMe_2Ph)(=NC_6H_3$-2,6-*i*$Pr_2$)) affords cyclic silyl ether **121**, which undergoes palladium-catalyzed cross-coupling with aryl iodide to give rise to alkenyl alcohol **122** [117].

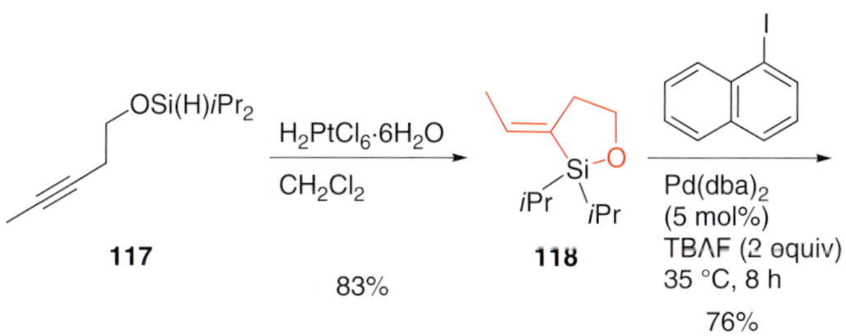

(9.43)

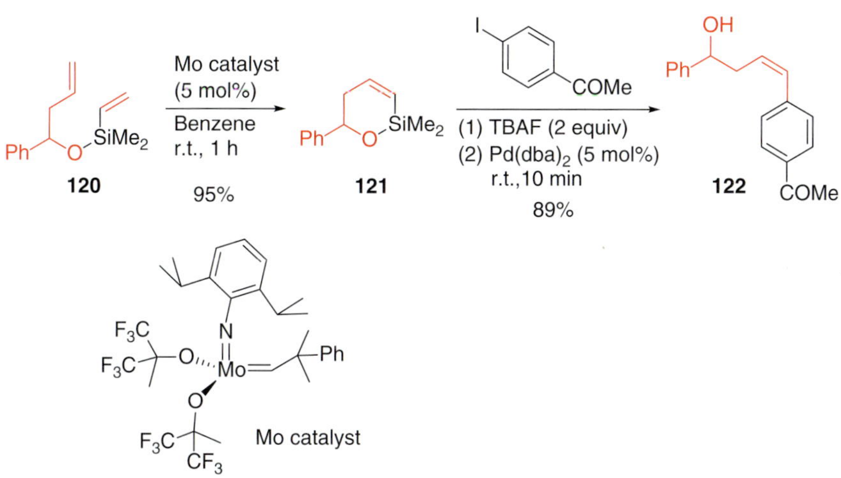

(9.44)

9.3 Cross-coupling of Silanols, Silanolates, Oligosiloxanes, and Polysiloxanes

9.3.1 Silanols and Silanolates

Cyclic oligosiloxanes such as D_3 (**123**, hexamethylcyclotrisiloxane) react with 3 equiv of organolithium RLi, as demonstrated by Sieburth and Mu, to give

3 equiv of lithium silanolate RMe_2SiOLi (**124**). An equimolar reaction of RLi with D_3 induces ring opening leading to linear trisiloxanes $R(SiMe_2O)_3Li$, which further reacts with 2 RLi to cleave an additional two Si-O bonds to give three molecules of **124**. Hydrolysis of **124** gives silanol **125**, whereas disiloxane **126** is produced by silylation with Me_3SiCl [118–120] (Eq. (9.45)). When a functionalized organolithium species such as aryl, heteroaryl, and alkenyl is introduced at the silicon atom, the resulting organosilanols are ready for the cross-coupling reaction.

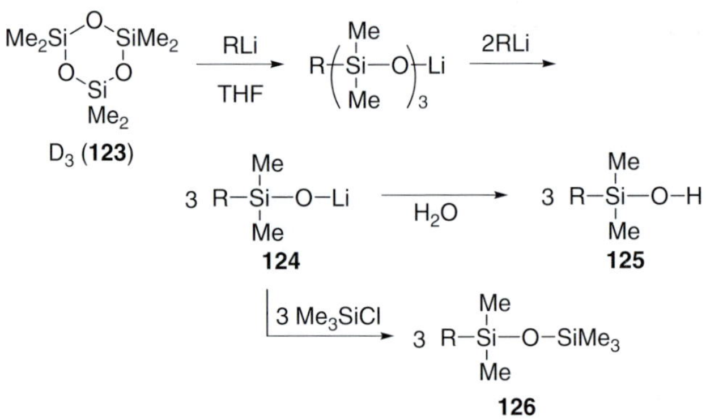

(9.45)

The use of organosilanols for the catalytic reaction was first shown by Mori and Hiyama in the oxidative Mizoroki–Heck-type reaction. The reaction of phenyldimethylsilanol **127** with methyl vinyl ketone (E = COMe) proceeds in the presence of a stoichiometric amount of $Pd(OAc)_2$ to undergo addition-elimination leading to **128**. The reaction proceeds also with a catalytic amount of palladium in the presence of $Cu(OAc)_2$/LiOAc as an oxidant [121]. Aryldimethylsilanol **127** also reacts with aryl iodide to give coupled unsymmetrical biaryl **129** catalyzed by palladium(0), where silver(I) oxide is revealed to serve as an effective additive [122] (Eq. (9.46)).

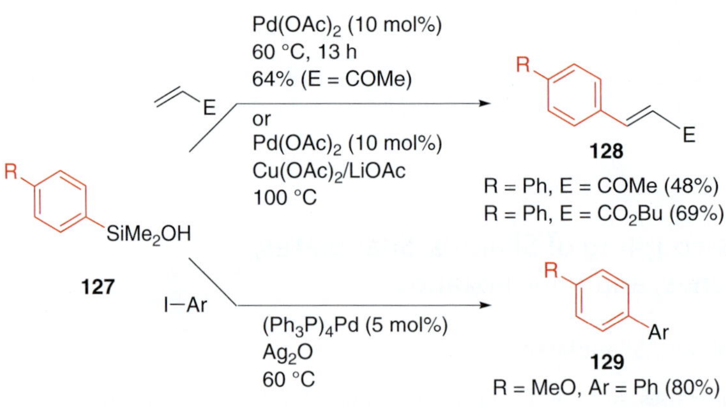

(9.46)

Silanediols and silanetriols also serve as organosilicon reagents for the smooth cross-coupling reactions. Silanediols **130** are readily prepared by hydrolysis of the related dichlorosilanes or alternatively by Pd/C-catalyzed hydrogenolysis of $PhMeSiH_2$ in aqueous solution (Eq. (9.47)), and are applicable not only to the cross-coupling reaction [123] but also to the Mizoroki–Heck-type reaction [124, 125], conjugate addition [124], hydroarylation of alkynes [126], and 1,2-addition to aldehydes [127] to give **131–136** by means of a rhodium or iridium catalyst (Figure 9.3).

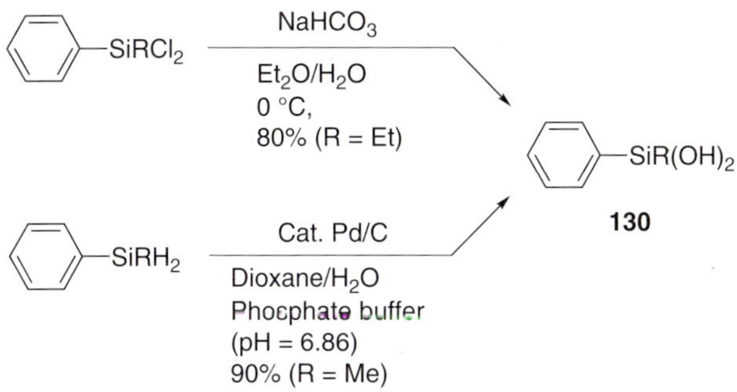

(9.47)

Denmark and Sweis employed silanols as a nucleophile for the cross-coupling reaction [128]. They extensively studied a wide range of cross-coupling-based organic transformations using various alkenyl, alkynyl, aryl, and heteroaryl silanols **137–145** [129–137] (Eq. (9.48)), demonstrating the high potential of the reaction by synthetic applications [138], total synthesis of natural products RK-397 (**146**) and (+)-papulacandin D (**147**) [139–141], and mechanistic studies [142–144].

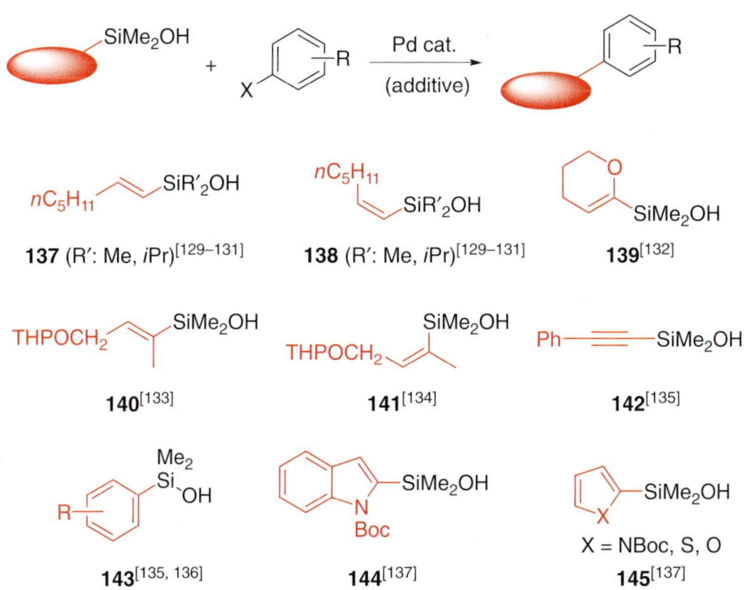

(9.48)

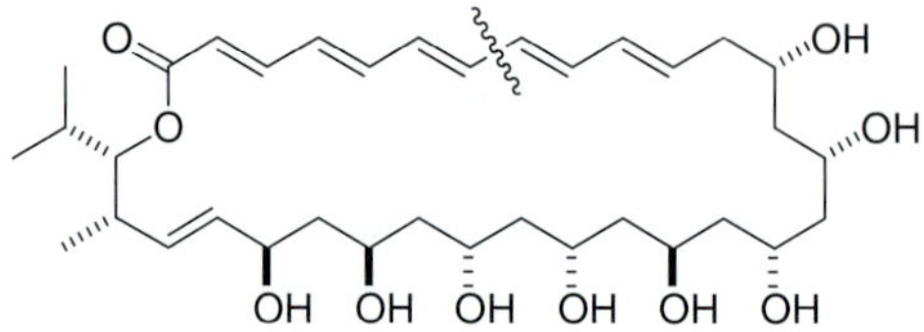

RK-397 (**146**)[139]

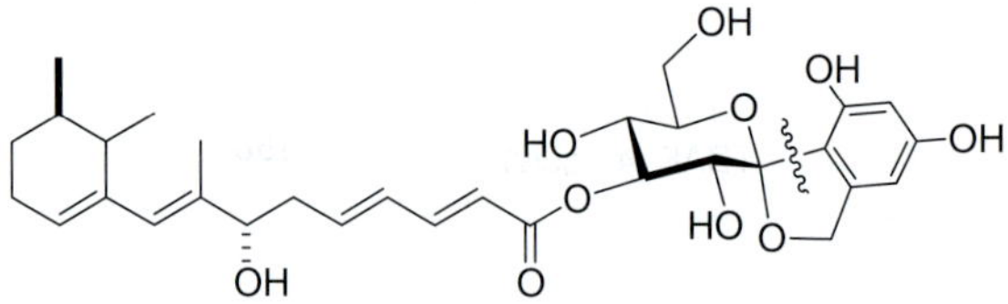

(+)-Papulacandin D (**147**)[140, 141]

Alkenylsilanols are deemed to be a silicon analog of allylic alcohols. Accordingly, Simmons–Smith cyclopropanation of alkenylsilanol **148** is accelerated by the directing effect of the hydroxy group to afford cyclopropylsilanol **149** [145, 146], which serves as a silane-coupling reagent for the cross-coupling reaction with bromobenzene to give 1,2-disubstituted cyclopropane **150** (Eq. (9.49)).

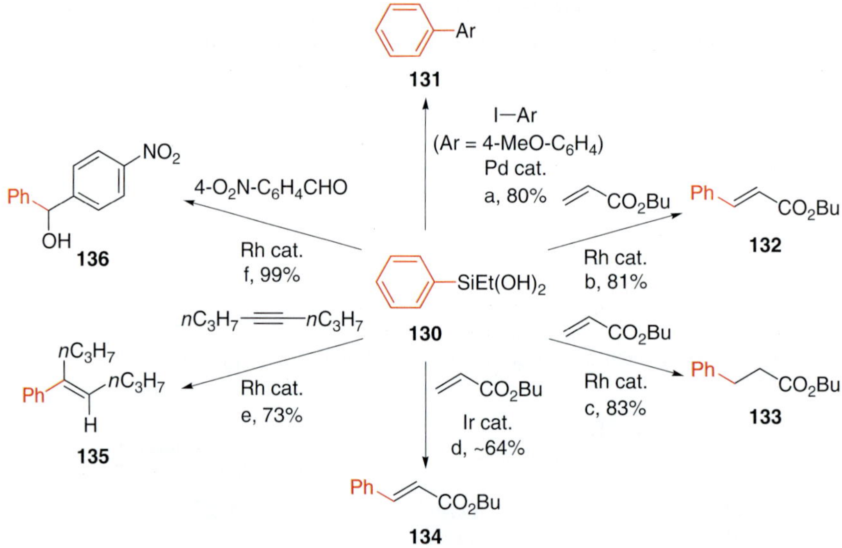

Figure 9.3 Synthetic transformations of silanediol **130**. (a) $(Ph_3P)_4Pd$ (5 mol%), tetrahydrofuran (THF), Ag_2O, 60 °C, 12 hours, (b) THF, $[Rh(OH)(cod)]_2$ (3 mol%), 70 °C, 24 hours, (c) THF-H_2O, $[Rh(OH)(cod)]_2$ (3 mol%), 70 °C, 24 hours, (d) toluene/(H_2O), $[Ir(OH)(cod)]_2$, (3 mol%), 120 °C, 24 hours, (e) toluene/H_2O (10 : 1), $[Rh(OH)(cod)]_2$ (3 mol%), 100 °C, 24 hours, and (f) THF, $[Rh(OH)(cod)]_2$ (3 mol%), 70 °C, 24 hours.

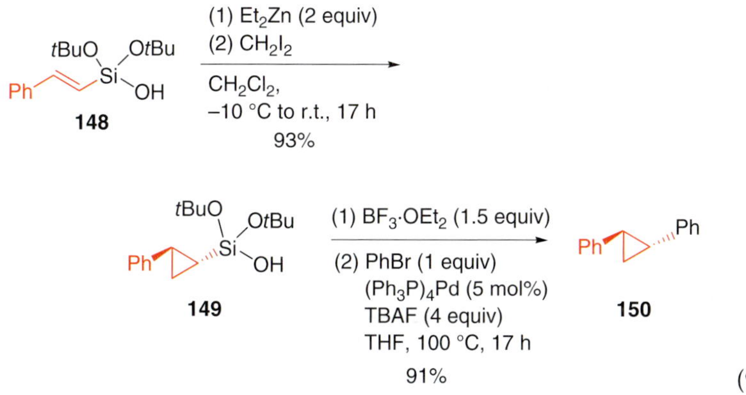

(9.49)

Organic silanolates are stable solid reagents which conveniently participate in the palladium-catalyzed cross-coupling with chloroarenes highly efficiently [147–151]. While a base like a fluoride ion is usually needed for the standard silicon cross-coupling, silanolates do not need any additional bases due to self-activation. Organic silanolates can couple with haloolefins and haloarenes under mild conditions in the presence of a palladium catalyst. Silanolates are readily prepared by treatment of the corresponding silanols with an alkali base. As shown in Eq. (9.50), 2,6-dimethylchlorobenzene successfully couples with potassium 1-heptenyldimethylsilanolate (**151**) catalyzed by a palladium catalyst and SPhos ($Cy_2[2\text{-}(2,6\text{-}(OMe)_2\text{-}C_6H_3)C_6H_4]P$) as a ligand to give alkenylated arene **152** in good yields [147]. When (Z)-alkenylsilanolates were used, (*Z*)-alkenyl products were obtained without isomerization. Mechanistic study on the cross-coupling using organic silanolates has been performed [152–154], suggesting two distinct pathways for transmetalation of vinylsilanolates through a neutral or an anionic palladium complex [154]. Aryl- and allyl silanolates **153** and **154** are applicable to the cross-coupling [149–151].

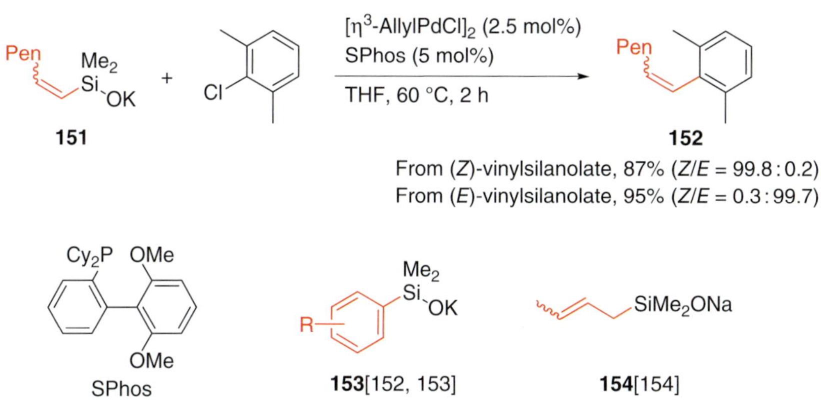

(9.50)

9.3.2 Disiloxanes, Oligosiloxanes, and Polysiloxanes

Disiloxanes are also activated by a fluoride ion (e.g. TBAF) to work as an organosilicon cross-coupling reagent [155, 156]. Hydrosilylation of pentamethyldisiloxane (**155**) with terminal alkynes **156** in the presence of a platinum catalyst gives *E*-alkene **157** stereoselectively, and the resulting alkenylsilane undergoes the cross-coupling reaction with iodobenzene (**158**) leading to **159** [155, 157]. When a iodorhodium(I) catalyst is employed for the hydrosilylation of alkyne **156**, either (*E*)-alkenylsilane **157** or (*Z*)-isomer **160** is produced in a stereoselective manner. Pretreatment of a silane reagent with a rhodium catalyst followed by addition of an alkyne delivers (*Z*)-alkenylsilane, while one-shot addition results in giving (*E*)-silane [18]. Subsequent cross-coupling with iodoarenes in the presence of a palladium catalyst leads to the corresponding (*E*)- and (*Z*)-alkenes, **159** and **161**, respectively (Eq. (9.51)).

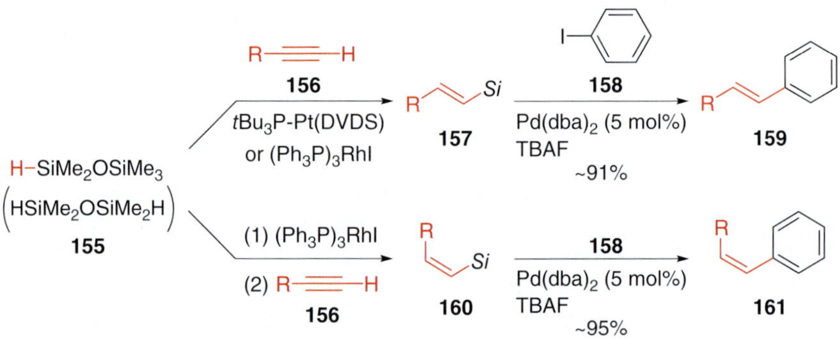

(9.51)

Disiloxanes exist in equilibrium with the corresponding silanolate species in the presence of a base like KOH [158]. This characteristic indicates that disiloxanes can be activated by a hydroxide ion toward the cross-coupling. Actually, (*p*-anisyl)disiloxane **162** undergoes cross-coupling with *p*-bromotoluene in the presence of KOH and H_2O as an effective disiloxane activator, thereby giving the corresponding coupled product, 4-methoxy-4′-methylbiphenyl [159] (Eq. (9.52)).

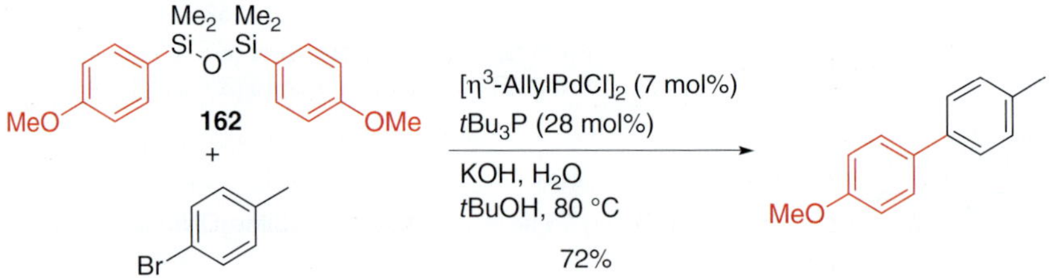

(9.52)

Disilane **163** is oxidized into disiloxane in the presence of a fluoride ion and CO_2 with the liberation of CO. Accordingly, a one-pot reaction of disilane **167** with aryl halide in the presence of a palladium/copper catalyst under an atmosphere of CO_2 achieves carbonylative cross-coupling to give diaryl ketone **164** [160] (Eq. (9.53)).

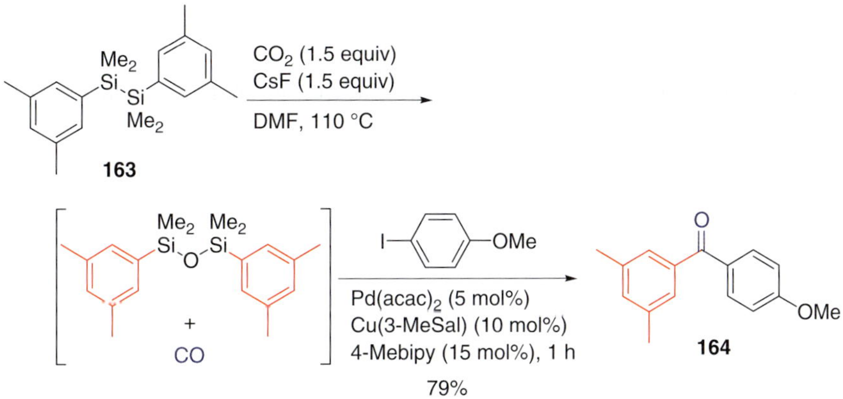

(9.53)

Cyclic and acyclic oligomeric siloxanes can also be employed as the silicon reagents for the cross-coupling reactions (Eq. (9.54)). Commercially available divinyldisiloxane (DVDS, **165**) undergoes vinylation of aryl halides in the presence of a palladium catalyst to give styrene derivatives [161]. Cyclic tetramer D_4^V (**166**) and trimer D_3^V (**167**), both commercially available from the silicone industry, give vinylated products under similar conditions. The cyclic siloxane trimer bearing a phenyl group D_3^{Ph} (**168**) also participates in the C—C bond-forming reactions. Cross-coupling with aryl halides is effected by a palladium catalyst [162, 163]. Arylsiloxanes **169** bearing a phenyl substituent are also prepared by hydrolytic condensation [163] of halosilanes or dehydrogenative functionalization of hydridosiloxanes D_4^H (**170**) [162] and have been employed for the coupling reactions. Poly(phenylmethylsiloxa ne) (**171**), which is available as thermally resistant silicone oil with various average molecular weights, also reacts with aryl halides to give coupled products. Polysiloxanes bearing substituted aryl and alkenyl groups **172** and **173** are similarly prepared by dehydrogenative arylation and hydrosilylation with poly(hydromethylsiloxane) PMHS (**174**) [162]. The use of palladium along with an electron-donating and bulky phosphine ligand for the reaction of **174** is effective for the cross-coupling reaction with unactivated aryl chlorides [164]. Conjugate addition of these siloxanes to α,β-unsaturated carbonyl compounds proceeds upon catalysis by a rhodium catalyst [165].

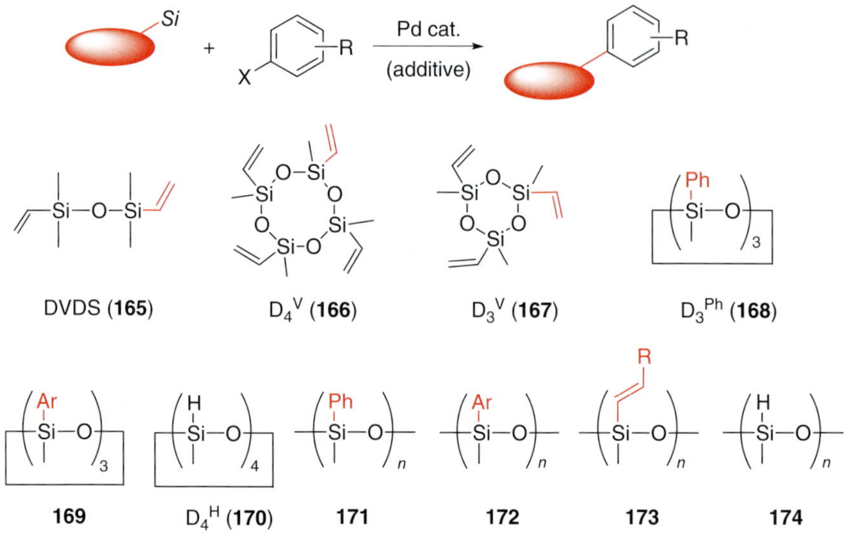

(9.54)

9.4 Cross-coupling of Allylsilane, Arylsilanes, and Trialkylsilanes

9.4.1 Silacyclobutyl, Allylsilanes, and Benzylsilanes

As described earlier, silicon reagents applicable to the cross-coupling should have electron-withdrawing heteroatom(s) on the silicon atom. This is because these heteroatoms enhance the Lewis acidity of the silicon center enough to form pentacoordinate silicates responsible for transmetalation. However, due to higher Lewis acidity of silicon, these silicon reagents are often less stable. Thus, tetraorganosilicon reagents are ideal for use. At present, nucleophiles containing trimethylsilyl, 1-methyl-1-silacyclobutyl, allyldimethylsilyl, aryldimethylsilyl, and benzyldimethylsilyl groups are applicable to cross-coupling. At length, trialkylsilyl groups such as triethylsilyl, *tert*-butyldimethylsilyl, and bulky triisopropylsilyl are found to be applicable to the cross-coupling reaction under suitable catalytic conditions. Hard bases such as TASF, TBAF, TMSOK, and CsF are potent to activate silicon reagents and mediate the cross-coupling reaction.

Silacyclobutanes were reported by Denmark and Choi to participate also in the reaction as the organosilicon reactant. They showed that 1-alkenyl-1-methylsilacyclobutanes bearing no heteroatom on the silicon atom cross-couple with aryl and alkenyl halides [166]. Both *E*- and *Z*-alkenylsilanes **175** and **176**, prepared according to Eq. (9.55), react with aryl halides in a stereospecific manner to give **177** and **178**, respectively, (Eq. (9.56)). 1-Aryl-1-silacyclobutane **179** also reacts with aryl halides in the presence of a palladium catalyst. The halogen atom on the silicon atom was found essential for successful coupling. Indeed, **179** (X = F or

Cl) coupled with 2-tolyl iodide using a Pd(II)/tBu$_3$P catalyst as well as 3 equiv of TBAF as an activator and gave biaryl **180** (X = F or Cl) [167] (Eq. (9.57)). Later, silanols derived from the silacyclobutanes were suggested to be the active species in these cases.

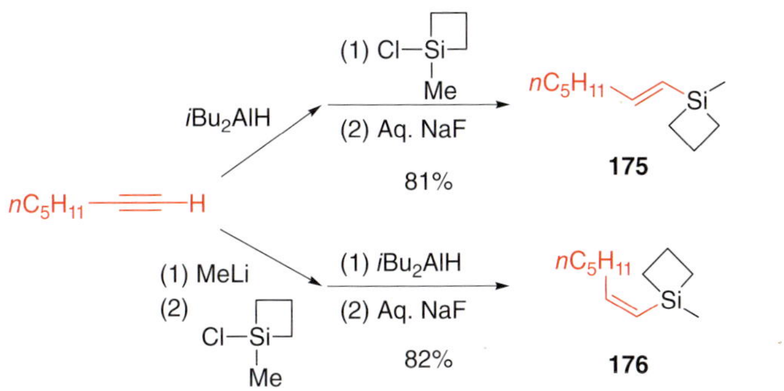

(9.55)

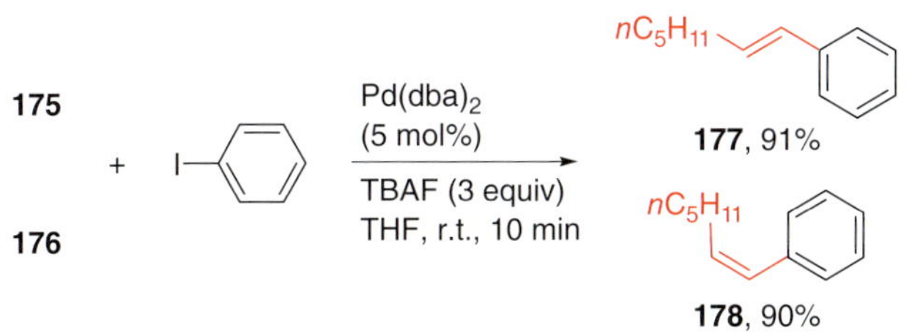

(9.56)

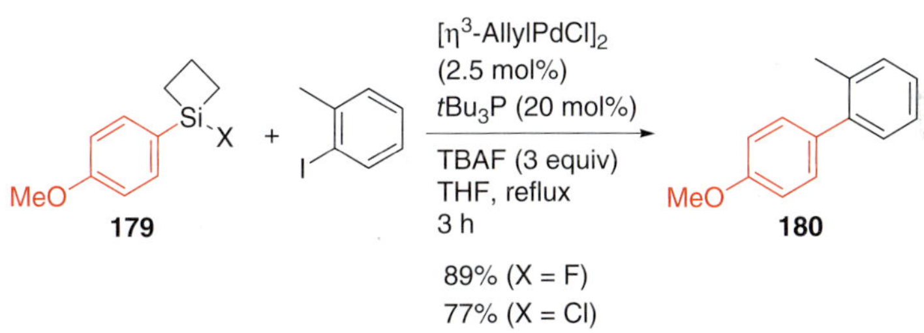

(9.57)

Phenyltriallylsilane (**181**) undergoes cross-coupling with chloroarenes under Pd/XPhos catalysis via activation by TBAF in excess, giving biaryls **182** [168] (Eq. (9.58)). Diphenyldiallylsilane also reacts with bromoarenes in the presence of PdCl$_2$/Cy$_3$P and TBAF to provide biaryls [169]. 2-Pyridyl(allyl)dimethylsilane **183** also effects the cross-coupling with iodoarenes in the presence of a palladium catalyst and Ag$_2$O [170] (Eq. (9.59)). Vinyl(β-methallyldimethyl)silanes such as **185** are employed in the cross-coupling with iodoarenes upon treatment with TBAF (2 equiv) to form corresponding products **186** [171] (Eq. (9.60)). The allylic moiety in these examples is assumed to be replaced by a fluorine (or hydroxy) before the cross-coupling reaction.

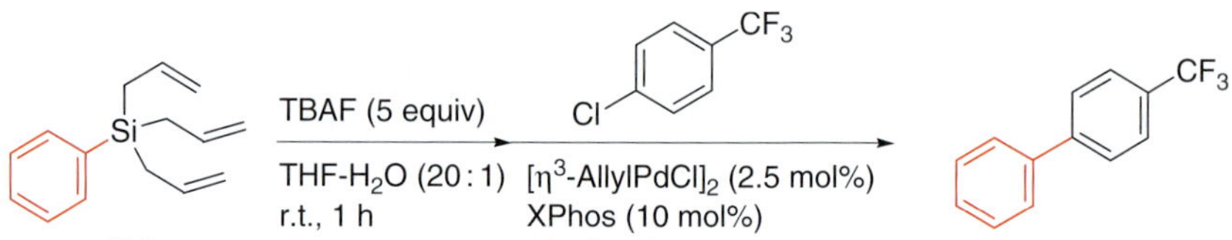

(9.58)

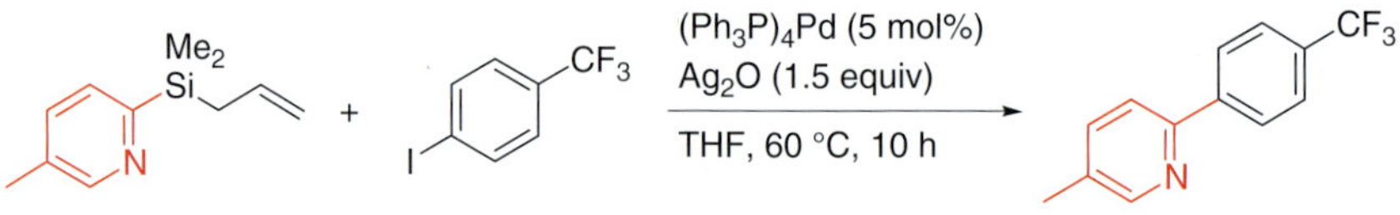

(9.59)

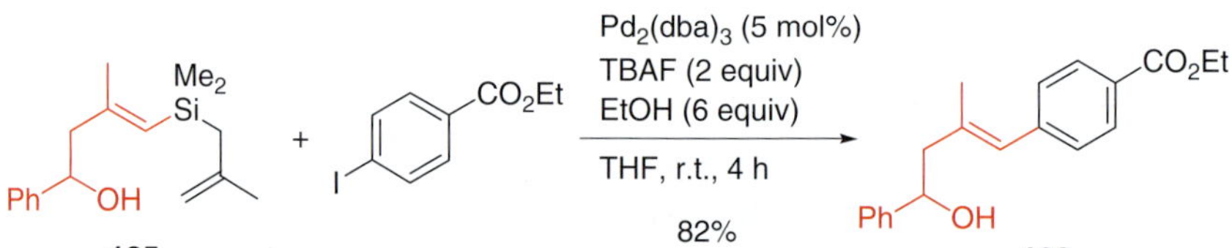

(9.60)

Organobenzylsilanes are also applicable to the palladium-catalyzed cross-coupling with haloarenes or haloalkenes [172, 173]. α-Alkylated vinyl(benzyl) silane **187** undergoes cross-coupling with sterically hindered *ortho*-iodotoluene after treatment with TBAF to form 1,1-disubstituted alkene **188** [172] (Eq. (9.61)). (*Z*)-*trans*-β-Benzylsilylstyrene **189** reacts with iodobenzene to give (*Z*)-diphenylethene (**190**) with a perfect stereoselectivity [173] (Eq. (9.62)). Herein, the benzyl group is assumed to be replaced by fluorine before the cross-coupling process. As shown in Figure 9.4, benzyldimethylsilyl groups are utilized as a wide variety of silicon nucleophiles. Examples are multi-substituted alkenylsilanes **191** [174–186], γ-CF_3-subsituted 1-propenylsilane **192** [187], silylmethylenecycles **193** [188], 1,3-dienylsilanes **194** [138, 189, 190], 1,3,5-trienylsilanes **195** [139], complex vinylsilanes such as **196** and **197** [191–194], and arylsilanes **198** [195].

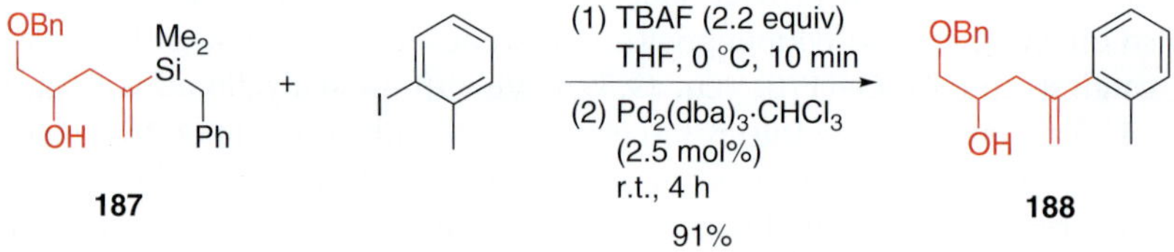

(9.61)

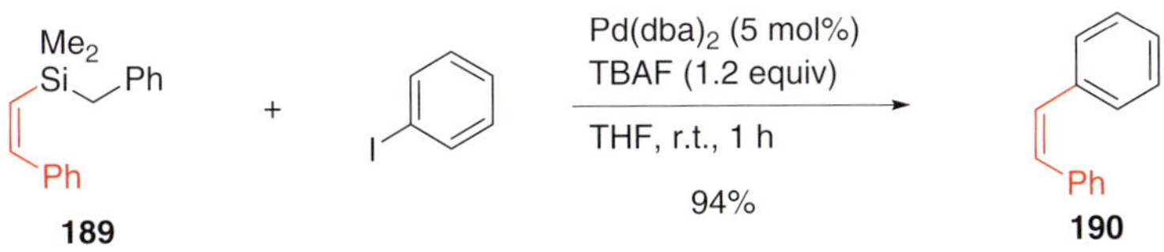

(9.62)

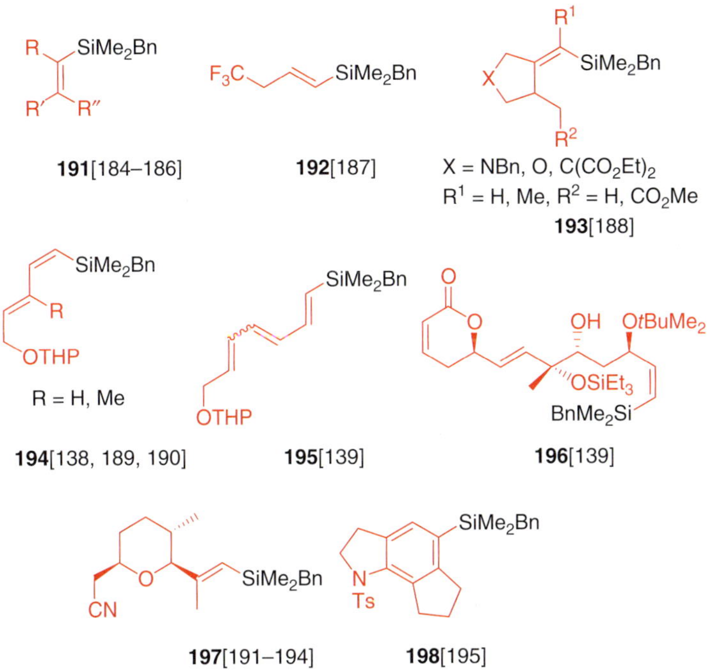

Figure 9.4 Various benzyldimethylsilyl-based nucleophiles.

Alkenyl-alkyl coupling is made possible using benzyldimethyl(vinyl)silanes, iodoalkanes, and a stoichiometric copper additive, giving internal alkenes. The example shown in Eq. (9.63) is achieved using (*E*)-1-butenylsilane **199** and 3-iodopropanol to give (*E*)-alkene **200** [196].

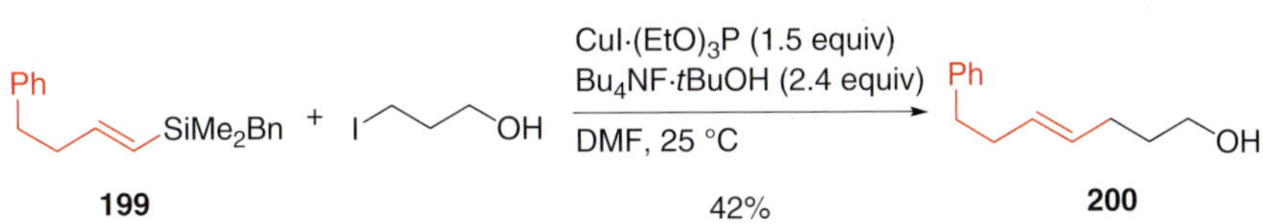

(9.63)

Alkenylsilanes **201** and **202** are prepared stereoselectively by hydrosilylation of terminal alkynes and hydrometalation of silylalkynes followed by protonation. The silane reagents **201** and **202** thus formed can be subjected to the metal-catalyzed coupling with organic electrophiles. Such tandem reactions have been applied to the synthesis of a retinoid vitamin A (**203**) and 11-*cis*-retinal (**204**) [189, 197] (Scheme 9.1).

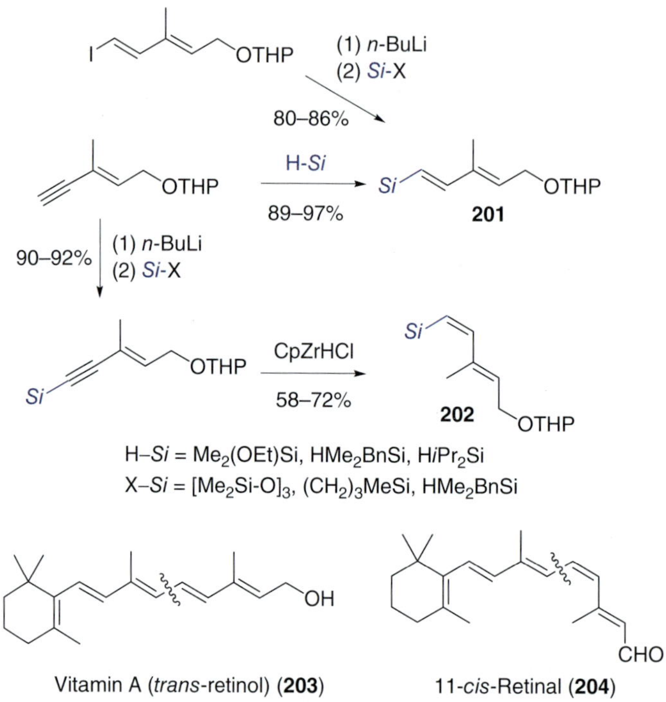

Scheme 9.1 Stereoselective synthesis of 1-silyl-butadienes and their synthetic targets.

9.4.2 Arylsilanes

Arylsilanes smoothly cross-couple with haloarenes in the presence of a palladium catalyst and an appropriate base. The phenyl group in di(2-furyl)(methyl)(phenyl)silane (**205**) selectively undergoes the cross-coupling with bromoarenes upon treatment with a palladium catalyst and TBAF, giving corresponding biaryls **206** [198] (Eq. (9.64)). Hereby, the furyl group is probably replaced by a fluorine before cross-coupling. Vinyl(thienyl)dimethylsilanes also cross-couple with iodoarenes under palladium catalysis and TBAF to form styrene derivatives [189, 199–203]. For example, (*Z*)-1-octenyldimethyl(2-thienyl)silane {(*Z*)-**207**} reacted with 1-iodo-4-trifluorobenzene to give (*Z*)-arylalkene with retention of configuration [199] (Eq. (9.65)). A 2-pyridyl group in lieu of the furyl and thienyl groups is also applicable to the cross-coupling [204–210]. Both vinyl- and arylpyridylsilanes **209** and **211** are transformed to stilbene (**210**) and 2-hydroxybiphenlyl (**212**) using a pertinent palladium catalyst and a base (Eqs. (9.66) and (9.67)).

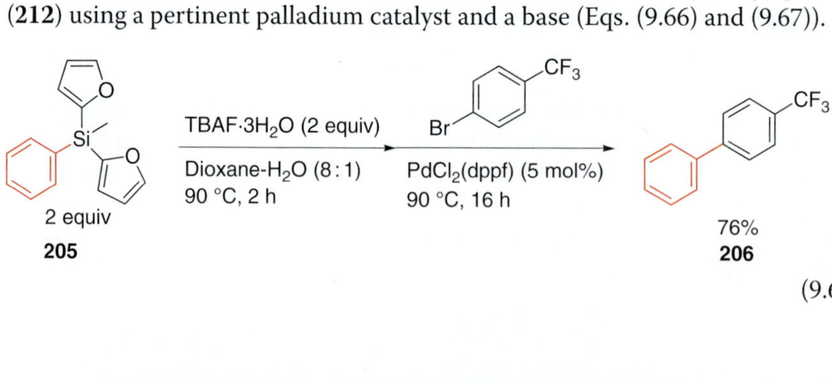

(9.64)

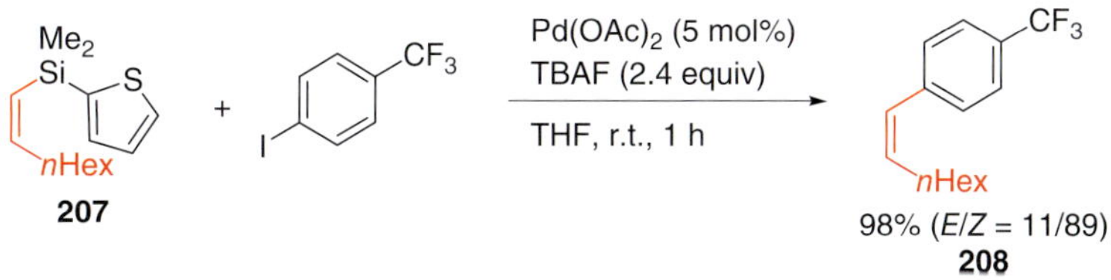

(9.65)

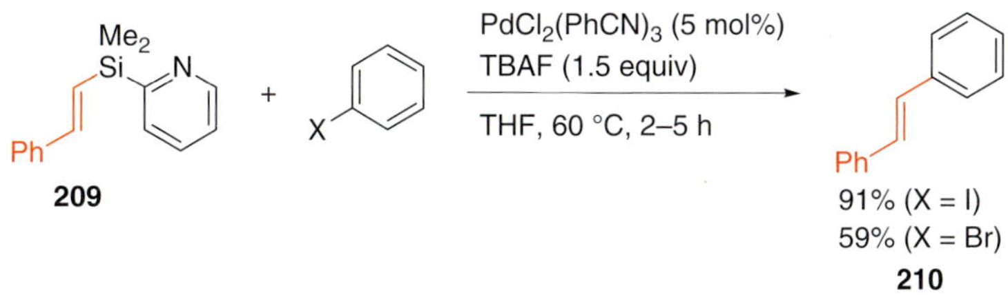

(9.66)

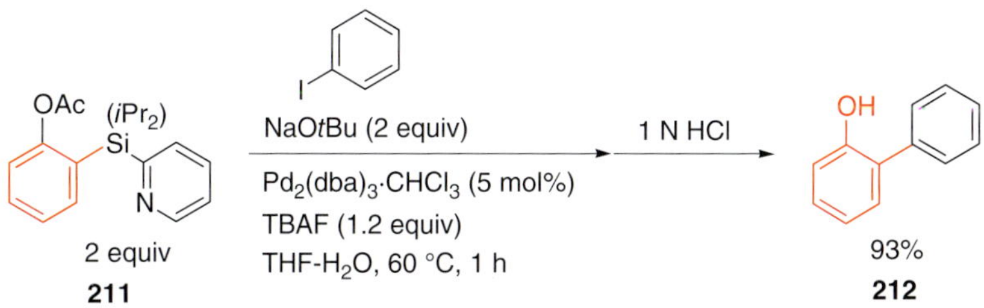

(9.67)

Combination of the Migita–Kosugi–Stille coupling using organotin and the pyridylsilane coupling allows a facile preparation of unsymmetric diarylmethanes. Reagent **213** bearing Si and Sn on a methylene carbon couples with substituted 4-iodoacetophenone **214** when catalyzed by palladium without a base to give selectively 4-acetylbenzylsilane **215**, which further couples with iodobenzene in the presence of Ag_2O and gives diarylmethane **216** [211] (Eq. (9.68)).

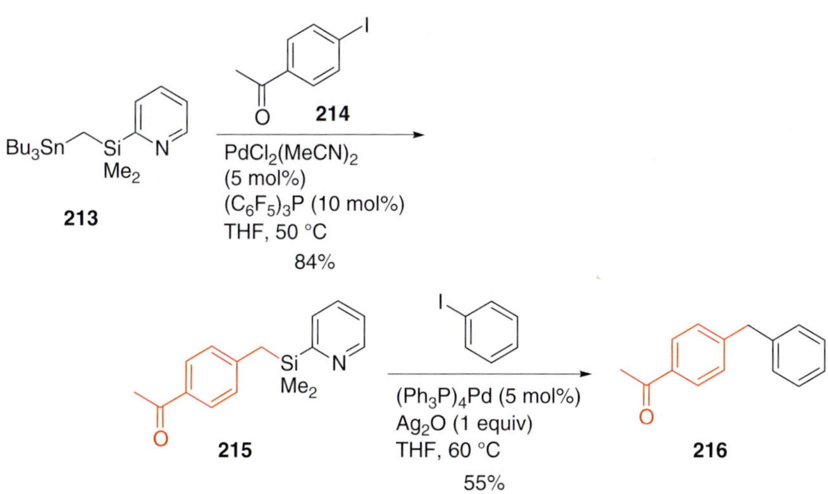

(9.68)

In addition to heteroaryl groups, electron-deficient aryl groups on silicon are potent to effect the cross-coupling. Pentafluoroaryl(vinyl)silanes **217** cross-couple with iodoarenes in the presence of palladium(0) and TBAF to give styrene derivatives **218** [212] (Eq. (9.69)). The 3,5-bis(trifluoromethyl)phenyl group also promotes the cross-coupling [213–218]. (*Z*)-Styryl{3,5-$(CF_3)_2C_6H_3$}silane **219** is transformed to (*Z*)-diarylethene **220** without any stereoisomerization [213] (Eq. (9.70)). Moreover, substrates with two {3,5-$(CF_3)_2C_6H_3$}silyl groups are applicable to the synthesis of polyphenylene(ethylene)s by sequential cross-coupling of diiodoarenes with bis(silylethenyl)arenes. As shown in Eq. (9.71), copolymerization of 2,5-dioctyloxy-1,4-diiodobenzene with bis(dimethyl-3,5-bis(trifluoromethyl) phenylsilyl-ethenyl)benzene (**221**) took place to give poly(phenylene-ethylene) **222** [215], the average molecular weight (M_n) being 5700 and the polydispersity index (M_w/M_n) 1.64. The stereochemistry is well retained under the polymerization conditions. 2-{3,5-$(CF_3)_2C_6H_3$}silylazulene **223** reacts with iodobenzene to form 2-phenylazulene **224** [218] (Eq. (9.72)).

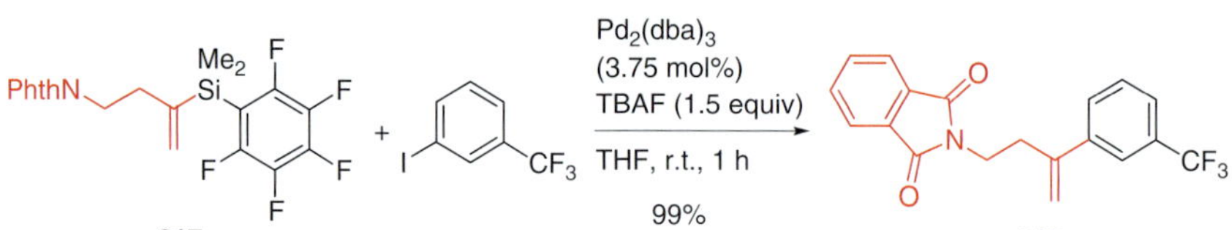

(9.69)

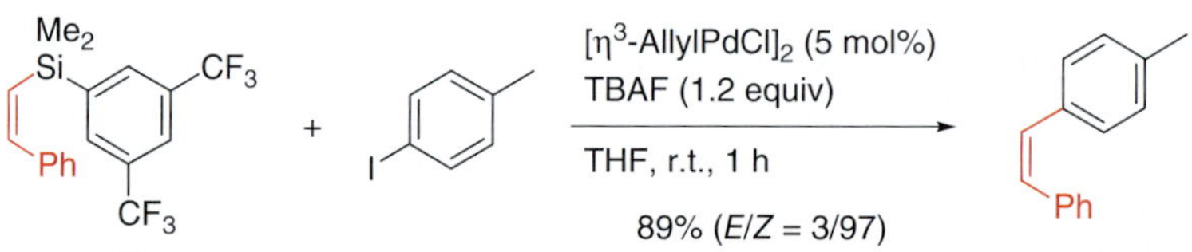

(9.70)

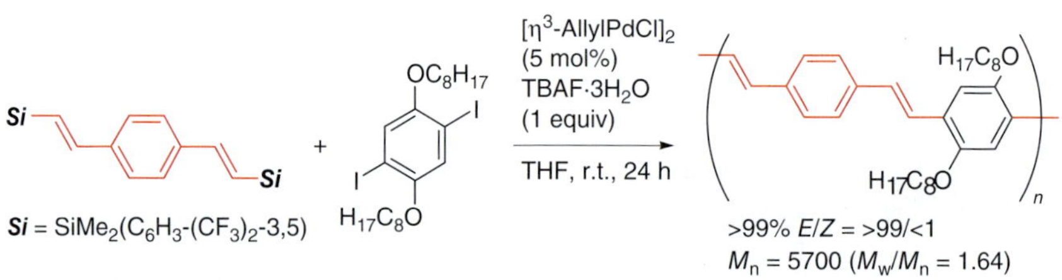

(9.71)

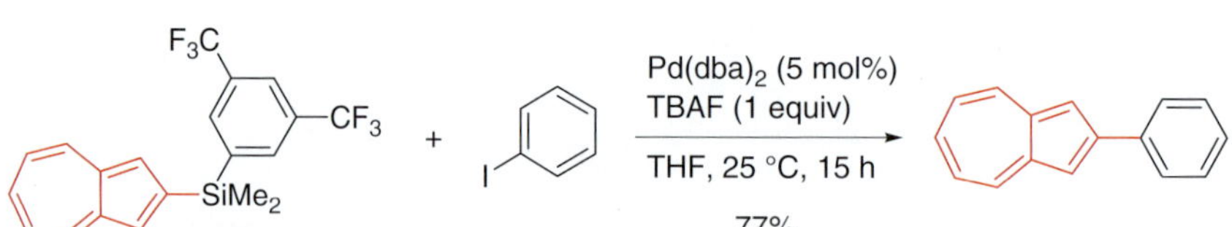

(9.72)

Vinylsilanes **225** having two phenyl groups on the silicon atom can cross-couple with iodoarenes with the aid of palladium/CuI catalyst and CsF to give α-fluorostyrene **226** [219, 220] (Eq. (9.73)). The phenyl group on the silicon atom is assumed to be replaced by fluorine to enhance the Lewis acidity of the silicon center to ease transmetalation. Although the phenyl group also may cross-couple to some extent (3%), the α-fluorovinyl group selectively participates in the cross-coupling in preference to the phenyl group. Moreover, alkenylsilane **227** also engages in selective cross-coupling upon treatment of 18-crown-6 and $KOSiMe_3$ [221, 222] (Eq. (9.74)). (*E*)-Vinylsilane **229** undergoes cross-coupling with iodobenzene with retention of the alkene stereochemistry in the presence of a palladium catalyst and TBAF [223] (Eq. (9.75)).

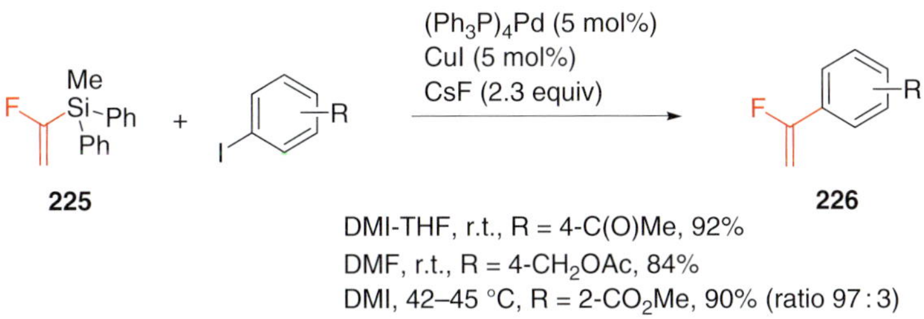

(9.73)

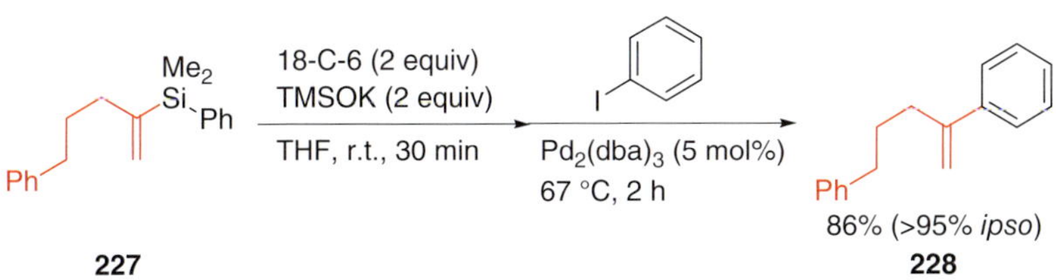

(9.74)

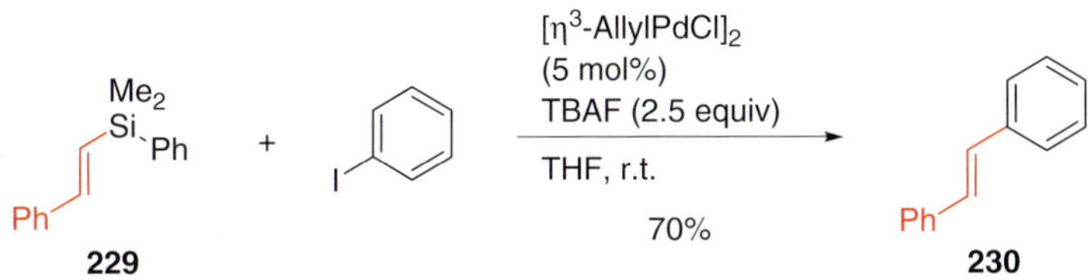

(9.75)

Silanecarboxylic acid is shown to couple with aryl halides in the presence of a palladium catalyst [224]. For example, $PhMe_2SiCO_2H$ (**231**) reacts with 4-iodoanisole smoothly at room temperature when Xantphos is employed as the ligand in the palladium catalyst to afford aromatic carboxylic acid **232**. Bromoarenes can also participate in the reaction with **231** to give product **232** but under harsh reaction conditions: higher temperatures (80 °C) and longer reaction time (19 hours) (Eq. (9.76)).

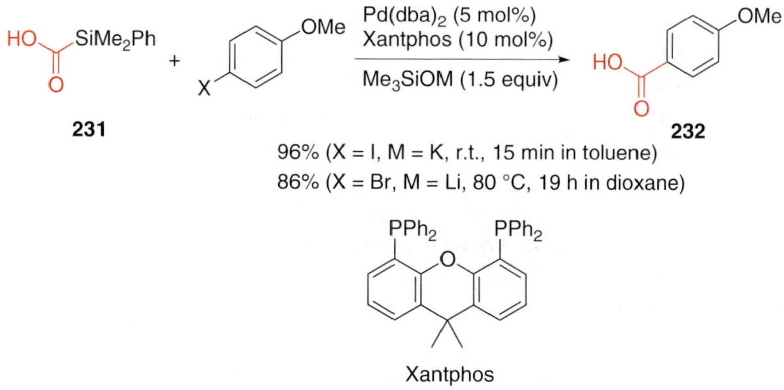

(9.76)

9.4.3 Trialkylsilanes

Among the organosilyl groups, simple trialkylsilyl groups such as trimethylsilyl, triethylsilyl, and triisopropylsilyl are ideal in view of stability, high solubility in various organic solvents, nontoxicity, easy handling, and high accessibility. However, their robustness makes them more difficult to utilize as simple cross-coupling nucleophiles. Thus, the organic group in organotrialkylsilanes is limited to heteroaryl groups, which enhance the Lewis acidity of the silicon center. 2-Pyridyl(trimethyl) silane **233** cross-couples with iodoarenes in the presence of a palladium catalyst, CuI cocatalyst, and TBAF to form 2-arylpyridines [225, 226] (Eq. (9.77)). Trimethyl(2-benzofuryl)silane (**235**) reacts with iodoarenes by a palladium catalyst, $AgNO_3$, and KF to provide 2-arylated benzofuranes [227] (Eq. (9.78)).

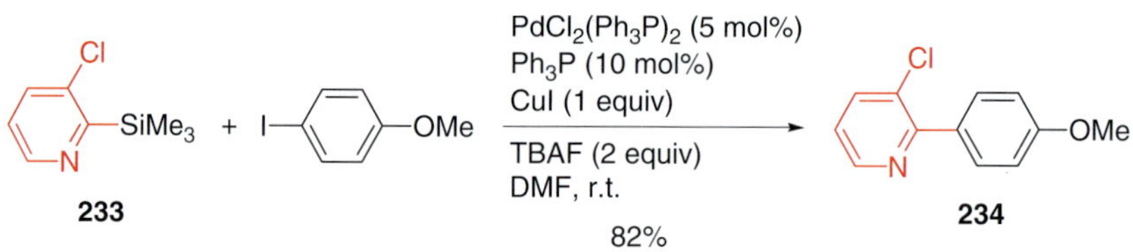

(9.77)

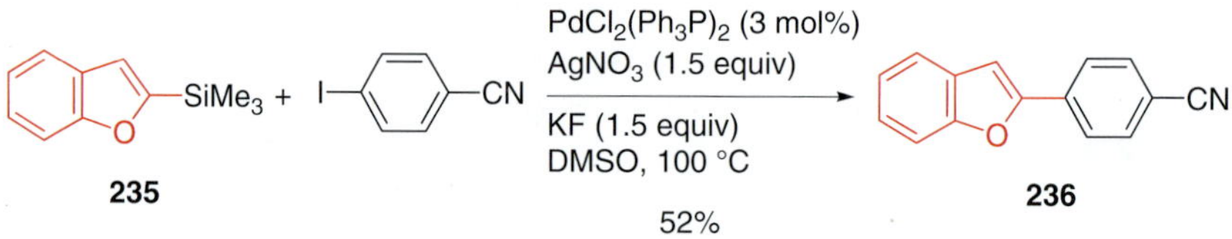

(9.78)

An intramolecular activating system for stable trialkylsilyl groups is an effective device for the cross-coupling reaction with haloarenes. γ-*cis*-Trimethylsilyl-substituted allylic alcohol **237** participates in the palladium-catalyzed cross-coupling with iodoarenes in the presence of CuCl and *t*BuOLi, thereby giving γ-*cis*-arylated allylic alcohols **238** [228] (Eq. (9.79)). The reaction is suggested to be triggered by the formation of allyloxycopper **239** followed by the generation of pentacoordinated silicate **240** by the intramolecular coordination of oxygen to silicon. Brook-type rearrangement then occurs to give vinylcopper **241**, which reacts with a palladium complex after oxidative addition of iodoarenes. Similarly, a combination of a palladium catalyst and a copper mediator utilizes 2-(3,4-epoxy-1-TBSO-butyl)-1-TMS (trimethylsilyl) benzene (**243**) as an aryl nucleophile precursor via epoxide opening and the Brook rearrangement [229]. Treatment of **243** with a dialkylcuprate generates monoprotected diol anion **244**, which undergoes double silicon migration and then reacts with a stoichiometric amount of CuI to give an aryl nucleophile **245**. The subsequent palladium-catalyzed reaction with vinyl bromides (or iodoarenes) produces cross-coupled products (Eq. (9.80)). This type of cross-coupling is reported to take place in the absence of a copper mediator. (*Z*)-β-(Trimethylsilyl)acrylic acid **247** is capable of palladium-catalyzed coupling with iodoarenes [230] (Eq. (9.81)). In this case, Cs_2CO_3 is basic enough to activate the carbon–silicon bond. 8-TBDMS (*t*-butyldimethylsilyl)-1-naphthol **249** cross-couples with iodobenzene in the presence of a palladium/Ph_3As catalyst and Cs_2CO_3, giving 8-aryl-1-OTBDMS-naphthalene **250** via a silyl-migration [231] (Eq. (9.82)). Attack of 1-hydroxy group to the silicon center mediated by Cs_2CO_3 may result in the formation of pentacoordinate silicate **251** followed by the activation of C8—Si bond.

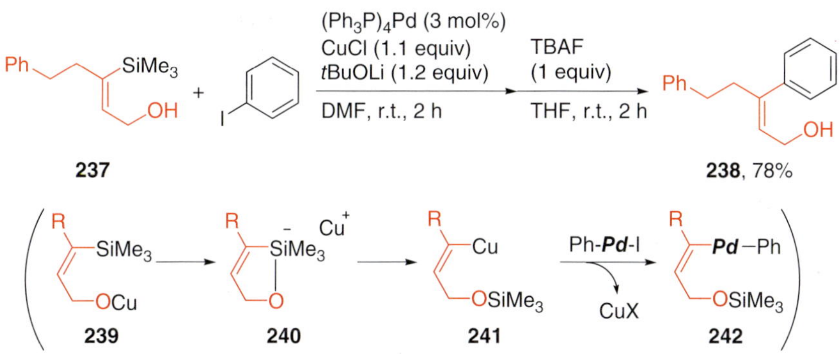

(9.79)

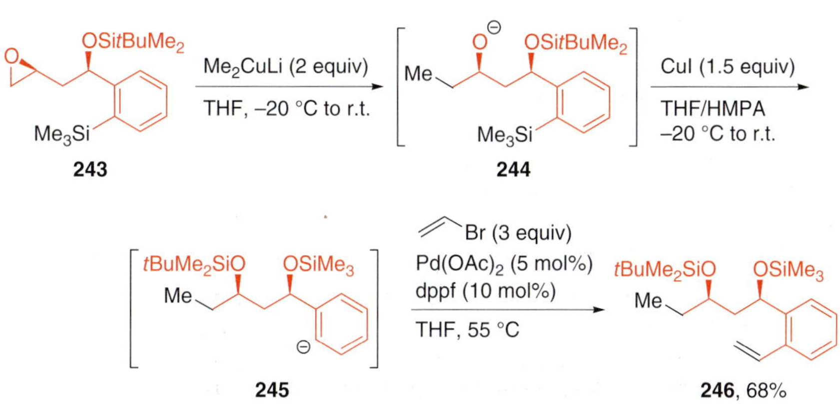

(9.80)

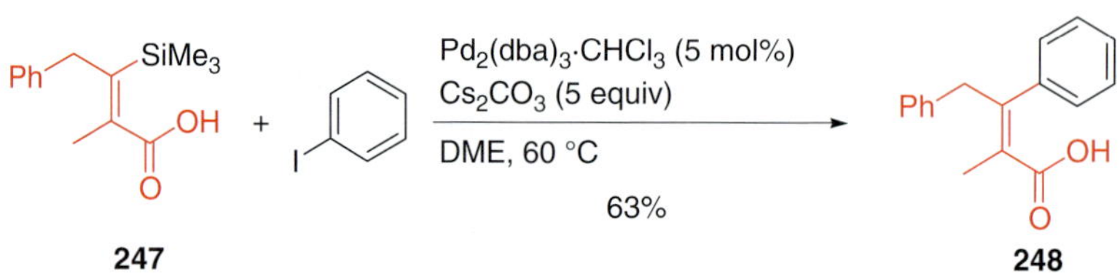

(9.81)

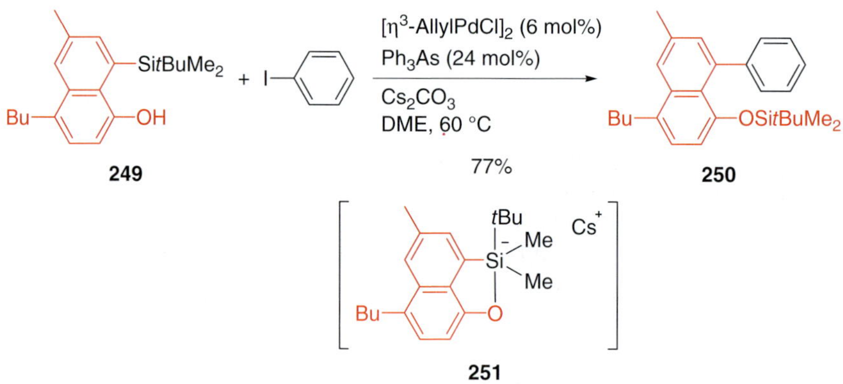

(9.82)

The examples shown in Eqs. (9.79) and (9.80) demonstrate that oxyanions attack a nearby silyl group intramolecularly to generate organocopper species, which then react with iodoarenes. These copper species have a nucleophilic character to react with alkyl electrophiles such as allyl chlorides, benzyl bromides, and alkyl iodides. For example, the trimethylsilyl group in *o*-trimethylsilylbenzyl alcohol **252** migrates upon exposure to CuO*t*Bu generated from CuI and LiO*t*Bu, and the resulting arylcopper intermediate reacts with 1-iodobutane to produce *o*-butylbenzyl alcohol **253** [232] (Eq. (9.83)). 2-*tert*-Butyldimethylsilylphenylketone **254** also is capable of this sort of copper-mediated alkylation sequence to give 2-alkylated product **255** [233] (Eq. (9.84)). The reaction is triggered by the formation of a copper enolate, which rearranges to an arylcopper through silyl migration.

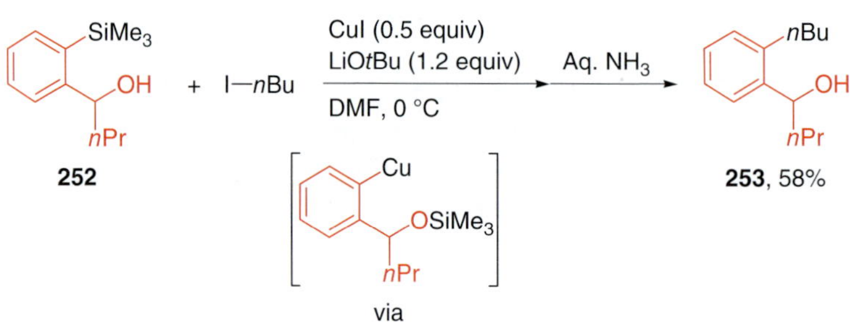

(9.83)

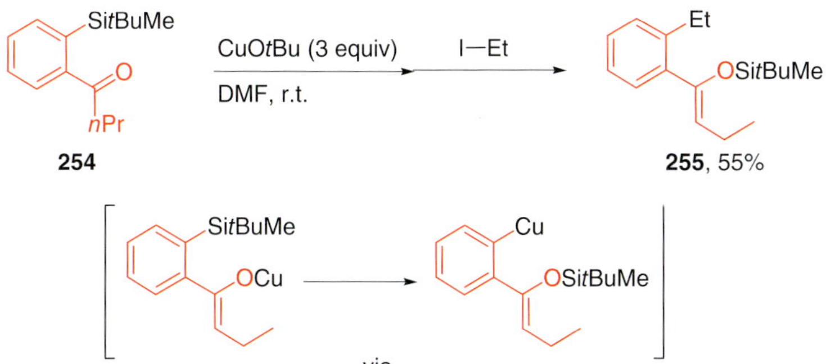

(9.84)

Copper(II) catalysis is effective for the cross-coupling with aryl- or heteroaryl-trialkylsilanes with various iodoarenes. For example, the reaction of 4,7-bis(triethyl)silyl-5,6-difluorobenzothiadiazole (**256**) with 1-bromo-4-iodobenzene in the presence of $CuBr_2$, Ph-DavePhos, and CsF (2.5 equiv) at 150 °C gives 4,7-diarylated double-coupling product **257** [234] (Eq. (9.85)). Disilylarenes such as 2,5-disilylthiophenes, 2,5-disilylfurans, and 1,4-disilyltetrafluorobenzene are applicable to this sort of double coupling. The fact that an aryl C—Br bond remains intact shows that the coupled products derived from 1-bromo-4-iodobenzene may be employed for further cross-coupling reactions. The key to the cross-coupling is a requisite combination of the arylsilanes, a Cu(II) catalyst, and a fluoride ion. Although the original Cu(II) catalytic conditions were not applicable to the cross-coupling with bromoarenes, a catalyst consisting of palladium(0), tris(2,4,6-trimethoxyphenyl)phosphine (TTMPP), and CuF_2 allows to use bromoarenes for similar cross-coupling with aryltrialkylsilanes. For example, benzothiophenes **258** having $SiEt_3$, $SiMe_3$, Si*t*$BuMe_2$, and even Si*i*Pr_3 undergo cross-coupling with *p*-bromoanisole to provide 2-anisylbenzothiophene (**259**) [235] (Eq. (9.86)).

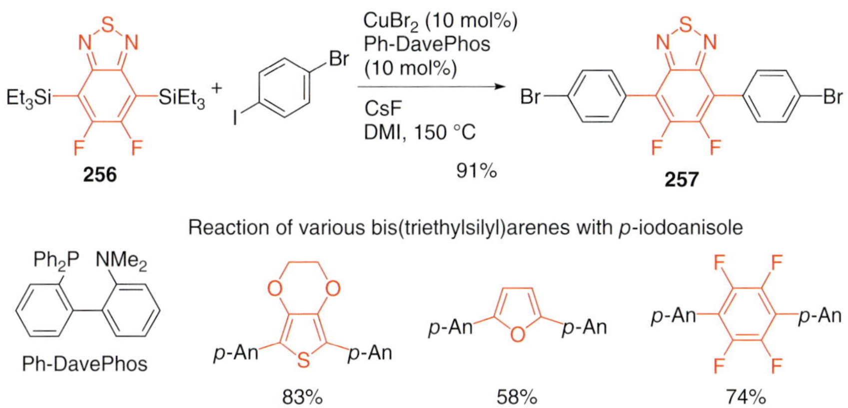

(9.85)

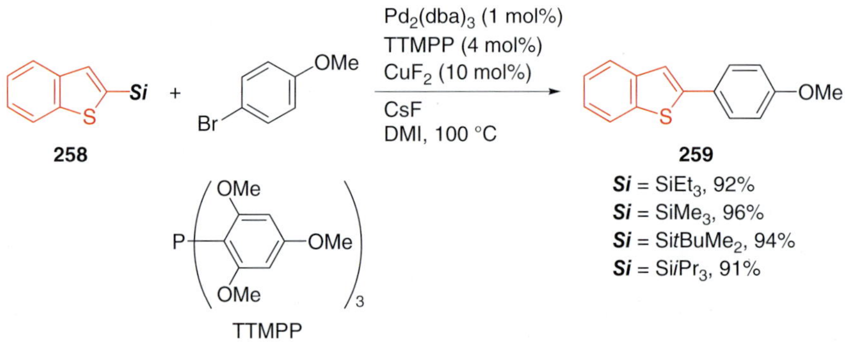

(9.86)

Alkynyltrimethylsilanes are directly applicable to the cross-coupling. A mixed catalyst system of a palladium and CuCl catalyzes the coupling reaction of trimethylsilylacetylene **260** with 4-acetylphenyl triflates in the absence of fluoride ions to give 1,2-diarylethyne **261** [236, 237] (Eq. (9.87)). Iodo- and chloroarenes are also used in the alkynylation. Relayed transmetalation from silicon to copper and then to palladium is considered to smoothly take place and achieve the cross-coupling. The alkynylation reaction proceeds without any palladium catalysts: CuCl/Ph_3P catalyst and potassium benzoate promote the cross-coupling of alkynylsilane **260** with 4-iodoanisole [238] (Eq. (9.88)). Co-use of $CuBr_2$/Ph-DavePhos catalyst and CsF also mediates similar transformations [234] (Eq. (9.89)).

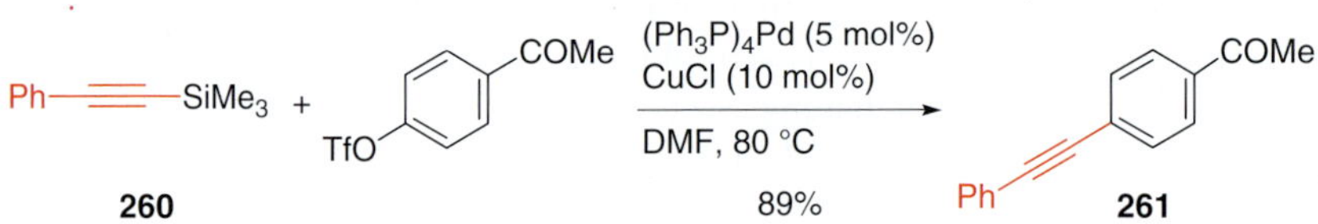

(9.87)

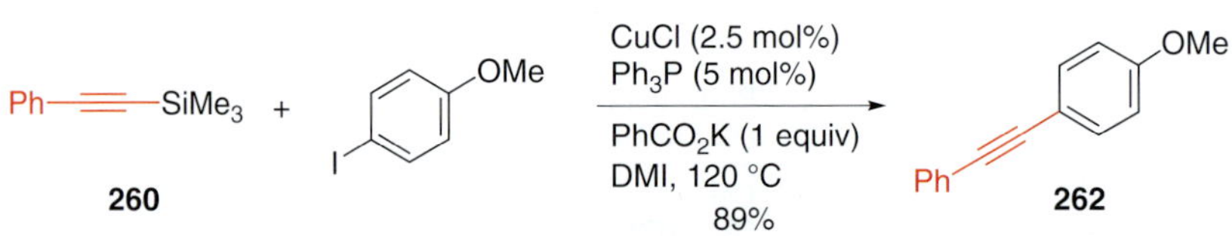

(9.88)

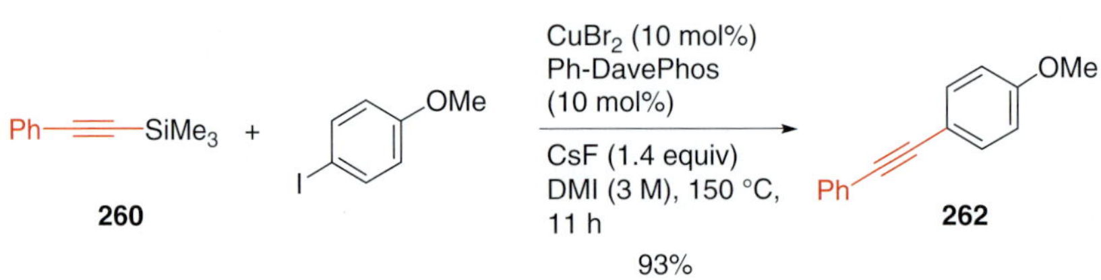

(9.89)

Triethyl(trifluoromethyl)silane (**263**) reacts with ethyl 4-iodobenzoate with a CuI/phen (1,10-phenanthroline) catalyst to give trifluoromethylarenes **264** [239] (Eq. (9.90)). Various trifluoromethyl copper reagents derived from trifluoromethylsilanes are generated and utilized for direct trifluoromethylation of iodoarenes, silylarenes, or aryldiazonium salts [240–243]. Similarly, difluoromethyltrimethylsilanes react with iodoarenes in a stoichiometric amount of CuI and CsF, forming difluoromethylarenes [244]. With a palladium catalyst, BrettPhos, and KF, trifluoromethylsilane **263** cross-couples with chloroarenes [245] (Eq. (9.91)). BrettPhos ligand is suggested to promote reductive elimination from an intermediate complex Ar–Pd–CF_3. *N,N*-Diethyl-α,α-difluoro-α-(trimethylsilyl)acetamide (**265**) can be used in the reaction with bromoarenes in the presence of $t$$Bu_2$CyP-ligated palladacycle catalyst **266**, giving α-aryl-α,α-difluoroacetamides **267** [246] (Eq. (9.92)). CuOAc catalyst is applicable to the cross-coupling of α,α-difluoro-α-(trimethylsilyl)acetamides with iodoarenes [247] (Eq. (9.93)).

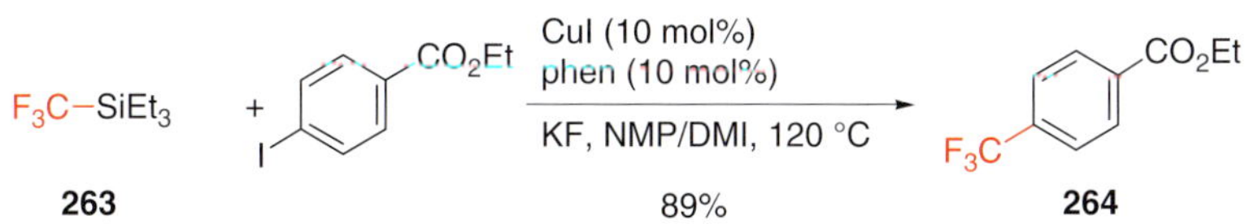

(9.90)

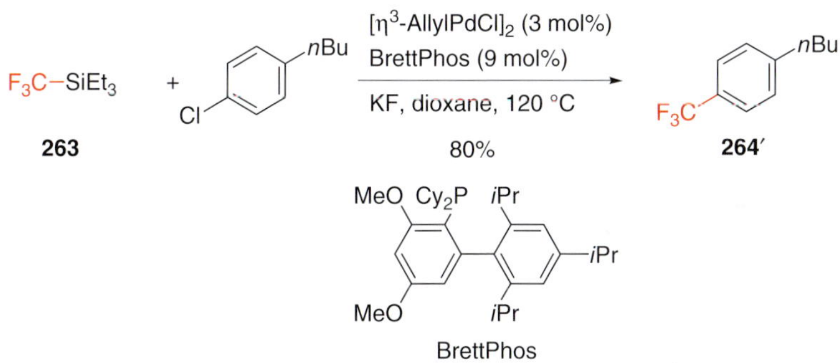

(9.91)

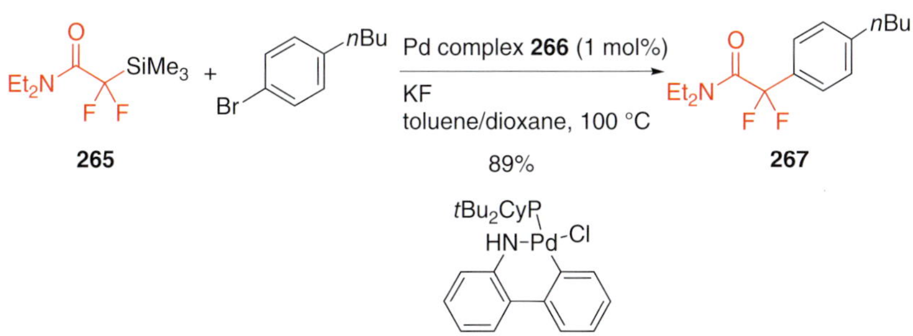

(9.92)

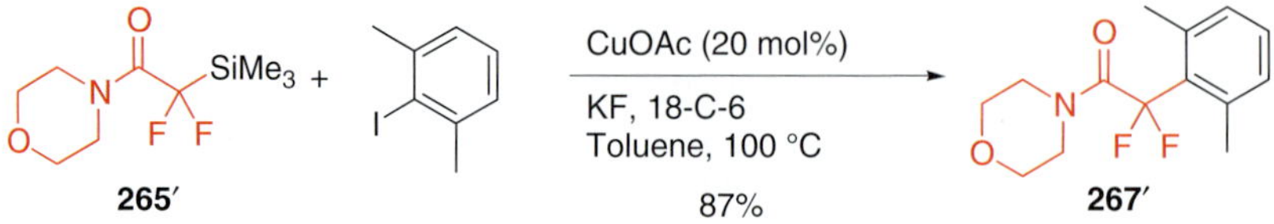

(9.93)

Polycondensed aromatic compounds react with phenyltrimethylsilane (**268**) by a palladium catalyst and a suitable oxidant such as copper(II) through C—H bond cleavage. For example, phenanthrene (**269**) cross-couples with **268** in the presence of $PdCl_2$ and $CuCl_2$ as an effective oxidant to give the phenylated product **270** [248] (Eq. (9.94)). Fluoranthene (**271**) and pyrene (**273**) are applicable to the direct arylation reaction to give **272** and **274**. This sort of reaction is considered to involve an electrophilic substitution on an electron-rich aromatic ring. Of note, in the absence of $CuCl_2$ catalyst, homocoupling of phenyltrimethylsilane is observed to form biphenyl even in the presence of naphthalene. The reaction of phenyltrimethylsilane (**268**) with 2-methylthiophene proceeds at the β-position in up to c. 90% selectivity in the presence of $PdCl_2(MeCN)_2$ (5 mol%) and $CuCl_2$ (2 equiv) to give 4-phenyl-2-methylthiophene (**275**) [249] (Eq. (9.95)). Fluoranthene (**271**) also undergoes C–H arylation leading to **272**, in which *o*-chloranil and AgOTf were employed as the oxidant [250] (Eq. (9.96)). The mechanism for the electrophilic aryl—Si bond activation by a palladium(II) is discussed using a DFT study [251]. The mechanism, regioselectivity, and role of palladium/*o*-chloranil catalytic reaction is also discussed in detail [252]. In addition, aryltrimethylsilanes undergo homocoupling at both *ipso*-positions by $Pd(MeCN)_4(BF_4)_2$ catalyst and *o*-chloranil [253] (Eq. (9.97)).

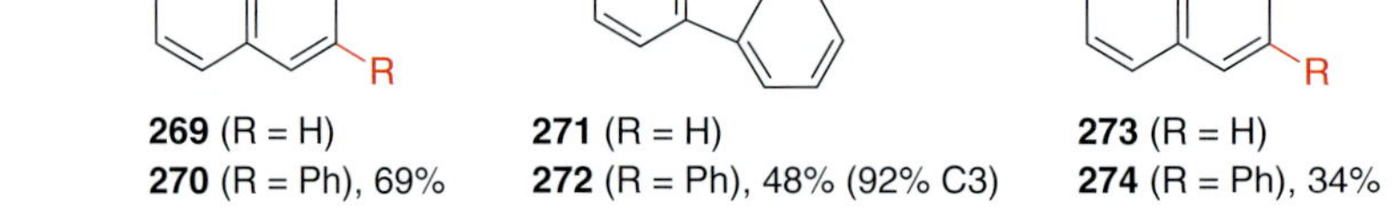

(9.94)

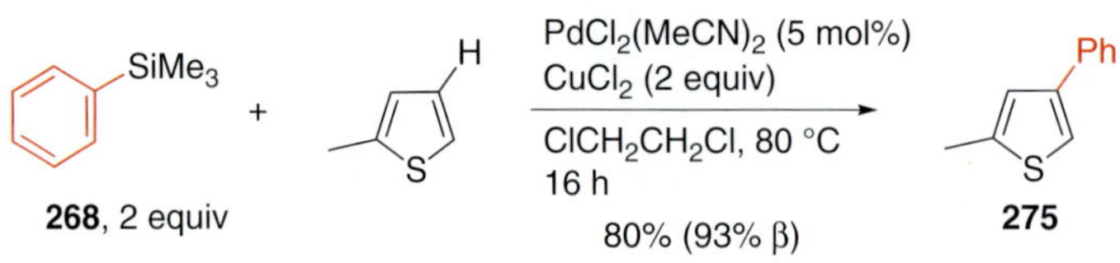

(9.95)

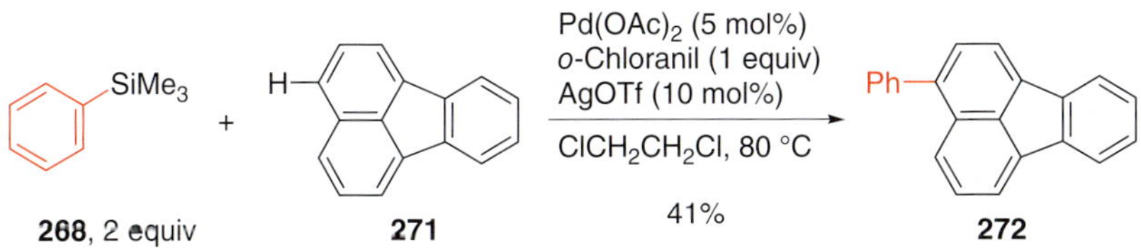

(9.96)

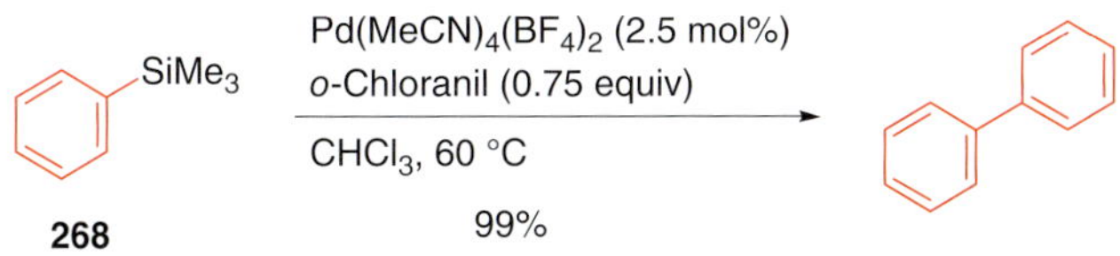

(9.97)

A combination of a gold(I) complex with a hypervalent iodine oxidant is effective for oxidative arylation of aryltrimethylsilanes. The $Ph_3PAuOTf$ catalyst along with $PhI(OAc)_2$ and 10-camphorsulfonic acid (CSA) in stoichiometric amounts can arylate electron-rich arenes using aryltrimethylsilanes to form biaryls [254] (Eq. (9.98)). Of note, a carbon–iodine bond, which is susceptible to oxidative addition to low-valent group 10 transition metals, is left unchanged. This reaction system is applicable to an intramolecular version [255]. For example, (2-benzylphenyl)trimethylsilane **278**, upon exposure to the above-mentioned conditions, gives fluorenes **279** (Eq. (9.99)). Moreover, aryltrimethylsilanes are methoxyethylated by ethylene and methanol in the presence of a Ph_3PAuCl catalyst and an IBA-OTf oxidant to give (2-methoxyethyl)benzene (**280**) [256] (Eq. (9.100)). Pyridine-based carbene-ligated AuCl complex {AuCl(PyC)} can be used in the oxidative C–H arylation of oxazole, indole, and benzothiophene using aryltrimethylsilanes in combination with IBX and CSA [257] (Eq. (9.101)).

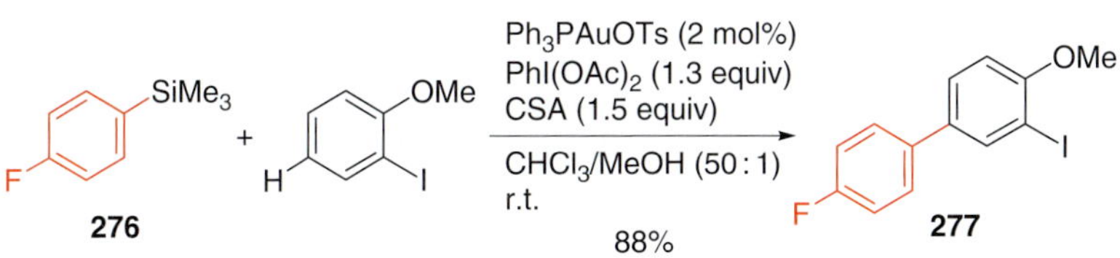

(9.98)

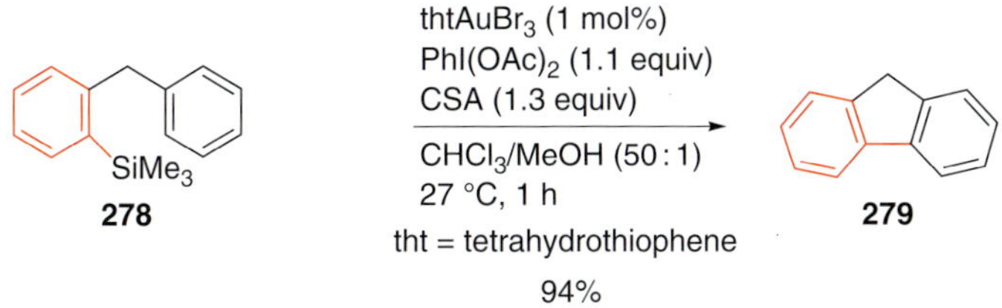

(9.99)

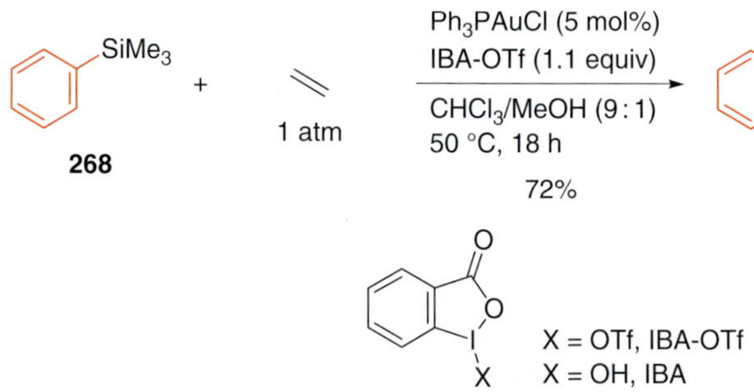

(9.100)

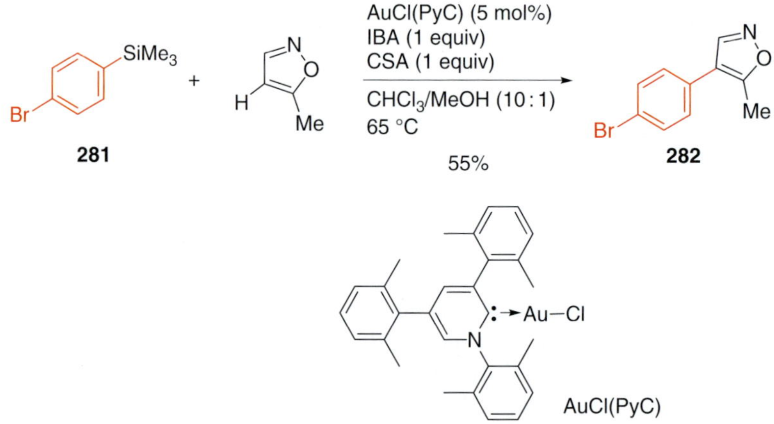

(9.101)

The reaction of 9-silafluorene proceeds with phenanthrene, pyrene, corannulene, and their analogs in the presence of a palladium(II) catalyst and *o*-chloranil [258, 259]. For example, phenanthrene underwent annulation with dimethyldibenzosilole **283** at the K-region selectively in the presence of $Pd(MeCN)_4(SbF_6)_2$ as a catalyst and *o*-chloranil, giving dibenzo[*g*,*p*]chrysenes [258] (Eq. (9.102)). Moreover, five-membered heteroaromatics such as benzothiophenes are applicable to the annulation reaction with 9-silafluorenes at *b* position [260] (Eq. (9.103)). The annulation reaction system using 9-silafluorenes can be employed in the formation of diarylphenanthrenes using diarylalkynes [261].

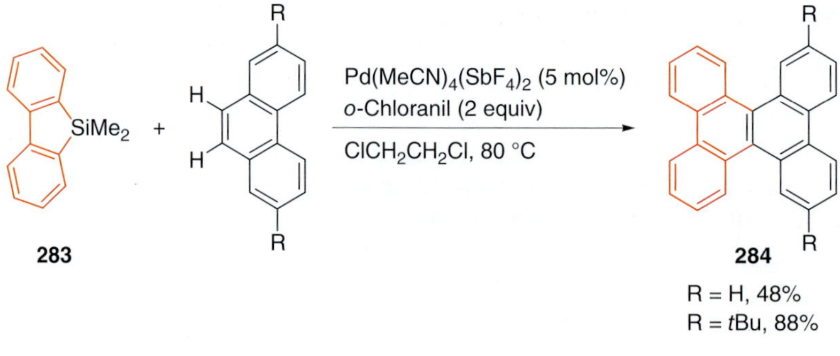

(9.102)

283 285 (9.103)

Desilylative acetoxylation of aryl(trimethyl)silanes takes place in acetic acid with $Pd(OAc)_2$ (5 mol%) in the presence of hypervalent iodine $PhI(OCOCF_3)_2$ to give acetoxyarene **287** [262] (Eq. (9.104)).

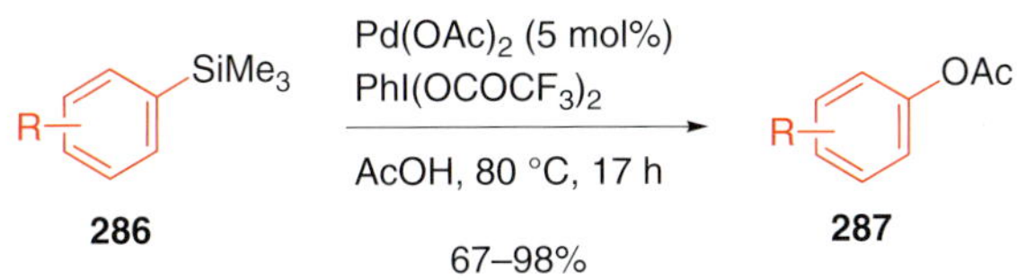

286 287 (9.104)

Trimethylsilyl- and dimethylphenylsilylalkenes **288** are acetylated by acetic anhydride in the presence of $[RhCl(CO)_2]_2$ to give enones **289** with retention of configuration [248] (Eq. (9.105)). The silicon reagents are activated electrophilically by a rhodium catalyst.

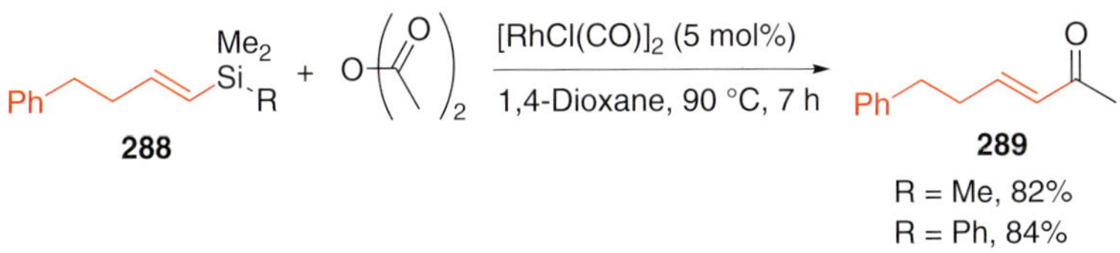

288 289 (9.105)

9.4.4 2-Hydroxymethylphenyl(dialkyl)silanes

Organosilanes substituted by a 2-(hydroxymethyl)phenyl group are highly convenient reagents for cross-coupling. The silyl group is activated mildly by the pendant hydroxy group mediated by a weak base such as potassium carbonate or potassium phosphate. The silyl functionality mediates the cross-coupling with bromoarenes in preference to boryl and stannyl groups. In addition, protection of the hydroxy group by conventional agents perfectly suppress the coupling.

The requisite silyl functionality, 2-(*h*ydroxy*m*ethyl)phenyldimethyl*si*lyl group, is named "HOMSi" for convenience. Typically, aryl-HOMSi reagents **291** are readily prepared by the reaction of cyclic silyl ether **290** with an aryllithium or an arylmagnesium halide [263] (Eq. (9.106)). Alternatively, palladium-catalyzed silylation of haloarenes with protected hydro-HOMSi **292** or disilane-type HOMSi **293** gives protected aryl-HOMSi **294** [264, 265] (Eq. (9.107)). Iridium-catalyzed direct silylation of heteroarenes using protected hydro-HOMSi **292** is also available, giving protected heteroaryl-HOMSi **295** [266] (Eq. (9.108)). The resulting coupled products, protected aryl-HOMSi **294** and heteroaryl-HOMSi **295**, are easily deprotected to give coupling-active (hetero)aryl-HOMSi reagents. In particular, vinyl-HOMSi reagents **297** are prepared by catalytic hydrosilylation of alkynes using protected hydro-HOMSi **292**, followed by deprotection [189, 263, 267] (Eq. (9.109)).

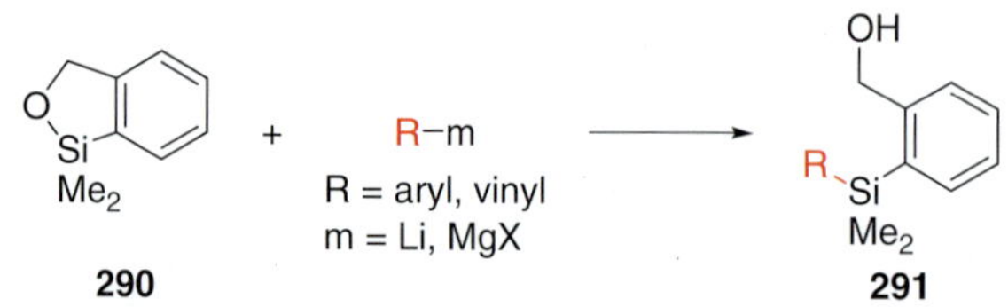

(9.106)

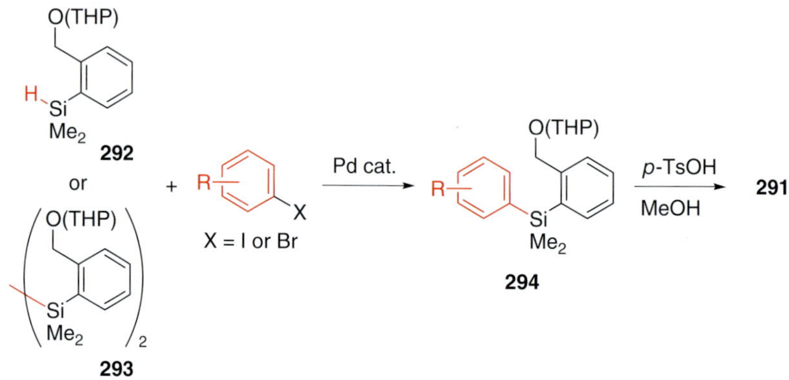

(9.107)

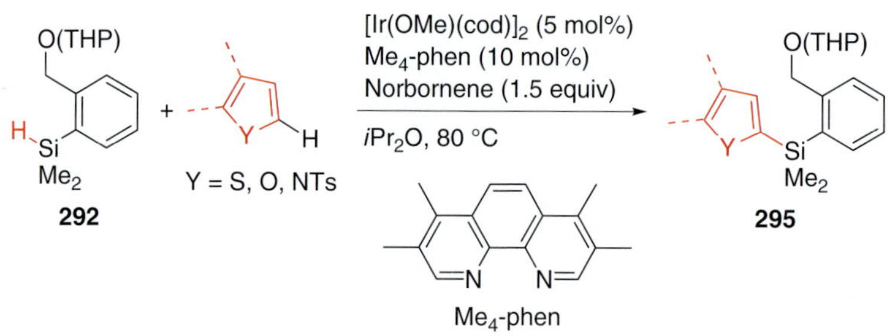

(9.108)

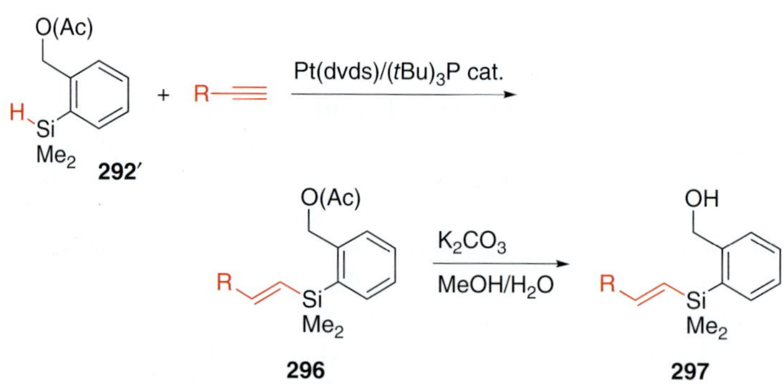

(9.109)

Organo-HOMSi reagents show high reactivity toward electrophiles such as haloarenes, sulfonylarenes, and vinyl halides [189, 248, 263–265, 268–272]. For example, (*Z*)-propenyl-HOMSi cross-couples with iodoarenes in the presence of a palladium/phosphine catalyst and K_2CO_3 to give the coupled product, (*Z*)-propenylarene **299** without stereoisomerization [248] (Eq. (9.110)). Coproduced cyclic silyl ether **290** is readily recovered by distillation and reconverted to the same or different HOMSi reagents. Aryl-HOMSi reagents cross-couple with (hetero)aryl bromides under Pd/Cu catalytic conditions, forming the corresponding biaryls [248, 268] (Eqs. (9.111) and (9.112)). Various functional groups are tolerated in the HOMSi-based cross-coupling. Of note, as shown in Eq. (9.112), once the hydroxy group in organo-HOMSi is simply protected, the reactivity for the cross-coupling is totally lost. This characteristic allows sequential synthesis of oligoarenes [268, 272]. As shown in Figure 9.5, various HOMSi reagents including mono- and disubstituted alkenyl-HOMSi, 1,3-dienyl-HOMSi, aryl-HOMSi, and heteroaryl-HOMSi reagents are used in the cross-coupling with many electrophiles to form biaryls, alkenylarenes, and polyenes [189, 248, 264, 272]. Arylene-bisHOMSi reagent **309**, for example, participates in the cross-coupling polymerization with dibromoarenes under the Pd/Cu catalytic conditions to furnish polyarylenes **310** [273] (Eq. (9.113)).

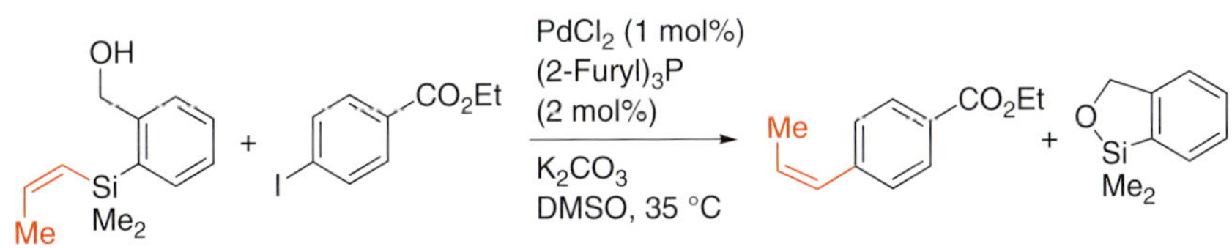

298, (*Z*/*E* = 94/6) **299**, 91% (*Z*/*E* = 94/6) **290**

(9.110)

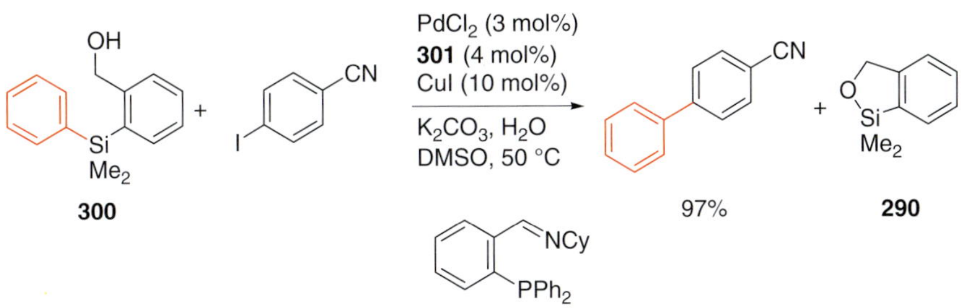

300 97% **290**

301

(9.111)

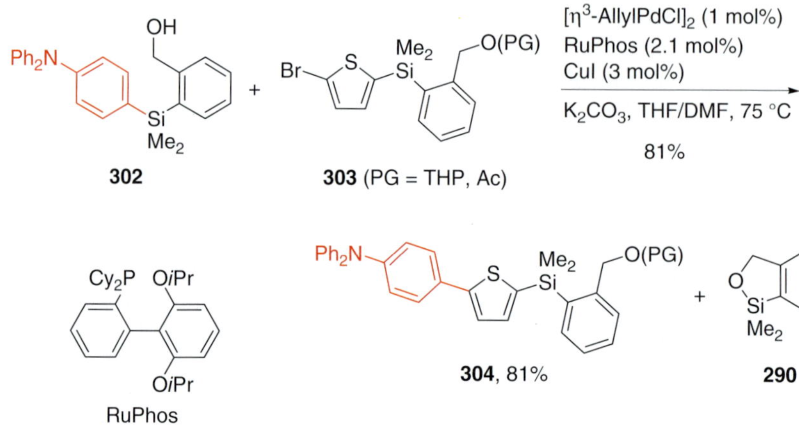

(9.112)

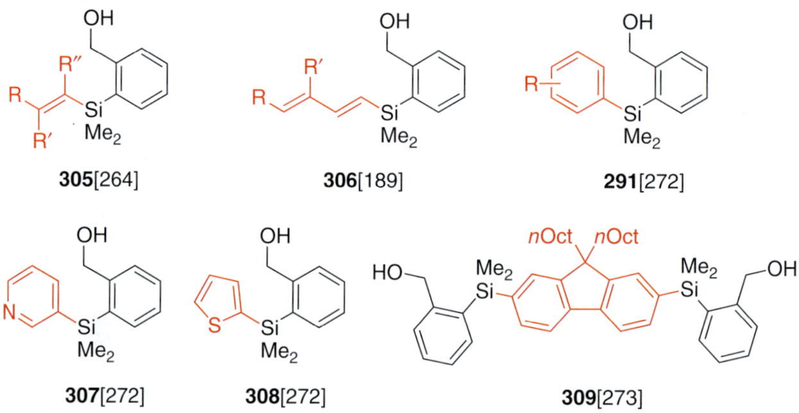

Figure 9.5 Typical organo-HOMSi reagents.

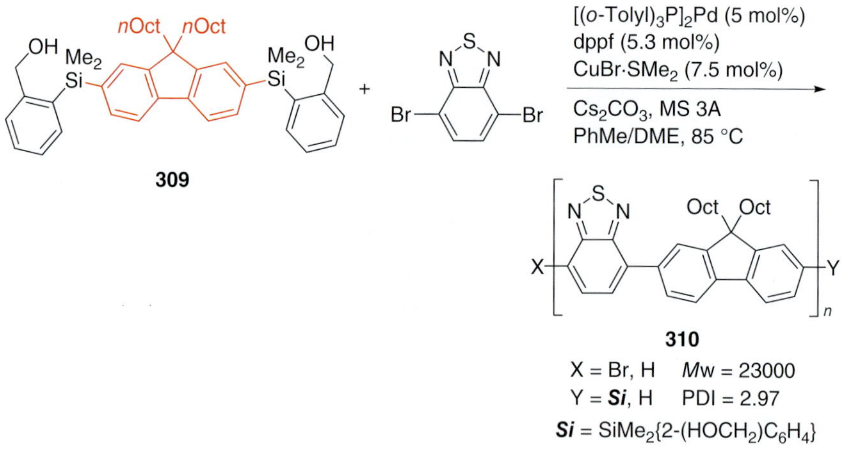

(9.113)

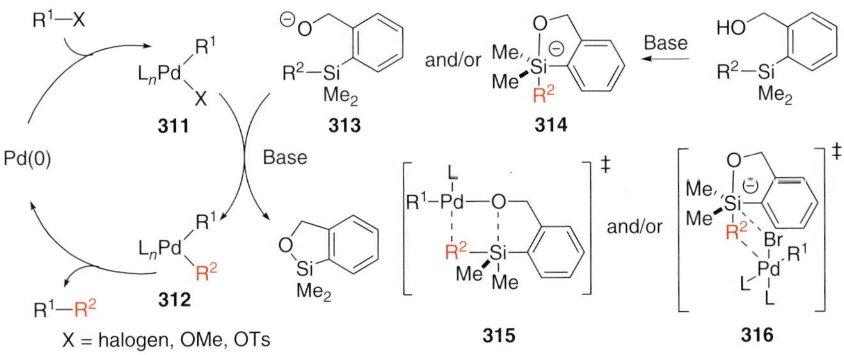

Figure 9.6 Plausible cross-coupling mechanism using organo-HOMSi reagents.

A plausible catalytic mechanism of the HOMSi-based cross-coupling is briefly shown in Figure 9.6. Oxidative addition of R^1–X to a palladium(0) complex generates a R^1–Pd(II)–X complex **311**. Also, organo-HOMSi reagents may be converted to alkoxide **313** and/or five-coordinated silicate **314** by a weak base. For subsequent transmetalation, two pathways are suggested: one involves interaction between palladium and oxygen in four-membered cyclic transition state **315** with R^2 equatorial in a trigonal bipyramidal silicate; another involves a four-membered cyclic transition state **316**, wherein R^2 occupies an axial position and the silicon atom becomes hexacoordinated. The copper cocatalyst most likely promotes the transmetalation step. Finally, reductive elimination from **312** produces the coupling products and regenerates the palladium(0) complex.

As long as allyl- or benzyl carbonates are used for the coupling reaction of organo-HOMSi reagents to give allylated or benzylated products (**318**, **319**), the use of a base is not needed [274] (Eqs. (9.114) and (9.115)). When oxidative addition of such carbonates to palladium(0) complex proceeds, decarboxylation accompanies to give the palladium alkoxide species, which can act as a base to activate the HOMSi reagents. In these cases, a copper cocatalyst is needed to promote the cross-coupling of aryl-HOMSi reagents.

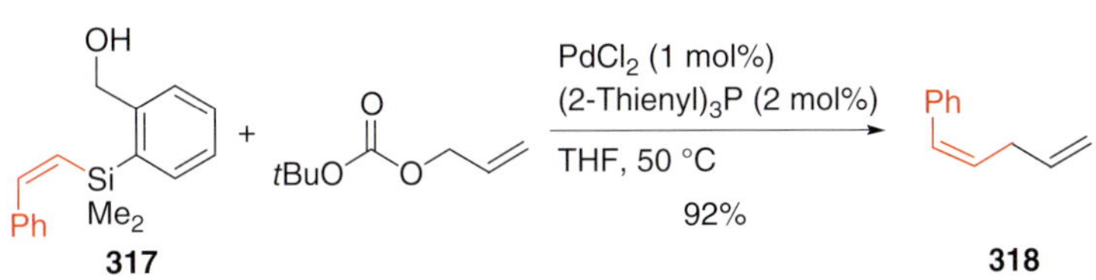

(9.114)

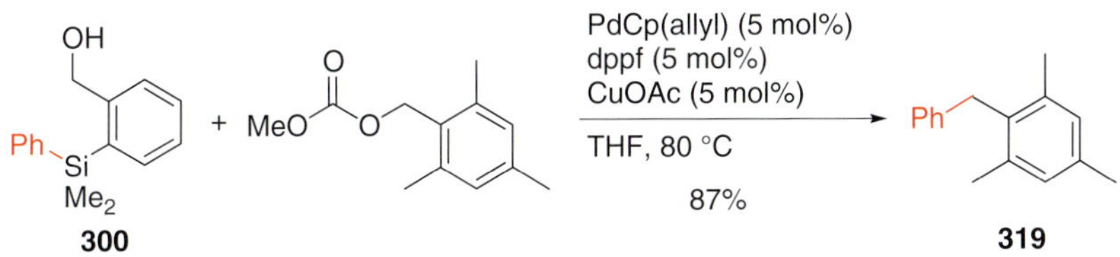

(9.115)

The HOMSi-based cross-coupling system is applicable to the alkyl-aryl cross-coupling reaction. Trimethylsilyl-HOMSi reagent **320** containing two methyl groups at the benzylic carbon can be used as a methylation reagent for chloroarenes. For example, **320** reacts with 4-chlorobenzonitrile in the presence of $Pd(OAc)_2$ and QPhos to give 4-methylbenzonitrile (**321**) along with cyclic silyl ether **322** [275] (Eq. (9.116)). This protocol allows employing various alkyl groups other than methyl group. Butyl-HOMSi reagent **323** can butylate 4-chlorobenzaldehyde in preference to two less reactive isopropyl groups, forming 4-butylbenzaldehyde **324** (Eq. (9.117)). Other *n*-alkylsilanes **325** containing alkenyl, hydroxy, siloxy, acetyl, and methoxycarbonyl are also introduced to aryl electrophiles. Moreover, a *sec*-alkyl group itself is also applicable to this reaction. Tri-*sec*-alkyl-2-(2-hydroxyprop-2-yl)phenylsilanes **326** react with bromoarenes under the same conditions, giving *sec*-alkylated arenes. This is the first example of the cross-coupling of *sec*-alkylsilanes.

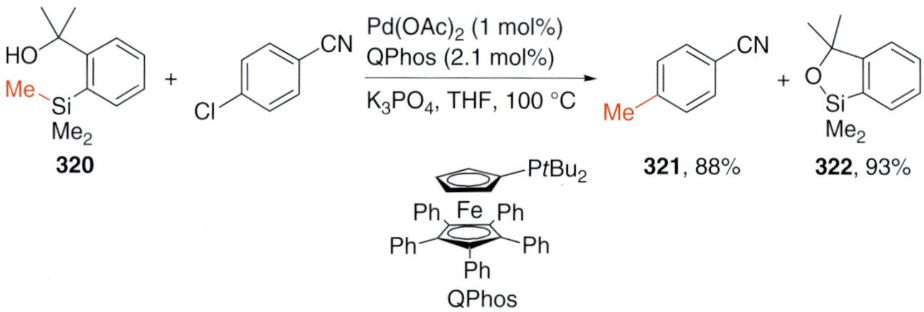

(9.116)

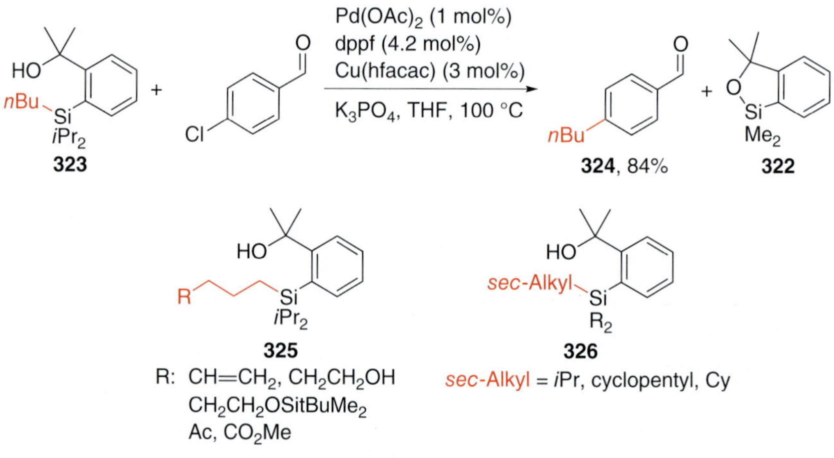

(9.117)

To improve the robustness of organo-HOMSi reagents without losing reactivity, the benzene moiety was converted to a saturated cyclohexane ring. The $C(sp^3)$—Si bond is apparently more stable than the original $C(sp^2)$—Si bond.

Cyclohexane-based aryl-HOMSi reagent **327** can be used in the nickel-catalyzed cross-coupling with chloroarenes or aryl tosylates, giving the corresponding biaryls **328** together with the cyclic silyl ether **329** [276, 277] (Eq. (9.118)). Benzonitriles can also be used as aryl electrophiles using the palladium catalyst [278]. Palladium/copper cooperative catalytic conditions allow the cross-coupling of cyclohexyl-based aryl-HOMSi reagents with a sterically demanding *ortho*-substituted aryl tosylate successfully to form biaryl **332** [279] (Eq. (9.119)).

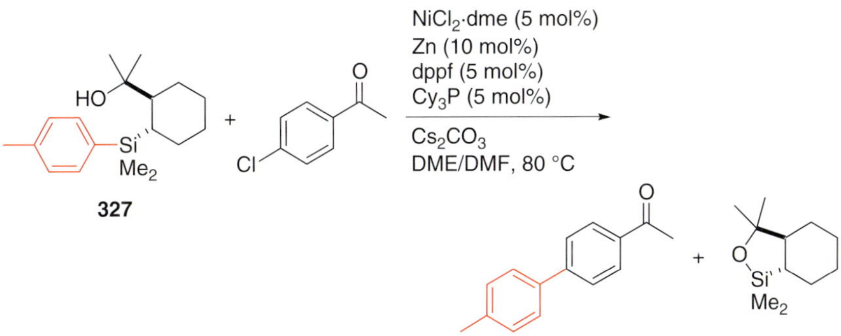

327 **328**, 85% **329**

(9.118)

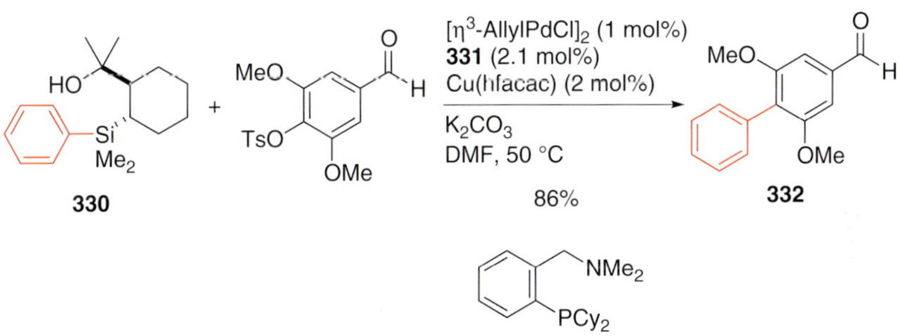

330 **332** **331**

(9.119)

Organo-HOMSi reagents are applicable to other cross-coupling reactions. Copper-mediated cyanation of aryl-HOMSi reagents takes place in the presence of ammonium iodide and dimethylformamide (DMF), giving aryl nitriles **334** [280] (Eq. (9.120)). Copper-catalyzed electrophilic amination of aryl-HOMSi reagent **300** proceeds using *O*-benzoyl-hydroxyamines and gives aniline derivatives **335** [281] (Eq. (9.121)). Three-component reaction among 1-aryl-1,3-butadiene (**336**), aldehyde, and phenyl-HOMSi reagent **300** occurs by a nickel catalysis, giving 1,5-diaryl-3-phenyl-4-penten-1-ol **337** [282] (Eq. (9.122)).

A rhodium complex catalyzes 1,4-addition of organo-HOMSi reagents to enones and 1,2-addition to alkynes and imines [283–286].

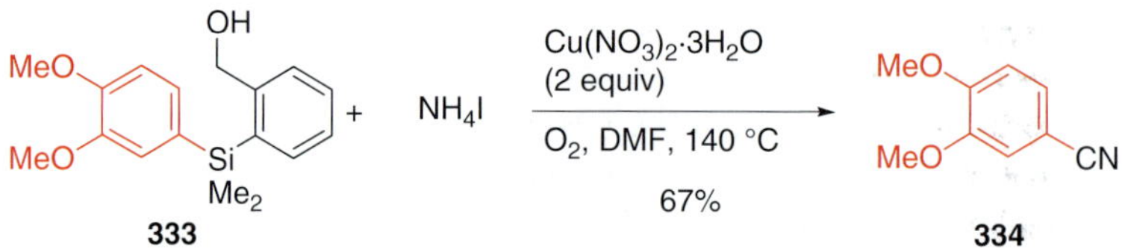

(9.120)

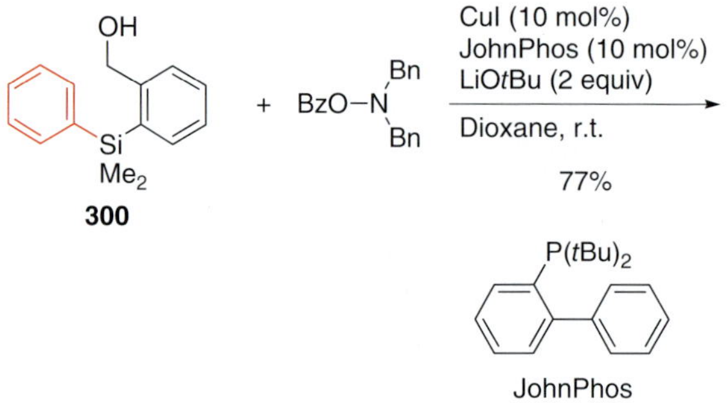

(9.121)

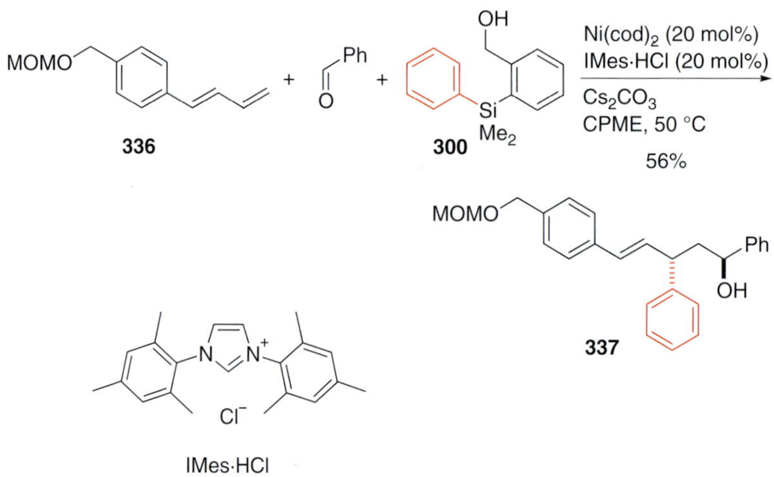

(9.122)

The cyclic silyl ether is readily converted to organo-{2-(lithioxymethyl) pheny}silanes by treatment of aryl- or vinyllithiums, which are *in situ* subjected to the palladium-catalyzed cross-coupling with haloarenes and haloalkenes [287–292]. This is the silicon-assisted Murahashi coupling applicable to various organolithium reagents. For example, geminally trifluoromethylated cyclic silyl ether **338** is a reusable, bench-stable, and highly reactive transfer agent and can be converted to lithiated aryl-HOMSi **339** with aryllithium, which furnishes biaryls by cross-coupling with chloroarenes in the presence of a palladium complex **340** and XPhos as a bulky biaryl ligand [291] (Eq. (9.123)). Various biaryls are prepared according to this procedure. Polymer-supported cyclic silyl ethers can be used as the

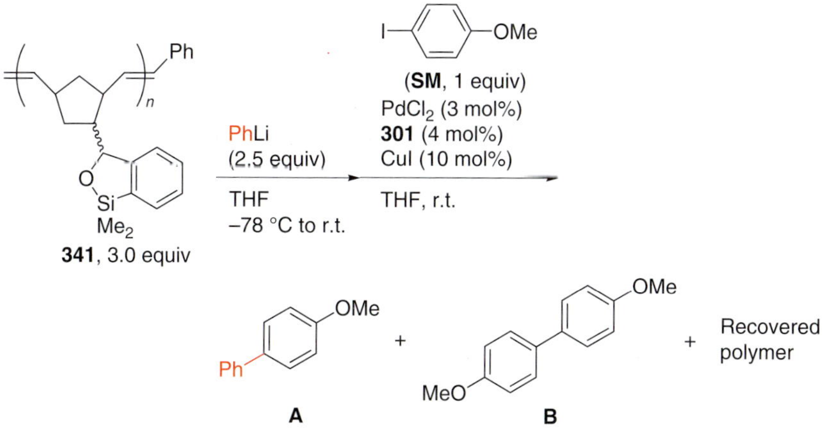

Cycle	Isolated yield of coupling product	^{1}H NMR results (**A**/**B**/**SM**)	Recovered polymer
0			(M_n = 63 500, PDI = 1.2)
1	98%	100 : nd : <1	98% (M_n = 73 000, PDI = 1.6)
2	91%	97 : nd : 3	99% (M_n = 104 400, PDI = 2.3)
3	81%	86 : 11 : 3	99% (M_n = 94 600, PDI, 2.6)

Figure 9.7 Polymer-supported cyclic silyl ether for cross-coupling.

cross-coupling mediator and easily recycled after the reaction [289, 290, 292]. As shown in Figure 9.7, polymer-supported silyl ether **341** can be recycled after the cross-coupling reaction [289].

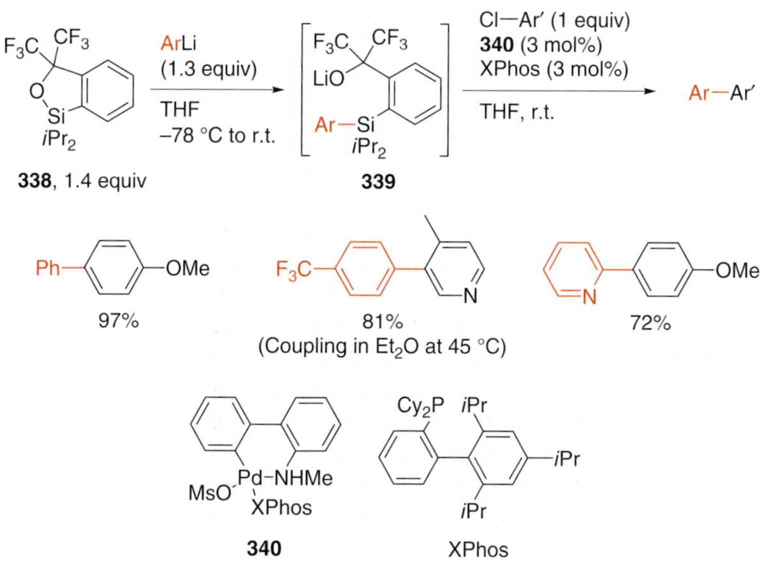

(9.123)

The Murahashi coupling modified by the cyclic silyl ether is applicable to the copper-catalyzed electrophilic amination reaction in a manner similar to the reaction in Eq. (9.121). Treatment of cyclic silyl ether **342** with aryllithium reagents affords lithiated aryl-HOMSi. Subsequent reaction with *O*-benzoyl-*N*,*N*-dibenzylhydroxyamine in the presence of a copper catalyst leads to the formation of aryl(dibenzyl)amine **343** [293] (Eq. (9.124)).

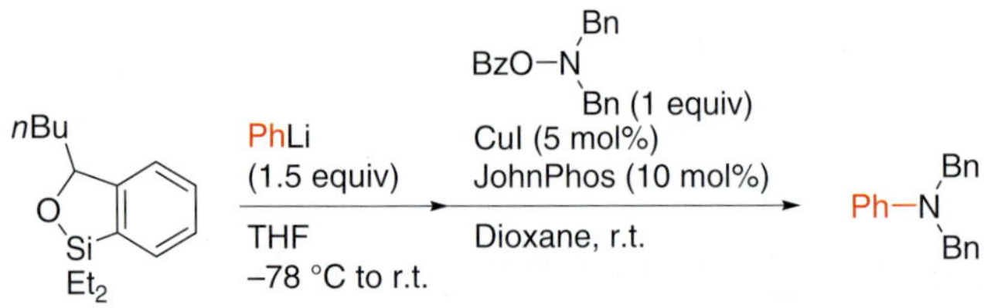

342 (1.8 equiv) **343**, 63% (9.124)

The concept of intramolecular activation works well with aryl- and alkenyl(3-hydroxypropyl)silanes **344**. These substrates are activated by a stoichiometric amount of CuO*t*Bu and $(EtO)_3P$ or (IPr)CuCl catalyst or *n*BuLi in a stoichiometric amount [294]. Styryl(3-hydroxypropyl)silane **344** is treated with BuLi followed by allyl chloride and (IPr)CuCl catalyst to form a 1,4-pentadiene product **345** (Eq. (9.125)). Similarly, aryl- or alkenyl-HOMSi reagents are applicable to the copper-catalyzed alkylation using allyl, benzyl, or alkyl halides. As shown in Eq. (9.126), biphenyl-HOMSi **346** reacts with allyl chloride in the presence of a CuI/$(EtO)_3P$ catalyst and LiO*t*Bu to produce an allylated biphenyl **347**.

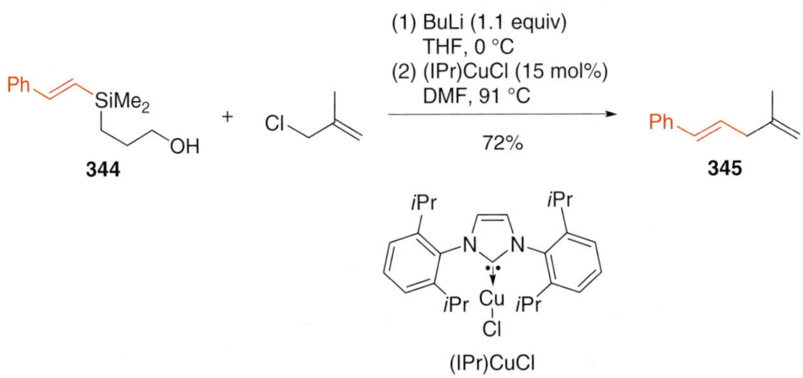

344 **345**

(9.125)

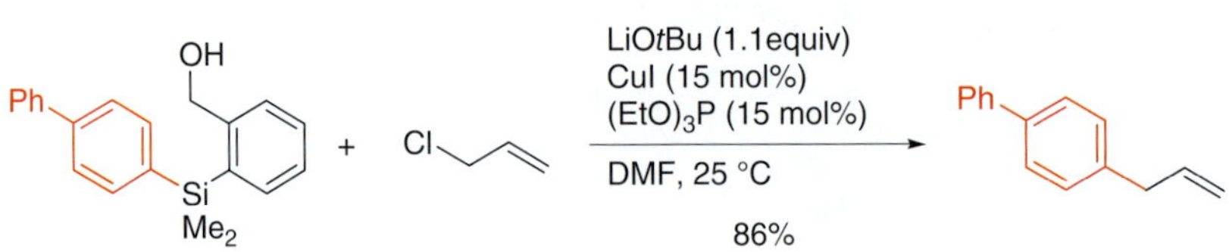

346 **347**

(9.126)

A similar cross-coupling strategy is effective with 1,2-oxasilabenzocyclopentene. Isomeric cyclic silyl ether **348**, upon treatment with aryllithium and CuI, affords a cross-coupling active silicon reagent **349**, which then cross-couples with iodonitrobenzene in the presence of a palladium catalyst to give biaryl **350** [295] (Eq. (9.127)).

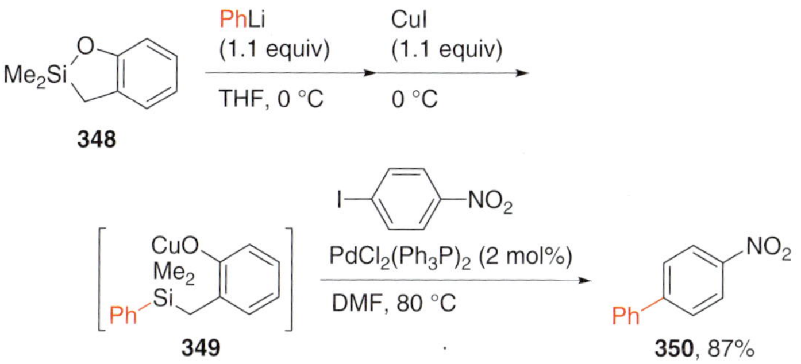

(9.127)

9.5 Summary

This chapter has focused on the recent progress in the silicon-based cross-coupling reaction. Historically, the cross-coupling active silyl groups were believed to be limited to those substituted by fluorine, chlorine, alkoxy, hydroxyl, or organic groups convertible *in situ* to one of these heteroatom groups. Later, hydroxy-linked carbonaceous groups such as 2-hydroxymethylphenyl on the silicon atom were introduced to stabilize the silicon reagents and attain activation readily by a weak base. Protection of the hydroxyl group by a conventional protective group makes the silicon center totally inactive. Presently, improvement of the reaction system and catalytic conditions led to use of trialkylsilyl groups as common and stable silicon coupling reagents. Actually, a number of approaches have been invented to achieve the cross-coupling using organotrialkylsilanes. Furthermore, a variety of metal complexes such as palladium, nickel, and copper are proposed as successful cross-coupling catalysts. Meanwhile, catalytic C–H activation reactions have been disclosed to be applicable to cross-coupling using organosilicon reagents. We believe that the future progress discussed herein will assist the progress in other fields of chemistry, e.g. pharmaceutical science and organic materials chemistry for a sustainable society. It is noteworthy that silicon-based cross-coupling reactions have already been reviewed in many media, focusing on potent silyl groups, catalysts, and protocols. Recent important reviews are listed at the end of the reference section, which will definitely help readers when they employ the silicon-based cross-coupling strategy [296–303].

References

1 Diederich, F. and Stang, P.J. (eds.) (1998). *Metal-Catalyzed Cross-Coupling Reactions*. Weinheim: Wiley-VCH.

2 de Meijere, A. and Diederich, F. (eds.) (2004). *Metal-Catalyzed Cross-Coupling Reactions*, 2e. Weinheim: Wiley-VCH.
3 de Meijere, A., Bräse, S., and Oestreich, M. (eds.) (2014). *Metal-Catalyzed Cross-Coupling Reactions and More*. Weinheim: Wiley-VCH.
4 Yoshida, J., Tamao, K., Yamamoto, H. et al. (1982). *Organometallics* 1: 542–549.
5 Hosomi, A., Kohra, S., and Tominaga, Y. (1988). *Chem. Pharm. Bull. (Tokyo)* 36: 4622–4625.
6 Hatanaka, Y. and Hiyama, T. (1988). *J. Org. Chem.* 53: 918–920.
7 Hatanaka, Y. and Hiyama, T. (1989). *J. Org. Chem.* 54: 268–270.
8 Matsuhashi, H., Asai, S., Hirabayashi, K. et al. (1997). *Bull. Chem. Soc. Jpn.* 70: 1943–1952.
9 Matsuhashi, H., Kuroboshi, M., Hatanaka, Y. et al. (1994). *Tetrahedron Lett.* 35: 6507–6510.
10 Matsuhashi, H., Asai, S., Hirabayashi, K. et al. (1997). *Bull. Chem. Soc. Jpn.* 70: 437–444.
11 Sugiyama, A., Ohnishi, Y., Nakaoka, M. et al. (2008). *J. Am. Chem. Soc.* 130: 12975–12985.
12 Amatore, C., Grimaud, L., Le Duc, G. et al. (2014). *Angew. Chem. Int. Ed.* 53: 6982–6985.
13 Correia, R. and DeShong, P. (2001). *J. Org. Chem.* 66: 7159–7165.
14 Mowery, M.E. and DeShong, P. (1999). *J. Org. Chem.* 64: 1684–1688.
15 Mowery, M.E. and DeShong, P. (1999). *Org. Lett.* 1: 2137–2140.
16 Manoso, A.S. and DeShong, P. (2001). *J. Org. Chem.* 66: 7449–7455.
17 Murata, M., Yoshida, S., Nirei, S. et al. (2006). *Synlett* 0118–0120.
18 Mori, A., Takahisa, E., Yamamura, Y. et al. (2004). *Organometallics* 23: 1755–1765.
19 Hagiwara, E., Gouda, K., Hatanaka, Y. et al. (1997). *Tetrahedron Lett.* 38: 439–442.
20 Wnuk, S.F., Garcia, P.I., and Wang, Z. (2004). *Org. Lett.* 6: 2047–2049.
21 Wang, Z., Pitteloud, J.-P., Montes, L. et al. (2008). *Tetrahedron* 64: 5322–5327.
22 Yuen, O.Y., So, C.M., Man, H.W. et al. (2016). *Chem. Eur. J.* 22: 6471–6476.
23 Clarke, M. (2005). *Adv. Synth. Catal.* 347: 303–307.
24 Lee, H.M. and Nolan, S.P. (2000). *Org. Lett.* 2: 2053–2055.
25 Gu, Z.-S., Shao, L.-X., and Lu, J.-M. (2012). *J. Organomet. Chem.* 700: 132–134.
26 Peñafiel, I., Pastor, I.M., and Yus, M. (2013). *Eur. J. Org. Chem.* 2013: 1479–1484.
27 Yang, J. and Wang, L. (2012). *Dalton Trans.* 41: 12031–12037.
28 Yang, J., Li, P., Zhang, Y., and Wang, L. (2014). *Dalton Trans.* 43: 7166–7175.
29 Ju, J., Nam, H., Jung, H.M. et al. (2006). *Tetrahedron Lett.* 47: 8673–8678.
30 Guo, M., Fu, L., Li, J. et al. (2016). *Molecules* 21: 987.
31 Guo, M., Qi, L., Zhang, Q. et al. (2014). *Org. Biomol. Chem.* 12: 7136–7139.
32 Lee, D.-H., Jung, J.-Y., and Jin, M.-J. (2010). *Chem. Commun.* 46: 9046–9048.
33 Alacid, E. and Nájera, C. (2006). *Adv. Synth. Catal.* 348: 945–952.
34 Alacid, E. and Nájera, C. (2006). *Adv. Synth. Catal.* 348: 2085–2091.
35 Alacid, E. and Nájera, C. (2008). *J. Org. Chem.* 73: 2315–2322.
36 Srimani, D., Sawoo, S., and Sarkar, A. (2007). *Org. Lett.* 9: 3639–3642.
37 Ranu, B.C., Dey, R., and Chattopadhyay, K. (2008). *Tetrahedron Lett.* 49: 3430–3432.

38 Premi, C. and Jain, N. (2013). *Eur. J. Org. Chem.* 2013: 5493–5499.
39 Dey, R., Chattopadhyay, K., and Ranu, B.C. (2008). *J. Org. Chem.* 73: 9461–9464.
40 Srimani, D., Bej, A., and Sarkar, A. (2010). *J. Org. Chem.* 75: 4296–4299.
41 Huang, S.-H., Liu, C.-H., and Yang, C.-M. (2014). *Green Chem.* 16: 2706–2712.
42 Yanase, T., Mori, S., Monguchi, Y. et al. (2011). *Chem. Lett.* 40: 910–912.
43 Yanase, T., Monguchi, Y., and Sajiki, H. (2012). *RSC Adv.* 2: 590–594.
44 Komáromi, A., Szabó, F., and Novák, Z. (2010). *Tetrahedron Lett.* 51: 5411–5414.
45 Diebold, C., Derible, A., Becht, J.-M. et al. (2013). *Tetrahedron* 69: 264–267.
46 Zhang, L., Li, P., Li, H. et al. (2012). *Catal. Sci. Technol.* 2: 1859–1864.
47 delPozo, J., Casares, J.A., and Espinet, P. (2016). *Chem. Eur. J.* 22: 4274–4284.
48 Traficante, C.I., Mata, E.G., and Delpiccolo, C.M.L. (2015). *RSC Adv.* 5: 26796–26800.
49 Spivak, D.A. (2003). Palladium-catalyzed cross-coupling of aryldiazonium salts with arylsilanes. US Patent US 20030120124 A1.
50 Cheng, K., Wang, C., Ding, Y. et al. (2011). *J. Org. Chem.* 76: 9261–9268.
51 Cheng, K., Zhao, B., Hu, S. et al. (2013). *Tetrahedron Lett.* 54: 6211–6214.
52 Zhang, L., Qing, J., Yang, P. et al. (2008). *Org. Lett.* 10: 4971–4974.
53 Zhang, L. and Wu, J. (2008). *J. Am. Chem. Soc.* 130: 12250–12251.
54 Cheng, K., Hu, S., Zhao, B. et al. (2013). *J. Org. Chem.* 78: 5022–5025.
55 Ogiwara, Y., Maegawa, Y., Sakino, D. et al. (2016). *Chem. Lett.* 45: 790–792.
56 Mehta, V.P., Sharma, A., and Van der Eycken, E. (2008). *Adv. Synth. Catal.* 350: 2174–2178.
57 Schmink, J.R. and Krska, S.W. (2011). *J. Am. Chem. Soc.* 133: 19574–19577.
58 Lee, J.-Y. and Fu, G.C. (2003). *J. Am. Chem. Soc.* 125: 5616–5617.
59 Strotman, N.A., Sommer, S., and Fu, G.C. (2007). *Angew. Chem. Int. Ed.* 46: 3556–3558.
60 Powell, D.A. and Fu, G.C. (2004). *J. Am. Chem. Soc.* 126: 7788–7789.
61 Dai, X., Strotman, N.A., and Fu, G.C. (2008). *J. Am. Chem. Soc.* 130: 3302–3303.
62 Gurung, S.K., Thapa, S., Vangala, A.S. et al. (2013). *Org. Lett.* 15: 5378–5381.
63 Gurung, S.K., Thapa, S., Shrestha, B. et al. (2014). *Synthesis* 46: 1933–1937.
64 Cornelissen, L., Lefrancq, M., and Riant, O. (2014). *Org. Lett.* 16: 3024–3027.
65 Takeda, T., Obata, R., Muramatsu, D. et al. (2014). *Chem. Commun.* 50: 15156–15158.
66 Takeda, T., Takeda, Y., and Tsubouchi, A. (2015). *Chem. Lett.* 44: 809–811.
67 Chen, K., Li, H., Li, Y. et al. (2012). *Chem. Sci.* 3: 1645–1649.
68 Katayev, D., Exner, B., and Gooßen, L.J. (2015). *ChemCatChem* 7: 2028–2032.
69 Corcé, V., Chamoreau, L.-M., Derat, E. et al. (2015). *Angew. Chem. Int. Ed.* 54: 11414–11418.
70 Goddard, J.-P., Ollivier, C., and Fensterbank, L. (2016). *Acc. Chem. Res.* 49: 1924–1936.
71 Lévêque, C., Chenneberg, L., Corcé, V. et al. (2016). *Org. Chem. Front.* 3: 462–465.
72 Lévêque, C., Chenneberg, L., Corcé, V. et al. (2016). *Chem. Commun.* 52: 9877–9880.
73 Chenneberg, L., Lévêque, C., Corcé, V. et al. (2016). *Synlett* 27: 731–735.
74 Lévêque, C., Corcé, V., Chenneberg, L. et al. (2017, 2017). *Eur. J. Org. Chem.* 2118–2121.

75 Jouffroy, M., Primer, D.N., and Molander, G.A. (2016). *J. Am. Chem. Soc.* 138: 475–478.
76 Patel, N.R., Kelly, C.B., Jouffroy, M. et al. (2016). *Org. Lett.* 18: 764–767.
77 Patel, N.R. and Molander, G.A. (2016). *J. Org. Chem.* 81: 7271–7275.
78 Jouffroy, M., Davies, G.H.M., and Molander, G.A. (2016). *Org. Lett.* 18: 1606–1609.
79 Lin, K., Wiles, R.J., Kelly, C.B. et al. (2017). *ACS Catal.* 7: 5129–5133.
80 Jouffroy, M., Kelly, C.B., and Molander, G.A. (2016). *Org. Lett.* 18: 876–879.
81 Carfagna, C., Musco, A., Sallese, G. et al. (1991). *J. Org. Chem.* 56: 261–263.
82 Hama, T., Liu, X., Culkin, D.A. et al. (2003). *J. Am. Chem. Soc.* 125: 11176–11177.
83 Guo, Y., Tao, G.-H., Blumenfeld, A. et al. (2010). *Organometallics* 29: 1818–1823.
84 Liu, X. and Hartwig, J.F. (2004). *J. Am. Chem. Soc.* 126: 5182–5191.
85 Huang, Z., Chen, Z., Lim, L.H. et al. (2013). *Angew. Chem. Int. Ed.* 52: 5807–5812.
86 Galarini, R., Musco, A., Pontellini, R. et al. (1992). *J. Mol. Catal.* 72: L11–L13.
87 Huang, Z., Liu, Z., and (Steve) Zhou, J. (2011). *J. Am. Chem. Soc.* 133: 15882–15885.
88 Kobayashi, K., Yamamoto, Y., and Miyaura, N. (2011). *Organometallics* 30: 6323–6327.
89 Aspin, S., López-Suárez, L., Larini, P. et al. (2013). *Org. Lett.* 15: 5056–5059.
90 Alberico, D., Scott, M.E., and Lautens, M. (2007). *Chem. Rev.* 107: 174–238.
91 Seregin, I.V. and Gevorgyan, V. (2007). *Chem. Soc. Rev.* 36: 1173–1193.
92 Chen, X., Engle, K.M., Wang, D.-H. et al. (2009). *Angew. Chem. Int. Ed.* 48: 5094–5115.
93 Lyons, T.W. and Sanford, M.S. (2010). *Chem. Rev.* 110: 1147–1169.
94 He, J., Wasa, M., Chan, K.S.L. et al. (2017). *Chem. Rev.* 117: 8754–8786.
95 Yang, S., Li, B., Wan, X. et al. (2007). *J. Am. Chem. Soc.* 129: 6066–6067.
96 Zhou, H., Xu, Y.-H., Chung, W.-J. et al. (2009). *Angew. Chem. Int. Ed.* 48: 5355–5357.
97 Zhao, S., Liu, B., Zhan, B.-B. et al. (2016). *Org. Lett.* 18: 4586–4589.
98 Hachiya, H., Hirano, K., Satoh, T. et al. (2010). *Angew. Chem. Int. Ed.* 49: 2202–2205.
99 Han, W., Mayer, P., and Ofial, A.R. (2011). *Chem. Eur. J.* 17: 6904–6908.
100 Liang, Z., Yao, B., and Zhang, Y. (2010). *Org. Lett.* 12: 3185–3187.
101 Li, W., Yin, Z., Jiang, X. et al. (2011). *J. Org. Chem.* 76: 8543–8548.
102 Bi, L. and Georg, G.I. (2011). *Org. Lett.* 13: 5413–5415.
103 Fan, H., Shang, Y., and Su, W. (2014). *Eur. J. Org. Chem.* 2014: 3323–3327.
104 Luo, H., Liu, H., Zhang, Z. et al. (2016). *RSC Adv.* 6: 39292–39295.
105 He, J., Takise, R., Fu, H. et al. (2015). *J. Am. Chem. Soc.* 137: 4618–4621.
106 Nareddy, P., Jordan, F., and Szostak, M. (2017). *Chem. Sci.* 8: 3204–3210.
107 Senthilkumar, N., Parthasarathy, K., Gandeepan, P. et al. (2013). *Chem. Asian J.* 8: 2175–2181.
108 Nareddy, P., Jordan, F., and Szostak, M. (2017). *Org. Biomol. Chem.* 15: 4783–4788.
109 Lu, M.-Z., Lu, P., Xu, Y.-H. et al. (2014). *Org. Lett.* 16: 2614–2617.

110 Lu, M.-Z., Wang, C.-Q., Song, S.-J. et al. (2017). *Org. Chem. Front.* 4: 303–307.

111 Lam, P.Y.S., Deudon, S., Averill, K.M. et al. (2000). *J. Am. Chem. Soc.* 122: 7600–7601.

112 Morstein, J., Kalkman, E.D., Cheng, C. et al. (2016). *Org. Lett.* 18: 5244–5247.

113 Kim, J., Ok, J., Kim, S. et al. (2014). *Org. Lett.* 16: 4602–4605.

114 Zhong, L., Su, Q., Xiao, J. et al. (2017). *Asian J. Org. Chem.* 6: 1072–1079.

115 Luo, H., Liu, H., Chen, X. et al. (2017). *Chem. Commun.* 53: 956–958.

116 Denmark, S.E. and Pan, W. (2001). *Org. Lett.* 3: 61–64.

117 Denmark, S.E. and Yang, S.-M. (2001). *Org. Lett.* 3: 1749–1752.

118 Sieburth, S.M. and Mu, W. (1993). *J. Org. Chem.* 58: 7584–7586.

119 Mori, A., Hishida, T., Soga, Y. et al. (1995). *Chem. Lett.* 24: 107–108.

120 Mori, A., Sato, H., Mizuno, K. et al. (1996). *Chem. Lett.* 25: 517–518.

121 Hirabayashi, K., Nishihara, Y., Mori, A. et al. (1998). *Tetrahedron Lett.* 39: 7893–7896.

122 Hirabayashi, K., Kawashima, J., Nishihara, Y. et al. (1999). *Org. Lett.* 1: 299–302.

123 Hirabayashi, K., Mori, A., Kawashima, J. et al. (2000). *J. Org. Chem.* 65: 5342–5349.

124 Mori, A., Danda, Y., Fujii, T. et al. (2001). *J. Am. Chem. Soc.* 123: 10774–10775.

125 Koike, T., Du, X., Sanada, T. et al. (2003). *Angew. Chem. Int. Ed.* 42: 89–92.

126 Fujii, T., Koike, T., Mori, A. et al. (2002). *Synlett* 298–300.

127 Fujii, T., Koike, T., Mori, A. et al. (2002). *Synlett* 295–297.

128 Denmark, S.E. and Sweis, R.F. (2002). *Acc. Chem. Res.* 35: 835–846.

129 Denmark, S.E., Wehrli, D., and Choi, J.Y. (2000). *Org. Lett.* 2: 2491–2494.

130 Denmark, S.E. and Wehrli, D. (2000). *Org. Lett.* 2: 565–568.

131 Denmark, S.E. and Sweis, R.F. (2001). *J. Am. Chem. Soc.* 123: 6439–6440.

132 Denmark, S.E. and Neuville, L. (2000). *Org. Lett.* 2: 3221–3224.

133 Denmark, S.E. and Pan, W. (2002). *J. Organomet. Chem.* 653: 98–104.

134 Chang, S., Yang, S.H., and Lee, P.H. (2001). *Tetrahedron Lett.* 42: 4833–4835.

135 Denmark, S.E. and Ober, M.H. (2003). *Org. Lett.* 5: 1357–1360.

136 Denmark, S.E. and Ober, M.H. (2004). *Adv. Synth. Catal.* 346: 1703–1714.

137 Denmark, S.E. and Baird, J.D. (2006). *Org. Lett.* 8: 793–795.

138 Denmark, S.E. and Tymonko, S.A. (2005). *J. Am. Chem. Soc.* 127: 8004–8005.

139 Denmark, S.E. and Fujimori, S. (2005). *J. Am. Chem. Soc.* 127: 8971–8973.

140 Denmark, S.E., Regens, C.S., and Kobayashi, T. (2007). *J. Am. Chem. Soc.* 129: 2774–2776.

141 Denmark, S.E., Kobayashi, T., and Regens, C.S. (2010). *Tetrahedron* 66: 4745–4759.

142 Denmark, S.E. and Sweis, R.F. (2004). *J. Am. Chem. Soc.* 126: 4876–4882.

143 Denmark, S.E., Sweis, R.F., and Wehrli, D. (2004). *J. Am. Chem. Soc.* 126: 4865–4875.

144 Denmark, S.E., Neuville, L., Christy, M.E.L. et al. (2006). *J. Org. Chem.* 71: 8500–8509.

145 Hirabayashi, K., Takahisa, E., Nishihara, Y. et al. (1998). *Bull. Chem. Soc. Jpn.* 71: 2409–2417.

146 Beaulieu, L.-P.B., Delvos, L.B., and Charette, A.B. (2010). *Org. Lett.* 12: 1348–1351.

147 Denmark, S.E. and Kallemeyn, J.M. (2006). *J. Am. Chem. Soc.* 128: 15958–15959.
148 Denmark, S.E. and Regens, C.S. (2011). *Tetrahedron Lett.* 52: 2165–2168.
149 Denmark, S.E., Smith, R.C., and Tymonko, S.A. (2007). *Tetrahedron* 63: 5730–5738.
150 Denmark, S.E., Smith, R.C., Chang, W.-T.T. et al. (2009). *J. Am. Chem. Soc.* 131: 3104–3118.
151 Denmark, S.E. and Werner, N.S. (2008). *J. Am. Chem. Soc.* 130: 16382–16393.
152 Denmark, S.E. and Smith, R.C. (2010). *J. Am. Chem. Soc.* 132: 1243–1245.
153 Tymonko, S.A., Smith, R.C., Ambrosi, A. et al. (2015). *J. Am. Chem. Soc.* 137: 6192–6199.
154 Tymonko, S.A., Smith, R.C., Ambrosi, A. et al. (2015). *J. Am. Chem. Soc.* 137: 6200–6218.
155 Sore, H.F., Boehner, C.M., Laraia, L. et al. (2011). *Org. Biomol. Chem.* 9: 504–515.
156 Frye, E.C., O'Connor, C.J., Twigg, D.G. et al. (2012). *Chem. Eur. J.* 18: 8774–8779.
157 Denmark, S.E. and Wang, Z. (2001). *Org. Lett.* 3: 1073–1076.
158 Sore, H.F., Boehner, C.M., MacDonald, S.J.F. et al. (2009). *Org. Biomol. Chem.* 7: 1068–1071.
159 Boehner, C.M., Frye, E.C., O'Connell, K.M.G. et al. (2011). *Chem. Eur. J.* 17: 13230–13239.
160 Lian, Z., Nielsen, D.U., Lindhardt, A.T. et al. (2016). *Nat. Commun.* 7: 13782.
161 Denmark, S.E. and Butler, C.R. (2008). *J. Am. Chem. Soc.* 130: 3690–3704.
162 Suguro, M., Yamamura, Y., Koike, T. et al. (2007). *React. Funct. Polym.* 67: 1264–1276.
163 Endo, M., Sakurai, T., Ojima, S. et al. (2007). *Synlett* 2007: 0749–0752.
164 Koike, T. and Mori, A. (2003). *Synlett* 1850–1852.
165 Koike, T., Du, X., Mori, A. et al. (2002). *Synlett* 2002: 0301–0303.
166 Denmark, S.E. and Choi, J.Y. (1999). *J. Am. Chem. Soc.* 121: 5821–5822.
167 Denmark, S.E. and Wu, Z. (1999). *Org. Lett.* 1: 1495–1498.
168 Sahoo, A.K., Nakao, Y., and Hiyama, T. (2004). *Chem. Lett.* 33: 632–633.
169 Sahoo, A.K., Oda, T., Nakao, Y. et al. (2004). *Adv. Synth. Catal.* 346: 1715–1727.
170 Nokami, T., Tomida, Y., Kamei, T. et al. (2006). *Org. Lett.* 8: 729–731.
171 Li, L. and Navasero, N. (2006). *Org. Lett.* 8: 3733–3736.
172 Trost, B.M., Machacek, M.R., and Ball, Z.T. (2003). *Org. Lett.* 5: 1895–1898.
173 Nishihara, Y., Saito, D., Tanemura, K. et al. (2009). *Org. Lett.* 11: 3546–3549.
174 Shindo, M., Matsumoto, K., and Shishido, K. (2005). *Chem. Commun.* 2477–2479.
175 Trost, B.M., Machacek, M.R., and Faulk, B.D. (2006). *J. Am. Chem. Soc.* 128: 6745–6754.
176 Morrill, C. and Mani, N.S. (2007). *Org. Lett.* 9: 1505–1508.
177 Junker, C.S., Welker, M.E., and Day, C.S. (2010). *J. Org. Chem.* 75: 8155–8165.
178 Rooke, D.A. and Ferreira, E.M. (2010). *J. Am. Chem. Soc.* 132: 11926–11928.
179 Lim, D.S.W. and Anderson, E.A. (2011). *Org. Lett.* 13: 4806–4809.

180 Takeda, T., Wasa, H., and Tsubouchi, A. (2011). *Tetrahedron Lett.* 52: 4575–4578.
181 Köpfer, A., Sam, B., Breit, B. et al. (2013). *Chem. Sci.* 4: 1876–1880.
182 McLaughlin, M.G., McAdam, C.A., and Cook, M.J. (2015). *Org. Lett.* 17: 10–13.
183 McAdam, C.A., McLaughlin, M.G., and Cook, M.J. (2015). *Org. Chem. Front.* 2: 510–514.
184 Trost, B.M., Hung, C.-I., Koester, D.C. et al. (2015). *Org. Lett.* 17: 3778–3781.
185 Miller, Z.D., Dorel, R., and Montgomery, J. (2015). *Angew. Chem. Int. Ed.* 54: 9088–9091.
186 Trost, B.M., Sharif, E.U., and Cregg, J.J. (2017). *Chem. Sci.* 8: 770–774.
187 Shimizu, R., Egami, H., Hamashima, Y. et al. (2012). *Angew. Chem. Int. Ed.* 51: 4577–4580.
188 Denmark, S.E. and Liu, J.H.-C. (2007). *J. Am. Chem. Soc.* 129: 3737–3744.
189 Bergueiro, J., Montenegro, J., Cambeiro, F. et al. (2012). *Chem. Eur. J.* 18: 4401–4410.
190 Takeda, T., Mori, A., Fujii, T. et al. (2014). *Tetrahedron* 70: 5776–5786.
191 Trost, B.M., Frederiksen, M.U., Papillon, J.P.N. et al. (2005). *J. Am. Chem. Soc.* 127: 3666–3667.
192 Zhang, Y. and Panek, J.S. (2007). *Org. Lett.* 9: 3141–3143.
193 Li, H., Xie, H., Zhang, Z. et al. (2015). *Chem. Commun.* 51: 8484–8487.
194 Trost, B.M., Biannic, B., Brindle, C.S. et al. (2015). *J. Am. Chem. Soc.* 137: 11594–11597.
195 Campbell, C.D., Greenaway, R.L., Holton, O.T. et al. (2014). *Chem. Commun.* 50: 5187–5189.
196 Takeda, T., Matsumura, R., Aasa, H. et al. (2014). *Asian J. Org. Chem.* 3: 838–841.
197 Montenegro, J., Bergueiro, J., Saá, C. et al. (2009). *Org. Lett.* 11: 141–144.
198 Murata, M., Ohara, H., Oiwa, R. et al. (2006). *Synthesis* 1771–1774.
199 Hosoi, K., Nozaki, K., and Hiyama, T. (2002). *Chem. Lett.* 138–139.
200 Hosoi, K., Nozaki, K., and Hiyama, T. (2002). *Proc. Jpn. Acad. Ser. B* 78: 154–160.
201 Vitale, M., Prestat, G., Lopes, D. et al. (2008). *J. Org. Chem.* 73: 5795–5805.
202 Liu, L. and Floreancig, P.E. (2010). *Angew. Chem. Int. Ed.* 49: 3069–3072.
203 Lorion, M.M., Matt, B., Alves, S. et al. (2013). *Chem. Eur. J.* 19: 12607–12612.
204 Itami, K., Nokami, T., and Yoshida, J. (2001). *J. Am. Chem. Soc.* 123: 5600–5601.
205 Itami, K., Nokami, T., Ishimura, Y. et al. (2001). *J. Am. Chem. Soc.* 123: 11577–11585.
206 Itami, K., Mitsudo, K., Nokami, T. et al. (2002). *J. Organomet. Chem.* 653: 105–113.
207 Kamei, T., Itami, K., and Yoshida, J.-i. (2004). *Adv. Synth. Catal.* 346: 1824–1835.
208 Itami, K., Ohashi, Y., and Yoshida, J.-i. (2005). *J. Org. Chem.* 70: 2778–2792.
209 Chernyak, N., Dudnik, A.S., Huang, C. et al. (2010). *J. Am. Chem. Soc.* 132: 8270–8272.
210 Huang, C., Chernyak, N., Dudnik, A.S. et al. (2011). *Adv. Synth. Catal.* 353: 1285–1305.
211 Itami, K., Mineno, M., Kamei, T. et al. (2002). *Org. Lett.* 4: 3635–3638.

212 Sore, H.F., Blackwell, D.T., MacDonald, S.J.F. et al. (2010). *Org. Lett.* 12: 2806–2809.

213 Katayama, H., Taniguchi, K., Kobayashi, M. et al. (2002). *J. Organomet. Chem.* 645: 192–200.

214 Katayama, H., Nagao, M., Ozawa, F. et al. (2006). *J. Org. Chem.* 71: 2699–2705.

215 Katayama, H., Nagao, M., Moriguchi, R. et al. (2003). *J. Organomet. Chem.* 676: 49–54.

216 Katayama, H., Nagao, M., Nishimura, T. et al. (2006). *Macromolecules* 39: 2039–2048.

217 Wakioka, M., Ikegami, M., and Ozawa, F. (2010). *Macromolecules* 43: 6980–6985.

218 Murai, M., Takami, K., Takeshima, H. et al. (2015). *Org. Lett.* 17: 1798–1801.

219 Hanamoto, T., Kobayashi, T., and Kondo, M. (2001). *Synlett* 281–283.

220 Hanamoto, T. and Kobayashi, T. (2003). *J. Org. Chem.* 68: 6354–6359.

221 Anderson, J.C., Anguille, S., and Bailey, R. (2002). *Chem. Commun.* 2018–2019.

222 Anderson, J.C. and Munday, R.H. (2004). *J. Org. Chem.* 69: 8971–8974.

223 Zhang, L., Hang, Z., and Liu, Z.-Q. (2016). *Angew. Chem. Int. Ed.* 55: 236–239.

224 Friis, S.D., Andersen, T.L., and Skrydstrup, T. (2013). *Org. Lett.* 15: 1378–1381.

225 Pierrat, P., Gros, P., and Fort, Y. (2005). *Org. Lett.* 7: 697–700.

226 Napier, S., Marcuccio, S.M., Tye, H. et al. (2008). *Tetrahedron Lett.* 49: 6314–6315.

227 Matsuda, S., Takahashi, M., Monguchi, D. et al. (2009). *Synlett* 12: 1941–1943.

228 Taguchi, H., Ghoroku, K., Tadaki, M. et al. (2002). *J. Org. Chem.* 67: 8450–8456.

229 Sanchez, L. and Smith, A.B. III (2012). *Org. Lett.* 14: 6314–6317.

230 Matsumoto, K. and Shindo, M. (2012). *Adv. Synth. Catal.* 354: 642–650.

231 Akai, S., Ikawa, T., Takayanagi, S. et al. (2008). *Angew. Chem. Int. Ed.* 47: 7673–7676.

232 Taguchi, H., Takami, K., Tsubouchi, A. et al. (2004). *Tetrahedron Lett.* 45: 429–432.

233 Tsubouchi, A., Matsuda, H., Kira, T. et al. (2009). *Chem. Lett.* 38: 1180–1181.

234 Komiyama, T., Minami, Y., and Hiyama, T. (2016). *Angew. Chem. Int. Ed.* 55: 15787–15791.

235 Komiyama, T., Minami, Y., Furuya, Y. et al. (2018). *Angew. Chem. Int. Ed.* 57: 1987–1990.

236 Nishihara, Y., Ikegashira, K., Mori, A. et al. (1997). *Chem. Lett.* 26: 1233–1234.

237 Nishihara, Y., Ikegashira, K., Hirabayashi, K. et al. (2000). *J. Org. Chem.* 65: 1780–1787.

238 Nishihara, Y., Noyori, S., Okamoto, T. et al. (2011). *Chem. Lett.* 40: 972–974.

239 Oishi, M., Kondo, H., and Amii, H. (2009). *Chem. Commun.* 1909–1911.

240 Dubinina, G.G., Furutachi, H., and Vicic, D.A. (2008). *J. Am. Chem. Soc.* 130: 8600–8601.

241 Morimoto, H., Tsubogo, T., Litvinas, N.D. et al. (2011). *Angew. Chem. Int. Ed.* 50: 3793–3798.

242 Danoun, G., Bayarmagnai, B., Grünberg, M.F. et al. (2013). *Angew. Chem. Int. Ed.* 52: 7972–7975.

243 Morstein, J., Hou, H., Cheng, C. et al. (2016). *Angew. Chem. Int. Ed.* 55: 8054–8057.

244 Fier, P.S. and Hartwig, J.F. (2012). *J. Am. Chem. Soc.* 134: 5524–5527.

245 Cho, E.J., Senecal, T.D., Kinzel, T. et al. (2010). *Science* 328: 1679–1681.

246 Ge, S., Arlow, S.I., Mormino, M.G. et al. (2014). *J. Am. Chem. Soc.* 136: 14401–14404.

247 Arlow, S.I. and Hartwig, J.F. (2016). *Angew. Chem. Int. Ed.* 55: 4567–4572.

248 Funaki, K., Kawai, H., Sato, T. et al. (2011). *Chem. Lett.* 40: 1050–1052.

249 Funaki, K., Sato, T., and Oi, S. (2012). *Org. Lett.* 14: 6186–6189.

250 Kawasumi, K., Mochida, K., Kajino, T. et al. (2012). *Org. Lett.* 14: 418–421.

251 Rauf, W. and Brown, J.M. (2017). *Org. Biomol. Chem.* 15: 8179–8185.

252 Shibata, M., Ito, H., and Itami, K. (2018). *J. Am. Chem. Soc.* 140: 2196–2205.

253 Shibata, M., Ito, H., and Itami, K. (2017). *Chem. Lett.* 46: 1701–1704.

254 Ball, L.T., Lloyd-Jones, G.C., and Russell, C.A. (2012). *Science* 337: 1644–1648.

255 Corrie, T.J.A., Ball, L.T., Russell, C.A. et al. (2017). *J. Am. Chem. Soc.* 139: 245–254.

256 Harper, M.J., Emmett, E.J., Bower, J.F. et al. (2017). *J. Am. Chem. Soc.* 139: 12386–12389.

257 Hata, K., Ito, H., Segawa, Y. et al. (2015). *Beilstein J. Org. Chem.* 11: 2737–2746.

258 Ozaki, K., Kawasumi, K., Shibata, M. et al. (2015). *Nat. Commun.* 6: 6251–6258.

259 Kato, K., Segawa, Y., and Itami, K. (2017). *Can. J. Chem.* 95: 329–333.

260 Ozaki, K., Matsuoka, W., Ito, H. et al. (2017). *Org. Lett.* 19: 1930–1933.

261 Ozaki, K., Murai, K., Matsuoka, W. et al. (2017). *Angew. Chem. Int. Ed.* 56: 1361–1364.

262 Gondo, K., Oyamada, J., and Kitamura, T. (2015). *Org. Lett.* 17: 4778–4781.

263 Nakao, Y., Imanaka, H., Sahoo, A.K. et al. (2005). *J. Am. Chem. Soc.* 127: 6952–6953.

264 Iizuka, M. and Kondo, Y. (2008). *Eur. J. Org. Chem.* 1161–1163.

265 Minami, Y., Shimizu, K., Tsuruoka, C. et al. (2014). *Chem. Lett.* 43: 201–203.

266 Minami, Y., Komiyama, T., and Hiyama, T. (2015). *Chem. Lett.* 44: 1065–1067.

267 Rooke, D.A. and Ferreira, E.M. (2012). *Org. Lett.* 14: 3328–3331.

268 Nakao, Y., Chen, J., Tanaka, M. et al. (2007). *J. Am. Chem. Soc.* 129: 11694–11695.

269 Shirbin, S.J., Boughton, B.A., Zammit, S.C. et al. (2010). *Tetrahedron Lett.* 51: 2971–2974.

270 Marcuccio, S.M., Epa, R., Moslmani, M. et al. (2011). *Tetrahedron Lett.* 52: 7178–7181.

271 Nakao, Y., Imanaka, H., Chen, J. et al. (2007). *J. Organomet. Chem.* 692: 585–603.

272 Chen, J., Tanaka, M., Sahoo, A.K. et al. (2010). *Bull. Chem. Soc. Jpn.* 83: 554–569.

273 Shimizu, K., Minami, Y., Nakao, Y. et al. (2013). *Chem. Lett.* 42: 45–47.

274 Nakao, Y., Ebata, S., Chen, J. et al. (2007). *Chem. Lett.* 36: 606–607.

275 Nakao, Y., Takeda, M., Matsumoto, T. et al. (2010). *Angew. Chem. Int. Ed.* 49: 4447–4450.

276 Tang, S., Takeda, M., Nakao, Y. et al. (2011). *Chem. Commun.* 47: 307–309.

277 Tang, S., Li, S.-H., Nakao, Y. et al. (2013). *Asian J. Org. Chem.* 2: 416–421.
278 Tang, S., Li, S.-H., and Yan, W.-b. (2012). *Tetrahedron Lett.* 53: 6743–6746.
279 Ohgi, A., Semba, K., Hiyama, T. et al. (2016). *Chem. Lett.* 45: 973–975.
280 Wang, Z. and Chang, S. (2013). *Org. Lett.* 15: 1990–1993.
281 Miki, Y., Hirano, K., Satoh, T. et al. (2013). *Org. Lett.* 15: 172–175.
282 Saito, N., Yamazaki, T., and Sato, Y. (2008). *Tetrahedron Lett.* 49: 5073–5076.
283 Nakao, Y., Chen, J., Imanaka, H. et al. (2007). *J. Am. Chem. Soc.* 129: 9137–9143.
284 Shintani, R., Ichikawa, Y., Hayashi, T. et al. (2007). *Org. Lett.* 9: 4643–4645.
285 Nakao, Y., Takeda, M., Chen, J. et al. (2008). *Chem. Lett.* 37: 290–291.
286 Nakao, Y., Takeda, M., Chen, J. et al. (2008). *Synlett* 774–776.
287 Smith, A.B. III, Hoye, A.T., Martinez-Solorio, D. et al. (2012). *J. Am. Chem. Soc.* 134: 4533–4536.
288 Martinez-Solorio, D., Hoye, A.T., Nguyen, M.H. et al. (2013). *Org. Lett.* 15: 2454–2457.
289 Nguyen, M.H. and Smith, A.B. III (2013). *Org. Lett.* 15: 4258–4261.
290 Nguyen, M.H. and Smith, A.B. III (2014). *Org. Lett.* 16: 2070–2073.
291 Martinez-Solorio, D., Melillo, B., Sanchez, L. et al. (2016). *J. Am. Chem. Soc.* 138: 1836–1839.
292 Nguyen, M.H., O'Brien, K.T., and Smith, A.B. III (2017). *J. Org. Chem.* 82: 11056–11071.
293 Nguyen, M.H. and Smith, A.B. III (2013). *Org. Lett.* 15: 4872–4875.
294 Tsubouchi, A., Muramatsu, D., and Takeda, T. (2013). *Angew. Chem. Int. Ed.* 52: 12719–12722.
295 Son, E.-C., Tsuji, H., Saeki, T. et al. (2006). *Bull. Chem. Soc. Jpn.* 79: 492–494.
296 Chang, W.-T.T., Smith, R.C., Regens, C.S. et al. (2011). *Organic Reactions*, vol. 75 (ed. S.E. Denmark), 213. Hoboken, New Jersey: Wiley-VCH.
297 Nakao, Y. and Hiyama, T. (2011). *J. Synth. Org. Chem Jpn.* 69: 1221–1230.
298 Nakao, Y. and Hiyama, T. (2011). *Chem. Soc. Rev.* 40: 4893–4901.
299 Sore, H.F., Galloway, W.R.J.D., and Spring, D.R. (2012). *Chem. Soc. Rev.* 41: 1845–1866.
300 Denmark, S.E. and Ambrosi, A. (2015). *Org. Process Res. Dev.* 19: 982–994.
301 Foubelo, F., Najera, C., and Yus, M. (2016). *Chem. Rec.* 16: 2521–2533.
302 Komiyama, T., Minami, Y., and Hiyama, T. (2017). *ACS Catal.* 7: 631–651.
303 Komiyama, T., Minami, Y., and Hiyama, T. (2017). *Synlett* 28: 1873–1884.

10

Lewis Base Activation of Silicon Lewis Acids

Sergio Rossi[1] *and Scott E. Denmark*[2]

[1] *Department of Chemistry, University of Milan, Via Golgi 19, 20133, Milan, Italy*
[2] *Department of Chemistry, University of Illinois, 600 S. Mathews Avenue, Urbana, IL 61801, USA*

10.1 Introduction

Hypercoordinate silicon compounds have been of great interest since their discovery in the nineteenth century, when J.L. Gay-Lussac and J.A. Thénard and J. Davy reported the existence of the $SiF_4{\cdot}2NH_3$ complex [1, 2]. Starting from this point, many *penta-*, *hexa-* and *hepta-*coordinated silicon species were identified and studied in the past two centuries [3]. Nowadays, the ability of silicon to form more bonds than the four predicted by the rules of primary valency is well documented and this particular extra-coordination has been involved in the development of many enantioselective reactions starting from different silicon-based Lewis acids. Typically, the silicon atom forms bonds with only four atoms to complete its electronic requirement for an outer shell octet. However, under specific circumstances, when a Lewis-acidic silicon species interacts with a donor molecule (a Lewis base), instead of generating a more stable acid–base adduct (as predicted by the general principle of Lewis acid–base theory), the formation of a new highly reactive adduct can occur, thus breaking the "Lewis–Langmuir octet rule."

Different theories have been advanced to explain this particular behavior, and one of the earliest involves a reorganization of the electron density around the silicon atom, as shown by Gutmann's semiempirical analysis (Scheme 10.1) [4, 5]. Gutmann recognized that the formation of the new dative bond between the Lewis base and the Lewis acidic silicon compound leads to an overall increase in the electron density on the acceptor fragment of the adduct. However, the distribution of this electron density is not equal among the constituent atoms and, indeed, an enhancement of both nucleophilic and electrophilic character is observed. This phenomenon results in the lengthening of the bonds in the newly formed adduct, with attendant expansion of the coordination sphere of the silicon atom. Because of this electronic redistribution, the new complex exhibits

Organosilicon Chemistry: Novel Approaches and Reactions, First Edition.
Edited by Tamejiro Hiyama and Martin Oestreich

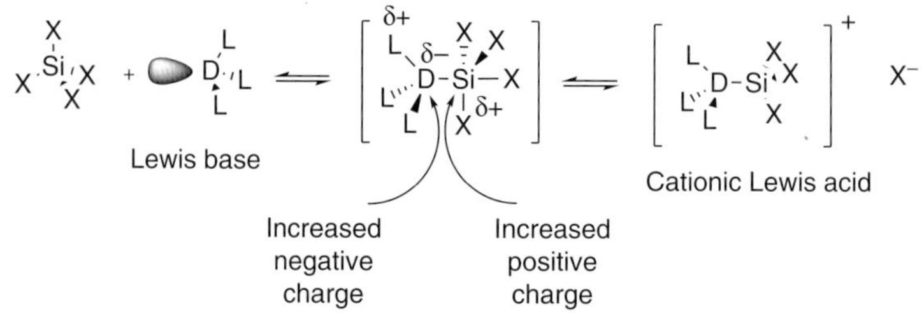

Scheme 10.1 Electronic redistribution resulting from Lewis acid–Lewis base complexation.

enhanced electrophilic reactivity at the central silicon atom and enhanced nucleophilic reactivity at the peripheral ligands.

In a very early rationalization of this behavior, the silicon atom is proposed to employ 3d orbitals, as for transition-metal complexation [6]. In the five-coordinate species, the silicon orbitals would have sp^3d hybridization (with trigonal-bipyramidal geometry), whereas in the *hexa*-coordinate species the hybridization would be sp^3d^2 (with octahedral geometry). The reduced s-character of the silicon orbitals in the hypercoordinate species would explain their increased Lewis acidity and the transfer of electron density to the ligands. However, silicon is not a transition metal and it is now generally accepted that the 3d orbitals on silicon are too diffuse to originate molecular bonds [7].

In a second theory, the silicon atom engages in hypervalent bonding (or in the expansion of the coordination sphere) [8], in which the silicon 3p orbitals are involved in the formation of one or two electron-rich, three-center four-electron bonds (hybrids first introduced by Pimentel building on early ideas of Rundle) [9–13]. According to this analysis, the formation of a *penta*- or *hexa*-coordinate silicon species would involve, respectively, one or two, three-center four-electron bonds, each formed by one silicon p orbital and two p orbitals of the donating ligands (Figure 10.1a). An important consequence of this fact is the nonequivalence of the ligand positions in five- and six-coordinated silicon species, wherein the σ-acceptor ligands prefer "hypervalent" bonds and the σ-donors preferentially forming normal covalent bonds with the sp^2 (for *penta*-coordinate compounds) or sp (for *hexa*-coordinate compounds) silicon orbitals.

The hypervalent bonds localize an electron-rich character on the peripheral ligands and an electron-deficient character at the silicon central atom. Mixing of the filled p-orbital on the acceptor with the filled p-orbital on the donor, according to molecular orbital (MO) theory, generates a pair of hybrid orbitals, in which the HOMO of this hybrid orbital (Ψ^2) contains a node at the central atom with the electron density localized on the peripheral atoms (Figure 10.1b).

In a more modern analysis, Weinhold and Landis applied the natural bond orbital (NBO) treatment to the classic Pimentel–Rundle model of three-center, four-electron bonding [14]. In their approach, the interaction between the Lewis base adduct composed as LB: (donor) + Si–X (acceptor) is described as a resonance limit with the form LB–Si + X:. As a consequence, the bond order of each pair of atoms can be calculated according to the natural resonance theory (NRT)

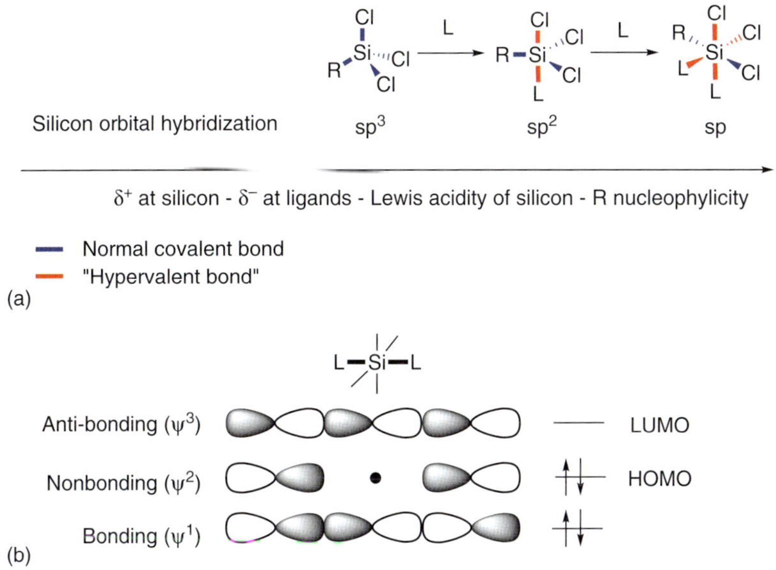

Figure 10.1 Hypervalent bonding molecular orbitals (linear combination of atomic orbitals). (a) Hypervalent bonding. (b) Molecular orbital diagram of three-center four-electron hybrids.

as the sum of a covalent contribution and of an ionic contribution. This unique resonance motif was classified as "hyperbonding" or "ω-bonding" character (Equations 1–3).

$$\mathrm{LB:} + \mathrm{Si-X} \leftrightarrow \mathrm{LB-Si} + \mathrm{X:} \tag{10.1}$$

$$b_{\mathrm{LB-Si}}^{(\mathrm{NRT})} = b_{\mathrm{LB-Si}}^{\mathrm{covalent}} + b_{\mathrm{LB-Si}}^{\mathrm{ion}} \tag{10.2}$$

$$b_{\mathrm{Si-X}}^{(\mathrm{NRT})} = b_{\mathrm{Si-X}}^{\mathrm{covalent}} + b_{\mathrm{Si-X}}^{\mathrm{ion}} \tag{10.3}$$

These hyperbonds possess typical characteristics such as a linear (or near-linear) bond geometry, increased bond lengths, high cationic character on the central atom, and enhanced anionic character of terminal atoms, as typically observed in the higher coordinate silicon compounds.

Most recently, Gillespie and a coworker (who rejects the wave function model as a molecular geometry explanation, because it is not an observable and measurable physical entity), asserts that the geometry of molecules can be explained using three complementary models: (i) the VSEPR (valence shell electron pair repulsion) model, (ii) the quantum theory of atoms in molecules (AIM) [15], and (iii) the LCP (ligand close-packing) model. In particular, Gillespie and a coworker criticizes the concept of the "hypervalency" as misleading because it arises from an overinterpretation of the octet rule. He asserts that the bonding in molecules with ligand–lone pair coordination numbers greater than four is no different from that of molecules in which the coordination number is exactly four, because bonding may range from a predominantly ionic to a predominantly covalent character [16]. The quantum theory of atoms in molecules has been used, for example, in the calculation of electron density distribution functions in substituted quasisilatranes [15], silatranes [17, 18], and also in several complexes of tris(pentafluorophenyl)silanes and other chlorosilanes [19] with neutral Lewis bases [20]. The AIM model has also been criticized for overemphasizing electron densities as determinants of molecular geometries and failing to explain chemical and physical proprieties of molecules [21].

Setting aside the dispute on which theory best describes the atypical behavior of silicon in LB–LA adducts [22], numerous experimental observations unambiguously demonstrate that the interaction of a Lewis base with a silicon Lewis acid generates a new silicon species with enhanced reactivity [23]. The focus of this chapter is to describe the use of these hypercoordinate silicon species in the promotion of catalytic processes. Generally, the hypercoordinate silicon species provide increased Lewis acidic character and, for this reason, are able to coordinate and activate many substrates. Depending on the substituents on the silicon atom, several different reaction pathways have been identified and are summarized in Scheme 10.2.

Upon formation of hypercoordinate silicon species **i** and **ii** by coordination of one or two Lewis bases (LBs) and in the absence of a nucleophilic substrate, the silicon adduct **ii** (more thermodynamically preferred than **i** [24, 25]) can react through a "direct transfer" of one peripheral ligand onto the substrate by an intermolecular process (path A, yellow box). This pathway is accessible only when R = H, and it is the activation method of $HSiCl_3$ in the reductions of imines and carbonyl compounds, wherein the hydride is transferred to the substrate, which is not coordinated to the silicon atom.

Alternatively, an expulsion of a ligand could occur, generating the cationic *penta*-coordinate silicon species **iii** (path B, red box). This species can then evolve following two different pathways. First, **iii** can coordinate the substrate (SB) at the silicon atom and generate a new hexacoordinate cationic species **iv** (path C1, green box). In the absence of an external nucleophile, one of the silicon substituents of species **iv** acts as the nucleophile and attacks the substrate through a six-membered, cationic transition state (path D1). This process is typically observed when tetrachlorosilane, allyltrichlorosilanes, and their derivatives are employed. Examples of this class of reactions are the Lewis base–catalyzed, $SiCl_4$-mediated opening of epoxides as well as the Lewis base–catalyzed allylation of aromatic aldehydes promoted by the presence of allyltrichlorosilane. Second, if species **iv** is formed in the presence of an external nucleophile that is unable to coordinate to the silicon atom, path D2 is available, wherein the substrate is activated to undergo nucleophilic attack from the external nucleophile through an intermolecular pathway. A typical reaction example is the allylation of aldehydes with allyltributyltin mediated by tetrachlorosilane activated by catalytic amounts of a Lewis base.

However, if an acidic nucleophile can coordinate the silicon atom, the cationic species **iii** will traverse to path C2 (blue box), leading to the neutral hexacoordinate species **v**. This newly formed species will evolve to hypercoordinate silicon species **vi** by simple expulsion of a silicon substituent and coordination of the substrate. The hypercoordinate species **vi** contains both the substrate and the

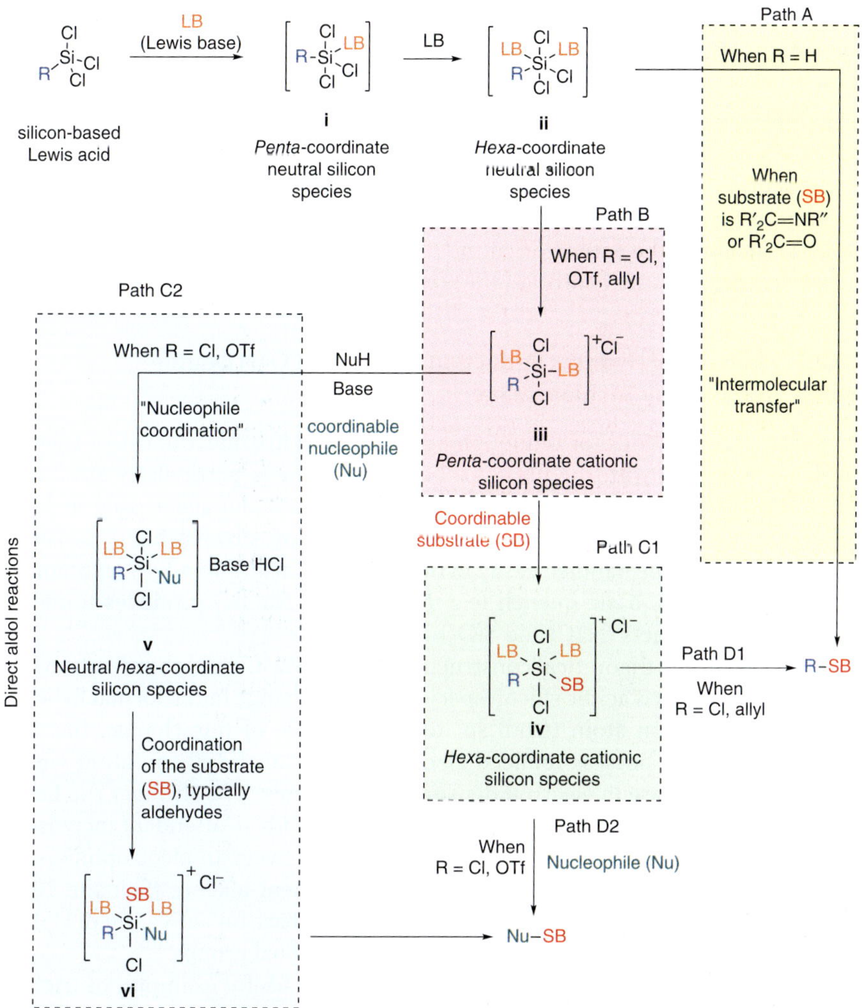

Scheme 10.2 Possible reaction pathways involving hypercoordinate silicon species.

nucleophile coordinated to the silicon atom, thus allowing formation of the new bond in an intramolecular manner under the guidance of the silicon atom. This pathway is formed in direct (double) aldol reaction performed by $SiCl_4$ or $SiCl_3OTf$, in which aldehydes react with activated enolates generated by deprotonation of ketones and esters by tertiary amines.

According to this classification, all of these processes involving hypercoordinate silicon species can be subdivided according to how the substrates and nucleophiles interact together and by the kind of process that will be involved in the formation of the new bond (intramolecular versus intermolecular pathway). All recent developments in these classes are discussed in detail in the following sections, with particular attention to the mechanisms involved.

10.2 Direct Transfer of a Silicon Ligand to a Substrate Not Coordinated to the Silicon Atom

The transfer of a nucleophilic silicon ligand to a substrate that is not coordinated to the silicon atom has received increasing attention in the past 20 years because this reaction mechanism is typical for the trichlorosilane-mediated reduction of carbon–oxygen and carbon–nitrogen double bonds. Coordination by the Lewis base is crucial for the activation of the reducing agent (which is otherwise quite unreactive) and for the control of the stereochemical course of the reduction.

10.2.1 Transfer of Hydride: Reduction of C=O and C=N Double Bonds Promoted by Trichlorosilane

Among all the methods for reduction by a nonmetal hydride, the use of trichlorosilane combined with the presence of a Lewis base is particularly attractive. Trichlorosilane is an inexpensive, stoichiometric reducing agent used in large quantities in the silicon industry. Even though trichlorosilane is highly sensitive to moisture, it can be handled easily in the laboratory using standard techniques. A simple, aqueous, basic quench (e.g. $NaHCO_3$ or NaOH) produces nontoxic inorganic by-products (NaCl and SiO_2).

According to the theoretical constructs described earlier, the coordination of a Lewis base to a Lewis acidic silicon species causes a change in the formal hybridization of the silicon atom (from sp^3 to sp). Because of this change, the new molecular orbitals possess reduced electron density at the silicon atom with a concomitant increase in electron density on the ligands. The stronger the Lewis base, the greater the electron density on the ligands with an attendant increase in the hydridic character of the hydrogen atom for delivery to electrophilic substrates such as carbonyls, imines, or electron deficient alkenes (Scheme 10.3). Moreover, the use of an enantiomerically pure Lewis base for activation of $HSiCl_3$ allows the enantioselective reduction of these functional groups.

In 1996, Kobayashi et al. reported one of the first useful examples of trichlorosilane activation by a Lewis base in the reductive amination of aldehydes and ketones catalyzed by DMF [26]. ^{29}Si NMR analysis allowed the identification of the hexacoordinate silicon species **2**, demonstrating the important role of the Lewis base in silicon activation (Scheme 10.4).

Under these mild reaction conditions, imine derivatives containing functional groups such as esters, nitros, nitriles, and sulfones react only at the C=N double bond, thus demonstrating exquisite selectivity in the reduction process. On the

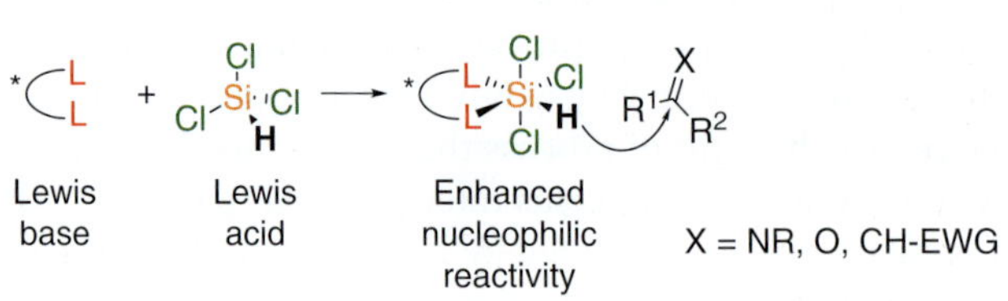

Scheme 10.3 Reduction of C=O, C=N, and C=C double bonds promoted by trichlorosilane.

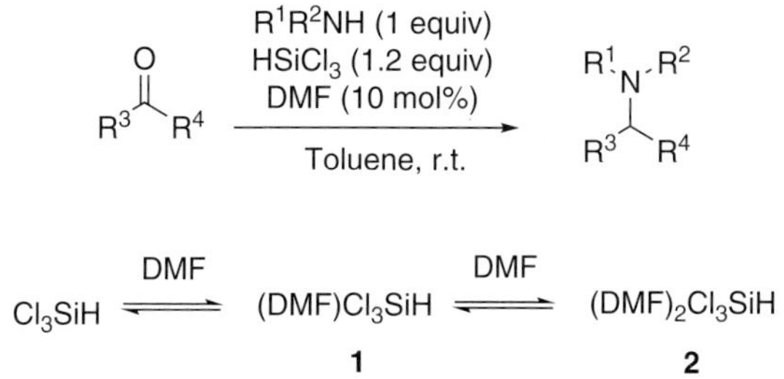

Scheme 10.4 Reductive amination of aldehydes promoted by $HSiCl_3$ and activated by DMF.

basis of this preliminary work, many other investigations followed, demonstrating that secondary amines and heterocycles are also compatible with the reductive elimination conditions. The use of $HSiCl_3$ as the reducing agent activated by a Lewis base was then extended to many other substrates including ketones, *N*-oxides, phosphine oxides, and sulfoxides, all of which resulted in the synthesis of the corresponding reduced products [27–32].

The use of chiral Lewis bases in combination with $HSiCl_3$ represents a logical extension of this process, providing the opportunity to control the stereochemical outcome of the process. This type of approach has been widely explored in recent years, leading to the development of many efficient catalysts.

The first example of an enantioselective, catalytic reduction with $HSiCl_3$ was reported in 1999 by Matsumura and coworkers who introduced *N*-formyl cyclic amines derived from proline (chiral DMF analogs) as activators in the reduction of ketones [33] and imines [34] in the presence of trichlorosilane (Scheme 10.5).

Although the enantioselectivity was modest (up to 77 : 23 er for alcohols and up to 83 : 17 er for amines, see Table 10.1 entries 1 and 2), this work set the stage for the development of many *N*-formyl-derived chiral catalysts, capable of delivering high yields and good enantioselectivities. Modification of the original Matsumura catalysts included the introduction of bulky substituents on the amide function [35, 36] and at the C(5) position of the pyrrolidine (catalyst **4**) [37, 38] as well as the introduction of other stereocenters (catalysts **6** and **7**) [35]. However, in the latter case, these catalysts still exhibited moderate enantioselectivity in the reduction of ketimines with trichlorosilane (Table 10.1, entries 4 and 5). Replacement of the proline core with a pipecolic acid or piperazine moiety led to an increased selectivity in reductive amination (entries 6 and 7) [39–42]; and, in this family of catalysts, the introduction of a second stereocenter provided an improvement in the enantioselectivity of the process (entries 8–12 [42–45]. Replacement of the *N*-formyl group with a different acyl group improves the enantioselectivity but lowers the yield [46, 47], as in the case of the presence of the trifluoroacetyl group (entry 13) [48].

Bidentate catalysts in which two proline units connected with achiral aliphatic linkers (**11**) [35], or with chiral linkers [49, 50] have also been developed, but no improvements in terms of enantioselectivity were observed (entries 14 and 15).

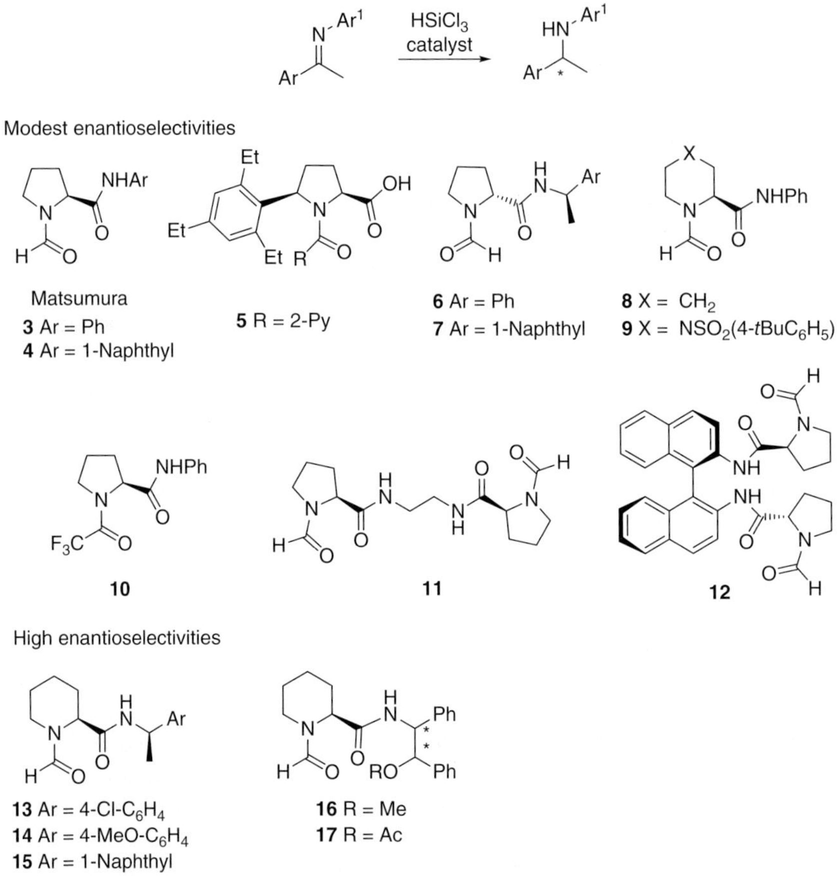

Scheme 10.5 *N*-formyl cyclic amines as catalyst for enantioselective reductions with $HSiCl_3$.

Malkov et al. disclosed an important breakthrough in 2004, when they developed the first highly selective catalyst derived from acyclic amino acids such as valine (Scheme 10.6) [51]. The commercially available *N*-methyl-(*S*)-valine-derived Lewis-basic catalysts (**18**) and (**19**) provide chiral amines with up to 97 : 3 er.

In the following years, Malkov et al. reported the syntheses of different substituted *N*-formyl amino acids (with either aromatic or aliphatic substituents) along with a detailed investigation of the reduction of aromatic imines with trichlorosilane (Scheme 10.7, Table 10.2) [48, 52–56]. This analysis provides important insights that helped clarify the reaction mechanism. These studies revealed that valine provides the highest enantioselectivities and also that the *N*-methyl formamide moiety is essential for high enantioselectivity (entries 1–6). Moreover, the presence of an NH as part of the anilide moiety is also a critical factor as tertiary amides failed to provide stereoselectivity (entries 7 and 9). Arene–arene interactions between the catalyst and the substrate are invoked as stabilizing the transition state during the hydride transfer (entry 8, see Scheme 10.6) [35, 48, 56]

Table 10.1 Enantioselective reduction of imine with trichlorosilane catalyzed by chiral formamides.

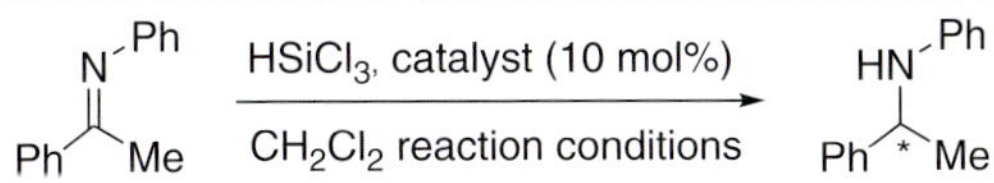

Entry	Catalyst	Temperature (°C)	Yield (%)	er
1	**3**	rt	91	77 : 23
2	**4**	rt	52	83 : 17
3	**5**	rt	92	88 : 12
4	**6**	0	94	86 : 14
5	**7**	−20	95	94 : 6
6	**8**	rt	47	82 : 18
7	**9**	rt	53	72 : 28
8	**13**	0	86	94.5 : 5.5
9	**14**	0	88	95 : 5
10	**15**	0	94	96 : 4
11	**16**	0	97	97 : 3
12	**17**	0	97	96 : 4
13	**10**	rt	57	92 : 8
14	**11**	0	93	88 : 12
15	**12**	0	82	65 : 35

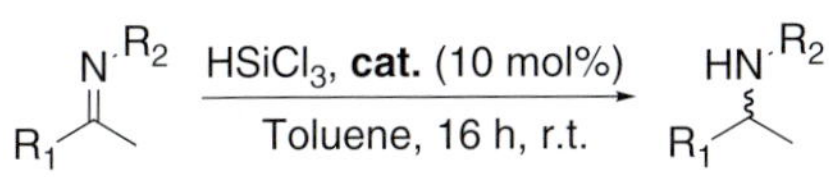

R_1 = Ph, 4-$MeOC_6H_4$, 2-naphth, 2-MeC_6H_4, cC_6H_{11}, 4-$CF_3C_6H_4$, *i*Pr, Ph−CH=CH
R_2 = Ph, 4-$MeOC_6H_4$, 3,5-$tBu_2C_6H_3$, 3-MeC_6H_4, 3,5-$Me_2C_6H_3$

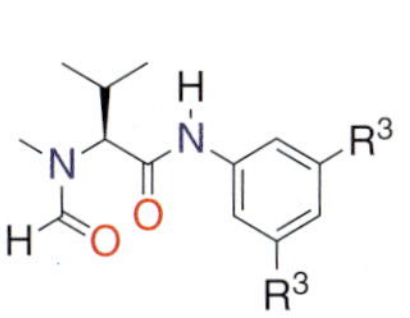

18 R^3 = Me up to 96 : 4 er
19 R^3 = *t*Bu up to 97.5 : 2.5 ee (cat. 5 mol%)

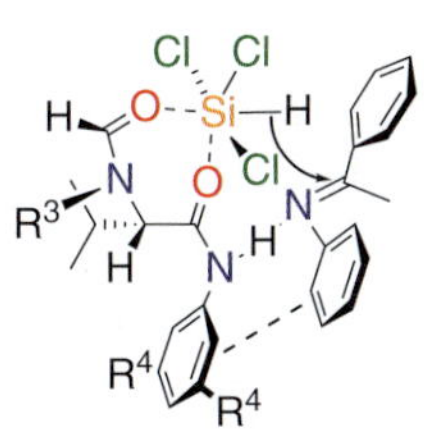

Proposed transition state

Scheme 10.6 *N*-Formyl derivatives of *N*-methyl-(*S*)-valine.

Interestingly, the absolute configuration of the product is primarily determined by the nature of the amino acid side chain; bulkier substituents in the 3 and 5 positions of the catalyst aromatic ring provide high levels of enantioselectivity for

18 R^1 = Me, R^2 = *i*Pr, R^3 = Me
19 R^1 = Me, R^2 = *i*Pr, R^3 = *t*Bu
20 R^1 = Me, R^2 = *t*Bu, R^3 = Me
21 R^1 = Me, R^2 = CH_2Ph, R^3 = Me

22 R^1 = H, R^2 = Me, R^3 = Me, R^4 = Ph
23 R^1 = H, R^2 = Me, R^3 = H, R^4 = Adamanthyl
24 R^1 = Ph, R^2 = Me, R^3 = H, R^4 = 3,5-$(CH_3)_2C_6H_3$

Scheme 10.7 Various substituted *N*-formylated amino acids.

Table 10.2 Enantioselective reduction of imine with trichlorosilane catalyzed by substituted *N*-formylated amino acids.

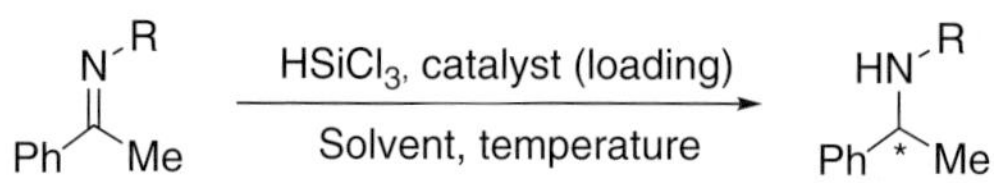

Entry	Catalyst (mol%)	R	Solvent (temperature)	Yield (%)	er
1	**18** (10)	4-$MeOC_6H_4$	toluene (rt)	85	95 : 5
2	**19** (5)	4-$MeOC_6H_4$	toluene (rt)	95	97 : 3
3	**19** (1)	Ph	toluene (rt)	92	96 : 4
4	**19** (5)	*c*-C_6H_{11}	toluene (rt)	86	92 : 8
5	**20** (10)	4-$MeOC_6H_4$	toluene (rt)	95	91 : 9
6	**21** (10)	4-$MeOC_6H_4$	toluene (rt)	84	74 : 26
7	**22** (10)	Ph	$CHCl_3$ (rt)	23	53 : 47
8	**23** (10)	Ph	$CHCl_3$ (0 °C)	89	77 : 33
9	**24** (10)	Ph	toluene (rt)	70	50 : 50

reduction of aromatic and nonaromatic ketimines. In the proposed transition state, the presence of a catalyst/substrate hydrogen bonding interaction and the coordination of the silicon atom by the two carboxy amide groups are suggested to play critical roles in determining the stereoselectivity of the catalyst (Scheme 10.6).

High enantioselectivities are also obtained with catalyst **19** (Sigamide). With only 5 mol% loading (Table 10.2, entries 1 and 2), aliphatic imines react smoothly to form the corresponding reduced products with high yields and good enantioselectivity (entry 3). Moreover, in contrast to all previous catalysts, Sigamide is very effective in the reduction of multifunctional ketimines bearing heterocyclic and aliphatic substrates (entry 4) and also in the reduction of α-chloroimines [56].

Replacing the isopropyl group of Sigamide with a *t*Bu group (catalyst **20**, entry 5) or with a benzyl group (catalyst **21**, entry 6) causes a decrease in terms of yield and enantioselectivity and nearly racemic products are obtained using catalyst **22** bearing a secondary amidic groups or catalyst **24** (entries 7–9).

Owing to the lower loadings of Sigamide required in combination with trichlorosilane to reduce a broad range of substrates at room temperature, it can be considered a viable alternative to the other reduction methods, such as catalytic hydrogenation or reduction with Hantzsch dihydropyridine ester catalyzed by chiral Lewis acids. Low enantioselectivities are observed only with imines derived from pyridines, but this may be ascribed to a competition in the coordination between the nitrogen of the catalysts and the pyridine moiety with trichlorosilane.

A further milestone in this field was the discovery that *N*-picolinoylpyrrolidines **25–28** are also very active compounds in the reduction of imines promoted by trichlorosilane demonstrating that the *N*-formyl group is not always essential (Scheme 10.8).

In 2006, Matsumura and coworkers reported the use of *N*-picolinoyl-(2*S*)-(diphenylhydroxymethyl) pyrrolidine **25** as a Lewis-basic activator in the reduction of aromatic imines to the corresponding amines mediated by trichlorosilane [57, 58]. High enantioselectivity is obtained using catalysts **25–28** (Table 10.3, entries 1–4). However, reactions performed using catalysts containing bulky substituents were less successful (entry 5) [59]. Replacing the proline with a 1,3-dioxane scaffold (**30**, entry 6) leads products with high yields and selectivities [60, 61] as do catalysts **31** and **32** derived from ephedrine (entries 7 and 8) [59, 62, 63]. With catalyst **32**, the pyridine ring, the free hydroxyl group, and the *N*-methyl substitution in the amino alcohol portion represent key structural elements to ensure high stereocontrol. This catalyst provides high chemical and stereochemical efficiency, with only 1 mol% loading

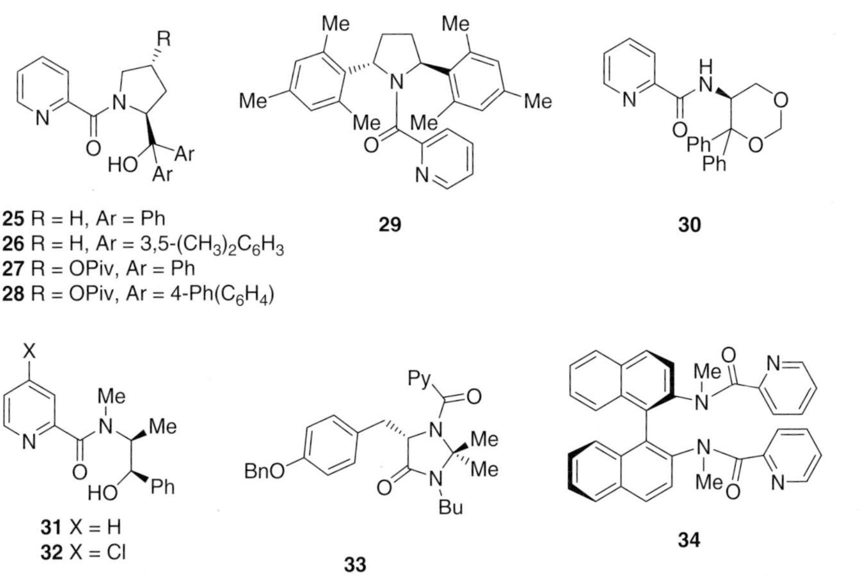

Scheme 10.8 *N*-Picolinoyl derivatives as enantioselective catalysts.

Table 10.3 Enantioselective reduction of imine with trichlorosilane catalyzed by substituted *N*-picolinoyl derivatives.

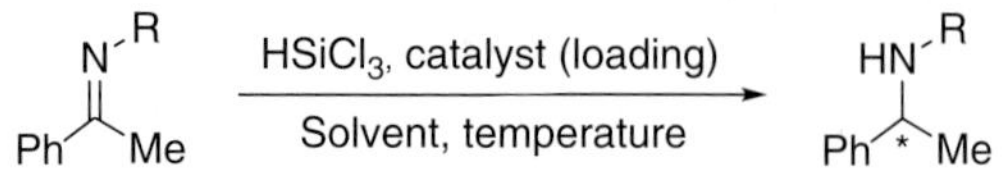

Entry	Catalyst (mol%)	R	Solvent	Temperature (°C)	Yield (%)	er
1	**25** (10)	Ph	CH_2Cl_2	rt	86	84 : 14
2	**26** (10)	Ph	CH_2Cl_2	rt	88	96 : 4
3	**27** (10)	4-$MeOC_6H_3$	CH_2Cl_2	−10	56	97 : 3
4	**28** (10)	4-$MeOC_6H_3$	CH_2Cl_2	rt	99	99 : 1
5	**29** (10)	Ph	CH_2Cl_2	rt	92	88 : 12
6	**30** (10)	4-$MeOC_6H_3$	$ClCH_2CH_2Cl$	−10	96	97 : 3
7	**31** (10)	Ph	CH_2Cl_2	0	98	88 : 12
8	**32** (10)	Ph	$CHCl_3$	−20	98	97.5 : 2.5
9	**32** (1)	Ph	$CHCl_3$	−20	98	94 : 6
10	**32** (10)	Bu	$CHCl_3$	0	98	95.5 : 4.5
11	**33** (10)	4-$MeOC_6H_3$	CH_2Cl_2	0	99	98.5 : 1.5
12	**33** (1)	4-$MeOC_6H_3$	CH_2Cl_2	0	99	98.5 : 1.5
12	**33** (0.1)	4-$MeOC_6H_3$	CH_2Cl_2	0	93	95 : 5
13	**33** (0.1)	4-$MeOC_6H_3$	CH_2Cl_2	−20	80	98.5 : 1.5
14	**33** (10)	Bu	CH_2Cl_2	0	80	98 : 2
14	**33** (10)	Bn	CH_2Cl_2	0	99	93 : 7
15	**33** (10)	Bu	CH_2Cl_2	0	99	92.5 : 7.5

N-Picolinoyl derivatives have also found applications in the reduction of α-, β-, and γ-imino esters [69–72], β-enamino esters [73, 74], benzoxazinones, quinoxalinones [75], and benzodiazepinones with good results [76].

(entry 9). Interestingly, with this catalyst, reduction of *N*-alkyl imines can be performed with high enantioselectivity as well (entry 10) [64]. Benaglia and coworkers have developed catalyst **33**, an *N*-picolinoyl derivative based on an imidazolidinone core [65] and catalyst **34** derived from the (*R*)-*N*,*N*′-(dimethylamino)binaphthalene scaffold. Both catalysts show remarkable activity in the reduction of *N*-aryl- and *N*-alkyl-substituted ketimines (entries 11 and 12) [66, 67]. Products are obtained in higher than 90% yields and enantioselectivity ranging from 95 : 5 to 99 : 1 er. Importantly, catalyst **33** still works efficiently at 1 mol% loading without any erosion of stereoselectivity. Furthermore, at 0.1 mol% loading, chiral amines can be obtained in 92% yield and 95 : 5 er at 0 °C, and in 80% yield and 98.5 : 1.5 er at −20 °C. It is worth mentioning that by performing the reaction with 0.1 mol% of the chiral base, the ACE (asymmetric catalyst efficiency) [68] of **33** is about 400, a value that very favorably compares with organometallic catalysts [27].

In 2006, Malkov et al. reported a novel, chiral Lewis-basic system derived from an oxazoline scaffold (Scheme 10.9). Catalyst **35** has found application in the reduction of aromatic ketones and imines, with trichlorosilane providing the products with modest yield but with higher enantioselectivity, although a 20 mol% catalyst loading is required (Table 10.4, entry 1) [77, 78].

Contemporaneously, Sun and coworkers developed a new class of catalysts featuring a sulfinamide group as the chiral element [79]. Catalyst **36** affords enantioselective ketimine reductions with results comparable to those performed using formamide catalysts when a 20 mol% catalyst loading is employed (entry 2). More efficient sulfinamides **37–39** were reported a few years later in which the presence of a sulfinyl group is combined with a proline [80, 81] or valine [82] scaffold. These new catalysts have been employed in the reduction of aromatic *N*-aryl and *N*-alkyl ketimines in the presence of trichlorosilane to

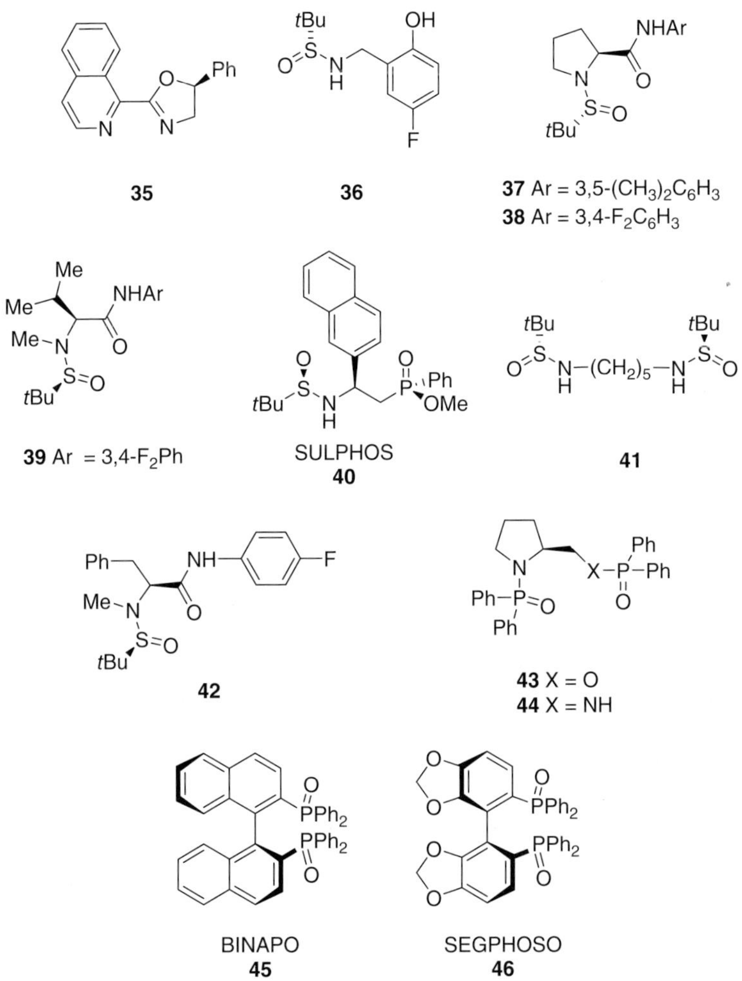

Scheme 10.9 Sulfinamides, phosphoramides, and oxazoline-derived organocatalysts.

Table 10.4 Enantioselective reduction of imine with trichlorosilane catalyzed by sulfinamides, phosphoramides, and oxazoline-derived organocatalysts.

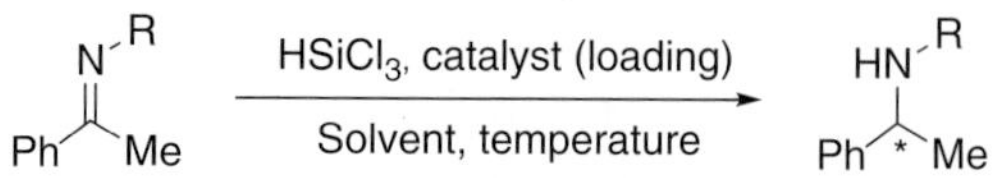

Entry	Catalyst (mol%)	R	Solvent (temperature)	Temperature (°C)	Yield (%)	er
1	**35** (20)	Ph	$CHCl_3$	−20	65	93 : 7
2	**36** (20)	Ph	CH_2Cl_2	−20	92	96 : 4
3	**37** (10)	Bn	toluene	0	98	98 : 2
4	**37** (10)	Allyl	toluene	0	82	96 : 4
5	**38** (10)	Bn	toluene	−20	86	97 : 3
6	**39** (20)	Bn	toluene	−40	95	96.5 : 3.5
7	**40** (10)	4-MeOPh	toluene	−78	96	99.5 : 0.5
8	**40** (20)	Ph	CH_2Cl_2	−20	92	96.5 : 3.5
9[a)]	**41** (10)	Ph	CH_2Cl_2	−20	91	98 : 2

a) Reaction performed in the presence of 30 mol% of 2,6-lutidine.

provide the corresponding amines in higher yields and comparable enantioselectivity (entries 3–6). More recently, the sulfinamide phosphinate **40** with three stereogenic centers was reported by Khiar and coworkers [83]. The naphthyl derivative SULPHOS (**40**) is efficient in asymmetric imine reduction, leading to a wide range of arylmethylamines in high yields with up to 99.5 : 0.5 er with 10 mol% catalyst loading (entries 7 and 8).

Other chiral sulfinimides have been employed in the reduction promoted by trichlorosilane. Catalyst **41** represents the first example of a bissulfinimide in which the two active sites are connected together with a five-methylene tether (entry 9) [84] and sulfonamide **42** based on a phenylalanine scaffold [85] that exhibits high enantioselectivity in the reduction of cyclic imines (Scheme 10.10).

Similarly, bisphosphine oxide derivatives have been investigated in trichlorosilane-mediated C=N reductions (Scheme 10.11). Catalysts **43** and **44**, based on the (*S*)-prolinol scaffold show high chemical efficiency but moderate enantioselectivity in the reduction of β-imino esters [86, 87]. Moreover, in 2009, Nakajima and coworkers demonstrated that β-amido enones can be converted

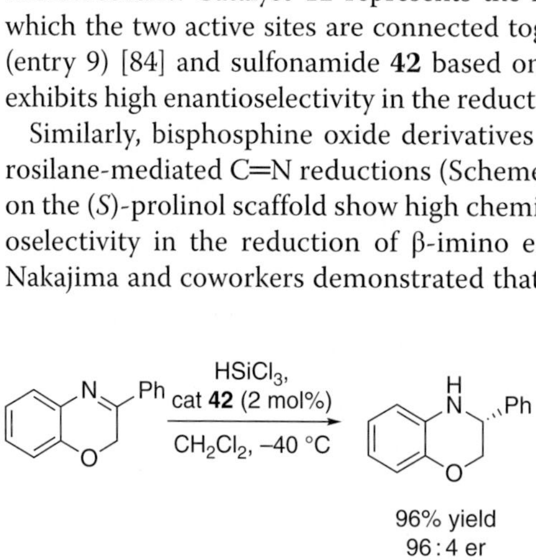

Scheme 10.10 Reduction of cyclic imines promoted by sulfonamide **42**.

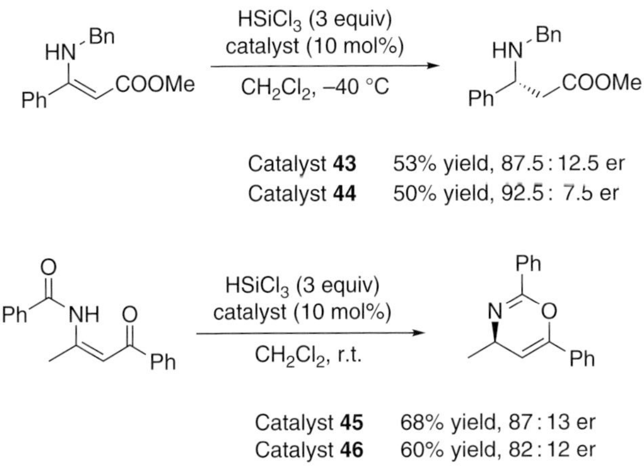

Scheme 10.11 Reduction of β-imino esters and β-amido enones performed with bisphosphinoxides.

in the corresponding enantioenriched 4*H*-1,3-oxazines when BINAPO (**45**) and SEGPHOSO (**46**) are employed as catalysts [88].

Despite the wide range of Lewis bases employed in stereoselective reduction reactions performed by trichlorosilane, removal of the catalysts from the reaction mixture is a common problem. To address this issue, and to facilitate recovery, the catalysts have been anchored onto different polymeric, solid supports (Scheme 10.12). In this way, simple filtration separates the catalyst from the reaction medium. However, because of the insolubility of these supported catalysts, a decrease in the enantioselectivity is generally observed.

Catalyst **18** has been supported onto different resins (Merrifield, extended Merrifield, Wang, TentaGel, Marshall), but catalyst **47** is the best, exhibiting 91 : 9 er when used in chloroform (Table 10.5, entry 1) [89]. Recycling studies show that catalyst **47** can be easily separated by filtration and reused at least five times without any loss of its chemical and stereochemical efficiency. The presence of a fluorous tag (**48**) or a third-generation dendron (**49**) [90] allows operation under homogeneous conditions, wherein comparable results with the untagged counterpart can be obtained. In the first case, catalyst **48** is recovered by filtration through a pad of fluorous silica, whereas catalyst **49** is recovered by precipitation with methanol (entries 2 and 3). Catalyst **18** is supported on gold nanoparticles [91] and on a polymethacrylate backbone [92]. Catalyst **50** shows lower enantioselectivities and reactivity after recycling, but catalyst **51** allows the reduction of imines under homogeneous conditions with good yields and higher enantioselectivity with only 7 mol% loading (entries 4 and 5). Interestingly, catalyst **51** can be removed by precipitation with methanol followed by centrifugation.

More recently, Benaglia and coworkers prepared the polystyrene-supported cinchona-based picolinamide **52** that promotes the C=N reduction of imines in high enantioselectivity, typically higher than 92 : 8 er and up to 95 : 5 er (entry 6). The applicability of this catalyst has been further extended in the development of

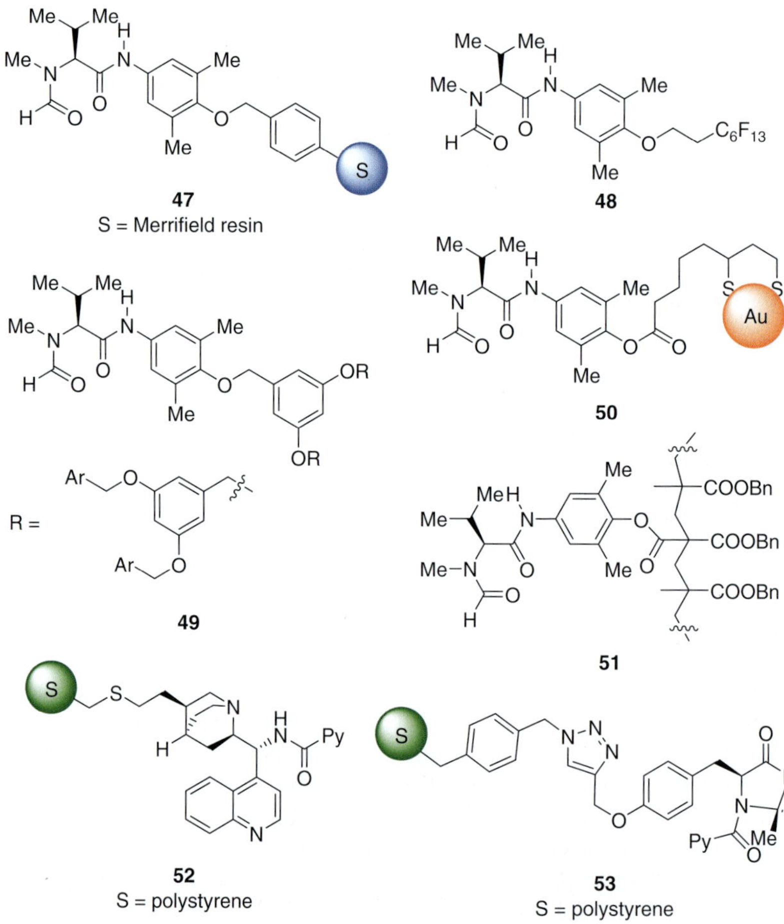

Scheme 10.12 Supported catalysts for the enantioselective reduction of imines with $HSiCl_3$.

a reactor containing **52** for the stereoselective reduction of imines with trichlorosilane under continuous flow conditions, albeit with a significant decrease in enantioselectivity (Scheme 10.13) [93]. Better results are obtained with the polystyrene-supported catalyst **53**. Even with a loading as low as 1 mol%, the reduction of imines is accomplished with stereoselectivities reaching between 95 : 5 and 98.5 : 1.5 er. The general scope of the method and the good recyclability of the immobilized catalyst have been demonstrated [94]. Moreover, the polystyrene-anchored chiral catalysts have been used to prepare packed bed reactors for the continuous-flow synthesis of chiral amines which can be obtained in excellent yields and enantioselectivities (Scheme 10.13).

In recent years, other types of Lewis base–catalyzed, $HSiCl_3$-mediated reductions have been reported. In 2016, Wang and coworkers developed a straightforward method for the synthesis of 1,1-disubstituted hydrazines starting from ketones or aldehydes and phenylhdrazines under catalysis with

Table 10.5 Enantioselective reduction of imine with trichlorosilane catalyzed by supported catalysts.

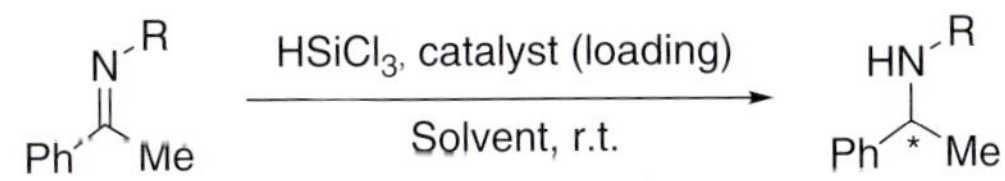

Entry	Catalyst (mol%)	R	Solvent	Yield (%)	er
1	**47** (25)	4-MeOPh	$CHCl_3$	81	91 : 9
2	**48** (10)	4-MeOPh	Toluene	90	95.5 : 4.5
3	**49** (10)	4-MeOPh	Toluene	89	94.5 : 4.5
4	**50** (20)	4-MeOPh	Toluene	90	92 : 8
5	**51** (7)	Ph	Toluene	86	93 : 7
6[a)]	**52** (20)	4-MeOPh	CH_2Cl_2	98	95 : 5
7	**53** (20)	4-MeOPh	CH_2Cl_2	94	98.5 : 1.5

a) Reaction run at 18 °C.

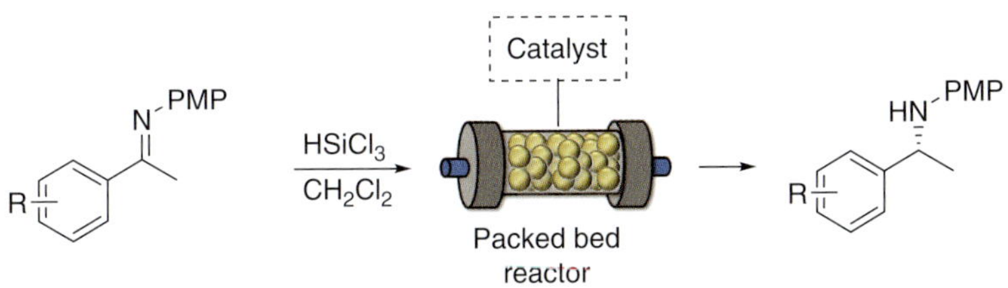

Catalyst 52 @ 0.02 ml/min res. time: 25 min   R = 4-NO_2 98% yield, 73:27 er
Catalyst 53 @ 0.4 ml/h, res. time: 30 min   R = H   92% yield, 92.5:7.5 er

Scheme 10.13 Continuous-flow reductions performed with supported catalysts.

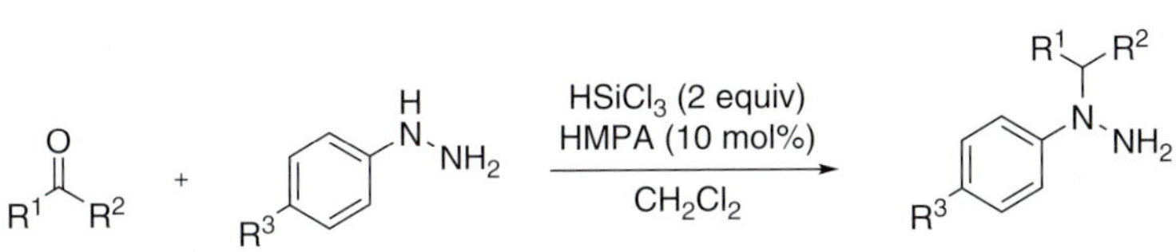

R^1 = aryl, alkl,
R^2 = H, aryl, alkyl
R^3 = H, F, Cl, NO_2

25 examples
72–98% yield

Scheme 10.14 Synthesis of 1,1-disubstituted hydrazines.

hexamethylphosphoramide (HMPA) (Scheme 10.14) [95]. Products are obtained in good to high yields independent of the electronic or steric characteristics of the substrates.

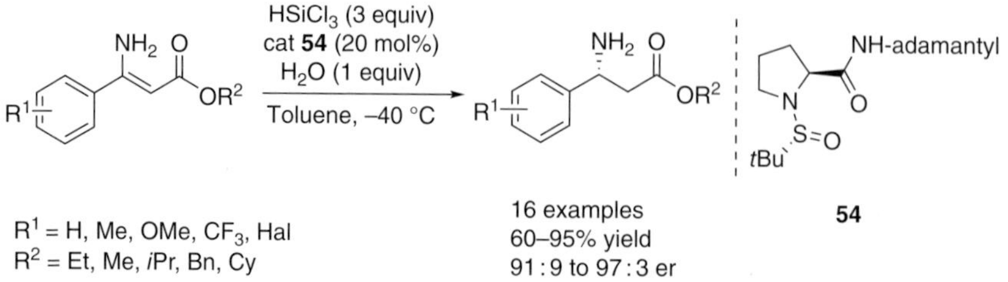

Scheme 10.15 Enantioselective reduction of unprotected β-enamino esters.

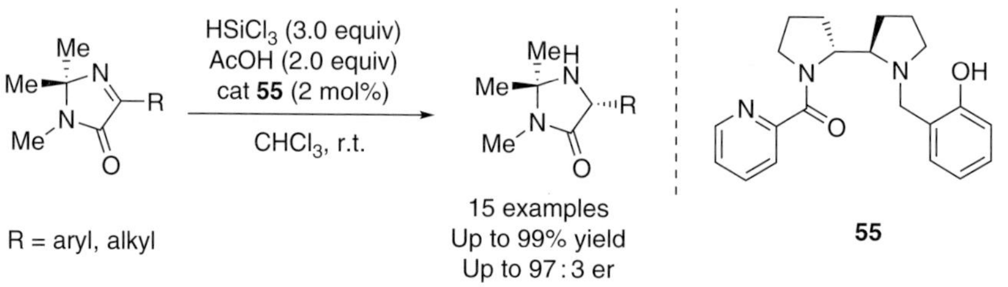

Scheme 10.16 Enantioselective reduction of imidazolinones.

Contemporaneously, these authors reported an unprecedented catalytic, enantioselective reduction of *N*-unsubstituted β-enamino esters using *N-tert*-butylsulfinyl-L-proline-derived amide **54**. A broad range of β-aryl-substituted β-amino esters can be synthetized in high yields and enantioselectivities [96]. This example illustrates that an *N*-protecting group is not necessary for the reduction of both electron-rich and electron-deficient β-enamino esters (Scheme 10.15).

In 2017, Kirsch and coworkers showed that imidazolinones can be reduced by trichlorosilane in a highly enantioselective manner when in combination with catalyst **55** [97]. Under mild reaction conditions and with 2 mol% catalyst loading, a broad range of chiral imidazolidinones bearing various structural motifs including sterically demanding substituents, alkyls and aryls can be obtained in high yields and excellent enantiomeric ratios (Scheme 10.16).

In concert with the development of enantioselective reduction processes, computational studies that help identify the factors responsible for high enantioselectivities have also been disclosed. High-level calculations (B3PW91/cc-pVDZ DFT theory) have been employed to model the imine reduction depicted in Scheme 10.17. In the absence of a Lewis base, the reduction process shows a large activation enthalpy $\Delta H^{\ddagger} = 40.0$ kcal/mol, and $\Delta G^{\ddagger} = 51.7$ kcal/mol; too high in energy to occur, as is experimentally observed. However, in the presence of an *N*-formyl prolinamide catalyst, chelation of $HSiCl_3$ leads to the formation of a hexacoordinate silicon species which lowers the energy of transition state **56** significantly ($\Delta H^{\ddagger} = 19.1$ kcal/mol, and $\Delta G^{\ddagger} = 47.1$ kcal/mol) [35]. This calculation supports the supposition that the imine substrate is not directly coordinated to

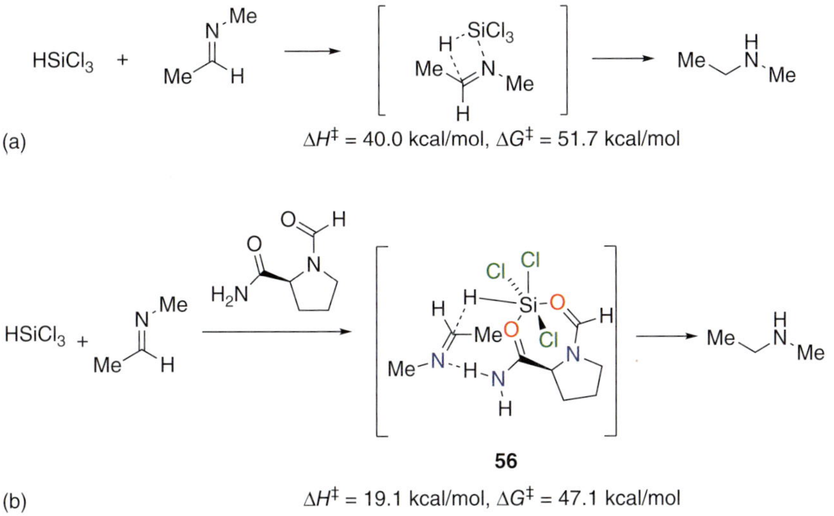

Scheme 10.17 DFT calculations of the uncatalyzed and catalyzed $HSiCl_3$ imine reduction. (a) Without Lewis base activation. (b) With Lewis base activation.

the silicon atom, but it is associated through a hydrogen bond to the NH group of the catalyst.

10.2.2 Reduction of Nitroaromatic Compounds by Trichlorosilane

Reduction of nitro groups represents one of the most straightforward entries to aliphatic and aromatic amines [98]. Among the various available methods, reduction by hydrogenation and transfer hydrogenation are the most commonly employed. However, these protocols sometimes lack functional-group compatibility, often requiring the presence of expensive transition-metal catalysts and harsh reaction conditions. Recently, Benaglia and coworkers reported that trichlorosilane, after activation with tertiary amines, is able to reduce aromatic nitro compounds to the corresponding anilines in high yields [99, 100]. Contrary to the interaction of a Lewis base with $HSiCl_3$ that leads to the formation of a more electrophilic hypercoordinate species, in the presence of a Brønsted base such as Et_3N or *i*$PrNEt_2$, a more nucleophilic silicon species is generated, which is competent for reduction of nitro compounds under mild reaction conditions (Scheme 10.18).

In most cases, complete conversion of the nitro derivative into the corresponding amine is observed. Nitriles, amides, ketones, alcohols, and carboxylic acid moieties are compatible, as are allylic and benzylic protecting groups. Moreover, halogenated nitro compounds can be converted to amines without any detectable traces of dehalogenated products [99]. Owing to the very mild reaction conditions, this method has found application in the total synthesis of a product of active pharmaceutical ingredient (API) as Aliskiren, in which an enantiopure aliphatic nitro compound with four stereocenters (one of which directly bears the

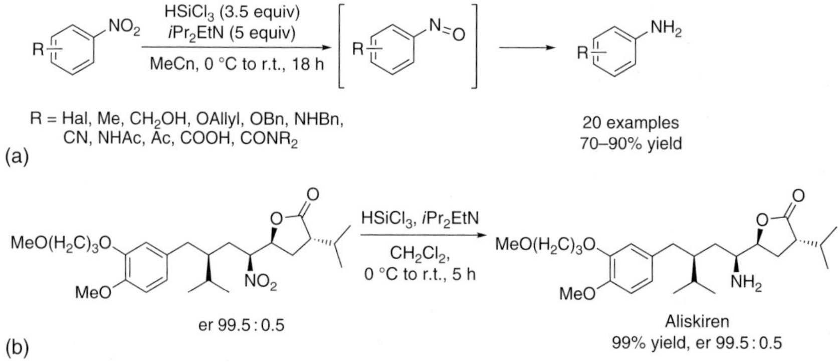

Scheme 10.18 Trichlorosilane-mediated reduction of nitro compounds to anilines. (a) NO_2 reduction promoted by $HSiCl_3$. (b) NO_2 reduction step in the total synthesis of Aliskiren.

nitro group) is reduced to the corresponding aliphatic amine in 99% yield without eroding the stereochemical integrity of the four stereogenic centers [101].

Mechanistic studies have provided some insight to the interaction of trichlorosilane with a Brønsted base. This type of interaction was already described in 1968 [102–106]. Initially, it was hypothesized that the combination of $HSiCl_3$ with a base could lead to the formation of a R_3NH^+/Cl_3Si^- ion pair, as illustrated in Scheme 10.19. However, Karsch proposed that this equilibrium may further evolve toward the formation of a dichlorosilylene species ($SiCl_2$) or in one of its stabilized forms (Scheme 10.19) [107].

To explain why $HSiCl_3$ reacts differently with a Brønsted or a Lewis base, the hard and soft bases and acids (HSAB) principle should be considered. Both the silicon and the hydrogen atom of $HSiCl_3$ are acidic; but since Lewis bases involved in $HSiCl_3$ activation are not strong Brønsted bases, they will preferentially interact with the soft silicon atom. On the other hand, a Brønsted base such as an aliphatic tertiary amine will prefer to interact with the harder acidic site in the molecule; the hydrogen atom. As result, the H atom in $HSiCl_3$ seems to be a proton rather than a hydride, unless activated by Lewis bases.

^{1}H, ^{13}C, and ^{29}Si NMR spectroscopic and computational analysis revealed that the reduction from a nitro to an amino group occurs in three steps [108]. At first, the $Me_3N–SiCl_2$ stabilized species **57** is generated (−20.6 kcal/mol). Next, the interaction with the nitro aromatic compound allows the formation of the corresponding nitroso derivate **58** (the addition of $Me_3N–SiCl_2$ to nitromethane occurs with an energy barrier of +15.8 kcal/mol). The transition state involves the addition of the negatively charged Si atom to the nitrogen of the nitro group.

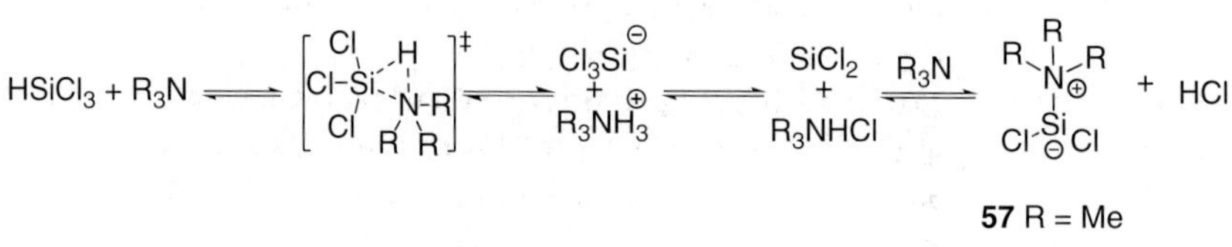

57 R = Me

Scheme 10.19 Interaction between $HSiCl_3$ and a tertiary amine.

Once the Si—N bond is formed, the intrinsic reaction coordinate (IRC) directly leads to the formation of a transient three-membered cycle involving the Si, N, and O atoms, which through electron rearrangement forms Me_3N–Si(O)Cl_2 and nitrosomethane **58**. This step represents the rate-determining step of the process [108]. The second step involves a further reduction of the nitroso group (performed in analogy with the first step) leading to the formation of compound **59**, in which a new molecule of Me_3N–$SiCl_2$ acts as a nucleophile toward the nitroso *N* atom. This process is faster than the previous addition with an energy barrier of −12.9 kcal/mol (when considering the dissociated reagents as a zero energy point). In the third and last step of the reduction, Me_3N–$SiCl_2$ attacks compound **59** by nucleophilic ring opening, breaking the N—O bond and leading to the formation of the silylated amine **60**, which upon workup generates the final amine product. The process has an energy barrier of −9.3 kcal/mol (Scheme 10.20) [108].

10.3 Direct Transfer of a Silicon Substituent to the Silicon-Coordinated Substrate

According to the general reaction mechanism illustrated in Scheme 10.2, the hexacoordinate hypervalent silicon species **ii** generated by interaction of the silicon Lewis acid with a Lewis base can express its Lewis acidity by coordinating a suitable substrate after the expulsion of a silicon ligand. The substrate bound to the silicon atom in the newly formed hypercoordinate silicon species **iv** is then activated toward a nucleophilic attack by an external nucleophile or by one of the silicon substituents (if an external nucleophile is absent).

Examples of silicon Lewis acid–mediated transformations performed in the absence of an external nucleophile are the opening of epoxides by chlorotrimethylsilane and tetrachlorosilane as well as the allylation of imines and aldehydes with allylic trichlorosilanes.

10.3.1 Opening of Epoxides

The facile ring opening of epoxides makes them extremely versatile intermediates, as demonstrated by their significant use in organic synthesis in the generation of 1,2-difunctionalized compounds [109–111]. One of the most common methods to activate epoxides toward nucleophilic opening is the polarization of the carbon–oxygen bond by the interaction of the electron pairs of the oxygen atom with a Lewis acid. Among the myriad methods available for the ring opening, Lewis base–catalyzed, Lewis acid–mediated reactions offer straightforward opportunities, especially for stereoselective applications.

10.3.1.1 Lewis Base–Catalyzed Epoxide Opening with Chlorotrimethylsilane

In one of the first examples, Andrews et al. found that epoxide ring opening with chlorotrimethylsilane could be very efficiently catalyzed by triphenylphosphine, leading to the formation of trimethylsilylated chlorohydrins (Scheme 10.21a) [112]. The insertion of TMSCl proceeds to completion in 15 minutes at −50 °C,

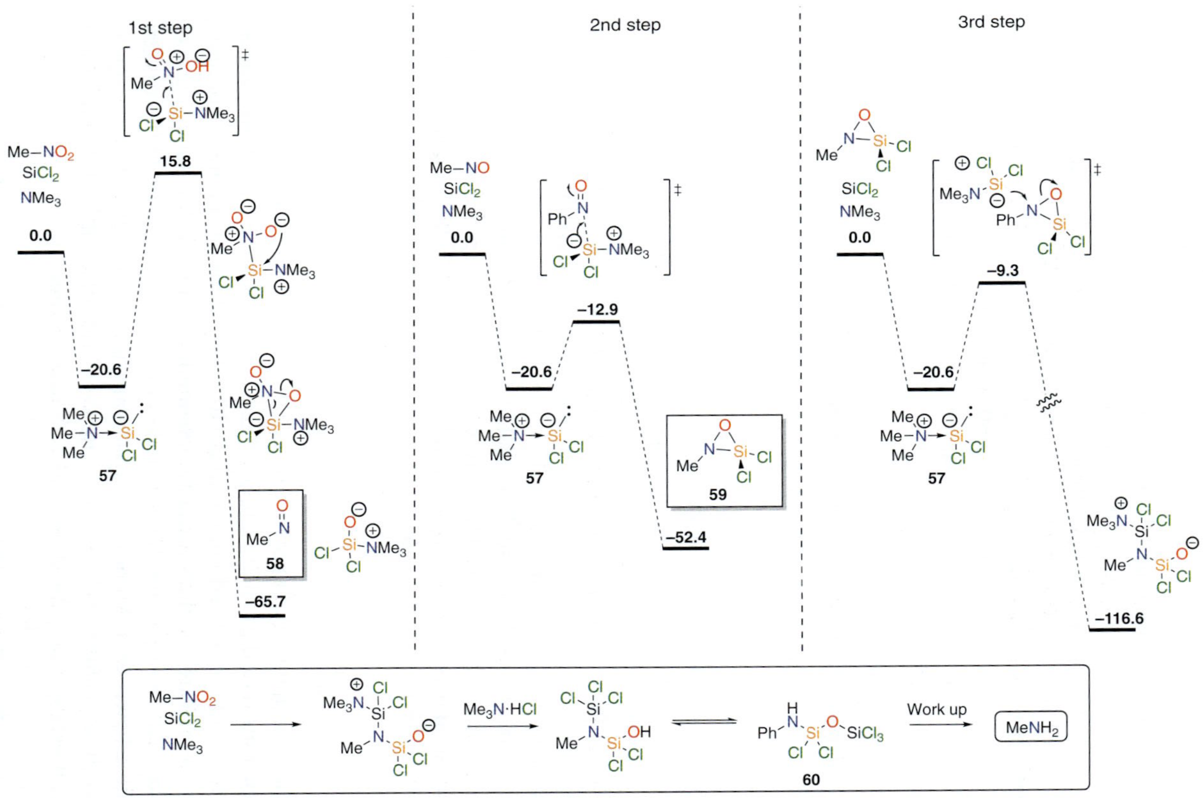

Scheme 10.20 Gibbs free energy profile for $HSiCl_3$-mediated reduction of $ArNO_2$ to $ArNH_2$. For calculation purposes, Ar group was replaced with Me group. MP2/6-311+G(d,p)//M06-2X/6-311++G(3df,3pd)[PCM] level of theory. The energies are in kcal/mol.

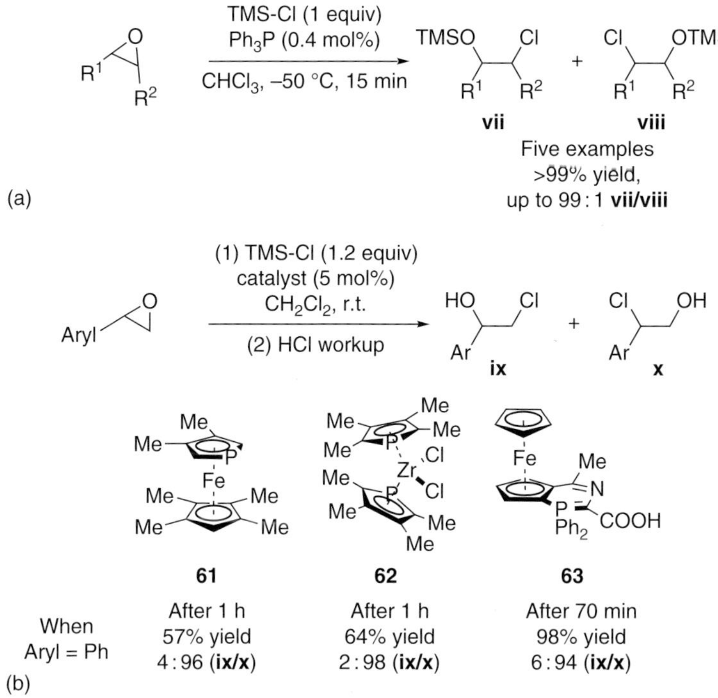

Scheme 10.21 Lewis base–catalyzed epoxide opening with chlorotrimethylsilane. (a) Triphenylphosphine-catalyzed opening of epoxides. (b) Metallocene-catalyzed opening of epoxides.

to produce the corresponding silylated chlorohydrins with high selectivity (up to 99) for the primary versus secondary chlorohydrin. Phosphametallocene complexes **61** and **62** containing Fe and Zr are also good activators for the selective ring opening of unsymmetrical and *meso*-epoxides (Scheme 10.21b) [113–116]. Catalyst **61** is more reactive than catalyst **62** for opening of *meso*-epoxides, but an opposite trend is observed if the reaction is performed with monosubstituted epoxides. Moreover, the selectivity toward primary (**vii-a**) and secondary (**vii-b**) chlorohydrins is highly substrate and catalyst dependent (generally favoring primary derivatives).

10.3.1.2 Lewis Base–Catalyzed Epoxide Opening with Silicon Tetrachloride

The first catalytic opening of *meso*-epoxides mediated by tetrachlorosilane involved the use of HMPA and was reported by Denmark et al. in 1998 (Scheme 10.22) [117]. This group reported using monophosphoramide **65** derived from (*R*)-*N*,*N*′-2,2′-dimethyl-1,1′-binaphthyldiamine.

Different *meso*-epoxides were subjected to this transformation, and it was found that higher level of enantioselectivities are usually obtained using acyclic substrates rather than cyclic ones (Table 10.6, entries 1 and 2). In 2001, Fu and coworkers showed that chiral *N*-oxides such as **65** are able to promote the opening of *meso*-epoxides with good enantioselectivities in the presence of $SiCl_4$ and $i$$Pr_2NEt$ [118], leading to the development of a large number of mono and bis

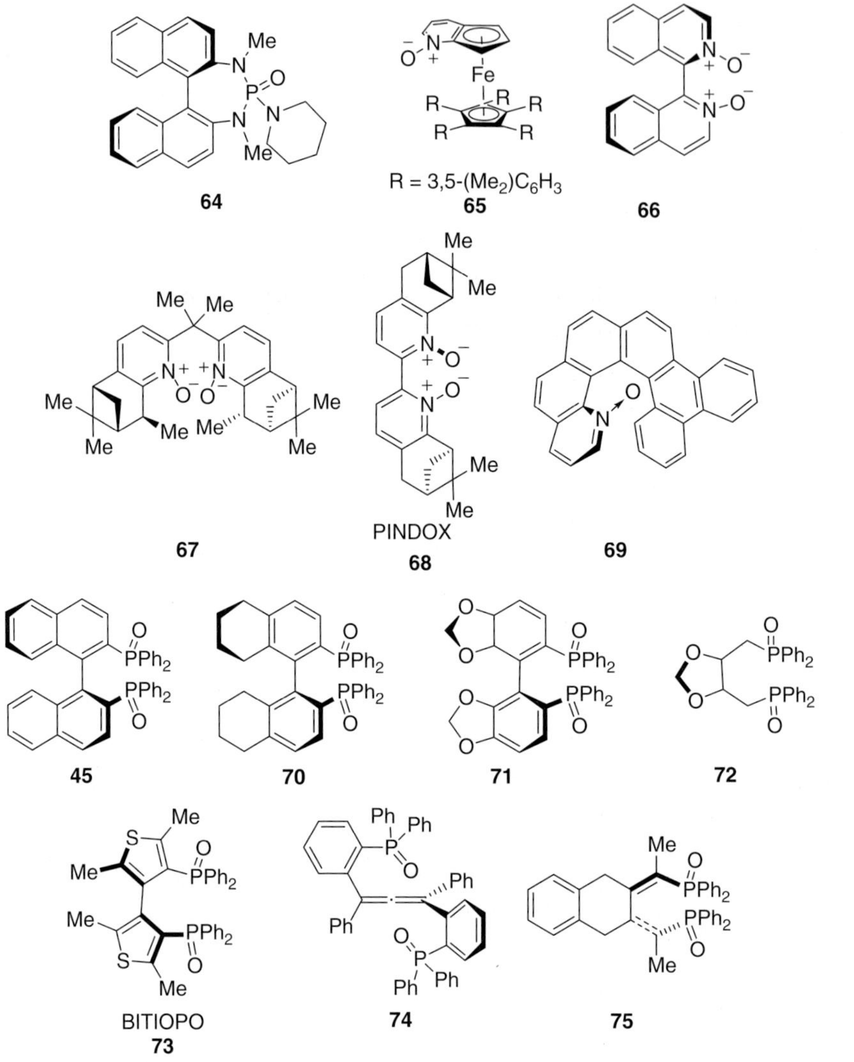

Scheme 10.22 Lewis base catalysts for opening of *meso*-epoxides.

N-oxide catalysts (66–69, entries 3–8). Nakajima et al. [119] reported that bis *N*-oxide **66** was able to efficiently promote the opening of acyclic epoxides leading to chlorohydrins with high enantioselectivities, but it was less effective in cyclic epoxides (entries 6 and 7).

A few years later, terpene-derived *N*-oxides were developed, such as catalyst **67** [120], and PINDOX *N*-oxide **68**, which are very efficient catalysts for the opening of *meso*-epoxides in macrocycles [121]. Takenaka et al. developed the azahelicene *N*-oxide **69** (characterized by the presence of a large aromatic surface) that provides chlorohydrins with high enantiomeric ratios (entry 10) [122].

Table 10.6 Opening of *meso*-epoxides using chiral Lewis bases.

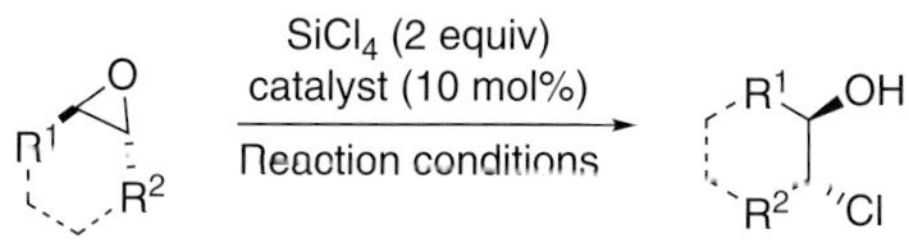

Entry	Catalyst	R^1, R^2	Reaction conditions	Yield (%)	er
1	HMPA	Ph, Ph	CH_2Cl_2, −78 °C, 120 min	96	—
2	HMPA	cyclohexyl	CH_2Cl_2, −78 °C, 20 min	89	—
3	**64**	Ph, Ph	CH_2Cl_2, −78 °C, 180 min	94	93.5 : 6.5
4	**64**	cyclohexyl	CH_2Cl_2, −78 °C, 20 min	90	75.8 : 24.2
5[a)]	**65**	Ph, Ph	CH_2Cl_2, rt, 120 min	88	97 : 3
6[a)]	**66**	Ph, Ph	CH_2Cl_2, −78 °C, 6 h	95	95 : 5
7[a)]	**66**	cyclohexyl	CH_2Cl_2, −78 °C, 6 h	83	50 : 50
8[a)]	**67**	Ph, Ph	CH_2Cl_2, −78 °C, 18 h	100	85 : 15
9[a)]	**68**	cyclodocedyl	CH_2Cl_2, −90 °C, 48 h	67	98 : 2
10[a)]	**69**	Ph, Ph	CH_2Cl_2, −78 °C, 48 h	77	97 : 3
11[a)]	**45**	Ph, Ph	CH_2Cl_2, −78 °C 4 h	94	95 : 5
12[a)]	**70**	Ph, Ph	CH_2Cl_2, −78 °C, 2 h	92	93 : 7
13[a)]	**71**	Ph, Ph	CH_2Cl_2, −78 °C, 2 h	93	94 : 6
14[a)]	**72**	Ph, Ph	CH_2Cl_2, −78 °C, 2 h	92	58.5 : 41.5
15[a)]	**73**	Ph, Ph	CH_2Cl_2, −78 °C, 12 h	99	90.5 : 9.5
16[a),b)]	**74**	Ph, Ph,	CH_2Cl_2, −78 °C, 12 h	97	97 : 3
17[a)]	**75**	Ph, Ph	CH_2Cl_2, −78 °C, 1 h	93	92 : 8

a) 1.5 equiv of iPr$_2$NEt was used.
b) 0.1 mol% as catalyst loading.

Later, Nakajima and coworkers showed that chiral, phosphine oxide catalyst **45** ((*S*)-BINAPO) leads to the formation of chlorohydrins with high enantioselectivity (up to 95 : 5 er, entry 11). Here again, iPr$_2$NEt is extremely important for high enantioselectivities likely to scavenge HCl, which induces a racemic background reaction [123, 124]. Following this approach, phosphine oxides **70–75** were developed to improve the stereoselectivity of the process. Catalysts **70** and **71** show activity comparable to BINAPO with cyclic and acyclic substrates, whereas (*R*,*R*)-DIOP-derived compound **72** is unselective in reactions involving acyclic *meso*-epoxides (entries 11–14) [123, 124]. The more electron-rich tetramethyl bis-(diphenylphosphino)-bithiophene oxide **73** (BITIOPO) leads to the formation of products in quantitative yield but in modest enantioselectivity (entry 15) [125]. Chiral allenic bisphosphine oxide **74** is an effective promoter of *meso*-epoxide opening reactions, showing high reactivity even when employed at 0.1 mol% loading [126]. In this case, chlorohydrins can be obtained in up to 97 : 3 er in almost quantitative yield (entry 16). Catalyst **75**, although presenting a particular conformationally rigid C_2-symmetric diene, shows only modest selectivity (entry 17) [127].

The development of new catalysts with increased efficiency in the epoxide opening was guided by mechanistic studies. Preliminary investigations with monofunctionalized *N*-oxides were performed by Fu and coworkers, who demonstrated that (i) the Lewis base–catalyzed epoxide opening mediated by $SiCl_4$ was zero order in $SiCl_4$ and in *i*Pr_2NEt; and (ii) a positive nonlinear effect was observed between the er of the catalyst and the er of the product [118]. By comparing the activity of pyridine *N*-oxides with one or two binding sites, Nakajima showed that the use of bis-*N*-oxides leads to the formation of products in higher yield and enantiomeric ratio compared to the mono-*N*-oxides.

This behavior was tentatively rationalized by hypothesizing the formation of a *penta*-coordinated silicon intermediate for mono-N-oxides and a more reactive, hexacoordinate silicon species, for bis-*N*-oxides [119]. However, a detailed study on the mechanism of Lewis base-catalyzed epoxide opening with $SiCl_4$ clarified in detail the role of reagents, the stoichiometry, the kinetics, and the mode of activation [117]. Among all the chlorosilanes species investigated, only silicon tetrachloride affords the product with a high level of stereocontrol; and to determine the stoichiometry of $SiCl_4$ required in the epoxide opening, conversions for the chlorination of *cis*-stilbene oxide catalyzed by HMPA at different loadings of $SiCl_4$ were evaluated. A linear correlation between conversion and the loading of the tetrachlorosilane was found, in accordance with the hypothesis that only one chloride per $SiCl_4$ is active under the reaction conditions and that the catalytically active species responsible for the ring opening does not change over the course of the reaction. Moreover, the enantioselectivity is independent of the catalyst loadings ranging from 100 to 4 mol%, but decreases when a 2 mol% loading of catalyst **64** is employed [117]. This behavior suggests that a single, highly stereoselective pathway is active under higher loading (favoring the formation of a 2 : 1 complex between the Lewis base and $SiCl_4$ – Scheme 10.23, cycle B) but that a less selective pathway operates at lower catalyst loadings, wherein only a single Lewis base·$SiCl_4$ complex is formed (Scheme 10.23, cycle A) [117]. To address this question, the opening of *cis*-stilbene oxide was studied using rapid injection NMR analysis [117, 128] (RINMR) in the presence of 1.1 equiv of $SiCl_4$ and HMPA (as the Lewis base) in deuterated dichloromethane at −78 °C [129]. The overall first-order behavior of the reaction was confirmed and the derived rate expression reveals a second-order dependence on HMPA and a zeroth-order dependence on $SiCl_4$ concentrations. This rate law supports the initial proposal that a 2 : 1 complex between the phosphoramide and $SiCl_4$ is the catalyst resting state and that the Lewis base is saturated with $SiCl_4$. Although kinetic studies performed with catalyst **64** encountered challenges of reproducibility [117], with the inability to determine the exact reaction order for this chiral Lewis base, the observed weak negative nonlinear effect supports the hypothesis that more than one catalyst molecule may be bound to $SiCl_4$ in the stereochemistry-determining transition structure. In addition, even if the resting state is an LB_2SiCl_4 species, it is not possible to conclude that this is also the kinetically competent species. An alternative catalytic cycle involving two silicon species, each with one phosphoramide bound, has been also proposed (Scheme 10.23, cycle C) [117] in which a cationic silicon species (**xi**) serves to activate the epoxide, while a second silicate

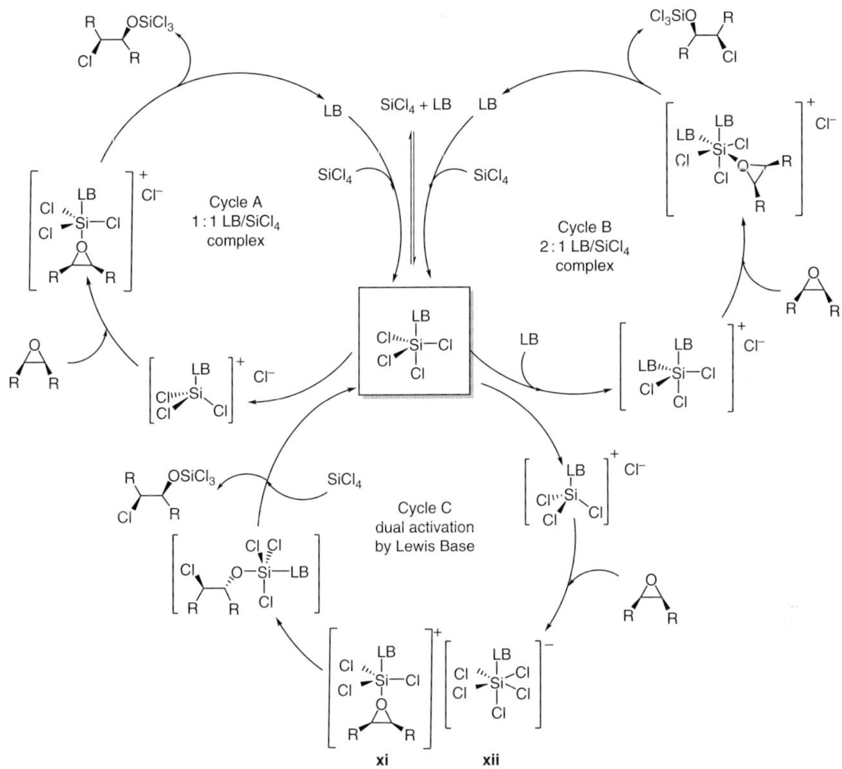

Scheme 10.23 Proposed catalytic cycles for epoxide opening in the presence of $SiCl_4$.

species (**xii**) activates a chloride toward nucleophilic attack. In this scenario, the chiral phosphoramide is intimately involved with the nucleophile, and, in principle, it is reasonable to assume that tethered Lewis-basic catalysts would increase the reaction rate and perhaps the enantioselectivity. However, studies conducted with bisphosphoramides failed to improve the enantioselectivity of the opening epoxides [117].

Unfortunately, these kinetic studies are unable to differentiate between the two mechanistic scenarios, and further mechanistic investigations are needed to identify the operative reaction pathway.

10.3.2 Allylation of Substrates Using Allylic Trichlorosilanes

10.3.2.1 Allylation of C=N Bonds

Enantiomerically enriched homoallylic amines are useful building blocks that have found applications in many synthetic processes since they can easily be converted into different functional groups. Despite their utility, only a few examples of enantioselective allylation of activated imines are known. In 1999, Kobayashi and Hirabayashi reported a synthesis of homoallylic amines by the addition of allyltrichlorosilane to benzoyl hydrazones (performed in DMF or HMPA as solvents) to afford the corresponding homoallylic benzoyl hydrazines in good to high yields (Table 10.7, entries 1–3) [130]. It was also shown that in

Table 10.7 Allylation of *N*-benzoylhydrazones with allyltrichlorosilane.

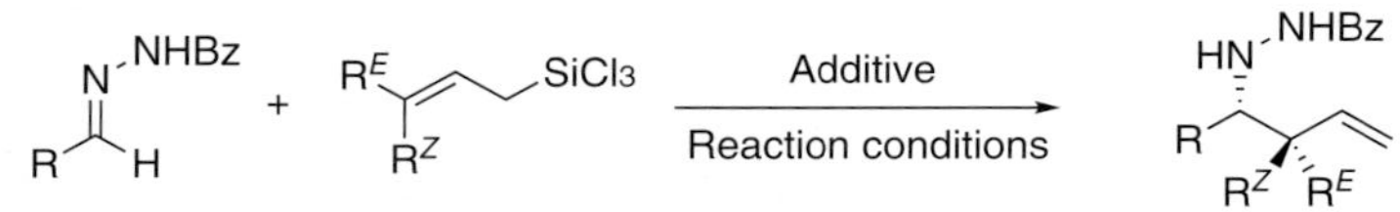

Entry	Additive (eq)	R	R^E, R^Z	Reaction conditions	Yield (%)	*E*:*Z*	er (major)
1	—	Ph	H, Me	DMF, 18 h, 0 °C	79	1 : 99	—
2	—	Ph	Me, H	DMF, 18 h, 0 °C	59	78 : 22	
3	—	Ph	H, H	HMPA, 2 h, 0 °C	91	—	—
4	DMSO (3)	Ph	H, H	CH_2Cl_2, 1 h, −78 °C	98		
5	**76** (1)	$PhCH_2CH_2$	H, H	CH_2Cl_2, 12 h, −78 °C	46	—	—
6[a)]	**77** (3)	$PhCH_2CH_2$	Me, H	CH_2Cl_2, 4 h, −78 °C	58	99 : 1	94.5 : 5.5
7[a)]	**77** (3)	$PhCH_2CH_2$	H, Me	CH_2Cl_2, 4 h, −78 °C	60	1 : 99	95.5 : 4.5
8[a)]	**78** (3)	*i*Pr	H, H	CH_2Cl_2, 1 h, −78 °C	>95	—	91 : 9
9[a)]	79•O (1.5)	Ph	H, H	CH_2Cl_2, 1 h, −78 °C	89	—	88 : 12
10[a)]	**79** (1.5)	Ph	H, H	CH_2Cl_2, 1 h, −78 °C	73	—	79 : 21
11[a)]	**80** (1.5)	*i*Pr	H, H	CH_2Cl_2, 20 min −78 °C	81	—	91 : 9
12[a)]	**81** (1)	*i*Pr	H, H	CH_2Cl_2, 72 h, −78 °C	70	—	5 : 95
13	**45** (2)	COOEt	H, H	CH_2Cl_2, 12 h, −78 °C	91	—	99 : 1
14[b)]	**82** (0.1)	COOEt	H, H	THF/H_2O, 9 : 1, 55 h, 0 °C	84	—	90.5 : 9.5

a) Reaction performed in the presence of 2-methyl-2-butene.
b) Reaction performed in the presence of 20 mol% of ZnF_2 and with allyl trimethoxysilane instead of allyl trichlorosilane.

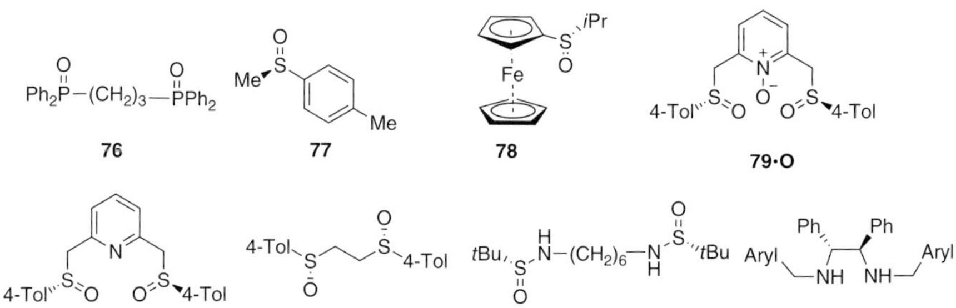

Scheme 10.24 Sulfoxides and sulfonamides as Lewis base additives for allylation of benzoyl hydrazones.

the crotylation with (*E*)- and (*Z*)-crotyltrichlorosilanes, *syn*- and *anti*-adducts were obtained stereospecifically, respectively [131, 132].

A few years later it was recognized that Lewis bases such as sulfoxides [133] and phosphine oxides [134] can act as activators for these transformations (entries 4 and 5, Scheme 10.24). An enantioselective version of allylation of benzoyl hydrazones was then developed using chiral mono- and bissulfoxides **77–80** (entries 6–11) [135–137], sulfinamide **81** (entry 12) [138], or phosphine oxides such as BINAPO (**45**) (entry 12) [139], leading to the formation of allylic amines in good yields. However, superstoichiometric amounts of these additives are necessary to achieve high stereoselectivities. Chiral diamine **82**, in the presence of zinc fluoride and allyltrimethoxysilane, effects the allylation of acylhydrazono esters under catalytic conditions, even if long reaction times are necessary to obtain the product in high yield (entry 13) [140].

10.3.2.2 Allylation of C=O Bonds

The allylation of carbonyl compounds is an important topic for synthetic applications, because it creates new C–C bonds with two consecutive stereocenters [141–144]. Early examples of Lewis base–promoted allylation reactions involving hypercoordinate allylsiliconates date back to the 1980s when Sakurai reported the addition of allyltrifluorosilanes and crotyltrifluorosilanes to aldehydes in the presence of a stoichiometric amount of CsF (Scheme 10.25) [145] or dilithium catecholates [146, 147]. A strong correlation between the geometry of the silane and the configuration of the allylic alcohol was observed: (*E*)-crotylsilane giving the *anti*-homoallyl alcohol and (*Z*)-crotylsilane giving the *syn*-adduct. This stereospecificity is consistent with a mechanism that involves a chair-like, six-membered transition state in which a hexacoordinate silicon atom is the organizational center [148, 149].

Several years later it was demonstrated that DMF, a Lewis-basic solvent, promoted the allylation of trichlorosilanes to aldehydes [150–153]. In this case as well, the correlation between the geometry of the crotylsilane and the relative configuration of the allylic alcohol was observed ($E \rightarrow$ *anti* and $Z \rightarrow$ *syn*) (Scheme 10.26). DFT calculations at the MP2/6-31G** level of theory revealed that the activation barrier of the allylation reaction between benzaldehyde and allyltrichlorosilane in the presence of DMF was considerably lower than that of

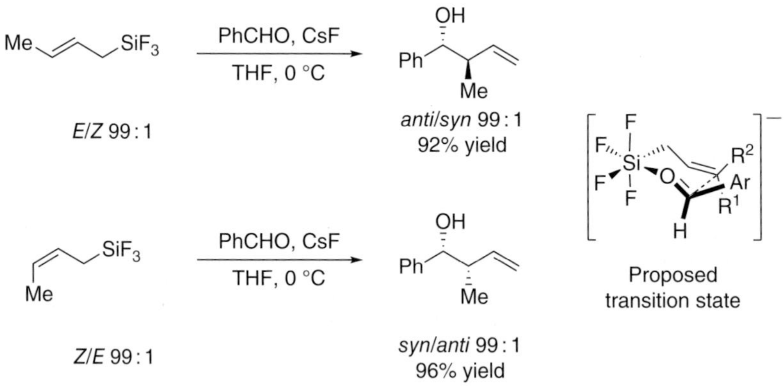

Scheme 10.25 Allylation of aldehydes performed with allyltrifluorosilanes.

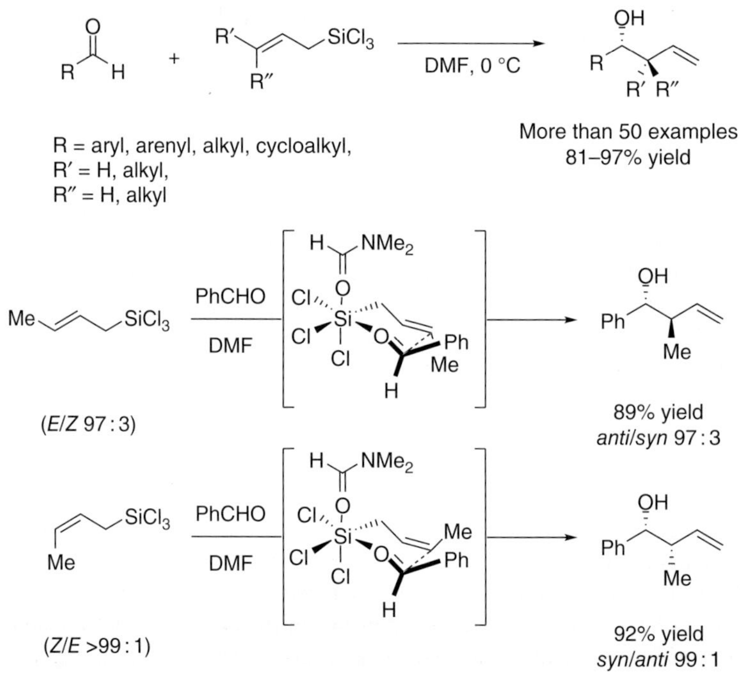

Scheme 10.26 Allylation of aldehydes performed with allyltrichlorosilane.

the reaction in the absence of the Lewis base (2.3 kcal/mol vs. 14.3 kcal/mol). Moreover, it was demonstrated that a hexacoordinate silicon atom possesses Lewis acidity significantly greater compared to that in a *penta*-coordinate structure [154].

In 1994, Denmark et al. reported the first example of enantioselective addition of allyltrichlorosilane to aldehydes promoted by phosphoramides **83–86** (Table 10.8 and Scheme 10.27) [155, 156]. Among them, only phosphoramides **83** and **86** showed useful enantioselectivity (entries 1 and 6). In the same context,

Table 10.8 Allylation of aldehydes with allyltrichlorosilane.

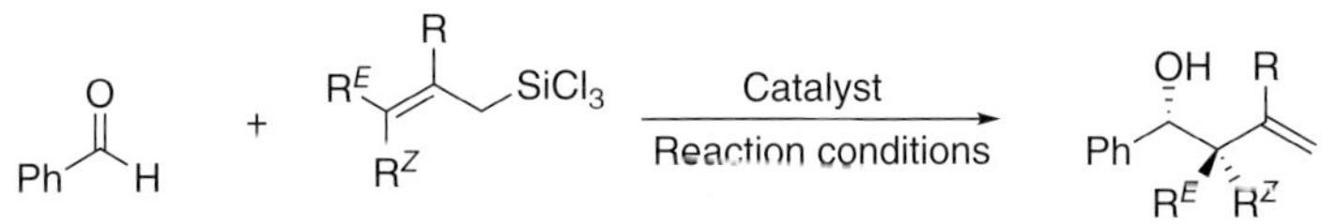

Entry	Catalyst (eq)	R^E, R^Z, R	Reaction conditions	Yield (%)	*anti:syn*	er (major)
1	**HMPA** (1)	H, H, H	CH_3CN, 4 min, 0 °C	86	—	—
2	**83** (1)	Me, H, H	CH_2Cl_2, 7.5 h, −78 °C	68	98 : 2	83 : 17
3	**83** (1)	H, Me, H	CH_2Cl_2, 7.5 h, −78 °C	72	2 : 98	80 : 20
4	**84** (1)	H, H, H	CH_2Cl_2, 6 h, −78 °C	43	—	60 : 40
5	**85** (1)	H, H, H	CH_2Cl_2, 6 h, −78 °C	81	—	80 : 20
6	**86** (1)	H, H, H	CH_2Cl_2, 12 h, −78 °C	59	—	86 : 14
7[a)]	**87** (0.1)	Me, H, H	CH_2Cl_2, 24 h, −40 °C	86	93 : 7	74 : 26
8[a)]	**87** (0.1)	H, Me, H	CH_2Cl_2, 24 h, −40 °C	87	1 : 99	63 : 37
9	**88** (0.1)	H, H, , H	CH_2Cl_2, 6 h, −78 °C	54	—	86 : 14
10	**89** (0.05)	Me, H, H	CH_2Cl_2, 6 h, −78 °C	82	99 : 1	93 : 7
11	**89** (0.05)	H, Me, H	CH_2Cl_2, 6 h, −78 °C	89	1 : 99	97 : 3
12	**90** (0.2)	H, H, H	Toluene, 48 h, −20 °C	93	—	93.5 : 6.5
13[b)]	**45** (0.05)	H, H, H	CH_2Cl_2, 24 h, rt	92	—	71.5 : 28.5
14[c)]	**73** (0.05)	H, H, H	CH_3CN, 40 h, 0 °C	85	—	96.5 : 3.5
15[c)]	**91** (0.02)	H, H, H	CH_3CN, 6 h, rt	93	—	21.5 : 78.5
16[b)]	**75** (0.01)	H, H, Me	CH_3CH_2CN, 9 h, −90 °C	88	—	95 : 5

a) Reaction performed in the presence of nBu$_4$NI (20 mol%).
b) Reaction performed in the presence of iPr$_2$NEt and Bu$_4$NI.
c) Reaction performed in the presence of iPr$_2$NEt.

Hamada reported the use of a catalytic amount of chiral phosphoramide **87** in the presence of nBu$_4$NI, necessary to accelerate the reaction [157, 158]. After an extensive study on the reaction mechanism by kinetic analysis, Denmark et al. reported that chiral bidentate phosphoramides **88** and **89** exhibited enhanced stereochemical efficiency leading to the final product in higher yield and enantiomeric ratio, in shorter reaction time and in lower loading [159–163]. Catalyst **89** efficiently promotes the allylation of aromatic aldehydes with only 5 mol% loading, affording the products in high yields and enantiomeric ratio (entries 10 and 11). Subsequently, Sun and coworker reported that BINOL-derived phosphoric acid **90** was able to also catalyze the addition of allyltrichlorosilane to benzaldehyde, leading to the formation of the corresponding homoallylic alcohol in 93% yield and high enantioselectivity (entry 12) [164]. It was suggested that the Lewis-basic P=O moiety of catalyst **90** coordinated the silicon atom, while, at the same time, the P—OH bond was involved in the activation of the aldehyde through a hydrogen bond. In 2005, it was demonstrated that phosphine oxides such as BINAPO (**45**) could act also as catalysts in the enantioselective addition

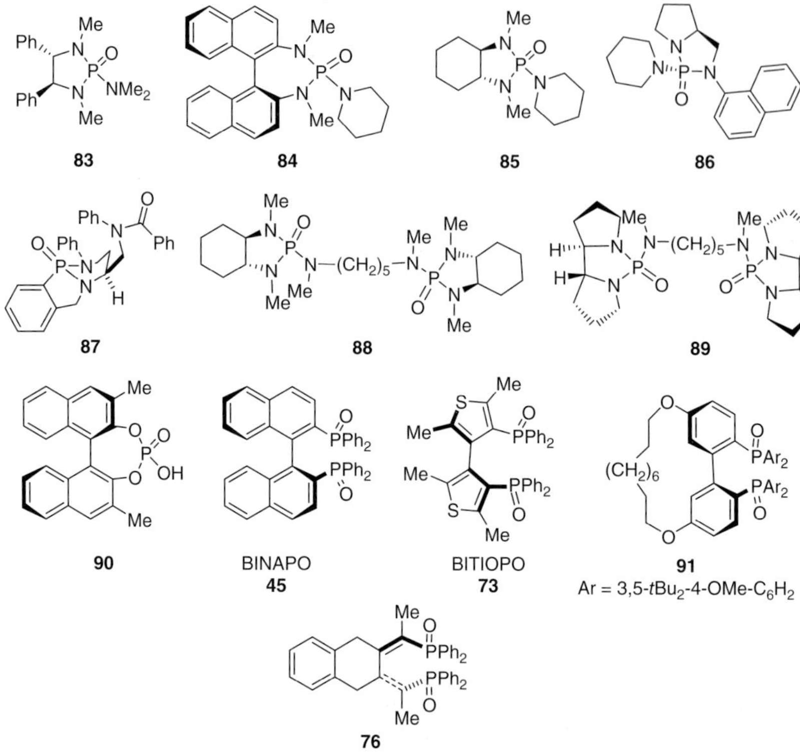

Scheme 10.27 Selected Lewis base catalyst for allylation of aldehydes.

of allyltrichlorosilane to aldehydes, albeit with lower enantioselectivity compared to bisphosphoramides (entry 13) [165]. An improvement in the catalytic efficiency of chiral phosphine oxides could be achieved using BITIOPO **73**, which promoted the allylation of benzaldehyde in 96.5 : 3.5 er (entry 14) [125, 166]. Since then, other bisphosphine oxides have been developed, such as bridged compound **91** which can promote the reaction with only 2 mol% loading [167], and the previously mentioned atropisomeric catalyst **76**, which is extremely reactive at lower loading and temperature (entry 16) [127].

The development of these efficient catalytic systems was made possible in part by the clarification of the allylation mechanism described in a series of detailed studies by Denmark and coworkers. Kinetic, stereochemical, and structural investigations combined with the observation of a nonlinear relationship between the enantiopurity of monophosphoramide **85** with the product revealed that the allylation proceeds through divergent reaction pathways [98]. Allylation of benzaldehyde with catalyst **85** in the presence of allyltrichlorosilane displays a 1.77 order dependence on the catalyst, clearly indicating two competing pathways involving both one and two phosphoramides. The simultaneous operation of both mono- and diphosphoramide pathways provides a unique explanation for the strong differences in enantioselectivities obtained in Lewis base–catalyzed allylations. The transition states for both pathways are shown in Scheme 10.28.

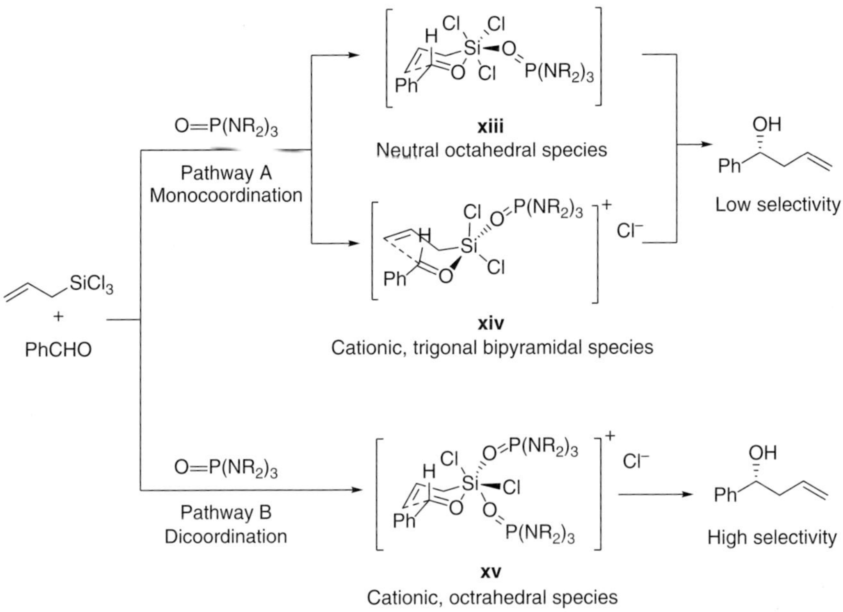

Scheme 10.28 Transition states coordinating one or two phosphoramides.

In the case of pathway A, in which only one phosphoramide is involved, the assembling of phosphoramide, allylic trichlorosilane, and aldehyde leads to the formation of either a hexacoordinate, octahedral, neutral transition state **xiii**, or, after the ionization of one chloride ion, a trigonal bipyramidal, *penta*-coordinate silicon species **xiv**. However, in both cases, because of the diminished influence of the singular chiral promoter in the former, only modest selectivity is expected. To date, it is not possible to distinguish which of the two species (**xiii** or **xiv**) is actively involved, but it is reasonable to presume that the cationic species **xiv** is kinetically competent, because the addition of ammonium salts enhances the reaction rate [157, 158].

On the other hand, the involvement of two phosphoramides in pathway B, leads to the formation of hexacoordinate, octahedral, cationic silicon species **xv** that results in the formation of products with higher enantioselectivities. This fact was experimentally confirmed by the observation of the high enantioselectivities obtained when bidentate catalyst **89** was employed in allylation reactions (Table 10.8, entry 10).

The stereochemical aspects of the catalytic activation of allylsilanes with bisphosphoramides have also been investigated by crystallographic studies and by ^{119}Sn and ^{31}P NMR analysis [160, 163]. The isolation of the complexes of **89** with $SnCl_4$, which forms more stable complexes than do silicon Lewis acids, confirms the octahedral geometry of the adducts, with a bidentate, cis-coordination of the ligand and a trans-arrangement of the chlorine atoms [162, 168]. Computational analysis of species **xii** derived from bidentate catalyst **89** allows a rationalization of the stereochemical outcome of the allylation reaction (Scheme 10.29) [162].

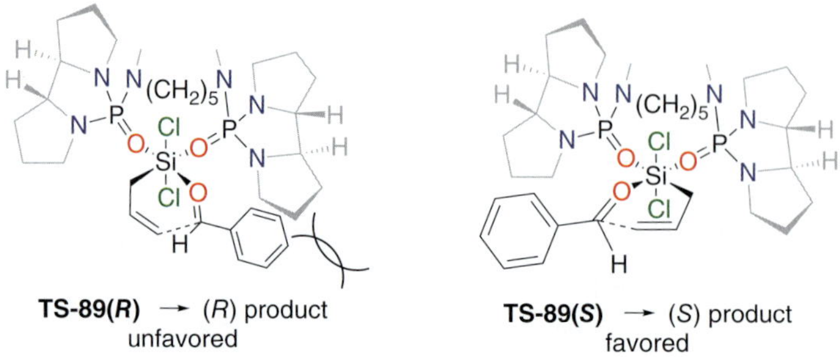

Scheme 10.29 Transition states of benzaldehyde allylation promoted by chiral phosphoramide **89**.

In the chair-like cyclic transition state TS-**89**(*R*), responsible for the formation of the allyl alcohol with (*R*)-configuration, the phenyl ring is located in an unfavorable position occupied by a forward-pointing pyrrolidine ring, creating destabilizing steric interactions. In the diastereomeric, chair-like arrangement in cyclic transition state TS-**89**(*S*), no unfavorable interactions are present, leading to the formation of the experimentally observed (*S*)-enantiomer.

It must be noted, however, that none of the catalysts previously described are able to perform the allylation with aliphatic aldehydes. Unfortunately, this behavior is related to the mechanism of the reaction rather than to the structure of the catalyst. It was demonstrated by ^{1}H-NMR spectroscopic analysis that in the reaction with aliphatic aldehydes, the chloride ion binds with the aldehyde resulting in the formation of an alkyl chloro silyl ether that precludes the addition of the allylating reagent (Scheme 10.30) [161].

Among Lewis-basic catalysts, the amine *N*-oxides represent another class of compounds that deserve special mention in allylation reactions (Scheme 10.31). The combination of the high Lewis basicity of the *N*-oxide oxygen atom, combined with the high affinity of silicon for oxygen, constitutes the key point for the further development of allyltrichlorosilane-mediated reactions. By analogy to the chelation model proposed for bisphosphoramides, both oxygens of the bis *N*-oxide catalysts are proposed to coordinate the silicon atom [169–171]. The first enantioselective allylation of aldehydes was achieved using C_2 symmetric biquinoline *N,N′*-dioxide **92** [172]. The reaction rate increases with the addition

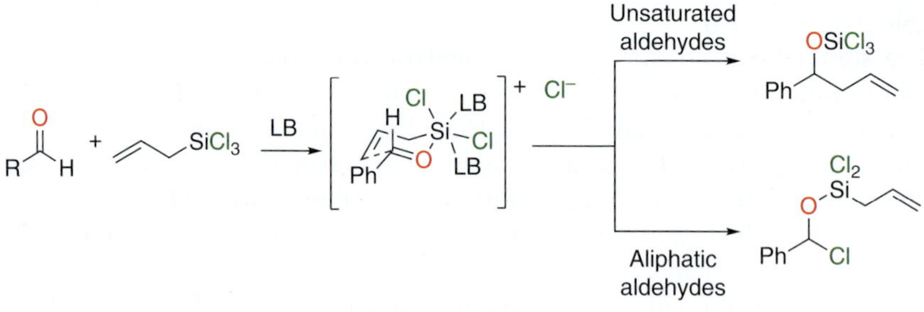

Scheme 10.30 Formation of a silyl chlorohydrin with aliphatic aldehydes.

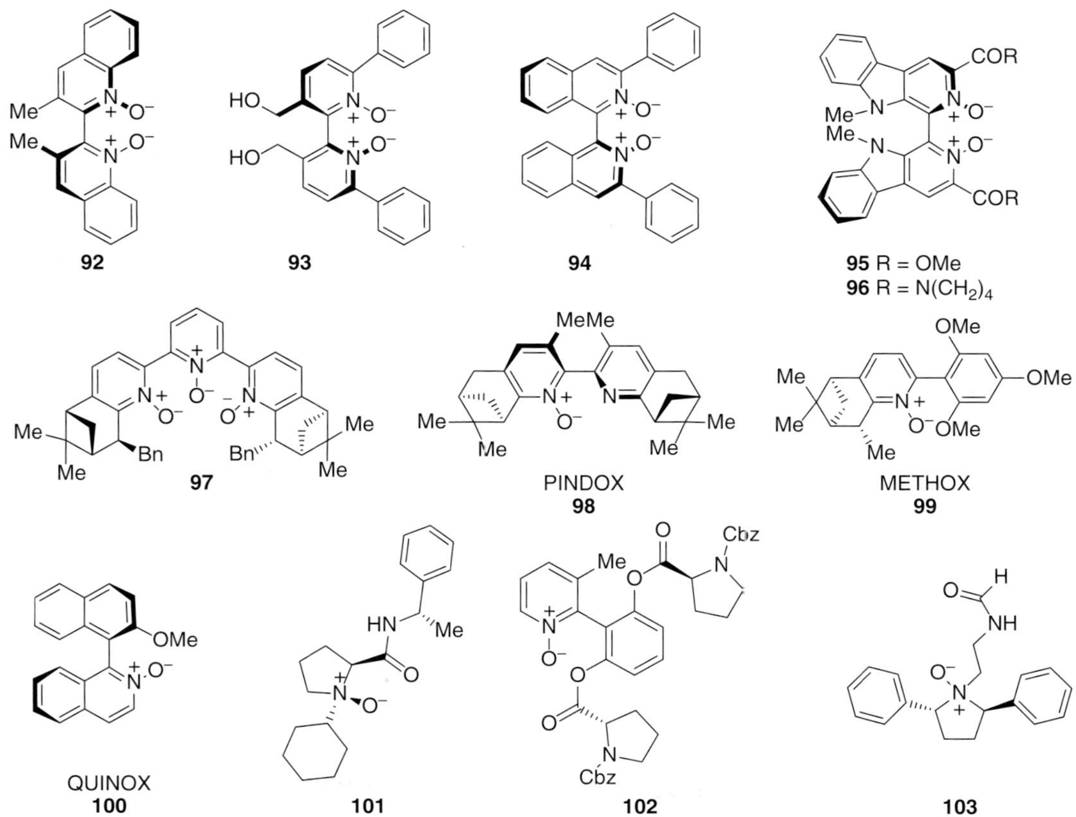

Scheme 10.31 Selected N-oxide-based catalysts for allylation of aldehydes.

of diisopropylethylamine to afford the homoallylic alcohol in high yield and enantioselectivity (Table 10.9, entry 1). Similar levels of enantioselection are obtained with axially chiral N-oxides **93** and **94** (entries 2–4) [169–171]. Catalyst **93** exhibited remarkable reactivity also at 0.1 mol% and at 0.01 mol% loading level, which makes this catalyst the most reactive to date. Biscarboline *N,N*-dioxides **95** and **96** represent another example of highly efficient catalysts, capable of operating at lower catalysts loading (1 mol%).

With these catalysts, aromatic, α,β-unsaturated and aliphatic aldehydes react well, giving the corresponding homoallylic alcohols in good yields and high enantiomeric ratios (entries 5–8) [173–175]. The use of catalysts bearing three *N*-oxide units have also been investigated (catalyst **97**), but in this case, their efficiency is lower compared to that of the other catalysts (entry 9) [175]. Another important contribution is the demonstration that two *N*-oxide groups are not necessary and replacing one *N*-oxide unit with a second coordination element. Different terpene-derived bipyridine mono *N*-oxides were developed, and the most successful derivative of this series, **98** (Me_2-PINDOX), led to the formation of products in up to 99 : 1 er (entry 10) [176, 177]. Unfortunately, this catalyst is not configurationally stable in solution; at room temperature, this catalyst loses

Table 10.9 Allylation of aldehydes with chiral *N*-oxides as catalysts.

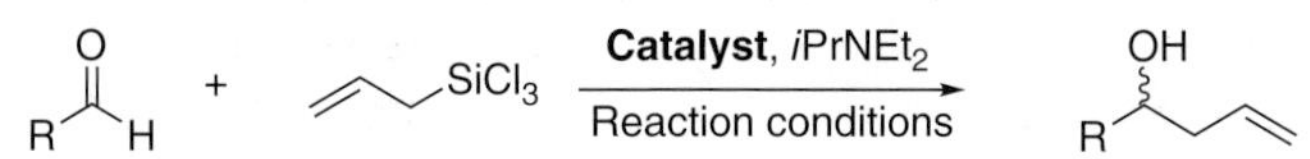

Entry	Catalyst (mol%)	R	Reaction conditions	Yield (%)	er
1	**92** (10)	Ph	CH_2Cl_2, 6 h, −78 °C	85	94 : 9
2	**93** (10)	Ph	CH_3CN, 2.5 h, 45 °C	96	92 : 8
3	**93** (1)	Ph	CH_3CN, 0.25 h, 45 °C	96	9.5 : 90.5
4	**94** (1)	Ph	CH_2Cl_2, 18 h, −40 °C	99	88.5 : 11.5
5	**95** (10)	Ph	CH_2Cl_2, 16 h, −80 °C	100	97.5 : 2.5
6	**95** (1)	Ph	CH_2Cl_2, 16 h, −80 °C	100	97.5 : 2.5
7	**95** (1)	Cyclohexyl	CH_2Cl_2, 16 h, −80 °C	53	98.5 : 1.5
8	**96** (1)	Ph	CH_2Cl_2, 20 h, −80 °C	100	93.5 : 6.5
9	**97** (10)	Ph	CH_2Cl_2, 2 h, 0 °C	89	87 : 13
10	**98** (10)	Ph	CH_2Cl_2, 12 h, −60 °C	72	99 : 1
11	**99** (10)	PhCH=CH−	CH_3CN, 5 days, 45 °C	75	94 : 6
12	**99** (5)	Ph	CH_3CN, 18 h, 45 °C	95	98 : 2
13	**100** (5)	4-CF_3−C_6H_4	CH_2Cl_2, 2 h, −40 °C	85	98 : 2
14	**101** (10)	Ph	$Cl(CH_2)_2Cl$, 24 h, 22 °C	82	93.5 : 6.5
15	**102** (10)	Ph	CH_3CN, 48 h, 0 °C	45	84 : 16
16	**103** (10)	Ph	CH_3CN, 48 h, 0 °C	51	90.5 : 9.5
17	**103** (30)	$PhCH_2CH_2$	CH_3CN, 48 h, 0 °C	47	90.5 : 9.5

its enantiomeric purity in a few days. Interestingly, it was shown that simple *N*-oxides such as the pinene-derived catalyst **9** [178–181] and the quinoline-type mono *N*-oxide **100** [182], devoid of a second coordinative element, are also good promoters for the allylation reaction (entries 11–13). Interestingly, the presence of axial chirality or the presence of two silicon coordinating units in the *N*-oxide structure are not absolute prerequisites for attaining high enantioselectivity in the allylation reaction. For example, Hoveyda and coworkers developed catalyst **101**, the first example of "non-pyridine-type" *N*-oxide bearing a stereogenic center at the nitrogen derived from proline (entry 14) [183]. Catalyst **101** promotes the allylation of aldehydes at room temperature with high enantioselectivity. In catalyst **102**, also derived from proline, the enantioselection is proposed to arise from coordination of the silicon atom to the pyridine *N*-oxide oxygen and the phenolic oxygen of one side arm [184]. A second example of chiral "non-pyridine-type" *N*-oxide, introduced by Benaglia et al., is a *trans*-2,5-diphenylpyrrolidine *N*-oxide **103** [185]. This compound catalyzes reactions of aliphatic aldehydes with allyltrichlorosilane to afford the homoallylic alcohol with good enantioselectivity (entries 16 and 17).

The stereochemical aspects of the catalytic activation of allyltrichlorosilanes with *N*-oxides were investigated by kinetic and computational studies. Initially, by analogy with the chelation model proposed by Denmark (Scheme 10.28), the PINDOX-derived transition state **xvi** was proposed [176, 177], wherein the allyl group is in a position trans to the *N*-oxide and the chlorine atoms are trans to each other (Scheme 10.32a). However, Wheeler and coworkers clearly demonstrated that the favored transition state was **xvii**, in which one of the chlorides and the aldehyde exchange places, leading to the formation of the homoallylic alcohol in good agreement with the experimental stereoselectivity [186, 187]. Next, the divergent behavior of METHOX catalyst **99** (compatible with both electron-rich and electron-poor aldehydes), and QUINOX catalyst **100** (more sensitive to the electronic properties of the substrate) was investigated. As in the case of phosphoramides, two different transition states are possible, one involving an ionic *penta*-coordinate silicon species **xviii** and one with a neutral hexacoordinate species **xix** (Scheme 10.32b) [182]. However, in both cases, only one molecule of the *N*-oxide catalyst is involved in the rate- and selectivity-determining step. The difference in the formation of transition state **xviii** or **xix** can be attributed to the increased steric congestion: less sterically hindered catalysts such as QUINOX prefer the hexacoordinate silicon species, whereas the bulkiest catalyst, METHOX, prefers a *penta*-coordinate species. This difference is also reflected in the solvent used; reactions performed with **100** are superior in dichloromethane, whereas catalyst **99** performs better in polar solvents such as acetonitrile [188, 189].

In addition to phosphoramides, phosphine oxides, and *N*-oxides, chiral sulfoxides [190–197] and sulfinamides [198] display modest to good enantioselectivities in the allylation of aldehydes promoted with allyltrichlorosilane. However, these types of Lewis bases are generally used in superstoichiometric amounts and are rarely recovered due to the reduction of the sulfoxide moiety, thus limiting their applicability. A few selected examples are reported in Scheme 10.33. Sulfonamides **110** and **111** work efficiently in catalytic activity, producing good

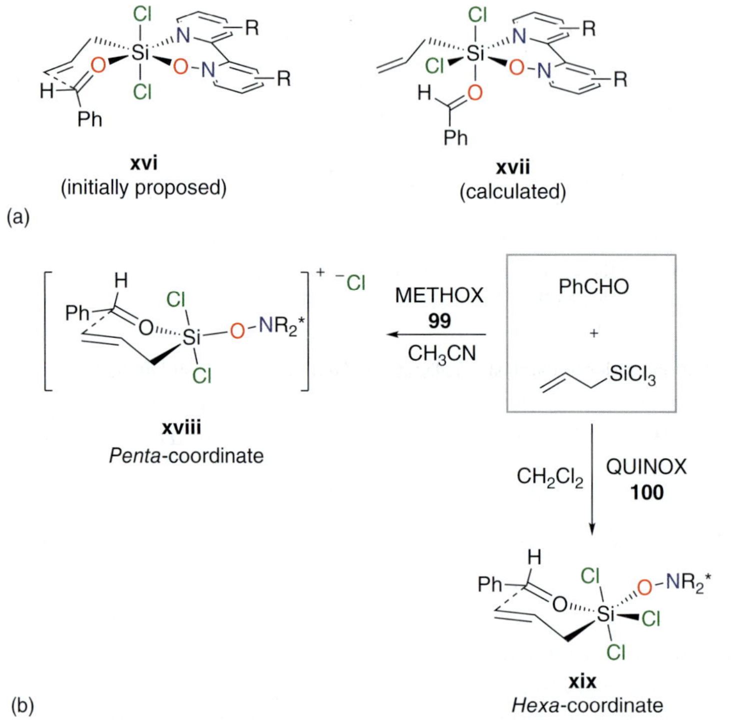

Scheme 10.32 Mechanism of allylation catalyzed by chiral pyridine N-oxides. (a) TS of allylation reaction perfomed with PINDOX. (b) TS of allylation reaction perfomed with METHOX and QUINOX.

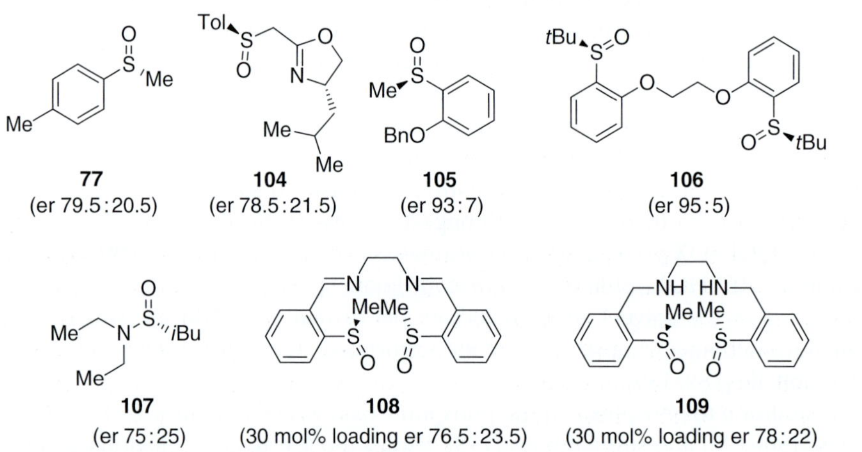

Scheme 10.33 Selected sulfoxides and sulfonamides as Lewis base promoters in the allylation of benzaldehyde with allyltrichlorosilane [190–193, 198].

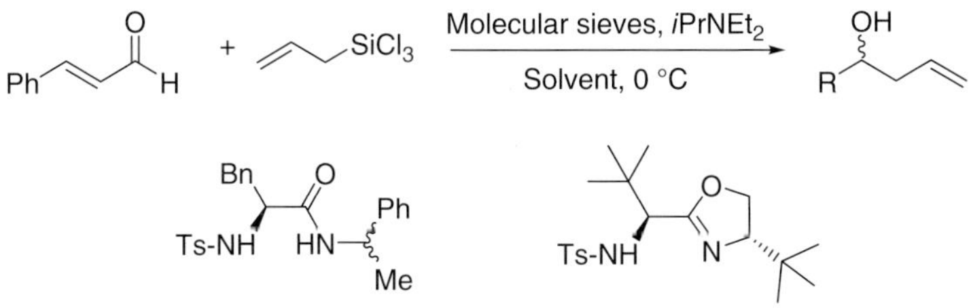

110
15 mol% loading,
CH_2Cl_2/THF, 7 : 3, 10 h
82% yield, 91.5 : 8.5

111
10 mol% loading,
CH_2Cl_2, 15 h
77% yield, 90:10

Scheme 10.34 Sulfonamides as catalysts in allylation of α,β-unsaturated aldehydes.

yields and enantioselectivities over a wide range of aromatic and α,β-unsaturated aldehydes (Scheme 10.34) [199, 200].

10.3.3 Aldol Reactions Involving Preformed Enoxysilane Derivatives

By analogy with the reaction between allyltrichlorosilanes and carbonyl compounds, the addition of enoxytrichlorosilane derivatives to aldehydes was evaluated. Here again, the silicon atom serves as an organizational center to bring together the nucleophile, electrophile, and the catalyst. However, on the basis of the higher nucleophilicity of enoxysilane derivatives, it was expected that these additions would be more facile [201, 202].

Trichlorosilyl derivatives of enolates are more nucleophilic compared to the corresponding allyltrichlorosilanes, and reaction with aldehydes (aldol addition) proceeds readily at room temperature without a catalyst [203]. In the absence of a Lewis base, the aldolization process provides the *syn* aldol product in high yield, according to a closed, boat-like transition structure (Table 10.10, entry 1). Nevertheless, the reaction can be accelerated by the presence of a Lewis base as reported by Denmark et al. in 1996, who developed a catalytic, enantioselective addition of trichlorosilyl ethers of achiral cyclic ketones [204] and achiral methyl ketones [205] to aldehydes catalyzed by mono and bisphosphoramides **112** and **113** (Scheme 10.35).

Surprisingly, small changes in the structure of the catalysts led to the generation of the opposite diastereomers. Catalyst **113a** promotes the addition of cyclohexanone-derived trichlorosilyl enol ether to benzaldehyde in 95% yield, 99 : 1 *anti/syn* ratio, and 96.5 : 3.5 er with 10 mol% loading; but when catalyst **113b** is used in the same reaction conditions, the final aldol product is obtained in 94% yield, 99 : 1 *syn/anti*-ratio and 76.5 : 23.5 er (entries 2 and 3).

Moreover, using the same catalyst **113a** but switching from cyclic trichorosilyl enol ether derivatives to acyclic nucleophiles, an inversion of the diastereoselection was again observed, with the predominant formation of *syn* diastereomer (entries 6–9). In both cases, diastereoselectivity is highly substrate dependent. Aldehydes bearing both electron-withdrawing and electron-donating substituents react in excellent yields and selectivities in short reaction

Table 10.10 Aldol addition between aldehydes and various trichlorosilyl enol ethers.

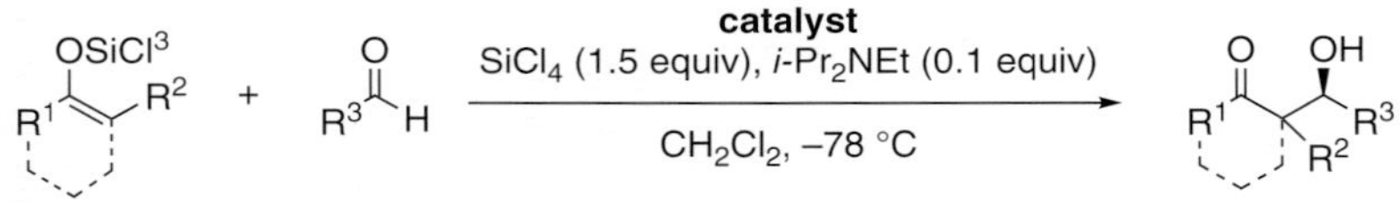

Entry	R^1	R^2	R^3	Catalyst (eq loading)	Time (h)	Yield (%)	*anti:syn*	er (major)
1[a]	$-(CH_2)_4-$	H	Ph	—	2	92	2 : 98	—
2	$-(CH_2)_4-$	H	Ph	**113a** (0.1)	2	95	99 : 1	96.5 : 3.5
3	$-(CH_2)_4-$	H	Ph	**113b** (0.1)	2	94	1 : 99	76.5 : 23.5
4	$-(CH_2)_4-$	H	1-Naphthyl	**113a** (0.1)	2	94	99 : 1	99.5 : 0.5
5	$-(CH_2)_4-$	H	(*E*)-CH=CHPh	**113a** (0.1)	2	94	99 : 1	98 : 2
6	Ph	Me	Ph	**113a** (0.15)	6	95	5 : 95	97.5 : 2.5
7	Ph	Me	4-Br-C_6H_4	**113a** (0.15)	6	89	8 : 92	98 : 2
8	Ph	Me	1-Naphthyl	**113a** (0.15)	6	96	25 : 75	92 : 8
9	Ph	Me	Ph	**113c** (0.1)	6	74	56.5 : 43.5	56.5 : 43.5
10	$-(CH_2)_4-$	H	Ph	**66** (0.03)	0.5	94	15 : 75	65 : 35
11	$-(CH_2)_4-$	Cl	Ph	**114** (0.03)	0.5	92	96 : 4	80 : 20
12	$-(CH_2)_4-$	H	Ph	**45** (0.1)	2	58	30 : 70	93 : 7
13	$-(CH_2)_4-$	H	4-$NO_2C_6H_4$	**45** (0.1)	0.25	90	4 : 96	98 : 2
14	Me	H	Ph	**113a** (0.05)	2	98	—	93.5 : 6.5
15	*n*Bu	H	Ph	**113a** (0.05)	2	98	—	92 : 8
16	*i*Pr	H	Ph	**113a** (0.05)	2	97	—	91 : 9
17	Ph	H	Ph	**113a** (0.05)	2	93	—	74.5 : 25.5
18	Et	Me	Ph	**113a** (0.1)	2	85	6 : 94	95 : 5
19	Me	Me	Ph	**113a** (0.1)	2	79	17 : 83	55 : 45
20	Me	Me	Ph	**112** (0.5)	2	85	33.5 : 66.5	50 : 50
21	H	*n*Pent	Ph	**112** (0.5)	2	92	1 : 99	95 : 5

a) Reaction conducted at 0 °C.

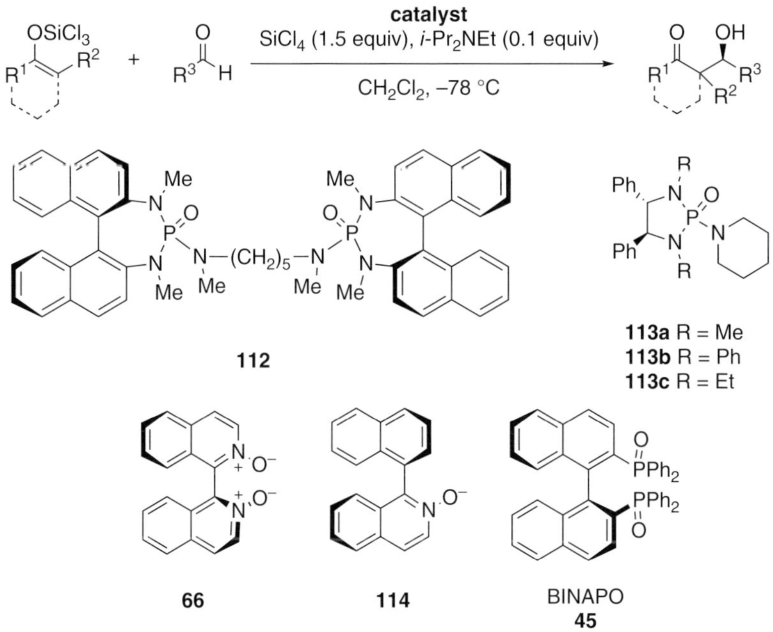

Scheme 10.35 Lewis bases active in the aldol reaction between aldehydes and silyl enol ethers.

times, as do olefinic aldehydes, but it was found that aliphatic aldehydes do not react with ether cyclic or acyclic substrates, presumably due to competitive enolization [206].

The high diastereo- and enantioselectivity of the aldolization process and the stereochemical response to the enolate geometry suggest that the transition state is a highly ordered, chair-like transition structure involving hexacoordinate siliconate species. Kinetic analysis of the divergence of reaction pathways in the chiral Lewis base–promoted aldol addition to trichlorosilyl enol ethers were conducted using rapid-injection NMR spectroscopy [207].

The uncatalyzed reaction displays a classic first-order dependence on each component; however, the aldol addition catalyzed by **113a** displays a second-order dependence on phosphoramide, thus providing direct evidence that two catalyzed pathways that depend on catalyst structure and concentration exist. More detailed studies [208, 209] explain the correlation between the mechanism and the structure of the Lewis base; sterically demanding phosphoramides such as **113b** bind to the enolate in a 1 : 1 manner and the resulting *penta*-coordinate cationic siliconate favors a boat-like arrangement (Scheme 10.36). Sterically less demanding phosphoramides such as **113a** bind the substrate in a 2 : 1 manner, and the resulting *hexa*-coordinate cationic siliconate favors a chair-like arrangement. According to these results, the predominant diastereoselection of the final product is dependent both on the geometry of the enolate and on the steric properties of the Lewis base. *E* enolates afford preferentially anti-products when employed with less hindered Lewis bases (and syn products with sterically

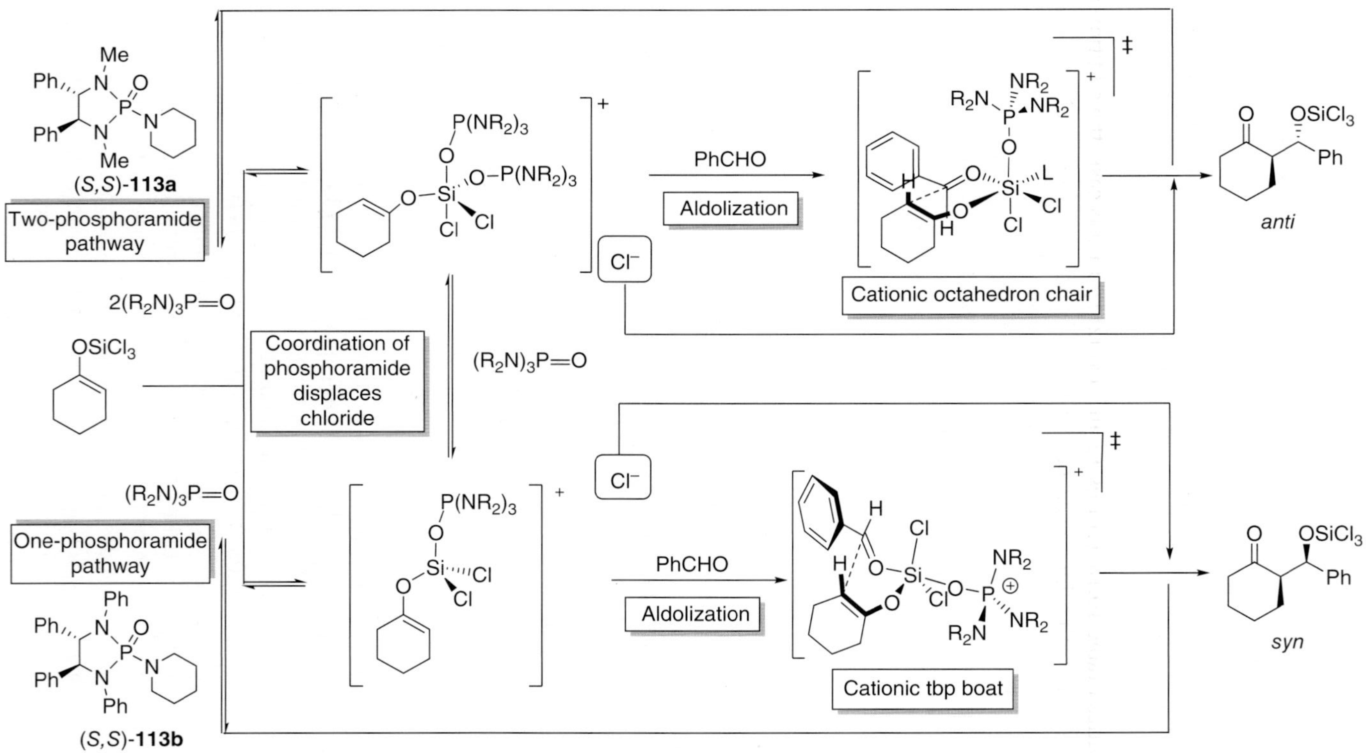

Scheme 10.36 Proposed mechanism and stereochemical rationale for aldol additions of trichlorosilyl enol ethers catalyzed by phosphoramides.

demanding Lewis bases), whereas aldol addition of *Z* enolates to aldehydes results in syn products if catalyzed with less hindered Lewis bases or in preferentially anti-products when bulky Lewis bases such as **113b** are employed.

Chiral *N,N'*-dioxides such as **66**, and monodentate *N*-oxide **114** were also demonstrated to be efficient catalysts in the catalytic, enantioselective aldol reactions of trichlorosilyl enol ethers, leading to the formation of the desired product in only 20 minutes with a 3 mol% loading (Table 10.10, entries 11 and 12) [210]. In addition, phosphine oxide BINAPO **45** was found to catalyze the addition of cyclohexanone-derived silyl enol ether with activated aromatic aldehydes, affording the corresponding *anti*-aldol adducts with high diastereo- and enantioselectivities (Table 10.10, entries 12 and 13) [211]. In this case, the addition of diisopropylethylamine increased both chemical and stereochemical efficiency. The authors proposed that the role of tertiary amine additive was both as an acid scavenger to neutralize the hydrogen chloride produced by the hydrolysis of trichlorosilyl enol ethers and as a promoter of the catalyst dissociation from the silicon atom at the end of the catalytic cycle.

Investigations of the aldol addition of trichlorosilyl enolates derived from methyl ketones were also performed (Table 10.10, entries 14–17) [212]. The size and electronic character of R^1 substituent had little role in the chemical efficiency of the process, providing high yields of aldol adducts when reacted with benzaldehyde. Only in the case of a phenyl substituent was lower enantioselectivity obtained. In addition, it was reported that high diastereoselectivities can be reached by the combination of either chiral enolates or chiral aldehydes with catalyst **113a**.

Z-Trichlorosilyl enolates of linear ketone enolates can be also employed in combination with catalyst **112**, leading to the formation of the final products in good yields but modest selectivities (Table 10.10, entries 18–21) [213]. Catalyst **112** results as more efficient in the reaction involving aldehyde-derived trichlorosilyl enol ethers, where aldol products can be obtained with high levels of diastereoselectivity, consistent with the reaction through a closed, chair-like transition structure [214].

10.4 Interaction of the Silicon-Activated Substrate with an External Non-Coordinated Nucleophile

As shown in Scheme 10.2, hypercoordinate cationic silicon species **iii** (with R = Cl), in the presence of an external nucleophile that is unable to be coordinated to the silicon atom, will evolve in the formation of hypercoordinate species **iv**, wherein the substrate is activated to undergo nucleophilic attack from the external nucleophile through an intermolecular pathway (pathway D2). A typical reaction example of this behavior is observed in Lewis base–catalyzed, $SiCl_4$-mediated reactions, such as the allylation of aldehydes with allyltributyltin or the aldol reactions performed with enoxysilane derivatives (Mukaiyama aldol).

10.4.1 Allylation of Aldehydes Mediated by Silicon Tetrachloride

In 2001, Denmark and Wynn reported the first example of a Lewis base–catalyzed, $SiCl_4$-mediated allylation using allyltributylstannane as the allylating reagent. Both mono- and bis-phosphoramides were investigated as the catalyst, and best results were obtained, respectively, with monophosphoramide **65** and bisphosphoramide **112** (Scheme 10.37) [215].

A variety of aromatic aldehydes with different electronic or steric contributions reacted smoothly, leading to the formation of the corresponding homoallylic alcohol in good yields, showing a great range of tolerance on the aldehyde substitution. Mechanistic studies demonstrated that the reaction obeys a second-order dependence in the Lewis base, explaining why better results are obtained using catalyst **112** instead of catalyst **65** [159, 207]. Unfortunately, aliphatic aldehydes are completely unreactive in this type of transformation, as is also observed with allyltrichlorosilanes owing to the formation of an alkyl chloro silyl ether. DFT calculations, followed by in situ IR and ^{1}H NMR spectroscopic studies using cyclohexanecarboxaldehyde, $SiCl_4$ and HMPA confirmed this hypothesis [203]. These mechanistic studies allowed the formulation of the catalytic cycle illustrated subsequently (Scheme 10.38). First, coordination of $SiCl_4$ by bisphosphoramide **112** occurs, causing ionization of a chloride ion and generation of a chiral, cationic trichlorosilyl species (**iii**). This species, after the coordination of the aldehyde, forms a reactive complex **iv** that undergoes intermolecular nucleophilic attack by allyltributylstannane. A new silicon species **xx** is then formed, which, after dissociation of the catalyst, leads to homoallylic alcohol protected as trichlorosilyl ether.

Several years later, the intermediacy of the hypothesized siliconium ion intermediate **iii** was confirmed by NMR spectroscopic analysis. A series of low-temperature ^{1}H, ^{31}P, and ^{29}Si NMR experiments conducted with different amounts of a Lewis base (HMPA) and $SiCl_4$ demonstrated the formation of several different silicon complexes, as shown in Figure 10.2 [24, 25].

Multinuclear NMR analysis and X-ray diffraction analysis of a 2 : 1 mixture of HMPA/$SiCl_4$ showed the presence of two neutral silicon species: **114**-*cis*

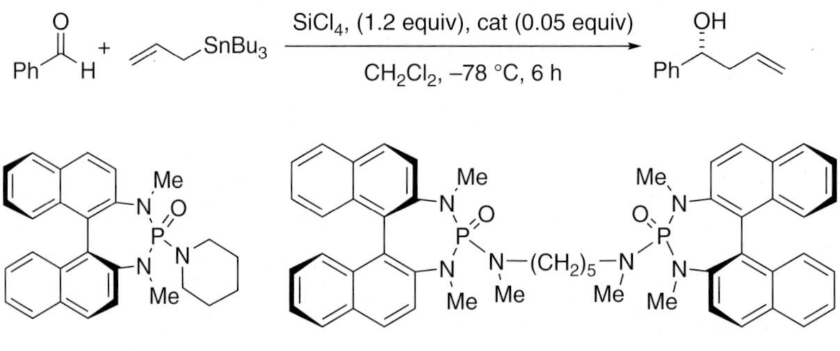

65
85% yield, er 89.5 : 10.5

112
91% yield, er 97 : 3

Scheme 10.37 Allylation of aldehydes promoted by $SiCl_4$ activated by phosphoramides.

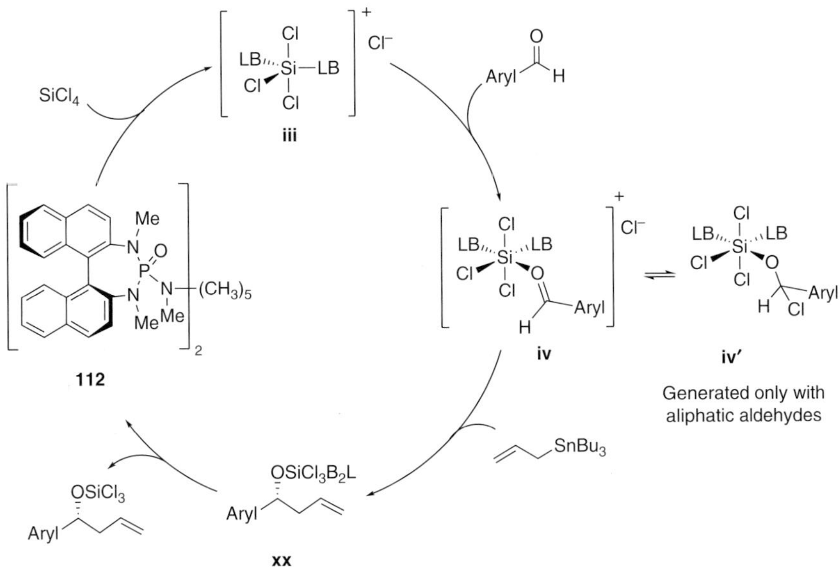

Scheme 10.38 Proposed mechanism for allylation reaction promoted by $SiCl_4$ and phosphoramides.

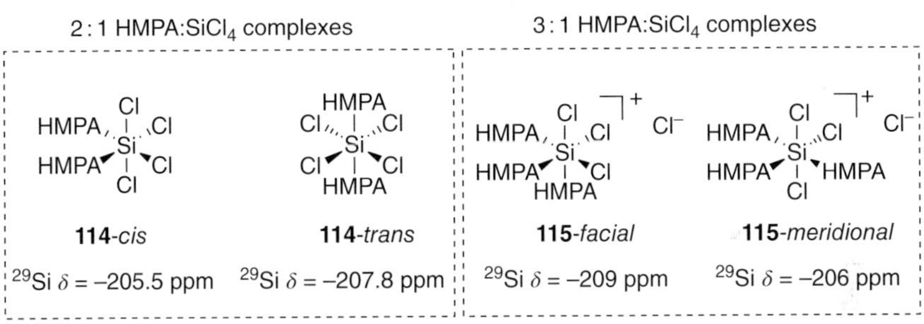

Figure 10.2 Silicon species identified by NMR analysis.

and **114**-*trans* corresponding to the two possible diastereomers. These two species disappear when at least three equivalents of HMPA are added, leading to the formation of **115**-*facial* and **115**-*meridional* silicon species. However, using bidentate phosphoramides, four other silicon species were identified (Figure 10.3) [24, 25].

Compounds **116**–**118** are neutral aggregates, whereas compound **119** is an ionic species. All these species are present in solution in varying amounts depending on the type of bisphosphoramide employed [129]. Kinetic experiments combined with RINMR and FTIR spectroscopic analysis revealed a zeroth-order dependence of the reaction on $SiCl_4$ and a 0.5 order for the bisphosphoramide. The first result indicates that $SiCl_4$ is saturated with the catalyst; the second reveals instead that all these dimeric species are not kinetically competent, but they constitute the resting states of the catalyst which must dissociate to form the kinetically active monomeric complex **iii**.

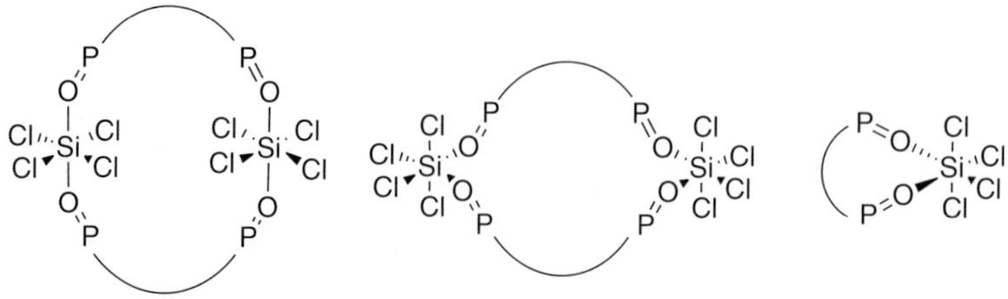

116
[*trans,trans*-$SiCl_4$-(**LB**)]$_2$
^{29}Si δ = −205.1 ppm (t), J = 15 Hz

117
[*cis,cis*-$SiCl_4$-(**LB**)]$_2$
^{29}Si δ = −205.5 ppm

118
[*cis,cis*-$SiCl_4$-(*R,R*)-**LB**]$_2$
^{29}Si δ = −185.5 ppm

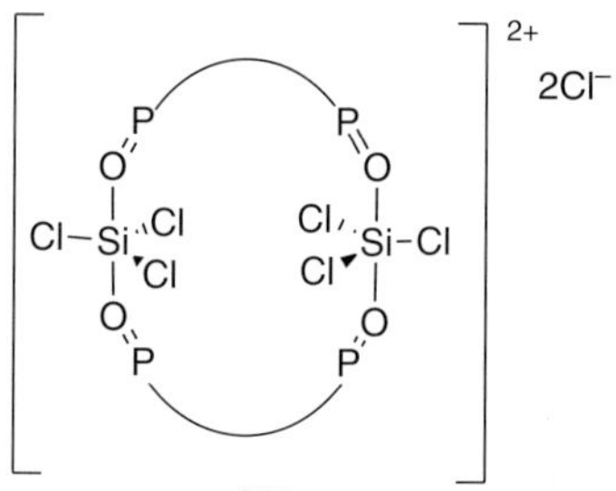

119
[*trans,trans*-$SiCl_3$-(**LB**)]$_2^{+2}$ $2Cl^-$
^{29}Si δ = −118.6 ppm

Figure 10.3 Bisphosphoramide-$SiCl_4$ complexation detected by ^{29}Si NMR analysis.

On the basis of the Hammond postulate that, "*species that are close in structure are also close in energy*" [217], neutral six-coordinate complex **118** could be considered the immediate precursor of species **iii** since it requires the least structural reorganization to form the desired kinetically active species and, indeed, the bisphosphoramide that forms this complex predominantly is also the fastest (but not the most selective).

10.4.2 Aldol Reactions Involving Trialkylsilyl Enol Derivatives

The venerable Mukaiyama aldol reaction revolutionized this process by preordaining which of the two carbonyl compounds would serve as the nucleophile and which would serve as the electrophile. Moreover, this variant introduced a conceptually distinct mode of catalysis by employing electrophilic activation of the acceptor through the action of Lewis acids. This advance naturally led to the development of highly efficient and selective chiral Lewis acids [203]. Accordingly, the use of $SiCl_4$, activated by Lewis bases, thus constituted a logical application of this novel type of Lewis acid.

10.4.2.1 Aldol Reactions Involving Trialkylsilyl Enol Ether Derivatives

Trialkylsilyl enol ethers of methyl ketones perform admirably as nucleophiles in aldol additions using $SiCl_4$. On the basis of the detailed investigation on the reaction mechanism reported previously [213], bisphosphoramides such as **112**

Table 10.11 Aldol addition between aldehydes and various silyl enol ethers.

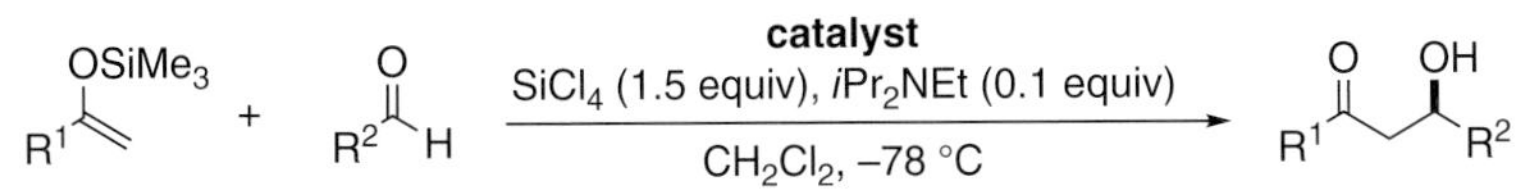

Entry	R^1	R^2	Catalyst (eq loading)	Time (h)	Yield (%)	er (major)
1	*n*Bu	C_6H_5	**112** (0.05)	2.5	86	99.5 : 0.5
2	*n*Bu	4-$CH_3OC_6H_4$	**112** (0.05)	4	97.5	99.5 : 0.5
3	*n*Bu	4-$CF_3C_6H_4$	**112** (0.05)	4	96	99.5 : 0.5
4	*n*Bu	(*E*)-PhCH=CH	**112** (0.05)	4	97.5	99.5 : 0.5
5	*n*Bu	2-Furyl	**112** (0.05)	6	88	99.5 : 0.5
6[a)]	H	Ph	**112** (0.15)	24	80	97.1 : 2.9

a) Product isolated as β-hydroxy dimethyl acetal.

were employed as efficient Lewis base catalysts for this type of transformation (Table 10.11) [218]. In this case, however, the trialkylsilyl enol ether acts as an external nucleophile able to attack the aldehyde activated by coordination to the silicon atom of tetrachlorosilane.

In the presence of 5 mol% of **112**, the addition of α-substituted trimethylsilyl enol ethers to a variety of aromatic, olefinic, and heteroaromatic aldehydes was accomplished in excellent yields (entries 1–5). Aldehydes bearing both electron-withdrawing and electron-donating substituents as well as heteroaromatic aldehydes react smoothly, allowing the formation of corresponding aldol products in high yields and selectivities in short reaction times. This method is also applicable to the crossed aldol reaction between aldehydes, by the addition of the trimethylsilyl enol ether of acetaldehyde to benzaldehyde (entry 6). In this case, increased catalytic loading and reaction time is necessary to improve the efficiency of the process. Moreover, for reasons of stability, a quenching with methanol is necessary, leading to the formation of the corresponding β-hydroxy dimethyl acetal in good yields. However, it must be noted that only trimethylsilyl enol ethers are effective substrates in crossed aldol reactions and no product is formed when more sterically hindered trialkylsilyl groups are used [219].

10.4.2.2 Aldol Reactions Involving Trialkylsilyl Ketene Acetals

The scope of this Lewis base–activated, $SiCl_4$-mediated process can be extended by engaging other π-nucleophiles of enhanced nucleophilicity such as silyl ketene acetals. In 2002, Denmark et al. reported a highly enantio- and diastereoselective addition of acetate- and propanoate-derived silyl ketene acetals to a variety of aromatic, olefinic, propargylic, and aliphatic aldehydes (Table 10.12) [216]. Except for aliphatic substrates (entry 3), which require long reaction times, the reactions catalyzed by **112** proceed rapidly at −78 °C, yielding the corresponding β-hydroxy esters in high chemical and stereochemical efficiency. Both electron-rich and electron-poor aromatic aldehydes react

Table 10.12 Aldol addition between aldehydes and various silyl ketene acetals.

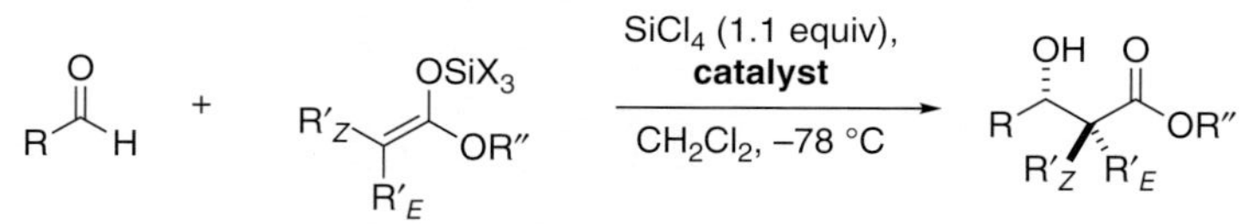

Entry	R	R'_E, R'_Z	R″	SiX_3	Catalyst (equiv loading)	Time (h)	Yield (%)	*syn:anti*	er (major)
1	Ph	H, H	Me	TBDMS	**112** (0.05)	0.25	86	—	99.5 : 0.5
2	PhCH=C(CH_3)	H, H	Me	TBDMS	**112** (0.05)	0.25	98	—	72.5 : 27.5
3	cyclohexyl	H, H	Me	TDBMS	**112** (0.05)	6	86	—	94.0 : 6.0
4	Ph	Me, H	Et	TBDMS	**112** (0.05)	3	78	98 : 2	88.8 : 11.2
5	Ph	H, Me	*t*Bu	TBDMS	**112** (0.05)	3	73	99 : 1	99.5 : 0.5
6	Ph	Me, Me	Me	TMS	**66** (0.1)	12	67	—	58.5 : 41.5
7	Ph	Me, Me	Me	TMS	**45** (0.1)	12	96	—	76 : 24
8	$PhCH_2CH_2$	Me, Me	Me	TMS	**45** (0.1)	12	nd	—	—

smoothly with different silyl ketene acetals. The reaction is diastereoconvergent, leading, uncharacteristically for a Mukaiyama aldol addition, to the formation of the *anti*-aldol addition product independent of the geometry of the carbon–carbon double bond (Table 10.12, entries 4 and 5). For this type of transformation, bisphosphoramides are the preferred catalysts, and it should be noted that other classes of catalysts such as bis *N*-oxide **66** [220] and phosphine oxides (BINAPO **45**) [221] are able to promote the addition of silyl ketene acetals to aromatic aldehydes only (entry 6), albeit with lower yields and diastereoselectivities.

Because the absolute configuration of the β-hydroxy esters obtained with catalyst **112** is independent of the double-bond geometry, (attack on the *Re* face of the aldehyde using (*S,S*)-**112**), it is reasonable to hypothesize that a common siliconium ion is involved for both pathways. Simple PM3 semiempirical calculations [24] reveal that the carbonyl oxygen of the aldehyde is coordinated *trans* to the phosphoramide P=O donor due to the presence of stabilizing π–π interaction between the aromatic ring of the aldehydes and one of the naphthyl rings of the catalyst in an edge-on manner [222]. In this conformation, the attack of the silyl ketene acetals onto the *Si* face is precluded by the chiral backbone of the catalysts, as shown in Figure 10.4.

Silyl ketene acetals are also employed as nucleophiles in the catalytic, enantioselective glycolate aldol addition to aldehydes for the reparation of stereodefined 1,2-diol units, in which the bond between the vicinal diol and both stereogenic centers are formed concomitantly as part of the process [223]. By kinetic enolization of a glycolate ester, *Z*-ketene acetals are formed exclusively with high geometrical purity. Addition of these nucleophiles to aldehydes performed in the presence of $SiCl_4$ and catalyzed by **112** proceeds with the formation of the corresponding products in high yields and diastereoselectivities (Table 10.13). However, it was found that this transformation was strongly influenced by all of the substituents on the double bond of the enoxysilane moiety. By varying the size of the substituents OR^1 and OR^2 on the silyl ketene acetal, both *syn*- and *anti*-diastereomers could be obtained with high stereoselectivity (Table 10.13).

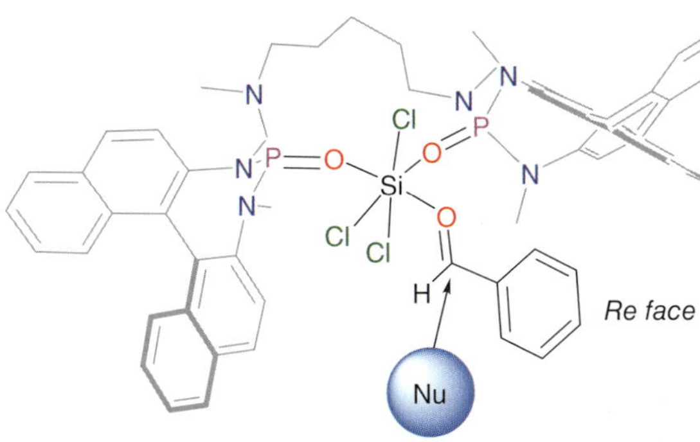

Figure 10.4 Stereochemical model for the addition to benzaldehyde using catalysts **112**.

Table 10.13 Glycolate aldol reaction with aldehydes.

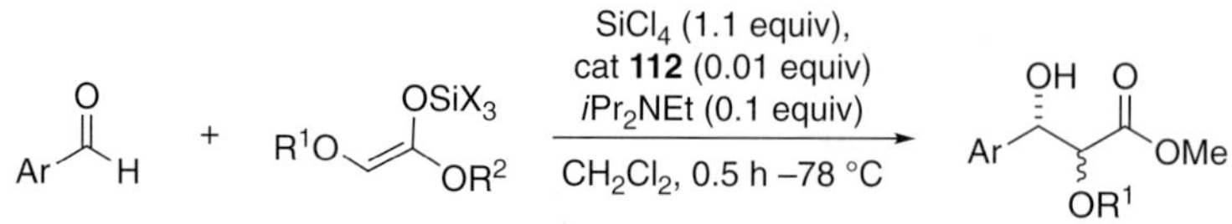

Entry	Ar	R^1	R^2	SiX_3	Yield (%)	*syn:anti*	er (major)
1	Ph	$Me_2(Ph)C$—	Me	TMS	87	99 : 1	96.6 : 3.4
2	1-Naphthyl	$Me_2(Ph)C$—	Me	TMS	94	>99 : 1	98.5 : 1.5
3	Ph	$Me_2(Et)C$—	Me	TBS	91	>1 : 99	95.1 : 4.9
4	1-Naphthyl	$Me_2(Et)C$—	Me	TBS	93	>1 : 99	97.4 : 2.6
5	$PhCH_2CH_2$	Bn	CH(*i*Pr)$_2$	TBS	82	2 : 98	96 : 4 : 3.6
6	PhCH=CH	Bn	CH(*i*Pr)$_2$	TBS	90	1 : 99	98 : 2
7	Ph	Me	Mr	TMS	98	57 : 43	73.5 : 26.5
8	Ph	*i*Pr	Me	TMS	95	86 : 14	80 : 20
9	Ph	*t*Bu	Me	TMS	93	99 : 1	93.5 : 6.5

To rationalize the stereochemical outcome, semiempirical computational analysis of the transition states was performed using a PM3 Hessian. These calculations show that the aldol addition proceeds through an open transition structure, wherein nonbonding steric interactions are the principal factor responsible for the experimentally observed differences in diastereoselectivity (Figure 10.5). Two different models of stereoselection were evaluated: model **xxi**, responsible for the formation of the *syn*-product, and model **xxvi** that leads to the formation of the *anti*-product.

Computational analysis suggests that model **xxii** engenders the fewest unfavorable steric interactions between the catalyst and the silyl ketene acetal (that adopts a so-called pinwheel conformation) [224]. In model **xxi**, the variation of the size of the OR^2 group has little influence on the reaction rate because the bulky *tert*-butyl group and the trimethylsilyl group of the silyl ketene acetal are located away from the catalyst complex, resulting in the formation of the *syn*-product. On the other hand, silyl ketene acetals bearing less sterically hindered ether groups prefer structure **xxiv**, which presents only a small steric interaction between the catalyst and the methoxy substituent of the silyl ketene acetal. In this case, *anti*-products were formed with high diastereoselectivity.

The complete catalytic cycle for the aldol reactions involving silyl ketene acetals is reported in Scheme 10.39 and it is analogous to that previously reported in Scheme 10.33 for the allylation reaction promoted by $SiCl_4$.

10.4.2.3 Vinylogous Aldol Addition

The addition of silyl dienolates to aldehydes represents the logical extension of the addition of silyl ketene acetals to aldehydes and was largely studied by different groups in the presence of phosphoramide-based catalysts and $SiCl_4$ as

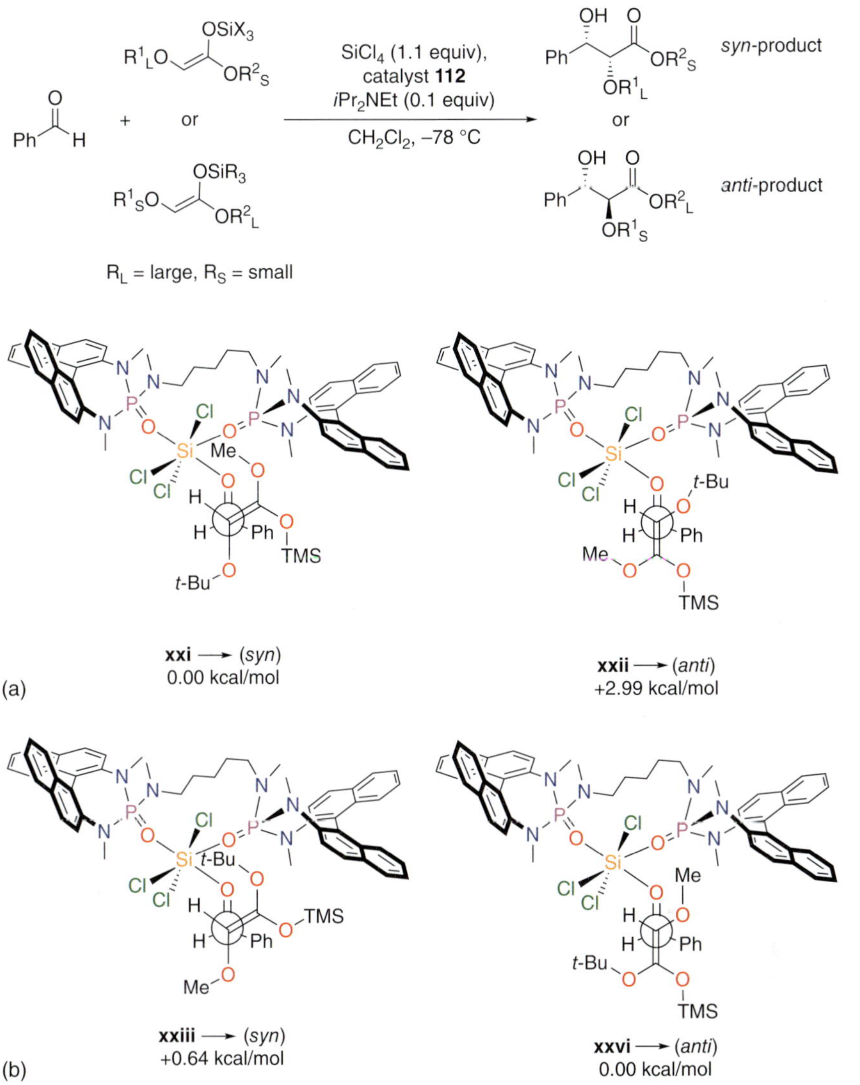

Figure 10.5 Models of stereoselection for catalytic, enantioselective addition of glycolate-derived silyl ketene acetals to aldehydes. (a) Transition states for glycolate aldol reactions with a bulky α-alkoxy group. (b) Transition states for glycolate aldol reactions with a bulky ester.

promoter [225–227]. Interestingly, despite the ambident nucleophilic characteristic of dienol ethers or vinyl ketene acetals, i.e. reaction at either the α- or the γ-carbon atom of the extended conjugated system, only γ-hydroxy enones with *E* configuration at the double bond are formed (Scheme 10.40).

In general, the reaction rate is dependent on the bulk of the substituents at γ-position of the dienol ether; the greater the steric hindrance, the lower the reactivity of the substrate in vinylogous aldol addition reactions. However, the γ-site selectivity and *E*-double-bond selectivity is independent of all substitution

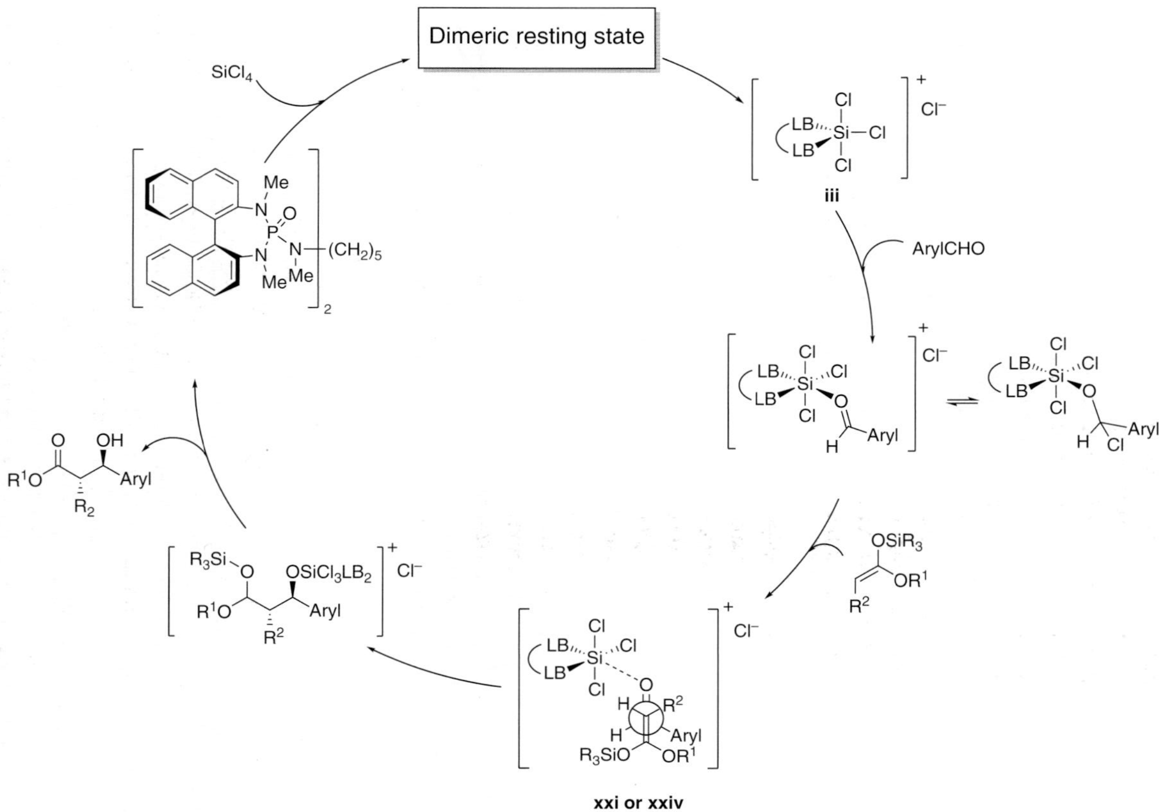

Scheme 10.39 Catalytic cycle for the addition of silyl ketene acetals to aldehydes.

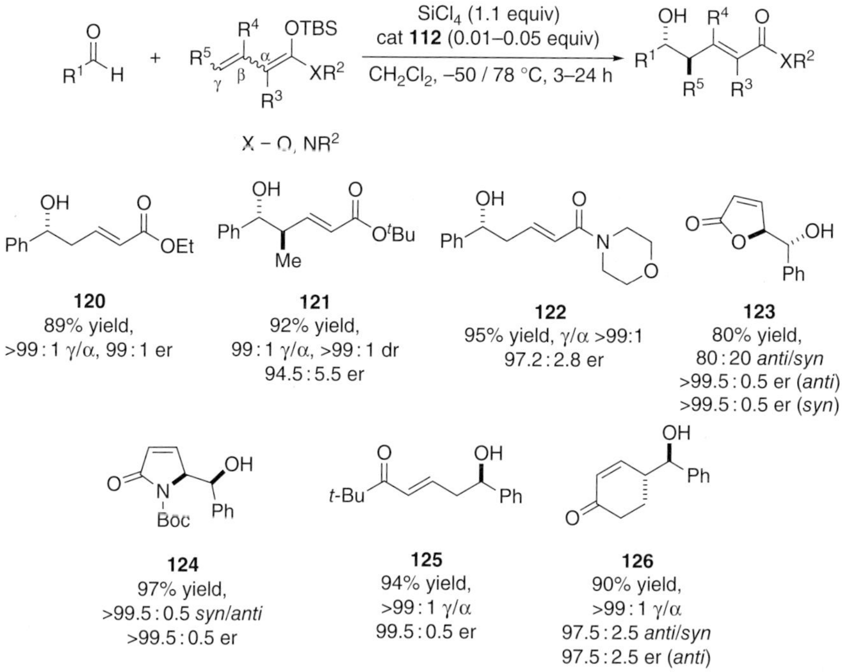

Scheme 10.40 Vinylogous aldol addition with aldehydes.

patterns on the dienolate and the reaction proceeds smoothly with aromatic, olefinic, and aliphatic aldehydes. Highly site-selective, vinylogous aldol additions involving conjugated *N,O*-silyl ketene acetals [228], 2-trimethylsilyloxyfurans [229], *N*-protected silyloxypyrroles [230], and silyl dienol ethers derived from simple α,β-unsaturated ketones [231] have been also successfully reported (compounds **122–126** of Scheme 10.40).

More recently, Curti et al. reported a bisphosphoramide-catalyzed, $SiCl_4$-mediated bisvinylogous and hypervinylogous aldol additions involving polyunsaturated silyloxy furans [232]. In this case, neutral, electron-rich, and electron-poor aromatic aldehydes afford the corresponding bisvinylogous aldol products in high yields, with complete ε-site selectivities and with excellent enantioselectivities in the range 97 : 3–99 : 1 er (Scheme 10.41). Unfortunately, in these cases, the control over the geometry of the exocyclic double bond is not very good and products with different *E/Z* ratios are formed.

10.4.3 Synthesis of Nitrile Derivatives from Silyl Ketene Imines

Mechanistic investigations of aldol reactions involving enoxysilane derivatives of aldehydes, ketones, and esters clearly illustrate that the electrophile (aldehyde) is the only species directly coordinated to silicon tetrachloride which, in turn, is activated by the chiral Lewis base. Because the facial differentiation of the substrate is controlled only by the configuration of the chiral Lewis base, other types of π-nucleophiles were investigated in Lewis base–catalyzed aldol

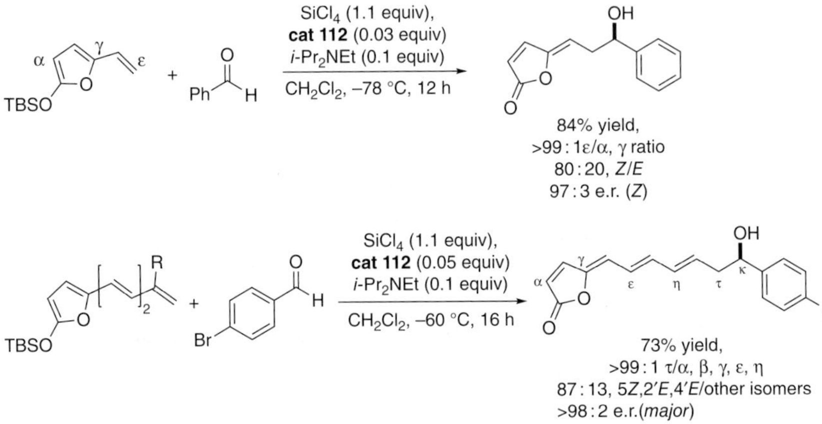

Scheme 10.41 Stereoselective bisvinylogous and hypervinylogous aldol additions promoted by $SiCl_4$ and catalyzed by **112**.

addition. Among them, silyl ketene imines have been successfully employed for the synthesis of enantioenriched β-hydroxy nitriles containing a quaternary stereogenic center. Using stable *N*-TBS ketene imines as substrates, the selective formation of nitrile derivatives proceeds in short reaction times, leading to the formation of products in good yield and with high levels of both diastereo- and enantioselectivity (Table 10.14) [233].

High yields and diastereoselectivities are observed with electron-rich aryl-substituted ketene imines; however, the best enantioselectivities are obtained with electron-poor aryl-substituted ketene imines (entries 1–3). Interestingly, triisopropylsilyl ketene imines are also reactive with aliphatic aldehydes; however, longer reaction times and $n\mathrm{Bu_4NI}$ are necessary to observe product formation (entry 4). The effect of $n\mathrm{Bu_4NI}$ is important because it increases the ionic strength of the solution, causing a modification of the equilibrium between hypercoordinate complexes **iv** and **iv′**) [234].

Table 10.14 Synthesis of nitrile derivatives from silyl ketene imines.

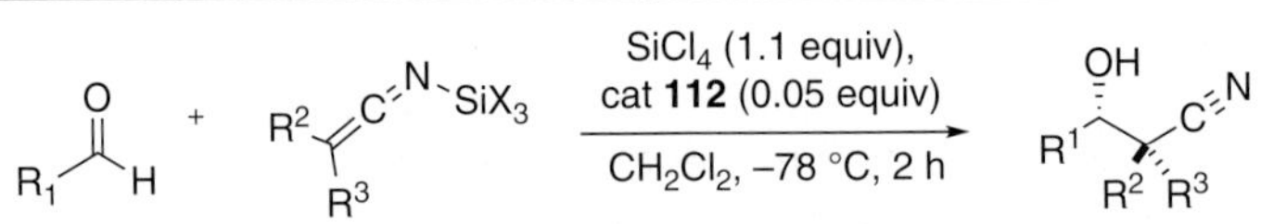

Entry	R^1	R^2	R^3	SiX_3	Yield (%)	dr	er (*major*)
1	C_6H_5	C_6H_5	Me	TBS	87	95 : 5	98.5 : 1.5
2	C_6H_5	4-CH_3O—C_6H_4	Me	TBS	90	99 : 1	99.1 : 0.9
3	C_6H_5	4-CF_3—C_6H_4	Et	TBS	73	97 : 3	99.5 : 0.5
4[a]	$CH_3(CH_2)_4$	C_6H_5	Me	TIPS	94	91 : 9	98 : 2

a) Reaction performed at −20 °C for 20 h with $n\mathrm{Bu_4NI}$ as additive.

Table 10.15 Addition of silyl ketene imines to aromatic enones.

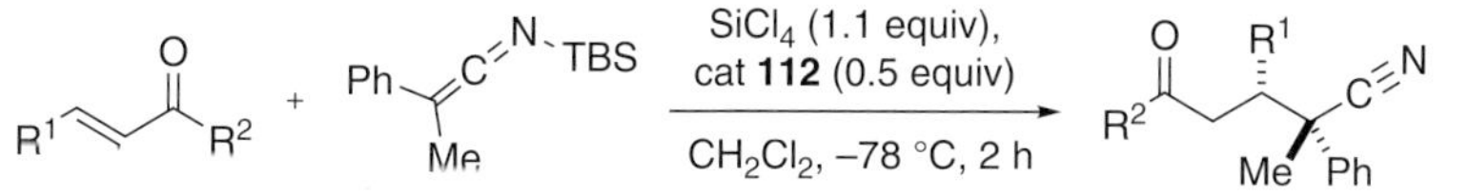

Entry	R_1	R_2	Yield (%)	1,4:1,2 addition	*syn:anti*	er (*syn*)
1	C_6H_5	CH_3	80	99 : 1	54 : 46	80 : 20
2	CH_3	H	74	92 : 8	84 : 16	51 : 49
3	C_6H_5	H	84	92 : 8	90 : 10	70 : 30
4	4-CH_3O—C_6H_4	H	79	95 : 5	91 : 9	72 : 28
5	2-Furyl	H	78	92 : 8	81 : 19	79 : 21

Other substrates such as α,β-unsaturated aldehydes and ketones are employed in the synthesis of nitrile derivatives from silyl ketene imines in a highly site-selective, Mukaiyama–Michael addition reaction (Table 10.15) [235, 236]. The combination of $SiCl_4$ with catalyst **112** promotes the addition of silyl ketene imines to a variety of aromatic enones and enals with high 1,4-selectivity and moderate diastereo- and enantioselectivity.

Highly site-selective 1,4-additions take place in good yield with α,β-unsaturated aldehydes, independent of the type of substitution on those substrates. However, poor diastereoselectivity is observed when the reaction is performed with α,β-unsaturated ketones.

A further extension of the scope of this reaction involves the use of *N*-silyl vinyl ketene imines that can be considered synthetic equivalents of allylic nitriles [237]. (*E*)-α,β-Unsaturated nitriles are preferentially formed in good yields and with good to excellent enantioselectivities, confirming that the *N*-silyl vinylketene imine undergoes addition onto the *Re* face of the aldehyde in an *s-trans* conformation. Various electron-rich, electron-poor, and hindered aromatic aldehydes are competent substrates in terms of yield, but only less sterically hindered electron-rich aromatic aldehydes exhibit the highest enantiomeric ratios (up to 98.5 : 1.5) (Table 10.16).

10.4.4 Passerini Reaction

The Passerini reaction is a three-component reaction between a carboxylic acid, a carbonyl compound such as a ketone or aldehyde, and an isocyanide that leads to the formation of α-acyloxy amides. Initially reported by Passerini in 1921 [238], the first example of catalytic, enantioselective version was developed over 80 years later. In 2003, Denmark and Fan reported the first example of catalytic and stereoselective synthesis of α-acyloxy amides using a phosphoramide-catalyzed and $SiCl_4$-mediated process (Scheme 10.42) [239, 240].

According to this protocol, α-hydroxy amides can be synthetized in high yields and high enantioselectivities, starting from aromatic, conjugated, and aliphatic aldehydes. Electron-donating or electron-withdrawing substituents

Table 10.16 Addition of N-silyl vinyl ketene imines to aromatic aldehydes.

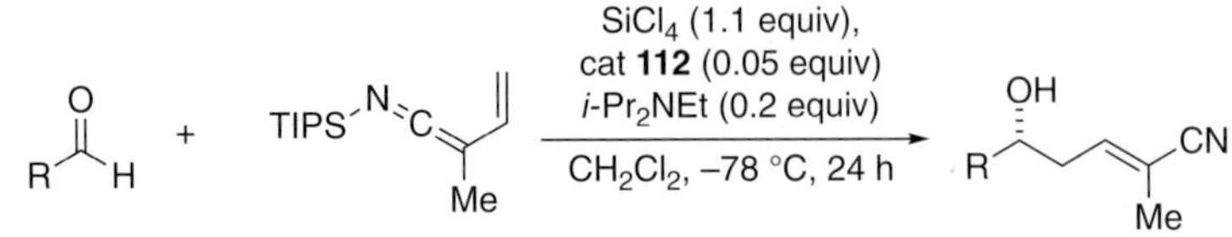

Entry	R	Yield (%)	*E*:*Z*	γ:α	er (*syn*)
1	Ph	97	99 : 1	98 : 2	93.5 : 6.5
2	C_6H_{11}	63	>98 : 2	>97 : 3	98.5 : 1.5
3	$(CH_3)_3CH_2CH_2$	82	>98 : 2	>97 : 3	96.5 : 3.5
4	4-CH_3O—C_6H_4	93	99 : 1	99 : 1	97 : 3
5	4-CF_3—C_6H_4	93	99 : 1	92 : 8	77 : 23

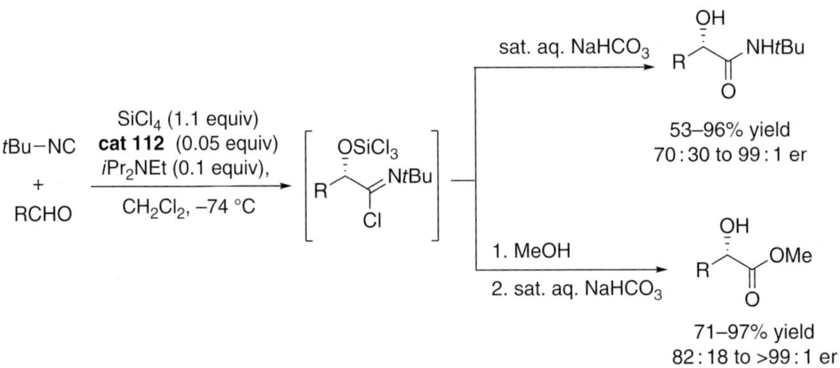

Scheme 10.42 Catalytic, enantioselective Passerini reaction promoted by $SiCl_4$.

on these substrates do not significantly influence the reaction and the desired product is formed in good yields and high enantioselectivities (Table 10.17, entries 1–3). On the other hand, enantioselectivities are dependent on the aldehyde structure, since sterically congested aromatic aldehydes cause a slight erosion of the enantioselection (entry 4), whereas aliphatic aldehydes afford lower enantiomeric ratios (entries 5 and 6).

Interestingly, by simply changing the quench procedure with the addition of MeOH before adding the saturated $NaHCO_3$ solution, it is possible to obtain carboxylic esters with comparable results. However, it must be noted that since the isocyanide itself is a strong Lewis base, the presence of $i Pr_2NEt$ is mandatory to suppress the self-catalyzed background reaction.

10.4.5 Phosphonylation of Aldehydes with Triethyl Phosphite

In addition to all of the carbon–carbon bond-forming transformations developed in the field of Lewis base–catalyzed, Lewis acid–mediated reactions, the

Table 10.17 Catalytic, enantioselective Passerini-type reaction.

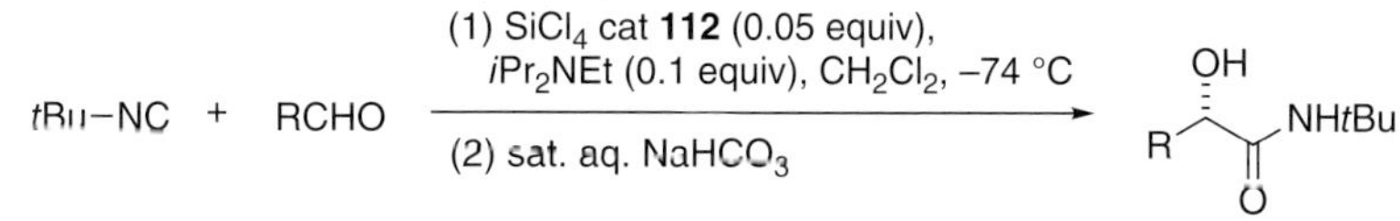

Entry	R	Yield (%)	er
1	Ph	96	>99 : 1
2	4-MeO—C_6H_4	89	98.3 : 1.7
3	4-CF_3—C_6H_4	89	96.5 : 3.5
4	(*E*)-PhCH=CH(CH_3)	86	67.4 : 32.6
5	$C_6H_5CH_2CH_2$	87	70 : 30
6	cyclohexyl	53	87.1 : 12.9

catalytic enantioselective Abramov-type phosphonylation of aldehydes with trialkyl phosphites stands out for the construction of carbon–phosphorus bonds (Scheme 10.43) [241, 242]. Although α-hydroxy phosphonates are obtained with modest enantioselectivity, this transformation allows access to α-substituted phosphonyl compounds that are useful synthetic intermediates. Different Lewis bases have been investigated, but only phosphine oxides afford enantioenriched products.

Reaction between benzaldehyde and triethyl phosphate performed in the presence of $SiCl_4$ and catalyst DMPP-DIOPO **127** provides the corresponding product in 90% yield and 78 : 28 er. Catalyst **127** is slightly more reactive than

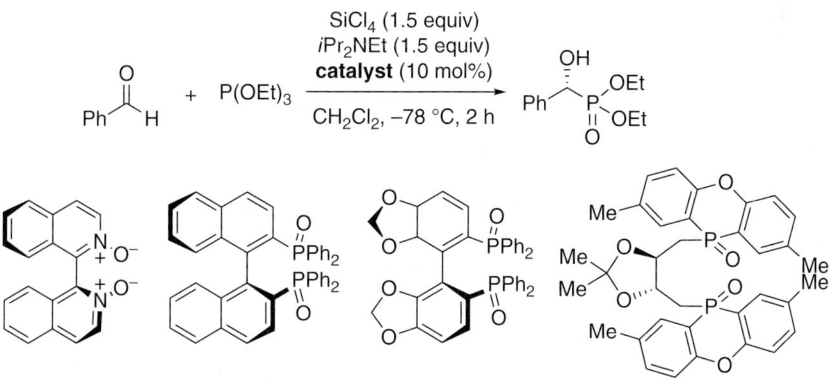

66	BINAPO **45**	SEGPHOSO **46**	DMPP-DIOPO **127**
60% yield 53.5:46.5 er	91% yield 70.5:29.5 er	95% yield 65.5:34.5 er	90% yield 72:28 er

Scheme 10.43 Phosphonylation of aldehydes with triethyl phosphite catalyzed by Lewis bases.

BINAPO **45**, since products are obtained with better yields; however, the enantioselectivities are comparable. Despite this, catalyst **127** is not influenced by the electronic characteristics of the aldehyde and is preferred to other phosphine oxide catalysts. The proposed mechanism involves the initial coordination of the aldehyde to a hypercoordinate silicon species derived by the interaction of $SiCl_4$ with the Lewis base. At this point, the trialkyl phosphite attacks the activated aldehyde to yield an α-trichlorosilyloxy phosphonate after Arbuzov-type dealkylation to form the corresponding alkyl chloride [243]. However, no kinetic or spectroscopic investigations have been performed and the details of this mechanism are still unclear.

10.5 Interaction of the Activated Substrate with an Externally Coordinated Nucleophile

As illustrated in Scheme 10.2, if species **iii** is generated in the presence of a nucleophile that can be coordinated to the silicon atom, pathway C2 (blue box) will be preferred. As a consequence, the formation of neutral hexacoordinate species **v** will occur. This new species generates species **vi** after the dissociation of a chlorine ion followed by the coordination of the substrate. Because this new species contains both the substrate and the nucleophile, the formation of a new carbon–carbon bond will occur in an intramolecular manner under the influence of the Lewis base.

10.5.1 Direct Aldol Reactions and Double Aldol Reaction

In 2009, Nakajima and coworkers developed an enantioselective aldol reaction of cyclic ketones with aromatic aldehydes using a hypercoordinate silicon complex generated from silicon tetrachloride and phosphine oxide **45** (Table 10.18, entries 1 and 2) [244]. This species activates the ketone to form a trichlorosilyl enol ether in situ by a deprotonation performed in the presence of diisopropylethylamine. Subsequent activation of the trichlorosilyl enol ether by the phosphine oxide promotes an aldol reaction with an aldehyde through a six-membered, chair-like transition state to afford the corresponding β-hydroxy ketones in moderate yields and good diastereo- and enantioselectivities. However, diastereo- and enantioselection can be improved by replacing $SiCl_4$ with the stronger Lewis acid $SiCl_3OTf$ (entry 3) [245]. An improved version of the direct aldol reaction between ketones and aldehydes has been developed by Benaglia and coworkers using the more electron-rich BITIOPO **73** and its modifications [246], in which aldol products are obtained in higher yields and with better diastereo- and enantioselectivities (entries 4 and 5) [247]. The use of aziridinyl phosphonate **129** instead results in lower yields [248].

^{1}H and ^{31}P NMR spectroscopic investigations have provided an explanation for the *anti*-diastereoselectivity observed and a catalytic cycle has been proposed (Scheme 10.44) [247]. According to this cycle, the formation of a six-membered, chair-like transition state is quite reasonable considering the predominant formation of the *anti*-diastereomer. Between the two limiting transition states for

Table 10.18 Direct aldol addition catalyzed by phosphine oxides in the presence of $SiCl_4$.

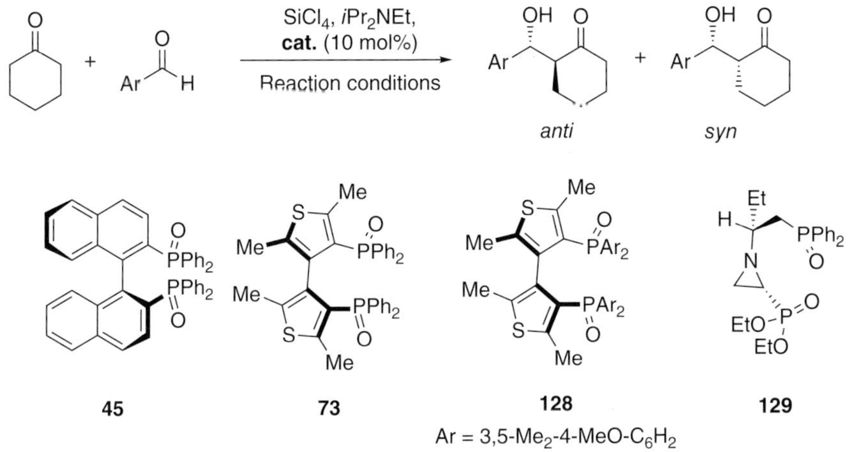

Entry	Catalyst	Ar	Reaction conditions	Yield (%)	*anti:syn*	er (*anti*)
1	**45**	Ph	CH_2Cl_2, −25 °C, 30 h	40	79 : 21	80 : 20
2	**45**	Ph	EtCN, 0 °C, 2 h	81	86 : 14	77 : 23
3[a)]	**45**	Ph	EtCN, −40 °C, 2 h	76	97 : 3	91.5 : 8.5
4	**73**	Ph	CH_2Cl_2, −25 °C, 30 h	70	93 : 7	91.5 : 8.5
5	**73**	4-$NO_2C_6H_5$	CH_2Cl_2, −25 °C, 15 h	71	87 : 13	96.5 : 3.5
6	**128**	Ph	CH_2Cl_2, 0 °C, 18 h	67	88 : 12	86 : 14
7	**129**	Ph	CH_2Cl_2, −20 °C, 21 h	29	92 : 8	68.5 : 31.5

a) Reaction performed with $SiCl_3OTf$ and Hex_2NMe.

the formation of the *anti*-diastereomer (**xxv** and **xxvi**, Scheme 10.43), model **xxvi** is favored because it does not cause steric repulsion between the diphenylphosphinoyl group of the Lewis base and the cyclohexanone-derived enol ether.

Unfortunately, aliphatic aldehydes are completely unreactive in direct aldol additions, indicating that the formation of the corresponding chlorohydrin probably occurs. On the other hand, aliphatic aldehydes can act as aldol donors by enolization with $SiCl_4$ with an amine base in a direct, crossed aldol addition, although the products are isolated in modest yields and lower enantioselectivities (Scheme 10.45a) [249].

Direct crossed aldol addition products can be obtained also starting from two aliphatic ketones in the presence of $SiCl_3OTf$ and a catalytic amount of chiral phosphine oxide **45**. In this case, both cyclic and acyclic ketones afforded the aldol products in very good yield and stereoselectivity (Scheme 10.45b) [243].

Aldol reactions performed with acyclic ketones in the presence of catalytic amounts of phosphine oxides provide branched [250–252] or linear [253] double aldol adducts (Scheme 10.46). By simple variation of the reaction stoichiometry, the combination of aromatic or aliphatic ketones with 2 equiv of aromatic

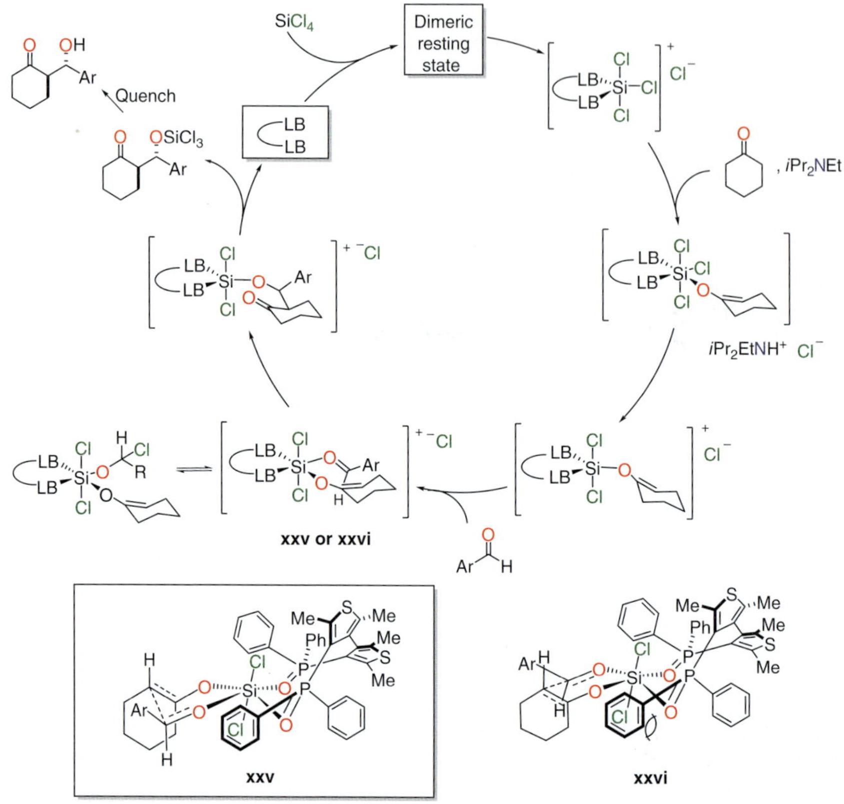

Scheme 10.44 Proposed catalytic cycle for direct aldol reaction catalyzed by BITIOPO **73**.

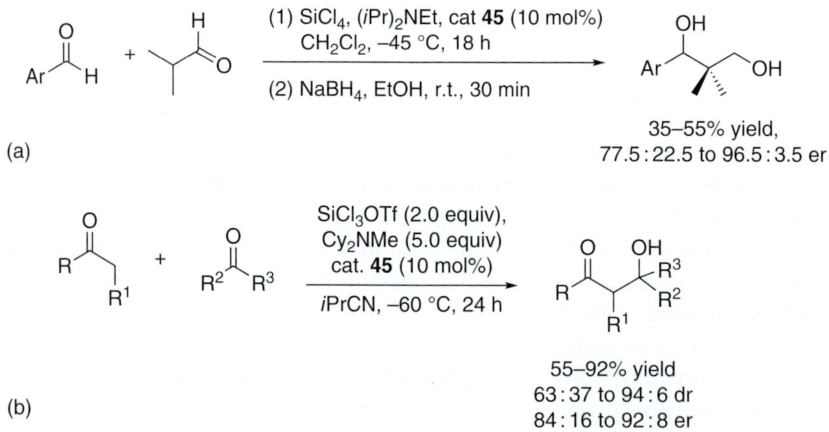

Scheme 10.45 Direct crossed aldol reaction promoted by BINAPO phosphine oxide. (a) Crossed aldol reactions of aldehydes. (b) Crossed aldol reaction of ketones.

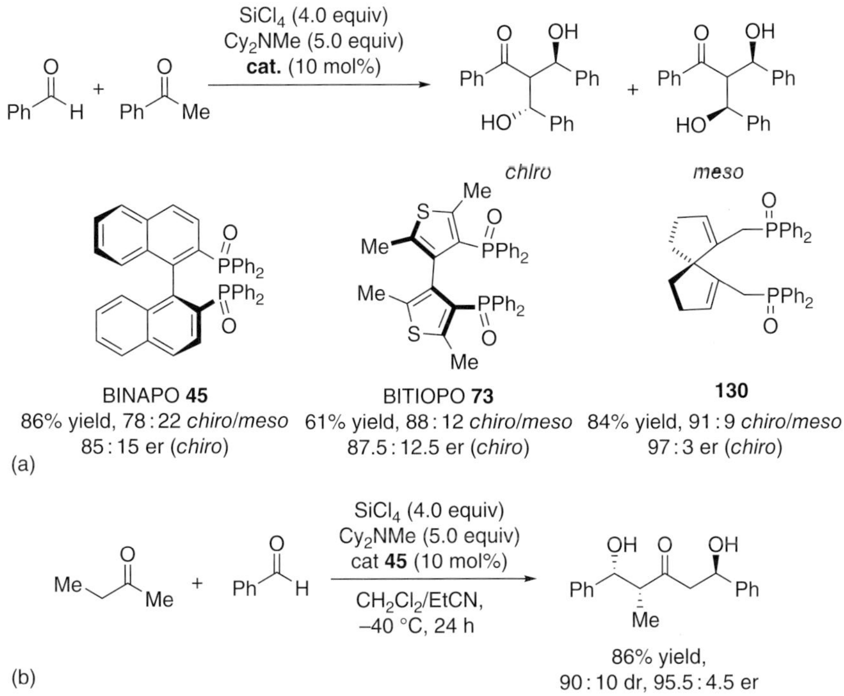

Scheme 10.46 Branched and linear double aldol addition promoted by phosphine oxides. (a) Branched, double aldol addition. (b) Linear, double aldol addition.

aldehydes leads to the formation of a mixture of two diastereomers in high yields with good diastereomeric ratios and high enantioselectivities. In the case of branched double aldol addition, the formation of two diastereoisomers is observed, in which only one of them is chiral (called "*chiro*"). The achiral isomer ("*meso*") can exist as two diastereomers because the central carbon is an achirotopic, stereogenic center. The products are obtained in good yield and high diastereo- and enantioselectivities using either electron-rich or electron-deficient aldehydes. This type of transformation has been studied using BINAPO **45** or BITIOPO **73**, a spiro[4,4]-1,6-nonadiene catalyst **130** developed by Ding and coworkers [254].

Direct, double aldol reaction can be also performed with enolizable ketones (Scheme 10.45b). In this case, the final product is a linear, double aldol adduct containing a 1,5-diol moiety with up to three stereogenic centers generated in a single operation. Both electron-rich and electron-deficient aldehydes react smoothly to generate corresponding products in good yield and high diastereo- and enantioselectivities. The relative configuration of the major diastereomer is 1,2-*syn*-1,5-*anti*. Interestingly, branched-type, double aldol reactions promoted by $SiCl_4$ and catalyzed by phosphine oxides have found application in the enantioselective syntheses of 2,3-dihydropyran-4-ones [255], whereas a linear double aldol addition protocol was employed in the synthesis of stereopentads [256] and in the total synthesis of (−)-ericanone [251, 257]. More interestingly, Kotani

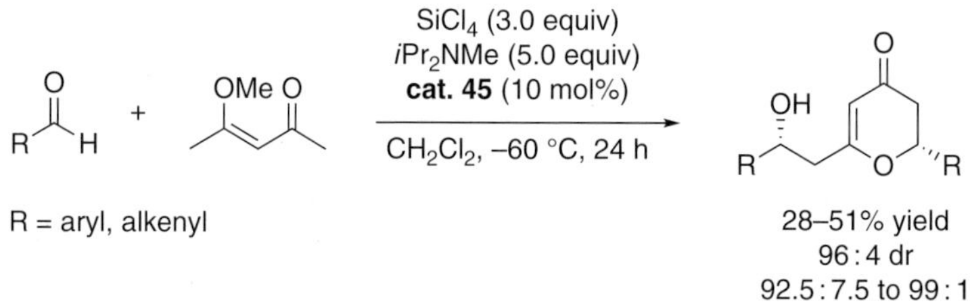

Scheme 10.47 Asymmetric aldol/vinylogous aldol/cyclization reaction catalyzed by BINAPO and promoted by $SiCl_4$.

and coworkers reported a cascade process for the synthesis of 2,6-disubstituted 2,3-dihydro-4-pyranones involving an asymmetric aldol reaction followed by a vinylogous aldol reaction with a consequent cyclization reaction using phosphine oxide catalysts. BINAPO **45** sequentially activates silicon tetrachloride and trichlorosilyl enol ethers to facilitate the asymmetric aldol/vinylogous aldol reaction of 4-methoxy-3-penten-2-one and conjugated aldehydes in a highly enantioselective manner. The subsequent cyclization produces 2,6-disubstituted 2,3-dihydro-4-pyranones in variable yields but in high enantioselectivities (Scheme 10.47) [258].

A modified version of the direct aldol addition was reported in 2010 by Sugiura et al., who described a diastereo- and enantioselective reductive aldol reaction with trichlorosilane using chiral Lewis bases as the catalysts [259]. Chiral Lewis base catalysts activate trichlorosilane to promote the tandem conjugate reduction/aldol reaction of α,β-unsaturated ketones with aldehydes to give enantiomerically enriched β-hydroxy ketones with good to high *syn* diastereo- and enantioselectivities. The reaction can engage α,β-unsaturated aldehydes owing to the chemoselective conjugate reduction of enones in the presence of enals (Scheme 10.48a). The same reaction has been implemented using a tertiary amine as the hydride donor in the presence of $SiCl_3OTf$ [260]. β-Hydroxy ketones

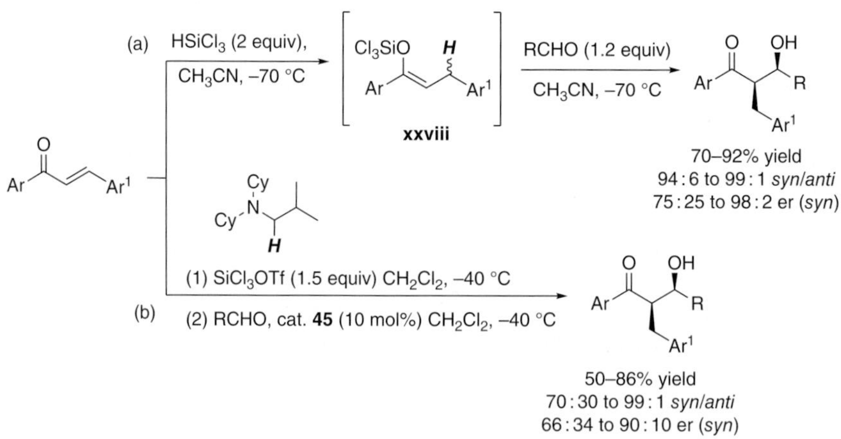

Scheme 10.48 Reductive aldol reaction catalyzed by BINAPO.

are obtained in modest diastereo- and enantioselectivities starting from both electron-rich and electron-deficient aldehydes under the action of phosphine oxide **45** (Scheme 10.48b). The proposed mechanism involves the formation of the (*Z*)-trichlorosilyl enol ethers **xxviii** which react via chair-like, six-membered ring transition states. Contrary to other Lewis base–catalyzed reactions discussed so far, aliphatic aldehydes react smoothly to afford the corresponding aldol adducts in good yields.

10.5.1.1 Direct Aldol Addition of Activated Thioesters

The scope of direct, stereoselective, catalytic aldol reactions has been extended by Rossi and Benaglia to the addition of *S*-phenyl thioesters [261] and trifluoroethyl thioesters [262] to aldehydes. Generally, esters are not considered good nucleophiles for these processes because of the lower acidity of the α-protons compared to those of ketones or aldehydes. However, the presence of a thiophenyl or trifluoromethyl moiety is able to decrease the pK_a of such protons to allow deprotonation by a tertiary amine. Reaction of properly activated trifluoroethyl thioesters with aromatic aldehydes, catalyzed by **73** in the presence of $SiCl_4$, affords β-hydroxy trifluoroethyl thioesters in moderate to good yields, up to 98 : 2 *syn*/*anti*-ratio and up to 94.5 : 5.5 enantiomeric ratio for the *syn* isomer (Table 10.15). The reaction can be performed also with *S*-phenyl thioesters, although the products are obtained with lower yields and diastereoselectivities. The observed inversion of the diastereoselectivity compared to the aldol reaction performed on ketones is correlated to the different configuration adopted by the *O*-silyl enolate **xxi** because of the bulky peripheral groups ("pinwheel" conformation, Scheme 10.49).

The origin of stereoselectivity has been elucidated with aid of DFT calculations which confirmed that the formation of the aldol product proceeds via a Zimmerman–Traxler, cationic, chair-like transition structure in which the aldehyde is coordinated and activated by the positively charged hypercoordinate silicon atom [261].

As shown in Table 10.19, electron-rich phosphine oxide **73** catalyzes the reaction in improved yields and enantioselectivities compared with BINAPO **45**. Proline-derived catalysts **43** and **44** instead are less effective in this type of

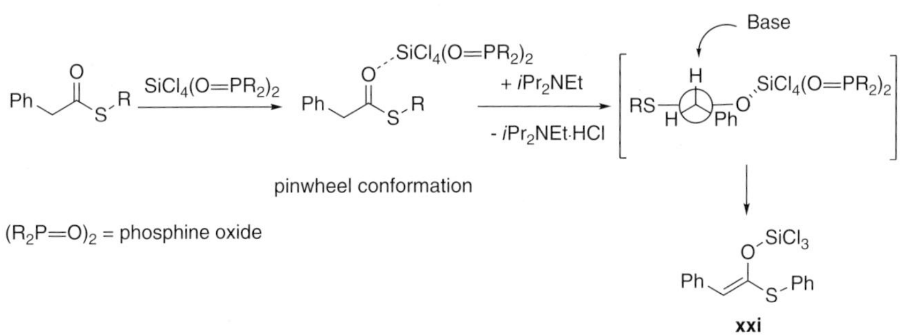

Scheme 10.49 In situ formation of silyl ketene thioacetal promoted by the presence of $SiCl_4$ and iPr_2NEt.

Table 10.19 Direct aldol addition catalyzed by phosphine oxides in the presence of $SiCl_4$.

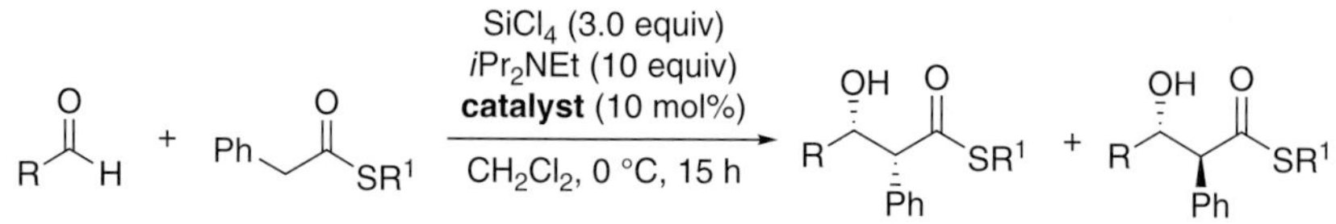

Entry	Catalyst	R	R^1	Yield (%)	*syn/anti*	er (*syn*)
1	**45**	Ph	CH_2CF_3	35	97 : 3	90.5 : 9.5
2	**45**	Ph	SPh	33	31 : 69	65 : 35
3	**73**	Ph	CH_2CF_3	80	98 : 2	94.5 : 5.5
4	**73**	Ph	SPh	62	71 : 29	87 : 13
5	**73**	2,6-$(MeO)_2C_6H_3$	SPh	61	92 : 8	95.5 : 4.5
6	**43**	Ph	CH_2CF_3	51	97 : 3	75.5 : 24.5
7	**44**	Ph	CH_2CF_3	40	94 : 6	73.5 : 26.5

transformation (entries 6 and 7) [86]. Unfortunately, β-hydroxy-activated thioesters are not amenable to direct, double aldol addition.

10.5.2 Enantioselective Morita–Baylis–Hillman Reaction

Hypercoordinate silicon species generated from a chiral phosphine oxide catalyst and silicon tetrachloride can promote the enantioselective Morita–Baylis–Hillman reaction. In this transformation, a chloride anion liberated from the hypervalent silicon complex smoothly generates a γ-chloro silyl enol ether that subsequently reacts with an aldehyde to afford the Baylis–Hillman adducts in good yields and with good enantioselectivities (Scheme 10.50) [263].

A particular case of this transformation is related to the use of α-substituted acroleins in combination with aromatic aldehydes and $SiCl_3OTf$. In such cases, the Morita–Baylis–Hillman product is not formed and the chlorinated aldol precursor can be isolated as a diol after reduction with $NaBH_4$ (Scheme 10.51) [264].

In general, the stereostructure of the final product cannot be predicted considering the type of substituents on the acrolein or aldehyde employed. C_1–C_3

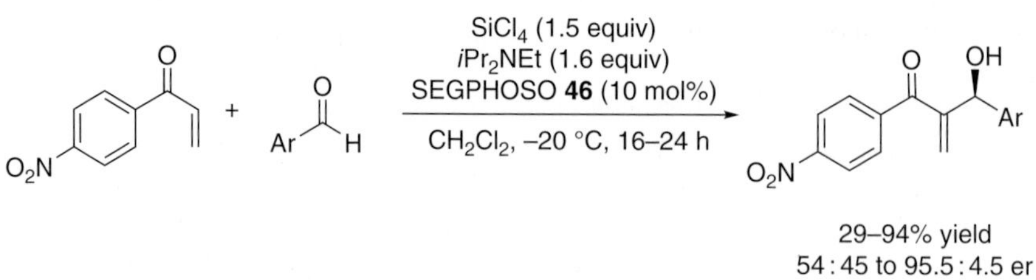

Scheme 10.50 Enantioselective MBH reaction catalyzed by SEGPHOSO **46**.

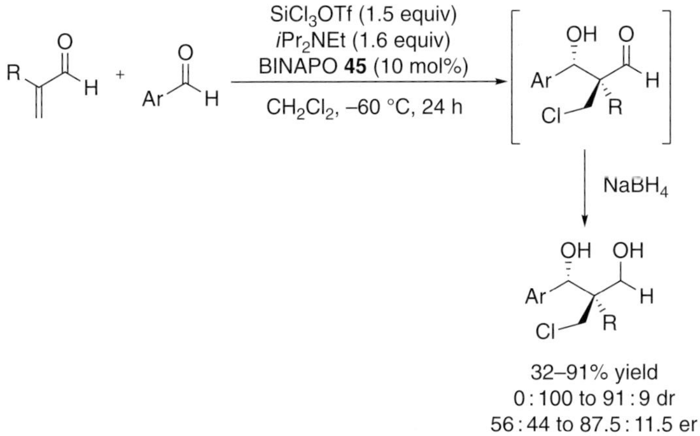

Scheme 10.51 Enantioselective chlorinative aldol reaction of α-substituted acroleins catalyzed by chiral phosphine oxides.

linearly substituted acroleins give the corresponding adducts with good enantioselectivities, but in lower diastereomeric ratios, in contrast with α-benzyl-substituted acrolein which allows the isolation of the corresponding product with almost complete diastereoselectivity, but in very low enantioselectivity. The use of electron-rich or electron-deficient aldehydes results in products with lower diastereoselectivities but higher enantioselectivities. However, hindered aldehydes destabilize the cyclic transition state, forcing the reaction to proceed by an acyclic pathway resulting in poorer diastereoselection.

10.5.3 Outlook and Perspective

Lewis base–catalyzed, Lewis acid–mediated reactions promoted by hypercoordinate silicon species are a powerful and useful alternative to reactions catalyzed by classical Lewis acids. Silicon tetrachloride is an inexpensive and readily available Lewis acid and can generate highly reactive, hypercoordinate species upon coordination with a chiral Lewis base. These newly formed hypercoordinate species can, in turn, activate different substrates to promote carbon–carbon and carbon–heteroatom bond formation. This new approach is very attractive, because the catalytically active species is generated only in the presence of the Lewis base and thus does not suffer from product inhibition and allows the use of the nascent Lewis acid in stoichiometric amounts. Indeed, the silicon moiety is frequently incorporated into the final product without interfering with catalytic turnover.

The high oxophilicity of silicon has enabled the activation of myriad functional groups for classical carbon–carbon bond-forming reactions. Future studies will evaluate the potential for this special class of Lewis acids to promote other reactions of carbonyl compounds such as cycloadditions, electrocyclic reactions, ene-reactions, pinacol-type rearrangements, and additions of other carbon nucleophiles.

Acknowledgment

We are grateful to the National Science Foundation and National Institutes of Health for generous financial support.

References

1 Gay-Lussac, J.L. and Thénard, J.A. (1809). Des propriétés de l'acide fluorique et surtout de son action sur le métal de la potasse. *Mem. Phys. Chem. Soc. Arcueil* 2: 317–331.
2 Davy, J. (1812). An account of some experiments on different combinations of fluoric acid. *Philos. Trans. R. Soc. London* 102: 352–369.
3 Chuit, C., Corriu, R.J.P., Reye, C., and Young, J.C. (1993). Reactivity of penta- and hexacoordinate silicon compounds and their role as reaction intermediates. *Chem. Rev.* 93: 1371–1448.
4 Gutmann, V. (1975). Empirical approach to molecular interactions. *Coord. Chem. Rev.* 15: 207–237.
5 Gutmann, V. (1978). *The Donor-Acceptor Approach to Molecular Interactions*. New York, NY: Plenum Press.
6 Kwart, H. and King, K. (1977). *Pentacovalency, in d-Orbitals in the Chemistry of Silicon, Phosphorus and Sulfur*. Berlin: Springer-Verlag.
7 Tandura, S.N., Voronkov, M.G., and Alekseev, N.V. (1986). Molecular and electronic structure of penta- and hexacoordinate silicon compounds. In: *Structural Chemistry of Boron and Silicon*, Topics in Current Chemistry, (editor Dr. Friedrich L.) vol. 131. Berlin, Heidelberg: Springer-Verlag.
8 Jensen, W.B. (2006). The origin of the term "hypervalent". *J. Chem. Educ.* 83: 1751.
9 Musher, J.I. (1969). The chemistry of hypervalent molecules. *Angew. Chem. Int. Ed. Engl.* 8: 54–68.
10 Curnow, O.J. (1998). A simple qualitative molecular-orbital/valence-bond description of the bonding in main group "hypervalent" molecules. *J. Chem. Educ.* 75: 910.
11 Gillespie, R. (2002). The octet rule and hypervalence: two misunderstood concepts. *Coord. Chem. Rev.* 233–234: 53–62.
12 Rundle, R.E. (1947). Halogen–halogen distances in halogen–polymer complexes. *J. Chem. Phys.* 15: 880–880.
13 Pimentel, G.C. (1951). The bonding of trihalide and bifluoride ions by the molecular orbital method. *J. Chem. Phys.* 19: 446–448.
14 Reed, A.E., Curtiss, L.A., and Weinhold, F. (1988). Intermolecular interactions from a natural bond orbital, donor-acceptor viewpoint. *Chem. Rev.* 88: 899–926.
15 Korlyukov, A.A., Lyssenko, K.A., Antipin, M.Y. et al. (2009). Si-fluoro substituted quasisilatranes (N→Si) $FYSi(OCH_2CH_2)_2NR$. *J. Organomet. Chem.* 694: 607–615.
16 Popelier, P.L.A. and Gillespie, R.J. (2001). *Chemical Bonding and Molecular Geometry from Lewis to Electron Densities*. New York, NY: Oxford University Press.

17 Korlyukov, A.A., Antipin, M.Y., Bolgova, Y.I. et al. (2010). Chemical bonding in the crystal structure of 1-hydrosilatrane. *Russ. Chem. Bull.* 58: 25–30.
18 Korlyukov, A.A., Antipin, M.Y., Buzin, M.I. et al. (2010). Molecular and crystal structure of two phases of 1-fluorosilatrane. Specific features of the electron density distribution. *J. Struct. Chem.* 50: 873–879.
19 Hagemann, M., Mix, A., Berger, R.J. et al. (2008). Strong intramolecular Si–N interactions in the chlorosilanes $Cl_{3-n}H_nSiOCH_2CH_2NMe_2$ (n = 1–3). *Inorg. Chem.* 47: 10554–10564.
20 Dilman, A.D., Levin, V.V., Korlyukov, A.A. et al. (2008). Complexation of tris(pentafluorophenyl)silanes with neutral Lewis bases. *J. Organomet. Chem.* 693: 1005–1019.
21 Frenking, G. (2003). Book review: chemical bonding and molecular geometry from Lewis to electron densities. By Ronald J. Gillespie and Paul L. A. Popelier. *Angew. Chem. Int. Ed.* 42: 143–147.
22 For an exhaustive discussion about Lewis Base-Lewis Acid Interaction see: Denmark, S.E. and Beutner, G.L. (2016). Principles, definitions, terminology, and orbital analysis of Lewis Base-Lewis acid interactions leading to catalysis. In: *Lewis Basis Catalysis in Organic Synthesis* (eds. E. Vedejs and S.E. Denmark), 31–54. Wiley-VCH.
23 Denmark, S.E. and Beutner, G.L. (2008). Lewis Base catalysis in organic synthesis. *Angew. Chem. Int. Ed.* 47: 1560–1638.
24 Denmark, S.E., Beutner, G.L., Wynn, T., and Eastgate, M.D. (2005). Lewis Base activation of Lewis acids: catalytic, enantioselective addition of silyl ketene acetals to aldehydes. *J. Am. Chem. Soc.* 127: 3774–3789.
25 Denmark, S.E. and Eklov, B.M. (2008). Neutral and cationic phosphoramide adducts of silicon tetrachloride: synthesis and characterization of their solution and solid-state structures. *Chemistry* 14: 234–239.
26 Kobayashi, S., Yasuda, M., and Hachiya, I. (1996). Trichlorosilane-dimethylformamide (Cl_3SiH-DMF) as an efficient reducing agent. Reduction of aldehydes and imines and reductive amination of aldehydes under mild conditions using hypercoordinate hydridosilicates. *Chem. Lett.* 25: 407–408.
27 Jones, S. and Warner, C.J. (2012). Trichlorosilane mediated asymmetric reductions of the C=N bond. *Org. Biomol. Chem.* 10: 2189–2200.
28 Rossi, S., Benaglia, M., Massolo, E., and Raimondi, L. (2014). Organocatalytic strategies for enantioselective metal-free reductions. *Catal. Sci. Technol.* 4: 2708–2723.
29 Benaglia, M., Guizzetti, S., and Rossi, S. (2011). Silicate-mediated stereoselective reactions catalyzed by chiral Lewis bases. In: *Catalytic Methods in Asymmetric Synthesis*, 579–624. Wiley.
30 Kočovský, P. and Malkov, A.V. (2016). Lewis bases as catalysts in the reduction of imines and ketones with Silanes ($n \rightarrow \sigma^*$). In: *Lewis Basis Catalysis in Organic Synthesis*, 1077–1112. Wiley-VCH.
31 Arceo, E. and Melchiorre, P. (2014). 8.03 Reduction of CN to CHNH by hydride delivery from C. In: *Comprehensive Organic Synthesis II*, 151–197. Elsevier Ltd.
32 Guizzetti, S. and Benaglia, M. (2010). Trichlorosilane-mediated stereoselective reduction of C=N bonds. *Eur. J. Org. Chem.* 5529–5541.

33 Iwasaki, F., Onomura, O., Mishima, K. et al. (1999). Catalytic activation of trichlorosilane for efficient and stereoselective reduction of ketones. *Tetrahedron Lett.* 40: 7507–7511.

34 Iwasaki, F., Onomura, O., Mishima, K. et al. (2001). First chemo- and stereoselective reduction of imines using trichlorosilane activated with N-formylpyrrolidine derivatives. *Tetrahedron Lett.* 42: 2525–2527.

35 Schreiner, P., Zhang, Z., Rooshenas, P., and Hausmann, H. (2009). Asymmetric transfer hydrogenation of ketimines with trichlorosilane: structural studies. *Synthesis* 2009: 1531–1544.

36 Baudequin, C., Chaturvedi, D., and Tsogoeva, S.B. (2007). Organocatalysis with chiral formamides: asymmetric allylation and reduction of imines. *Eur. J. Org. Chem.* 2007: 2623–2629.

37 Onomura, O., Kirira, P.G., Tanaka, T. et al. (2008). Diastereoselective arylation of l-proline derivatives at the 5-position. *Tetrahedron* 64: 7498–7503.

38 Matsumura, Y., Ogura, K., Kouchi, Y. et al. (2006). New efficient organic activators for highly enantioselective reduction of aromatic ketones by trichlorosilane. *Org. Lett.* 8: 3789–3792.

39 Wang, Z., Cheng, M., Wu, P. et al. (2006). L-piperazine-2-carboxylic acid derived N-formamide as a highly enantioselective Lewis basic catalyst for hydrosilylation of N-aryl imines with an unprecedented substrate profile. *Org. Lett.* 8: 3045–3048.

40 Wu, P., Wang, Z., Cheng, M. et al. (2008). Development of highly enantioselective new Lewis basic N-formamide organocatalysts for hydrosilylation of imines with an unprecedented substrate profile. *Tetrahedron* 64: 11304–11312.

41 Collados, J.F., Quiroga-Feijóo, M.L., and Alvarez-Ibarra, C. (2009). New chiral Lewis bases derived from L-pipecolinic acid showing stereocontrol highly dependent on the catalyst design in the hydrosilylation of N-phenyl ketimines. *Eur. J. Org. Chem.* 2009: 3357–3367.

42 Wang, Z., Wang, C., Zhou, L., and Sun, J. (2013). L-pipecolinic acid derived Lewis base organocatalyst for asymmetric reduction of N-aryl imines by trichlorosilane: effects of the side amide group on catalytic performances. *Org. Biomol. Chem.* 11: 787–797.

43 Wang, Z., Ye, X., Wei, S. et al. (2006). A highly enantioselective Lewis basic organocatalyst for reduction of N-aryl imines with unprecedented substrate spectrum. *Org. Lett.* 8: 999–1001.

44 Zhou, L., Wang, Z., Wei, S., and Sun, J. (2007). Evolution of chiral Lewis basic N-formamide as highly effective organocatalyst for asymmetric reduction of both ketones and ketimines with an unprecedented substrate scope. *Chem. Commun.* 2977–2979.

45 Xiao, Y.C., Wang, C., Yao, Y. et al. (2011). Direct asymmetric hydrosilylation of indoles: combined Lewis base and Bronsted acid activation. *Angew. Chem. Int. Ed.* 50: 10661–10664.

46 Kanemitsu, T., Umehara, A., Haneji, R. et al. (2012). A simple proline-based organocatalyst for the enantioselective reduction of imines using trichlorosilane as a reductant. *Tetrahedron* 68: 3893–3898.

47 Bonsignore, M., Benaglia, M., Raimondi, L. et al. (2013). Enantioselective reduction of ketoimines promoted by easily available (S)-proline derivatives. *Beilstein J. Org. Chem.* 9: 633–640.

48 Malkov, A.V., Stončius, S., MacDougall, K.N. et al. (2006). Formamides derived from N-methyl amino acids serve as new chiral organocatalysts in the enantioselective reduction of aromatic ketimines with trichlorosilane. *Tetrahedron* 62: 264–284.

49 Wang, Z., Wei, S., Wang, C., and Sun, J. (2007). Enantioselective hydrosilylation of ketimines catalyzed by Lewis basic C_2-symmetric chiral tetraamide. *Tetrahedron: Asymmetry* 18: 705–709.

50 Wen, Y., Xiong, Y., Chang, L. et al. (2007). Chiral bisformamides as effective organocatalysts for the asymmetric one-pot, three-component strecker reaction. *J. Org. Chem.* 72: 7715–7719.

51 Malkov, A.V., Mariani, A., MacDougall, K.N., and Kocovsky, P. (2004). Role of noncovalent interactions in the enantioselective reduction of aromatic ketimines with trichlorosilane. *Org. Lett.* 6: 2253–2256.

52 Malkov, A.V., Figlus, M., Stoncius, S., and Kocovsky, P. (2007). Organocatalysis with a fluorous tag: asymmetric reduction of imines with trichlorosilane catalyzed by amino acid-derived formamides. *J. Org. Chem.* 72: 1315–1325.

53 Malkov, A.V., Stoncius, S., and Kocovsky, P. (2007). Enantioselective synthesis of 1,2-diarylaziridines by the organocatalytic reductive amination of alpha-Chloro ketones. *Angew. Chem. Int. Ed.* 46: 3722–3724.

54 Malkov, A.V., Stoncius, S., Vrankova, K. et al. (2008). Dynamic kinetic resolution in the asymmetric synthesis of beta-amino acids by organocatalytic reduction of enamines with trichlorosilane. *Chem. Eur. J.* 14: 8082–8085.

55 Malkov, A.V., Vranková, K., Sigerson, R.C. et al. (2009). New organocatalysts for the asymmetric reduction of imines with trichlorosilane. *Tetrahedron* 65: 9481–9486.

56 Malkov, A.V., Vrankova, K., Stoncius, S., and Kocovsky, P. (2009). Asymmetric reduction of imines with trichlorosilane, catalyzed by sigamide, an amino acid-derived formamide: scope and limitations. *J. Org. Chem.* 74: 5839–5849.

57 Onomura, O., Kouchi, Y., Iwasaki, F., and Matsumura, Y. (2006). New organic activators for the enantioselective reduction of aromatic imines with trichlorosilane. *Tetrahedron Lett.* 47: 3751–3754.

58 Zheng, Y., Xue, Z., Liu, L. et al. (2013). Highly enantioselective hydrosilylation of N-(1,2-diarylethylidene)arylamines. *Org. Biomol. Chem.* 11: 412–415.

59 Zheng, H.J., Chen, W.B., Wu, Z.J. et al. (2008). Highly enantioselective synthesis of beta-amino acid derivatives by the Lewis base catalyzed hydrosilylation of beta-enamino esters. *Chemistry* 14: 9864–9867.

60 Jiang, Y., Liu, L.X., Yuan, W.C., and Zhang, X.M. (2012). Lewis base organocatalyzed enantioselective hydrosilylation of 1,4-benzoxazines. *Synlett* 23: 1797–1800.

61 Chen, X., Hu, X.Y., Shu, C. et al. (2013). Synthesis of a series of novel chiral Lewis base catalysts and their application in promoting asymmetric hydrosilylation of beta-enamino esters. *Org. Biomol. Chem.* 11: 3089–3093.

62 Zheng, H., Deng, J., Lin, W., and Zhang, X. (2007). Enantioselective hydrosilylation of ketimines with trichlorosilane promoted by chiral N-picolinoyl amino alcohols. *Tetrahedron Lett.* 48: 7934–7937.
63 Guizzetti, S., Benaglia, M., Cozzi, F., and Annunziata, R. (2009). Chiral Lewis base promoted trichlorosilane reduction of ketimines. An enantioselective organocatalytic synthesis of chiral amines. *Tetrahedron* 65: 6354–6363.
64 Guizzetti, S. and Benaglia, M. (2009). European Patent Application EP07023240.0, 22 September 2008.
65 Brenna, D., Porta, R., Massolo, E. et al. (2017). A new class of low-loading catalysts for highly enantioselective, metal-free imine reduction of wide general applicability. *ChemCatChem* 9: 941–945.
66 Guizzetti, S., Benaglia, M., Cozzi, F. et al. (2009). Enantioselective catalytic reduction of ketoimines with trichlorosilane promoted by readily available chiral Lewis bases. *Chirality* 21: 233–238.
67 Guizzetti, S., Benaglia, M., and Celentano, G. (2009). 1,1′-binaphthyldiamine-based Lewis bases as readily available and efficient organocatalysts for the reduction of N-aryl and N-alkyl ketimines. *Eur. J. Org. Chem.* 2009: 3683–3687.
68 El-Fayyoumy, S., Todd, M.H., and Richards, C.J. (2009). Can we measure catalyst efficiency in asymmetric chemical reactions? A theoretical approach. *Beilstein J. Org. Chem.* 5: 67.
69 Guizzetti, S., Benaglia, M., and Rossi, S. (2009). Highly stereoselective metal-free catalytic reduction of imines: an easy entry to enantiomerically pure amines and natural and unnatural alpha-amino esters. *Org. Lett.* 11: 2928–2931.
70 Xue, Z.-Y., Jiang, Y., Yuan, W.-C., and Zhang, X.-M. (2010). The first highly enantioselective Lewis base organocatalyzed hydrosilylation of α-imino esters. *Eur. J. Org. Chem.* 2010: 616–619.
71 Jiang, Y., Chen, X., Zheng, Y. et al. (2011). Highly diastereoselective and enantioselective synthesis of alpha-hydroxy beta-amino acid derivatives: Lewis base catalyzed hydrosilylation of alpha-acetoxy beta-enamino esters. *Angew. Chem. Int. Ed.* 50: 7304–7307.
72 Xue, Z.-Y., Liu, L.-X., Jiang, Y. et al. (2012). Highly enantioselective Lewis base organocatalyzed hydrosilylation of γ-imino esters. *Eur. J. Org. Chem.* 2012: 251–255.
73 Zhang, X.-M., Shu, C., Hu, X.-Y. et al. (2014). Lewis Base catalyzed asymmetric hydrosilylation of α-substituted β-enamino esters: facile access to enantioenriched β2-amino esters via dynamic kinetic resolution. *Synlett* 25: 1879–1882.
74 Weickgenannt, A. and Oestreich, M. (2011). A new motif for Lewis Base catalysis: asymmetric reduction of β-enamino esters. *ChemCatChem* 3: 1527–1529.
75 Xue, Z.-Y., Jiang, Y., Peng, X.-Z. et al. (2010). The first general, highly enantioselective Lewis base organo- catalyzed hydrosilylation of benzoxazinones and quinoxalinones. *Adv. Synth. Catal.* 352: 2132–2136.
76 Chen, X., Zheng, Y., Shu, C. et al. (2011). Enantioselective synthesis of 4-substituted4,5-dihydro-1H-[1,5]benzodiazepin-2(3H)-ones by the Lewis base-catalyzed hydrosilylation. *J. Org. Chem.* 76: 9109–9115.

77 Malkov, A.V., Liddon, A.J., Ramirez-Lopez, P. et al. (2006). Remote chiral induction in the organocatalytic hydrosilylation of aromatic ketones and ketimines. *Angew. Chem. Int. Ed.* 45: 1432–1435.

78 Malkov, A.V., Stewart-Liddon, A.J., McGeoch, G.D. et al. (2012). Catalyst development for organocatalytic hydrosilylation of aromatic ketones and ketimines. *Org. Biomol. Chem.* 10: 4864–4877.

79 Pei, D., Wang, Z., Wei, S. et al. (2006). S-Chiral sulfinamides as highly enantioselective organocatalysts. *Org. Lett.* 8: 5913–5915.

80 Wang, C., Wu, X., Zhou, L., and Sun, J. (2008). A highly enantioselective organocatalytic method for reduction of aromatic N-alkyl ketimines. *Chemistry* 14: 8789–8792.

81 Wu, X., Li, Y., Wang, C. et al. (2011). Chiral Lewis base catalyzed highly enantioselective reduction of N-alkyl *beta*-enamino esters with trichlorosilane and water. *Chem. Eur. J.* 17: 2846–2848.

82 Wang, C., Wu, X., Zhou, L., and Sun, J. (2015). L-Valine derived chiral N-sulfinamides as effective organocatalysts for the asymmetric hydrosilylation of N-alkyl and N-aryl protected ketimines. *Org. Biomol. Chem.* 13: 577–582.

83 Chelouan, A., Recio, R., Borrego, L.G. et al. (2016). Sulfinamide phosphinates as chiral catalysts for the enantioselective organocatalytic reduction of imines. *Org. Lett.* 18: 3258–3261.

84 Pei, D., Zhang, Y., Wei, S. et al. (2008). Rationally-designed, S-chiral bissulfinamides as highly enantioselective organocatalysts for reduction of ketimines. *Adv. Synth. Catal.* 350: 619–623.

85 Liu, X.W., Wang, C., Yan, Y. et al. (2013). An organocatalyst bearing stereogenic carbon and sulfur centers as an efficient promoter for enantioselective hydrosilylation of 1,4-benzooxazines. *J. Org. Chem.* 78: 6276–6280.

86 Bonsignore, M., Benaglia, M., Cozzi, F. et al. (2012). Readily available (S)-proline-derived organocatalysts for the Lewis acid-mediated Lewis base-catalyzed stereoselective aldol reactions of activated thioesters. *Tetrahedron* 68: 8251–8255.

87 Benaglia, M., Bonsignore, M., Annunziata, R., and Celentano, G. (2011). New, readily available organocatalysts for the enantioselective reduction of α-imino- and β-imino esters. *Synlett* 1085–1088.

88 Sugiura, M., Kumahara, M., and Nakajima, M. (2009). Asymmetric synthesis of 4H-1,3-oxazines: enantioselective reductive cyclization of N-acylated *beta*-amino enones with trichlorosilane catalyzed by chiral Lewis bases. *Chem. Commun.* 3585–3587.

89 Malkov, A.V., Figlus, M., and Kocovsky, P. (2008). Polymer-supported organocatalysts: asymmetric reduction of imines with trichlorosilane catalyzed by an amino acid-derived formamide anchored to a polymer. *J. Org. Chem.* 73: 3985–3995.

90 Figlus, M., Caldwell, S.T., Walas, D. et al. (2010). Dendron-anchored organocatalysts: the asymmetric reduction of imines with trichlorosilane, catalysed by an amino acid-derived formamide appended to a dendron. *Org. Biomol. Chem.* 8: 137–141.

91 Malkov, A.V., Figlus, M., Cooke, G. et al. (2009). Organocatalysts immobilised onto gold nanoparticles: application in the asymmetric reduction of imines with trichlorosilane. *Org. Biomol. Chem.* 7: 1878–1883.

92 Malkov, A.V., Figlus, M., Prestly, M.R. et al. (2009). Soluble polymer-supported organocatalysts: asymmetric reduction of imines with trichlorosilane catalyzed by an amino acid derived formamide anchored to a soluble polymer. *Chem. Eur. J.* 15: 9651–9654.
93 Fernandes, S.D., Porta, R., Barrulas, P.C. et al. (2016). Stereoselective reduction of imines with trichlorosilane using solid-supported chiral picolinamides. *Molecules* 21: https://doi.org/10.3390/molecules21091182.
94 Porta, R., Benaglia, M., Annunziata, R. et al. (2017). Solid supported chiral N-picolylimidazolidinones: recyclable catalysts for the enantioselective, metal- and hydrogen-free reduction of imines in batch and in flow mode. *Adv. Synth. Catal.* 359: 2375–2382.
95 Wang, T., Di, X., Wang, C. et al. (2016). Reductive hydrazination with trichlorosilane: a method for the preparation of 1,1-disubstituted hydrazines. *Org. Lett.* 18: 1900–1903.
96 Ye, J., Wang, C., Chen, L. et al. (2016). Chiral Lewis base-catalyzed, enantioselective reduction of unprotected β-Enamino esters with trichlorosilane. *Adv. Synth. Catal.* 358: 1042–1047.
97 Wagner, C., Kotthaus, A.F., and Kirsch, S.F. (2017). The asymmetric reduction of imidazolinones with trichlorosilane. *Chem. Commun.* 53: 4513–4516.
98 Orlandi, M., Brenna, D., Harms, R. et al. (2018). Recent developments in the reduction of aromatic and aliphatic nitro compounds to amines. *Org. Process Res. Dev.* https://doi.org/10.1021/acs.oprd.6b00205.
99 Orlandi, M., Tosi, F., Bonsignore, M., and Benaglia, M. (2015). Metal-free reduction of aromatic and aliphatic nitro compounds to amines: a $HSiCl_3$-mediated reaction of wide general applicability. *Org. Lett.* 17: 3941–3943.
100 Porta, R., Puglisi, A., Colombo, G. et al. (2016). Continuous-flow synthesis of primary amines: metal-free reduction of aliphatic and aromatic nitro derivatives with trichlorosilane. *Beilstein J. Org. Chem.* 12: 2614–2619.
101 Rossi, S., Benaglia, M., Porta, R. et al. (2015). A stereoselective catalytic nitroaldol reaction as the key step in a strategy for the synthesis of the renin inhibitor aliskiren. *Eur. J. Org. Chem.* 2015: 2531–2537.
102 Skell, P.S. and Harris, R.F. (1969). Reactions of triatomic carbon with alcohols. *J. Am. Chem. Soc.* 91: 699–701.
103 Benkeser, R.A. and Smith, W.E. (1969). Forming carbon-silicon bonds. Reductive silylation of carbonyl compounds. *J. Am. Chem. Soc.* 91: 1556–1557.
104 Benkeser, R.A. and Smith, W.E. (1968). Trichlorosilane-tertiary amine combinations as reducing agents for polyhalo compounds. Potential analogies with phosphorus chemistry. *J. Am. Chem. Soc.* 90: 5307–5309.
105 Benkeser, R.A., Gaul, J.M., and Smith, W.E. (1969). Silylation of organic halides. New method of forming the carbon–silicon bond. *J. Am. Chem. Soc.* 91: 3666–3667.
106 Benkeser, R.A., Foley, K.M., Gaul, J.M. et al. (1969). Silylation of acid chlorides. Forming the carbon–silicon bond. *J. Am. Chem. Soc.* 91: 4578–4579.
107 Roy, S., Stollberg, P., Herbst-Irmer, R. et al. (2005). Carbene-dichlorosilylene stabilized phosphinidenes exhibiting strong intramolecular charge transfer transition. *J. Am. Chem. Soc.* 137: 150–153.

108 Orlandi, M., Benaglia, M., Tosi, F. et al. (2016). $HSiCl_3$-mediated reduction of nitro-derivatives to amines: is tertiary amine-stabilized $SiCl_2$ the actual reducing species? *J. Org. Chem.* 81: 3037–3041.

109 Jacobsen, E.N. (2000). Asymmetric catalysis of epoxide ring-opening reactions. *Acc. Chem. Res.* 33: 421–431.

110 Wilson, T.W. and Denmark, S.E. (2016). Reactions of epoxides ($N \rightarrow \sigma^*$). In: *Lewis Base Catalysis in Organic Synthesis*, 1113–1152. Wiley-VCH.

111 Yudin, A.K. (2006). *Aziridines and Epoxides in Organic Synthesis*. Wiley-VCH.

112 Andrews, G.C., Crawford, T.C., and Contillo, L.G. (1981). Nucleophilic catalysis in the insertion of silicon halides into oxiranes: a synthesis of O-protected vicinal halohydrins. *Tetrahedron Lett.* 22: 3803–3806.

113 Garrett, C.E. and Fu, G.C. (1997). π-Bound phosphorus heterocycles as catalysts: ring opening of epoxides with TMSCl in the presence of a phosphaferrocene. *J. Org. Chem.* 62: 4534–4535.

114 Wang, L.S. and Hollis, T.K. (2003). Demonstration of a phosphazirconocene as a catalyst for the ring opening of epoxides with TMSCl. *Org. Lett.* 5: 2543–2545.

115 Hwang, G.-H., Ryu, E.-S., Park, D.-K. et al. (2001). A new class of planar-chiral ferrocenes: serendipitous formation of 1,2-ferrocenediylazaphosphinines. *Organometallics* 20: 5784–5787.

116 Kim, T.-J., Paek, S.H., Shim, S.C., and Cho, C.S. (2003). 1,2-ferrocenediylazaphosphinines: a new class of nucleophilic catalysts for ring-opening of epoxides. *Synlett* 849–851.

117 Denmark, S.E., Barsanti, P.A., Beutner, G.L., and Wilson, T.W. (2007). Enantioselective ring opening of epoxides with silicon tetrachloride in the presence of a chiral Lewis base: mechanism studies. *Adv. Synth. Catal.* 349: 567–582.

118 Tao, B., Lo, M.M.C., and Fu, G.C. (2001). Planar-chiral pyridine-N-oxides, a new family of asymmetric catalysts: exploiting an η^5-C_5Ar_5Ligand to achieve high enantioselectivity. *J. Am. Chem. Soc.* 123: 353–354.

119 Nakajima, M., Saito, M., Uemura, M., and Hashimoto, S. (2002). Enantioselective ring opening of meso-epoxides with tetrachlorosilane catalyzed by chiral bipyridine N,N′-dioxide derivatives. *Tetrahedron Lett.* 43: 8827–8829.

120 Chelucci, G., Baldino, S., Pinna, G.A. et al. (2008). Chiral pyridine N-oxides derived from monoterpenes as organocatalysts for stereoselective reactions with allyltrichlorosilane and tetrachlorosilane. *Tetrahedron* 64: 7574–7582.

121 Malkov, A.V., Gordon, M.R., Stoncius, S. et al. (2009). Desymmetrization of cyclic *meso*-epoxides with silicon tetrachloride catalyzed by PINDOX, a chiral bipyridine mono-N-oxide. *Org. Lett.* 11: 5390–5393.

122 Takenaka, N., Sarangthem, R.S., and Captain, B. (2008). Helical chiral pyridine N-oxides: a new family of asymmetric catalysts. *Angew. Chem. Int. Ed.* 47: 9708–9710.

123 Tokuoka, E., Kotani, S., Matsunaga, H. et al. (2005). Asymmetric ring opening of *meso*-epoxides catalyzed by the chiral phosphine oxide BINAPO. *Tetrahedron: Asymmetry* 16: 2391–2392.

124 Kotani, S., Furusho, H., Sugiura, M., and Nakajima, M. (2013). Facile synthesis of chiral 1,2-chlorohydrins via the ring-opening of *meso*-epoxides catalyzed by chiral phosphine oxides. *Tetrahedron* 69: 3075–3081.

125 Simonini, V., Benaglia, M., and Benincori, T. (2008). Novel chiral biheteroaromatic diphosphine oxides for Lewis base activation of Lewis acids in enantioselective allylation and epoxide opening. *Adv. Synth. Catal.* 350: 561–564.

126 Pu, X., Qi, X., and Ready, J.M. (2009). Allenes in asymmetric catalysis: asymmetric ring opening of meso-epoxides catalyzed by allene-containing phosphine oxides. *J. Am. Chem. Soc.* 131: 10364–10365.

127 Ogasawara, M., Kotani, S., and Nakajima, H. (2013). Atropisomeric chiral dienes in asymmetric catalysis: C(2)-symmetric (Z,Z)-2,3-bis[1-(diphenylphosphinyl)ethylidene]tetralin as a highly active Lewis base organocatalyst. *Angew. Chem. Int. Ed.* 52: 13798–13802.

128 Denmark, S.E., Williams, B.J., Eklov, B.M. et al. (2010). Design, validation, and implementation of a rapid-injection NMR system. *J. Org. Chem.* 75: 5558–5572.

129 Denmark, S.E., Eklov, B.M., Yao, P.J., and Eastgate, M.D. (2009). On the mechanism of Lewis base catalyzed aldol addition reactions: kinetic and spectroscopic investigations using rapid-injection NMR. *J. Am. Chem. Soc.* 131: 11770–11787.

130 Kobayashi, S. and Hirabayashi, R. (1999). Highly stereoselective synthesis of homoallylic amines based on addition of allyltrichlorosilanes to benzoylhydrazones under neutral conditions. *J. Am. Chem. Soc.* 121: 6942–6943.

131 Hirabayashi, R., Ogawa, C., Sugiura, M., and Kobayashi, S. (2001). Highly stereoselective synthesis of homoallylic amines based on addition of allyltrichlorosilanes to benzoylhydrazones. *J. Am. Chem. Soc.* 123: 9493–9499.

132 Ogawa, C., Sugiura, M., and Kobayashi, S. (2002). Stereoselective synthesis of bothsyn-andanti-N-tert-alkylamines using highly stereospecific crotylation of ketone-derived acylhydrazones with crotyltrichlorosilanes. *J. Org. Chem.* 67: 5359–5364.

133 Kobayashi, S., Ogawa, C., Konishi, H., and Sugiura, M. (2003). Chiral sulfoxides as neutral coordinate-organocatalysts in asymmetric allylation of N-acylhydrazones using allyltrichlorosilanes. *J. Am. Chem. Soc.* 125: 6610–6611.

134 Ogawa, C., Konishi, H., Sugiura, M., and Kobayashi, S. (2004). Phosphine oxides as efficient neutral coordinate-organocatalysts for stereoselective allylation of N-acylhydrazones. *Org. Biomol. Chem.* 2: 446–448.

135 Fernandez, I., Valdivia, V., Gori, B. et al. (2005). The isopropylsulfinyl group: a useful chiral controller for the asymmetric aziridination of sulfinylimines and the organocatalytic allylation of hydrazones. *Org. Lett.* 7: 1307–1310.

136 García-Flores, F., Flores-Michel, L.S., and Juaristi, E. (2006). Asymmetric allylation of N-benzoylhydrazones promoted by novel C_2-symmetric bissulfoxide organocatalysts. *Tetrahedron Lett.* 47: 8235–8238.

137 Fernandez, I., Valdivia, V., Leal, M.P., and Khiar, N. (2007). C_2-symmetric bissulfoxides as organocatalysts in the allylation of benzoyl hydrazones: spacer and concentration effects. *Org. Lett.* 9: 2215–2218.

138 Fernandez, I., Alcudia, A., Gori, B. et al. (2010). Chiral sulfur derivatives in the allylation of acyl hydrazones: C_2-symmetric *bis*-sulfinamides as enhanced chiral organic promoters. *Org. Biomol. Chem.* 8: 4388–4393.

139 Ogawa, C., Sugiura, M., and Kobayashi, S. (2004). Stereospecific, enantioselective allylation of *alpha*-hydrazono esters by using allyltrichlorosilanes with BINAP dioxides as neutral-coordinate organocatalysts. *Angew. Chem. Int. Ed.* 43: 6491–6493.

140 Yamamoto, H. and Wadamoto, M. (2007). Silver-catalyzed asymmetric allylation: allyltrimethoxysilane as a remarkable reagent. *Chem. Asian J.* 2: 692–698.

141 Benaglia, M., Guizzetti, S., and Pignataro, L. (2008). Stereoselective reactions involving hypervalent silicate complexes. *Coord. Chem. Rev.* 252: 492–512.

142 Oestreich, M. and Rendler, S. (2005). Hypervalent silicon as a reactive site in selective bond-forming processes. *Synthesis* 2005: 1727–1747.

143 Orito, Y. and Nakajima, M. (2006). Lewis Base catalyzed asymmetric reactions involving hypervalent silicate intermediates. *Synthesis* 2006: 1391–1401.

144 Malkov, A.V. and Kočovský, P. (2016). Lewis base-catalyzed reactions of SiX_3-based reagents with C=O, C=N ($n \rightarrow \sigma^*$). In: *Lewis Basis Catalysis in Organic Synthesis* (eds. E. Vedejs and S.E. Denmark), 1011–1038. Wiley-VCH.

145 Kira, M., Kobayashi, M., and Sakurai, H. (1987). Regiospecific and highly stereoselective allylation of aldehydes with allyltrifluorosilane activated by fluoride ions. *Tetrahedron Lett.* 28: 4081–4084.

146 Kira, M., Sato, K., and Sakurai, H. (1988). Chemistry of organosilicon compounds. 240. Pentacoordinate allylsilicates: characterization and highly stereoselective reaction with aldehydes. *J. Am. Chem. Soc.* 110: 4599–4602.

147 Kira, M., Sato, K., and Sakurai, H. (1990). Highly stereoselective allylation of aldehydes with pentacoordinate allylsilicates in hydroxylic media. Discrimination between linear and *alpha*-branched alkanals. *J. Am. Chem. Soc.* 112: 257–260.

148 Kira, M., Hino, T., and Sakurai, H. (1989). Stereohomogeneous synthesis of (*E*)- and (*Z*)-crotyltrifluorosilanes and highly stereoselective allylation of aldehydes. *Tetrahedron Lett.* 30: 1099–1102.

149 Kira, M., Hino, T., and Sakurai, H. (1991). Regiospecific and diastereoselective crotylation of aldimines with crotyltrifluorosilanes activated by fluoride ions. *Chem. Lett.* 20: 277–280.

150 Kobayashi, S. and Nishio, K. (1993). Facile and highly stereoselective allylation of aldehydes using allyltrichlorosilanes in DMF. *Tetrahedron Lett.* 34: 3453–3456.

151 Kobayashi, S. and Nishio, K. (1994). Facile and highly stereoselective synthesis of homoallylic alcohols using organosilicon intermediates. *J. Org. Chem.* 59: 6620–6628.

152 Kobayashi, S. and Nishio, K. (1994). One-pot synthesis of homoallylic alcohols from 1,3-dienes: tandem vicinal difunctionalization of 1,3-dienes by hydride addition-aldehyde coupling sequence via organosilicon intermediates. *Synthesis* 1994: 457–459.

153 Ogawa, C., Sugiura, M., and Kobayashi, S. (2003). Polymer-supported formamides as reusable organocatalysts for allylation of aldehydes with allyltrichlorosilane. *Chem. Commun.* 192–193.

154 Sakata, K. and Fujimoto, H. (2010). Quantum-chemical study of neutral Lewis base catalyzed allylation of aldehydes with allyltrichlorosilanes. *Organometallics* 29: 1004–1011.

155 Denmark, S.E., Coe, D.M., Pratt, N.E., and Griedel, B.D. (1994). Asymmetric allylation of aldehydes with chiral Lewis bases. *J. Org. Chem.* 59: 6161–6163.

156 Denmark, S.E., Fu, J., Coe, D.M. et al. (2006). Chiral phosphoramide-catalyzed enantioselective addition of allylic trichlorosilanes to aldehydes. Preparative and mechanistic studies with monodentate phosphorus-based amides. *J. Org. Chem.* 71: 1513–1522.

157 Nemoto, T., Hitomi, T., Nakamura, H. et al. (2007). Synthesis of novel P-stereogenic phenylphosphonamides and their application to Lewis base-catalyzed asymmetric allylation of benzaldehyde. *Tetrahedron: Asymmetry* 18: 1844–1849.

158 Short, J.D., Attenoux, S., and Berrisford, D.J. (1997). Additive effects on ligand activated allylation of aldehydes by allyltrichlorosilane. *Tetrahedron Lett.* 38: 2351–2354.

159 Denmark, S.E. and Fu, J. (2000). On the mechanism of catalytic, enantioselective allylation of aldehydes with chlorosilanes and chiral Lewis bases. *J. Am. Chem. Soc.* 122: 12021–12022.

160 Denmark, S.E. and Fu, J. (2001). Catalytic, enantioselective addition of substituted allylic trichlorosilanes using a rationally-designed 2,2'-bispyrrolidine-based bisphosphoramide. *J. Am. Chem. Soc.* 123: 9488–9489.

161 Denmark, S.E. and Fu, J. (2002). Asymmetric construction of quaternary centers by enantioselective allylation: application to the synthesis of the serotonin antagonist LY426965. *Org. Lett.* 4: 1951–1953.

162 Denmark, S.E. and Fu, J. (2003). Understanding the correlation of structure and selectivity in the chiral-phosphoramide-catalyzed enantioselective allylation reactions: solution and solid-state structural studies of bisphosphoramide•$SnCl_4$ complexes. *J. Am. Chem. Soc.* 125: 2208–2216.

163 Denmark, S.E., Fu, J., and Lawler, M.J. (2006). Chiral phosphoramide-catalyzed enantioselective addition of allylic trichlorosilanes to aldehydes. Preparative studies with bidentate phosphorus-based amides. *J. Org. Chem.* 71: 1523–1536.

164 Cheng, K., Fan, T., and Sun, J. (2011). Chiral phosphoric acid catalyzed enantioselective allylation of aldehydes with allyltrichlorosilane. *Chin. J. Chem.* 29: 1669–1671.

165 Nakajima, M., Kotani, S., Ishizuka, T., and Hashimoto, S. (2005). Chiral phosphine oxide BINAPO as a catalyst for enantioselective allylation of aldehydes with allyltrichlorosilanes. *Tetrahedron Lett.* 46: 157–159.

166 Benaglia, M. and Rossi, S. (2010). Chiral phosphine oxides in present-day organocatalysis. *Org. Biomol. Chem.* 8: 3824–3830.

167 Chen, J., Liu, D., Fan, D. et al. (2013). Synthesis of axially chiral C_{10}-bridgePHOS oxides and their use as organocatalysts in enantioselective allylations of aldehydes. *Tetrahedron* 69: 8161–8168.

168 Denmark, S.E. and Fu, J. (2003). Catalytic enantioselective allylation with chiral Lewis bases. *Chem. Commun.* 167–170.

169 Shimada, T., Kina, A., Ikeda, S., and Hayashi, T. (2002). A novel axially chiral 2,2'-bipyridine *N,N'*-dioxide. Its preparation and use for asymmetric allylation of aldehydes with allyl(trichloro)silane as a highly efficient catalyst. *Org. Lett.* 4: 2799–2801.

170 Shimada, T., Kina, A., and Hayashi, T. (2003). A new synthetic route to enantiomerically pure axially chiral 2,2'-bipyridine *N,N'*-dioxides. Highly efficient catalysts for asymmetric allylation of aldehydes with allyl(trichloro) silanes. *J. Org. Chem.* 68: 6329–6337.

171 Kina, A., Shimada, T., and Hayashi, T. (2004). A new approach to axially chiral bipyridineN,N′-dioxides bearing aromatic substituents and their use for catalytic asymmetric allylation of aldehydes with allyl(trichloro)silane. *Adv. Synth. Catal.* 346: 1169–1174.

172 Nakajima, M., Saito, M., Shiro, M., and Hashimoto, S.-i. (1998). (S)-3,3′-dimethyl-2,2'-biquinolineN,N'-dioxide as an efficient catalyst for enantioselective addition of allyltrichlorosilanes to aldehydes. *J. Am. Chem. Soc.* 120: 6419–6420.

173 Bai, B., Shen, L., Ren, J., and Zhu, H.J. (2012). Chiral biscarboline N,N′-dioxide derivatives: highly enantioselective addition of allyltrichlorosilane to aldehydes. *Adv. Synth. Catal.* 354: 354–358.

174 Bai, B., Zhu, H.-J., and Pan, W. (2012). Structure influence of chiral 1,1′-biscarboline-*N,N′*-dioxide on the enantioselective allylation of aldehydes with allyltrichlorosilanes. *Tetrahedron* 68: 6829–6836.

175 Deng, Y., Pan, W., Pei, Y.-N. et al. (2013). Addition of aldehydes with allyltrichlorosilane catalyzed by chiral *bis*-N–O secondary amides. *Tetrahedron* 69: 10431–10437.

176 Malkov, A.V., Orsini, M., Pernazza, D. et al. (2002). Chiral 2,2'-bipyridine-type N-monoxides as organocatalysts in the enantioselective allylation of aldehydes with allyltrichlorosilane. *Org. Lett.* 4: 1047–1049.

177 Malkov, A.V., Bell, M., Orsini, M. et al. (2003). New Lewis-basic N-oxides as chiral organocatalysts in asymmetric allylation of aldehydes. *J. Org. Chem.* 68: 9659–9668.

178 Malkov, A.V., Kabeshov, M.A., Barlog, M., and Kocovsky, P. (2009). Enantioselective and catalytic method for *alpha*-crotylation of aldehydes with a kinetic self-refinement of stereochemistry. *Chem. Eur. J.* 15: 1570–1573.

179 Malkov, A.V., Bell, M., Castelluzzo, F., and Kocovsky, P. (2005). METHOX: a new pyridine *N*-oxide organocatalyst for the asymmetric allylation of aldehydes with allyltrichlorosilanes. *Org. Lett.* 7: 3219–3222.

180 Malkov, A.V., Bell, M., Vassieu, M. et al. (2003). New pyridine-derived N-oxides as chiral organocatalysts in asymmetric allylation of aldehydes. *J. Mol. Catal. A: Chem.* 196: 179–186.

181 Malkov, A.V., Barlog, M., Jewkes, Y. et al. (2011). Enantioselective allylation of *alpha,beta*-unsaturated aldehydes with allyltrichlorosilane catalyzed by METHOX. *J. Org. Chem.* 76: 4800–4804.

182 Malkov, A.V., Dufkova, L., Farrugia, L., and Kocovsky, P. (2003). Quinox, a quinoline-type N-oxide, as organocatalyst in the asymmetric allylation of

aromatic aldehydes with allyltrichlorosilanes: the role of arene–arene interactions. *Angew. Chem. Int. Ed.* 42: 3674–3677.
183 Traverse, J.F., Zhao, Y., Hoveyda, A.H., and Snapper, M.L. (2005). Proline-based N-oxides as readily available and modular chiral catalysts. Enantioselective reactions of allyltrichlorosilane with aldehydes. *Org. Lett.* 7: 3151–3154.
184 Pignataro, L., Benaglia, M., Annunziata, R. et al. (2006). Structurally simple pyridine N-oxides as efficient organocatalysts for the enantioselective allylation of aromatic aldehydes. *J. Org. Chem.* 71: 1458–1463.
185 Benaglia, M., Simonini, V., Pignataro, L. et al. (2008). A new class of chiral Lewis basic metal-free catalysts for stereoselective allylations of aldehydes. *Synlett* 2008: 1061–1065.
186 Lu, T., Porterfield, M.A., and Wheeler, S.E. (2012). Explaining the disparate stereoselectivities of *N*-oxide catalyzed allylations and propargylations of aldehydes. *Org. Lett.* 14: 5310–5313.
187 Lu, T., Zhu, R., An, Y., and Wheeler, S.E. (2012). Origin of enantioselectivity in the propargylation of aromatic aldehydes catalyzed by helical N-oxides. *J. Am. Chem. Soc.* 134: 3095–3102.
188 Malkov, A.V., Stoncius, S., Bell, M. et al. (2013). Mechanistic dichotomy in the asymmetric allylation of aldehydes with allyltrichlorosilanes catalyzed by chiral pyridine N-oxides. *Chem. Eur. J.* 19: 9167–9185.
189 Malkov, A.V., Ramirez-Lopez, P., Biedermannova, L. et al. (2008). On the mechanism of asymmetric allylation of aldehydes with allyltrichlorosilanes catalyzed by QUINOX, a chiral Isoquinoline N-oxide. *J. Am. Chem. Soc.* 130: 5341–5348.
190 Massa, A., Malkov, A.V., Kočovský, P., and Scettri, A. (2003). Asymmetric allylation of aldehydes with allyltrichlorosilane promoted by chiral sulfoxides. *Tetrahedron Lett.* 44: 7179–7181.
191 Massa, A., Acocella, M.R., Sio, V.D. et al. (2009). A catalytic asymmetric allylation of aldehydes with allyl trichlorosilane activated by a chiral tetradentate *bis*-sulfoxide. *Tetrahedron: Asymmetry* 20: 202–204.
192 De Sio, V., Massa, A., and Scettri, A. (2010). Chiral sulfoxides as activators of allyl trichlorosilanes in the stereoselective allylation of aldehydes. *Org. Biomol. Chem.* 8: 3055–3059.
193 Monaco, G., Vignes, C., De Piano, F. et al. (2012). Chiral sulfoxides in the enantioselective allylation of aldehydes with allyltrichlorosilane: a kinetic study. *Org. Biomol. Chem.* 10: 9650–9659.
194 Zhao, H., Kayser, M.M., and Wang, Y. (2008). Mesoporous silica modified with enantiopure sulfoxide as catalyst in allylation of aldehydes. *Microporous Mesoporous Mater.* 116: 196–203.
195 Rowlands, G.J. and Kentish Barnes, W. (2003). Chiral sulfoxides in the enantioselective allylation of aldehydes with allyltrichlorosilane. *Chem. Commun.* 2712–2713.
196 Wang, P., Chen, J., Cun, L. et al. (2009). Aryl *tert*-butyl sulfoxide-promoted highly enantioselective addition of allyltrichlorosilane to aldehydes. *Org. Biomol. Chem.* 7: 3741–3747.
197 Sio, V.D., Acocella, M.R., Villano, R., and Scettri, A. (2010). New chiral imino- and amino-sulfoxides as activators of allyl trichlorosilane in the asymmetric allylation of aldehydes. *Tetrahedron: Asymmetry* 21: 1432–1435.

198 Fulton, J.R., Kamara, L.M., Morton, S.C., and Rowlands, G.J. (2009). The sulfinyl moiety in Lewis base-promoted allylations. *Tetrahedron* 65: 9134–9141.
199 Ghosh, D., Sahu, D., Saravanan, S. et al. (2013). Synthetically amenable amide derivatives of tosylated amino acids as organocatalysts for enantioselective allylation of aldehydes: computational rationale for enantioselectivity. *Org. Biomol. Chem.* 11: 3451–3460.
200 Ghosh, D., Sadhukhan, A., Maity, N.C. et al. (2014). Oxazoline derivatives tagged with tosylated amino acids as recyclable organocatalysts for enantioselective allylation of aldehydes. *RSC Adv.* 4: 12257.
201 Denmark, S.E. and Wynn, T. (2001). Lewis base activation of Lewis acids: catalytic enantioselective allylatio and propargylation of aldehydes. *J. Am. Chem. Soc.* 123: 6199–6200.
202 Reich, H.J. (2016). Mechanism of C—Si bond cleavage using Lewis bases (n → σ*). In: *Lewis Base Catalysis in Organic Synthesis* (eds. E. Vedejs and S.E. Denmark), 233–280. Wiley-VCH.
203 Beutner, G.L. and Denmark, S.E. (2013). Lewis base catalysis of the Mukaiyama directed aldol reaction: 40 years of inspiration and advances. *Angew. Chem. Int. Ed.* 52: 9086–9096.
204 Denmark, S.E., Winter, S.B.D., Su, X., and Wong, K.-T. (1996). Chemistry of trichlorosilyl enolates. 1. New reagents for catalytic, asymmetric aldol additions. *J. Am. Chem. Soc.* 118: 7404–7405.
205 Denmark, S.E. and Stavenger, R.A. (2000). Asymmetric catalysis of aldol reactions with chiral Lewis bases. *Acc. Chem. Res.* 33: 432–440.
206 Denmark, S.E., Stavenger, R.A., Wong, K.-T., and Su, X. (1999). Chiral phosphoramide-catalyzed aldol additions of ketone enolates. Preparative aspects. *J. Am. Chem. Soc.* 121: 4982–4991.
207 Denmark, S.E. and Pham, S.M. (2000). Kinetic analysis of the divergence of reaction pathways in the Chiral Lewis base promoted aldol additions of trichlorosilyl enol ethers: a rapid-injection NMR study. *Helv. Chim. Acta* 83: 1846–1853.
208 Denmark, S.E., Su, X., and Nishigaichi, Y. (1998). The chemistry of trichlorosilyl enolates. 6. Mechanistic duality in the Lewis base-catalyzed aldol addition reaction. *J. Am. Chem. Soc.* 120: 12990–12991.
209 Denmark, S.E., Pham, S.M., Stavenger, R.A. et al. (2006). Chiral phosphoramide-catalyzed aldol additions of ketone trichlorosilyl enolates. Mechanistic aspects. *J. Org. Chem.* 71: 3904–3922.
210 Nakajima, M., Yokota, T., Saito, M., and Hashimoto, S. (2004). Enantioselective aldol reactions of trichlorosilyl enol ethers catalyzed by chiral N,N′-dioxides and monodentate N-oxides. *Tetrahedron Lett.* 45: 61–64.
211 Nakajima, M., Kotani, S., and Hashimoto, S. (2006). Enantioselective aldol reactions of trichlorosilyl enol ethers catalyzed by the chiral phosphine oxide BINAPO. *Synlett* 2006: 1116–1118.
212 Denmark, S.E., Wong, K.T., and Stavenger, R.A. (1997). The chemistry of trichlorosilyl enolates .2. Highly-selective asymmetric aldol additions of ketone enolates. *J. Am. Chem. Soc.* 119: 2333–2334.
213 Denmark, S.E. and Pham, S.M. (2003). Stereoselective aldol additions of achiral ethyl ketone-derived trichlorosilyl enolates. *J. Org. Chem.* 68: 5045–5055.

214 Denmark, S.E. and Ghosh, S.K. (2001). The first catalytic, diastereoselective, and enantioselective crossed-aldol reactions of aldehydes. *Angew. Chem. Int. Ed.* 40: 4759–4762.

215 Denmark, S.E. and Wynn, T. (2001). Lewis base activation of Lewis acids: catalytic enantioselective allylation and propargylation of aldehydes. *J. Am. Chem. Soc.* 123: 6199–6200.

216 Denmark, S.E., Wynn, T., and Beutner, G.L. (2002). Lewis base activation of Lewis acids. Addition of silyl ketene acetals to aldehydes. *J. Am. Chem. Soc.* 124: 13405–13407.

217 Hammond, G.S. (1955). A correlation of reaction rates. *J. Am. Chem. Soc.* 77: 334–338.

218 Denmark, S.E. and Heemstra, J.R. Jr. (2003). Lewis base activation of Lewis acids. Catalytic enantioselective addition of silyl enol ethers of achiral methyl ketones to aldehydes. *Org. Lett.* 5: 2303–2306.

219 Denmark, S.E. and Bui, T. (2005). Lewis base catalyzed enantioselective aldol addition of acetaldehyde-derived silyl enol ether to aldehydes. *J. Org. Chem.* 70: 10190–10193.

220 Denmark, S.E. and Fan, Y. (2002). Catalytic, enantioselective aldol additions to ketones. *J. Am. Chem. Soc.* 124: 4233–4235.

221 Shimoda, Y., Tando, T., Kotani, S. et al. (2009). Enantioselective aldol reaction of silyl ketene acetals promoted by a Lewis base-activated Lewis acid catalyst. *Tetrahedron: Asymmetry* 20: 1369–1370.

222 Salonen, L.M., Ellermann, M., and Diederich, F. (2011). Aromatic rings in chemical and biological recognition: energetics and structures. *Angew. Chem. Int. Ed.* 50: 4808–4842.

223 Denmark, S.E. and Chung, W.J. (2008). Lewis base activation of Lewis acids: catalytic enantioselective glycolate aldol reactions. *Angew. Chem. Int. Ed.* 47: 1890–1892.

224 Wilcox, C.S. and Babston, R.E. (1984). Thermodynamic results for geometrical isomerism in silyl ketene acetals. *J. Org. Chem.* 49: 1451–1453.

225 Denmark, S.E. and Beutner, G.L. (2003). Lewis base activation of Lewis acids. Vinylogous aldol reactions. *J. Am. Chem. Soc.* 125: 7800–7801.

226 Denmark, S.E., Heemstra, J.R., and Beutner, G.L. (2005). Catalytic, enantioselective, vinylogous aldol reactions. *Angew. Chem. Int. Ed.* 44: 4682–4698.

227 Casiraghi, G., Battistini, L., Curti, C. et al. (2011). The vinylogous aldol and related addition reactions: ten years of progress. *Chem. Rev.* 111: 3076–3154.

228 Denmark, S.E. and Heemstra, J.R. Jr. (2006). Lewis base activation of Lewis acids. Vinylogous aldol addition reactions of conjugated *N,O*-silyl ketene acetals to aldehydes. *J. Am. Chem. Soc.* 128: 1038–1039.

229 Palombi, L., Acocella, M.R., Celenta, N. et al. (2006). Highly enantioselective vinylogous addition of 2-trimethylsilyloxyfuran to aldehydes promoted by the $SiCl_4$/chiral Lewis base system. *Tetrahedron: Asymmetry* 17: 3300–3303.

230 Curti, C., Sartori, A., Battistini, L. et al. (2009). Asymmetric, catalytic, vinylogous aldol reactions using pyrrole-based dienoxy silanes. Enantioselective synthesis of α,β-unsaturated γ-butyrolactam synthons. *Tetrahedron Lett.* 50: 3428–3431.

231 Denmark, S.E. and Heemstra, J.R. (2004). Lewis base activation of Lewis acids: vinylogous aldol additions of silyl dienol ethers to aldehydes. *Synlett* 2411–2416.

232 Curti, C., Battistini, L., Sartori, A. et al. (2011). Catalytic, asymmetric hypervinylogous Mukaiyama aldol reactions of extended furan-based silyl enolates. *Org. Lett.* 13: 4738–4741.

233 Denmark, S.E., Wilson, T.W., Burk, M.T., and Heemstra, J.R. Jr. (2007). Enantioselective construction of quaternary stereogenic carbons by the Lewis base catalyzed additions of silyl ketene imines to aldehydes. *J. Am. Chem. Soc.* 129: 14864–14865.

234 Denmark, S.E., Wilson, T.W., and Burk, M.T. (2014). Enantioselective construction of quaternary stereogenic carbon atoms by the Lewis base catalyzed additions of silyl ketene imines to aldehydes. *Chem. Eur. J.* 20: 9268–9279.

235 Denmark, S. and Wilson, T. (2010). Construction of quaternary stereogenic carbon centers by the Lewis base catalyzed conjugate addition of silyl ketene imines to α,β-unsaturated aldehydes and ketones. *Synlett* 2010: 1723–1728.

236 Denmark, S.E. and Wilson, T.W. (2010). N-Silyl oxyketene imines are underused yet highly versatile reagents for catalytic asymmetric synthesis. *Nat. Chem.* 2: 937–943.

237 Denmark, S.E. and Wilson, T.W. (2012). Lewis base catalyzed enantioselective additions of An N-silyl vinylketene imine. *Angew. Chem. Int. Ed.* 51: 3236–3239.

238 Passerini, M.T. (1921). Sopra gli isonitrili (I). Composto del p-isonitril-azobenzolo con acetone ed Acido acetico. *Gazz. Chim. Ital.* 51: 126–129.

239 Denmark, S.E. and Fan, Y. (2005). Catalytic, enantioselective α-additions of isocyanides: Lewis base catalyzed Passerini-type reactions. *J. Org. Chem.* 70: 9667–9676.

240 Denmark, S.E. and Fan, Y. (2003). The first catalytic, asymmetric α-additions of isocyanides. Lewis-base-catalyzed, enantioselective Passerini-type reactions. *J. Am. Chem. Soc.* 125: 7825–7827.

241 Nakanishi, K., Kotani, S., Sugiura, M., and Nakajima, M. (2008). First asymmetric Abramov-type phosphonylation of aldehydes with trialkyl phosphites catalyzed by chiral Lewis bases. *Tetrahedron* 64: 6415–6419.

242 Ohmaru, Y., Sato, N., Mizutani, M. et al. (2012). Synthesis of aryl group-modified DIOP dioxides (Ar-DIOPOs) and their application as modular Lewis base catalysts. *Org. Biomol. Chem.* 10: 4562–4570.

243 Aoki, S., Kotani, S., Sugiura, M., and Nakajima, M. (2012). Trichlorosilyl triflate-mediated enantioselective directed cross-aldol reaction between ketones using a chiral phosphine oxide as an organocatalyst. *Chem. Commun.* 5524–5526.

244 Kotani, S., Shimoda, Y., Sugiura, M., and Nakajima, M. (2009). Novel enantioselective direct aldol-type reaction promoted by a chiral phosphine oxide as an organocatalyst. *Tetrahedron Lett.* 50: 4602–4605.

245 Kotani, S., Aoki, S., Sugiura, M., and Nakajima, M. (2011). Trichlorosilyl triflate for enantioselective direct-type aldol reaction with chiral phosphine oxide. *Tetrahedron Lett.* 52: 2834–2836.

246 Rossi, S., Benaglia, M., Cirilli, R., and Benincori, T. (2015). Synthesis of novel chiral bithiophene based phosphine oxides as Lewis bases in organocatalytic stereoselective reactions. *Asymmet. Catal.* 2: 17–25.

247 Rossi, S., Benaglia, M., Genoni, A. et al. (2011). Biheteroaromatic diphosphine oxides-catalyzed stereoselective direct aldol reactions. *Tetrahedron* 67: 158–166.

248 Dogan, Ö. and Tan, D. (2015). Enantioselective direct aldol reactions promoted by phosphine oxide aziridinyl phosphonate organocatalysts. *Tetrahedron: Asymmetry* 26: 1348–1353.

249 Rossi, S., Benaglia, M., and Genoni, A. (2014). Organic reactions mediated by tetrachlorosilane. *Tetrahedron* 70: 2065–2080.

250 Shimoda, Y., Kotani, S., Sugiura, M., and Nakajima, M. (2011). Enantioselective double aldol reaction catalyzed by chiral phosphine oxide. *Chem. Eur. J.* 17: 7992–7995.

251 Nakajima, M., Kotani, S., and Sugiura, M. (2014). Enantioselective double aldol reactions involving the sequential activation of silicon tetrachloride by chiral phosphine oxides. *Synlett* 25: 631–640.

252 Genoni, A., Benaglia, M., Rossi, S., and Celentano, G. (2013). Enantiomerically pure bithiophene diphosphine oxides as catalysts for direct double aldol reactions. *Chirality* 25: 643–647.

253 Shimoda, Y., Kubo, T., Sugiura, M. et al. (2013). Stereoselective synthesis of multiple stereocenters by using a double aldol reaction. *Angew. Chem. Int. Ed.* 52: 3461–3464.

254 Zhang, P., Han, Z., Wang, Z., and Ding, K. (2013). Spiro[4,4]-1,6-nonadiene-based diphosphine oxides in Lewis base catalyzed asymmetric double-aldol reactions. *Angew. Chem. Int. Ed.* 52: 11054–11058.

255 Kotani, S., Miyazaki, S., Kawahara, K. et al. (2016). Stereoselective synthesis of highly functionalized 2,3-dihydro-4-pyranones using phosphine oxide as catalyst. *Chem. Pharm. Bull.* 64: 189–192.

256 Kotani, S., Kai, K., Sugiura, M., and Nakajima, M. (2017). Sequential catalysis of phosphine oxide for stereoselective synthesis of stereodyads. *Org. Lett.* 19: 3672–3675.

257 Kotani, S., Kai, K., Shimoda, Y. et al. (2016). Concise asymmetric construction of C_2-symmetric 1,9-diarylnonanoids using a hypervalent silicon complex: total synthesis of (−)-ericanone. *Chem. Asian J.* 11: 376–379.

258 Alim, N.R., Miyazaki, S., Shimoda, Y. et al. (2017). Asymmetric aldol/vinylogous aldol/cyclization reaction using phosphine oxide catalysts. *Chem. Pharm. Bull.* 65: 989–993.

259 Sugiura, M., Sato, N., Sonoda, Y. et al. (2010). Diastereo- and enantioselective reductive aldol reaction with trichlorosilane using chiral Lewis bases as organocatalysts. *Chem. Asian J.* 5: 478–481.

260 Osakama, K., Sugiura, M., Nakajima, M., and Kotani, S. (2012). Enantioselective reductive aldol reaction using tertiary amine as hydride donor. *Tetrahedron Lett.* 53: 4199–4201.

261 Rossi, S., Annunziata, R., Cozzi, F., and Raimondi, L. (2015). Phosphine oxide catalyzed, tetrachlorosilane-mediated enantioselective direct aldol reactions of thioesters. *Synthesis* 47: 2113–2124.

262 Rossi, S., Benaglia, M., Cozzi, F. et al. (2011). Organocatalytic stereoselective direct aldol reaction of trifluoroethyl thioesters. *Adv. Synth. Catal.* 353: 848–854.

263 Kotani, S., Ito, M., Nozaki, H. et al. (2013). Enantioselective Morita–Baylis–Hillman reaction catalyzed by a chiral phosphine oxide. *Tetrahedron Lett.* 54: 6430–6433.

264 Kotani, S., Hanamure, T., Nozaki, H. et al. (2017). Enantioselective chlorinative aldol reaction of α-substituted acroleins catalyzed by chiral phosphine oxides. *Tetrahedron: Asymmetry* 28: 282–287.

11

Hydrosilylation Catalyzed by Base Metals

Yusuke Sunada[1] *and Hideo Nagashima*[2]

[1] *The University of Tokyo, Institute of Industrial Science, 4-6-1 Komaba, Meguro-ku, Tokyo 153-8580, Japan*
[2] *Kyushu University, Institute for Materials Chemistry and Engineering, 6-1 Kasugakoen, Kasuga, Fukuoka 816-8580, Japan*

11.1 Introduction

Hydrosilylation is the term used for addition reactions of Si—H bonds to unsaturated molecules containing C=C, C≡C, C=O, C=N or C≡N bonds [1]. Precious metal catalysts such as those including Pt, Pd, Rh, Ir, and Ru play key roles in such reactions, demonstrating high catalytic activity and high selectivity. Hydrosilylation of carbonyl compounds is a method to reduce C=O bonds, and a number of synthetic protocols for alcohols and amines including their asymmetric syntheses have been achieved by the appropriate choice of hydrosilanes and catalysts [2]. In contrast, hydrosilylation of alkenes and alkynes has been utilized for the preparation of organosilicon compounds. In particular, platinum-catalyzed hydrosilylation of alkenes is a key technology in the worldwide silicone industry, producing various organosilicon compounds including silane coupling reagents, silicone fluids, silicone rubbers, and silicone resins [3]. A report in 2007 by Howell disclosed that "the silicone industry worldwide used around 5.6 tons of platinum," which corresponded to US$235 million in 2007 [4]. The author also commented that the price of platinum was reportedly US$42 per gram in 2007. Furthermore, it almost doubled from the start of 2007 to the middle of 2008. This situation is a typical example of the scarcity of chemical elements reported by Nakamura and Sato; some elements (i.e. the rare metals) face deficiencies in supply by rapid worldwide economic growth [5]. This is one of the motivations to find substitutes for platinum catalysts for the hydrosilylation of alkenes by inexpensive non-precious metal catalysts.

The first-row transition-metal elements are promising candidates as platinum substitutes. Iron is the most abundant transition metal in the Earth's crust, is environmentally friendly, and safe for humans [6]. This is followed by cobalt and nickel, which are much more cost-effective than platinum. Studies of organometallic complexes containing these elements have provided another strong

Organosilicon Chemistry: Novel Approaches and Reactions, First Edition.
Edited by Tamejiro Hiyama and Martin Oestreich

motivation to investigate catalyst designs using iron, cobalt, and nickel for the hydrosilylation of alkenes. Like hydrogenation reactions, hydrosilylation is believed to proceed by two-electron mechanisms [7]. The famous Chalk–Harrod cycle [7a,c,d] and the modified Chalk–Harrod cycle [7b] are typical examples of two-electron mechanisms, which involve oxidative addition of a Si—H bond to a metal center accompanied by an increase in the oxidation state of the metal by two. In contrast, iron is a typical element that easily promotes one-electron redox processes like those involving Fe(II) and Fe(III). One-electron chemistry is also often seen in cobalt and nickel. Thus, one of the main issues to overcome is the achievement of two-electron chemistry of Fe (and Co, Ni) by appropriate design of the catalyst. Ligand design is the key to solving this problem, as summarized by two recent accounts by Chirik [8a] and Nagashima [8b]. There are currently two strategies: one is the introduction of strong-field ligands to make the catalytic intermediate low-spin, and the other is the use of redox-active ligands. It should be noted that complexes that prefer one-electron chemistry are often paramagnetic, and thus it is difficult to investigate their structures and reactions in solution. Many such complexes are also very unstable to air and moisture. Calculations, which are now powerful for studies of low-spin, precious metal catalysis, are not well suited to the study of catalytic cycles involving paramagnetic intermediates. In other words, challenges still remain in how to conduct research to develop non-precious metal catalysts.

This chapter explores the fundamental chemistry of organoiron, cobalt, and nickel complexes, which show good catalytic performance for the hydrosilylation of alkenes. The topics are primarily focused on the hydrosilylation of alkenes with alkoxyhydrosilanes and hydrosiloxanes, including silicone polymers having Si–H groups, using organometallic complexes with redox-active ligands, and those with strong-field ligands such as CO, CNR, and N-heterocyclic carbenes (NHC). In many of the papers cited here, the studies were supported by characterization of the organometallic complexes as the catalyst precursor, which is useful for improved catalyst design.

11.2 Base-Metal Catalysts for Hydrosilylation of Alkenes with Alkoxyhydrosilanes and Hydrosiloxanes

Base-metal catalysts for the hydrosilylation of alkenes have received considerable attention since the early stage of hydrosilylation chemistry. Examples are iron [9] and cobalt [10] carbonyl complexes and cobalt [11] and nickel [12] phosphine complexes. As noted in the introduction, these 3d late transition metals tend to prefer one-electron redox reactions, while catalytic hydrosilylation is typically a two-electron redox process. This contradiction was solved by a series of work by Chirik, who summarized in his account in 2015 [8a] that two-electron chemistry in base-metal catalysis can be achieved by metals surrounded by weak fields but with redox-active ligands or those stabilized by strong-field ligands. In the following sections, iron and cobalt catalysts having redox-active ligands are first described, followed by iron and cobalt catalysts bearing CO, CNR, and NHC as strong-field ligands. Features of nickel catalysts are summarized in the final section. In all cases, the catalysis is focused on the hydrosilylation of alkenes with alkoxyhydrosilanes or hydrosiloxanes.

11.2.1 Iron and Cobalt Catalysts

11.2.1.1 Catalysts Bearing Tridentate Nitrogen Redox-Active Ligands and Related Catalysts

Cationic Fe(II) complexes with bis(imino)pyridine (PDI) ligands behave as efficient olefin polymerization catalysts [13]. A dinitrogen complex of iron bearing a $PDI-Ar^{iPr}$ ($PDI-Ar^{iPr}$ = 2,6-[2,6-$^iPr_2$—C_6H_3N=CMe]$_2C_5H_3N$) ligand (**1a**) [14a] prepared by Chirik and coworkers is a five-coordinate Fe complex with three canonical structures including "$(PDI-Ar^{iPr})Fe^0$," "$(PDI-Ar^{iPr})^{1-}Fe^I$," and "$(PDI-Ar^{iPr})^{2-}Fe^{II}$," as shown in Figure 11.1 and Scheme 11.1 [15]. Among them, the anti-ferromagnetic low-spin state of the iron center of the "$(PDI-Ar^{iPr})^{2-}Fe^{II}$" species is suitable for the activation of H—H and H—SiR_3 bonds on the iron center. This is similar to low-spin precious metals that undergo oxidative addition of H_2 and R_3SiH with ease.

Complex **1a** acts as an excellent catalyst for hydrosilylation of alkenes with $PhSiH_3$, showing a turnover frequency (TOF) exceeding 360 h^{-1} [14a]. In contrast, the attempted hydrosilylation of alkenes with tertiary hydrosilanes by **1a**, which has a bulky 2,6-diisopropylphenyl group on the imino-nitrogen, was not successful. However, the collaboration of Chirik's group and researchers from Momentive Performance Co., Ltd. yielded a solution, and hydrosilylation of

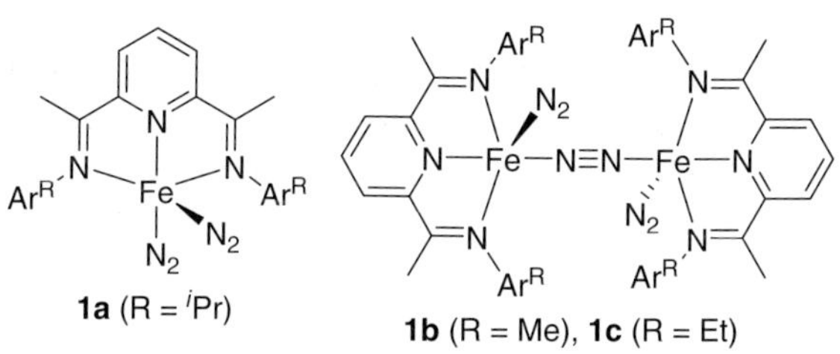

Figure 11.1 Dinitrogen complexes of iron bearing a $PDI-Ar^{iPr}$ ligands (**1a**–**1c**), where the 2,6-$R_2C_6H_3$– group of the imino moiety is defined as Ar^R (R = Me, Et, ^{i}Pr).

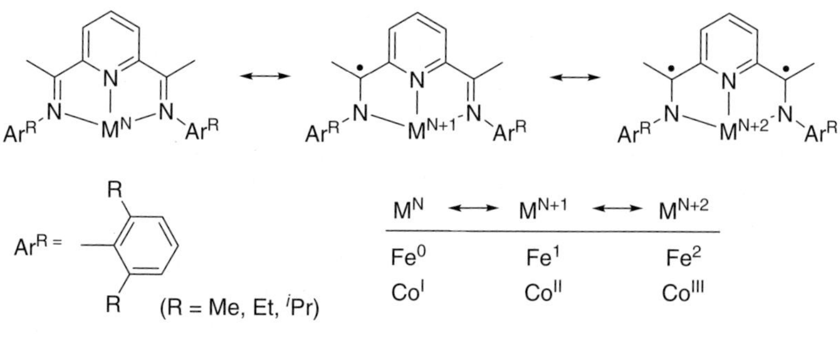

M^N	M^{N+1}	M^{N+2}
Fe^0	Fe^1	Fe^2
Co^I	Co^{II}	Co^{III}

Scheme 11.1 Redox activity of bis(imino)pyridine ligands in transition-metal complexes.

alkenes with trialkylsilanes, alkoxyhydrosilanes, and hydrosiloxanes was achieved by **1b** and **1c** having less bulky substituents on the imino groups [14b,c]. In particular, **1b** was extremely active for the hydrosilylation of hex-1-ene with either 1,1,1,3,5,5,5-heptamethyltrisiloxane (MD′M) or $(EtO)_3SiH$ at room temperature in minutes with TON > 25 000 (TON = turnover number), and was effective for silicone curing, namely, cross-linking of $Me_3SiO(Me_2SiO)_m(MeHSiO)_nSiMe_3$ with $CH_2{=}CHSiMe_2O(Me_2SiO)_nSiMe_2CH{=}CH_2$ in the presence of 500 ppm of **1b**. The regioselectivity of the Si—H addition was completely controlled in an anti-Markovnikov manner, even for the reaction with styrene, which often affords the Markovnikov product. The catalyst was useful for the hydrosilylation of allyl ethers including a methyl-capped oligoethylene glycol unit, although relatively higher catalytic loading was necessary. Unique selectivity, which cannot be achieved with conventional platinum catalysts, was observed in the hydrosilylation of 1,2,4-trivinylcyclohexane with alkoxyhydrosilanes [14d].

A drawback of bis(imino)pyridine iron(0) complexes for practical applications is their high sensitivity toward air and moisture. One approach to mitigate this is the use of bis[(trimethylsilyl)methyl]iron(II) complexes bearing PDI ligands [16a]. Although the details are not clear yet, mechanisms to generate catalytically active species likely involve reductive cleavage of the Fe—C bonds by the hydrosilanes. As reported in the same paper, simple 2,2′;6,2″-terpyridine behaved as an alternative to the bis(imino)pyridine ligand. The trimethylsilylmethyl iron complexes **2** and **3**, shown in Figure 11.2, demonstrated moderate catalytic activity toward hydrosilylation of oct-1-ene with tertiary hydrosilanes including MD′M and $(EtO)_3SiH$ [16a]. Trials to generate active species similar to those from bis(imino)pyridine iron(0) complexes, **1a**–**1c**, were studied by EtMgBr or alkoxides as the activator [16e,f].

The redox activity of bis(imino)pyridine cobalt(I) complexes was investigated theoretically [17]. Similar to bis(imino)pyridine iron(0) complexes, electron transfer from the cobalt center to the bis(imino)pyridine ligand provides three canonical forms as shown in Scheme 11.1, which are described as Co(I)[bis(imino)

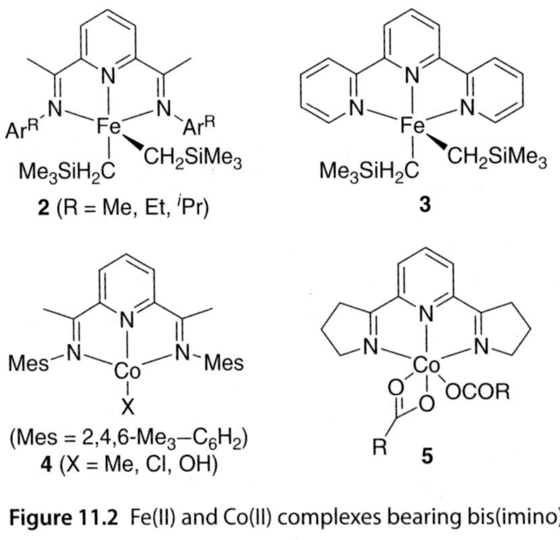

Figure 11.2 Fe(II) and Co(II) complexes bearing bis(imino)pyridine ligands.

pyridine], Co(II)[bis(iminopyridine)]$^-$, and Co(III)[bis(imino)pyridine]$^{2-}$. Co(0) analogs of **1a**–**1c** [18a] and a series of Co(I) complexes bearing a bis(imino)pyridine ligand (PDI)Co(X) **4**, where X = Me, Cl, or OH, were prepared and characterized [18b]. The complex bearing mesityl groups on the Ar groups of the PDI ligand showed interesting catalytic performance for alkene hydrosilylation with MD′M. Although the catalytic activity was much lower than that in the bis(imino) pyridine iron(0) complexes **1a**–**1c**, dehydrogenative silylation occurred to form a 1 : 1 mixture of allylic silane and alkane. Dehydrogenative silylation is commonly observed in the iron-carbonyl-catalyzed reactions of alkenes with tertiary hydrosilanes; however, the formed silicon products were a mixture of allyl and vinyl silanes. Selective formation of allyl silanes is a characteristic feature of **4**.

A modification to the cobalt bis(imino)pyridine complex **4** was made on the basis of the idea that sterically less bulky substituents on the imino groups would effectively open the coordination sphere for a hydrosilane and an alkene. As shown in Figure 11.2, complex **5** was prepared by the reaction of a less sterically hindered bis(imino)pyridine ligand with Co(2-ethylhexanoate)$_2$. While **4** achieved selective dehydrogenative silylation to form allylic silanes, **5** was a catalytic precursor affording the products in high (>98 : 2) hydrosilylation selectivity when alkoxyhydrosilanes and hydrosiloxanes were used as the hydrosilane. The catalyst was effective for the hydrosilylation of alkenes containing other functional groups such as ketones, amides, amines, and epoxides. The reaction of $(EtO)_3SiH$ with allyl glycidyl ether on a 10-g scale was carried out with a low catalyst loading (0.025 mol%, 50 ppm Co wt), which achieved TON = 4000 at 45 °C in 20 minutes. Catalyst **5** was also effective for silicone curing. Since **5** and the catalyst system composed of metal carboxylates and isocyanide ligands, described subsequently, can be prepared by air-stable precursors, they were the most practical catalyst precursors for the hydrosilylation of hydrosiloxanes and/or alkoxyhydrosilanes in reports until 2017.

11.2.1.2 Catalysts Containing CO, CNR, and NHC Ligands

Iron and cobalt complexes bearing redox-active ligands are a clear solution for catalysts active for hydrosilylation. In sharp contrast, appropriate low-spin designs of base-metal catalysts bearing strong π-acceptor ligands such as CO and CNR have also led to the discovery of effective catalyst systems toward the hydrosilylation of alkenes. It is known that iron carbonyls such as $Fe(CO)_5$ and $Fe_3(CO)_{12}$ act as catalysts for the hydrosilylation of alkenes under photoirradiation or high temperatures. The active species is proposed to be "$Fe(CO)_3$" [9]. However, the dehydrogenative silylation reaction occurs concomitantly, giving a 1 : 1 mixture of $RCH{=}CHSiR'_3$ and RCH_2CH_3. Hydrosilylation catalyzed by $Co_2(CO)_8$ proceeds under slightly milder conditions with higher selectivity compared with that by the iron carbonyls [10]. Isomerization of terminal alkenes to inactive internal alkenes occurred concomitantly. Recently, Marciniec et al. reported the reaction of vinylsilanes with hydrosiloxanes by the $Fe(CO)_3$(divinylsiloxane) catalyst, which was applicable to silicone-curing reactions. However, these reactions were accompanied by dehydrogenative silylation as a side reaction [19].

In 2013, Nagashima and coworkers reported the design and synthesis of an iron dicarbonyl complex having a disilaferracyclic structure **6** (Figure 11.3) [20].

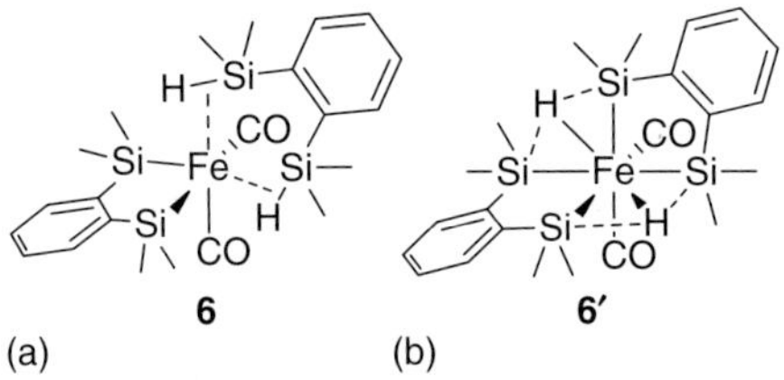

Figure 11.3 Proposed (a) and optimized (b) structure of **6**.

Complex **6** efficiently catalyzed hydrosilylation of ethylene with tertiary hydrosilanes including 1,1,1,3,3-pentamethyldisiloxane (PMDS), MD′M, Me_2PhSiH, Et_3SiH, and $(EtO)_3SiH$, and the highest TON reached 1000 in the reaction with PMDS. Hydrosilylation of substituted alkenes such as oct-1-ene and cyclopentene with PMDS was also realized by **6**. It should be noted that hydrosilylation of oct-2-ene with PMDS also took place under the same reaction conditions to give 1-organosilyloctane, a hydrosilylated product identical to that formed from oct-1-ene. The reaction involved extensive alkene migration and subsequent hydrosilylation of the resulting isomerized alkenes. An interesting feature of the catalysis by **6** is that hydrosilylation selectively proceeded in an anti-Markovnikov manner and no by-products derived from dehydrogenative silylation were formed from the alkenes except in the case with styrene.

DFT calculations revealed that **6** has a unique structure with four Fe—Si and two Fe—H bonds with two Si···H···Si interactions, so-called SISHA (secondary interaction between silicon and hydrogen atoms) [20d]. Detailed mechanistic studies for the hydrogenation of an alkene catalyzed by **6** suggested that all of the catalytic intermediates and transition states were in the low-spin state, and were effectively stabilized by two strongly π-accepting CO ligands, two Fe—Si σ-bonds, and the octahedral coordination geometry of the iron catalyst. Hydrosilylation of alkenes by **6** would proceed through similar mechanisms.

Studies on **6** as the hydrosilylation catalyst led to the discovery of the in situ generation of catalytically active iron species having isocyanide ligands. Combinatorial catalyst screening identified catalyst systems consisting of (η^4-COT)$_2$Fe (COT = 1,3,5,7-cyclooctatetraene) and open ferrocene, (MPDE)$_2$Fe (MPDE = η^5-3-methylpentadienyl), with π-accepting isocyanide ligands that behaved as efficient catalysts for the hydrosilylation of alkenes with PMDS [21]. The combination of (COT)$_2$Fe (0.05 mol%) and isocyanoadamantane (CNAd; Ad = adamant-1-yl) (0.1 mol%) catalyzed the hydrosilylation of styrene at 50 °C. Selective anti-Markovnikov addition occurred to give 1-organosilyl-2-phenylethane. This result is in stark contrast to that reported by Murai's catalysis mediated by $Fe_3(CO)_{12}$, which selectively afforded products derived from dehydrogenative silylation [9c]. The (COT)$_2$Fe/CNAd catalyst system is tolerant toward styrene derivatives having several functional groups including F, Cl, and CO_2Et, and the highest TON reached 5000. It is noteworthy that the modification of poly(dimethyl)siloxane, $Me_2HSi(OSiMe_2)_nSiHMe_2$ (n = c. 27), with styrene was also accomplished to give the corresponding polydimethylsiloxane bearing β-phenethyl groups at the polymer ends in quantitative yield.

By virtue of catalyst screening by the combinatorial approach, the same research group discovered new catalyst systems composed of iron or cobalt carboxylates, such as metal pivalates [$Fe(OPv)_2$ and $Co(OPv)_2$ (OPv = pivalate)], with isocyanide ligands, which were active for the hydrosilylation of various alkenes with hydrosiloxanes [22]. Interestingly, the utilities of $Fe(OPv)_2$ and $Co(OPv)_2$ were complementary. A 1 : 2 mixture of $Fe(OPv)_2$ and CNAd was useful for selective hydrosilylation of styrene derivatives and allylic ethers with PMDS or Me_2PhSiH to form the corresponding anti-Markovnikov adducts in high yields. The highest TON reached 9700. In contrast, a 1 : 3 mixture of $Co(OPv)_2$ and CNAd was useful for hydrosilylation of alkenes including α-methylstyrene derivatives, 2-norbornene, oct-1-ene, and oct-2-ene, and allylic ethers with PMDS, MD′M or Me_2PhSiH. Hydrosilylation of oct-1-ene was preceded by rapid alkene migration to a mixture of internal octenes, which subsequently reacted with the hydrosilanes to afford 1-silylated octane as a single product. The cobalt-catalyzed reaction of 2-norbornene with PMDS resulted in 2-silylated norbornane, whose stereochemistry was exclusively exo. The highest TON reached was 1885.

Catalytically active species were generated by the activation of $Fe(OPv)_2$ or $Co(OPv)_2$ by hydrosilanes in the presence of isocyanide. Pretreatment of the $Fe(OPv)_2$ or $Co(OPv)_2$/CNR catalyst system by reducing reagents resulted in efficient generation of catalytically active species. Typical reducing reagents were alkoxyhydrosilanes such as $(EtO)_3SiH$ and $(EtO)_2MeSiH$ and hydroboranes such as pinacolborane. These hydrides were much less reactive than hydrosiloxanes, and hydrosilylation with alkoxyhydrosilanes or hydroboration with pinacolborane did not compete with the hydrosilylation with hydrosiloxanes. For instance, the catalyst preactivation accelerated the hydrosilylation of 3-(2-methoxyethoxy) prop-1-ene or $CH_2{=}CHCH_2(OCH_2CH_2)_nOMe$ (n = 8) catalyzed by the $Co(OPv)_2$/CNAd/$(EtO)_2MeSiH$ catalyst system to form the corresponding product in high yield. These reactions were slow in the absence of the activator.

Modification of silicone fluid, $Me_2HSiO(SiMe_2O)_nSiHMe_2$ (n = c. 27), was accomplished with α-methylstyrene by the $Co(OPv)_2$/CNAd/$(EtO)_3SiH$ catalyst system and that with styrene by the $Fe(OPv)_2$/CNAd/$(EtO)_2MeSiH$ catalyst system. Approximately, 500 ppm of the catalyst system was enough to perform the modification in quantitative yield. The metal residue was easily removed by passage through a short plug of alumina, and the metal content in the modified silicone was lower than 1 ppm Cross-linking of $CH_2{=}CHSiMe_2O(SiMeO)_nSiMe_2CH{=}CH_2$ (n = c. 47) and $Me_3SiO(Si(H)MeO)_mSiMe_3$ (m = c. 8) was performed with 195 ppm of the $Co(OPv)_2$/CNAd catalyst at 120 °C, which successfully gave the insoluble silicone gel.

In 2017, Deng and a coworker reported the hydrosilylation of various alkenes with $(EtO)_3SiH$ catalyzed by $(NHC)Co[N(SiMe_3)_2]_2$, a cobalt(II) amide complex bearing the NHC ligand [23]. They found that the product distribution in the reaction of oct-1-ene with $(EtO)_3SiH$ varied depending on the NHC ligand used (Figure 11.4). When the reaction was performed with $(IMes)Co[N(SiMe_3)_2]_2$ (**7a**), isomers of oct-1-ene were formed as the main products. In contrast, a cobalt analog having a sterically small NHC ligand, $(IMesMe)Co[N(SiMe_3)_2]_2$ (**7b**), afforded the corresponding hydrosilylated product in high yield with high

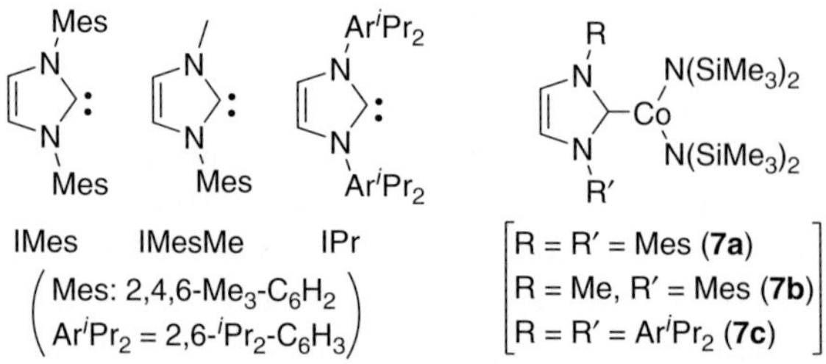

Figure 11.4 Structure of the NHC ligands and (NHC)Co[N$(SiMe_3)_2$]$_2$ (**7**).

selectivity. The reactions catalyzed by **7b** were tolerant to esters, epoxides, and secondary amines. This catalyst system was sensitive toward the steric environment around the "Si–H" group of the hydrosilane. The hydrosilylation of alkenes with $(EtO)_3SiH$ effectively proceeded to afford the corresponding products in moderate to high yields. However, reactions with $(EtO)_2MeSiH$ or Ph_3SiH gave the products in lower yields, while no reaction took place with Et_3SiH and MD′M. Further studies to seek catalytically active species yielded a two-coordinate Co(I) complex, (IPr)Co[N$(SiMe_3)_2$] (**8c**), from (IPr)Co[N$(SiMe_3)_2$]$_2$ (**7c**) and $(EtO)_3SiH$. Treatment of **7c** with $(EtO)_3SiH$ in benzene afforded a Co(I) hydride complex (IPr)Co(H)(η^6-C_6H_6) (**9c**) (Scheme 11.2). The authors observed the formation of products due to dehydrogenative silylation, and proposed that a Co(I)-silyl species may be formed by the reaction of (NHC)Co(I) hydride species and $(EtO)_3SiH$, which acts as the catalytically active species.

A novel cobalt catalyst, (^{DIPP}CCC)CoN_2 (**10**), having the bis(NHC)-based pincer type ligand [^{DIPP}CCC = bis(diisopropylphenyl-imidazol-2-ylidene)], was synthesized by Fout and coworkers [15e]. The catalyst **10** showed excellent functional-group compatibility in the hydrosilylation of various functionalized alkenes with tertiary hydrosilanes. Alkenes bearing aldehyde, ketone, ester, hydroxy, epoxide, allyl ether, nitrile, and amine groups effectively underwent the reaction with Me_2PhSiH or MD′M in the presence of 5 mol% of **10** in benzene at

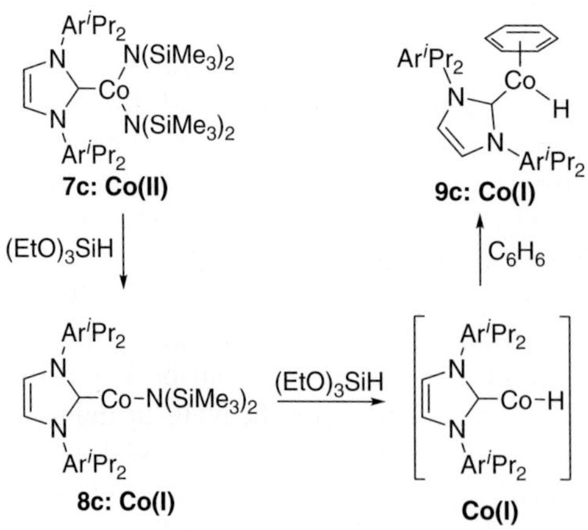

Scheme 11.2 Reactions with (IPr)cobalt amide complexes with $(EtO)_3SiH$.

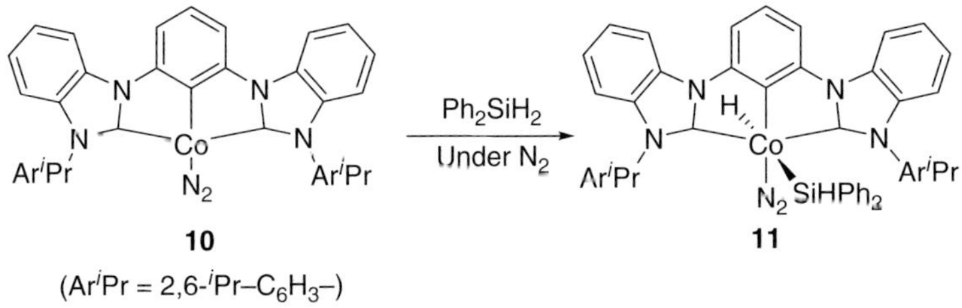

Scheme 11.3 Reaction of (^{DIPP}CCC)CoN_2 (**10**) with Ph_2SiH_2 to afford **11**.

room temperature with these functional groups remaining intact. Mechanistic studies based on the stoichiometric reactions revealed formation of a Co(III)-silyl-hydride complex, (^{DIPP}CCC)Co($HSiPh_2$)(H)(N_2) (**11**) from Ph_2SiH_2 and **10** (Scheme 11.3). Insertion of the alkene into either the Co—H or Co—silyl bond in **11** would be followed by reductive elimination of the hydrosilylated product.

11.2.1.3 Miscellaneous

β-Ketiminate cobalt(I) complexes **12a–12c** were recently investigated by Holland and coworkers as catalysts for the hydrosilylation of alkenes with $(EtO)_3SiH$ or $PhSiH_3$ [24]. Their catalytic performance was surveyed by the hydrosilylation of hex-1-ene with $(EtO)_3SiH$ at 60 °C under neat conditions. In this study, the degree of the steric bulk of the supporting ligand on each complex was quantified using the *G* parameter [25] (Figure 11.5). They found that catalysis by **12a** and **12b** with higher *G*-values gave the hydrosilylated products with low selectivity (12–23%), whereas internal hexenes were formed as the major product (57–63%). In sharp contrast, catalysis by **12c** with a lower *G* parameter showed good selectivity and activity to give the corresponding product in 89% yield. Hydrosilylation of functionalized alkenes bearing silyl ether, halide, ester, tertiary amine, and amide substituents with $PhSiH_3$ was catalyzed by **12c** at room temperature, whereas those with $(EtO)_3SiH$ required slightly harsher reaction conditions (four hours at 60 °C). In the reaction with 4,8-dimethylnona-1,7-diene, the sterically less-hindered double bond was selectively hydrosilylated in the anti-Markovnikov manner.

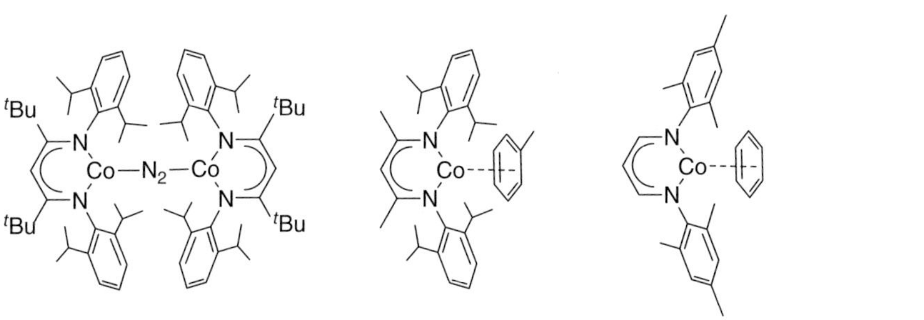

Figure 11.5 β-Ketiminate cobalt (I) complexes with different *G*-values, where *G*-value represents the percentage of a metal's coordination sphere shielded by the ligands [25].

A recent report by Lee discusses the catalysis of (aminomethyl)pyridine cobalt(II) dihalide for hydrosilylation reactions relevant to industrial processes to produce silicones [26]. Treatment of vinylsiloxanes or alkoxy vinylsilanes with MD′M, (EtO)Me_2SiH or Me_2PhSiH at 22 °C gave the corresponding hydrosilylated products in high yields (70–99% yield) with complete anti-Markovnikov selectivity. As the catalyst, 0.25 mol% of (aminomethyl)pyridine cobalt(II) dihalide complexes activated by 0.75 mol% of $LiCH_2SiMe_3$ was used. The same catalyst was applied to silicone curing. In a typical example, cross-linking of polymeric vinyl- and hydrosiloxanes occurred under neat conditions at 22 °C. With 50 ppm of the catalyst, the mixture provided a silicone gel within four to five minutes, whereas the gel was formed within 20 seconds using 200 ppm of the catalyst.

It is worth pointing out that Dow Chemicals applied for patents for the hydrosilylation of alkenes catalyzed by a mixture of metal amides, ligands, and organometallic activators. Examples of execution included the hydrosilylation of hex-1-ene with MD′M with catalysts containing Fe and Co amides [27].

11.2.2 Nickel Catalysts

A redox-active ligand-based nickel catalyst was developed by Chirik, and the most appropriate ligand for nickel was found to be an α-diimine [28]. It is known that cationic α-diimine nickel complexes behave as good catalysts for olefin polymerization catalysts [29]. As shown in Scheme 11.4, the combination of an air-stable and hydrocarbon-soluble nickel carboxylate, $Ni(2\text{-}EH)_2$ (2-EH = 2-ethylhexanoate) (**13**) (1 mol%), with an α-diimine ligand, {^{iPr}DI = $[ArN{=}C(Me)]_2$; Ar = 2,6-$^iPr_2$–C_6H_3}, generated catalytically active species in contact with hydrosilanes, which promoted the anti-Markovnikov hydrosilylation of oct-1-ene with tertiary alkoxy- and siloxy-substituted silanes. Cross-linking of a silicone containing vinyl groups with that containing Si–H moieties was also achieved by a combination of (^{iPr}DI) and $Ni(2\text{-}EH)_2$ in the presence of $(EtO)_3SiH$ as the activator. A stoichiometric reaction of a mixture of (^{iPr}DI) and $Ni(2\text{-}EH)_2$ with $(EtO)_3SiH$ resulted in the formation of a dinuclear nickel(I) complex having bridging hydrides $[(^{iPr}DI)NiH]_2$ (**14a**). Deuterium labeling and kinetic studies suggested a mechanism involving fragmentation of the dimeric structure of **14a**

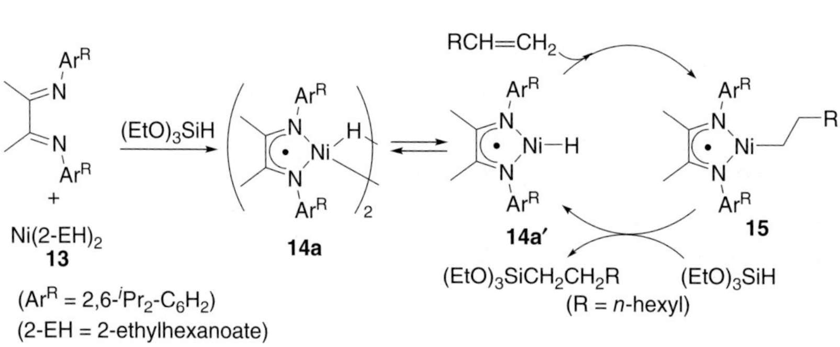

Scheme 11.4 Proposed mechanism for the hydrosilylation of oct-1-ene with $(EtO)_3SiH$ and **14a**.

to a monomeric nickel hydride ($(^{iPr}DI)NiH$ (**14a′**). Fast and reversible alkene insertion into the Ni—H bond in **14a′** took place to form **15**. Reaction of **15** with $(EtO)_3SiH$ proceeded through either σ-bond metathesis or a sequence of oxidative addition of the Si—H bond to the nickel center and subsequent reductive elimination to afford the hydrosilylated product. It is noteworthy that the formal oxidation state of the nickel center of **14a**, **14a′**, and **15** is Ni(I), but calculated electronic structures of **14a′** and **15** are best described as two Ni(II) centers with anionic radical α-diimine ligands. One-electron transfer occurs from the nickel center to the α-diimine ligand.

11.3 Hydrosilylation of Alkenes with Primary and Secondary Hydrosilanes by Base-Metal Catalysts

As described in the earlier sections, the history of hydrosilylation of alkenes started with platinum-catalyzed hydrosilylation with chlorohydrosilanes [3a]. In the early stage, hydrosilylation research was performed using more easily handled trialkylsilanes, which contributed to the understanding of the nature of the catalysts and their mechanisms. While alkoxyhydrosilanes and hydrosiloxanes are practically important for the production of silane coupling reagents and silicone materials in the silicone industry, primary and secondary hydrosilanes have no industrial usage, and therefore, have not attracted attention from organosilicon chemists. This situation changed in the early 2000s, when several groups reported that primary and secondary hydrosilanes were more reactive than tertiary hydrosilanes. In recent studies, $PhSiH_3$ and Ph_2SiH_2 have often been used for benchmark tests to check the activity and selectivity of new base-metal catalysts [30–33]. As shown in Scheme 11.5, eq. 1, the three Si—H bonds in $RSiH_3$ are capable of reacting with three alkenes. The reactions occur stepwise, and, in general, the reaction rate of the first, second, and third hydrosilylations decreased in this order, and under appropriate conditions, $k_1 > k_2 > k_3$, selective monohydrosilylation is achievable. In certain catalysts with sterically hindered ligands that allow limited space for coordination of a hydrosilane and an alkene, only $RSiH_3$ with its small size is useful as the hydrosilane. This explains our comment in the former part of this chapter that catalysts realizing reactions with primary and secondary hydrosilanes are not always active for hydrosilylation of more sterically bulky tertiary hydrosilanes including hydrosiloxanes and alkoxyhydrosilanes.

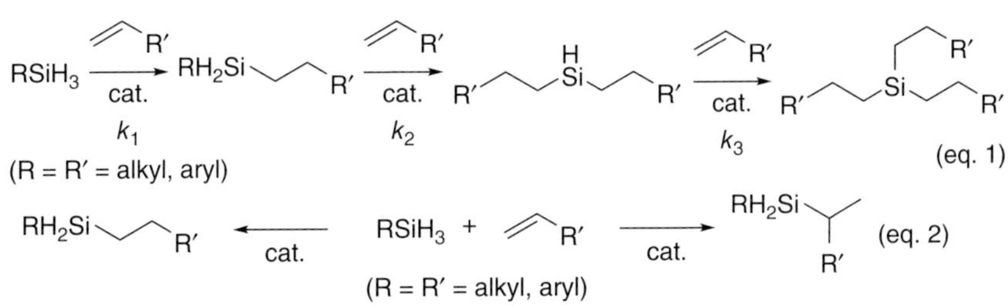

Scheme 11.5 General schemes for the hydrosilylation of alkenes with primary hydrosilanes.

The advantage of hydrosilylations with $RSiH_3$ and R_2SiH_2 in organic synthesis is the facile chemical modification of the products. In particular, the Tamao–Fleming oxidation realizes conversion of the $RSiH_2$ or R_2SiH moiety to a hydroxy group. Thus, this two-step reaction, hydrosilylation of alkenes followed by the Tamao–Fleming oxidation, is a method for producing alcohols from alkenes similar to a process including hydroboration and subsequent treatment with H_2O_2 [34]. Markovnikov and anti-Markovnikov addition of a Si—H bond, shown in Scheme 11.5, eq. 2, is the focus of recent base-metal-catalyzed hydrosilylation, which provides a strategy to control the reaction course by the choice of catalysts. The Markovnikov addition triggered the exploration of asymmetric hydrosilylation by chiral base-metal complexes leading to synthetic methods for chiral alcohols. In addition to the reaction with alkenes, hydrosilanes behave as good reducing reagents for carbonyl compounds by catalysis of transition metals [2c]. Primary and secondary hydrosilanes are often used as efficient reducing reagents. This raises the question of whether hydrosilylation of alkenes can be achieved with carbonyl functionalities remaining intact. This section reviews base-metalcatalyzed hydrosilylation of alkenes with $RSiH_3$ and R_2SiH_2 from the viewpoint of synthetic organic chemistry.

Most iron and cobalt catalysts active for the hydrosilylation of alkenes with $RSiH_3$ and R_2SiH_2 have tridentate ligands (Figure 11.6). As described in the previous sections, pioneering work by Chirik and coworkers in 2004 revealed a bis(imino)pyridine Fe(0) complex bearing a 2,6-diisopropylphenyl group on the imino group as an excellent catalyst [14a]. In the presence of 0.3 mol% of the complex **1a**, reaction of oct-1-ene with $PhSiH_3$ resulted in anti-Markovnikov addition at room temperature to form 1-$PhSiH_2$(*n*-octyl) quantitatively. The TOF reached $364\,h^{-1}$. Selective hydrosilylation of a terminal C=C bond in limonene was achieved at ambient temperature, and a secondary hydrosilane was obtained as a single product. As a variation of bis(imino)pyridine, substituted terpyridine was investigated as a ligand by both Chirik and coworkers [16b] and Nakazawa and coworkers [16b,c]. Nakazawa and coworkers established the method for in situ generation of active species by treatment of (substituted terpyridine)$FeBr_2$ **16** with borohydrides [16b]. This stimulated studies to avoid the use of extremely air-sensitive Fe(0) catalysts, providing dialkyl Fe(II) catalysts [16a] and catalysts composed of dihalo-Fe(II) precursors **17** and

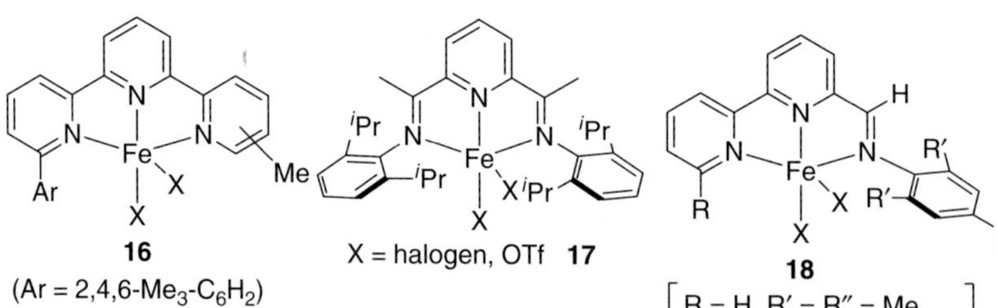

16
(Ar = 2,4,6-Me_3-C_6H_2)

X = halogen, OTf **17**

18

R = H, R′ = R″ = Me
R = H, R′ = ⁱPr, R″ = H
R = Me, R′ = R″ = Me
R = Me, R′ = ⁱPr, R″ = H

Figure 11.6 Fe(II) complexes as catalyst precursors for the hydrosilylation.

reducing reagents [16e,f]. In the latest paper, a modification of **16** provided a more efficient catalyst **18** [16d]. One of the derivatives of **18** activated by $NaBH_4$ showed high activity toward the hydrosilylation of oct-1-ene with $PhSiH_3$ at room temperature, in which the TON reached over 1.2×10^4 after 24 hours. Unfortunately, the products were a mixture of 1-$PhSiH_2$(*n*-octyl) and PhSiH(*n*-octyl)$_2$. It should be mentioned that a TON of 700 was achieved for MD′M by the same catalyst system. As described in the previous section, bis(imino)pyridine cobalt complexes behaved as hydrosilylation and dehydrogenative silylation catalysts. Dihalo- or dicarboxylato-(PDI)CoX_2 [X = halogen, carboxylate] complexes exist as Co(II), whereas monoalkyl (PDI)Co complexes are Co(I). The active species is Co(I), which is generated by reduction of (PDI)CoX_2.

Besides N^N^N-tridentate ligands such as bis(imino)pyridines and terpyridines [16a–c, 35], iron and cobalt P^N^N complexes were reported, as shown in Figure 11.7. Rauchfuss and coworkers reported the preparation and detailed characterization of iron and cobalt complexes, **19** and **20**, and their derivatives [36]. Both **19** and **20** activated by $NaBEt_3H$ catalyzed the hydrosilylation of oct-1-ene with Ph_2SiH_2 at room temperature. The activity (TOF) was as high as 2000 and 800 h^{-1}, respectively. Interestingly, the authors observed selective anti-Markovnikov addition in the iron-catalyzed reactions, but significant amounts of the products from Markovnikov addition were obtained when some of the derivatives of **20**, e.g. **20′**, were used as the catalyst. Iron and cobalt P^N^N complexes reported by Huang and coworkers demonstrated the excellent properties of these catalysts on Markovnikov/anti-Markovnikov selectivity and functional-group compatibility. The iron complex **21** activated by $NaHBEt_3$ catalyzed the hydrosilylation of terminal alkenes having ketone, ester, and amide groups with $PhSiH_3$ or Ph_2SiH_2 to give the corresponding anti-Markovnikov products [37]. The alkene hydrosilylation took place with carbonyl groups in the molecule remaining intact. The iron complex **21′** is a more robust catalyst than **21**, with improved catalyst efficiency. Interestingly, the iron complex **21′** promoted anti-Markovnikov addition with $PhSiH_3$, while selective Markovnikov addition was achieved using **22** as the cobalt homolog of **21′** [38]. A recent paper from Ge's research group realized switching of Markovnikov/anti-Markonikov selectivity simply by changing the cobalt ligand [39]. Using Co(acac)$_2$ (acac = acetylacetonato) as the catalyst precursor, selective Markovnikov addition took place for styrene derivatives by addition of Xantphos, whereas anti-Markovnikov addition

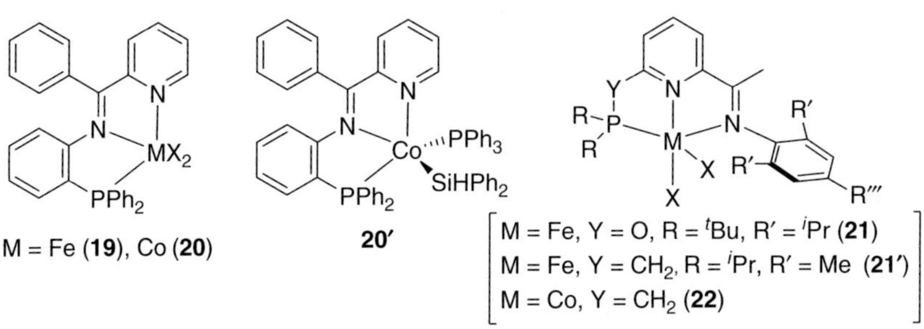

Figure 11.7 Iron and cobalt complexes having P^N^N ligands.

was accomplished in the presence of 1,1′-bis(diphenylphosphino)ferrocene (dppf). Similar Markovnikov to anti-Markovnikov switching was achieved for aliphatic alkenes by changing the ligand from bis(imino)pyridine to Xantphos (Xantphos = 4,5-bis(diphenylphosphino)-9,9-dimethylxanthene).

Catalytic asymmetric synthesis is a recent topic in organic chemistry. Two papers from Lu's research group suggested a bright future for base-metal-catalyzed hydrosilylation of alkenes in this field [40–42]. As shown in Scheme 11.6, eq. 1, a chiral version of the bis(imino)pyridine iron complex **23** catalyzed the hydrosilylation of 1,1-disubstituted alkenes at room temperature [40]. With 5 mol% of **23** activated by $NaHBEt_3$, various 1-alkyl-1′-arylethylenes underwent the hydrosilylation to form the corresponding silicon compounds having a substituent at the β-position. Efficient asymmetric induction occurred to give the product in over 90% ee in most cases. Reactions of 1,1′-dialkylethylene were also examined, which proceeded smoothly, but asymmetric induction was less efficient. Selective Markovnikov addition by cobalt catalysts described earlier was applied to asymmetric synthesis. As shown in Scheme 11.6, eq. 2, the complex **24** as the cobalt homolog of **23** (1 mol%) activated by NaO^tBu exhibited good catalytic activity for Markovnikov addition of $ArSiH_3$ to alkenes at room temperature [42]. The process could be applied to hydrosilylation of not only aryl or heteroarylethylene but also terminal aliphatic alkenes to give the desired branched products with over 98% selectivity with excellent enantioselectivity. The reactions were tolerant toward functional groups such as halides, esters, ketones, and imides.

As noted earlier, most iron and cobalt catalysts that activated primary and secondary hydrosilanes have N^N^N or P^N^N ligands. The only exception was Deng's NHC–Co complex showing high activity toward hydrosilylation of

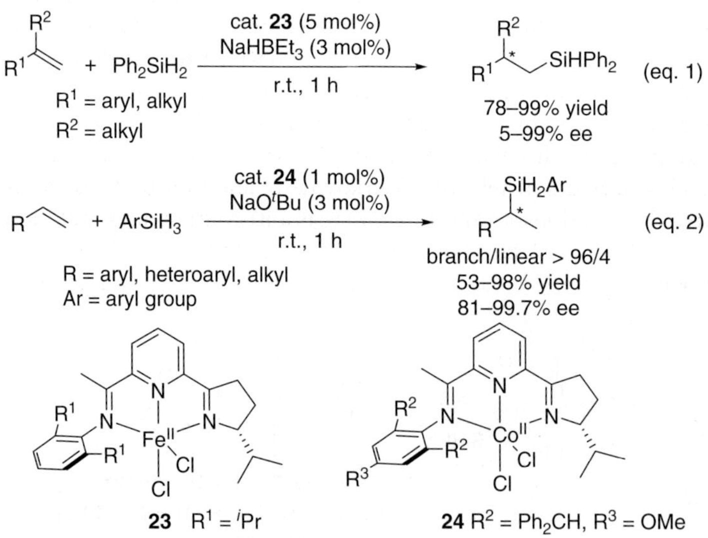

Scheme 11.6 Asymmetric hydrosilylation catalyzed by iron or cobalt complexes. "*" expresses an asymmetric carbon.

oct-1-ene to give the anti-Markovnikov product with >80% selectivity [43]. Nickel catalysts differ from iron and cobalt catalysts in that a variety of species from pincer-type complexes to nanoparticles are considered to be active species. Ni(II)–phosphine complexes have been investigated from the early stages of catalytic alkene hydrosilylation studies. After 2000, indenylnickelphosphine complexes related to styrene polymerization were investigated and found to have some catalytic activity toward the Markovnikov addition of styrene [44, 45]. Analogous to the catalysts for Shell higher olefin process (SHOP), several Schiff-base catalysts for the hydrosilylation of Et_2SiH_2 were reported by Nakajima and coworkers [46].

Hu and coworkers reported that nickel(II) complexes bearing a N^N^N pincer ligand showed good catalytic activity toward the hydrosilylation of various alkenes with Ph_2SiH_2 and alkoxyhydrosilanes. In their first paper, a bis(amino)amide Ni(II) complex **25** shown in Figure 11.8 catalyzed the chemoselective anti-Markovnikov hydrosilylation of various alkenes with Ph_2SiH_2 with 1 mol% of the catalyst at room temperature in THF [47]. Alkenes having ester, keto, NH_2, formyl, epoxy, and secondary amide groups could be used as the substrates without solvents. The high catalytic activity was exemplified by a solvent-free reaction of oct-1-ene with Ph_2SiH_2 in the presence of **25** (0.025 mol%) at room temperature, which was complete within 2.5–3 minutes, giving a TOF of about 83 000 h^{-1}. The amount of the loaded catalyst could be reduced to 0.01 mol% and the reaction finished within seven to eight minutes with a TON of 10 000.

The same research group later discovered that a different N^N^N-pincer complex **26** was a catalyst for the hydrosilylation of alkenes with alkoxyhydrosilanes in the presence of NaO^tBu (Figure 11.8) [48]. It should be noted that the reaction did not afford the expected alkoxysilane products as shown in Scheme 11.7. Reactions of terminal alkenes with $Me(MeO)_2SiH$ proceeded smoothly in THF at room temperature by Ni complex **26** (2.5 mol%)/NaO^tBu (5 mol%) to produce dimethylalkylsilanes in high yields (Scheme 11.7, eq. 1). Similar reactions of alkenes with $(EtO)_2MeSiH$ catalyzed by **26** activated by NaO^tBu gave methyldialkylsilanes, whereas those with $(MeO)_3SiH$ afforded dialkylsilanes (Scheme 11.7, eqs. 2 and 3). The reactions showed high functional-group compatibility with ether, epoxide, *tert*-butyldimethylsilyl-, protected alcohol, halide, amine, and acetal groups. Alkoxyhydrosilanes such as $Me(MeO)_2SiH$, $(EtO)_2MeSiH$, and $(MeO)_3SiH$ acted as surrogates for gaseous Me_2SiH_2, $MeSiH_3$, and SiH_4. The products obtained by eqs. 1–3 corresponded to those formed by the hydrosilylation of alkenes with Me_2SiH_2, $MeSiH_3$, and SiH_4, respectively. Preliminary mech-

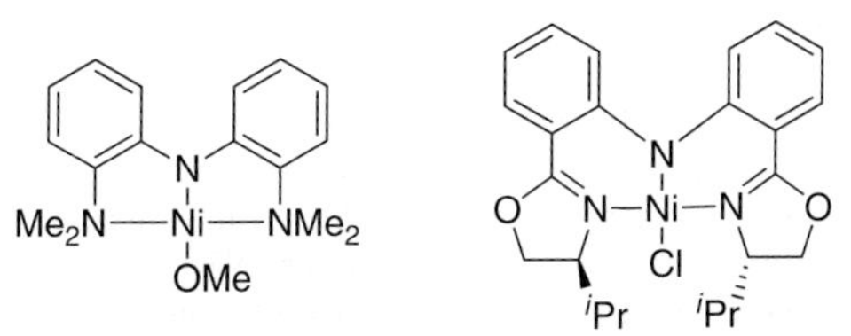

25 **26**

Figure 11.8 Nickel complexes having N^N^N-pincer ligand.

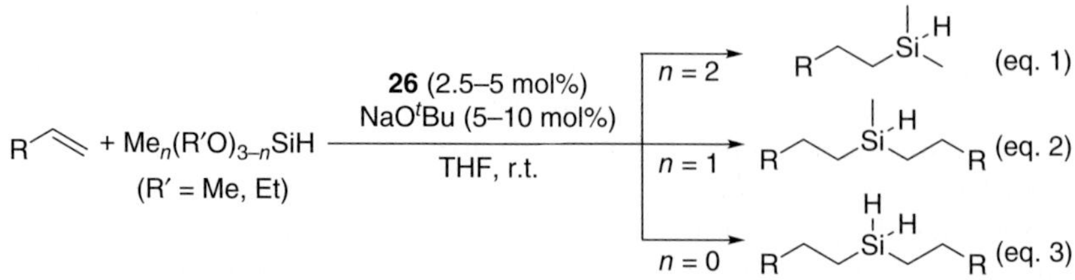

Scheme 11.7 Hydrosilylation of alkenes with alkoxyhydrosilanes catalyzed by **26** in the presence of NaO^{t}Bu, where alkenes are terminal functionalized alkenes bearing ether, epoxide, *tert*-butyldimethylsiloxy, halide, amine, and acetal groups.

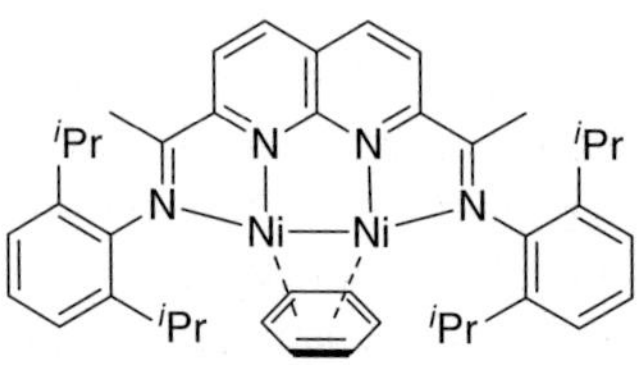

27

Figure 11.9 Dinuclear nickel complex **27** having a tetradentate nitrogen ligand.

anistic studies revealed that NaO^{t}Bu catalyzed the disproportionation of these alkoxysilanes to the hydrosilanes.[1]

Use of a tetradentate nitrogen ligand for the nickel species resulted in the formation of a dinuclear nickel complex **27** having a Ni^I—Ni^I bond (Figure 11.9) [49]. Bimetallic activation of R_2SiH_2 by the dinickel moiety in **27** was investigated, since it may be involved in the catalytic cycle. Despite interest in the mechanism, however, the catalytic activity for the hydrosilylation of oct-1-ene with Ph_2SiH_2 was not high.

One unique feature of nickel is that compounds without strong ligands like those containing nitrogen and phosphorus sometimes exhibit hydrosilylation activity. In 2012, Lipschutz and Tilley reported that a simple nickel amide, Ni[N(SiMe$_3$)(2,6-^{i}Pr$_2$C$_6$H$_3$)]$_2$, catalyzed the hydrosilylation of oct-1-ene with Ph_2SiH_2 [50]. This is in contrast to the high catalytic activity of Fe[N(SiMe$_3$)$_2$]$_2$ toward hydrosilylation of ketones and aldehydes with primary and secondary hydrosilanes [51]. The active catalytic species for the catalysis of these metal amides has not been discussed in the literature, but Ni(O^{t}Bu)$_2$ as an insoluble solid showed alkene hydrosilylation activity, and the authors assumed the

1 *Caution:* Alkoxyhydrosilanes are attractive reactants for the hydrosilylation of alkenes because the resulting alkylalkoxysilanes are useful as silane coupling reagents. However, they are also reactive with bases, leading to disproportionation to form pyrophoric and explosive SiH_4, $MeSiH_3$, and Me_2SiH_2. Formation of these silanes in the experiments was clearly observed by NMR spectroscopy and other methods reported in the literature. Special care is needed when new base-metal catalysts composed of metal halides with bases such as KO^{t}Bu, NaO^{t}Bu, and NaOMe are applied to the hydrosilylation with alkoxyhydrosilanes on large scales. Catalysts that do not contain a strong base as a component are desirable [16g, 48].

catalytically active species to be nickel nanoparticles [52]. Three other simple nickel catalysts, $[(\eta^6\text{-arene})Ni(\eta^3\text{-allyl})]^+$ [53], $Ni(OAr)_2$ in which the OAr group is derived from 9-hydroxyphenalenone [54], and $Ni(acac)_2$ [55] showed moderate catalytic activity toward the hydrosilylation of alkenes with primary or secondary hydrosilanes. In the reactions with the former catalyst, activation of a Si—H bond is assumed to be assisted by the coordinated allyl ligand, although both η^3-allyl and η^6-arene ligands are weakly coordinating ligands and could be eliminated during the catalysis. The latter two catalysts required the addition of reducing reagents to exhibit catalytic activity. The authors of Ref. [52] proposed radical mechanisms involving the single-electron transfer from dianionic Ni^{II} species. Mechanisms for these simple systems await further investigation.

The results so far described have focused on the hydrosilylation of 1-alkenes. Different features of the base-metal catalysis are obtained by changing the substrate from 1-alkenes to internal alkenes, dienes, allenes, and alkynes. The hydrosilylation of alkenes is often accompanied by alkene isomerization. This is explained by addition and elimination of metal-hydride intermediate species to form C=C bonds. This mechanism also explains the generally observed result that hydrosilylation of internal alkenes affords the same product, 1-silylalkane, as that of terminal alkenes. A recent paper discloses reactions in which a silyl group is selectively introduced at the benzylic position [56]. As shown in Scheme 11.8, eq. 1, (2,9-diaryl-1,10-phenanthroline)$FeCl_2$ (**28**) activated by EtMgBr catalyzed the hydrosilylation of alkenes having an aryl group at the terminal position with $PhSiH_3$, leading to the formation of a silyl product having a $PhSiH_2$ group at the

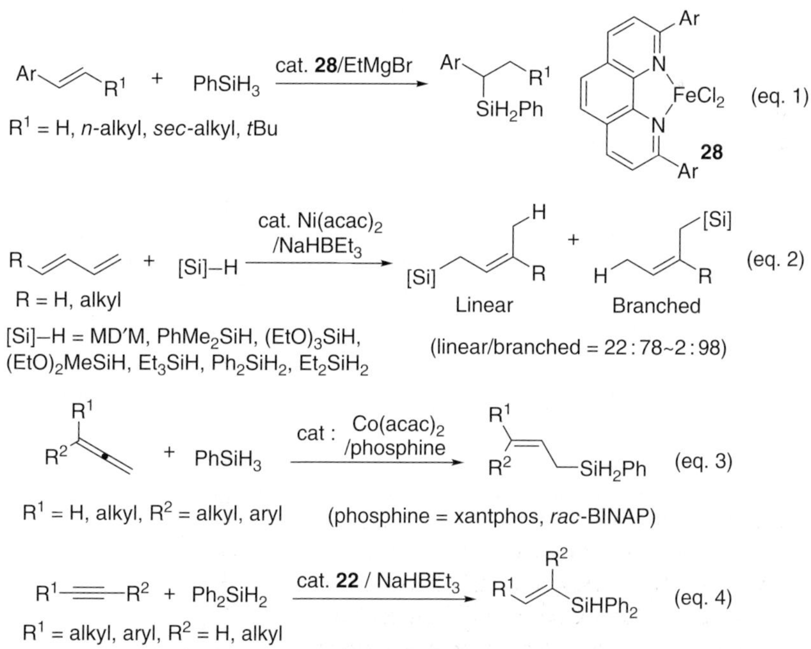

Scheme 11.8 Hydrosilylation of internal alkenes, dienes, allenes, and alkynes.

benzyl position. The same catalyst promoted Markovnikov addition to styrene (R = H in eq. 1), whereas anti-Markovnikov addition preferably took place for the reaction of alkenes without aryl substituents. The authors proposed that this unique benzylic selectivity arose from π–π-interactions between the aryl group in the substrate and delocalized π-bonds on the ligand. The hydrosilylation of dienes was investigated by bis(imino)pyridine cobalt complexes [57] and $Ni(acac)_2$ [55] as the catalysts (Scheme 11.8, eq. 2). A mixture of products due to 1,2-addition and 1,4-addition was obtained, and selective preparation of one of the isomers was investigated by changing the catalyst and the reaction conditions. Hydrosilylation of allenes [58] and alkynes [59] was also investigated by cobalt catalysts, in which unique selectivity affording the product having *Z*-alkenes was reported (Scheme 11.8, eqs. 3 and 4).

11.4 Conclusion and Future Outlook

Hydrosilylation reactions occur via two-electron redox processes involving the oxidative addition of a Si—H bond to a metal center and reductive elimination of the hydrosilylated product. A disadvantage of base-metal catalysts is their preference for one-electron redox processes over two-electron redox catalysis. Thus, strategies are needed to design base-metal catalysts to make two-electron processes possible. As described in this chapter, two approaches to catalytically active base-metal complexes have been developed: one is the use of redox-active ligands like bis(imino)pyridine derivatives, and the other is the introduction of strong-field ligands including CO, CNR, and NHC. One of the interests in base-metal-catalyzed hydrosilylation is as a possible substitute for the current platinum-catalyzed reactions of alkenes with alkoxyhydrosilanes and hydrosiloxanes, which is relevant to industrial modification of silicone fluid and silicone curing. As discussed in the first part of this chapter, several iron and cobalt catalysts showing high catalytic activity and high selectivity appear in the literature. Although primary and secondary hydrosilanes are not attractive from the industrial point of view, some reactions using them present excellent synthetic methods for organic synthesis, including asymmetric synthesis of chiral alcohols by regioselective hydrosilylation of terminal alkenes with $PhSiH_3$ and subsequent oxidative transformation of the resulting silanes to alcohols. Rapid growth of this field has provided unexpectedly fruitful aspects as potential substitutes for platinum in industrially important synthetic processes of silicones, new synthetic approaches in organic synthesis, and design of base-metal catalysts based on mechanistic studies with well-defined catalyst precursors. However, this is only the beginning, and more investigation is needed to achieve highly efficient and selective catalytic hydrosilylations of a variety of alkenes.

References

1 (a) Marcinec, J.G.B., Urbaniak, W., and Kornetka, Z.W. (1992). *Comprehensive Handbook on Hydrosilylation*. Oxford: Pergamon. (b) Matisons, J. and

Marciniec, B. (2009). *Hydrosilylation A Comprehensive Review on Recent Advances*. Dordrecht: Springer Netherlands.

2 (a) Larson, G.L. and Fry, J.L. (2009). *Ionic and Organometallic-Catalyzed Organosilane Reductions*, vol. 81. Wiley. (b) Addis, D., Das, S., Junge, K., and Beller, M. (2011). *Angew. Chem. Int. Ed.* 50: 6004–6011. (c) Nagashima, H. (2015). *Synlett* 26: 866–890. (d) Pesti, J. and Larson, G.L. (2016). *Org. Process Res. Dev.* 20: 1164–1181.

3 (a) Speier, J.L. (1979). *Adv. Organomet. Chem.* 17: 407–447. (b) Troegel, D. and Stohrer, J. (2011). *Coord. Chem. Rev.* 255: 1440–1459. (c) Nakajima, Y. and Shimada, S. (2015). *RSC Adv.* 5: 20603–20616.

4 Holwell, A.J. (2008). *Platinum Met. Rev.* 52: 243–246.

5 Nakamura, E. and Sato, K. (2011). *Nat. Mater.* 10: 158–161.

6 Thayer, A. (2005). *Chem. Eng. News* 83: 55–58.

7 (a) Chalk, A.J. and Harrod, J.F. (1965). *J. Am. Chem. Soc.* 87: 1133–1135. (b) Seitz, F. and Wrighton, M.S. (1988). *Angew. Chem. Int. Ed.* 27: 289–291. (c) Sakaki, S., Mizoe, N., Sugimoto, M., and Musashi, Y. (1999). *Coord. Chem. Rev.* 190: 933–960. (d) Roy, A.K. and Taylor, R.B. (2002). *J. Am. Chem. Soc.* 124: 9510–9524. (e) Meister, T.K., Riener, K., Gigler, P. et al. (2016). *ACS Catal.* 6: 1274–1284.

8 (a) Chirik, P.J. (2015). *Acc. Chem. Res.* 48: 1687–1695. (b) Nagashima, H. (2017). *Bull. Chem. Soc. Jpn.* 90: 761–775.

9 (a) Nesmeyanov, A.N., Freidlina, R.K., Chukovskaya, E.C. et al. (1962). *Tetrahedron* 17: 61–68. (b) Schroeder, M.A. and Wrighton, M.S. (1977). *J. Organomet. Chem.* 128: 345–358. (c) Kakiuchi, F., Tanaka, Y., Chatani, N., and Murai, S. (1993). *J. Organomet. Chem.* 456: 45–47. (d) Naumov, R.N., Itazaki, M., Kamitani, M., and Nakazawa, H. (2012). *J. Am. Chem. Soc.* 134: 804–807.

10 (a) Chalk, A.J. and Harrod, J.F. (1967). *J. Am. Chem. Soc.* 89: 1640–1647. (b) Reichel, C.L. and Wrighton, M.S. (1980). *Inorg. Chem.* 19: 3858–3860. (c) Takeshita, K., Seki, Y., Kawamoto, K. et al. (1987). *J. Org. Chem.* 52: 4864–4868. (d) Brookhart, M. and Grant, B.E. (1993). *J. Am. Chem. Soc.* 115: 2151–2156. (e) Chatani, N., Kodama, T., Kajikawa, Y. et al. (2000). *Chem. Lett.* 29: 14–15.

11 (a) Archer, N.J., Haszeldine, R.N., and Parish, R.V. (1971). *J. Chem. Soc. D* 524–525. (b) Archer, N.J., Haszeldine, R.N., and Parish, R.V. (1979). *J. Chem. Soc., Dalton Trans.* 695.

12 (a) Kiso, Y., Kumada, M., Maeda, K. et al. (1973). *J. Organomet. Chem.* 50: 311–318. (b) Kiso, Y., Kumada, M., Tamao, K., and Umeno, M. (1973). *J. Organomet. Chem.* 50: 297–310.

13 (a) Small, B.L., Brookhart, M., and Bennett, A.M.A. (1998). *J. Am. Chem. Soc.* 120: 4049–4050. (b) Britovsek, G.J.P., Bruce, M., Gibson, V.C. et al. (1999). *J. Am. Chem. Soc.* 121: 8728–8740.

14 (a) Bart, S.C., Lobkovsky, E., and Chirik, P.J. (2004). *J. Am. Chem. Soc.* 126: 13794–13807. (b) Russell, S.K., Darmon, J.M., Lobkovsky, E., and Chirik, P.J. (2010). *Inorg. Chem.* 49: 2782–2792. (c) Tondreau, A.M., Atienza, C.C., Weller, K.J. et al. (2012). *Science* 335: 567–570. (d) Hojilla Atienza, C.C., Tondreau, A.M., Weller, K.J. et al. (2012). *ACS Catal.* 2: 2169–2172.

15 (a) Bart, S.C., Chlopek, K., Bill, E. et al. (2006). *J. Am. Chem. Soc.* 128: 13901–13912. (b) Bart, S.C., Lobkovsky, E., Bill, E. et al. (2007). *Inorg. Chem.* 46:

7055–7063. (c) Darmon, J.M., Stieber, S.C.E., Sylvester, K.T. et al. (2012). *J. Am. Chem. Soc.* 134: 17125–17137. (d) Stieber, S.C.E., Milsmann, C., Hoyt, J.M. et al. (2012). *Inorg. Chem.* 51: 3770–3785. (e) Ibrahim, A.D., Entsminger, S.W., Zhu, L., and Fout, A.R. (2016). *ACS Catal.* 6: 3589–3593.

16 (a) Tondreau, A.M., Atienza, C.C.H., Darmon, J.M. et al. (2012). *Organometallics* 31: 4886–4893. (b) Kamata, K., Suzuki, A., Nakai, Y., and Nakazawa, H. (2012). *Organometallics* 31: 3825–3828. (c) Hayasaka, K., Kamata, K., and Nakazawa, H. (2016). *Bull. Chem. Soc. Jpn.* 89: 394–404. (d) Toya, Y., Hayasaka, K., and Nakazawa, H. (2017). *Organometallics* 36: 1727–1735. (e) Greenhalgh, M.D., Frank, D.J., and Thomas, S.P. (2014). *Adv. Synth. Catal.* 356: 584–590. (f) Challinor, A.J., Calin, M., Nichol, G.S. et al. (2016). *Adv. Synth. Catal.* 358: 2404–2409. (g) Docherty, J.H., Peng, J., Dominey, A.P., and Thomas, S.P. (2017). *Nat. Chem.* 9: 595–600.

17 Knijnenburg, Q., Hetterscheid, D., Kooistra, T.M., and Budzelaar, P.H.M. (2004). *Eur. J. Inorg. Chem.* 1204–1211.

18 (a) Bowman, A.C., Milsmann, C., Atienza, C.C.H. et al. (2010). *J. Am. Chem. Soc.* 132: 1676–1684. (b) Atienza, C.C., Diao, T., Weller, K.J. et al. (2014). *J. Am. Chem. Soc.* 136: 12108–12118. (c) Schuster, C.H., Diao, T., Pappas, I., and Chirik, P.J. (2016). *ACS Catal.* 6: 2632–2636.

19 (a) Marciniec, B., Kownacka, A., Kownacki, I., and Taylor, R. (2014). *Appl. Catal. A: Gen.* 486: 230–238. See also (b) Kownacka, A., Kownacki, I., Kubicki, M. et al. (2014). *J. Organomet. Chem.* 750: 132–139. (c) Marciniec, B. and Majchrzak, M. (2000). *Inorg. Chem. Commun.* 3: 371–375.

20 (a) Sunada, Y., Tsutsumi, H., Shigeta, K. et al. (2013). *Dalton Trans.* 42: 16687–16692. (b) Sunada, Y. and Nagashima, H. (2017). *Dalton. Trans.* 46: 7644–7655. (c) Sunada, Y. and Nagashima, H. (2017). *J. Synth. Org. Chem. Jpn.* 75: 1253–1263. (d) Tahara, A., Tanaka, H., Sunada, Y. et al. (2016). *J. Org. Chem.* 81: 10900–10911. (e) Hoshi, K., Tahara, A., Sunada, Y. et al. (2017). *Bull. Chem. Soc. Jpn.* 90: 613–626. (f) Sunada, Y., Soejima, H., and Nagashima, H. (2014). *Organometallics* 33: 5936–5939.

21 Sunada, Y., Noda, D., Soejima, H. et al. (2015). *Organometallics* 34: 2896–2906.

22 Noda, D., Tahara, A., Sunada, Y., and Nagashima, H. (2016). *J. Am. Chem. Soc.* 138: 2480–2483.

23 Liu, Y. and Deng, L. (2017). *J. Am. Chem. Soc.* 139: 1798–1801.

24 Chen, C., Hecht, M.B., Kavara, A. et al. (2015). *J. Am. Chem. Soc.* 137: 13244–13247.

25 Guzei, I.A. and Wendt, M. (2006). *Dalton Trans.* 3991–3999.

26 Lee, K.L. (2017). *Angew. Chem. Int. Ed.* 56: 3665–3669.

27 (a) Brandstadt, K., Cook, S., Nguyen, B.T. et al. (2011). WO Patent Application 2013043846 A1 to Dow Corning Corporation, priority date 20 September 2011; See also (b) Brandstadt, K., Cook, S., Nguyen, B.T. et al. (2011). WO Patent Application 2013043912 A2 to Dow Corning Corporation, priority date 20 September 2011. (c) Brandstadt, K., Cook, S., Nguyen, B.T. et al. (2011). WO Patent Application 2013043783 A2 to Dow Corning Corporation, priority date 20 September 2011.

28 Pappas, I., Treacy, S., and Chirik, P.J. (2016). *ACS Catal.* 6: 4105–4109.

29 Johnson, L.K., Killian, C.M., and Brookhart, M. (1995). *J. Am. Chem. Soc.* 117: 6414–6415.
30 Du, X. and Huang, Z. (2017). *ACS Catal.* 7: 1227–1243.
31 Zhang, M. and Zhang, A. (2010). *Appl. Organomet. Chem.* 24: 751–757.
32 Sun, J. and Deng, L. (2016). *ACS Catal.* 6: 290–300.
33 Nakajima, Y., Sato, K., and Shimada, S. (2016). *Chem. Rec.* 16: 2379–2387.
34 Jones, G.R. and Landais, Y. (1996). *Tetrahedron* 52: 7599–7602.
35 Gorczynski, A., Zaranek, M., Witomska, S. et al. (2016). *Catal. Commun.* 78: 71–74.
36 (a) Gilbert-Wilson, R., Chu, W.-Y., and Rauchfuss, T.B. (2015). *Inorg. Chem.* 54: 5596–5603. (b) Chu, W.-Y., Gilbert-Wilson, R., Rauchfuss, T.B. et al. (2016). *Organometallics* 35: 2900–2914.
37 Peng, D., Zhang, Y., Du, X. et al. (2013). *J. Am. Chem. Soc.* 135: 19154–19166.
38 Du, X., Zhang, Y., Peng, D., and Huang, Z. (2016). *Angew. Chem. Int. Ed.* 55: 6671–6675.
39 Wang, C., Teo, W.J., and Ge, S. (2017). *ACS Catal.* 7: 855–863.
40 Chen, J., Cheng, B., Cao, M., and Lu, Z. (2015). *Angew. Chem. Int. Ed.* 54: 4661–4664.
41 Nikonov, G.I. (2015). *ChemCatChem* 7: 1918–1919.
42 Cheng, B., Lu, P., Zhang, H. et al. (2017). *J. Am. Chem. Soc.* 139: 9439–9442.
43 Mo, Z., Liu, Y., and Deng, L. (2013). *Angew. Chem. Int. Ed.* 52: 10845–10849.
44 Fontaine, F.-G., Nguyen, R.-V., and Zargarian, D. (2003). *Can. J. Chem.* 81: 1299–1306.
45 Gareau, D., Sui-Seng, C., Groux, L.F. et al. (2005). *Organometallics* 24: 4003–4013.
46 Srinivas, V., Nakajima, Y., Ando, W. et al. (2015). *Catal. Sci. Technol.* 5: 2081–2084.
47 Buslov, I., Becouse, J., Mazza, S. et al. (2015). *Angew. Chem. Int. Ed.* 54: 14523–14526.
48 Buslov, I., Keller, S.C., and Hu, X. (2016). *Org. Lett.* 18: 1928–1931.
49 Steiman, T.J. and Uyeda, C. (2015). *J. Am. Chem. Soc.* 137: 6104–6110.
50 Lipschutz, M.I. and Tilley, T.D. (2012). *Chem. Commun.* 48: 7146–7148.
51 Yang, J. and Tilley, T.D. (2010). *Angew. Chem. Int. Ed.* 49: 10186–10188.
52 Buslov, I., Song, F., and Hu, X. (2016). *Angew. Chem. Int. Ed.* 55: 12295–12299.
53 Mathew, J., Nakajima, Y., Choe, Y.K. et al. (2016). *Chem. Commun.* 52: 6723–6726.
54 Vijaykumar, G., Pariyar, A., Ahmed, J. et al. (2018). *Chem. Sci.* 9: 2817–2825.
55 Srinivas, V., Nakajima, Y., Ando, W. et al. (2016). *J. Organomet. Chem.* 809: 57–62.
56 Hu, M.-Y., He, Q., Fan, S.-J. et al. (2018). *Nat. Commun.* 9: 221.
57 Raya, B., Jing, S., Balasanthiran, V., and RajanBabu, T.V. (2017). *ACS Catal.* 7: 2275–2283.
58 Wang, C., Teo, W.J., and Ge, S. (2017). *Nat. Commun.* 8: 2258.
59 Du, X., Hou, W., Zhang, Y., and Huang, Z. (2017). *Org. Chem. Front.* 4: 1517–1521.

12

Silylenes as Ligands in Catalysis

Yu-Peng Zhou and Matthias Driess

Technische Universität Berlin, Department of Chemistry: Metalorganics and Inorganic Materials, Sekr. C2, Strasse des 17. Juni 135, D-10623 Berlin, Germany

12.1 Introduction

NHCs and their metal complexes have been recognized to act as versatile (pre)catalysts for various types of catalytic organic transformations [1]. Silylenes, the silicon analogs of carbenes, are no longer laboratory curiosities but valuable building blocks for the synthesis of new functional silicon compounds with high potential exceeding the Lewis donor properties of NHCs and phosphines [2]. The σ-donor and π-acceptor ability, ligand-to-metal charge transfer, and steric parameters of several Si(II) compounds have been calculated by Szilvási and a coworker [3], who indicate that silylene ligands can compete with or even exceed the favorable features of widely used NHC and phosphine ligands regarding these parameters.

12.2 Applications of Silylene Ligands in Catalysis

Since the first report on Suzuki cross-coupling using the monodentate silylene–Pd(0) complex as (pre)catalyst in 2001, substantial progress has been made in this field. Silylene ligands have demonstrated their excellence in various catalytic transformations, including carbon–carbon bond-formation, carbon–heteroatom bond-formation, and reduction reactions. Silylene ligands are not a simple isoelectronic replacement for phosphine and carbene ligands but facilitate catalytic efficiencies and selectivities by taking advantage of the unique electronic features and the cooperative effects of silylene ligands.

12.2.1 Carbon–Carbon Bond-Forming Reactions

In 2001, the silylene dinuclear Pd complex **1** was introduced to catalysis by Fürstner et al. (Figure 12.1) [4]. Complex **1** bearing two bridged N-heterocyclic silylene (NHSi) ligands was synthesized by the reaction of $(Ph_3P)_4Pd$ and 1 M

Organosilicon Chemistry: Novel Approaches and Reactions, First Edition.
Edited by Tamejiro Hiyama and Martin Oestreich

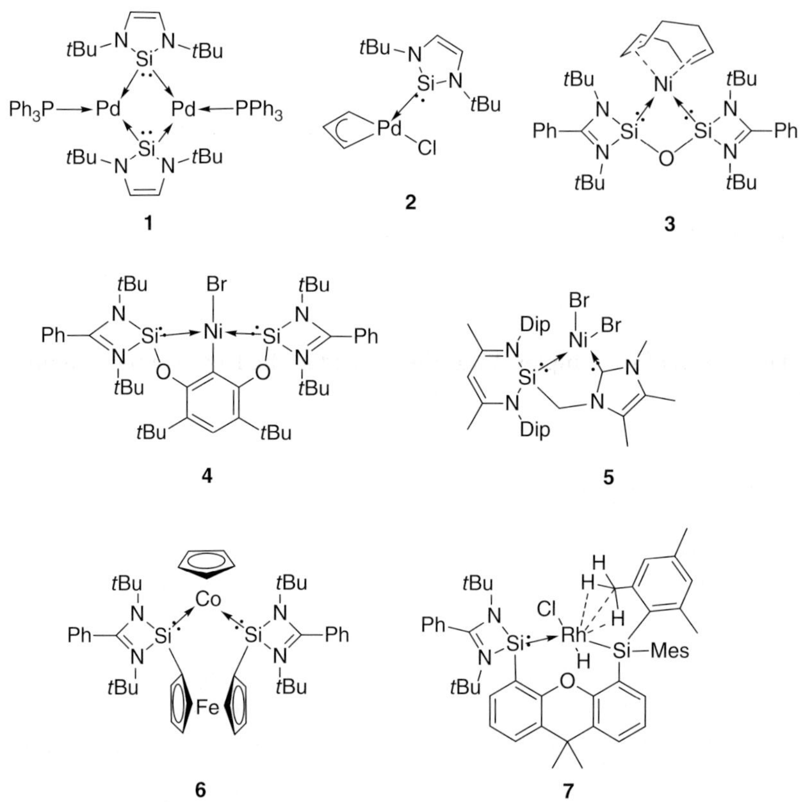

Figure 12.1 Silylene–metal complexes which were applied in catalytic carbon–carbon bond-forming reactions.

equivalents of the corresponding silylene. Catalytic tests indicated that silylene complex **1** was an active catalyst in the Suzuki coupling reaction of aryl boronic acids with aryl bromides (Scheme 12.1).

In 2008, Roesky and coworkers prepared another catalytically active Pd complex **2**, which is a mononuclear (η^3-C_3H_5)–Pd(II) complex bearing one NHSi

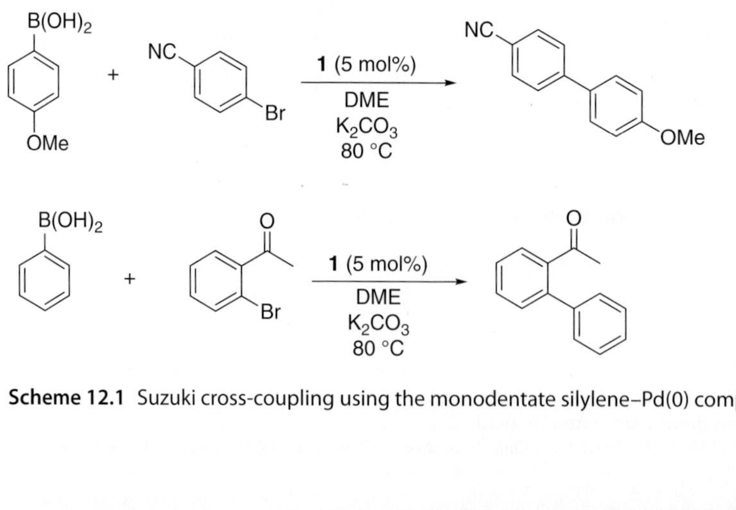

Scheme 12.1 Suzuki cross-coupling using the monodentate silylene–Pd(0) complex **1**.

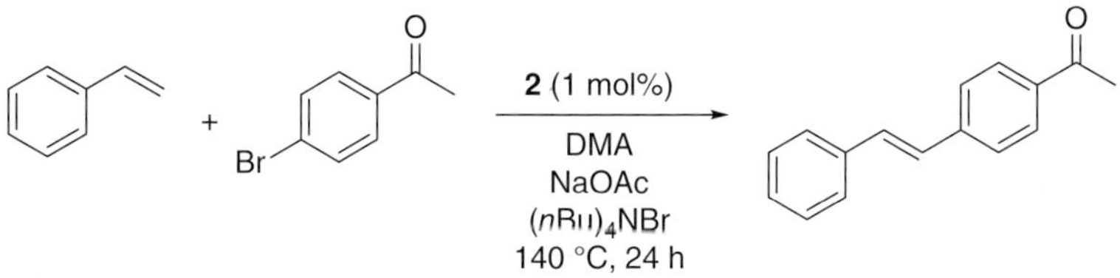

Scheme 12.2 Heck coupling of bromoacetophenone and styrene using silylene–Pd(II) complex **2**.

ligand (Figure 12.1) [5]. Complex **2** was then tested in the Heck coupling of bromoacetophenone and styrene, giving almost quantitative yield after 24 hours at 140 °C (Scheme 12.2).

In 2010, the first chelating bis-silylene ligand [LSi:–O–Si:L] (L = PhC[N*t*Bu]$_2$) was developed by our group through the dehydrochlorination reaction of the amidinate disiloxane precursor [LSiH(Cl)–O–SiH(Cl)L] with [$(CH_3)_3Si]_2NLi$ [6]. The corresponding Ni(0) complex **3** was generated by the reaction of the bis-silylene with $Ni(COD)_2$ in toluene (Figure 12.1). Three years later, Inoue and coworkers employed complex **3** as a precatalyst for carbon–carbon cross-coupling reactions [7]. It was shown that complex **3** efficiently catalyzes carbon–carbon bond formation of aryl halides with organozinc reagents or Grignard reagents (Schemes 12.3 and 12.4, respectively).

Reported by Driess and coworkers in 2013, the Ni(II) complex **4** was prepared by the reaction of the corresponding pincer-type silylene ligand and $NiBr_2$(dme) in the presence of excess Et_3N (Figure 12.1) [8]. Complex **2** was found to be an active

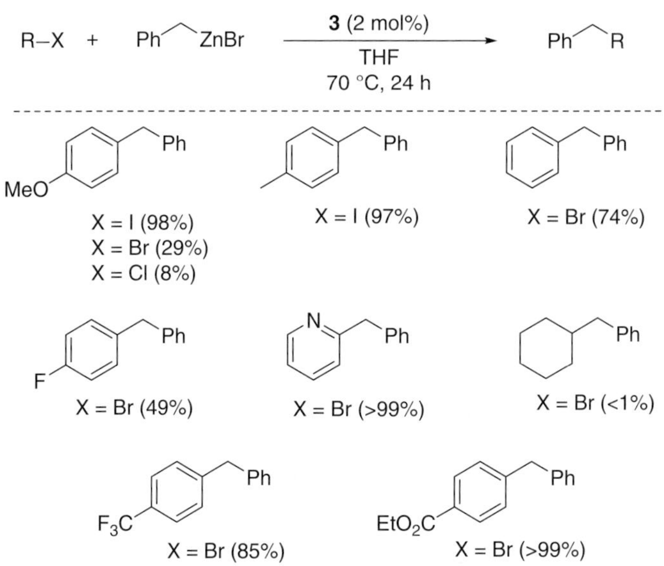

Scheme 12.3 Nickel-catalyzed carbon–carbon bond formation of benzylzinc bromide with various aryl halides using **3** as precatalyst.

(pre)catalyst for the Sonogashira coupling of phenylacetylene with (*E*)-1-iodooct-1-ene. Further mechanism study suggested that the Si(II) center plays an important role in stabilizing the heterobimetallic Ni/Cu intermediate **4a**, which further undergoes oxidative addition yielding the Ni(IV) intermedia **4b** (Scheme 12.5).

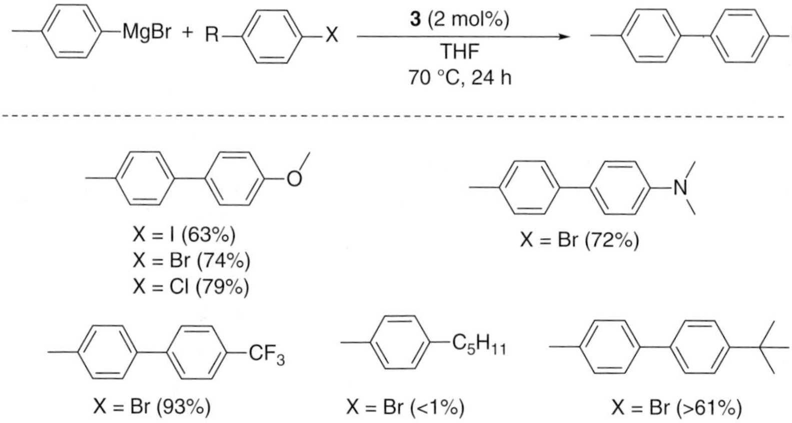

Scheme 12.4 Carbon–carbon bond formation of Grignard reagents with different aryl halide derivatives using the **3** precatalyst.

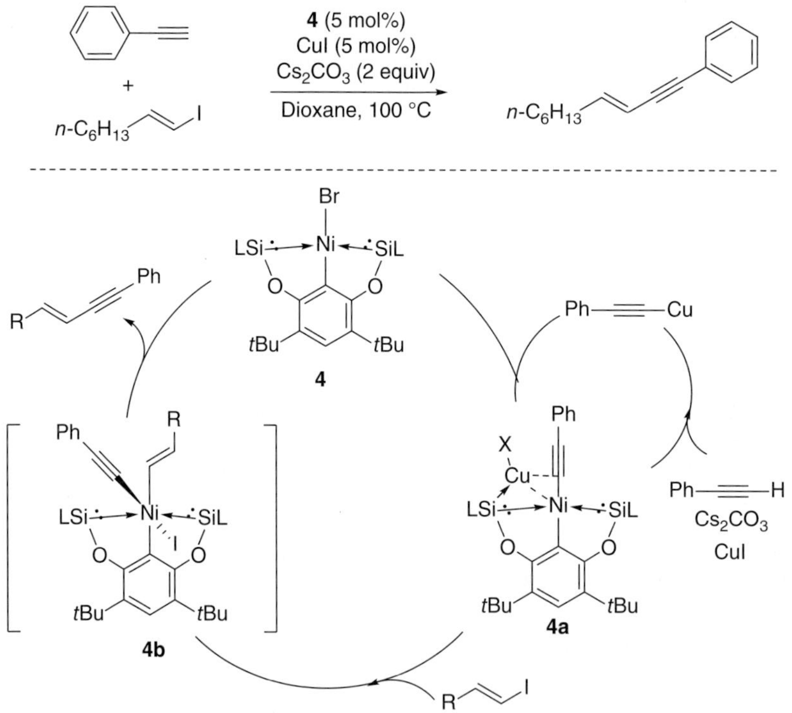

Scheme 12.5 Sonogashira cross-coupling reaction catalyzed by silylene–Ni complex **4** and the proposed mechanism. L = PhC(N^{t}Bu)$_2$.

In 2015, a mixed bidentate NHSi–NHC ligand was developed by our group. Its Ni(II) complex **5** was synthesized by the reaction of $NiBr_2$(dme) with N-heterocyclic silylcarbene LSi(H)(CH_2)NHC (L = CH{C=CH_2}(CMe)(NAr)$_2$, Ar = 2,6-*i*$Pr_2C_6H_3$, NHC = 3,4,5-trimethylimidazol-2-ylidene) in toluene at room temperature (Figure 12.1) [9]. **5** was subsequently applied successfully in Kumada–Corriu coupling reactions with promising catalytic performances (up to 99%, Scheme 12.6).

In addition to coupling reactions, silylene ligands have also been applied in cyclotrimerization and hydroformylation reactions. In 2012, the catalytically active Co(I) complex **6** bearing the ferrocene bridged bis-silylene ligand was synthesized by reducing the $CoBr_2$ coordinated silylene with KC_8 in the presence of NaCp (Figure 12.1) [10]. Complex **6** was investigated as a (pre)catalyst for the [2+2+2] cyclotrimerization reaction of phenylacetylene which gave triphenylbenzene as a product. Moreover, the cyclotrimerization of the same substrate in acetonitrile solutions resulted in the formation of substituted pyridines (Scheme 12.7).

In 2017, rhodium-catalyzed hydroformylation of styrene using silylene ligands was reported by Schomäcker and coworkers. The hydroformylation of styrene at 30 bar CO/H_2 pressure in the presence of [HRh(CO)(PPh_3)$_3$] with an excess of the bis-silylene ligand **L-3** resulted in superior catalytic activity [11]. In contrast, the reactions with excess of the monodentate silylene ligands **L-1** and **L-2** as well as phosphine/germylene analogs **L-4**, **L-5**, and **L-6** under the same reaction

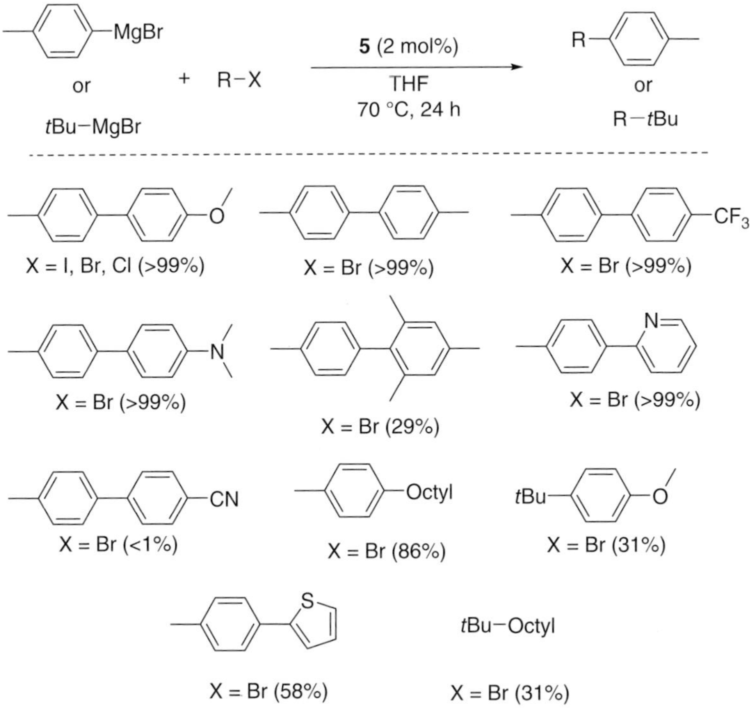

Scheme 12.6 Kumada–Corriu-type cross-coupling reactions catalyzed by silylene–Ni complex **5**.

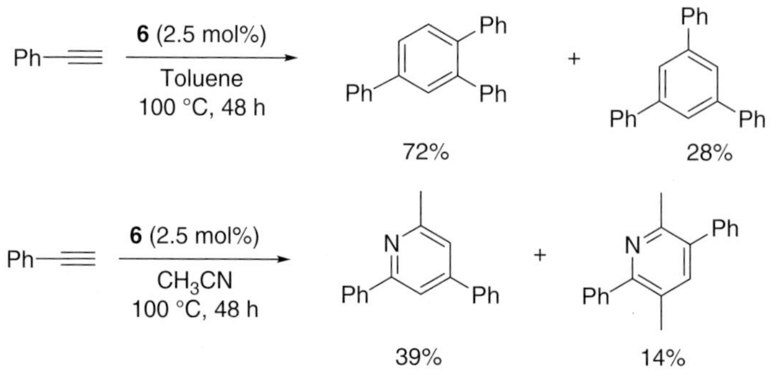

Scheme 12.7 Catalytic cyclotrimerization using silylene-Co(I) complex **6** as a precatalyst.

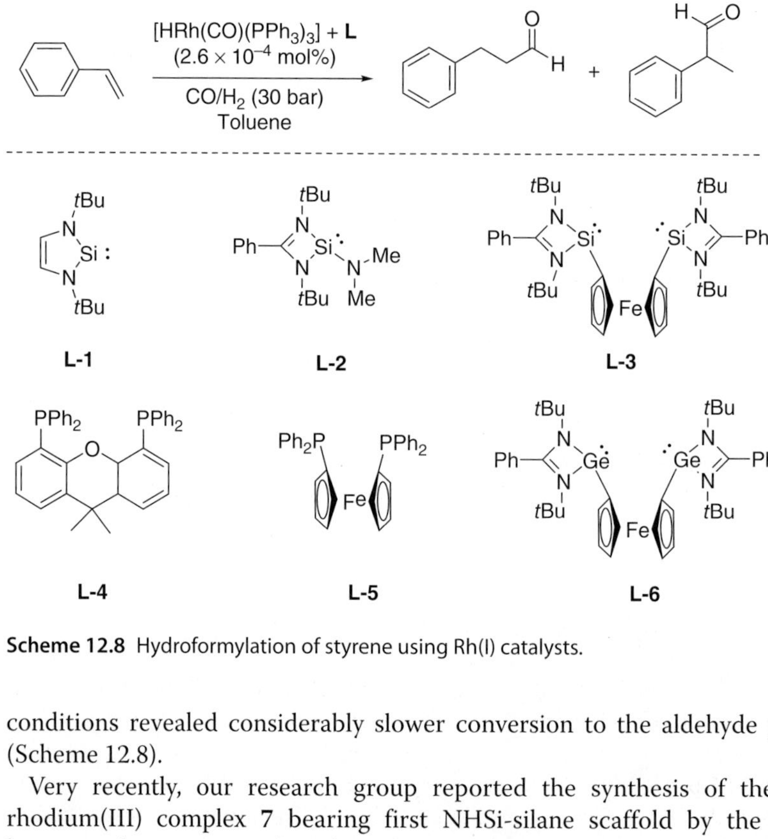

Scheme 12.8 Hydroformylation of styrene using Rh(I) catalysts.

conditions revealed considerably slower conversion to the aldehyde products (Scheme 12.8).

Very recently, our research group reported the synthesis of the unique rhodium(III) complex **7** bearing first NHSi-silane scaffold by the reaction between the corresponding ligand and 0.5 M equivalents of $[Rh(coe)_2Cl]_2$ (coe = cyclooctene) (Figure 12.1). Rh complex **7** acted as highly efficient precatalysts in the C–H functionalization of pyridine and pyrazole derivatives with alkenes or alkynes under mild reaction conditions in good yields of 65–96% (Scheme 12.9). The reaction mechanism was further calculated by DFT calculations, and these suggest that the agostic bis(silyl) hydridorhodium(III) complex is likely to act as an intermediate in the catalytic cycle.

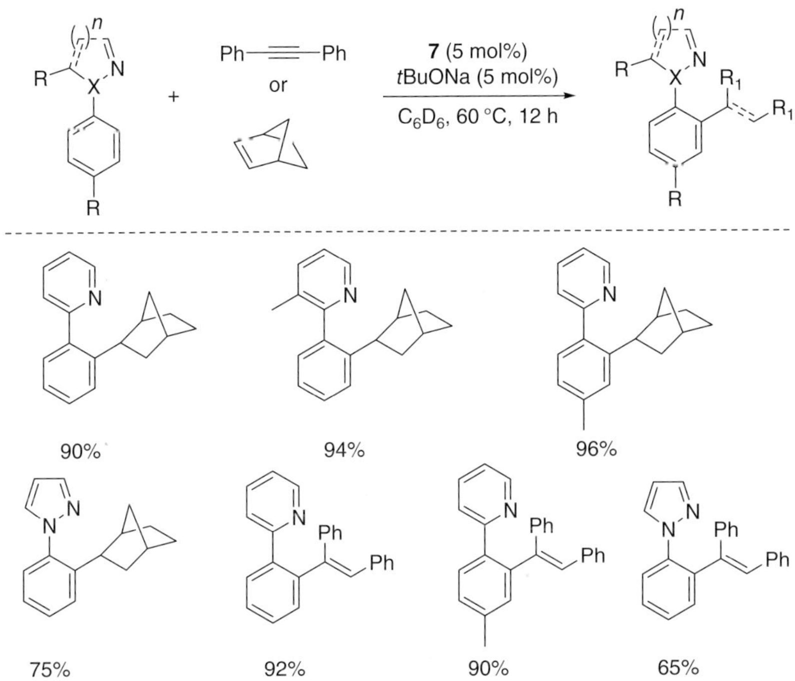

Scheme 12.9 C—H bond functionalizations catalyzed by rhodium catalyst **7**.

12.2.2 Carbon–Heteroatom Bond-Forming Reactions

In 2012, Hartwig and coworkers reported the first SiCSi pincer silylene ligand [12]. Iridium complex **8** was synthesized by reacting the corresponding ligand with 0.5 M equivalents of [IrCl(COE)$_2$] (Figure 12.2). Complex **8** was found to be an active precatalyst for the borylation of arenes using pinacolborane. The surpassing catalytic performances of silylenes against its phosphine and germylene analogs suggested that the silylene ligand has stronger donor capacity, enabling a promoting catalytic efficiency (Scheme 12.10).

In 2017, Cabeza et al. prepared a new mesitylsilylene [13]. Iridium(III) complexes **9a** and **9b** were synthesized by the reactions of the silylene ligand with the corresponding iridium precursors at room temperature, which led to C–H activation and oxidative addition of C–H bonds to iridium centers (Scheme 12.11). Both iridium complexes were tested in metal-mediated catalysis. The deuteration reactions of styrene and 1,3,5-trimethoxybenzene were successfully achieved using **9a** as a (pre) catalyst and CD_3OD as a deuterium source (Scheme 12.11). Furthermore, dehydrogenative borylation of benzene and toluene was tested using complex **9b** as (pre) catalyst (Scheme 12.12). The reactions went fast in the presence of COE (71% yield for benzene in 3 hours and 31% yield for toluene in 5 hours) but slowed down without COE (56% yield for benzene in 24 hours and 52% yield for toluene in 48 hours).

In the same year, the cobalt(II) complex (**10**) bearing an SiNSi pincer silylene ligand was reported by Cui and coworkers by the reaction of the silylene ligand with $CoBr_2$ in THF (Figure 12.2) [14]. Complex **10** enabled the regioselective

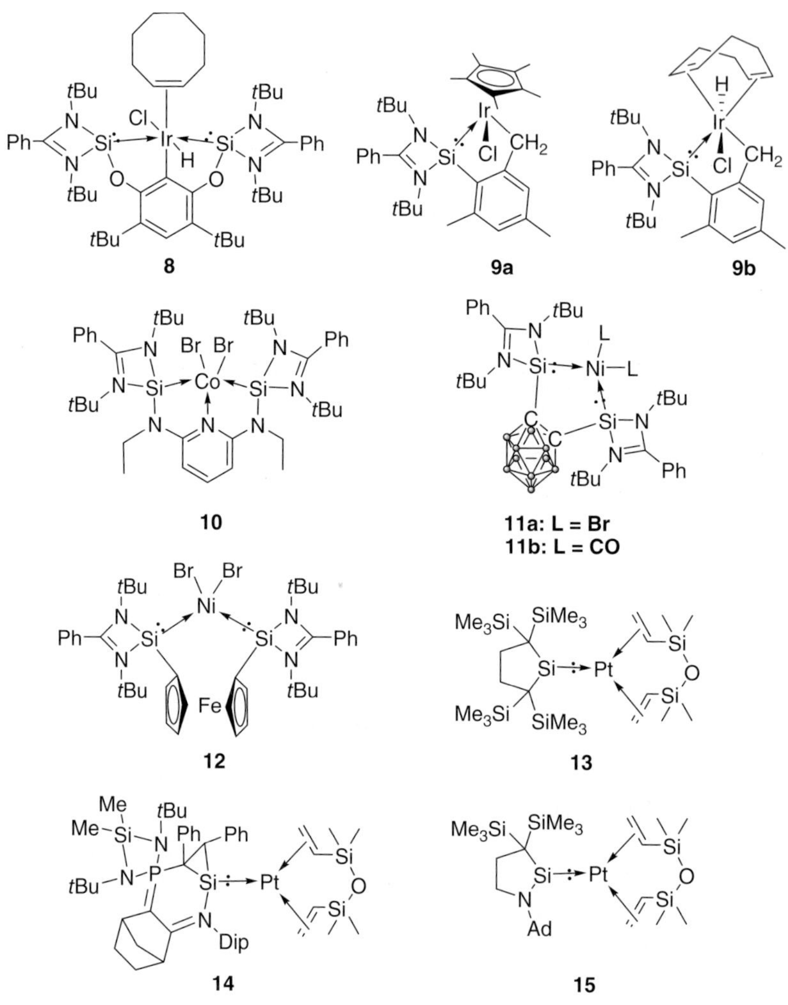

Figure 12.2 Silylene–metal complexes which were applied in catalytic carbon–heteroatom bond-formation reactions.

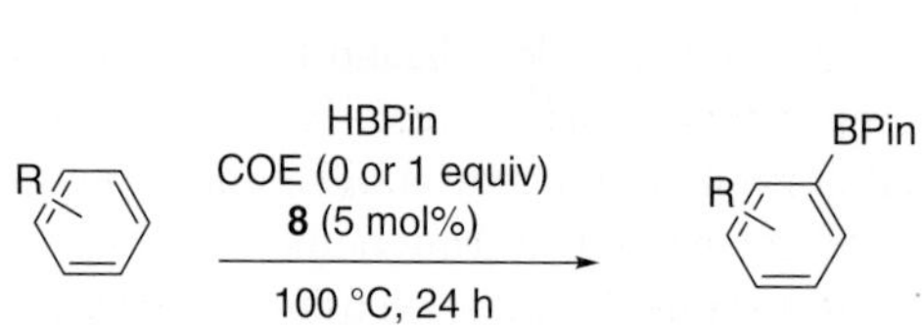

Scheme 12.10 Catalytic performances of arene borylation catalyzed by Ir complexes **8**.

C–H borylation of different heterocycles and fluorinated arenes. The borylation of 2,6-lutidine gave the product with substitution in the 4-position in high yield (98%). However, the reaction of 2-methoxy-6-methylpyridine was much slower and yielded only 71% of the product. The borylation of 2,3-lutidine selectively borylated at the 5-position, whereas 2-benzylpyridine yielded a mixture of 4- and

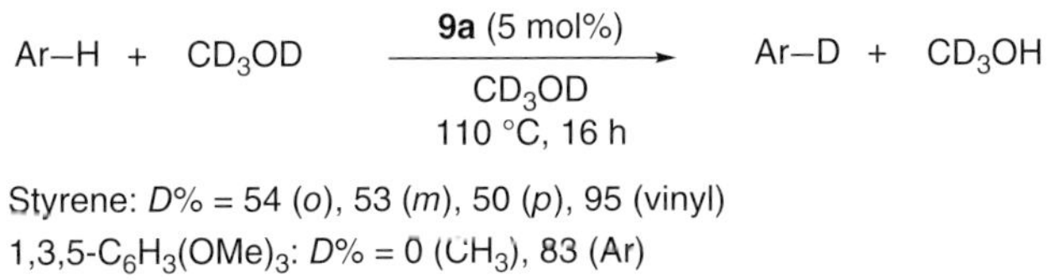

Scheme 12.11 Catalytic H/D exchange reaction using silylene–Ir(III) complex **9a** as precatalyst.

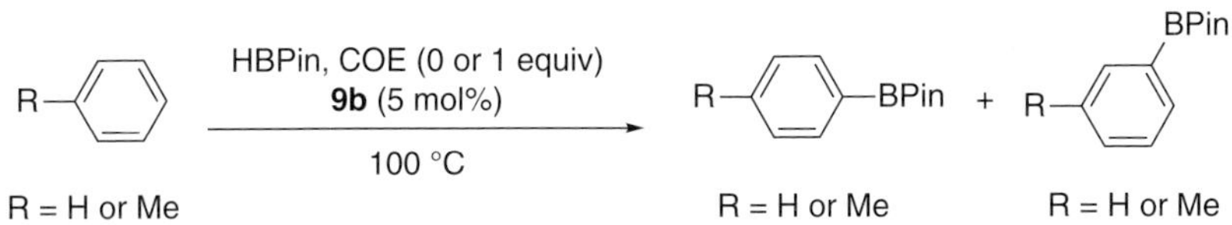

Scheme 12.12 Catalytic dehydrogenative borylation of arenes catalyzed by silylene–Ir(III) complex **9b**.

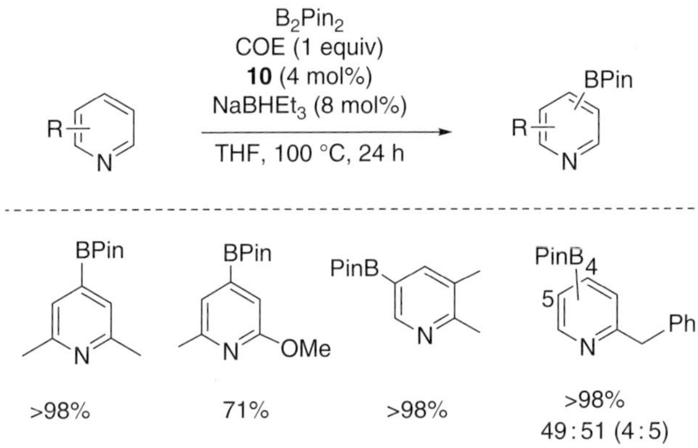

Scheme 12.13 Borylation of pyridine derivatives catalyzed by pincer-type silylene–Co(II) complex **10**.

5-borylated isomers (Scheme 12.13). In contrast to pyridine derivatives, the borylation of furan analogs was carried out without cyclohexene and gave 2,5-diborylation products, whereas 2-methylfuran was selectively borylated at the 5-position with 0.5 M equivalent of B_2Pin_2 (Scheme 12.14).

Catalytic borylations of fluoro- and trifluoromethylarenes employing cobalt(II) complex **10** were also tested. Disubstituted trifluoromethylbenzene derivatives underwent sterically controlled borylation at the least hindered position. Borylation of 1,2-disubstituted benzene yielded the 4-borylated product as a major regioisomer. 1,3-Disubstituted benzene predominantly gave the 5-borylated product giving up to 97% yield. Borylation of 1,3-difluorobenzenes and 1,2,3-trifluorobezene proceeded at the 5-position. Fluorobenzene yielded the 3-borylated regioisomer as the major product. Borylation of 2-methylfluorobenzene occurred at the 4-position with 72% selectivity. The bis-borylation of naphthalene yielded

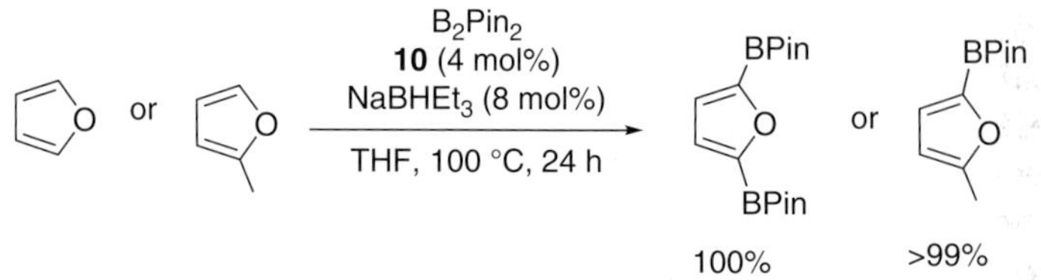

Scheme 12.14 Borylation of furan and its derivatives catalyzed by pincer-type silylene–Co(II) complex **10**.

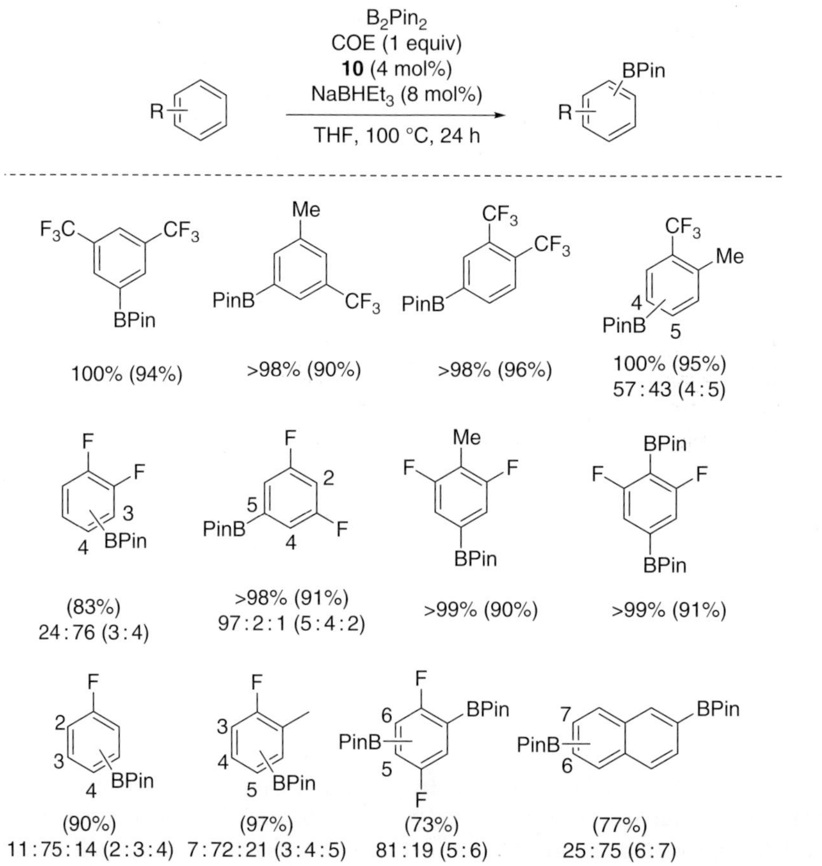

Scheme 12.15 Borylation of fluoro- and trifluoroarenes catalyzed by pincer-type silylene–Co(II) complex **10**.

the 6- and 7-borylated products in a molar ratio of 1 : 3, whereas borylation of pyrene selectively gave the 2,7-borylated product (Scheme 12.15).

In 2016, the new chelating bis-silylene-substituted *o*-carborane ligand was synthesized by our group. Coordination of the extraordinarily strong donor ligand to nickel(II) yielded the corresponding complex **11a**, which undergoes reduction with KC_8 in the presence of CO to afford the tetrahedral nickel(0) complex **11b**

(Figure 12.2). The CO stretching vibration modes of **11b** indicate that the Si(II) atoms in bis-silylene are even stronger σ donors than the P(III) atoms in phosphines and C(II) atoms in N-heterocyclic carbene ligands. Compared with the ferrocendiyl-bridged bis-silylene analog **12** and the related bis(phosphanyl)-substituted *o*-carborane-containing complexes, complex **11a** shows a superior catalytic activity in Buchwald–Hartwig amination reactions of aryl halides with secondary amines. Screening and optimization experiments revealed that the catalytic amination of various aryl halides succeeds already with 0.5 mol% of the complex **11a** precatalyst in the presence of $AgBPh_4$ and 1.2 equivalents of KO*t*Bu in excellent yields. The use of **12** or the phosphine analogs resulted in significantly lower yields of the desired product, indicating the superior steering role of the bis-silylene to facilitate metal-mediated carbon–nitrogen coupling reactions (Scheme 12.16) [15].

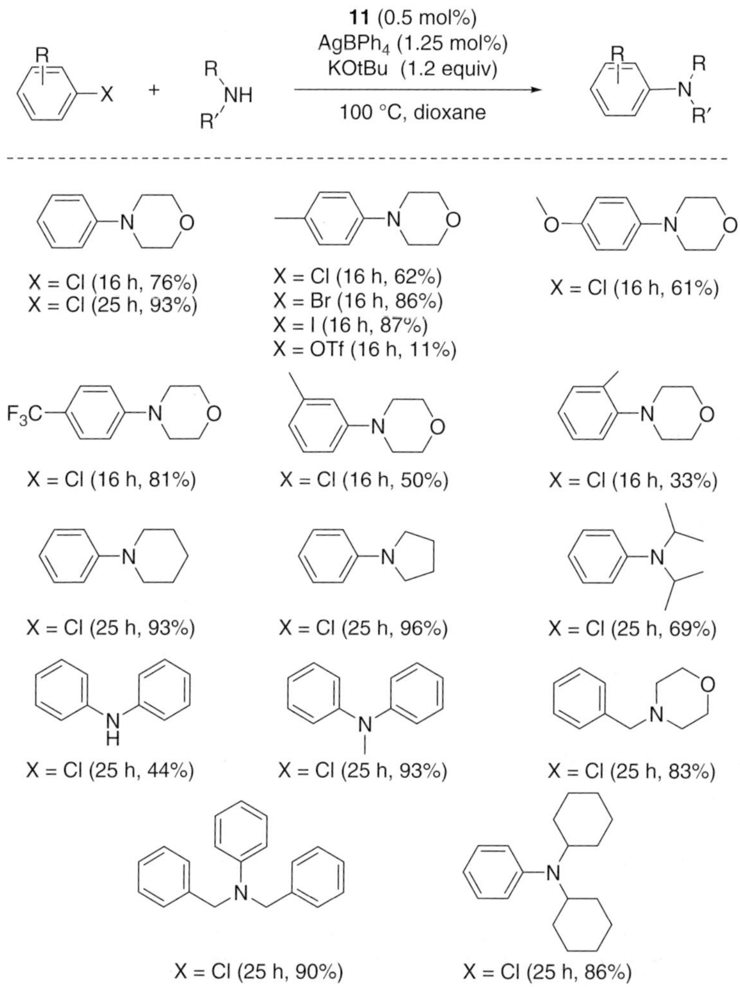

Scheme 12.16 Buchwald–Hartwig amination catalyzed by silylene nickel(II) complex **11**.

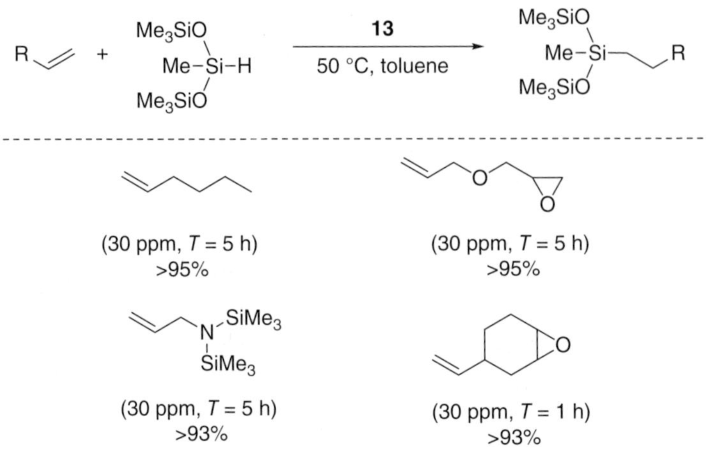

Scheme 12.17 Hydrosilylation of alkenes catalyzed by silylene–Pt(0) complex **13**.

In 2016, Iwamoto and coworkers reported the olefin hydrosilylation reactions catalyzed by platinum(0) complex **13** bearing a dialkylsilylene as ligand (Figure 12.2) [16]. The catalytic activity of complex **13** was assessed for the hydrosilylation of $(Me_3SiO)_2SiMeH$ and hex-1-ene in the presence of 30 ppm (w/w) of **13** at 50 °C for five hours. In addition, **13** also catalyzes the hydrosilylation of terminal alkenes bearing different functional groups with very high conversion rates (Scheme 12.17).

In the same year, Baceiredo and coworkers for the first time applied the base-stabilized silacyclopropylidene complexes in catalysis [17]. The reaction of silacyclopropylidene at room temperature with 0.5 equiv of Karstedt complex $[Pt_2(dvtms)_3]$ (dvtms = Divinyltetramethyldisiloxane) afforded air-stable platinum(0) complex **14** (Figure 12.2). Hydrosilylation of oct-1-ene catalyzed by complex **14** (30 ppm) is as fast as that with the Karstedt catalyst (>95% conversion = 20 versus 15 minutes) and proceeds in a more selective manner (isolated yield = 91% versus 78%) (Scheme 12.18).

One year later, cyclic alkyl(amino)silylene was also employed as ligand in platinum-catalyzed alkene hydrosilylation by Iwamoto and coworkers [18]. The platinum(0) complex **15** bearing a cyclic alkyl(amino) silylene was synthesized using the same method (Figure 12.2). The catalytic performance of **15** in the catalytic hydrosilylation of $(Me_3SiO)_2MeSiH$ with various terminal alkenes that contain functional groups was comparable to platinum(0) complex **14** bearing a dialkylsilylene (Scheme 12.19).

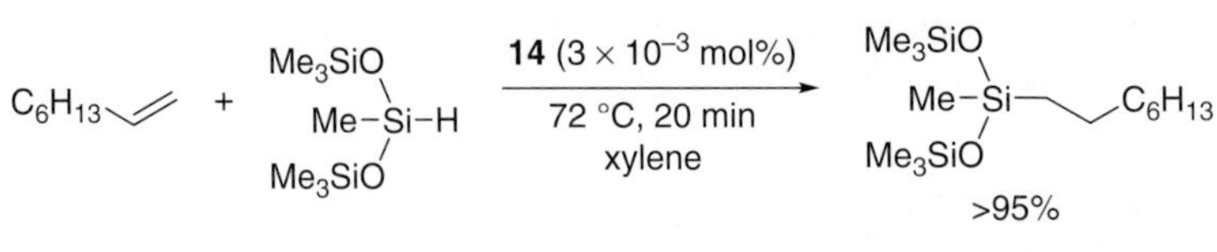

Scheme 12.18 Hydrosilylation of alkenes employing silylene–Pt(0) complex **14** as precatalyst.

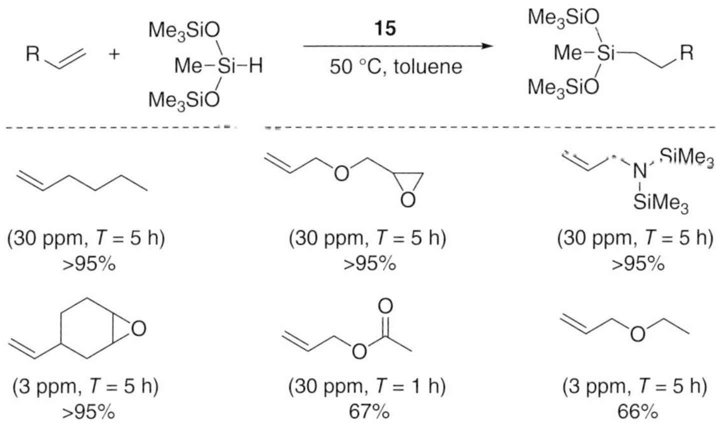

Scheme 12.19 Hydrosilylation of alkenes suing silylene–Pt(0) complex **15** as precatalyst.

12.2.3 Reduction Reactions

In 2012, our group reported the silylene–Fe(0) complexes **16–18** bearing *N,N*-di(*tert*-butyl)amidinato silylene ligands (Figure 12.3) [19]. The catalytic activity of complex **17** was then checked in the hydrosilylation of carbonyl compounds. The corresponding alcohols were obtained in high yields after the workup process (Scheme 12.20). Further experimental and theoretical results reveal a ketone-mediated 1,2-H migration from Si(II) to Fe(0), which corroborates that the NHSi ligand plays a key role in the catalytic cycle (Scheme 12.21).

In 2014, our group reported iron complex **19** based on the tridentate SiNSi ligand (Figure 12.3) [20], which was prepared by the reaction of corresponding pincer-type silylene ligand with $(Me_3P)_4Fe$ or by the reaction of the corresponding $FeCl_2$–silylene complex bearing the same ligand with KC_8 in the presence of excess amount of Me_3P. Catalytic carbonyl hydrosilylation reactions were also checked using complex **19**. It was found that the catalyst showed good functional-group tolerance for different substituents. The mechanism of hydrosilylation catalyzed by complex **19** has been investigated by Driess and coworkers in 2015 [21]. The catalytically active species **19a** was isolated through the stoichiometric reaction of precatalyst complex **19** and $(Et_3O)_3SiH$. Interestingly, the further experimental and theoretical analysis suggests that the iron center of complex **19a** is not directly involved in the catalytic cycle due to steric effects. The C=O activation occurs at the Si(IV) Lewis acid center of complex **19a** and intermediate **19b** is generated. One $(Et_3O)_3SiH$ molecule then coordinates to **19b** and gives the intermediate **19c**, which further releases the hydrosilylation product (Scheme 12.22).

In addition, copper(I) complex **20** bearing base-stabilized silacyclopropylidene has also been used as a catalyst for the hydrosilylation of bulky ketones such as adamant-1-yl methyl ketone (Figure 12.3) [17a]. Complex **20** is considerably more active than the corresponding Ph_3P–CuI complex and is

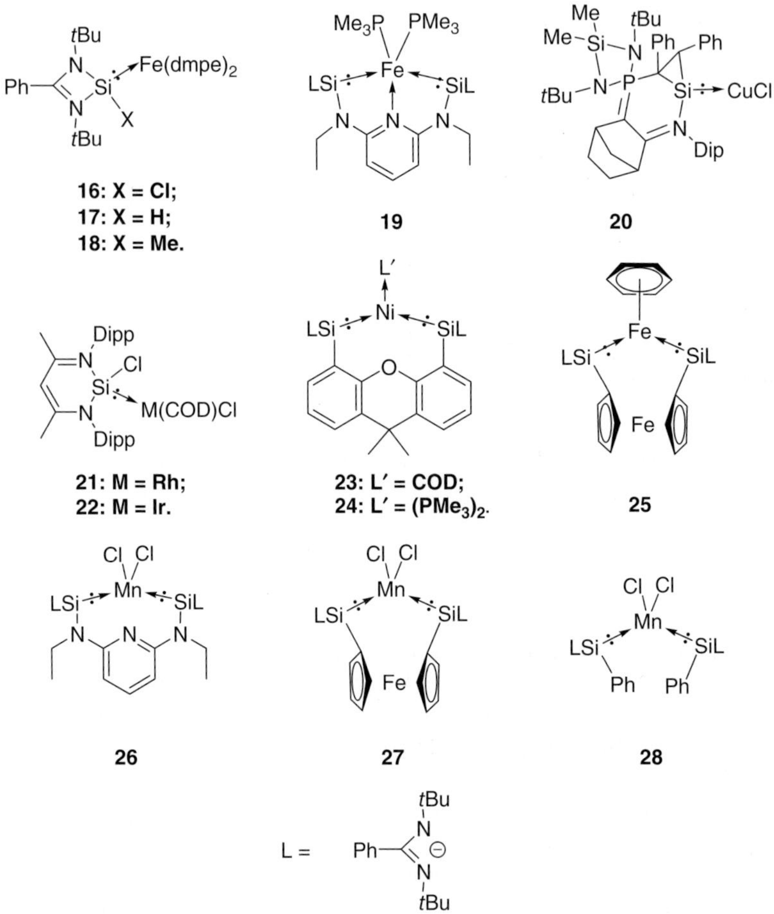

Figure 12.3 Silylene–metal complexes which were applied in catalytic reduction reactions.

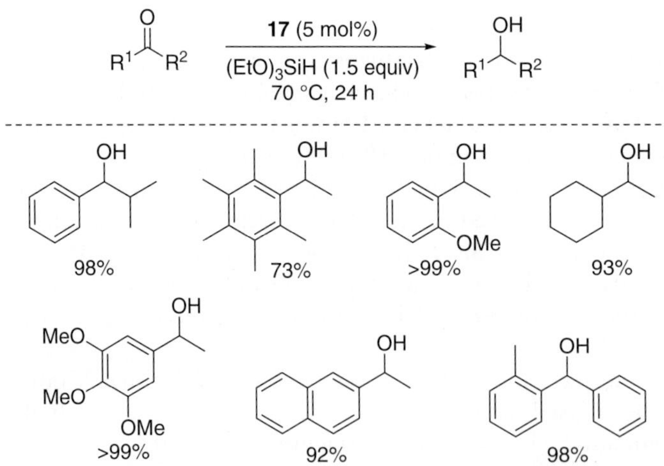

Scheme 12.20 Hydrosilylation of ketones catalyzed by silylene–Fe(0) complex **17**.

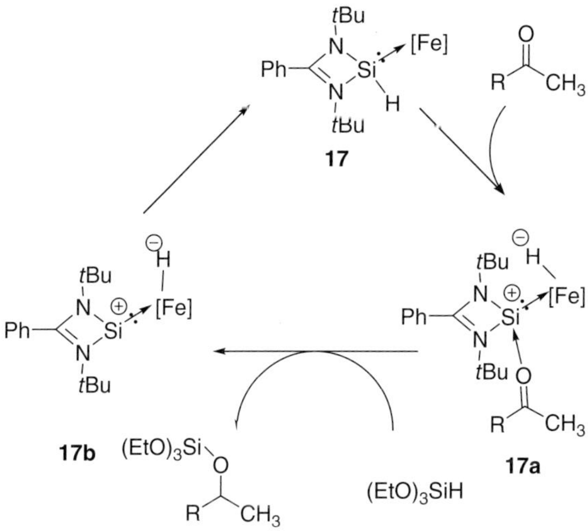

Scheme 12.21 Proposed mechanism for hydrosilylation of ketones catalyzed by Fe–silylene complex **17**.

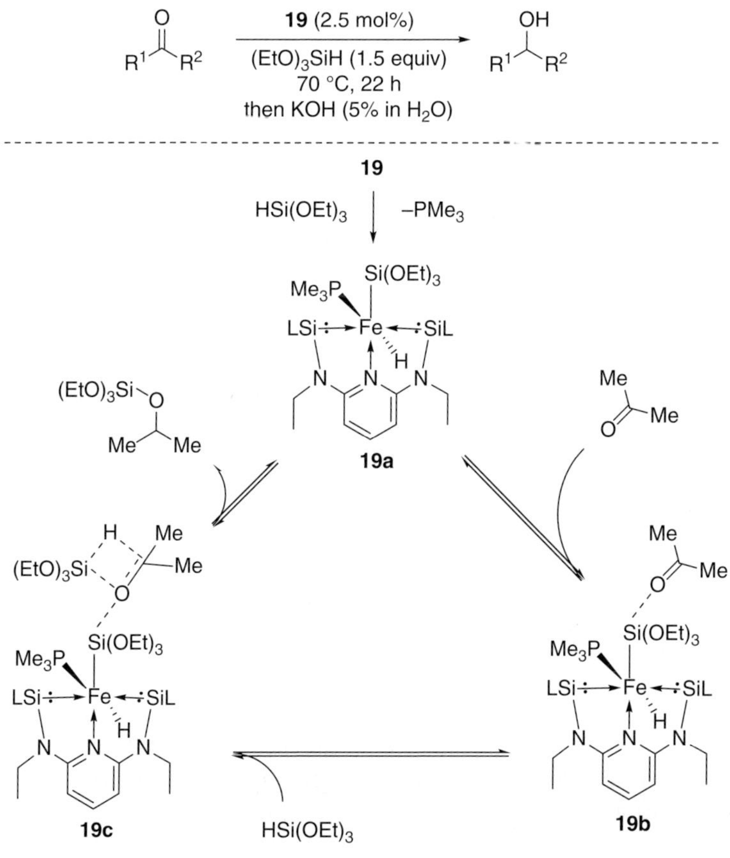

Scheme 12.22 Peripheral mechanism for hydrosilylation of ketones.

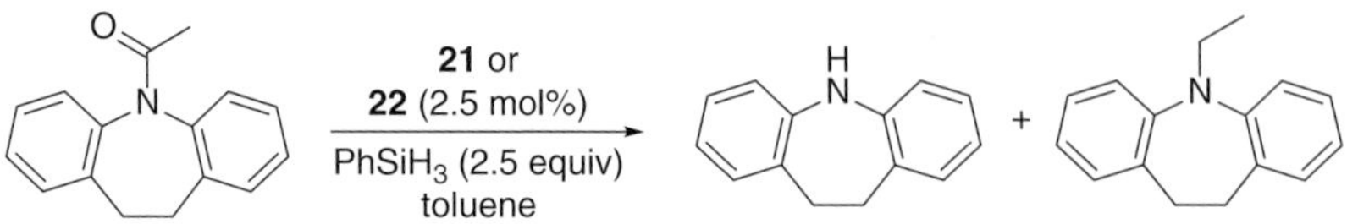

Scheme 12.23 Reduction of dibenzoazepine using complex **21** or **22** as a precatalyst.

as efficient as those with an extremely bulky bowl-shaped phosphine ligand developed by Tsuji et al. [17b].

Silylene–Rh(I) and –Ir(I) complexes **21** and **22** bearing β-diketiminate silylene ligand were applied in the reduction of amide substrates (Figure 12.3) [22]. The carbon–oxygen bond cleavage product was observed in 61% yield when the complex **21** was applied. However, a mixture of carbon–oxygen and carbon–nitrogen bond cleavage products was observed using complex **22** as precatalyst (Scheme 12.23).

In 2017, our group prepared a new bis-silylene ligand based on the xanthene backbone [23]. The corresponding nickel(0) complex **23** bearing COD ligand was found to be a very active catalyst in the hydrogenation of alkenes under mild conditions. Nickel(0) complex **24** bearing Me_3P ligands was also found to be active but less efficient. Functional groups such as OMe, CF_3, CN, a pyridine ring, and a carbonyl group were tolerated. Dienes gave the corresponding alkanes almost quantitatively after 36–84 hours. Turnover number (TON) and turnover frequency (TOF) of dodec-1-ene and styrene reached up to 1000 and $250\,h^{-1}$, respectively (Scheme 12.24). Moreover, a unique Si(II)-assisted H_2 activation mode, in which the low-valent silicon centers are involved in the key step of dihydrogen cleavage and transfer, were suggested by DFT calculations (Scheme 12.25).

In the same year, η^6-arene iron(0) complex **25** was also prepared by Driess and coworkers through reduction of the corresponding Fe(II) dihalide precursors in the presence of benzene (Figure 12.3) [24]. The catalytic hydrogenation of various ketones and an aldehyde were carried out using complex **25**. The scope experiment suggested that the hydrogenation of aldehyde and ketone derivatives gave moderate to good yields of corresponding alcohols. Remarkably, the high chemoselectivity of C=O over C=C reduction was achieved using allylacetone and chalcone as substrates (Scheme 12.26).

In 2018, manganese complexes **26–28** stabilized by a pincer-type bis(NHSi)-pyridine ligand, a bidentate bis(NHSi)-ferrocene ligand, and a monodentate NHSi ligand, respectively, were synthesized and reported by our group (Figure 12.3) [25]. Complexes **26–28** represent the first time reported well-defined manganese complexes catalyzing the (transfer) hydrogenation of unsaturated hydrocarbons. Manganese complex **26** gave the best performance in stereoselective transfer semi-hydrogenation of alkynes. The catalytic transfer semi-hydrogenation of different alkynes using ammonia-borane (NH_3BH_3) as H-source yielded (*E*)-alkenes as main products (Scheme 12.27).

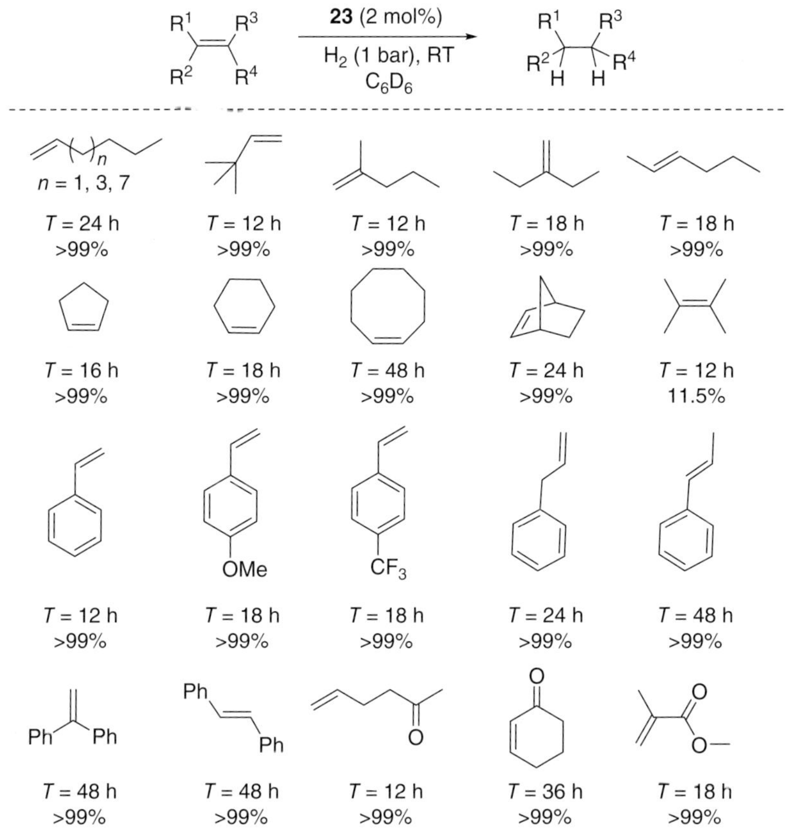

Scheme 12.24 Hydrogenation of alkenes catalyzed by silylene–Ni(0) complex **23**.

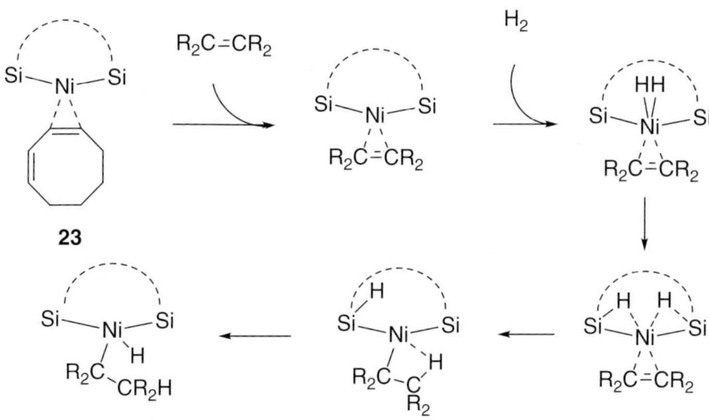

Scheme 12.25 DFT-calculated mechanism for silicon-assisted H_2 activation on the Ni(0) center of complex **23** (only key steps are shown).

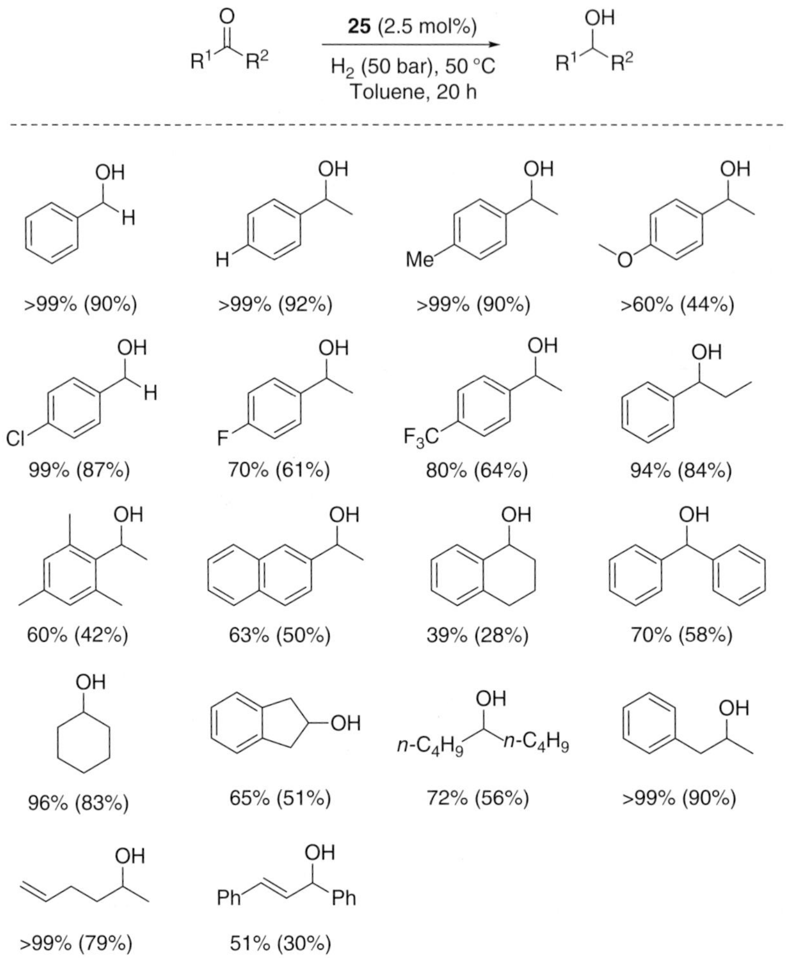

Scheme 12.26 Hydrogenation of carbonyl compounds catalyzed by silylene–Fe(0) complex **25**.

12.3 Summary and Outlook

In conclusion, silylene ligands have demonstrated their superior role in different transition-metal-mediated catalytic transformations including carbon–carbon bond formation, carbon–heteroatom bond formation, and reduction reactions.

A large number of silylenes have been isolated and reported in recent years. However, compared to the great numbers of NHC and phosphine ligands used in catalysis, the silylene ligands used in catalysis are still scarce. One of the main challenges is to develop new multidentate-silylenes to bring about the discovery of new catalytic performances (e.g. enantioselective catalysis using chiral silylene ligands). We hope that silylene ligands will become a preferred option for synthetic chemists in the future.

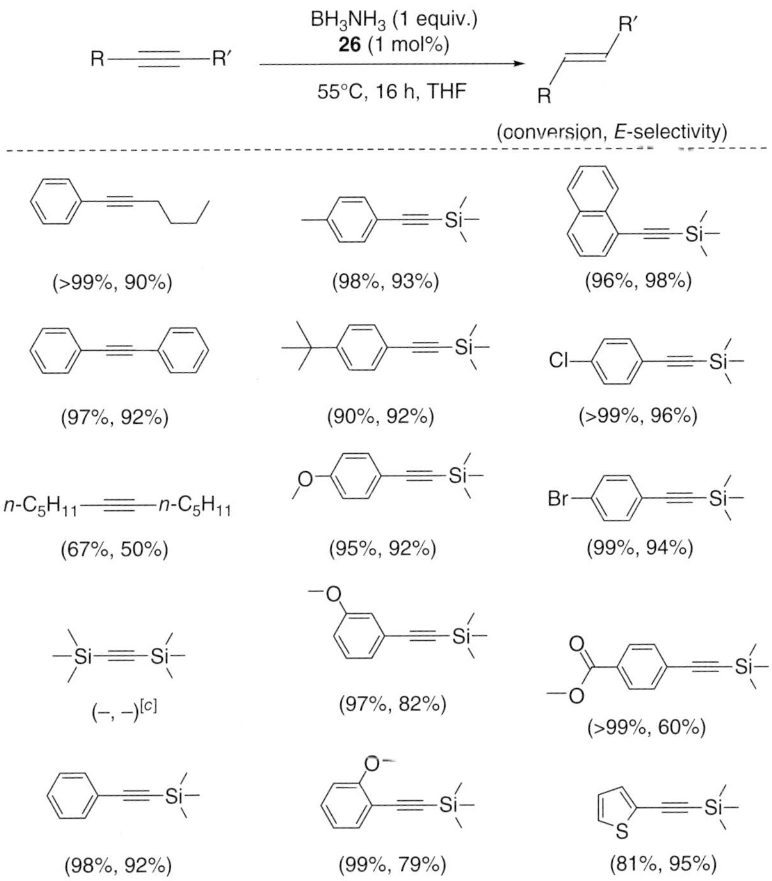

Scheme 12.27 Stereoselective transfer semi-hydrogenation of alkynes catalyzed by Mn–silylene complex **26**.

Acknowledgment

We are grateful to the Deutsche Forschungsgemeinschaft (Cluster of Excellence "UniCat" and DR 226-18/1) and the Berlin International Graduate School of Natural Science and Engineering (BIG-NSE, Y.P.Z.) for financial support.

References

1 (a) Peris, E. (2017). *Chem. Rev.* https://doi.org/10.1021/acs.chemrev.6b00695. (b) Díez-González, S., Marion, N., and Nolan, S.P. (2009). *Chem. Rev.* 109: 3612–3676. (c) Hahn, F.E. and Jahnke, M.C. (2008). *Angew. Chem. Int. Ed.* 47: 3122–3172.

2 (a) Raoufmoghaddam, S., Zhou, Y.P., Wang, Y., and Driess, M. (2017). *J. Organomet. Chem.* 829: 2–10. (b) Gallego, D., Blom, B., Tan, G., and Driess, M. (2015). *Reference Module in Chemistry, Molecular Sciences and Chemical Engineering*. Elsevier https://doi.org/10.1016/B978-0-12-409547-2.11449-0. (c) Blom, B., Gallego, D., and Driess, M. (2014). *Inorg. Chem. Front.* 1: 134–148.
3 Benedek, Z. and Szilvasi, T. (2015). *RSC Adv.* 5: 5077–5086.
4 Fürstner, A., Krause, H., and Lehmann, C.W. (2001). *Chem. Commun.* 2372–2373.
5 Zhang, M., Liu, X., Shi, C. et al. (2008). *Z. Anorg. Allg. Chem.* 634: 1755–1758.
6 Wang, W., Inoue, S., Yao, S., and Driess, M. (2010). *J. Am. Chem. Soc.* 132: 15890–15892.
7 Someya, C.I., Haberberger, M., Wang, W. et al. (2013). *Chem. Lett.* 42: 286–288.
8 Gallego, D., Brück, A., Irran, E. et al. (2013). *J. Am. Chem. Soc.* 135: 15617–15626.
9 Tan, G., Enthaler, S., Inoue, S. et al. (2015). *Angew. Chem. Int. Ed.* 54: 2214–2218.
10 Wang, W., Inoue, S., Enthaler, S., and Driess, M. (2012). *Angew. Chem. Int. Ed.* 51: 6167–6171.
11 Schmidt, M., Blom, B., Szilvási, T. et al. (2017). *Eur. J. Inorg. Chem.* 2017: 1284–1291.
12 Brück, A., Gallego, D., Wang, W. et al. (2012). *Angew. Chem. Int. Ed.* 51: 11478–11482.
13 Cabeza, J.A., Garcia-Alvarez, P., and Gonzalez-Alvarez, L. (2017). *Chem. Commun.* 53: 10275–10278.
14 Ren, H., Zhou, Y.P., Bai, Y. et al. (2017). *Chem. Eur. J.* 23: 5663–5667.
15 Zhou, Y.P., Raoufmoghaddam, S., Szilvási, T., and Driess, M. (2016). *Angew. Chem. Int. Ed.* 55: 12868–12872.
16 Iimura, T., Akasaka, N., and Iwamoto, T. (2016). *Organometallics* 35: 4071–4076.
17 (a)Troadec, T., Prades, A., Rodriguez, R. et al. (2016). *Inorg. Chem.* 55: 8234–8240. (b) Fujihara, T., Semba, K., Terao, J., and Tsuji, Y. (2010). *Angew. Chem., Int. Ed.* 49: 1472–1446.
18 Iimura, T., Akasaka, N., Kosai, T., and Iwamoto, T. (2017). *Dalton Trans.* 46: 8868–8874.
19 Blom, B., Enthaler, S., Inoue, S. et al. (2013). *J. Am. Chem. Soc.* 135: 6703–6713.
20 Gallego, D., Inoue, S., Blom, B., and Driess, M. (2014). *Organometallics* 33: 6885–6897.
21 Metsanen, T.T., Gallego, D., Szilvasi, T. et al. (2015). *Chem. Sci.* 6: 7143–7149.
22 Stoelzel, M., Präsang, C., Blom, B., and Driess, M. (2013). *Aust. J. Chem.* 66: 1163–1170.
23 Wang, Y., Kostenko, A., Yao, S., and Driess, M. (2017). *J. Am. Chem. Soc.* 139: 13499–13506.
24 Luecke, M.P., Porwal, D., Kostenko, A. et al. (2017). *Dalton Trans.* 46: 16412–16418.
25 Zhou, Y.P., Mo, Z., Luecke, M.P., and Driess, M. (2018). *Chem. Eur. J.* 24: 4780–4784.

13

Enantioselective Synthesis of Silyl Ethers Through Catalytic Si—O Bond Formation

Amir H. Hoveyda and Marc L. Snapper

Merkert Chemistry Center, Boston College, 2609 Beacon Street, Chestnut Hill, MA, 02467-3860, USA

13.1 Introduction

Strategies in chemical synthesis have often relied on protection of certain polar functionalities so that transformations may be performed selectively at other specific sites within a multifunctional molecule. From early on, alkyl- or aryl-substituted silyloxy groups have been considered promising protecting groups for alcohols. A key requirement was finding an appropriate balance of robustness under a variety of reaction conditions, in addition to facile installation and removal of the protecting group. Along these lines, Corey and Venkateswarlu speculated that the *t*-butyldimethylsilyl group (TBDMS) should be robust enough to survive typical transformations ranging from with organometallic reagents to Jones oxidation [1]. The challenge they faced in protecting an alcohol unit was low reactivity between *t*-butyldimethylsilyl chlorides and alcohols. Fortunately, their efforts led identifying imidazole in dimethylformamide (DMF) as an effective combination for converting a secondary alcohol to its corresponding silyl ether. They also showed that silyl removal may be achieved under mild conditions with a fluoride reagents. While this landmark contribution ushered in the use of TBDMS functionality as a practical hydroxyl protecting group, *the fact that the reaction was shown to be accelerated (i.e. catalyzed) by imidazole in DMF offered the first enticing indication that a corresponding enantioselective catalytic silylation of an alcohol might be feasible.* In light of this longstanding evidence, it is interesting to note that there were Nonetheless, no efforts were reported towards this end for nearly three decades (vide infra) [2].

Additional support for *catalytic* silylation of alcohols did emerge however. Chaudhary and Hernandez demonstrated an improved selectivity for the installing a TBDMS unit on a primary hydroxy group in the presence of 4-dimethylaminopyridine (DMAP) (versus imidazole catalyst [3]. It was suggested that the

Organosilicon Chemistry: Novel Approaches and Reactions, First Edition.
Edited by Tamejiro Hiyama and Martin Oestreich

greater selectivity that the reaction with DMAP displays for the protection of primary versus secondary alcohol supports its close interaction with the silyl chloride, thus inhibiting subsequent interactions particularly with the secondary hydroxyl groups.

More recently, in other transformations involving silylated nucleophiles (i.e. allyl silanes, silylenol ethers, etc.), the role of Lewis base activation of Lewis acids has become more apparent [4]. This key concept provided the fundamental understanding of how a Lewis base, such as imidazole or DMAP might function as an effective catalyst in *accelerating* silyl ether formation. With this experimental and theoretical background, the stage was set for the introduction of methods for catalytic enantioselective alcohol silylation [5].

13.2 Lewis Base–Catalyzed Enantioselective Silylations of Alcohols

Concurrent with advancements in hydroxyl group protection was the development of enantioselective group transfer reactions on alcohols. Of this class of reactions, perhaps the most thoroughly studied are the enantioselective acylations of alcohols [6]. Impressive solutions for resolving and desymmetrizing a broad range of hydroxyl-containing substrates are now available using either enzymes or small-molecule catalysts (i.e. Lewis bases) to provide highly enantioenriched acylated products. While this strategy has provided important access to enantioenriched building blocks and products, it is limited by the nature of the acyl group that is used to desymmetrize or resolve the alcohol. For example, the resulting esters are typically base and acid sensitive and do not often serve as a suitable hydroxyl protecting group. That being the case, additional steps are often required to suitably protect the hydroxyl group or ester for subsequent transformations elsewhere on the molecule [7]. Clearly, the ability to directly silylate a particular hydroxyl group enantioselectively could overcome this limitation. Unlike enzyme-mediated acylations, however, Nature has not provided us with any direct guidance.

13.2.1 Early Lewis Base–Mediated Enantioselective Silylations of Alcohols

The first reported attempt of catalyzing an enantioselective silylation of a secondary hydroxyl group was reported in 2001 by Ishikawa and coworkers (Eqs. (13.1) and (13.2)) [8]. In this case, the authors examined a range of chiral Lewis-basic guanidines to effect the partial resolution of indan-1-ol. The selectivity for the enantioselective formation of a TBDMS ether group was modest and the rate of reaction was quite slow (Eq. (13.1)). As shown in Eq. (13.2), bicyclic guanidines were slightly more selective for the formation of a triisopropylsilyl ether (TIPS).

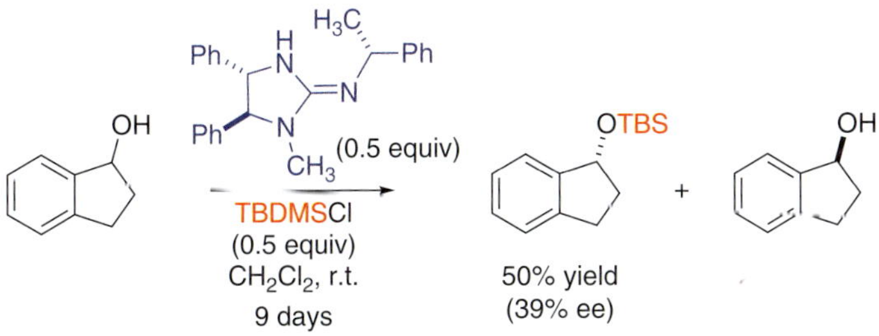

(13.1)

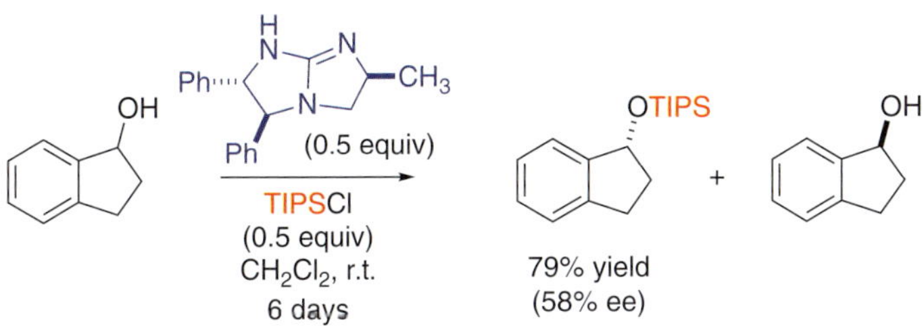

(13.2)

The Ishikawa group was able to recover and recycle their chiral guanidines; however, they were not able to demonstrate any catalytic turnover for these additives. Based on preliminary NMR studies, the authors suggested that the selection mechanism might proceed through a chiral silylguanidinium salt. While of limited synthetic value, this work provided strong support for the development of a catalytic enantioselective silylation of alcohols.

13.2.2 Lewis- and Brønsted Base-Catalyzed Enantioselective Silylations of Polyols

In 2006, the Hoveyda/Snapper team reported a catalytic enantioselective silylation of *meso*-1,2- and 1,3-diols using an imidazole-based catalyst **1** and readily available silyl chlorides [9]. Although catalyst loadings were high (i.e. 30 mol%) and the reaction rates were slow (two to five days), the yields and enantioselectivities of the mono-silylated diols were good. Scheme 13.1, A shows representative cyclic and acyclic meso-1,2-diols that can be desymmetrized with *t*-butyldimethyl silyl chloride.

The substrate scope beyond 1,2-diols proved to be limited. For example, isolated secondary alcohols were not viable substrates. Some *meso*-1,3-cyclic diols, on the other hand, could be silylated under these and related conditions, albeit in lower yields (due to over-silylation) and enantioselectivities. In addition, other common silylating agents, such as triethylsilyl choride (TESCl) or TIPSCl, were shown to be competent electrophiles. These results are illustrated in Scheme 13.1, B.

Based on extensive structural modifications to the catalyst aimed at improving the reaction rate, a mechanistic model emerged describing how the catalyst might function in this reaction. In particular, three regions of the imidazole catalyst **1** were identified as necessary for reactivity and selectivity. Figure 13.1 shows the initially proposed transition state model for this asymmetric silylation.

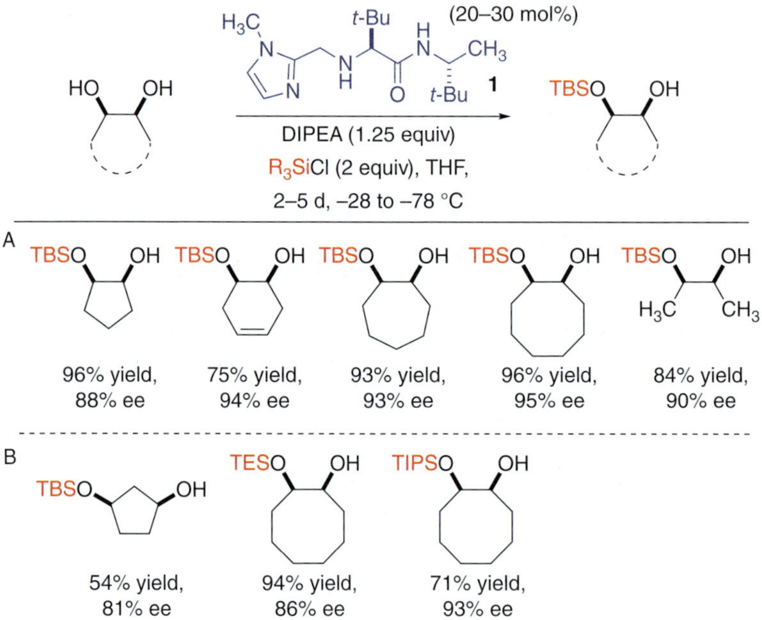

Scheme 13.1 Catalytic *meso*-diol desymmetrization.

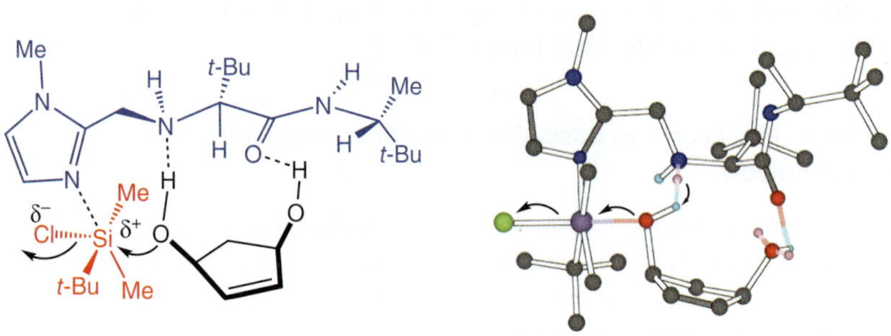

Figure 13.1 The initial transition state model for catalytic enantioselective silyl ether formation.

The imidazole group was thought to coordinate to the silyl group causing a Lewis-basic activation of the silicone atom toward nucleophilic attack by the hydroxyl group [4]. The secondary amine was thought to serve as a general base, deprotonating the hydroxyl group as it displaces the silyl chloride. Finally, the amide functionality was believed to provide a second point of interaction with the catalyst through a hydrogen bond with the diol substrate. Later studies (discussed subsequently), however, suggest that the catalyst might not be functioning in this manner.

This initial report on the desymmetrization of diols with the Hoveyda/Snapper imidazole-containing peptidyl catalyst was then extended to effect the kinetic resolution of a range of racemic vicinal *syn*-diols [10]. Under optimized reaction conditions, the less hindered hydroxyl group of the reactive enantiomer is usually preferentially silylated with selectivity factors (*s*) ranging from 8 to 50 (data not shown). This kinetic resolution allows for the easy separation of the unreacted diol from the regioselectively mono-silylated product, both in an enantioenriched manner. The results, where the imidazole catalyst differentiates between enantiotopic hydroxyl groups adjacent to a methyl substituent versus a vicinal hydroxyl group next to larger substituents, are high lighted in Scheme 13.2.

In a similar manner, the imidazole catalyst was also able to silylate selectively terminal 1,2-diols. Here, the primary hydroxyl group of one enantiomer reacts preferentially with the silyl chloride. The 2-hydroxyl group can be either secondary or tertiary. In some sterically congested cases, the selectivity factors (*s*) can approach 200, allowing for either the product or the starting material to be isolated with high enantiomeric excess, depending on the degree of conversion. These results are summarized in Scheme 13.3.

In cases where the site selectivity for the silylation of the vicinal diols are low, Rodrigo et al. showed that these compounds could be resolved through an enantioselective regiodivergent reaction [11]. In this case, when running the reaction to full conversion, each *syn*-vicinal diol enantiomer becomes selectively silylated on a regiotopic hydroxyl group. That being the case, the two silylated regioisomers can then be isolated in an enantiomerically enriched manner. These results are described in Scheme 13.4. While the catalyst **1** can differentiate between a methyl and an ethyl group next to the hydroxyl functionality (as shown earlier), it cannot distinguish readily between an sp_3 methylene and an sp_2 methine group. The first three examples show that the pseudo-enantiotopic hydroxyl groups of these cyclic alkenes all react at a competitively similar rate. Likewise, the catalyst

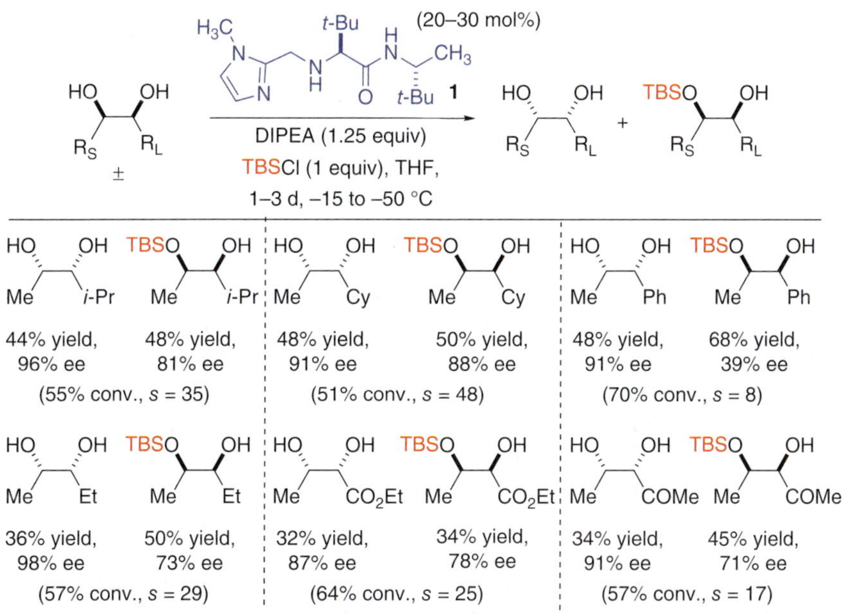

Scheme 13.2 Kinetic resolution of diols.

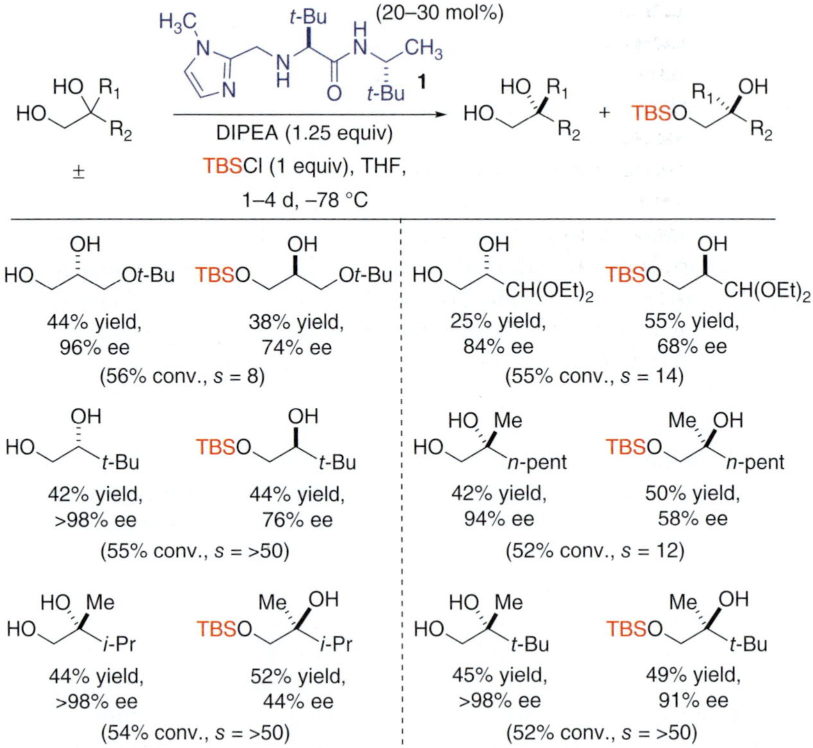

Scheme 13.3 Kinetic resolution of 1,2-diols.

does not distinguish between longer alkyl groups either; an alcohol next to an ethyl substituent reacts similarly to one next to a propyl group.

The utility of this strategy was highlighted in the first enantioselective synthesis of sapinofuranone A. As shown in Scheme 13.5, Sharpless' enantioselective dihydroxylation on 1,3-cyclohexadiene provides the desired diol enantiomer **2** with modest selectivity (32% ee). When this enantioselective reaction is combined with the regiodivergent silylation, the desired silylated regioisomer **3** is obtained in 64% yield (greater than the 50% theoretical yield if using racemic diol) and with enhanced enantiomeric excess (97% versus 88% ee). Oxidative cleavage of the double bond, followed by a *cis*-selective Wittig olefination and tetra-*n*-butylammonium fluoride (TBAF) desilylation delivers the target natural product in a good overall yield.

Subsequent studies extended the scope of the approach to cyclic and acyclic *meso*-1,2,3-triols [12]. Depending on the reactivity of the substrate, either TBSCl or TESCl could be used to desymmetrize a wide range of cyclic and acyclic *meso*-1,2,3-triols in the presence of 30 mol% of their imidazole catalyst **1**. These findings are shown in Scheme 13.6. In many cases, the enantioselectivities are exceptional. This is because of an effective secondary kinetic resolution where to which the initially generated silyl ether is subjected. In this case, the minor silylated enantiomer, still containing a reactive diol

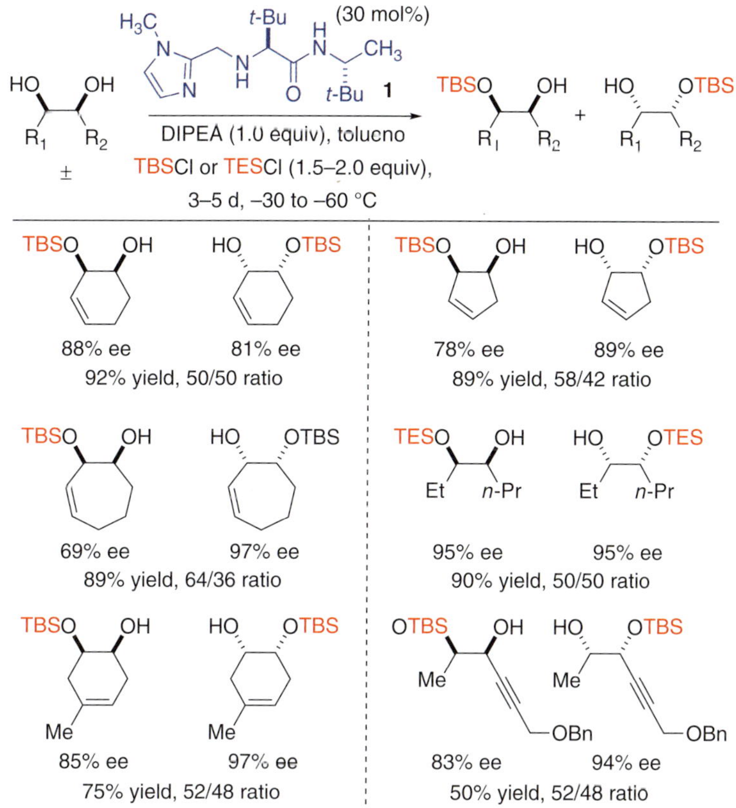

Scheme 13.4 Regiodivergent resolution of 1,2-diols.

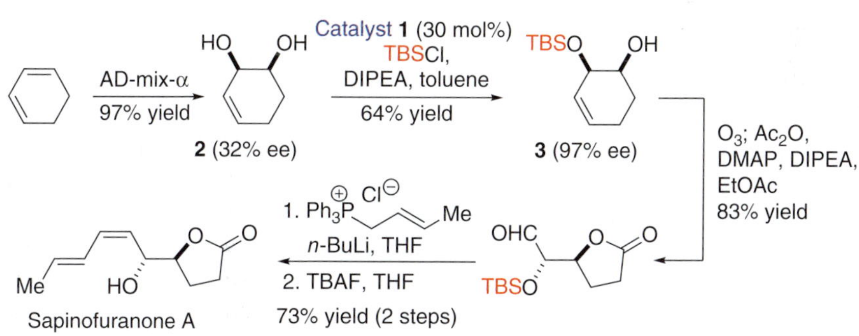

Scheme 13.5 Enantioselective synthesis of sapinofuranone A.

functionality, can undergo another catalytic silyl ether formation more rapidly than the major mono-silylated product. This serves to enhance the enantiopurity of the major product at the expense of its yield and the presence of a 1,3-*bis*-silylated by-product.

The absolute stereochemistry of the catalyzed silylation is substrate dependent. The 1,2,3-triols with primary hydroxyl groups lead to enantioselectivity

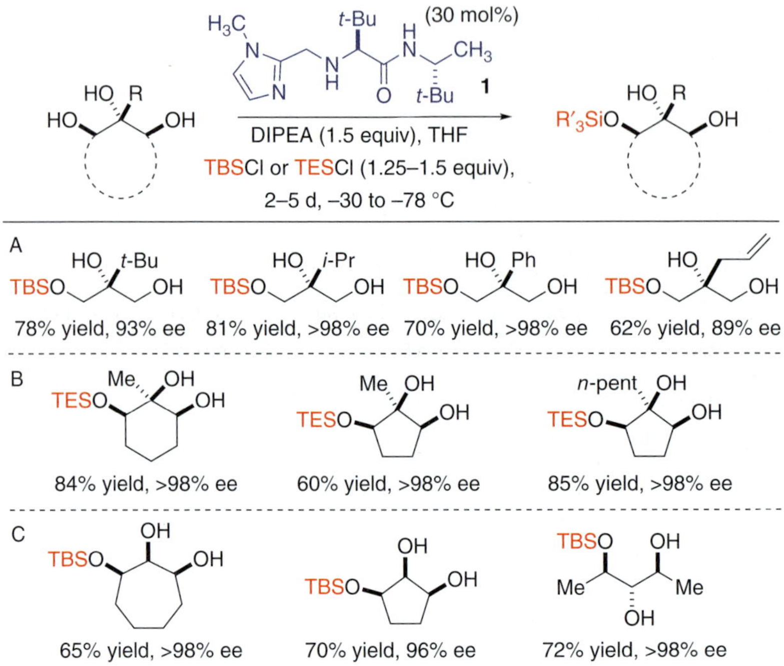

Scheme 13.6 Catalytic desymmetrization of *meso* triols.

opposite that of the substrates where the outer hydroxyl groups are secondary (Scheme 13.6, B and C). While the initially proposed transition state model was used to explain this switch in selectivity, minor modifications are likely necessary given the revised mechanistic model discussed subsequently. In any case, as the results in Scheme 13.6, C show, the catalyst can carry out a site-selective silylation on substrates with three similar secondary alcohols.

Catalytic triol desymmetrization was shown later to play a critical role in total syntheses of cleroindicins D, F, and C. The substrate for the desymmetrization could be prepared readily in five steps through an oxidative functionalization of a commercially available *para*-substituted phenol derivative. The requisite tetraol acetal system **4** was then desymmetrized with 2.2 equiv of TESCl. The excess silylating reagent was necessary due to the rapid background reaction of the primary hydroxyl group with the TESCl. Once the primary hydroxyl group was silylated, the catalyst then carried out a site-selective silylation of one of the enantiotopic secondary hydroxyl groups. This provided the *bis*-silylated product **5** in 83% yield and excellent enantiomeric purity. To complete the total synthesis, the remaining secondary alcohol was activated with mesyl chloride and then, in one step under acidic conditions, the cyclohexylenone was unmasked and reacted through an acidic hydrolysis of the dimethyl acetal, elimination of the mesylate, hydrolysis of the hydroxyl triethylsilyl (TES) groups, and conjugate addition of the terminal hydroxyl group into the enone. This sequence provided

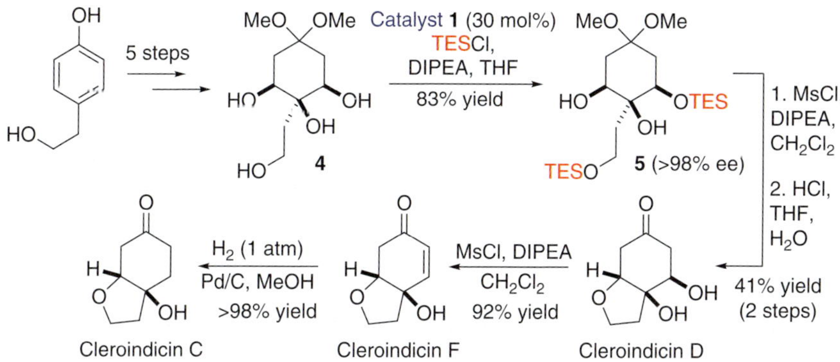

Scheme 13.7 Enantioselective total synthesis of cleroindicin D, C, and F.

cleroindicin D in 45% yield for that final acid-mediated step. Cleroindicin D could then be converted to cleroindicins F and C through straightforward manipulations (Scheme 13.7).

In 2013, some of the most significant shortcomings of the initially reported methodology were addressed [13]. Specifically, it was demonstrated that a transformation may be substantially accelerated when a tetrazole cocatalyst is present. The inclusion of 7.5–20 mol% of commercially available 5-ethylthio-tetrazole to the enantioselective silylation with a concurrent reduction of the recyclable chiral imidazole catalyst from 30 to 20 mol% (or less) allowed for reactions that, without the tetrazole, required up to five days to go to completion, to now proceed to completion in approximately one hour or less. Moreover, some substrates that failed to react under the initial reaction conditions were viable under these modified conditions. A direct comparison of the desymmetrization of *meso*-diols under the original and updated reaction conditions are shown in Scheme 13.8.

It was also shown that the loading of the chiral imidazole catalyst could be reduced to 5 mol%, which extended the reaction time to eight hours in one case. As shown in Scheme 13.9, a similar reaction rate enhancement was also observed for the kinetic resolution of racemic 1,2-diols. Reactions that required one to three days now were complete within one hour with comparable (or improved) selectivity factors.

The positive impact of tetrazole as co-catalyst originated from computational studies on the role of the imidazole catalyst **1** in the silylation transition state. Preliminary computational studies did not the initially reported transition state model. It was perceived that the "bifunctional" catalyst is unable to achieve the geometry necessary for both nucleophilic activation of the silicon, while also selectively deprotonating the nucleophilic hydroxyl group. Kinetic studies were inconclusive for this highly concentrated, heterogeneous reaction; nevertheless, it was speculated that the reaction is bimolecular in an imidazole catalyst. One imidazole activates the silyl group, while another serves as a chiral Brønsted base. Accordingly it was speculated that a less hindered silyl activator might be preferred. After screening a number of

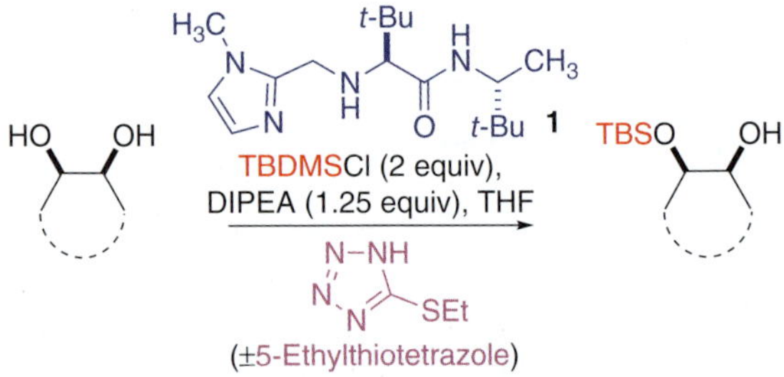

Entry	Product	Cat. **1** (mol%)	5-Ethylthiotetrazole (mol%)	Time (h)	Temp. (°C)	Yield (%)	ee (%)
(1)	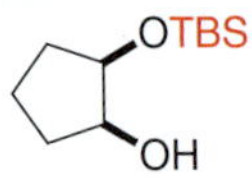	30	—	**69**	−40	96	88
		20	7.5	1	−40	97	91
(2)	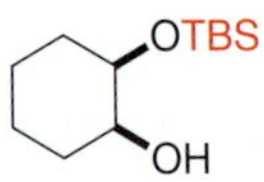	30	—	**120**	−28	82	92
		20	7.5	1	−40	93	94
(3)	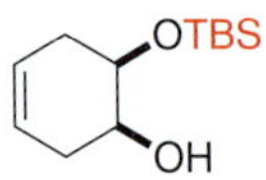	30	—	**120**	−40	75	94
		20	7.5	1	−40	82	95
(4)	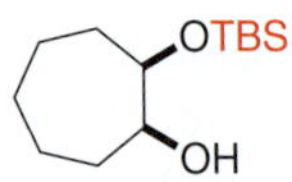	30	—	**120**	−40	93	93
		20	7.5	1	−40	93	93
(5)	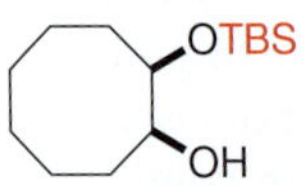	30	—	**120**	−40	96	95
		20	7.5	1	−40	93	95
(6)	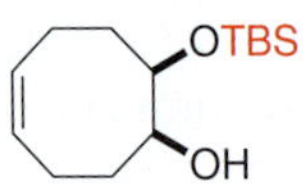	30	—	**120**	−40	96	95
		20	7.5	1	−40	93	95
(7)	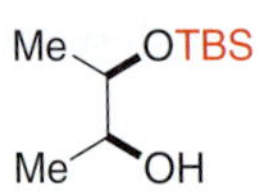	20	—	**120**	−28	84	90
		20	7.5	1	−40		93

Scheme 13.8 Improved conditions for the catalytic desymmetrization of *meso*-diols.

nitrogen-containing heterocyclic additive, tetrazoles were shown to provide a substantial rate enhancement without significant erosion in enantioselectivity. Based on these new findings, the transition state model for this reaction was outlined.

Catalyst **1** was thus demonstrated to be a highly effective Brønsted base, where the basicity of the imidazole is communicated and combined with the basicity of the secondary amine through a hydrogen bond network involving the second, non-nucleophilic hydroxyl group. It is only when this second hydroxyl group is positioned does the secondary amine become basic enough to deprotonate the hydroxyl group becoming silylated. The generality of this mechanistic hypothesis has not yet been tested.

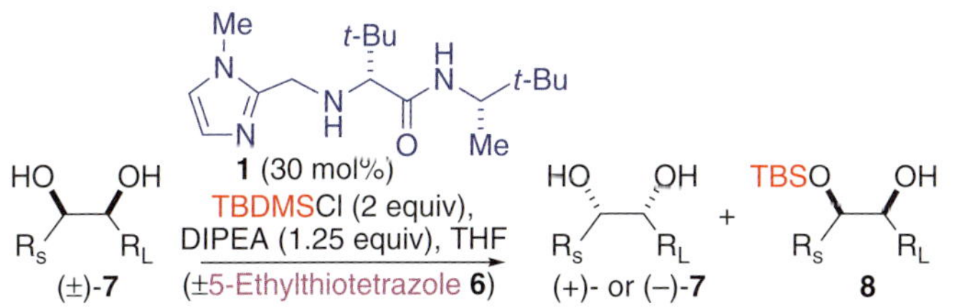

Entry	Substrate	6 (mol%)	Time (h)	Temp. (°C)	Recovered 7 (yield%)	Recovered 7 (ee%)	Product 8 (yield%)	Product 8 (ee%)	*s*
(1)	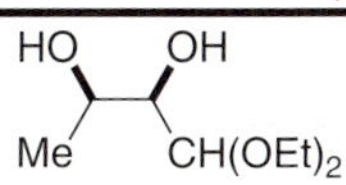	—	**24**	−30	44	98	52	80	***>50***
		5	1	−40	53	79	47	90	***46***
(2)	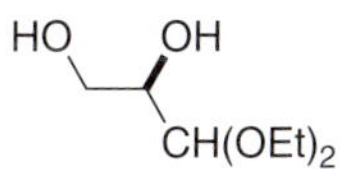	—	**24**	−78	25	84	55	68	***14***
		7.5	1	−78	41	82	49	70	***14***
(3)	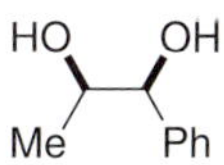	—	**72**	−15	30	96	68	39	***8***
		7.5	1	−40	47	92	52	86	***43***
(4)	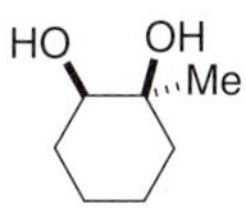	—	**24**	−30	—	—	<5	—	—
		20	12	−40	62	57	36	90	***39***

Scheme 13.9 Improved conditions for kinetic resolution of 1,2-diols.

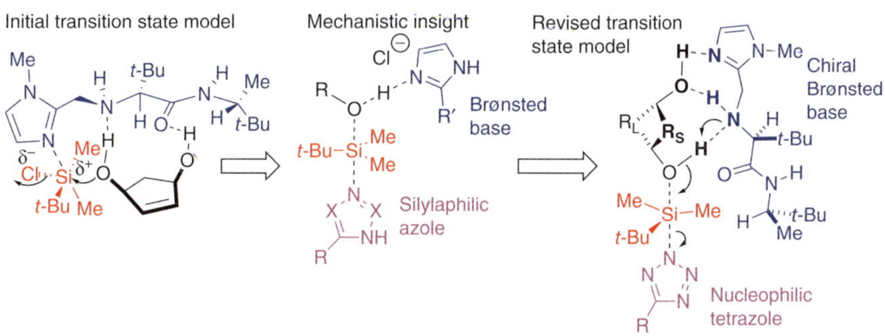

Figure 13.2 Revised transition state model for catalytic enantioselective silyl ether formation.

13.2.3 Directed Lewis Base-Catalyzed Enantioselective Silylations of Polyols

The Tan group used a temporary covalent interaction between their "scaffolding catalyst" **9** and *meso*-diol substrates to control the absolute stereochemistry of the silylation step [14]. This strategy is illustrated in Scheme 13.10. Key to the success of this approach is the incorporation of the methoxyoxazolidine component in their catalyst that serves as a reversible alcohol binding functionality. Under mildly acidic conditions, the methoxy group can be exchanged for a variety of other hydroxyl-containing compounds including the diol substrate. When the diol is bound through one of its enantiotopic hydroxyl groups, in one case, its

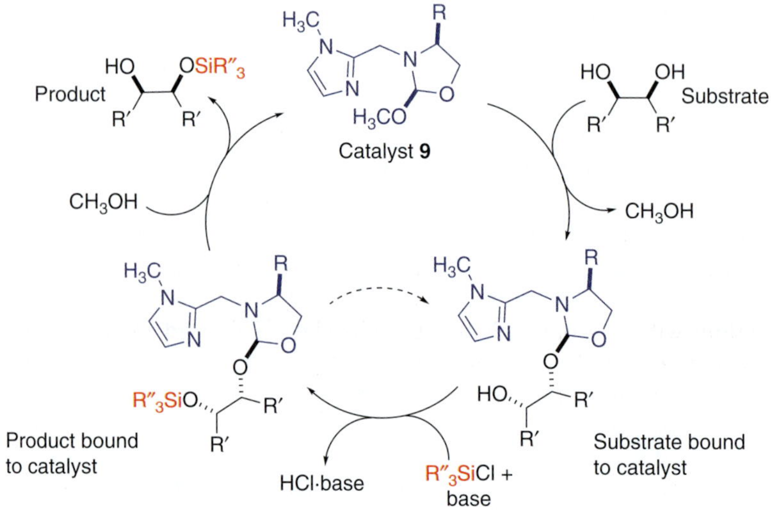

Scheme 13.10 Catalytic cycle for "scaffolding catalyst" **9** resulting in *meso*-diol desymmetrization.

remaining free hydroxyl group is found to be in close proximity to the catalytically active imidazole functionality, which can serve to selectively silylate that group. An acid-mediated exchange of alcohols can then return the starting catalyst (if non-silylated methanol is still available) or connect a new diol substrate directly to the catalyst, releasing the mono-silylated product.

The application of this unique catalytic strategy in the desymmetrization of several *meso*-1,2-diols using TBSCl is shown in Scheme 13.11. With 20 mol% of bifunctional catalyst **9** and some acidic buffer additives to support hydroxyl group exchange on the catalyst, both cyclic and acyclic *meso*-1,2-diols were shown to be good substrates for this selective silylation (i.e. ee's as high as 95%, data not shown). Limiting the scope of the methodology, but supporting the

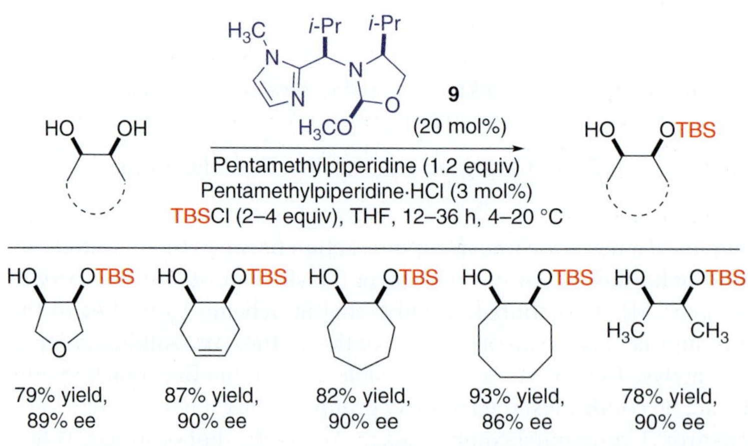

Scheme 13.11 *meso*-diol Desymmetrization with TBSCl.

proposed mechanism, both *trans*-1,2-diols and *meso*-1,3-diols were shown to be nonviable substrates in this enantioselective silylation.

With adjustments to reaction times and concentrations, other silylating agents, such as TESCl, *t*-butyldiphenylsilyl chloride (TBDPSCl), and DMPSCl, were also shown to be effective agents in the desymmetrization of *meso*-1,2-diols. Using the smaller TESCl reagent, in particular, allowed the Tan team to extend the substrate scope to include acyclic *meso*-diols that were slow to react with TBSCl. For example, *meso*-hydrobenzoin could be desymmetrized in 82% yield and 90% ee with TESCl, whereas this substrate displayed inferior reactivity with TBSCl.

In a follow-up study, the scaffolding catalyst's ability to make reversible covalent interactions with appropriately functionalized diols allowed Tan and coworkers to use their catalyst to override the inherent substrate silylation preferences to give a highly catalyst-controlled silylation result [15]. For example, methyl imidazole–catalyzed silylation of a 1,2-diol, where one of hydroxyl groups is primary and the other is secondary, occurs nearly exclusively on the primary alcohol. With **9** or **10**, however, the silylation results become enantiomer specific. That is, in this catalyst-controlled reaction, one enantiomer of a racemic 1,2-diol mixture is silylated preferentially on its more sterically hindered secondary hydroxyl, whereas the other enantiomer is silylated on its primary hydroxyl group. This enantiospecific silylation allows for easy separation of the regioisomerically functionalized enantiomers in high yields (the maximum yield of a single enantiomer is 50%) and good enantiomeric purities. These results for a range of racemic terminal 1,2-diols are summarized in Scheme 13.12.

As the data indicate, the enantioenriched isomers with the primary hydroxyl group silylated are isolated in somewhat lower enantiomeric purities than are the slower reacting isomers with the secondary alcohols silylated. In a follow-up study, it was demonstrated that by minimizing background silylation of the primary hydroxyl groups, improved selectivities for the primary silylated 1,2-diol products may be achieved [16]. These selectivities were improved using a slower introduction of a limited amount of TESCl and diisopropylethylamine (DIPEA). Instead of achieving a regiodivergent reaction where both enantiomers react as mentioned, this protocol was more like a traditional kinetic resolution, where the faster reacting enantiomer was isolated. The results of this kinetic resolution are shown in Scheme 13.13.

The Tan group also demonstrated an effective desymmetrization of glycerol through the use this reversible covalent catalytic silylation strategy [17]. As summarized in Eq. (13.3), with a variety of bulky silyl chlorides in their catalytic system, highly enantiomerically enriched mono-silylated glycerols can be synthesized. The enantioselectivities reported for the desymmetrizations of glycerol are notably higher that the desymmetrizations of the *meso*-1,2-diols [14]; however, the yields are somewhat lower. In addition, appreciable amounts of the *meso*-bis-silylated glycerols were also observed in this transformation. Initial silylation selectivities were similar to those of the diols; however, on further reaction, the undesired enantiomer was an effective substrate for the catalyst, so it was silylated preferentially to give the *bis*-silylated product. This secondary kinetic resolution on the mono-silylated glycerol enantiomers

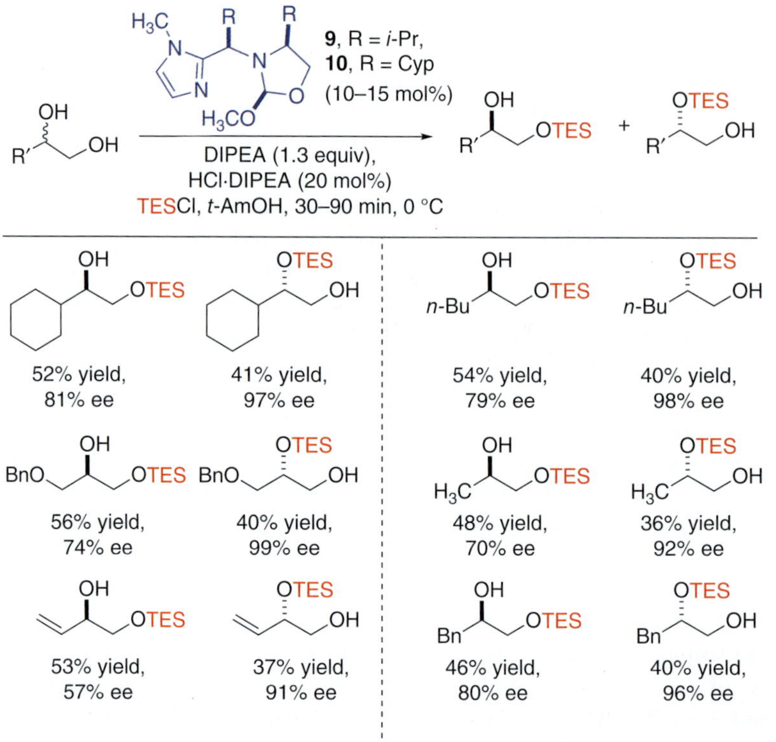

Scheme 13.12 Regiodivergent resolution of 1,2-diols.

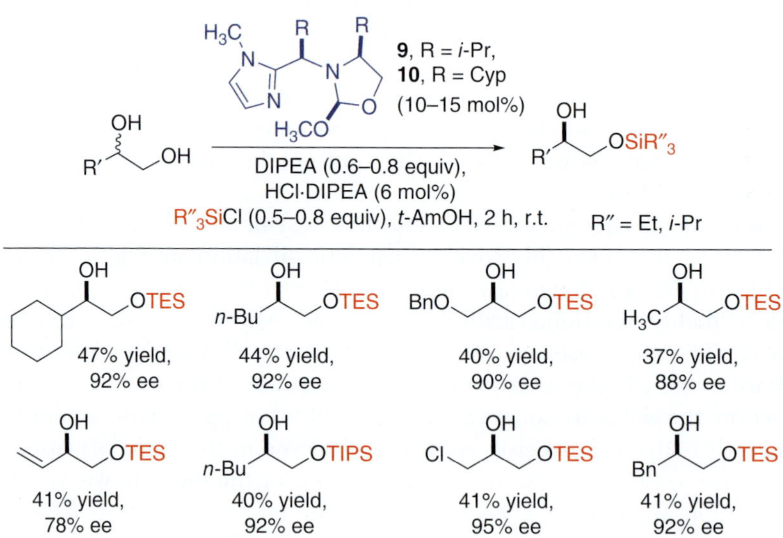

Scheme 13.13 Kinetic resolution of 1,2-diols.

serves to enhance enantioselectivity of the final product at a small expense of overall yield.

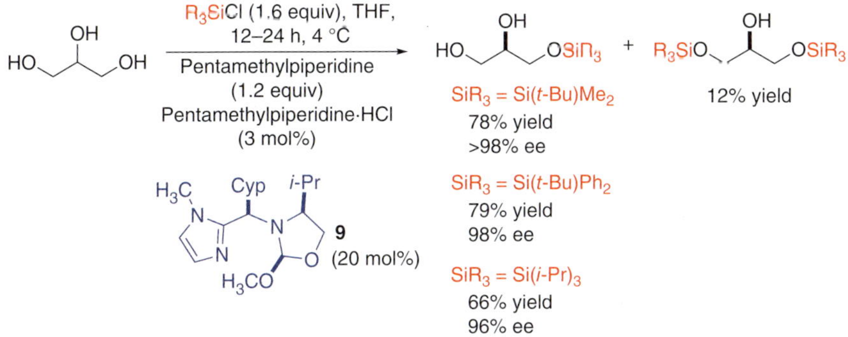

(13.3)

13.2.4 Lewis Base-Catalyzed Enantioselective Silylations of Alcohols

The Wiskur group's first contribution in the area of enantioselective hydroxyl group silylation arose from their efforts to develop an enantioselective Mukaiyama aldol process [18]. In the presence of a methylated cinchona alkaloid salt **11** (30 mol%), a silylketene acetal was shown to react with benzaldehyde to produce an aldol product that, after desilylation, was shown to be racemic. On closer inspection, however, it was noted that product **13** was formed in low enantiomeric excess. Subsequent studies supported the conclusion that the chiral ammonium salt **11** was promoting the kinetic resolution of the alkoxide enantiomers generated in the aldol reaction, albeit with low selectivity factors. The results of this early study are summarized in Scheme 13.14.

In a subsequent study, Wiskur identified reaction conditions, including a Lewis-basic heterocyclic catalyst **14**, which, for the first time, resulted in the

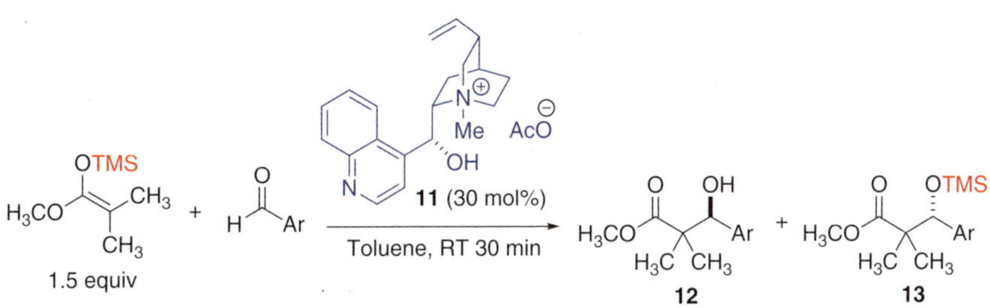

Ar	Yield (**12** + **13**)	ratio **12** : **13**	ee **12** (%)	*s*
C_6H_5	85	1 : 6	68	2.1
2-F-C_6H_4	78	1 : 3.5	84	3.7
4-Br-C_6H_4	83	1 : 5	78	2.7

Scheme 13.14 Mukaiyama aldol/silylative kinetic resolution.

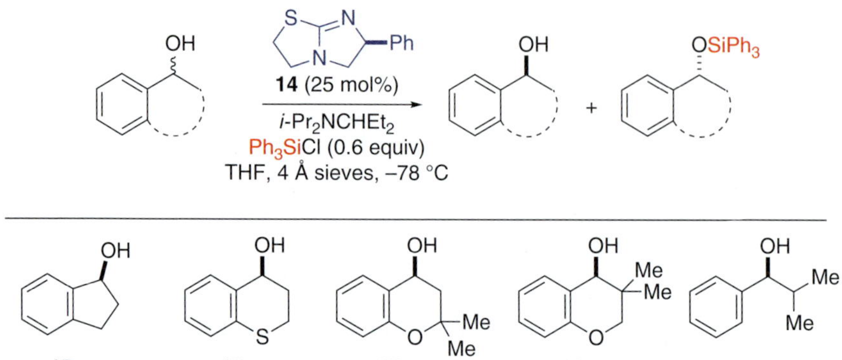

15	16	17	18	19
60% conv., 1 h	52% conv., 8 h	52% conv., 8 h	<5% conv., 24 h	29% conv., 48 h
84% ee, *s* = 8.6	88% ee, *s* = 25	88% ee, *s* = 25		26% ee, *s* = 5.6

Scheme 13.15 Kinetic resolution of isolated secondary alcohols.

silylative kinetic resolution of isolated secondary alcohols [19]. As shown in Scheme 13.15, the isothiourea (−)-tetramisole **14** was shown to be an effective catalyst at 25 mol% loading for the silylative kinetic resolution of several types of benzylic alcohols. Unhindered bicyclic benzylic alcohols, such as the first three examples (**15–17**), were particularly well suited for this resolution. Hindered and acyclic benzylic alcohols, such as **18** and **19**, were less reactive and selective in this transformation.

The silylating agent in this transformation is noteworthy. While bulky trialkylsilyl chlorides were unreactive, optimal reactivity and selectivity were obtained with triphenylsilyl chloride. Wiskur suggested that the chiral propeller-like conformation of the triphenyl groups on the silyl chloride might play a role in the differentiation between the secondary alcohol enantiomers.

In a follow-up study, Wiskur and co-workers expanded the substrate scope of their methodology to include α-hydroxy lactones and lactams [20]. While similar to the abovementioned strategy, key differences necessary for the success of these type of substrates included using the isothiourea (−)-benzotetramisole **20** as the catalyst, increasing the concentration of the reaction and switching to readily available Hünig's base from the esoteric trialkylamine used previously. As summarized in Scheme 13.16, under these conditions, α-hydroxy-γ-butyrolactone could be resolved with modest efficiency (s = 5.8). The presence of additional steric bulk near (i.e. β-position) the reactive hydroxyl group, however, increases the selectivity of the resolution significantly. The β,β-diethyl α-hydroxy-γ-butyrolactone, for example, can be resolved with a selectivity factor of 100. Not surprisingly, the reactivity rate for this more hindered substrate, however, decreases substantially. A similar trend is seen with the α-hydroxy lactams as well, but these compounds are generally less selective. Nonetheless, the authors indicate that this is the first reported nonenzymatic resolution of this class of compounds.

The strategy was then extended to acyclic α-hydroxy esters and amides substrates, as well. As the representative example illustrated in Scheme 13.17

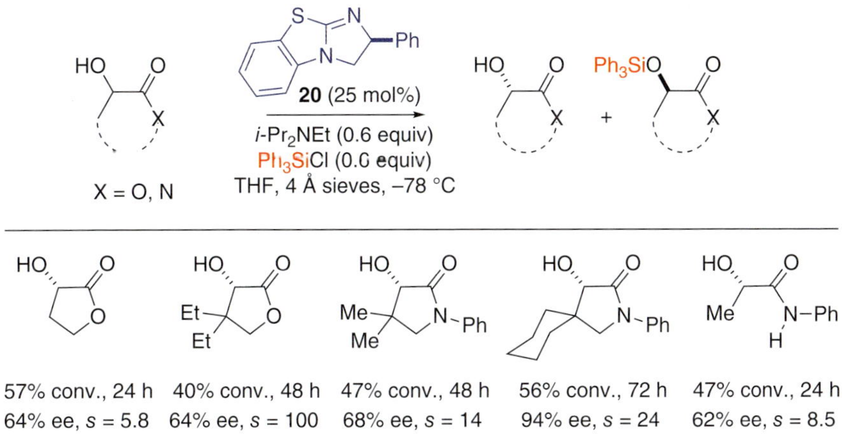

Scheme 13.16 Kinetic resolution of α-hydroxy lactones and lactams.

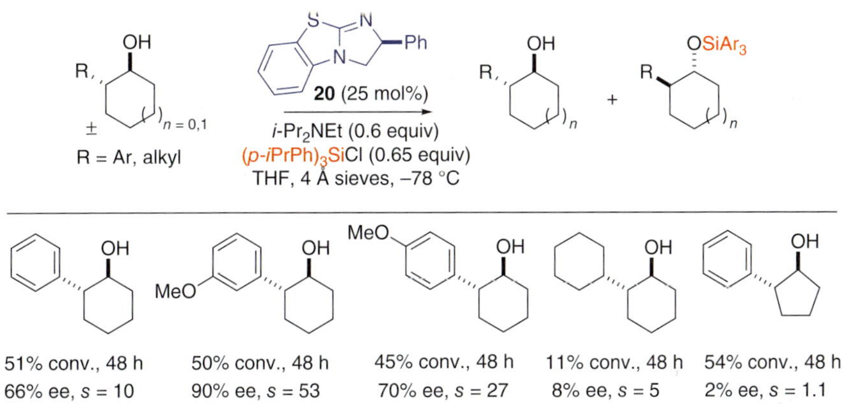

Scheme 13.17 Kinetic resolution of *trans*-2-arylcyclohexanols [21].

indicates, regardless of the neighboring steric environment, the selectivity factors were again modest with this class of substrates.

In broaden substrate Wiskur et al. examined 2-substituted cycloalkanols as potential substrates. To increase selectivity, they used *para*-isopropyl-substituted triarylsilyl chloride as their silylating agent. It was found that cyclic alcohols with an adjacent *cis*-substituent were unreactive under these reaction conditions. For cyclohexanols, It was argued that *cis*-2-substituted alcohols were unreactive because the hydroxyl group resides in the more hindered axial conformation. In support of this hypothesis, they showed that some *trans*-2-arylcyclohexanols, where the hydroxyl group is equatorial, were particularly good substrates for effecting the resolution. As shown in Scheme 13.17, *trans*-2-phenyl cyclohexanol provided sufficient reactivity and modest selectivity ($s = 10$) in the kinetic resolution. Moreover, the *trans*-2-*o*-methyoxyphenyl cyclohexanol was much more selective ($s = 53$) in this transformation. *trans*-2-Alkyl substituted cyclohexanols and *trans*-2-phenyl-substituted cyclopentanols emerged as less effective substrates showing either low selectivity or reactivity. Currently, it is unclear why minor modifications to the substrates lead to these significant reactivity and selectivity differences.

As shown in Eq. (13.4), the diastereo- and enantioselectivity displayed by catalyst **20** allowed Wiskur to reduce a racemic 2-aryl substituted cyclohexanone to produce, non-selectively, four alcohol isomers (two enantiomers for both the *cis*- and *trans*-diastereomeric isomers) Without purification, one of the four alcohol isomers present in the mixture was converted to its corresponding silyl ether. That is, only one enantiomer of the *trans*-diastereomeric mixture reacts, not affecting either enantiomer of the *cis*-diastereomer. This allowed for convenient isolation of that *trans*-isomer in high enantiomeric purity.

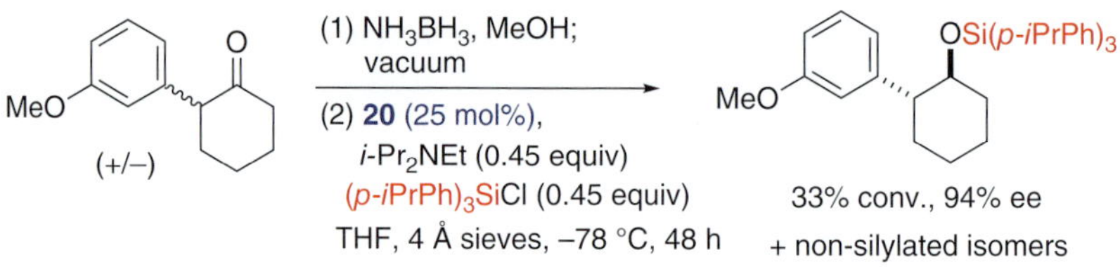

(13.4)

An immobilized, recyclable triaryl silyl chloride was subsequently introduced [22]. The polystyrene-bound triphenylsilyl chloride was used in the presence of their chiral isothiourea catalysts (**14**, **20**) to trap preferentially the more reactive alcohol enantiomer. The results are summarized in Scheme 13.18. In these cases, the ease of purification through precipitation and filtration of the polystyrene-bound silyl reagent was somewhat mitigated by the lower selectivity and reactivity (i.e. conversion) displayed in these reactions. Nevertheless, the polystyrene-bound triphenylsilyl chloride could be

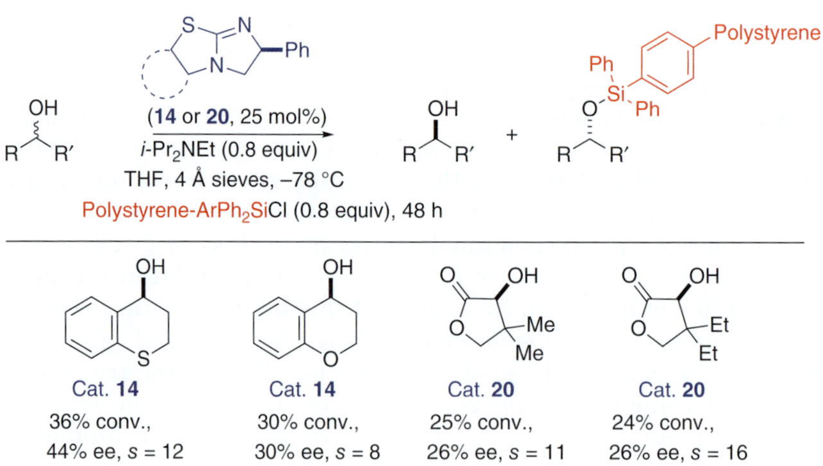

Scheme 13.18 Kinetic resolution with an immobilized triarylsilyl chloride.

regenerated and reused, as well as provide the more reactive alcohol isomer in an enantioenriched manner.

For additional insight, the electronic and steric effects of *para*-substituted triaryl silyl chlorides on silylative kinetic resolution of secondary alcohols was investigated [23]. The electronic nature of the substituents was determined to be the most significant factor. It was thus established that electron-donating groups on the *para*-substituted triaryl silyl chlorides lead to slower, more selective transformations via a later transition state, while electron-withdrawing substituents accelerated the reactions, but were less selective due to relatively late earlier transition state. The redistribution of positive charge in the rate-determining step was consistent with an S_N2-type transition state, where the incoming alcohol attacks the triarylsilyl group and displaces the chiral tetramisole catalyst **14**. This transition state is illustrated in Scheme 13.19. Of particular note, this interaction also provides some rationale of how the chiral catalyst could influence the propeller-like conformation triarylsilyl chlorides effectively enhancing the diastereomeric differences of the two alcohol enantiomers in their respective transition states [24].

Nakata and coworkers have used a similar strategy for kinetic resolution of a variety of racemic 1-indanols using a chiral guanidine catalyst [25]. Building upon Wiskur's observations, Nakata and coworkers showed that their optimized guanidine catalyst **21** could effect the silylative kinetic resolution with higher selectivities and lower catalyst loading. Like Wiskur, Nakata's system worked best with triphenylsilyl chloride as the electrophilic silylating agent. Representative examples of this transformation are shown in Scheme 13.20.

In general, substituents on the aromatic portion of the indanol ring system were all well tolerated, undergoing the silylative kinetic resolution with useful selectivity factors between 30 and 89. In addition, substituents on the aliphatic portion of the indanol ring system were also tolerated as long as they were not adjacent to the reactive hydroxyl group.

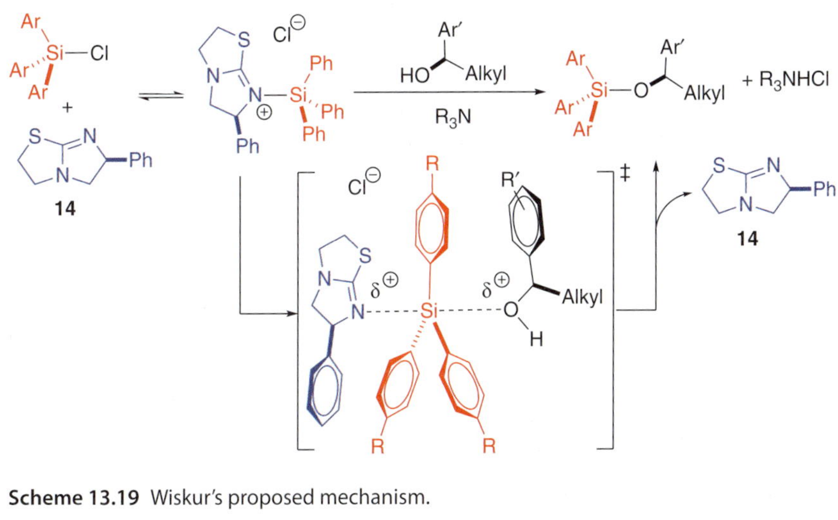

Scheme 13.19 Wiskur's proposed mechanism.

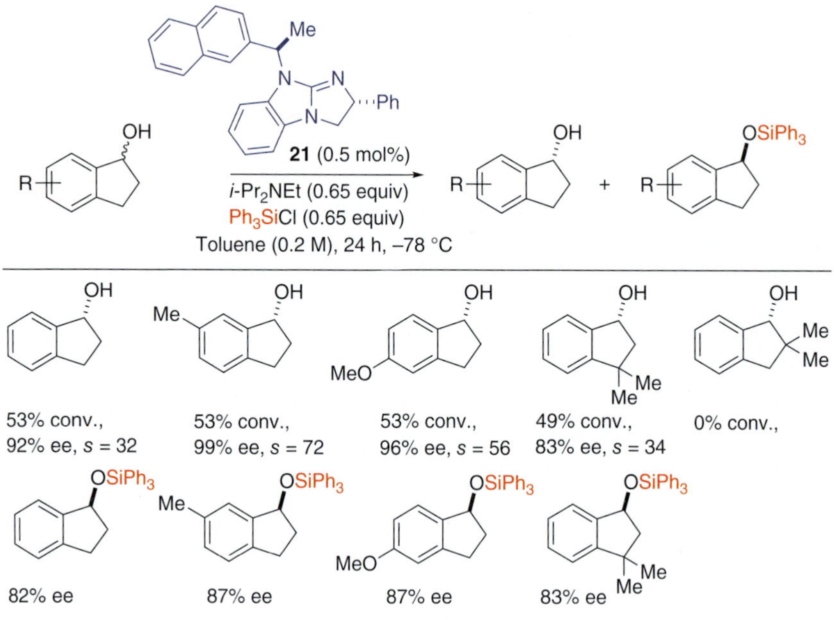

Scheme 13.20 Nakata's kinetic resolution of 1-indanols.

Given their similarities, it is interesting to speculate on the reactivity differences between Wiskur's isothiourea catalyst **20** and Nakata's related guanidine catalyst **21**. The additional N-substituent on Nakata's guanidine catalyst likely plays some role here. While there is evidence for a matched/mismatched case regarding the stereogenic center on the N-substituent, it does not control the overall stereochemistry of the reaction. More likely, the bulk of this substituent and its resulting greater interaction with the triphenyl silyl group in the transition state described by Wiskur probably accounts for the higher selectively of this catalyst. One might expect that this greater bulk could also lead to a decrease in reactivity of the guanidine relative to the isothiourea catalyst; however, this could be more than offset by the great Lewis basicity of the guanidine system. In short, the greater steric bulk combined with the additional Lewis basicity of the guanidine system **21** combines to provide a highly effective catalyst for the silylative kinetic resolution of these secondary alcohols.

13.2.5 Lewis Base–Mediated Enantioselective Desilylations of Mono-Alcohols

Adopting a different strategy, the Song group developed a novel desilylative kinetic resolution of racemic secondary trimethylsilyl alcohols [26]. With 20 mol% of a chiral crown ether **22**, which can chelate potassium, it was found that KF in dioxane can preferentially cleave one trimethylsilyl alcohol enantiomer. Representative cases are shown in Scheme 13.21. Although the reaction is

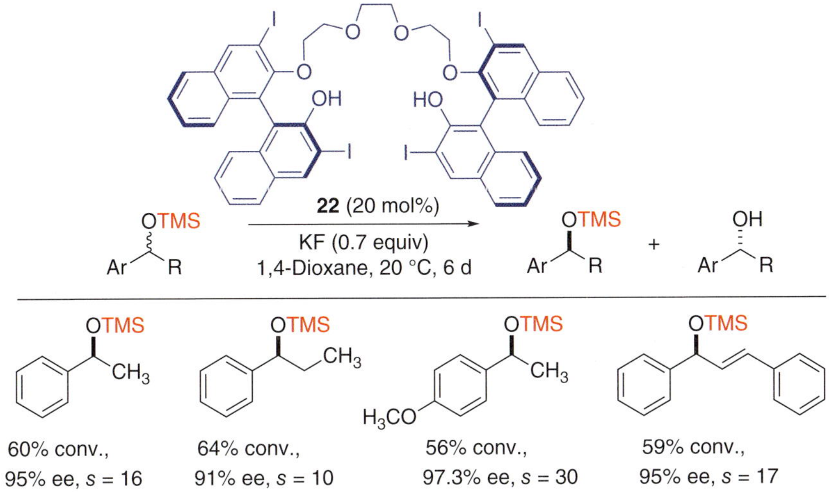

Scheme 13.21 Song's desilylative kinetic resolution of secondary silyl alcohols.

sluggish, the substrate scope is reasonably broad. Some insight into the role of the chiral crown either is provided through an X-ray crystallographic structure of the crown ether bound to KF.

13.3 Brønsted Acid-Catalyzed Enantioselective Silylation of Alcohols

The List group reported that a Brønsted acid catalyst could be used to effect the silylative desymmetrization of diols [27]. Whereas the efforts described earlier used chiral bases interacting with electrophilic silylating agents to achieve an asymmetric silylation of the hydroxyl group, this work feature enantioselective a chiral Brønsted acid catalyst. As illustrated in Scheme 13.22, the chiral acid **23** (i.e. STRIP) in a nonpolar solvent, such as cyclohexane, is used to protonate a basic silyl transfer reagent, such as hexamethyldisilazide (HMDS). The resulting chiral ion pair would then transfer selectively a TMS group to a meso-diol acceptor molecule.

Two examples of the transformation are illustrated in Eqs. (13.5) and (13.6). The meso-1,2-diphenylethane-1,2-diol illustrated in Eq. (13.5) is representative of the majority of diaryl-ethane-1,2-diols examined in this transformation. In these examples, the reported selectivities and yields were of similar synthetically useful levels. Cyclohexane-1,2-diol, as shown in Eq. (13.6), provided monosilylated product in an excellent yield, however, with slightly lower levels of selectivity. In nearly all cases, the reaction times were fairly long (1–10 days) to achieve these levels of conversion and selectivity. Nevertheless, this contribution represents a unique and underexplored strategy to effect enantioselective silylation.

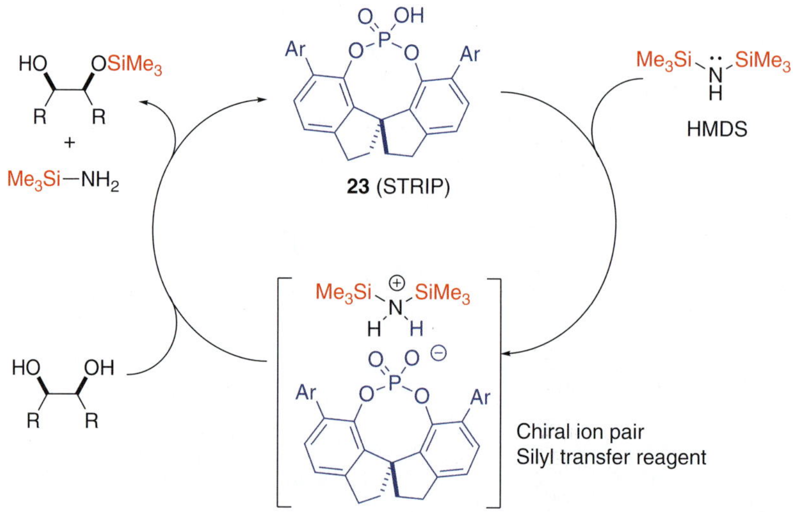

Scheme 13.22 Proposed catalytic cycle for the Brønsted acid-catalyzed silylation of a hydroxyl group.

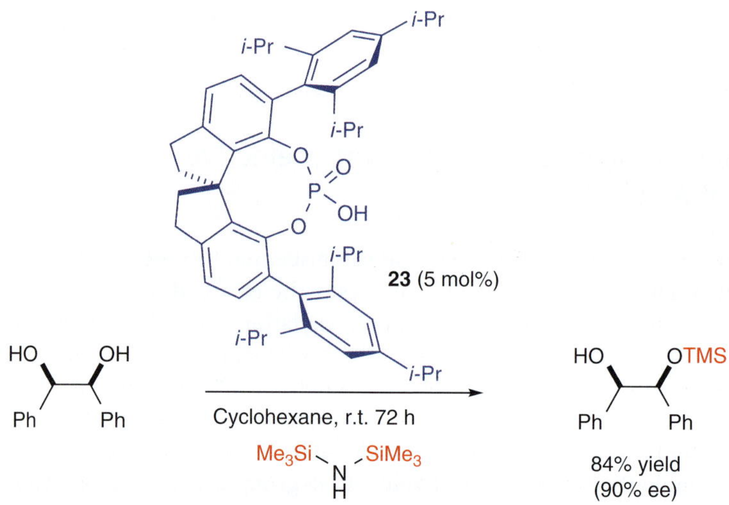

(13.5)

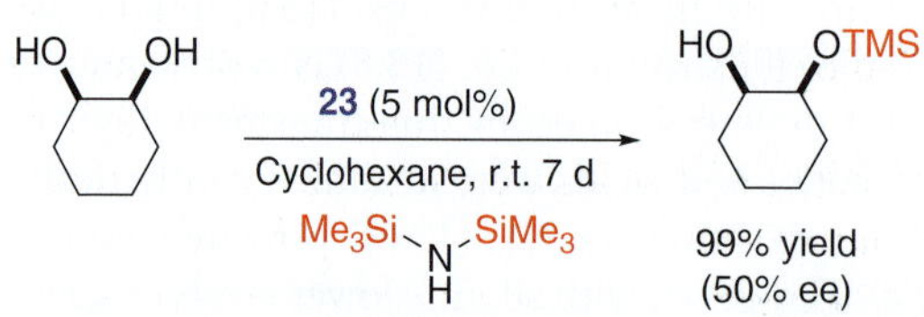

(13.6)

In the same year, the Song group reported that their chiral crown ether **22** used for desilylations [26] could be used to carry out silylative kinetic resolution of secondary alcohols [28]. Their results are summarized in Scheme 13.23. In this approach, the phenolic hydrogen on the catalyst serves as a chiral Brønsted acid

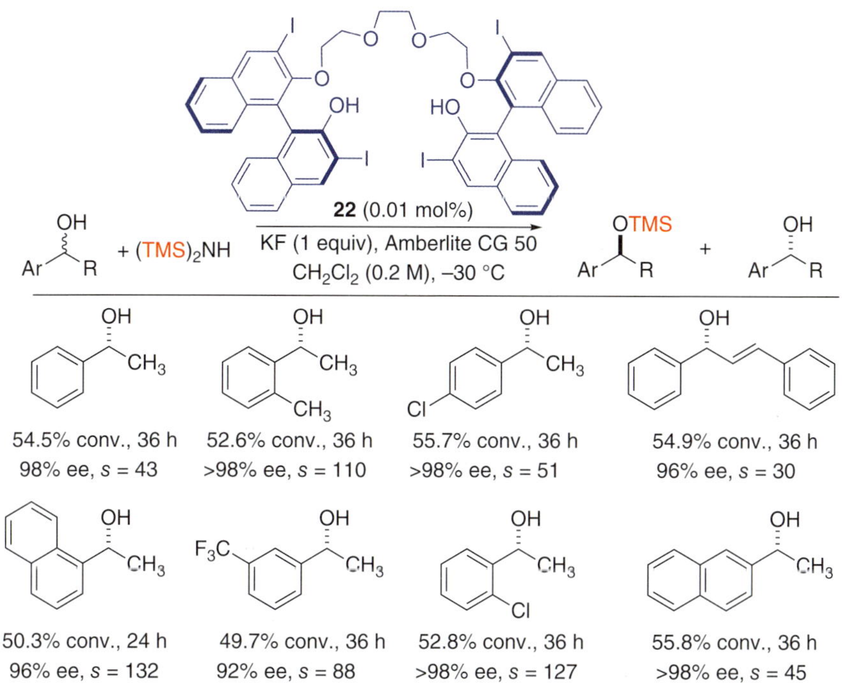

Scheme 13.23 Song's silylative kinetic resolution of secondary alcohols.

protonating HMDS, which serves as the silylating source for the secondary alcohol. Song suggests that the ether oxygens of the catalyst also play a role by interacting with the secondary alcohol's proton. As shown, a relatively broad range of benzylic secondary alcohols can be resolved with good to excellent selectivities with notably low catalyst loadings.

The key to achieving high catalyst efficiency is the introduction of fluoride and a proton source to the reaction. Song and his team speculated that silylation of one of the catalytic phenolic groups was one of the catalyst's deactivation mechanisms in this reaction. They also reasoned that the phenolic trimethylsilyl group on the catalyst would be much more sensitive to fluoride than the silyl ether product. Thus, with the introduction of a fluoride (KF) and proton source (Amberlite CG 50), the catalyst's turnover numbers and frequencies increased significantly.

As shown in Scheme 13.24, this highly effective catalytic silylation could also be extended to desymmetrization of diols. Specifically, a several *meso*-hydrobenzoins were desymmetrized through the use of this protocol with slightly higher catalysts loadings (1 mol%) and with good selectivity.

13.4 Hydroxyl Group Silylations with Organometallic Complexes

From early on, an alternative enantioselective silylation strategy has been under development by the Oestreich group. Instead of using silyl chlorides as electrophiles

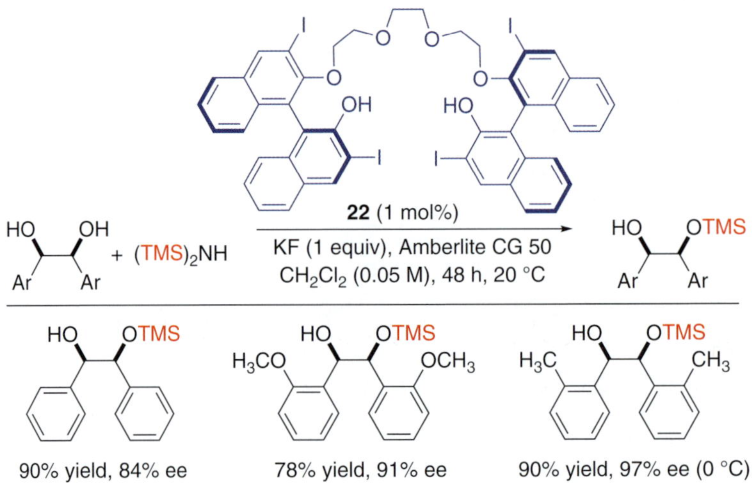

Scheme 13.24 Song's silylative desymmetrization of *meso*-diols.

in the silylation of alcohols, they employed a metal-catalyzed oxidative coupling between functionalized silanes and hydroxy groups. In addition to generating the silyl ether, hydrogen (H_2) is the major by-product of this transformation.

13.4.1 Directed, Catalytic Enantioselective Hydroxyl Group Silylations with Chiral Silanes

Oestreich and coworkers showed that enantiomerically enriched, bulky cyclic silanes may be used to resolve some secondary alcohols [29]. In general, the reactions can proceed with useful levels of selectivity; however, the need for a pendant ligand on the alcohol substrate to ensure the desired kinetic selectivity did limit the overall utility and scope of this transformation. Shown here (Scheme 13.25) are illustrative examples of this unique approach. The yield for the alkyne substrate is slightly lower due to a competitive *Z*-selective reduction of the alkyne functionality.

Given the precious nature of their chiral silyl reagent, Oestreich and coworkers demonstrated that the silane **24** could be recycled through a reductive cleavage of the silicon–oxygen bond of the silylated substrate with complete retention of its stereochemical integrity (Eq. (13.7)). This allows for reuse of the reagent **24** in this auxiliary-based kinetic resolution strategy.

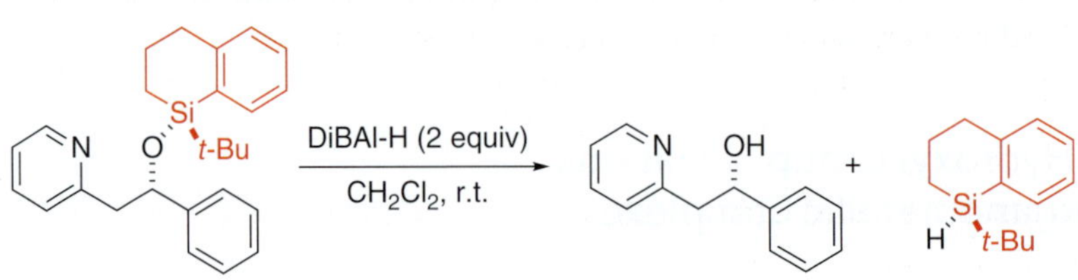

(86 : 14 dr) (78% yield, 71% ee) **24** (98% yield, 96% ee) (13.7)

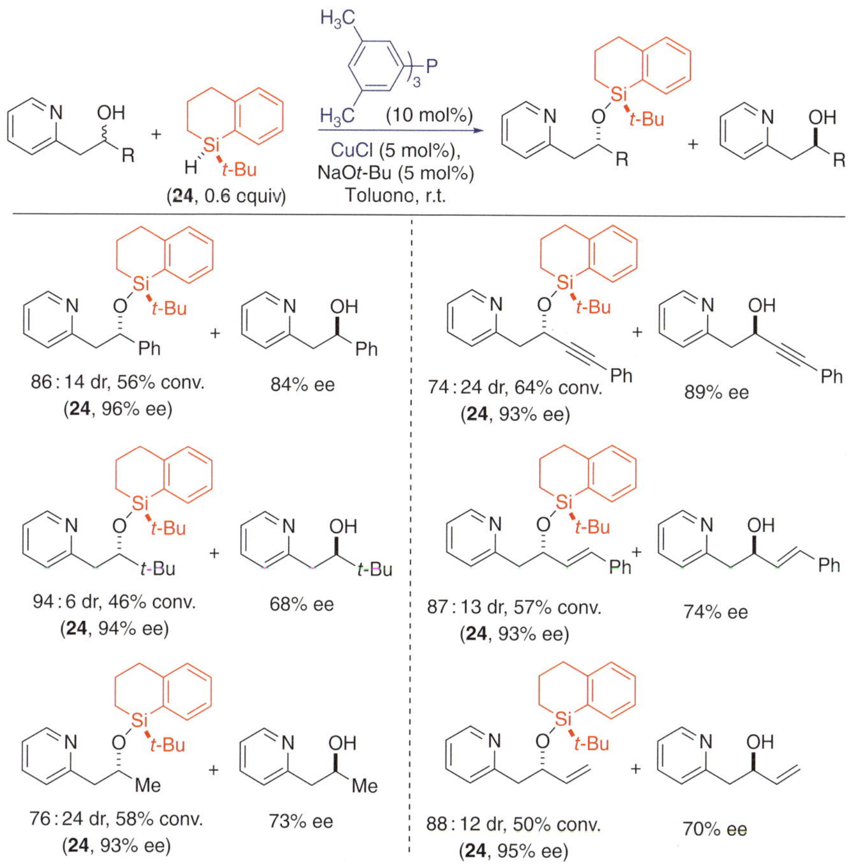

Scheme 13.25 Kinetic resolution of pyridine-containing secondary alcohols with a Cu-based catalyst and chiral silane **24**.

Access to chiral reagent **24** in an enantiomerically enriched manner is possible through several strategies [30]. Shown in Scheme 13.26 is a typical resolution. Preparation of the racemic silane **24** is achieved in one step (45% yield) by treating *tert*-butyltrichlorosilane with the corresponding Grignard reagent prepared from of 1-bromo-2-(3-bromopropyl)benzene, followed by $LiAlH_4$ reduction of the silylhalogen bond (Si—X where X = Cl and Br). Treatment of racemic **24** with chlorine gas then provides the racemic chlorosilane **25** in 99% yield. Subjecting racemic chlorosilane **25** to deportonated (−)-menthol provides a separable mixture of diastereomeric silylethers that are epimeric at silicon. After crystallographic and chromatographic separation, the individual diastereomers can then be reduced with DiBAl-H to provide the cyclic silane enantiomers **24** in high enantiomeric purity.

Oestreich have provided additional information on the scope and mechanism of the above silylation [31]. Computational studies suggest that the silylation step occurs through a four-center, σ-bond metathesis between the chiral silane **24** and pyridine-chelated copper alkoxide. The proposed catalytic cycle involving a

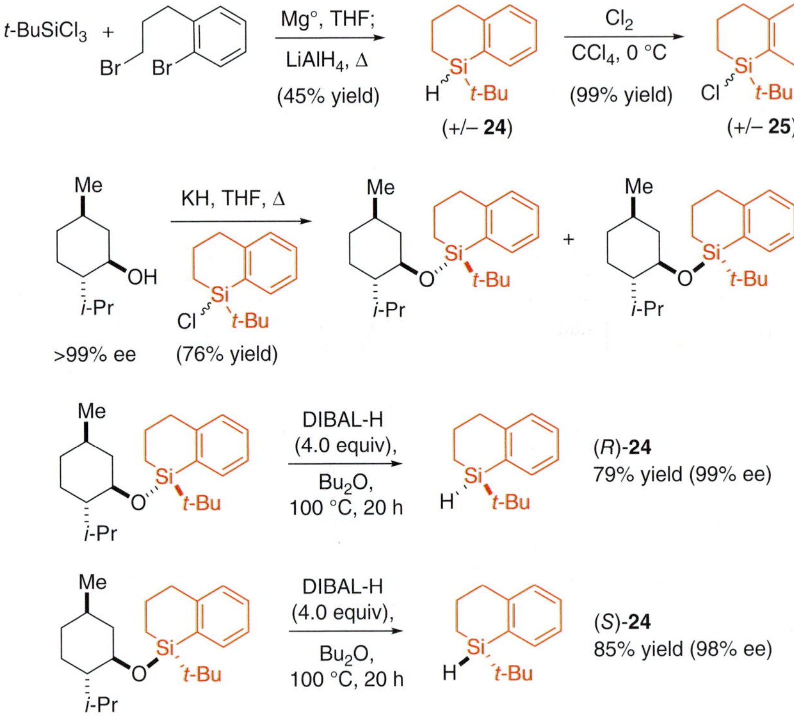

Scheme 13.26 Preparation of chiral silane.

σ-bond metathesis step can be found in Scheme 13.27. The stereoretentive nature of the silyl transfer, coupled with the stereoretentive DiBAl-H reduction, allows for recycling of the chiral silane reagent.

Through further catalyst and ligand screening, the Oestreich team was able to identify more selective, rhodium-catalyzed conditions that provided the kinetic resolution of these secondary alcohols with significantly higher selectivity factors (*s* values) [32]. They found that N-heterocyclic carbene ligands with Rh(I) led to a highly diastereostereoselective secondary alcohol silylation with racemic (or highly enantiomerically enriched) chiral silanes. Scheme 13.28 illustrates some of the substrates resolved with their enantiomerically enriched silane **24**. It is impressive to note that the selectivity factor (*s* values) for at least one of these examples appears to be approximately 900.

For more hindered, pyridine-containing tertiary propargyl alcohols, the standard chiral silane reagent **24** proved to be insufficiently reactiive [33]. Likewise, the rhodium-catalyzed conditions were also ineffective on these substrates. For these more hindered tertiary systems, a more strained cyclic chiral silane **26** was perceived to be more effective resolving agent. Considering the the transition state, it was argued that the strained cyclic chiral silane would facilitate this transformation through its increased Lewis acidity [4]. This strategy proved successful. Examples involving the more reactive chiral silane **26** are

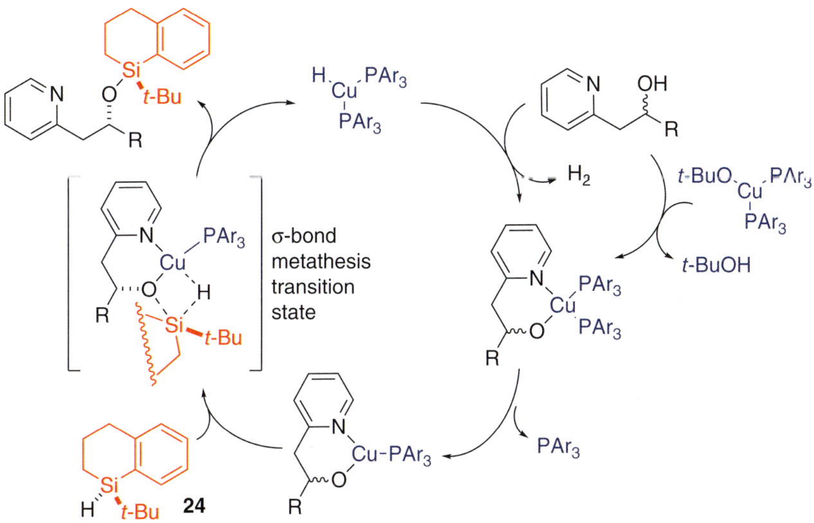

Scheme 13.27 Proposed catalytic cycle for the copper-catalyzed kinetic resolution of pyridine-containing secondary alcohols using chiral silanes.

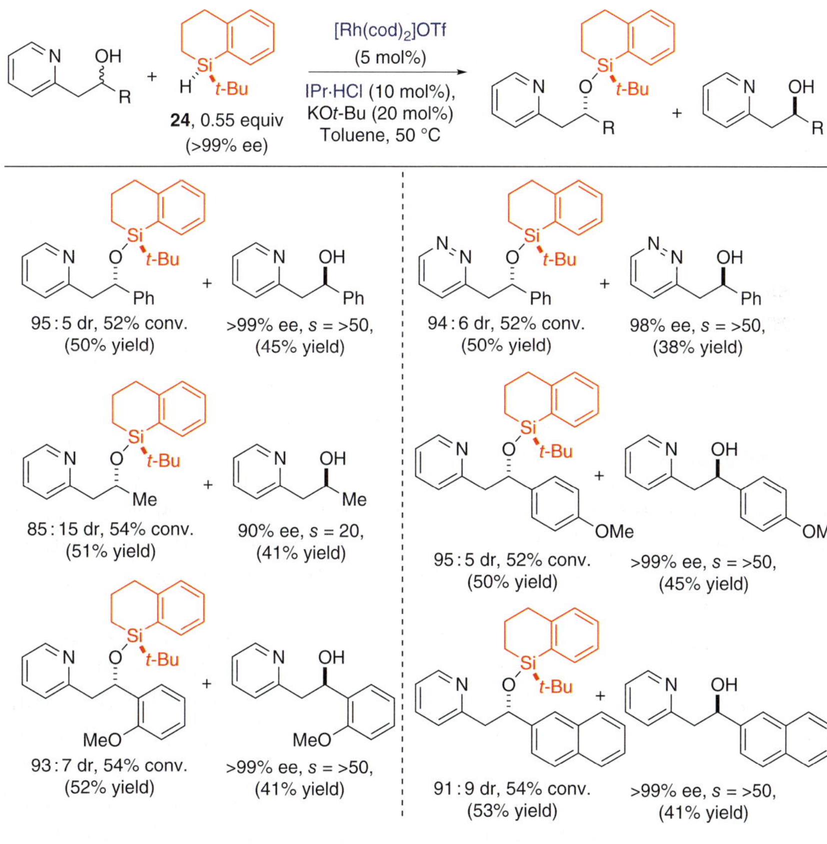

Scheme 13.28 Oestreich's rhodium-catalyzed kinetic resolution of pyridine-containing secondary alcohols using chiral silanes.

Table 13.1 Oestreich's copper-catalyzed kinetic resolution of pyridine-containing tertiary alcohols using a strained chiral silane **26**.

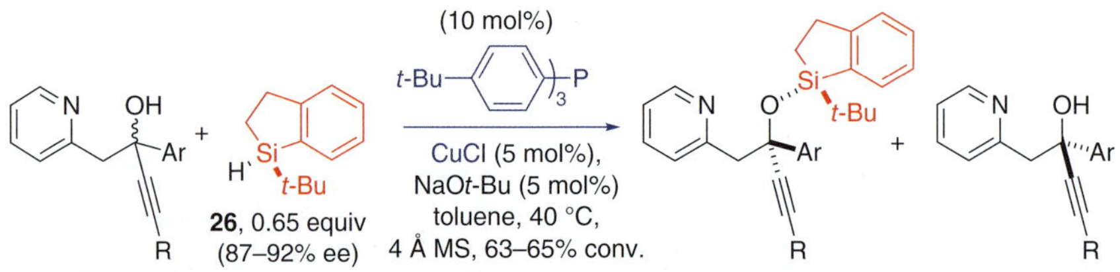

26, 0.65 equiv (87–92% ee)

R = TMS	Ar = Ph	**26** 92% ee	*s* = 8	58% yield, 76 : 24 dr	34% yield, 88% ee
R = TMS	Ar = 4-tolyl	**26** 90% ee	*s* = 7	58% yield, 78 : 22 dr	34% yield, 84% ee
R = Ph	Ar = Ph	**26** 92% ee	*s* = 9	62% yield, 78 : 22 dr	25% yield, 92% ee

shown in Table 13.1. While limited in substrate scope and selectivity, this contribution represents one of the limited number of kinetic resolutions of tertiary alcohols at the time.

To expand the scope of their kinetic resolution strategy, Oestreich prepared and resolved trifluoromethyl-containing secondary alcohols with various pendant N-chelating groups [34]. Several examples are presented in Scheme 13.29. The considerable selectivity and reactivity of these trifluoromethyl substituted carbinols, compared with alkyl- and aryl-substituted carbinols, suggested a possible favorable interaction between a fluorine and silicon atom in the transition state. Supporting this hypothesis were control experiments where the pendant chelating pyridine ring was substituted with a phenyl group. The selectivity and reactivity in this CF_3-substituted system was higher than those of the corresponding methyl-substituted substrate, an observation that supported a favorable interaction of the trifluoromethyl group in the transition state.

13.4.2 Metal-Catalyzed Enantioselective Hydroxy Group Silylations with Chiral Silanes

In 2008, Bellemin-Laponnaz and coworkers examined polymeric silanes in the kinetic resolution of secondary alcohols through a copper-catalyzed dehydrogenative coupling [35]. One potential advantage of this reagent is the ease of separation of the faster reacting, polymer-bound alcohol enantiomer from the soluble, slower reacting enantiomer. Various chiral phosphine ligands on copper were screened to control the selectivity in the kinetic resolution, however, without much success. Building on Oestreich's earlier observation [29], it was noted that polymeric silanes prefunctionalized with a chiral alcohol additive provide modest selectivities in the kinetic resolutions of racemic secondary alcohols, independent of the nature of the

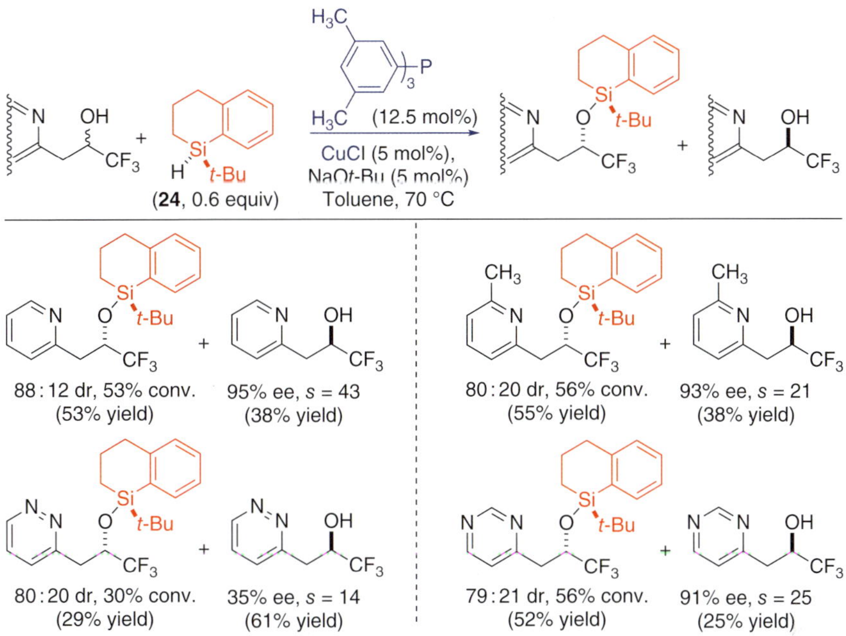

Scheme 13.29 Kinetic resolution of CF_3-substituted secondary alcohols with a Cu-based catalyst and chiral silane **24**.

phosphine ligand used on copper. This transformation is illustrated in Eq. (13.8). It was concluded that a reagent-controlled strategy involving chiral silanes would be more effective in controlling the selectivity in the silylative kinetic resolution of secondary alcohols than a chiral catalyst-controlled process.

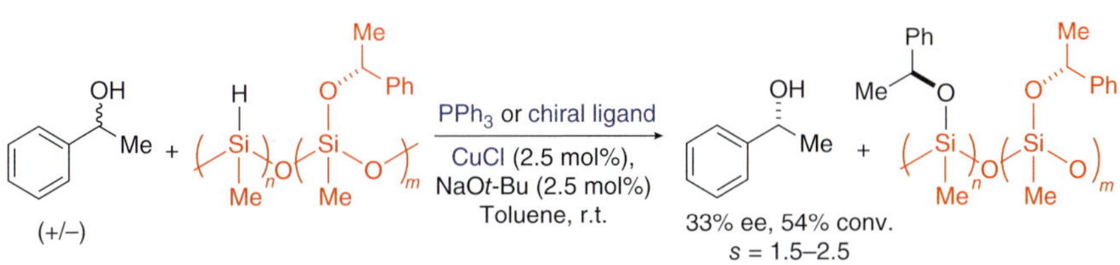

(13.8)

13.4.3 Directed, Enantioselective Catalytic Hydroxy Group Silylations with Achiral Silanes

A noteworthy advance in silylation of alcohols through the catalytic silane oxidation strategy was reported in 2010 [36]. Through extensive catalyst screening, the Oestreich team was able to move from a reagent-controlled process to a potentially more cost effective, chiral catalyst-controlled kinetic resolution of secondary alcohols containing a donor group.

In particular, they identified taddol-based ligands (i.e. **27**) with copper chloride and cesium carbonate that promoted the dehydrogenative couple of the secondary alcohol with achiral silanes with useful selectivity. The nature of the substituents on the silane proved critical toward balancing the reactivity and selectivity of this catalytic system. While trialkylsilanes were unreactive, triaryl silanes were reactive, but minimally selective. The mixed diarylalkyl silane **28**, on the other hand, provided the desired mix of reactivity and selectivity to achieve selective silylation. Several examples of this transformation are shown in Scheme 13.30. While the requirement of the second copper ligating functionality still limits the substrate scope of the reaction, the catalyst-controlled nature of the transformation represents a substantial advancement in this general approach.

13.4.4 Enantioselective Catalytic Hydroxy Group Silylations with Achiral Silanes

Another major advance is the development of dehydrogenative coupling of hydrosilanes, promoted by Cu-based catalysts for a more broadly applicable kinetic resolution of secondary alcohols [37]. Whereas earlier work was limited to alcohols with a copper ligating functionality, switching to a chiral bidentate ligand (i.e. **29**) and further optimization largely addressed the latter limitation. One key to the success to this strategy was the use of trialkylsilanes, such as *n*-Bu_3SiH, as a coupling partner.

Illustrative examples of this dehydrogenative coupling process are shown in Scheme 13.31. As indicated, a broad range of benzylic secondary alcohols can be resolved with moderate to outstanding selectivities, depending on the substrate substituents. Acyclic and cyclic allylic alcohols are also viable substrates if a substoichometric amount of sacrificial styrene is added to the reaction to circumvent olefin reduction of the allylic substrate.

Oestreich has also demonstrated that the enantioselective coupling process can be applied to an assortment of hydroxy oxime esters [38]. Representative examples are presented in Scheme 13.32. Since the selectivity factors for the *E*- and the *Z*-oxime isomers were similar, it does not appear that the oxime nitrogen needs to serve as a copper liganding atom in this transformation.

The utility of was highlighted by converting the optically enriched α-hydroxy oxime ethers to a variety of other useful functionalities. Some of those transformations are summarized in Scheme 13.33.

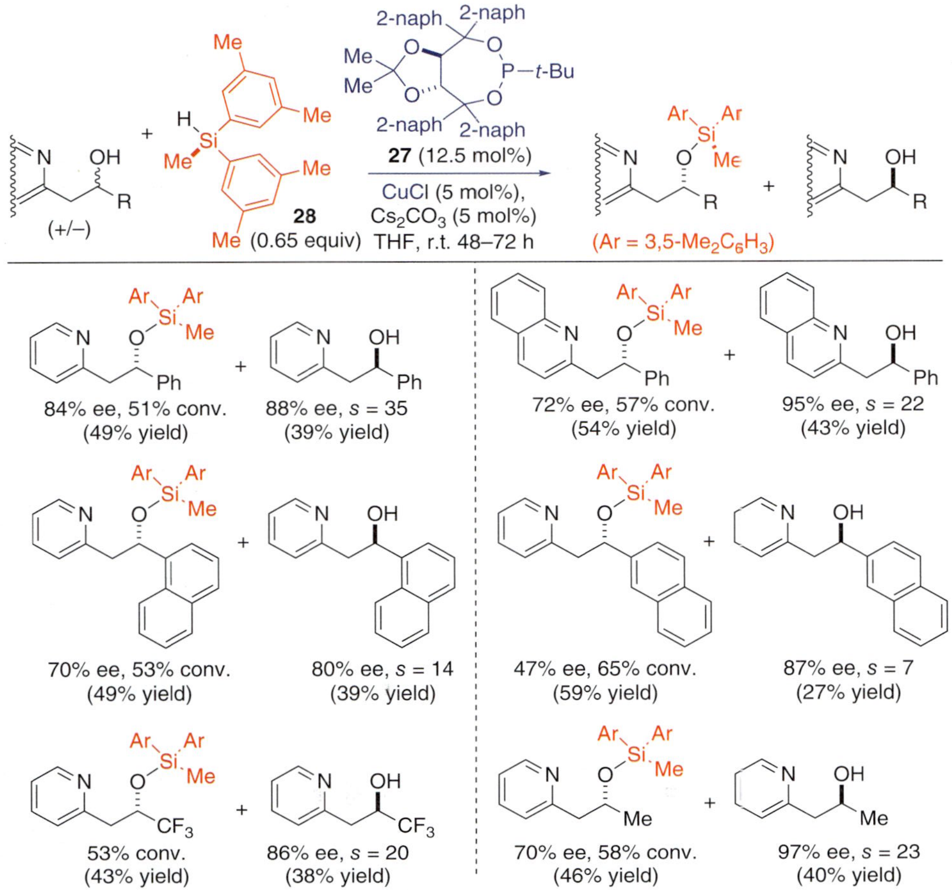

Scheme 13.30 Catalytic kinetic resolution of secondary alcohols with donor groups with diarylmethyl silanes.

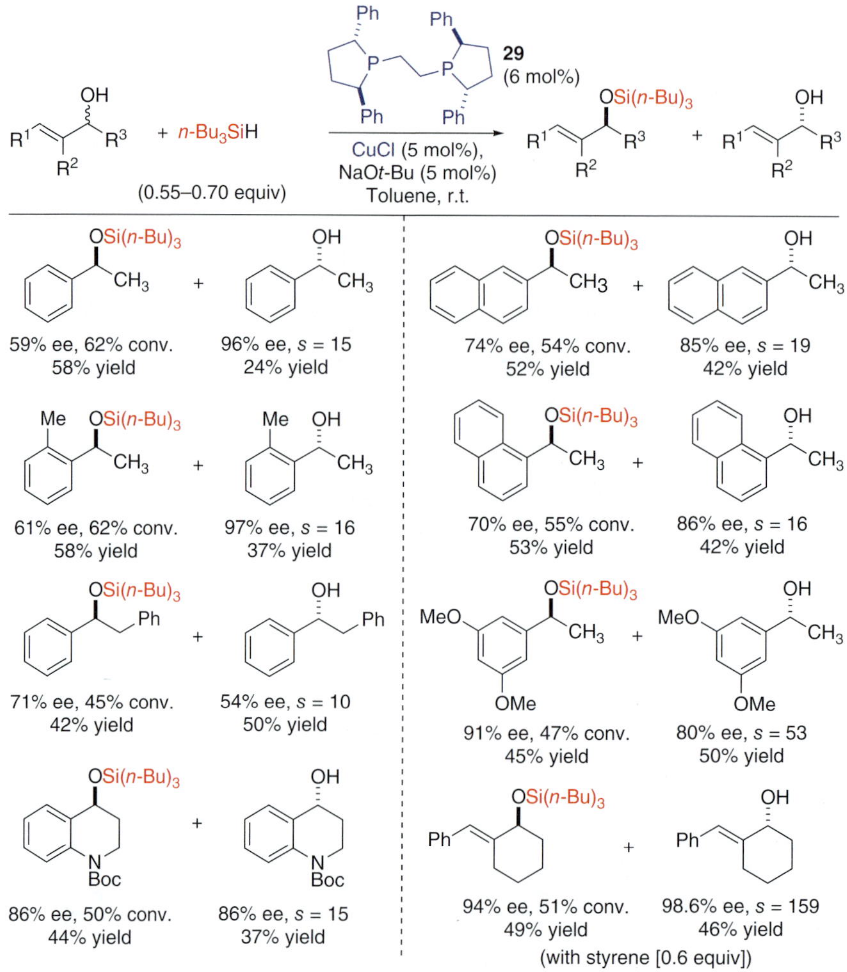

Scheme 13.31 Catalytic kinetic resolution of secondary alcohols.

13.5 Conclusions

The importance of silyl ethers as protecting groups in complex molecule total synthesis has been well established. Described earlier were assortments of complementary strategies that serve to extend considerably the synthetic utility of this important functionality. The new chiral catalysts and reagents that have been developed within the past two decades now offer novel asymmetric access to a variety of alcohols or silyl ethers, through either a silylative or desilylative process. Still, there are numerous limitations to be addressed, but considerable progress has already been made. It is clear that exciting new breakthroughs regarding catalysts and substrate selectivity and scope will continue to be reported in the future.

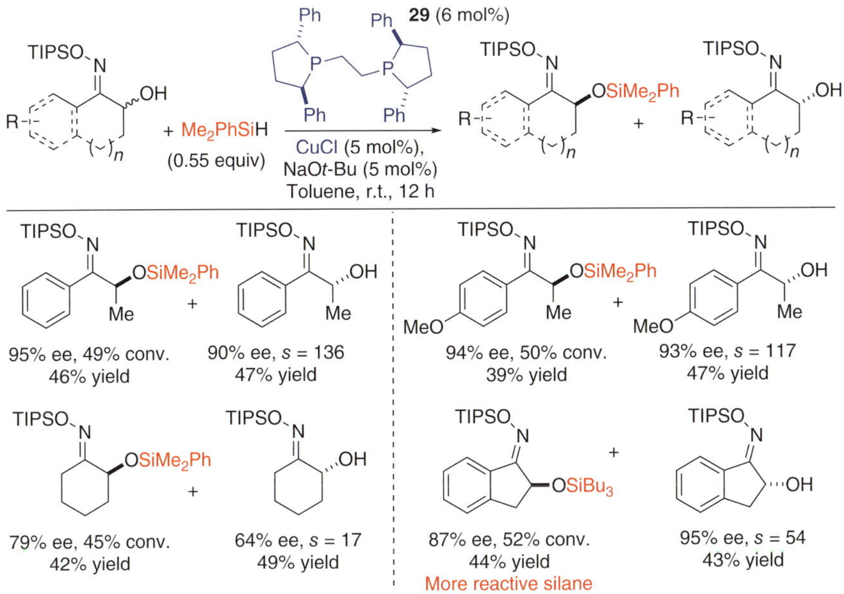

Scheme 13.32 Oestreich's catalytic asymmetric kinetic resolution of α-hydroxy oxime ethers.

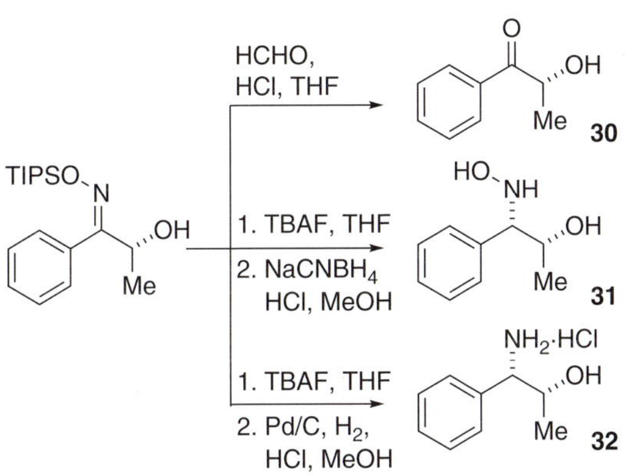

Scheme 13.33 α-Hydroxy oximes as precursors to other useful functionalities (e.g. **30**–**32**).

References

1 Corey, E.J. and Venkateswarlu, A. (1972). *J. Am. Chem. Soc.* 94: 6190–6191.

2 For alternative catalytic approaches to the same functionality (e.g. asymmetric hydrosilylation), see: (a)Nishiyama, H., Sakaguchi, H., Nakamura, T. et al. (1989). *Organometallics* 8: 846–848.(b) Sawamura, M., Kuwano, R., and Ito, Y. (1994). *Angew. Chem. Int. Ed.* 33: 111–113.(c) Tao, B. and Fu, G.C. (2002). *Angew. Chem. Int. Ed.* 41: 3892–3894.(d) Gade, L.H., Cesar, V., and Bellemin-Laponnaz, S. (2004). *Angew. Chem. Int. Ed.* 43: 1014–1017.(e) Zhu, G., Terry, M., and Zhang, X. (1997). *J. Organomet.*

Chem. 547: 97–101.(f) Nishibayashi, Y., Takei, I., Uemura, S., and Hidai, M. (1998). *Organometallics* 17: 3420–3422.(g) Chianese, A.R. and Crabtree, R.H. (2005). *Organometallics* 24: 4432–4436.(h) Malacea, R., Poli, R., and Manoury, E. (2010). *Coord. Chem. Rev.* 254: 729–752.(i) Díez-González, S. and Nolan, S.P. (2007). *Org. Prep. Proced. Int.* 39: 523–559.(j) Marciniec, B. (ed.) (2009). *Hydrosilylation: A Comprehensive Review on Recent Advances*. Netherlands: Springer.

3 (a) Chaudhary, S.K. and Hernandez, O. (1979). *Tetrahedron Lett.* 2: 99–102.For a more recent study, see: (b)Patschinski, P., Zhang, C., and Zipse, H. (2014). *J. Org. Chem.* 79: 8348–8357.

4 (a) Denmark, S.E., Jacobs, R.T., Dai-Ho, G., and Wilson, S. (1990). *Organometallics* 9: 3015–3019.(b) Myers, A.G., Kephart, S.E., and Chen, H. (1992). *J. Am. Chem. Soc.* 114: 7922–7923.(c) Denmark, S.E.;.B., Griedel, B.D., and Coe, D.M. (1993). *J. Org. Chem.* 58: 988–990.(d) Matsumoto, K., Oshima, K., and Utimoto, K. (1994). *J. Org. Chem.* 59: 7152–7155.(e) Zhang, X., Houk, K.N., and Leighton, J.L. (2005). *Angew. Chem. Int. Ed.* 44: 938–941.

5 For reviews, see: (a)Xu, L.-W., Chen, Y., and Lu, Y. (2015). *Angew. Chem. Int. Ed.* 54: 9456–9466.(b) Weickgenannt, A., Mewald, M., and Oestreich, M. (2010). *Org. Biomol. Chem.* 8: 1497–1504.(c) Rendler, S. and Oestreich, M. (2008). *Angew. Chem. Int. Ed.* 47: 248–250.

6 For example, see: Oriyama, T. (2011). *Acylation of Alcohols and Amines in Science of Synthesis, Stereoselective Synthesis*, vol. 3 (eds. J.G. De Vries, G.A. Molander and P.A. Evans), 829–849.

7 For example, see: Myers, A.G., Hammond, M., and Wu, Y. (1996). *Tetrahedron Lett.* 37: 3083–3086.

8 Isobe, T., Fukuda, K., Araki, Y., and Ishikawa, T. (2001). *Chem. Commun.* 243–244.

9 Zhao, Y., Rodrigo, J., Hoveyda, A.H., and Snapper, M.L. (2006). *Nature* 443: 67–70.

10 Zhao, Y., Mitra, A.W., Hoveyda, A.H., and Snapper, M.L. (2007). *Angew. Chem. Int. Ed.* 46: 8471–8474.

11 Rodrigo, J.M., Zhao, Y., Hoveyda, A.H., and Snapper, M.L. (2011). *Org. Lett.* 13: 3778–3781.

12 You, Z., Hoveyda, A.H., and Snapper, M.L. (2009). *Angew. Chem. Int. Ed.* 48: 547–550.

13 Manville, N., Alite, H., Haeffner, F. et al. (2013). *Nat. Chem.* 5: 768–774.

14 (a) Sun, X., Worthy, A.D., and Tan, K.L. (2011). *Angew. Chem. Int. Ed.* 50: 8167–8171.(b) Tan, K.L., Sun, X., and Worthy, A.D. (2012). *Synlett* 23: 321–325.

15 Worthy, A.D., Sun, X., and Tan, K.L. (2012). *J. Am. Chem. Soc.* 134: 7321–7324.

16 Sun, X., Worthy, A.D., and Tan, K.L. (2013). *J. Org. Chem.* 78: 10494–10499.

17 Giustra, Z.X. and Tan, K.L. (2013). *Chem. Commun.* 49: 4370–4372.

18 Patel, S.G. and Wiskur, S.L. (2008). *Tetrahedron Lett.* 50: 1164–1166.

19 Sheppard, C.I., Taylor, J.L., and Wiskur, S.L. (2011). *Org. Lett.* 13: 3794–3797.

20 Clark, R.W., Deaton, T.M., Zhang, Y. et al. (2013). *Org. Lett.* 15: 6132–6135.

21 Wang, L., Akhani, R.K., and Wiskur, S.L. (2015). *Org. Lett.* 17: 2408–2411.

22 Akhani, R.K., Clark, R.W., Yuan, L. et al. (2015). *ChemCatChem* 7: 1527–1530.

23 Akhani, R.K., Moore, M.I., Pribyl, J.G., and Wiskur, S.L. (2014). *J. Org. Chem.* 79: 2384–2396.

24 Wang, L., Zhang, T., Redden, B.K. et al. (2016). *J. Org. Chem.* 81: 8187–8193.
25 Yoshimatsu, S., Yamada, A., and Nakata, K. (2018). *J. Org. Chem.* 83: 452–458.
26 Yan, H., Jang, H.B., Lee, J.-W. et al. (2010). *Angew. Chem. Int. Ed.* 49: 8915–8917.
27 Hyodo, K., Gandhi, S., van Gemmeren, M., and List, B. (2015). *Synlett* 26: 1093–1095.
28 Park, S.Y., Lee, J.-W., and Song, C.E. (2015). *Nat. Commun.* 6: 7512.
29 Rendler, S., Auer, G., and Oestreich, M. (2005). *Angew. Chem. Int. Ed.* 44: 7620–7624.
30 Rendler, S., Auer, G., Keller, M., and Oestreich, M. (2006). *Adv. Synth. Catal.* 348: 1171–1182.
31 Rendler, S., Plefka, O., Karatas, B. et al. (2008). *Chem. Eur. J.* 14: 11512–11528.
32 Klare, H.F.T. and Oestreich, M. (2007). *Angew. Chem. Int. Ed.* 46: 9335–9338.
33 Karatas, B., Rendler, S., Fröhlich, R., and Oestreich, M. (2008). *Org. Biomol. Chem.* 6: 1435–1440.
34 Steves, A. and Oestreich, M. (2009). *Org. Biomol. Chem.* 7: 4464–4469.
35 Issenhuth, J., T; Dagorne, S., and Bellemin-Laponnaz, S. (2008). *J. Mol. Catal. A* 286: 6–10.
36 Weickgenannt, A., Mewald, M., Muesmann, T.W.T., and Oestreich, M. (2010). *Angew. Chem. Int. Ed.* 49: 2223–2226.
37 Dong, X., Weickgenannt, A., and Oestreich, M. (2017). *Nat. Commun.* 8: 15547.
38 Dong, X., Kita, Y., and Oestreich, M. (2018). *Angew. Chem. Int. Ed.* 57: 10728–10731.

14

Chiral Silicon Molecules

Kazunobu Igawa and Katsuhiko Tomooka

Kyushu University, Institute for Materials Chemistry and Engineering, 6-1 Kasuga-koen, Kasuga, Fukuoka, 816-8580, Japan

14.1 Introduction

14.1.1 General Background of Chiral Silicon Molecules

Chirality is one of the most important structural properties of a molecule [1]. Therefore, a large number of fundamental and applied studies on chiral molecules have been performed. The main focus of these studies has been chiral molecules having a chiral carbon center, synonymous with "asymmetrically substituted carbon" (hereafter described as "chiral carbon molecules"), which are ubiquitous in nature and synthesized in enantioenriched form from natural sources or by various asymmetric synthetic methods [2]. On the other hand, chiral molecules having a chiral silicon center, synonymous with "asymmetrically substituted silicon" (hereafter described as "chiral silicon molecule"), are also available by synthesis but are nonexistent in nature [3]. Chiral silicon molecules have chiral functionality quite different from chiral carbon molecules owing to their different structural and electronic properties (Figure 14.1) [4]. For example, the bond radii of silicon are larger than those of carbon; thus, the Si—C(sp^3) bond (c. 1.87 Å) is c. 1.2 times longer than the C—C(sp^3) bond (c. 1.54 Å) [5], which results in a unique chiral environment around the silicon atom. Moreover, other properties such as the higher electropositivity of silicon compared to carbon, instability of silicon multiple bonds, stronger Si—O bonds than C—O bonds, and formation of conjugated systems with Si—C σ bond orbitals also result inherent chiral characteristics on chiral silicon molecules.

In the past decades, studies of chiral silicon molecules have gradually expanded. In this chapter, the preparation, transformation, and application of enantioenriched chiral silicon molecules are reviewed with a focus on recent studies.

Organosilicon Chemistry: Novel Approaches and Reactions, First Edition.
Edited by Tamejiro Hiyama and Martin Oestreich

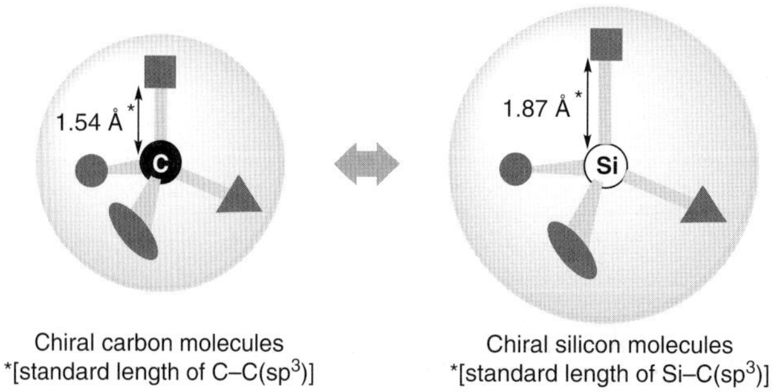

Figure 14.1 Chiral carbon molecules and chiral silicon molecules.

14.1.2 History of Chiral Silicon Molecules

In 1907, Kipping demonstrated that sodium salts of siloxane **1** show optical rotations of $[\alpha]_D$ +3.3° or $[\alpha]_D$ −4.5° when prepared from its salts of *d*- or *l*-methylhydrindamine, respectively (Figure 14.2) [6]. This result is recognized as the first report of a chiral silicon molecule.

Half a century after Kipping's discovery, Sommer et al. synthesized alkoxysilane **2** having a chiral silicon center along with a (−)-menthoxy group and separated the diastereomers, epimeric at the silicon atom, by fractional crystallization (Figure 14.3) [7], which allowed the preparation of a variety of 1-NpPhMeSi-X-type chiral silicon molecules in enantioenriched form. Sommer's group carried out extensive stereochemical studies on their nucleophilic substitution reactions (vide infra) [8].

Around 1970, pioneering works on the asymmetric synthesis of chiral silicon molecules were independently developed by three groups: the Klebe, Corriu, and Kumada groups. From then to now, this research field has received some attention, in particular after the 1990s [9–11]. The recent study of chiral silicon mol-

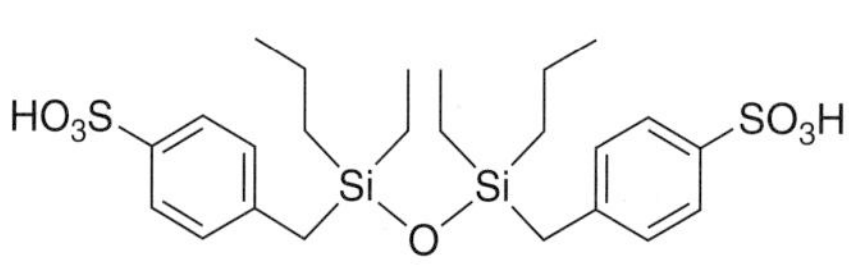

1

Figure 14.2 Kipping's chiral siloxane **1**.

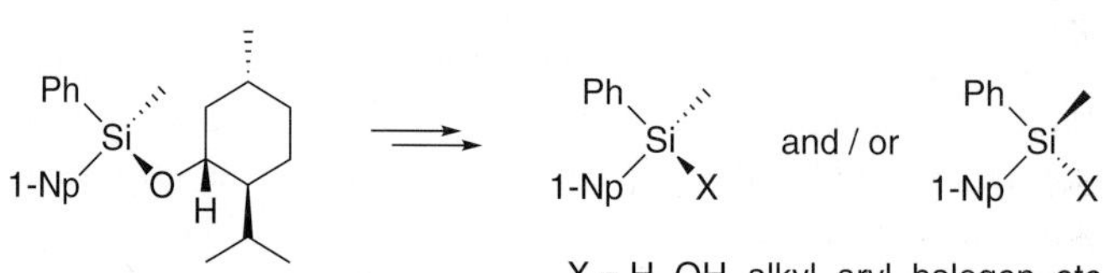

(R_{Si})-**2**

Figure 14.3 Sommer's (−)-menthoxysilane **2**.

ecules was prompted by the advancements in analytical methods for chiral molecules, the separation of enantiomers by high-performance liquid chromatography (HPLC) or gas chromatography (GC) using a chiral stationary phase, the determination of absolute stereochemistry by single-crystal X-ray crystallographic analysis, the development of efficient asymmetric synthetic methods, and the deeper understanding of basic organosilicon chemistry.

14.2 Preparation of Enantioenriched Chiral Silicon Molecules

14.2.1 Classification of Preparation Methods for Enantioenriched Chiral Silicon Molecules

Preparation methods for enantioenriched chiral silicon molecules are classified into separation of stereoisomers, either enantiomers or diastereomers epimeric at the silicon atom (X and/or Y has other chiral element) (Figure 14.4a) and asymmetric synthesis with stereoselective bond formation of molecules having prochiral silicon atoms (hereafter described as prochiral silicon molecules, which includes molecules having other chiral elements, such as chiral carbon center, axial chirality, and planar chirality) (Figure 14.4b) in a way similar to enantioenriched chiral carbon molecules.

However, special consideration is necessary for the asymmetric synthesis of chiral silicon molecules because of the significant differences between the fundamental characteristics of silicon and carbon. In the case of chiral carbon molecules, approaches based on the desymmetrization of sp^2-hybidized S_1 symmetric carbon atoms as in alkenes, aldehydes, ketones, and imines (Eq. (14.1a)) as well as the desymmetrization of sp^3-hybidized S_1 symmetric carbon atoms (Eq. (14.1b)) are efficient and well developed. In sharp contrast, the desymmetrization of sp^2-hybridized silicon atoms is unsuitable for the stereochemical control

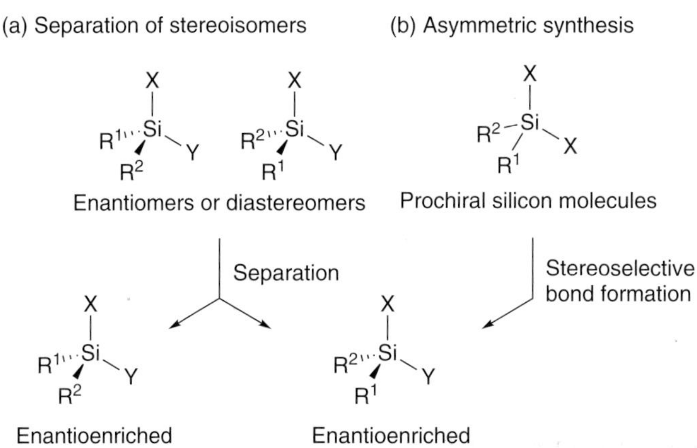

Figure 14.4 Classification of preparation methods for enantioenriched chiral silicon molecules. (a) Separation of stereoisomers and (b) asymmetric synthesis.

of asymmetry at silicon because of the difficulty in handling labile sp^2 silicon molecules as substrates (Eq. (14.2a)). In contrast, molecules containing sp^3-hybridized silicon atoms are reasonably stable and reactive, thus the desymmetrization of sp^3-hybidized S_1 symmetric silicon atoms is the most appropriate approach for the stereochemical control of chirality at silicon (Eq. (14.2b)). In this section, typical examples of the asymmetric synthesis and separation of stereoisomers of chiral silicon molecules are described.

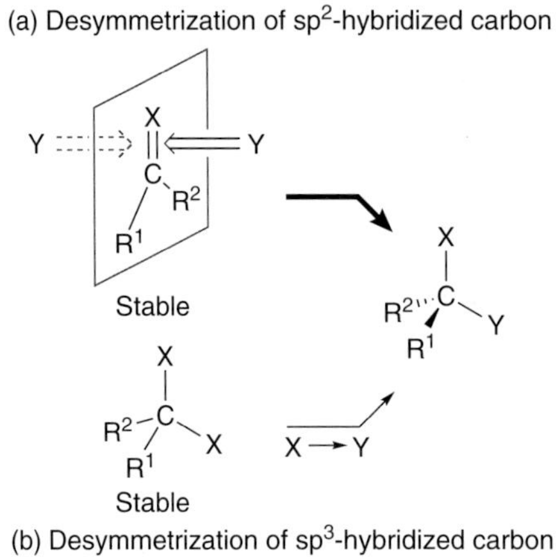

(14.1)

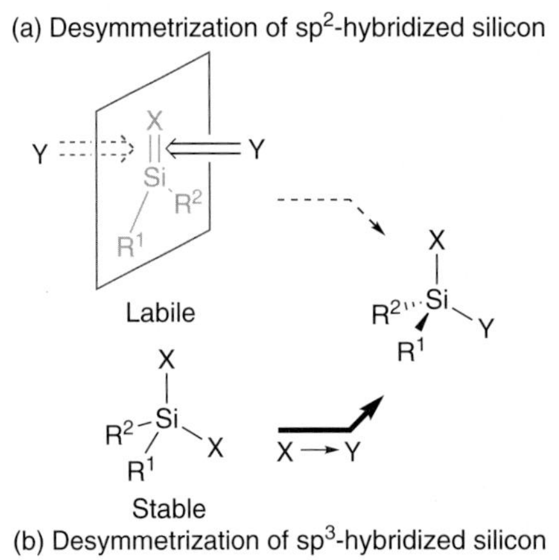

(14.2)

14.2.2 Separation of Stereoisomers of Chiral Silicon Molecules

14.2.2.1 Classification of Separation Methods for Stereoisomers of Chiral Silicon Molecules

Direct separation of a chiral silicon molecules **A** and *ent*-**A** was difficult several decades ago before the development of high-performance chiral stationary-phase or kinetic resolution methods (Eq. (14.3)). Therefore, a detour route involving the conversion of **A** and *ent*-**A** into diastereomers **B** and epi_{Si}-**B** epimeric at the silicon atom by the introduction of an enantioenriched other chiral element such as chiral carbon center was developed (Eq. (14.4)), for example,

Sommer's (−)-menthoxysilane **2** [7]. **B** and *epi*$_{Si}$-**B** can be separated by fractional crystallization and/or silica gel column chromatography, and they can be converted into "chiral-carbon-free" enantioenriched chiral silicon molecules by suitable transformation reactions. As mentioned, with the recent development of chiral technologies, **A** and *ent*-**A** have been directly separated by HPLC using chiral stationary phase [12], as well as kinetic resolution by asymmetric reactions. These methods are discussed in Sections 14.2.2.2 and 14.2.2.3.

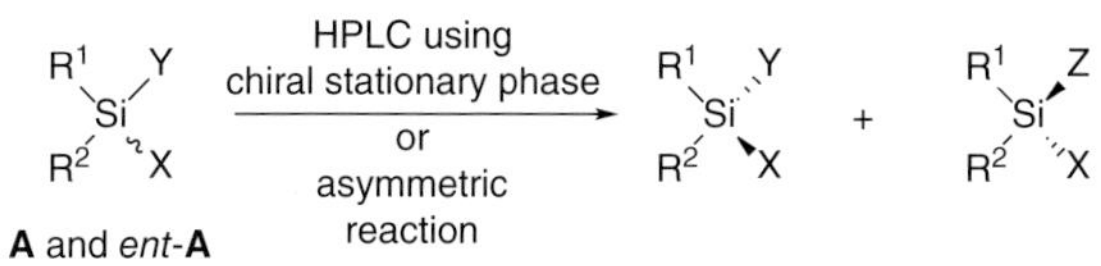

(14.3)

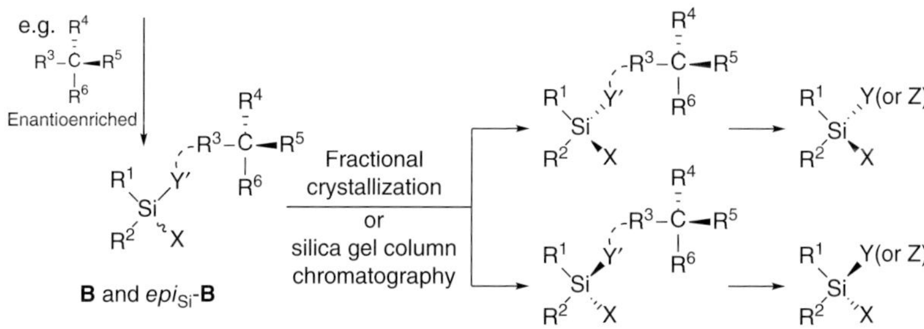

(14.4)

14.2.2.2 Separation of Silicon Epimers of Chiral Silicon Molecules

Silicon epimers are prepared by the introduction of other chiral element as chiral auxiliary or chiral resolving agent to a racemic mixture of chiral silicon molecules. Tacke and coworkers obtained almost enantiopure 2-aminoethylsilane (S_{Si})-**3a–c** by fractional crystallization after salt formation with tartaric acid derivative (*S*,*S*)-**4**, followed by treatment with aq. NaOH (Eq. (14.5)) [13]. On the other hand, (R_{Si})-**3a–c** were obtained by fractional crystallization of the salt with (*R*,*R*)-**4**.

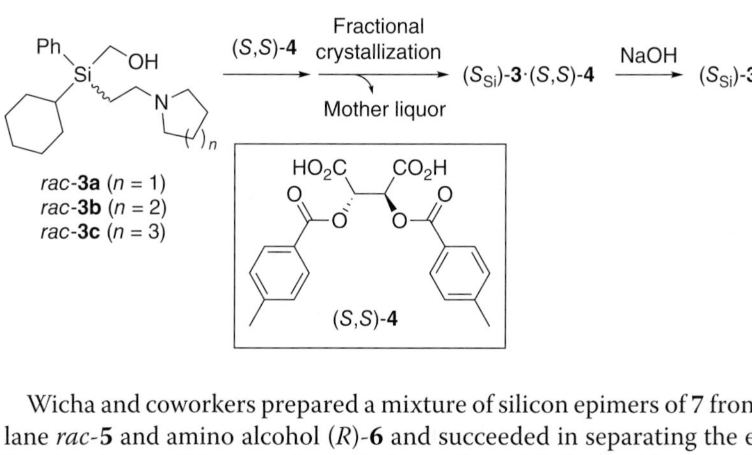

(14.5)

Wicha and coworkers prepared a mixture of silicon epimers of **7** from chlorosilane *rac*-**5** and amino alcohol (*R*)-**6** and succeeded in separating the epimers by fractional crystallization after salt formation with hydrochloric acid to afford crystals of (R_{Si},R)-7·HCl (Eq. (14.6)) [14]. Thus, they obtained crystalline (R_{Si},R)-7·HCl and (S_{Si},R)-7 from the mother liquor, which were converted into enantioenriched hydrosilanes (S_{Si})-**8** and (R_{Si})-**8**, respectively, by reduction with $LiAlH_4$.

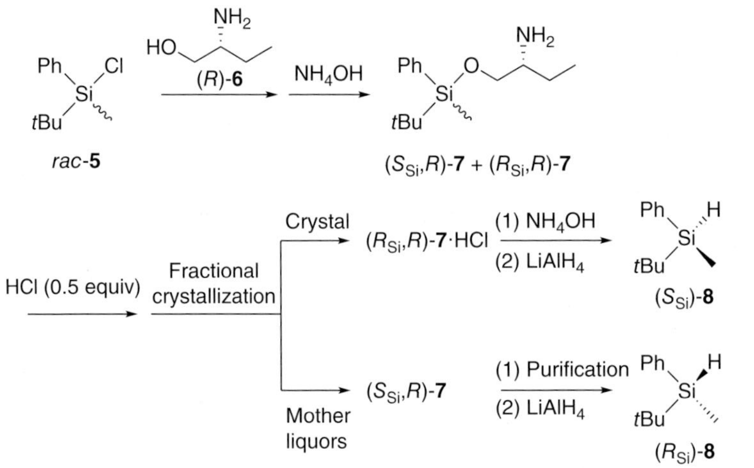

(14.6)

Similarly, the silicon epimers of a variety of chiral compounds **9–12** can be separated by silica gel column chromatography (Table 14.1) [15–18]. The thus-obtained enantioenriched α-hydroxyethylsilane **9** can be converted into acylsilane **13** (Table 14.1, entry 1). Furthermore, enantioenriched alkoxysilanes **10–12** can be converted into enantioenriched hydrosilanes **14–16** by reduction with $LiAlH_4$ or diisobutylaluminum hydride (DIBAL) (Table 14.1, entries 2–4).

In contrast, Tamao, Kawachi, and coworkers reported the dynamic kinetic resolution of silicon epimers of fluorosilane **17** in the presence of a catalytic amount of AgF with the expectation of the epimerization of the chiral silicon center via silicate **18**, which has a prochiral silicon atom (Eq. (14.7)) [19]. As a result, (R_{Si})-**17** was selectively precipitated and isolated in 68% yield with 98% dp (diastereomeric purity: the mole fraction of major diastereomer). Klebe and Finkbeiner and Corriu and Lanneau also reported the dynamic kinetic resolution of chiral chlorosilane and chiral silaoxazolidinone, respectively [9, 20].

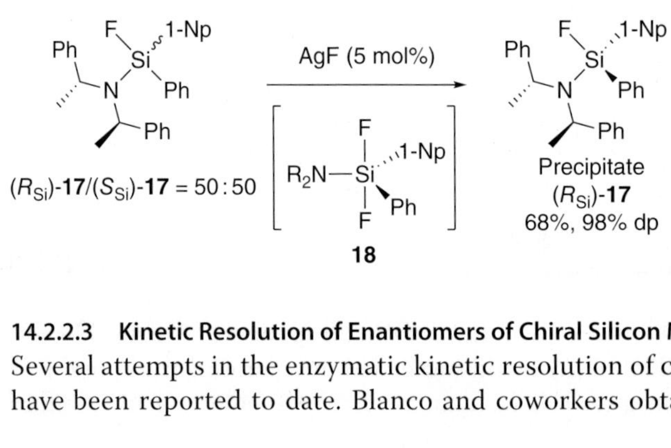

(14.7)

14.2.2.3 Kinetic Resolution of Enantiomers of Chiral Silicon Molecules

Several attempts in the enzymatic kinetic resolution of chiral silicon molecules have been reported to date. Blanco and coworkers obtained enantioenriched

Table 14.1 Separable silicon epimers with silica gel column chromatography.

Entry	Silicon epimer	Reagents for transformation after optical resolution	Obtained chiral silicon molecules
1	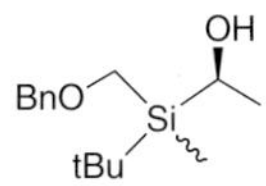 **9**	$(COCl)_2$, DMSO Et_3N	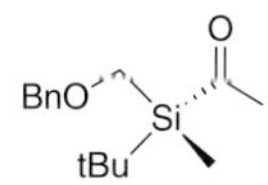 (R_{Si})-**13** and (S_{Si})-**13** (up to 92±2% ee)[a]
2	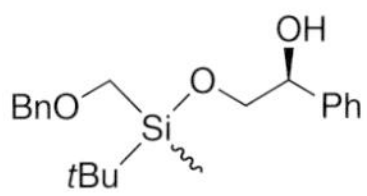 **10**	$LiAlH_4$	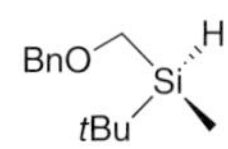 (S_{Si})-**14** and (R_{Si})-**14** (94% ee)
3	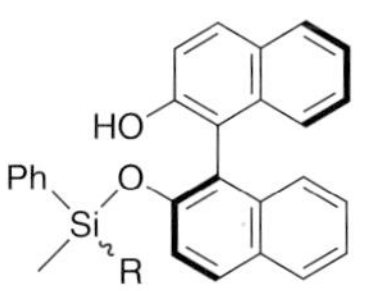 **11a** (R = *t*Bu) **11b** (R = *i*Pr)	$LiAlH_4$	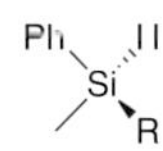 (R_{Si})-**15a** and (S_{Si})-**15a** (R = *t*Bu, >98% ee) (R_{Si})-**15b** and (S_{Si})-**15b** (R = *i*Pr, >98% ee)
4	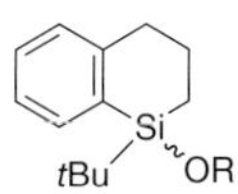 **12a** [R = (−)-Men] **12b** [R = (*S*)-phenethyl]	DIBAL	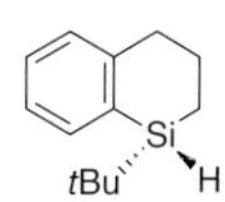 (S_{Si})-**16** and (R_{Si})-**16** (up to 99% ee)

ee = enantiomeric excess; DMSO = dimethyl sulfoxide.
a) Enantiopurity of **9** before oxidation.

hydroxymethylsilane **19** (15% ee) along with enantioenriched ester **21** (20% ee) from the *rac*-**19** by lipase-promoted esterification with oxime ester **20** (Eq. (14.8)) [21].

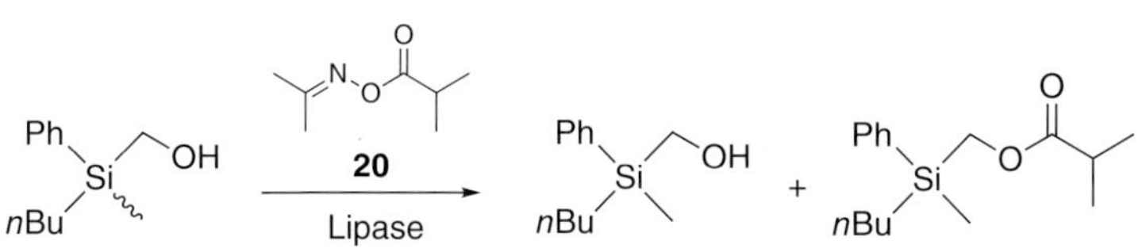

rac-**19** — **19** 15% ee — **21** 40%, 20% ee (14.8)

Tacke and Heinrich obtained enantioenriched ester (R_{Si})-**24** in 68% ee from *rac*-**22** by papain-promoted esterification with carboxylic acid **23** (Eq. (14.9)) [22]. Moreover, hydrolysis of enantioenriched (R_{Si})-**24** (68% ee) and another round of papain-promoted esterification of the resulting (R_{Si})-**22** afforded enantiopure (R_{Si})-**24**.

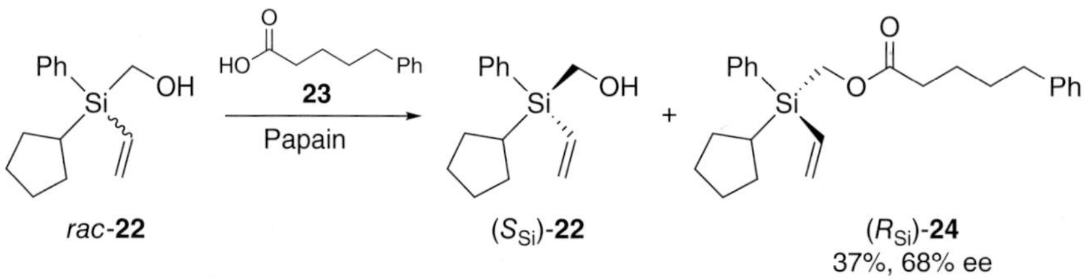

(14.9)

Yamamoto and coworkers attempted kinetic resolution of 1-cyclohexenylsilanol **25** by Katsuki–Sharpless epoxidation and found that the reaction using cyclododecyl tartrate (*S*,*S*)-**26** as a chiral ligand afforded enantiopure (R_{Si})-**25** along with epoxidation products (S_{Si},*S*,*S*)-**27** [(S_{Si},*S*,*S*)-**27**/(S_{Si},*R*,*R*)-**27** = 95 : 5] (Eq. (14.10)) [23].

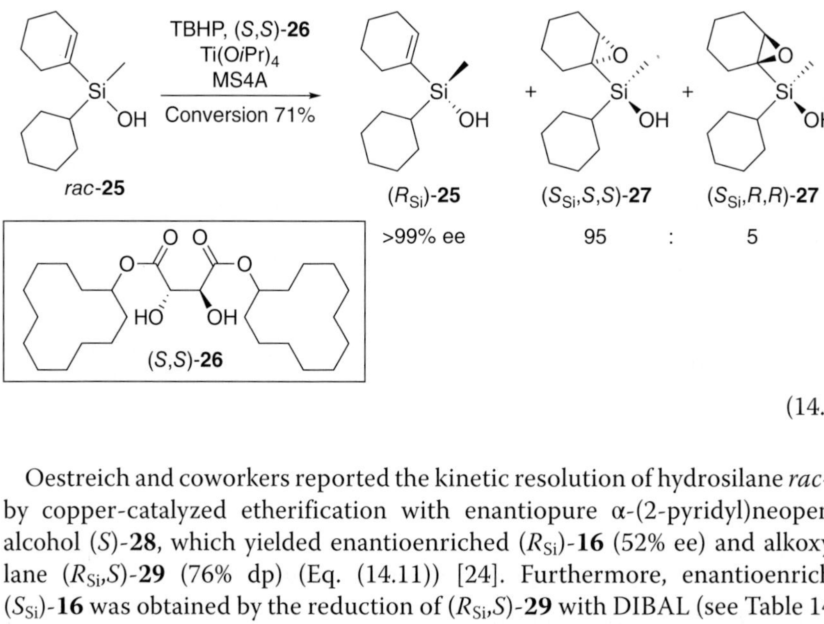

(14.10)

Oestreich and coworkers reported the kinetic resolution of hydrosilane *rac*-**16** by copper-catalyzed etherification with enantiopure α-(2-pyridyl)neopentyl alcohol (*S*)-**28**, which yielded enantioenriched (R_{Si})-**16** (52% ee) and alkoxysilane (R_{Si},*S*)-**29** (76% dp) (Eq. (14.11)) [24]. Furthermore, enantioenriched (S_{Si})-**16** was obtained by the reduction of (R_{Si},*S*)-**29** with DIBAL (see Table 14.1, entry 4).

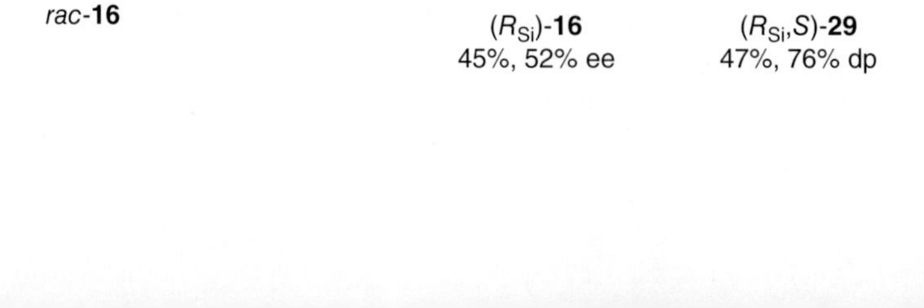

(14.11)

14.2.3 Asymmetric Synthesis of Chiral Silicon Molecules

14.2.3.1 Classification of Asymmetric Synthetic Methods for Chiral Silicon Molecules

As mentioned, all of the reported asymmetric synthetic methods are based on the desymmetrization of sp^3-hybridized S_1 symmetric silicon atoms. They can be classified into four types of reactions (Eq. (14.12)). Three of the four methods are transformation with Si—X bond cleavage. These can be divided by the difference in the X substituent: heteroatom substituents (X = NR_2, OR, Cl), hydrogen (X = H), and carbon substituents (X = R_C). The fourth method involves transformation reaction without Si—X bond cleavage, that is transformation on the carbon substituent R_C. These four methods are discussed in Sections 14.2.3.2–14.2.3.5.

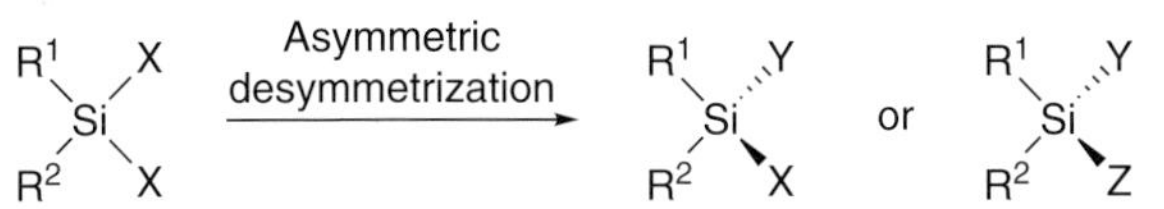

Section 14.2.3.2: X = NR_2, OR, Cl
Section 14.2.3.3: X = H
Section 14.2.3.4: X = R_C
} with Si–X bond cleavage
Section 14.2.3.5: X = R_C without Si–X bond cleavage (14.12)

14.2.3.2 Desymmetrization of Prochiral Silicon Atoms by Substitution of a Heteroatom Substituent

The asymmetric substitution of one of two amino or alkoxy groups or chlorine atoms at the silicon atom has been investigated (Eq. (14.13)). The stereochemistry of the reactions is controlled by asymmetric induction from a chiral reagent or stereodirecting group attached to the prochiral silicon substrate (Table 14.2).

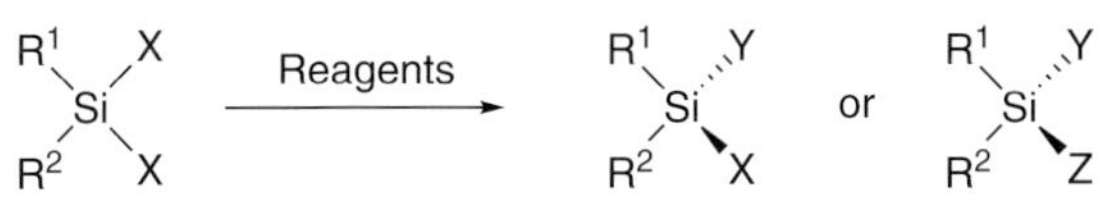

Y = NR_2, OR, Cl (14.13)

14.2.3.2.1 Diastereoselective Approach

Asymmetric Induction from Chiral Nucleophile Pioneering work on the diastereoselective substitution reactions of prochiral silicon molecules with chiral nucleophile was reported by Klebe and Finkbeiner in 1966 [9]. The substitution reaction of achiral diamidosilane **30** and enantiopure *N*-phenyl phenylalanine (**31**) afforded a diastereomeric mixture of silaoxazolidinone **32** as a kinetic product (Table 14.2, entry 1). However, one diastereomer of **32** was produced by thermodynamic control during the recrystallization process to provide **32** in >99% dp. The thus-obtained enantiopure **32** can be converted into enantioenriched dialkoxysilane **33** by sequential substitution of the *N*-phenyl phenylalanine ligand with naphth-1-ol and methanol (Eq. (14.14)).

Table 14.2 Desymmetrization with substitution of heteroatom substituent.

Entry	Substrate	Reagents	Product	Yield stereoselectivity
Diastereoselective approach				
Asymmetric induction from chiral nucleophile				
1	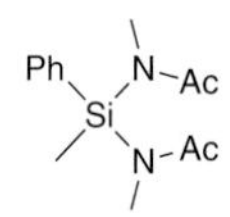 **30**	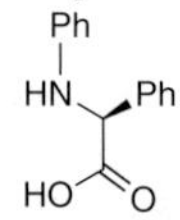 **31**	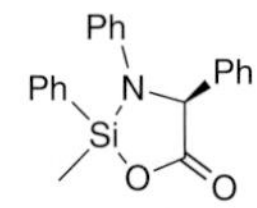 **32**	92% >99% dp
2	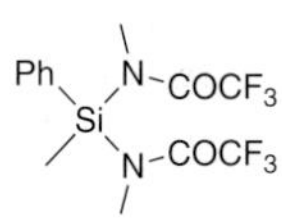 **34**	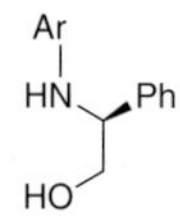 **35**	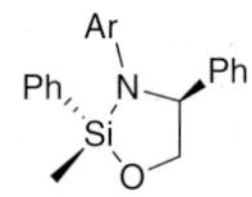 **36**	90%[a)] 78% dp[a)]
3	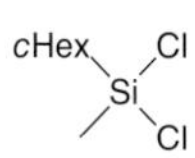 **37**	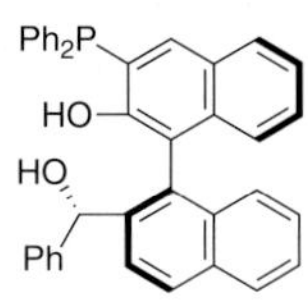 **38** + 2,6-lutidine	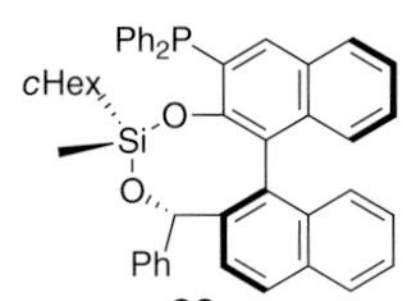 **39**	92% >99% dp
Asymmetric induction from chirality of substituent(s) on silicon				
4	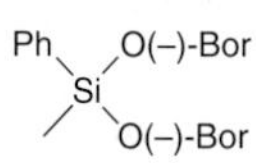 **40**	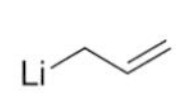	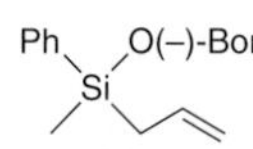 **41**	20% 60% dp
5	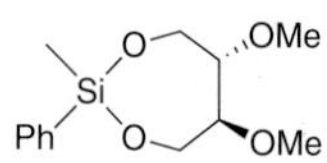 **42**	*n*PrMgBr then BSTFA	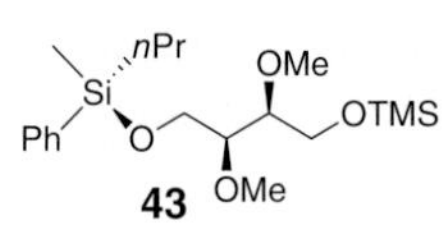 **43**	91% dp
6	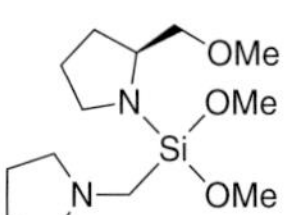 **44**	*t*BuLi	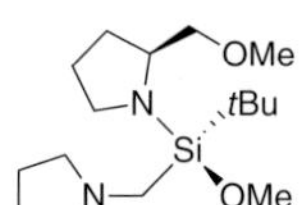 **45**	80% >99% dp
7	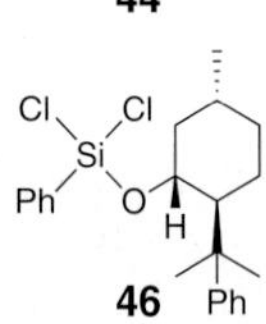 **46**	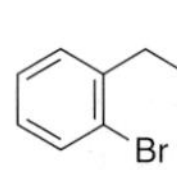 + Mg **47**	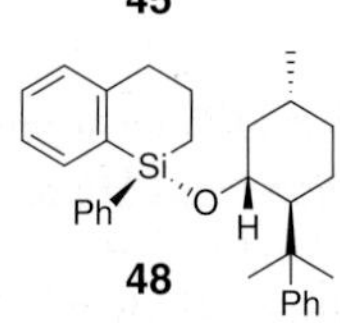 **48**	37% 60% dp
Enantioselective approach				
8	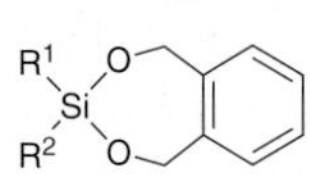 **49**	R^3Li + 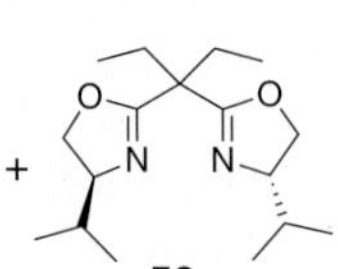 **50**	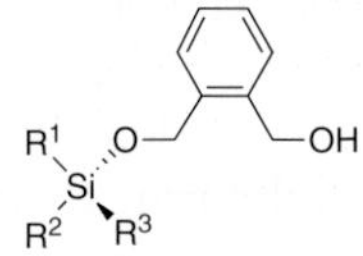 **51**	92%[b)] 84% ee[b)]

a) Ar = 4-MeOC_6H_4.

b) R^1 = Me, R^2 = *c*Hex, R^3 = *t*Bu. (–)-Bor = 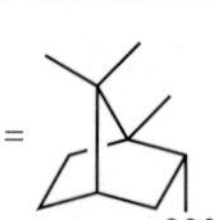 BSTFA = *N*,*O*-bis(trimethylsilyl) trifluoroacetamide.

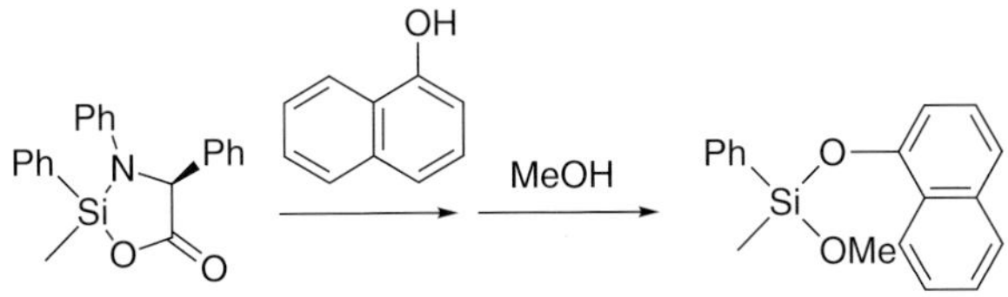

32
>98% dp

33
Enantioenriched

(14.14)

Wada, Oka and coworkers found that a substitution of achiral diamidosilane **34** and amino alcohol **35** proceeds with kinetic control to afford silaoxazolidine **36** with high diastereomeric excess (**36a** [Ar = 4-$MeOC_6H_4$] in 89% dp) (Table 14.2, entry 2) [25]. In contrast to Klebe's study of **32** [9], the dp of **36** was decreased to 60% at 90 °C in toluene due to an unfavorable thermodynamic equilibrium. The thus-obtained silane **36** was converted into hydrosilanes **15** almost without loss of enantiopurity by treatment with Grignard reagents, followed by Boc protection of the resulting amine and subsequent DIBAL reduction (Eq. (14.15)).

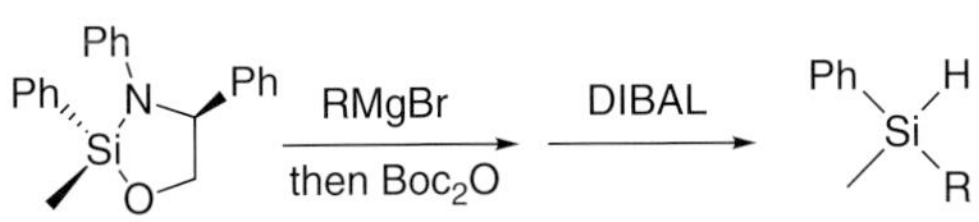

(R_{Si},*S*)-**36**
>99% dp

15
Up to 96% ee

(14.15)

On the other hand, Xu and coworkers obtained dialkoxysilane **39** in a highly diastereoselective manner from dichlorosilane **37** and chiral binaphthyl diol **38** (Table 14.2, entry 3) [26].

Asymmetric Induction at Silicon by Chiral Substituent(s) Kawakami and coworkers reported that the reaction of prochiral silicon molecule **40** having two enantiopure 2-bornyloxy groups [(−)-BorO] as chiral leaving groups (LGs) with allyllithium proceeds with moderate diastereoselectivity to afford **41** in 60% dp (Table 14.2, entry 4) [27]. On the other hand, Masuda and coworkers designed seven-membered cyclic dialkoxysilane **42** prepared from dichlorosilane and C_2-symmetric diol as a chiral substrate for the substitution reaction at the prochiral silicon atom and its reaction with *n*PrMgBr afforded monoalkoxysilanes **43** in 91% dp (Table 14.2, entry 5) [28].

Bauer and Strohmann achieved a substitution reaction with dimethoxysilane **44**, which has a proline-derived chiral stereodirecting group, with *t*BuLi which provided monoalkoxysilane **45** as a single stereoisomer (Table 14.2, entry 6) [29]. Oestreich and coworkers found that the reaction of dichlorosilane **46**, which has a (−)-menthoxy group as a chiral stereodirecting group, with a dianionic nucleophile prepared from dibromide **47** and magnesium by a Barbier-type reaction provided cyclic monoalkoxysilane **48** in 60% dp (Table 14.2, entry 7) [18].

14.2.3.2.2 Enantioselective Approach

Tomooka, Igawa and coworkers developed an enantioselective nucleophilic substitution reaction of achiral dialkoxysilane **49** with alkyllithium in the presence of chiral coordinating agent (*S*,*S*)-**50**, which provided enantioenriched monoalkoxysilane **51** (up to 84% ee) (Table 14.2, entry 8) [30]. This is the first example of the stereoselective synthesis of a chiral silicon molecule via enantioselective Si—C bond formation. The thus-obtained alkoxysilane **51** can be converted into enantioenriched silanol **52** by Birch reduction without loss of enantiopurity (Eq. (14.16)).

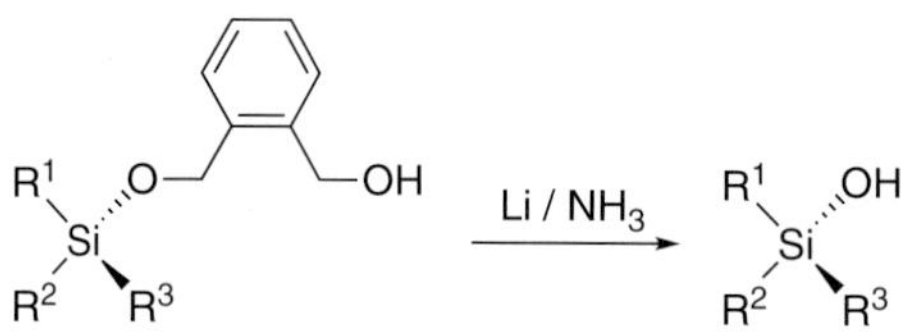

51 **52** (14.16)

14.2.3.3 Desymmetrization of Dihydrosilane

Hydrosilanes can be converted into a variety of functionalized silicon molecules via Si—H bond cleavage and Si—X bond formation (e.g. X = OR, NR_2, or R_C) using transition-metal-catalyzed reactions; thus, the asymmetric desymmetrization of achiral dihydrosilane **53** is a reasonable approach to obtain enantioenriched chiral silicon molecules (Eq. (14.17)). In this section, the reactions are categorized by the type of bond formation: Si—O/Si—N bond formation or Si—C bond formation via a Si—H bond cleavage.

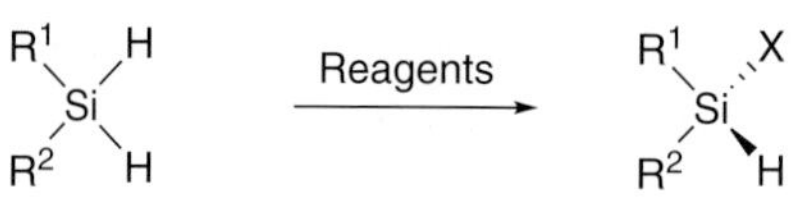

53 (14.17)

14.2.3.3.1 Stereoselective Si—O or Si—N Bond Formation

Asymmetric Induction by a Chiral Catalyst Enantioselective dehydrogenative etherification of dihydrosilane **53a** and an achiral alcohol was developed by Corriu and Moreau using a chiral rhodium catalyst prepared in situ from $[RhCl(coe)]_2$ (coe = cyclooctene) and (*R*,*R*)-**55** (Figure 14.5) [10]. In the case of the reaction with alcohol **54**, alkoxyhydrosilane **56a** was obtained in 19% ee (Table 14.3, entry 1). On the other hand, Kumada, Hayashi and Yamamoto reported the asymmetric desymmetrization of **53a** by the hydrosilylation of benzophenone (**57**), which was catalyzed by (R_P)-**58** (Figure 14.5) to afford alkoxysilane (S_{Si})-**56a** in 28% ee (Table 14.3, entry 2) [11]. Guan and Li developed dehydrogenative amination of dihydrosilane **53a** with benzylamine (**59**) promoted by a chiral yttrium catalyst (*R*,*R*)-**60** (Figure 14.5) to afford aminohydrosilane **61** in 23% ee (Table 14.3, entry 3) [31].

Double Asymmetric Induction from a Chiral Alcohol and a Chiral Catalyst To improve the stereoselectivity of Si—O bond formation of dihydrosilanes, Corriu and Moreau used the double asymmetric induction approach for the dehydrogenative etherification of dihydrosilane **53a** with chiral alcohols such as (−)-menthol and a chiral rhodium catalyst prepared from (*R*,*R*)-**55** and [RhCl (coe)]$_2$ to afford

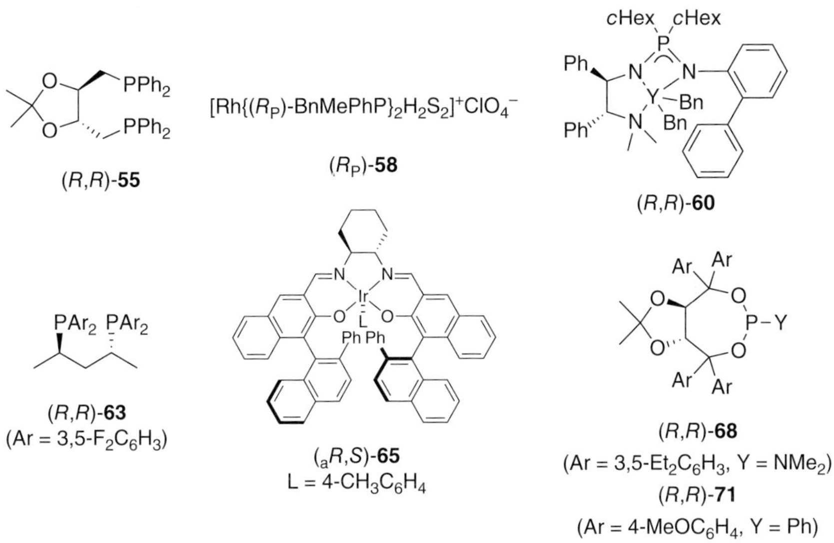

Figure 14.5 Chiral reagents in Table 14.3.

alkoxyhydrosilane **56b** in 74% dp (Table 14.3, entry 4) [10]. Furthermore, Leighton and coworkers developed the dehydrogenative etherification of dihydrosilane **53b** with double asymmetric induction from homopropargylic alcohol (*S*)-**62** and a chiral copper catalyst prepared from (*R*,*R*)-**63** (Figure 14.5), CuCl and NaO*t*Bu to afford alkoxyhydrosilane (R_{Si})-**56c** in 97% dp (Table 14.3, entry 5) [32].

14.2.3.3.2 Enantioselective Si—C Bond Formation

Katsuki and coworkers succeeded in enantioselective carbene insertion into the Si—H bond of dihydrosilanes **53c** promoted by a chiral salen iridium catalyst ($_{a}R$,*S*)-**65** (Figure 14.5) with α-diazoester **64** to afford hydrosilane **66** in a highly diastereo- and enantioselective manner (up to 99% dp, 99% ee) (Table 14.3, entry 6) [33]. Nishihara, Yamanoi and coworkers developed enantioselective arylation of dihydrosilanes with aryliodide promoted by a chiral catalyst prepared from chiral phosphoramide (*R*,*R*)-**68** (Figure 14.5) and $Pd_2(dba)_3$ (dba = dibenzylideneacetone), which afforded arylsilane **69** in 76% ee from **53d** and **67** (Table 14.3, entry 7) [34]. Tomooka, Igawa and coworkers developed the asymmetric desymmetrization of dihydrosilane by the hydrosilylation of an alkyne promoted by a chiral platinum catalyst prepared from chiral phosphonite (*R*,*R*)-**71** (Figure 14.5) and $Pt(dba)_3$, which afforded alkenylhydrosilane (S_{Si})-**72** in 82% ee from dihydrosilane **53e** with hex-3-yne (**70a**: R = Et) (Table 14.3, entry 8) [35a]. A similar enantioselective hydrosilylation of alkyne with a chiral cobalt catalyst was reported by Huang and coworkers [35b].

14.2.3.4 Desymmetrization of Prochiral Silicon Atoms by Enantioselective Substitution of a Carbon Substituent

Enantioselective transformations of achiral silicon substrates via Si—C bond cleavage have also been developed (Eq. (14.18)) (Table 14.4).

Table 14.3 Desymmetrization of dihydrosilanes.

Entry	Substrates		Reagents	Product	Yield stereoselectivity
Stereooselective Si—O or Si—N bond formation					
Asymmetric induction from chiral catalyst					
1	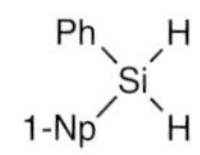 **53a**	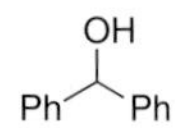 **54**	(−)-(*R*,*R*)-**55** $[RhCl(coe)]_2$	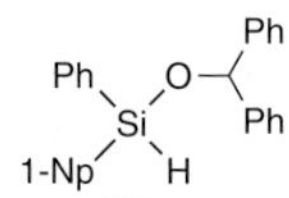 **56a**	– 19% ee
2	**53a**	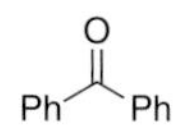 **57**	(R_P)-**58** (79% ee)	(S_{Si})-**56a**	– 28% ee
3	**53a**	$H_2N{-}Bn$ **59**	(*R*,*R*)-**60**	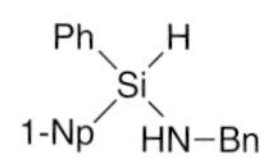 **61**	71%[a] 23% ee[a]
Double asymmetric induction from chiral alcohol and chiral catalyst					
4	**53a**	(−)-menthol	(*R*,*R*)-**55** $[RhCl(coe)]_2$	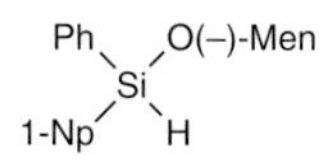 **56b**	– 48% dp
5	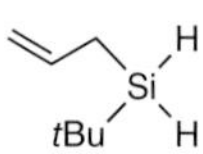 **53b**	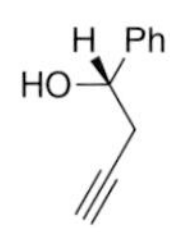 (*S*)-**62**	(*R*,*R*)-**63** CuCl NaO*t*Bu	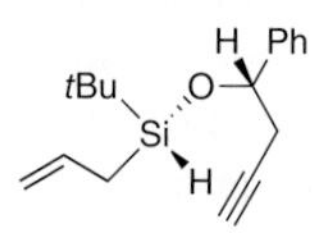 (R_{Si})-**56c**	54% 97% dp
Enantioselective Si—C bond formation					
6	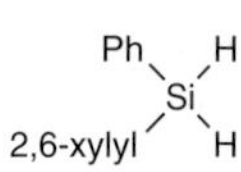 **53c**	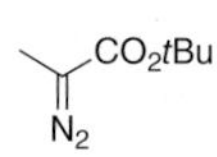 **64**	($_aR$,*S*)-**65**	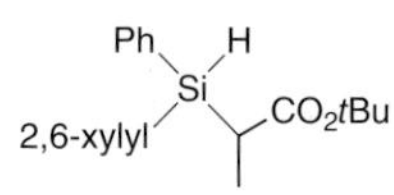 **66**	69% 99% dp 99% ee
7	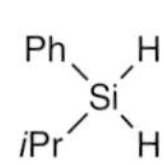 **53d**	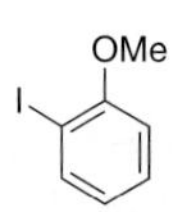 **67**	(*R*,*R*)-**68** $Pd_2(dba)_3$	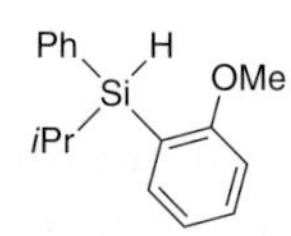 **69**	73% 76% ee
8	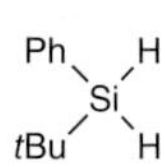 **53e**	R—≡—R **70**	(*R*,*R*)-**71** $Pt(dba)_3$	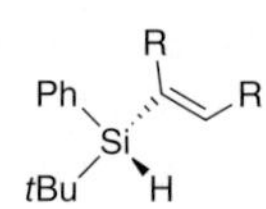 (S_{Si})-**72**	47%[b,c] 81%[b,d] 82% ee[b]

a) Determined after cHex$_2$BCl treatment.
b) R = Et.
c) Isolated yield.
d) Yield based on recovered starting material.

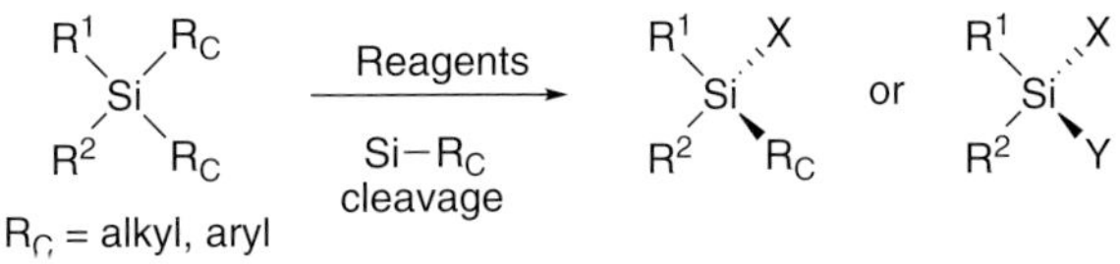

(14.18)

14.2.3.4.1 Diastereoselective Approach

Tomooka, Nakazaki and a coworker synthesized chiral silicon dialkoxysilane **74** from diphenylsilane **73** and benzyl alcohol via montmorillonite K10- and MS

Table 14.4 Desymmetrization with substitution of a carbon substituent.

Entry	Substrates	Reagents conditions	Product	Yield stereoselectivity
Diastereoselective approach				
1	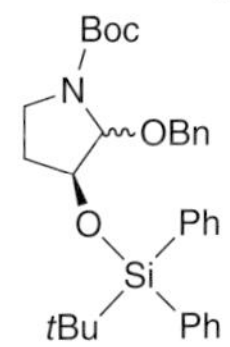 **73** BnOH	K10 MS4A	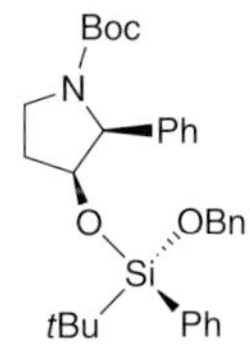 **74**	51% >95% dp
2	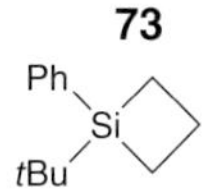 **76** 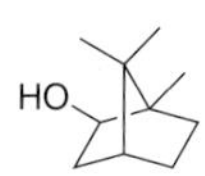	UV light (254 nm)	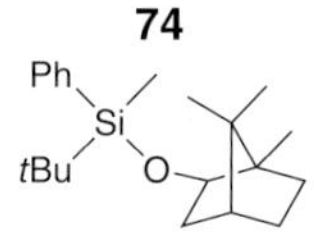 **77**	70% 65% dp
Enantioselective approach				
Si—C bond cleavage of silacyclobutane				
3	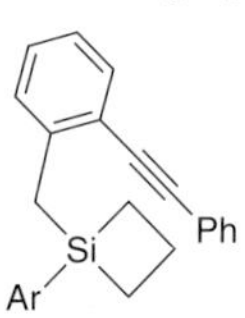 **79** –	($_a$S,S,S)-**80** PdCp(η^3-C_3H_5)	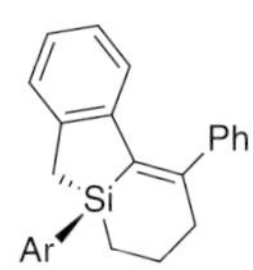 **81**	91%[a)] 95% ee[a)]
4	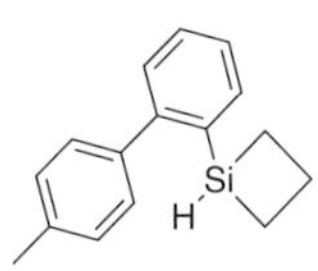 **82** 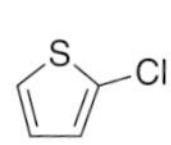 **83**	($_a$R)-**84** Rh(cod)Cl	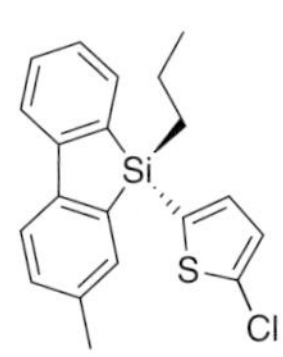 **85**	60% 86% ee
Si—Ar bond cleavage				
5	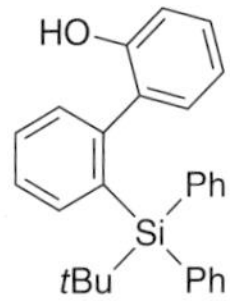 **86** –	(S,S)-**87** $[Rh(OH)(coe)_2]_2$	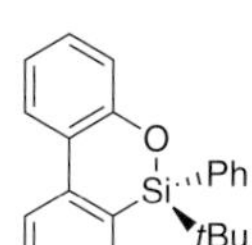 (R_{Si})-**88**	87% 91% ee

(Continued)

Table 14.4 (Continued)

Entry	Substrates		Reagents conditions	Product	Yield stereoselectivity
6	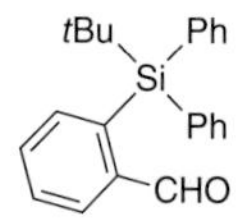 **89**	–	(R,R)-**90** Ni(cod)$_2$ NaOtBu	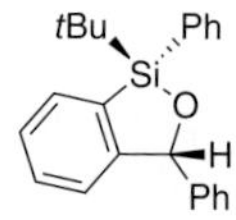 (S_{Si})-**91**	99% 95% ee
Si–alkyny bond cleavage					
7	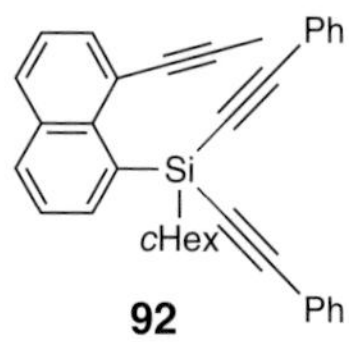 **92**	–	($_a$$R$,$S$,$S$)-**93** $[RhCl(C_2H_4)_2]_2$ $NaBAr^F{}_4$	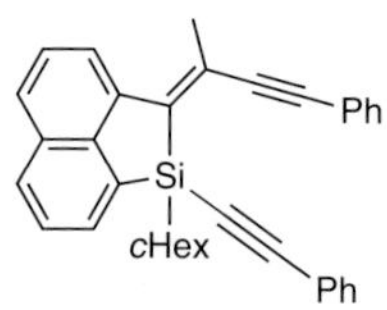 (+)-**94**	62% 94% ee
8	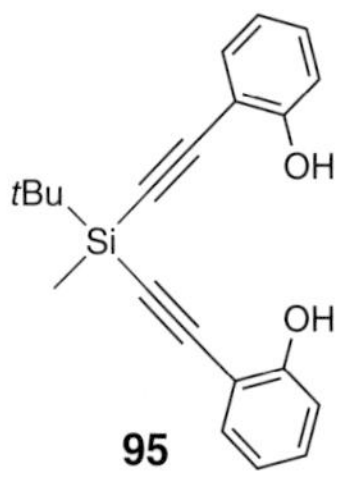 **95**	–	($_a$$S$)-**96** $[Rh(cod)_2]BF_4$	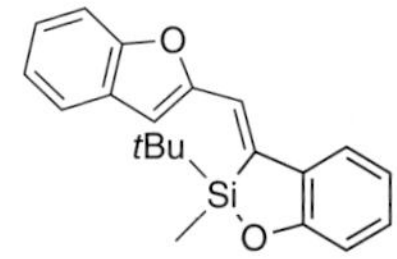 **97**	70% 65% ee
Si–Me bond cleavage					
9	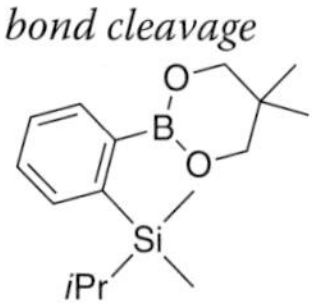 **100**	 **101**	(S_P,S_P)-**102** $[RhCl(C_2H_4)_2]_2$ Na_2CO_3	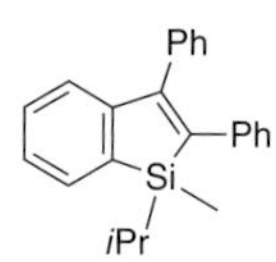 **103**	46% 98% ee

Ar^F = 3,5-$(CF_3)_2C_6H_3$.
a) Ar = 3,5-$(MeO)_2C_6H_3$.

4A-promoted 1,4-phenyl migration from silicon to carbon via Si—C bond cleavage and Si—O bond formation with high diastereoselectivity (>95% dp) (Table 14.4, entry 1) [36]. The thus-obtained dialkoxysilane **74** can be converted into chiral alkoxysilanol **75** via β-elimination of the siloxy moiety almost without loss of enantiopurity (Eq. (14.19)).

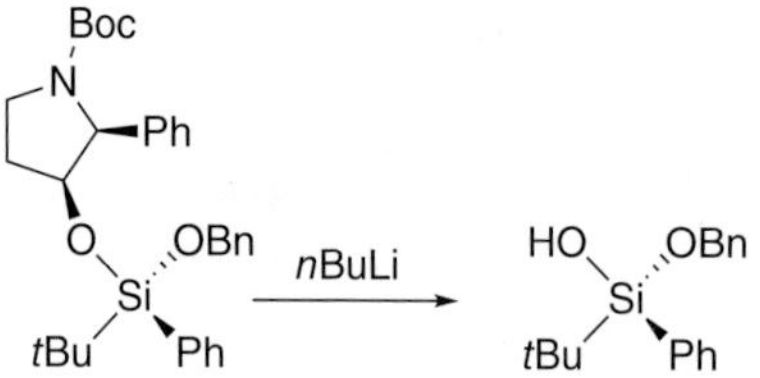

74 >95% dp → **75** 87%, 95% ee (14.19)

Bertrand and coworkers found that the photochemical reaction of achiral silacyclobutane **76** with (−)-isoborneol provided alkoxysilane **77** in 65% dp (Table 14.4, entry 2) [37]. In this reaction, silene **78** was proposed as the reactive species (Eq. (14.20)).

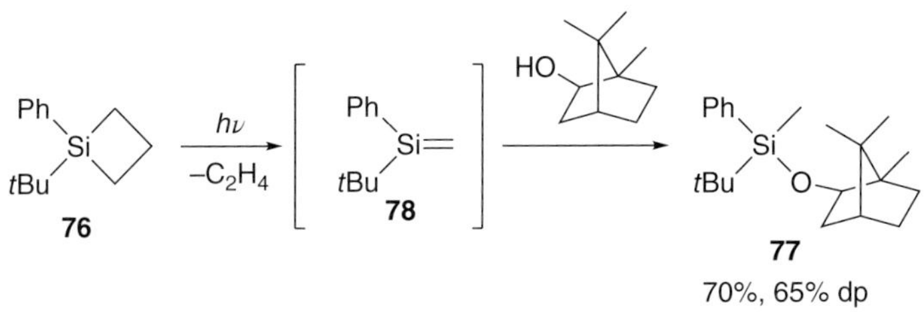

(14.20)

14.2.3.4.2 *Enantioselective Approach*

Si—C Bond Cleavage of Silacyclobutane In addition to photochemically induced Si—C bond cleavage of silacyclobutanes, transition-metal-catalyzed bond cleavage has been developed. The groups of Hayashi and He discovered an enantioselective transformation of silacyclobutanes using chiral transition-metal catalysts. Hayashi, Shintani and a coworker reported enantioselective alkyne insertion into the Si—C bond of silacyclobutane **79a** [Ar = 3,5-$(MeO)_2C_6H_3$] promoted by a chiral palladium catalyst prepared from ($_a$*S*,*S*,*S*)-**80** (Figure 14.6) and PdCp(η^3-C_3H_5) to afford tricyclic chiral silicon molecule **81a** [Ar = 3,5-$(MeO)_2C_6H_3$] in 95% ee (Table 14.4, entry 3) [38]. He and coworkers later developed an enantioselective ring-opening reaction of silacyclobutane **82** with 2-chlorothiophene (**83**) promoted by a chiral rhodium catalyst prepared from ($_a$*R*)-**84** (Figure 14.6) and Rh(cod)Cl (cod = cycloocta-1,5-diene) to afford chiral dibenzosilole **85** in 86% ee (Table 14.4, entry 4) [39].

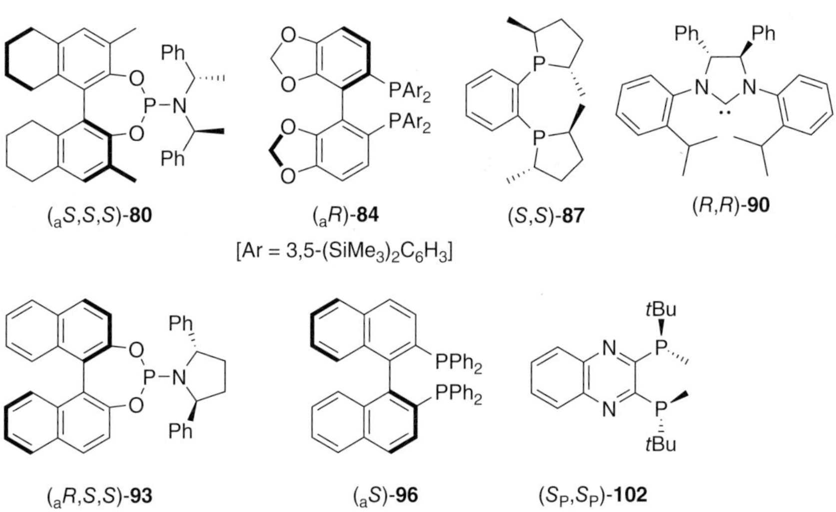

Figure 14.6 Chiral ligands in Table 14.4.

Si—Ar Bond Cleavage Hayashi, Shintani and coworkers developed an enantioselective substitution reaction of the aryl group on achiral diarylsilane **86** with a chiral rhodium catalyst prepared from (*S*,*S*)-**87** (Figure 14.6) and [Rh(OH)(coe)$_2$]$_2$ to afford chiral dibenzooxazoline **88** in 91% ee via intramolecular silicate formation with a hydroxy group (Table 14.4, entry 5) [40]. Ogoshi, Hoshimoto and coworkers developed the 1,4-aryl migration of diphenylsilane **89** promoted by a chiral nickel catalyst prepared from (*R*,*R*)-**90** (Figure 14.6) and Ni(cod)$_2$ in the presence of NaO*t*Bu to afford benzoxasilole (S_{Si})-**91** in 95% ee (Table 14.4, entry 6) [41].

Si—Alkynyl Bond Cleavage Nozaki, Shintani and a coworker reported an enantioselective intramolecular alkynylsilylation of bisalkynylsilane **92** with a chiral rhodium catalyst prepared from ($_a$*R*,*S*,*S*)-**93** (Figure 14.6) and [RhCl(C$_2$H$_4$)]$_2$ to afford siladihydroacenaphthylene (+)-**94** in 94% ee (Table 14.4, entry 7) [42]. On the other hand, Tanaka and coworkers reported the enantioselective transformation via the Si—C bond cleavage of bisalkynylsilane **95** promoted by a chiral rhodium catalyst prepared from (*S*)-**96** (Figure 14.6) and [Rh(cod)$_2$]BF$_4$ to provide benzofuranylmethylidene-benzoxasiloles **97** in 65% ee (Table 14.4, entry 8) [43]. They proposed that the reaction proceeded via 1,2-silyl migration on one of the two alkynyl groups (**95** to **98**), followed by 1,3-alkynyl migration from silicon to carbon with Si—O bond formation (**98** to **99**) and oxycyclization on the alkynyl group (**99** to **97**) (Eq. (14.21)).

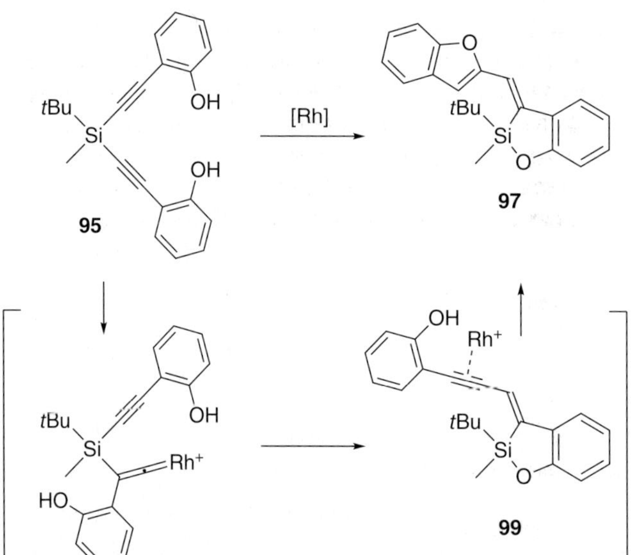

(14.21)

Si—Me Bond Cleavage Chatani, Tobisu and coworkers found a highly enantioselective insertion of alkyne **101** into the Si—Me bond of dimethylsilane **100** using a chiral rhodium catalyst prepared from (S_P,S_P)-**102** (Figure 14.6) and [RhCl(C$_2$H$_4$)$_2$]$_2$. This reaction afforded chiral benzosilole **103** in 98% ee (Table 14.4, entry 9) [44].

14.2.3.5 Desymmetrization of Prochiral Silicon Atoms by Transformations of Carbon Substituent(s) without Si—C Bond Cleavage

The transformation of one of two identical substituents connected to S_1 symmetric silicon atoms by Si—C bonds without Si—C bond cleavage is also an efficient approach to enantioenriched chiral silicon molecules (Eq. (14.22)) (Table 14.5).

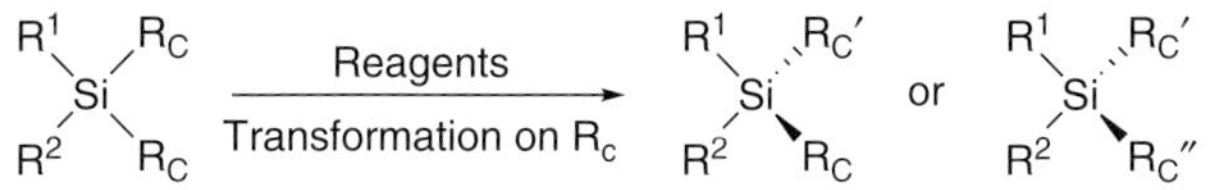

R_C: carbon substituent (14.22)

14.2.3.5.1 Reactions Promoted by Lipase or a Chiral Base

Blanco and Djerourou performed the desymmetrization of bis(hydroxymethyl) silane **104** by lipase-promoted asymmetric esterification with oxime ester **20**, affording monoester **105** in 76% ee (Table 14.5, entry 1) [45]. Xu and coworkers reported a similar lipase-catalyzed asymmetric esterification of bis(hydoroxymethylaryl)silane **106** that afforded chiral monoester **107** in 78% ee (Table 14.5, entry 2) [46]. Tomooka, Igawa and coworkers reported the enantioselective β-elimination of silacyclopentene oxide **108** to afford silacyclopentenol **110** in up to >98% ee using a new chiral lithium amide **109** in combination with 1,8-diazabicyclo[5.4.0]undec-7-ene, (DBU) (Table 14.5, entry 3) [47].

14.2.3.5.2 Chiral Transition-metal-Catalyst–promoted Reactions

Nozaki, Shintani and coworkers reported the enantioselective [2+2+2] cyclization of bisethynyl silane **111** with isocyanate **112** promoted by a rhodium catalyst prepared from ($_a$*R*)-**113** (Figure 14.7) and $[RhCl(C_2H_4)_2]_2$, which afforded pyridinone-fused benzosilole **114** in 91% ee (Table 14.5, entry 4) [48]. Nishiyama, Naganawa and coworkers reported the enantioselective intramolecular hydrosilylation of bisalkenylsilane **115** with a chiral rhodium catalyst prepared from ($_a$*S*)-**116** (Figure 14.7) and $[RhOMe(cod)]_2$, yielding chiral silane **117** in 80% ee (Table 14.5, entry 5) [49]. Hayashi, Shintani and coworkers reported the enantioselective biaryl coupling of biphenylsilane **118** promoted by a chiral palladium catalyst prepared from (*R*,$_p$*S*)-**119** (Figure 14.7) and $Pd(OAc)_2$ to afford dibenzosilole

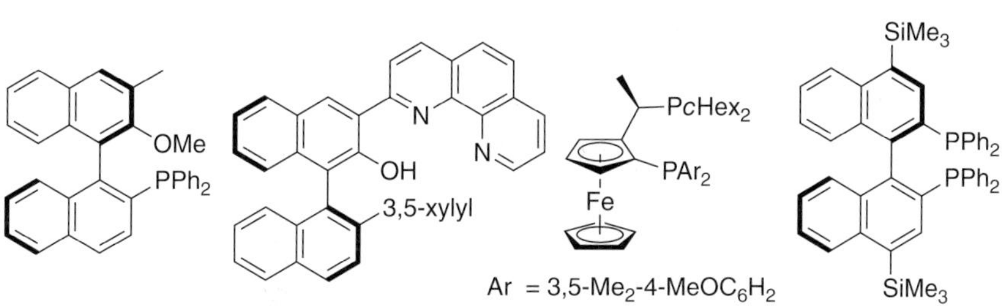

($_a$*R*)-**113** ($_a$*S*)-**116** (*R*,$_p$*S*)-**119** ($_a$*R*)-**122**

Figure 14.7 Chiral ligands in Table 14.5.

Table 14.5 Desymmetrization with transformation of carbon substituent(s).

Entry	Substrate	Reagents	Product	Yield stereoselectivity
Lipase– or chiral base–promoted reaction				
1	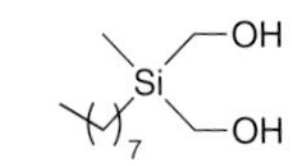 **104**	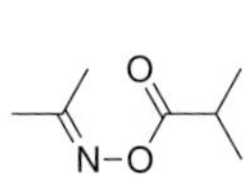 **20** lipase	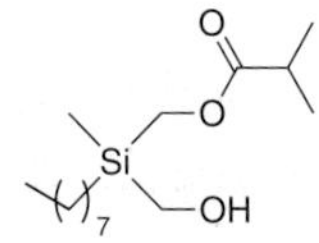 **105**	70% 76% ee
2	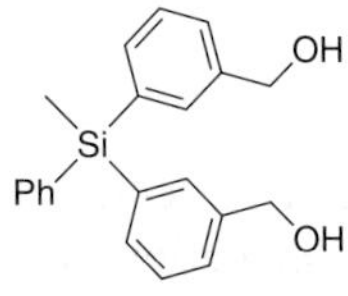 **106**	Ac_2O lipase	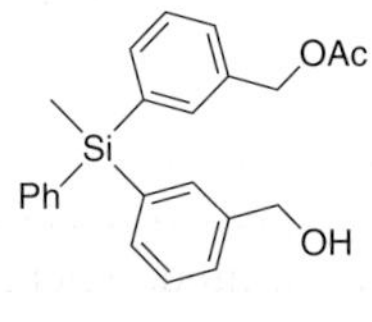 **107**	90% 78% ee
3	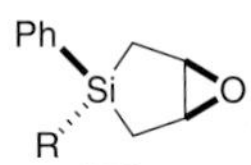 **108**	none 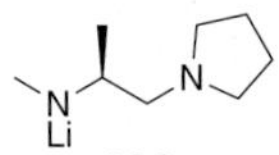 **109** DBU	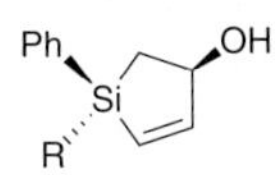 **110**	>99%[a] >98% ee[a]
Chiral transition-metal-catalyst-promoted reactions				
4	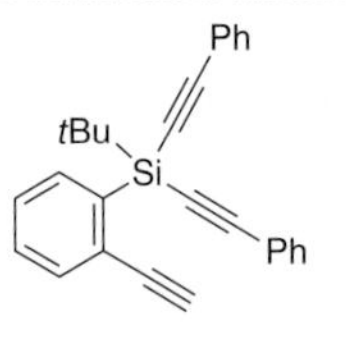 **111**	 **112** ($_a$*R*)-**113** $[RhCl(C_2H_4)_2]_2$ $NaBAr_4$	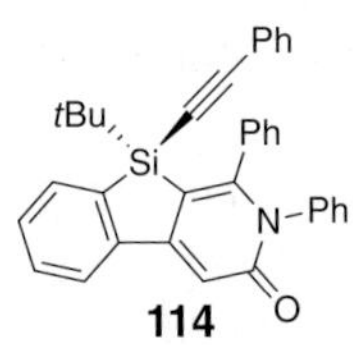 **114**	85% 91% ee
5	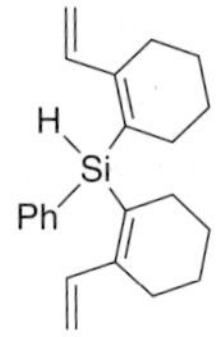 **115**	none ($_a$*S*)-**116** $[RhOMe(cod)]_2$	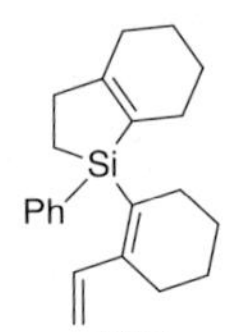 **117**	>99% 80% ee
6	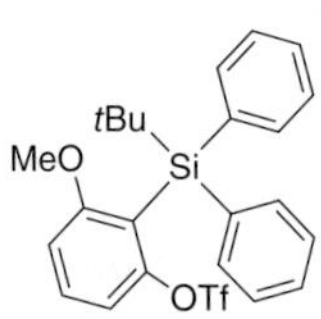 **118**	none (*R*,$_p$*S*)-**119** $Pd(OAc)_2$ Et_2NH	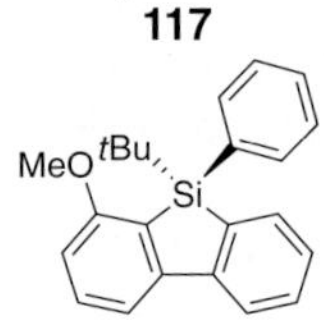 (*S*$_{Si}$)-**120**	92% 94% ee
7	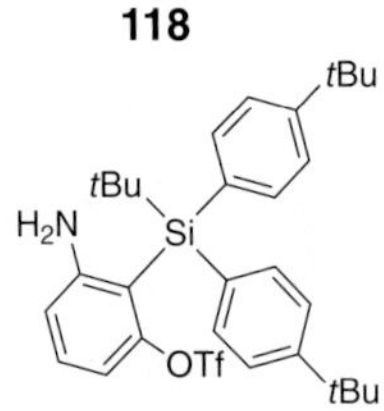 **121**	none ($_a$*R*)-**122** $Pd(OAc)_2$ Et_3N	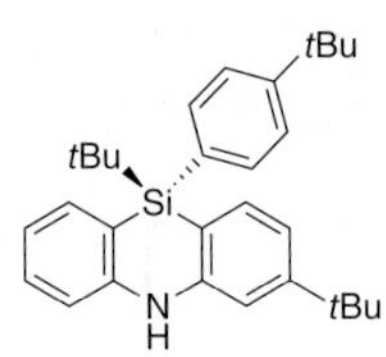 (*S*$_{Si}$)-**123**	78% 98% ee

a) R = Me.

(S_{Si})-**120** in 94% ee (Table 14.5, entry 6) [50]. They also developed the enantioselective *N*-arylation of silane **121** with a chiral palladium catalyst prepared from ($_aR$)-**122** (Figure 14.7) and $Pd(OAc)_2$ to afford 5,10-dihydrophenazasilane (S_{Si})-**123** in 98% ee (Table 14.5, entry 7) [51].

14.3 Stereoselective Transformation of Enantioenriched Chiral Silicon Molecules

14.3.1 Classification of Stereoselective Transformation of Chiral Silicon Molecules

Although a number of asymmetric synthesis and separation methods for chiral silicon molecules have been developed, the structural variety of chiral silicon molecules is quite limited. Therefore, stereoselective transformation is important to increase the diversity of chiral silicon molecules. Thus, stereoselective nucleophilic substitution, electrophilic substitution, and oxidation of enantioenriched chiral silicon molecules have been studied extensively. In this section, reported reactions are summarized and classified into the three categories. Moreover, multistep functionalization for the synthesis of highly functionalized chiral silicon molecules is also described.

14.3.2 Nucleophilic Substitution at a Chiral Silicon Center

As silicon is an electropositive atom, nucleophilic substitution is a reasonable approach for the transformation of silicon molecules. However, the stereochemical course of nucleophilic substitution reactions at chiral silicon center changes depending on inversion or retention of configuration via a silicate intermediate. That is, differences in the kind of nucleophile (Nu^-), LG, staying groups (R^1, R^2, and R^3), and reaction conditions (e.g. solvents, additive, and reaction temperature) affect the stereochemical course (Eq. (14.23)). This is in sharp contrast to the reactions of carbon centers, which provide inversion products via the Walden inversion (S_N2) or a 50 : 50 mixture of inversion and retention products via a carbocation intermediate (S_N1) (Eq. (14.24)).

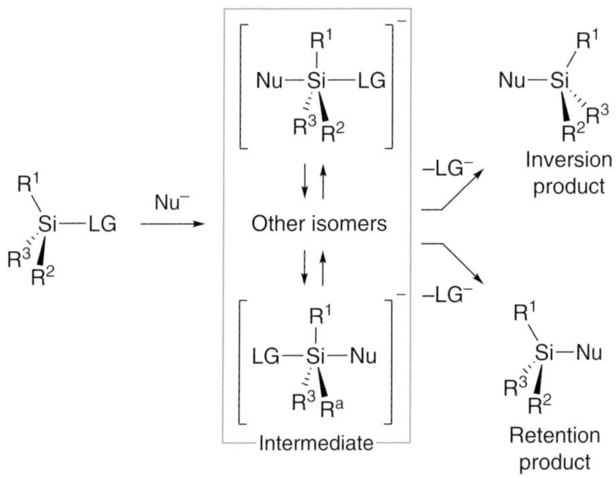

(14.23)

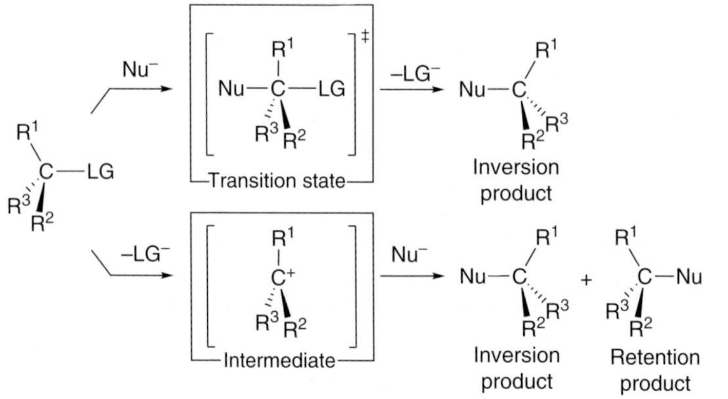

(14.24)

Stereochemical course of nucleophilic substitution in the enantioenriched 1-NpMePhSi-LG system with a variety of nucleophiles has been systematically investigated [8], but the sense of stereoselectivity was not clearly defined. For example, the reactions of methoxysilane **124** with ethyllithium, 4-methoxyphenyllithum, and ethynyllithium mainly provided substitution products with retention of configuration (Table 14.6, entries 1–3) [52a]. However, the reactions with allyllithium and benzyllithium mainly provided inversion products (Table 14.6, entries 4 and 5) [52a]. Furthermore, the reactions of **124** with DIBAL provided retention products both in *n*-hexane and diethyl ether (Table 14.6, entries 6 and 7); in contrast, the reactions of fluorosilane **125** provided the substituted product

Table 14.6 Stereochemical course of nucleophilic substitution on asymmetric silicon.

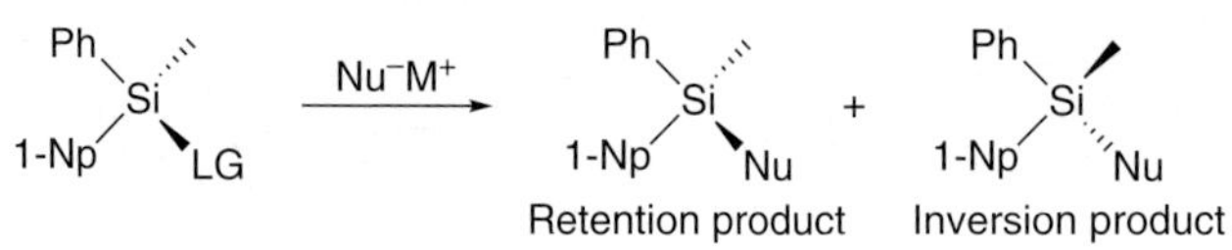

Entry	Substrate	LG	Nu^-M^+	Stereoselectivity
1	**124**	OMe	EtLi	74% retention
2	**124**	OMe	*o*-MeC_6H_4Li	97% retention
3	**124**	OMe	Ph—≡—Li	>99% retention
4	**124**	OMe	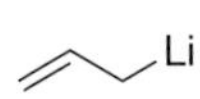	58% inversion
5	**124**	OMe	$PhCH_2Li$	79% inversion
6[a)]	**124**	OMe	DIBAL	>99% retention
7[b)]	**124**	OMe	DIBAL	99% retention
8[a)]	**125**	F	DIBAL	98% retention
9[b)]	**125**	F	DIBAL	90% inversion

a) Reaction in hexane.
b) Reaction in Et_2O.

with retention or inversion product in hexane or diethyl ether, respectively (Table 14.6, entries 8 and 9) [52b].

To control the stereochemical course of nucleophilic substitution at the chiral silicon center, Strohmann, Bauer and coworkers reported stereocontrol using a staying group effect [29, 53]. For example, the reaction of chiral benzyl silane (S_{Si})-**126** (98% ee), substituted by a 1-piperidinylmethyl, with *n*BuLi provided *n*Bu-substituted silane (R_{Si})-**127** (62% ee) via elimination of the benzyl group with 63% inversion of configuration (Eq. (14.25)) [53]. Density functional theory (DFT) calculations suggest that the reaction of **126** proceeds via transition state **128**, in which the piperidinyl nitrogen atom coordinates to the lithium of *n*BuLi at the back of the benzyl group, thus resulting in the inversion product via silicate formation.

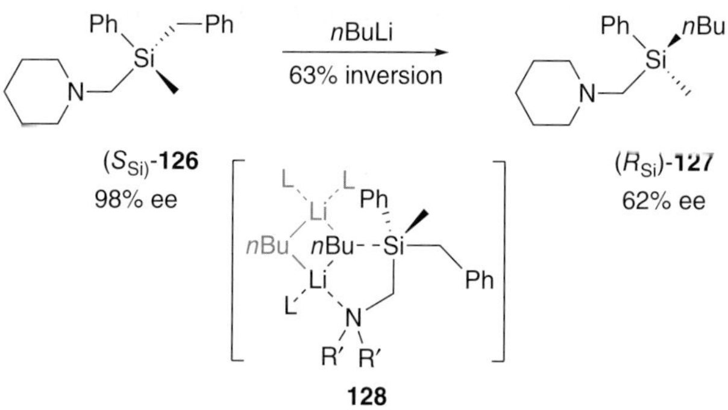

(14.25)

On the other hand, Kawakami and Suzuki reported substantial additive and temperature effects on the stereochemical course of the nucleophilic substitution of fluorosilane (R_{Si})-**129** with silyllithium **130** (Eq. (14.26)) [54]. The reaction in tetrahydrofuran (THF) at room temperature provided disilane **131** in 80% retention of configuration. On the addition of hexamethylphosphoramide (HMPA) (1.0 equiv), the % retention in the product increased to >99%. In sharp contrast, the presence of LiBr (3.0 equiv) or lowering the reaction temperature to −78 °C caused a dramatic change in the stereochemistry: 93% inversion resulted in both cases.

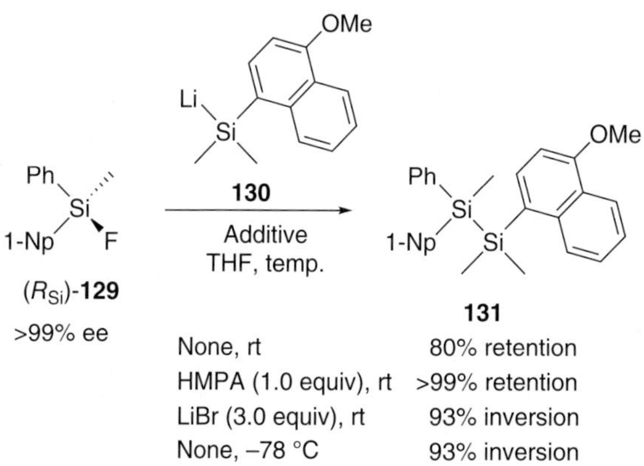

(14.26)

With regard to intramolecular nucleophilic substitution reactions, Tomooka, Nakazaki and a coworker developed the retro-[1,4]-Brook rearrangement of chiral allyloxylsilane (S_{Si})-**132**, which provided enol silyl ether (R_{Si})-**133** with perfect retention of configuration at the chiral silicon center (Eq. (14.27)) [55]. In this reaction, the deprotonation of **132** provides allylic lithium **134** to afford lithium enolate **136** with Si—C bond formation at the γ-position via transition state **135**.

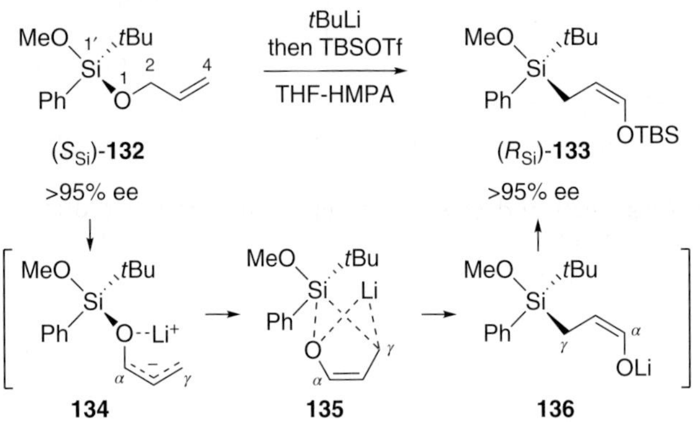

(14.27)

14.3.3 Electrophilic Substitution at Chiral Silicon Center

It is known that the stereochemistry of silyl anion is generally more stable than that of carbanion [56, 57]; thus, enantioenriched chiral silylanion can be transformed into functionalized chiral silicon molecules with high stereoselectivity by proper reaction conditions with an electrophile. Enantioenriched chiral silylanion have been prepared from the corresponding chiral stannylsilanes and disilanes by reductive Si—Sn or Si—Si bond cleavage, respectively.

Kawakami and coworkers prepared enantioenriched silyllithium **138** from stannylsilane (R_{Si})-**137** (90% ee) by treatment with MeLi in THF at −78 °C, and hydrosilane (S_{Si})-**139** (>85% ee) was obtained by the protonation of **138** with retention of configuration (Eq. (14.28)) [58]. This result means that the stereochemistry of **138** was mostly retained during the lithiation and protonation sequence. However, a reduction in the enantiopurity of (S_{Si})-**139** to 50% ee was observed when **138** was kept at 0 °C for one hour before the protonation.

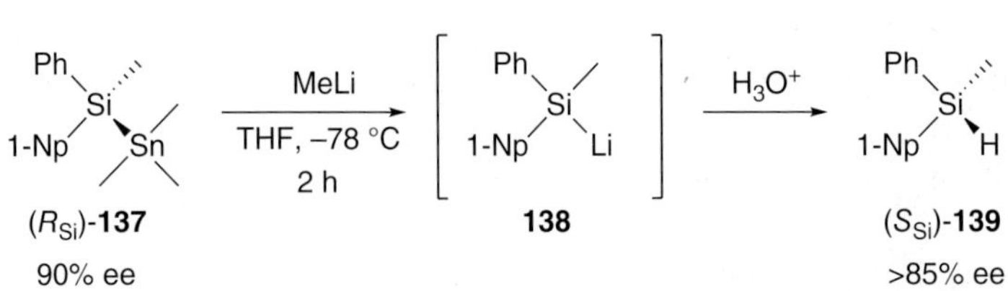

(14.28)

Strohmann et al. prepared enantioenriched silyllithium **141a** (M = Li) from disilane (R_{Si})-**140** (>98% ee) by reduction with lithium metal in THF and obtained disilane (R_{Si})-**142** (>98% ee) by treatment with Me_3SiCl with perfect retention of configuration (Eq. (14.29)) [59]. Similar with silyllithium **138**, the enantiopurity of **141a** was reduced to 53% ee after two hours at 20 °C. Interestingly, in the case of **141b** (M = MgX), which was prepared by the transmetalation of **141a** with $MgBr_2(thf)_4$, no racemization occurred in two hours at 20 °C [59a].

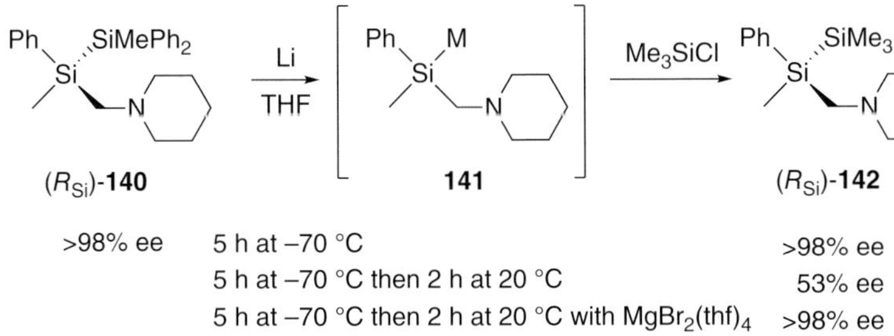

(14.29)

On the other hand, Oestreich and coworkers reported that the reductive lithiation of enantioenriched chlorosilane (R_{Si})-**143** (98% ee) with lithium di-*tert*-butylbiphenylide (LDBB) followed by protonation yielded hydrosilane (S_{Si})-**145** in 48% ee with partial retention of configuration (Eq. (14.30)) [60]. The decrease of the enantiopurity resulted from the formation of an achiral dichlorosilicate **146** from part of (R_{Si})-**143** with in situ–generated LiCl. They evaluated the proposed mechanism by the individual reactions of enantioenriched **143** with LiCl in THF, which showed a substantial decrease in enantiomeric purity.

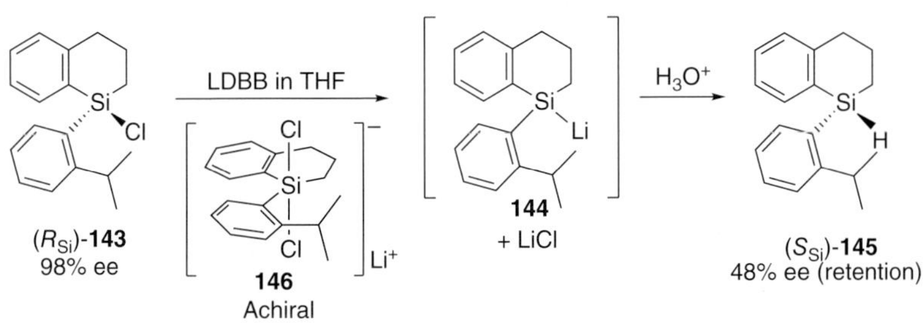

(14.30)

14.3.4 Oxidation at Chiral Silicon Center

A variety of oxidation reactions of enantioenriched chiral hydrosilanes to silanol have been reported. The stereochemical course of the reactions shows a trend: soluble oxidants mainly provide retention products, but insoluble solid oxidants mainly provide inversion products (Table 14.7). The oxidation of enantioenriched chiral alkenylsilane is also described in this section.

14.3.4.1 Oxidation of Hydrosilane

14.3.4.1.1 Oxidation with Soluble Oxidants

The reactions of hydrosilane **139** with a stoichiometric amount of oxaziridine **147** or dimethyldioxirane (DMDO) stereoselectively afforded silanols **52** with retention of configuration (Table 14.7, entries 1,2) [61, 62]. Oxidation with a catalytic amount of $MeReO_3$ [63], Mes_2Te [64], or $(TBA)_4[\gamma\text{-}SiW_{10}O_{34}(H_2O)_2]$ [65] (TBA = tetrabutylammonium) with a reoxidizing agent also afforded silanols

Table 14.7 Stereoselectivity in oxidation of chiral hydrosilane.

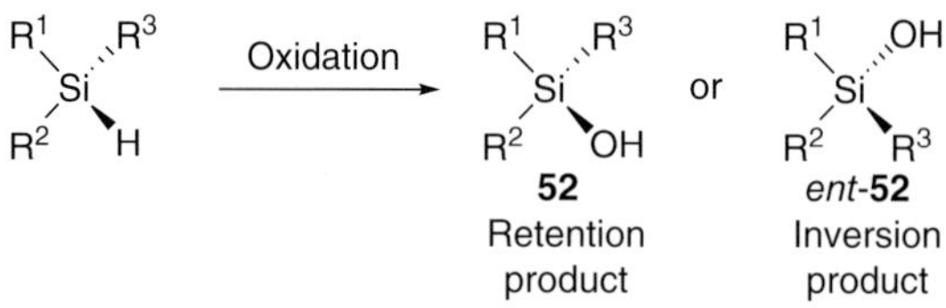

Entry	Substrate	Reagents	Yields	Stereoselectivity
Oxidation with soluble oxidant in reaction solvent				
1	**139**	Oxaziridine **147**	—	>99% retention
2	**139**	DMDO	>98%	>95% retention
3	**139**	$MeReO_3$ Urea hydroperoxide	96%	94% retention
4	**139**	Mes_2Te, air, $h\nu$ Hematoporphyrin	79%	93% retention
5	**139**	$(TBA)_4[\gamma\text{-}SiW_{10}O_{34}(H_2O)_2]$ H_2O_2	>95%	93% retention
6	**139**	$[RuCl_2(p\text{-cymene})]_2$, air	97%	74% inversion
Oxidation with insoluble solid oxidant				
7	**139**	Raney nickel, H_2O	97%	94% inversion
8	**148**	Pd/C-$100H_{ox}$, H_2O	96%	81% inversion
9	**149**	RuHAP, O_2	—	99% inversion

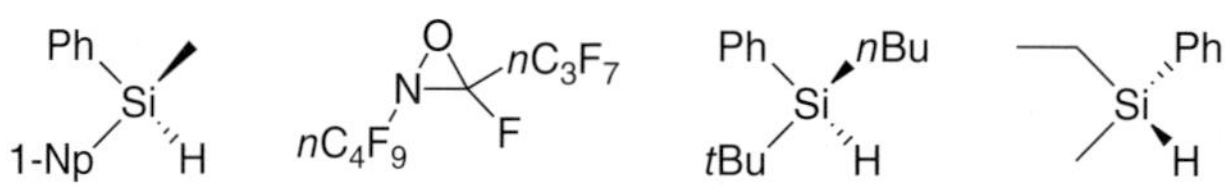

139 **147** **148** **149**

with retention of configuration (Table 14.7, entries 3–5). Exceptionally, the reaction with a catalytic amount of soluble $[RuCl_2(p\text{-cymene})]_2$ in combination with air as a reoxidizing agent afforded a silanol with inversion of configuration (Table 14.7, entry 6) [66].

14.3.4.1.2 Oxidation with Insoluble Solid Oxidants

Oxidation of hydrosilanes **139** and **148** with Raney nickel or Pd/C-$100H_{OX}$ with H_2O produced inversion products (Table 14.7, entries 7, 8) [67, 68]. Yoshizawa, Kamachi and coworkers performed DFT calculations and proposed the reaction mechanism on Pd/C-$100H_{OX}$, in which the dissociative adsorption of hydrosilane forms Si—Pd bonds on the metal surface with retention of configuration, followed by the backside approach of H_2O to the Si—Pd bond with inversion of configuration, thus yielding the total inversion product [68]. Oxidation of **149** using a hydroxyapatite-bound Ru complex (RuHAP) with O_2 also afforded the corresponding inverted silanol with a high degree of stereoselectivity (Table 14.7, entry 9) [69].

14.3.4.2 Oxidation of Alkenylsilane

Tomooka, Igawa and coworkers found that ozone cleaves the Si—C(sp^2) bond and forms Si—O bonds stereoselectively. For example, ozone oxidation of chiral alkenylsilane (S_{Si})-**150** (53% ee), which was prepared by the enantioselective hydrosilylation of alkyne (see Table 14.3, entry 8), provided chiral peroxysilane (R_{Si})-**151**. This was converted into silanol (R_{Si})-**52a** (44% ee) by treatment with a base, resulting in 83% retention of configuration from (S_{Si})-**150** (Eq. (14.31)) [35a].

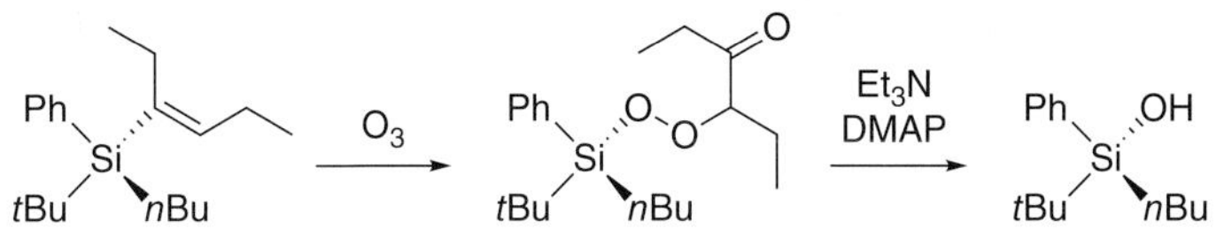

(S_{Si})-**150** 53% ee   (R_{Si})-**151**   (R_{Si})-**52a** 88%, 44% ee   (14.31)

14.3.5 Multistep Functionalization of Chiral Silicon Molecules

To explore novel biochemistry and materials chemistry based on chiral silicon molecules, multistep transformations from simple enantioenriched chiral silicon molecules without loss of the enantiomeric purity are required. Pioneering work on the multistep transformations of chiral silicon molecule was reported by Tacke and Heinrich. They demonstrated the synthesis of silicon analogs of bioactive chiral carbon molecules. For example, enantiomerically pure ammonium salt (R_{Si})-**154** was synthesized from (R_{Si})-**22** via trimethylsilyl (TMS) protection of the hydroxymethyl group on the silicon atom, aminolithiation of the vinyl group, removal of the TMS group, and treatment with MeI (Eq. (14.32)) [22]. (R_{Si})-**154** shows significant bioactivity as a antimuscarinic agent (see Section 14.4.4).

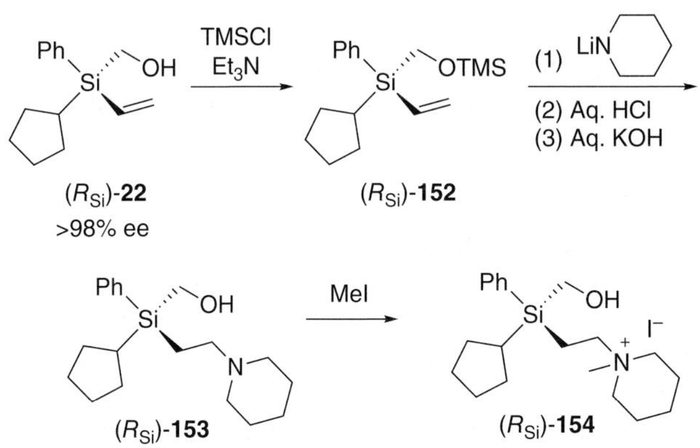

(R_{Si})-**22** >98% ee   (R_{Si})-**152**

(R_{Si})-**153**   (R_{Si})-**154** >98% ee   (14.32)

Enantioenriched chiral carboxylic acids with chiral carbon at the α-position are common structures in biological compounds that are important for basic life processes. Accordingly, their silicon congeners are highly attractive as a new class of bioactive compounds because of their unique reactivity and physical properties. To this end, Tomooka, Igawa and a coworker synthesized silacarboxylic acid **158** from enantioenriched silanol (S_{Si})-**52a** (R = *n*Bu) (>98% ee) and (S_{Si})-**52b** (R = Me) (>98% ee) in four steps, (i) methyl etherification, (ii) $LiAlH_4$ reduction, (iii) radical chlorination, and (iv) reductive lithiation followed by CO_2 treatment (Eq. (14.33)) [70]. The thus-obtained (S_{Si})-**158a** and (S_{Si})-**158b** maintained the level of enantiopurity almost the same as the starting silanols **52** [(S_{Si})-**158a**: >98% ee, (S_{Si})-**158b**: 96% ee]. In sharp contrast to the significant racemization of enantioenriched chlorosilane **143** by reductive lithiation reported by Oestreich and coworkers (see Eq. (14.30)), that of chlorosilanes **156** proceeded in a highly stereoretentive manner, probably because of the steric hindrance of the *t*Bu group.

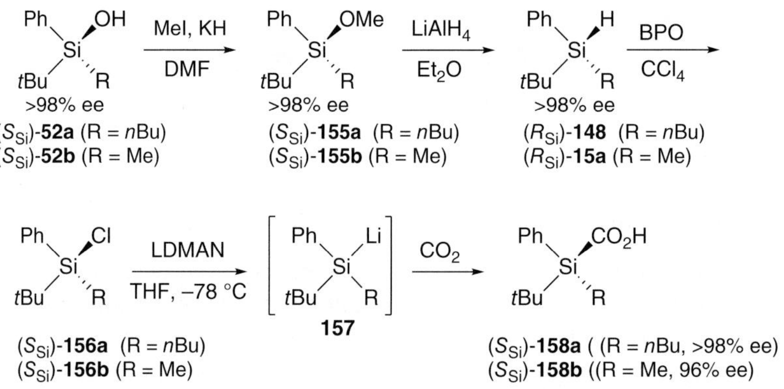

(14.33)

Of note, the unique character of silacarboxylic acid was observed in its esterification reactions. For example, the *N*,*N*′-dicyclohexylcarbodiimide (DCC) condensation reaction of silacarboxylic acids **158** provides substantial amounts of silyl ethers **159** and silyl silacarboxylic acid esters **160**, both resulting from decarbonylation (Eq. (14.34), condition A). These results mean that nucleophilic acyl substitution of electrophilically activated silacarboxylic acid is slower than nucleophilic substitution on the electropositive silicon atom. In sharp contrast, the Mitsunobu reaction of **158** provided the desired esters **161** in good yields, in which silacarboxylic acid reacted as nucleophile (Eq. (14.34), condition B).

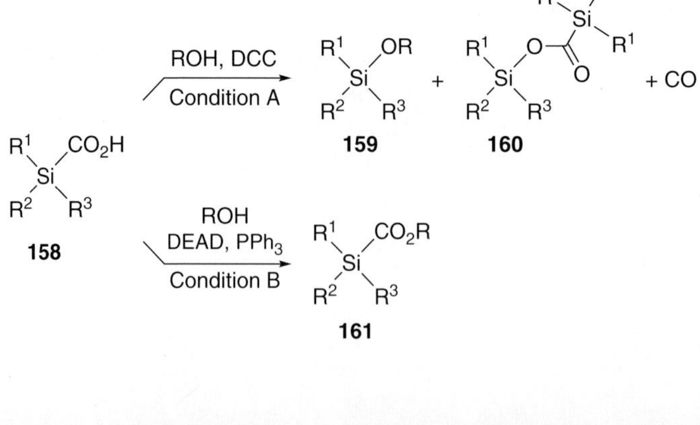

(14.34)

Furthermore, Tomooka, Igawa and coworkers performed the stereoselective introduction of several oxygen functionalities to silacyclopentenol **110**; that is, silacyclopentane triols **165** were synthesized as the sole stereoisomers via epoxidation with *meta*-chloroperoxybenzoic acid (mCPBA), C2-selective addition of allyl alcohol, and hydrolysis of the allyl ether moiety by isomerization to an enol ether with the Wilkinson's catalyst [$RhCl(PPh_3)_3$] in the presence of 1,4-diazabicyclo[2.2.2]octane (DABCO), followed by acidic hydrolysis (Eq. (14.35)) [47, 71]. The thus-obtained triol **165** shows unique biological activity (see Section 14.4.4).

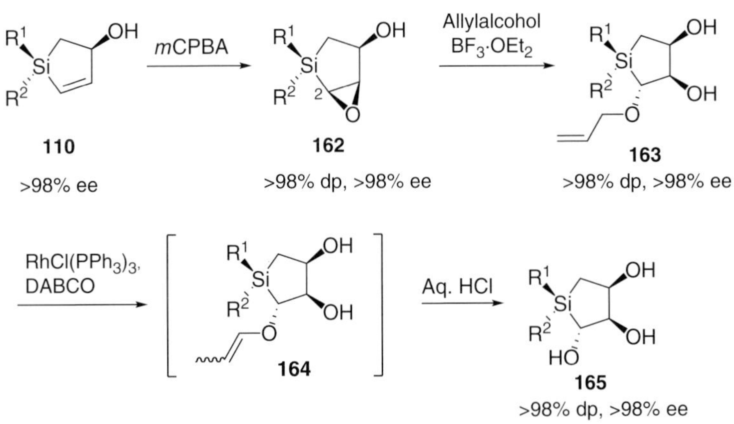

(14.35)

14.4 Application of Enantioenriched Chiral Silicon Molecules

14.4.1 Classification of Applications of Chiral Silicon Molecules

As mentioned earlier in this chapter, the applications of chiral silicon molecules are relatively unexplored compared with those of chiral carbon molecules, and only a few pioneering works have been reported so far. The applications can be classified into three categories; chiral reagents, chiral materials, and bioactive molecules. The reported studies are described in this order.

14.4.2 Application as Chiral Reagents

Bienz and coworkers reported stereoselective transformations of ketones by using chiral silicon moieties as chiral stereodirecting group [15, 16, 72]. For example, the nucleophilic addition of PhLi to acyl silane (S_{Si})-**13** (92 ± 2% ee) afforded tertiary alcohol (R_{Si},*R*)-**166**, which was converted into ether (R_{Si},*R*)-**167** via the [1,2]-Brook rearrangement. Finally, (*R*)-α-phenethyl alcohol (**168**) (88 ± 2% ee) was obtained from (R_{Si},*R*)-**167** by tetra-*n*-butylammonium fluoride (TBAF) treatment (Eq. (14.36)) [15].

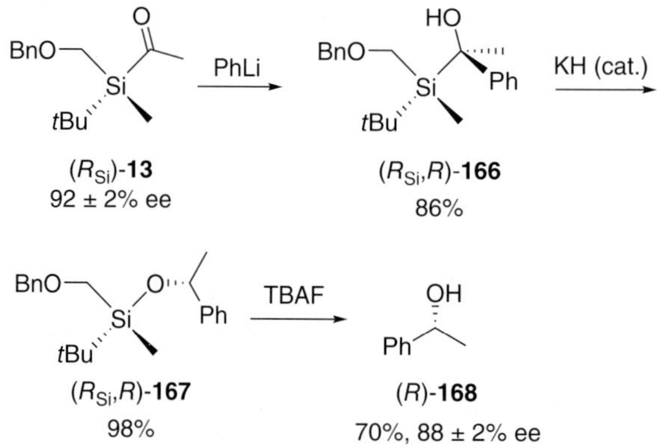

(14.36)

Leighton and coworkers reported a high level of 1,5-remote stereocontrol of chiral diol **169** using the intramolecular silylformylation–allylsilylation reaction of homopropargylic alcohol derivative (R_{Si},*S*)-**56c**, which was prepared by the enantioselective dehydrogenative etherification of dihydrosilane (see Table 14.3, entry 5) [32]. In this reaction, bicyclic intermediate **171** was generated via the stereoselective intramolecular Hosomi–Sakurai reaction of β-silylenal **170** (Eq. (14.37)).

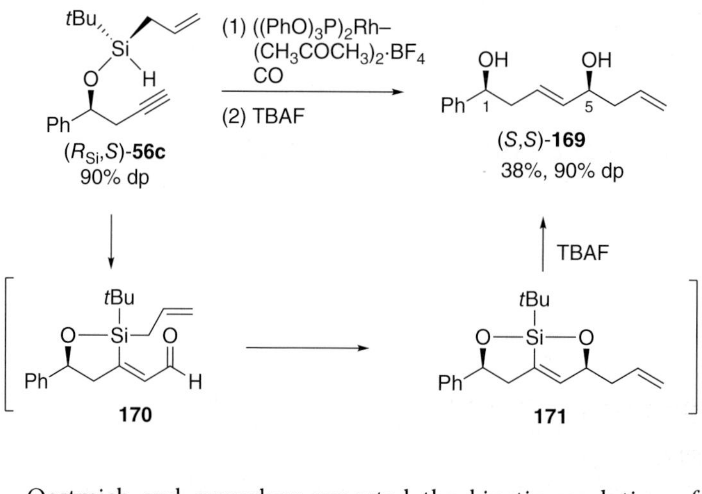

(14.37)

Oestreich and coworkers reported the kinetic resolution of chiral alcohol *rac*-**172**, which has a 2-pyridyl group, by copper-catalyzed etherification with enantioenriched hydrosilane (R_{Si})-**16** (96% ee). The reaction afforded silyl ether (S_{Si},*S*)-**173** (86% dp) along with **172**, which was recovered in 84% ee (*R*) with 56% conversion (Eq. (14.38)) [73].

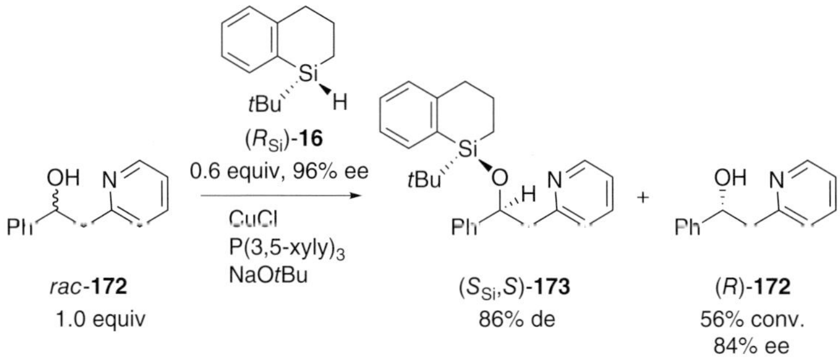

(14.38)

They also developed the highly stereoselective hydrosilylation of alkenes using chiral hydrosilane **16**. For example, the reaction of norbornene (**174**) with (R_{Si})-**16** (85% ee) in the presence of cationic palladium complex **175** afforded silane (S_{Si},*R*)-**176** in >99% dp and much higher enantiopurity than that of **16** (93% ee) (Eq. (14.39)) [74]. This result means that the hydrosilylation proceeded with a positive nonlinear effect in terms of the enantiopurity of hydrosilane **16**.

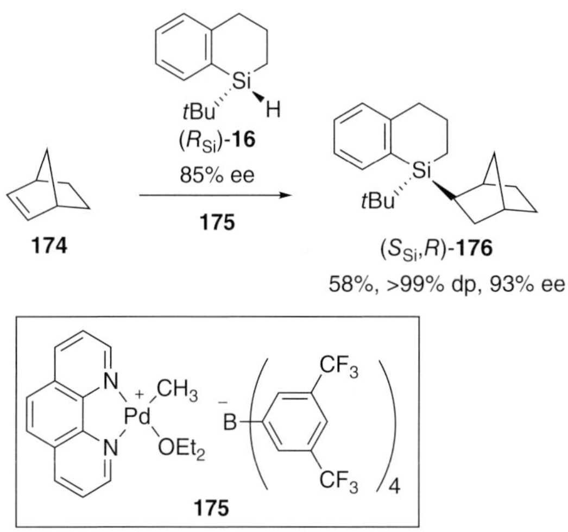

(14.39)

14.4.3 Application as Chiral Materials

14.4.3.1 Chiral Silicon Polymer

Kawakami and coworkers synthesized a variety of chiral-silicon-containing polymers in a stereocontrolled manner using enantioenriched chiral silicon monomers [27, 75]. For example, the reaction of allylhydrosilane (R_{Si})-**177** (60% ee) with Karstedt's catalyst [Pt(dvds)] (dvds = tetramethyldivinyldisiloxane) (5 mmol%)

afforded polycarbosilanes (**178**, M_n = 11 000, M_w/M_n = 2.3) with a population of partial structures in 52% isotactic polymer (*I*-**178**), 32% heterotactic polymer (*H*-**178**), and 16% syndiotactic polymer (*S*-**178**) (Eq. (14.40)) [27]. In contrast, the similar polymerization of *rac*-**177** afforded a polymer containing only 25% isotactic polymer, with 50% heterotactic polymer, and 25% syndiotactic polymer.

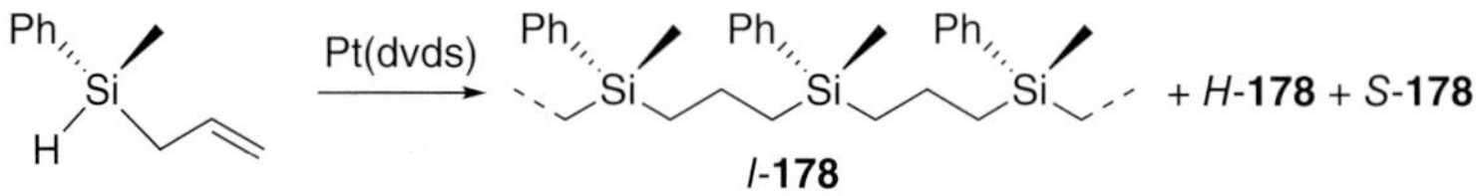

	I-**178**		
(R_{Si})-**177** (60% ee)	52%[a]	32%[a]	16%[a]
rac-**177**	25%[a]	50%[a]	25%[a]

a) Calculated triad population. (14.40)

Kawakami and coworkers also reported the ring-opening polymerization of enantioenriched cyclic silanes. The reaction of siloxane (S_{Si})-**179** (>98% ee) with PhLi (1.7 mol%) proceeded at Si3 via selective nucleophilic addition to afford optically active isotactic polymer *I*-(S_{Si})-**180** (M_n = 11 300, M_w/M_n = 1.12, $[\alpha]^{25}_D$ −8.0) (Eq. (14.41)) [75a]. Furthermore, the reaction of silacyclobutane (+)-**181** with Et_3SiH in the presence of Karstedt's catalyst (1 mol%) afforded optically active isotactic polymer *I*-(−)-**182** (M_n = 35 600, M_w/M_n = 1.73, $[\alpha]^{25}_D$ −33.5) (Eq. (14.42)) [75b].

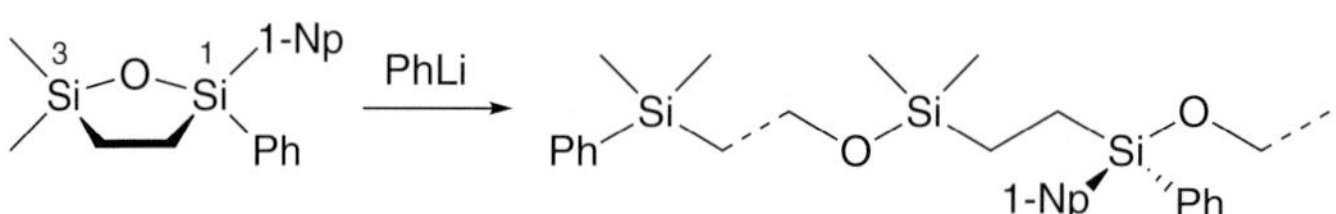

(S_{Si})-**179**
>98% ee

I-(S_{Si})-**180** (14.41)

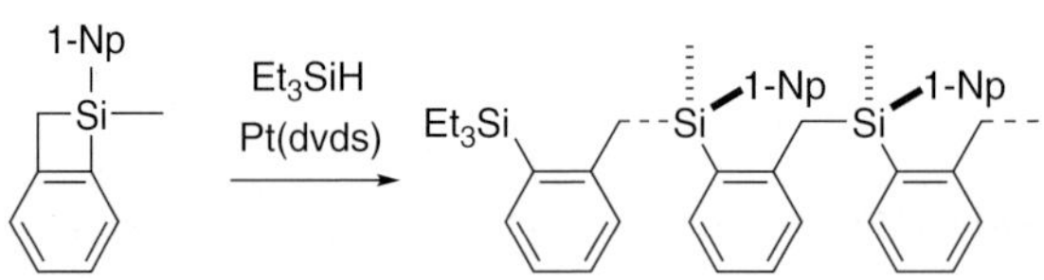

(+)-**181**
>98% ee

I-(−)-**182** (14.42)

Nozaki, Shintani and coworkers performed the polymerization of enantioenriched (R_{Si})-**183** (99% ee) by Sonogashira coupling between iodobenzene- and terminal alkyne moieties to afford isotactic polymer *I*-(R_{Si})-**184** (M_n = 19 000, $[\alpha]^{25}_D$ −282) (Eq. (14.43)) [48b].

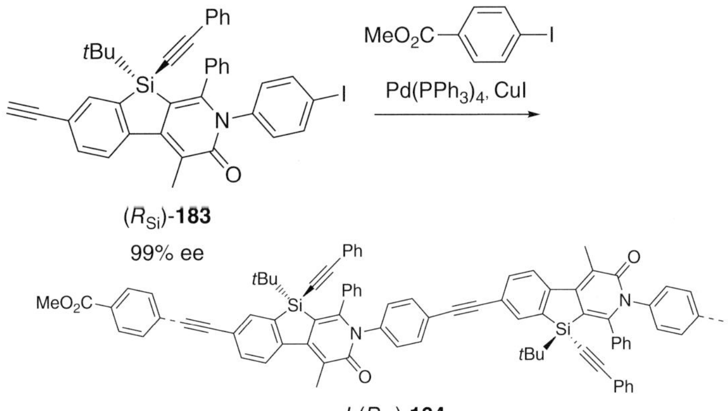

(14.43)

14.4.3.2 Circular Polarized Luminescence of Chiral Silicon Molecules

Nishihara, Yamanoi and coworkers evaluated circular polarized luminescence (CPL) properties of enantioenriched arylsilanes and found that pyrene-substituted (S_{Si}) **185** has a substantial g_{lum} value (-1.6×10^{-3}) (Figure 14.8) [34b]. Interestingly, its doubly substituted chiral silicon molecules having silicon-containing derivative (S_{Si},S_{Si})-**186** exhibits a larger g_{lum} value with the opposite sign ($+8.0 \times 10^{-3}$) than that of (S_{Si})-**185**.

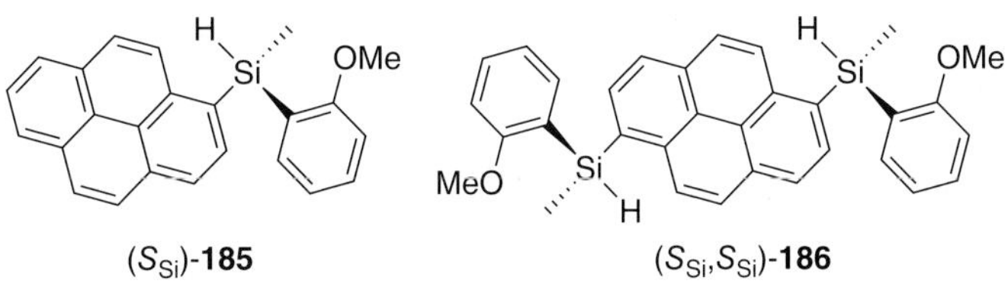

Figure 14.8 Chiral silicon molecules having CPL property.

14.4.4 Applications as Bioactive Molecules

How would the stereochemistry of a chiral silicon center be recognized within an organism? Biofunctionality is one of the most attractive and challenging subjects in chiral silicon chemistry and would make up a new class of medicinal chemistry. Pioneering works in this field were carried out by Tacke and coworkers. They designed novel antimuscarinic agent **188** (E = C) based on the structure of trihexyphenidyl (**187**), which is a typical antiparkinsonian agent owing to its antimuscarinic activity (Figure 14.9) [22, 76]. They also synthesized the chiral silicon congener **154** (E = Si) in an enantioenriched form (Eq. (14.32)). Biological studies with muscarinic receptors obtained from human NB-OK showed that the *R*-isomers of **188** and **154** have higher activity than the corresponding *S*-isomers. Moreover, the subtype selectivity of silicon congener **154** is higher than that of **188**.

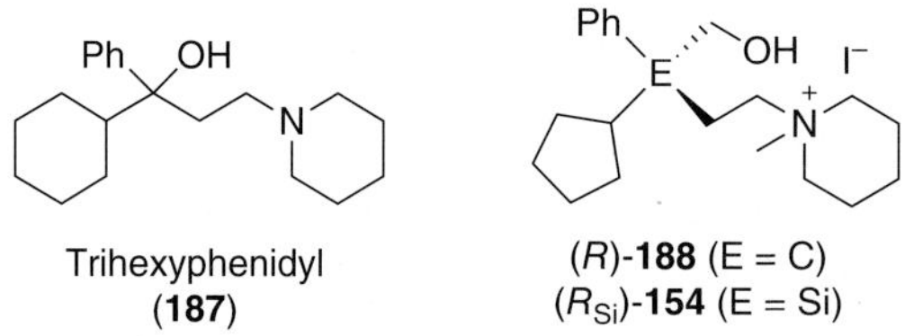

Figure 14.9 Structures of antimuscarinic agents.

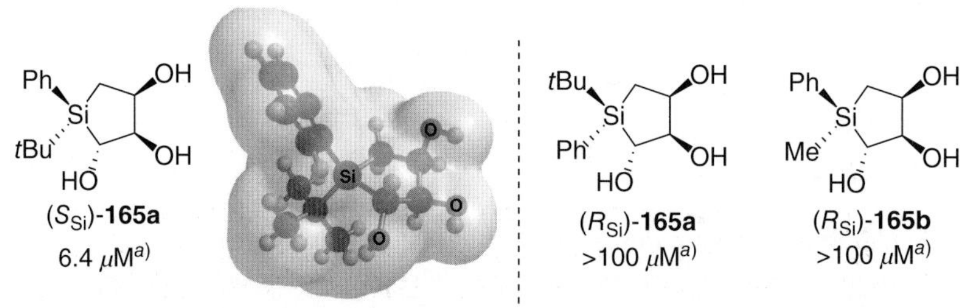

Figure 14.10 Binding activity of silacyclopentane triols **165** toward 5-HT_{2b} serotonin acceptor, and space-filling model of (S_{Si})-**165a**. a) IC_{50} value for the specific binding to 5-HT_{2b} with 1.20 nM [^{3}H]-lysergic acid diethylamide as the ligand.

Recently, Tomooka, Igawa and coworkers found that the bioactive silacyclopentane derivative silacyclopentane triol (S_{Si})-**165a** shows substantial binding activity toward the 5-HT_{2b} serotonin acceptor [47]. In sharp contrast, the silicon epimer (R_{Si})-**165a** and analog (R_{Si})-**165b** having Me groups instead of the *t*Bu groups of **165a** show no meaningful activity (Figure 14.10). Notably, **165a** has a bulky phenyl and *t*Bu group on the silicon in the five-membered ring. In contrast, the carbon congener is difficult to synthesize owing to the steric repulsion of the *t*Bu and Ph groups around the smaller chiral environment of the carbon atom.

14.5 Summary and Conclusions

Now, over 110 years have passed since the first report of chiral silicon molecules by Kipping in 1907, which was only 33 years after Van't Hoff and Le Bel independently propounded the tetrahedral configuration of carbon in 1874. However, studies on chiral silicon molecules have been quite limited compared with that of chiral carbon counterparts. Recently, the number of reports concerning chiral silicon molecules has been increasing gradually, particularly in terms of their asymmetric synthesis. The dynamic study on chiral sila-drugs and chiral sila-materials will open a new horizon in the chiral technology based on silicon chemistry.

References

1 (a) Riehl, J.P. (2010). *Mirror-Image Asymmetry: An Introduction to the Origin and Consequences of Chirality*. Hoboken, NJ: Wiley. (b) Eliel, E.L. and Wilen, S.H. (1994). *Stereochemistry of Organic Compounds*. New York, NY: Wiley.

2 (a) Morrison, J.D. (ed.) (1983–1985). *Asymmetric Synthesis*, vol. 1–5. New York, NY: Academic Press. (b) Carreira, E.M. and Yamamoto, H. (eds.) (2012). *Comprehensive Chirality*, vol. 1–9. Amsterdam: Elsevier.

3 Reviews on chiral silicon molecules, see: (a) Corriu, R.J.P., Guerin, C., and Moreau, J.J.E.(eds. E.L. Eliel, S.H. Wilen and N.L. Allinger) (1984). *Topics in Stereochemistry*, vol. 15, 43–199. New York, NY: Wiley. (b) Oestreich, M. (2007). *Synlett* 1629–1643. (c) Xu, L.-W., Li, L., Lai, G.-Q., and Jiang, J.-X. (2011). *Chem. Soc. Rev.* 40: 1777–1790. (d) Bauer, J.O. and Strohmann, C. (2016). *Eur. J. Inorg. Chem.* 2868–2881. (e) Shintani, R. (2015). *Asian J. Org. Chem.* 4: 510–514. (f) Shintani, R. (2018). *Synlett* 29: 388–396.

4 Brook, M.A. (2000). *Silicon in Organic, Organometallic, and Polymer Chemistry*. New York, NY: Wiley.

5 (a) Kaftory, M., Kapon, M., and Botoshansky, M. (1998). *The Chemistry of Organic Silicon Compounds*, vol. 2 (eds. Z. Rappoport, Y. Apeloig and S. Patai), 181–266. Chichester: Wiley. (b) Sheldric, W.S. (1989). *The Chemistry of Organic Silicon Compounds* (eds. S. Patai and Z. Rappoport), 227–304. Chichester: Wiley.

6 (a) Kipping, F.S. (1907). *J. Chem. Soc. Trans.* 91: 209–240. (b) Kipping, F.S. (1908). *J. Chem. Soc. Trans.* 93: 457–473. (c) Challenger, F. and Kipping, F.S. (1910). *J. Chem. Soc. Trans.* 97: 755–773.

7 (a) Sommer, L.H. and Frye, C.L. (1959). *J. Am. Chem. Soc.* 81: 1013. (b) Sommer, L.H., Frye, C.L., Parker, G.A., and Michael, K.W. (1964). *J. Am. Chem. Soc.* 86: 3271–3276.

8 (a) Sommer, L.H. (1965). *Stereochemistry, Mechanism, and Silicon. An Introduction to the Dynamic Stereochemistry and Reaction Mechanism of Silicon Centers*. New York, NY: McGraw-Hill. (b) Corriu, R.J.P., Guerin, C., and Moreau, J.J.E. (1989). *The Chemistry of Organic Silicon Compounds* (eds. S. Patai and Z. Rappoport), 305–370. Chichester: Wiley. (c) Holmes, R.R. (1990). *Chem. Rev.* 90: 17–31.

9 (a) Klebe, J.F. and Finkbeiner, H. (1966). *J. Am. Chem. Soc.* 88: 4740–4741. (b) Klebe, J.F. and Finkbeiner, H. (1968). *J. Am. Chem. Soc.* 90: 7255–7261.

10 Corriu, R.J.P. and Moreau, J.J.E. (1973). *Tetrahedron Lett.* 14: 4469–4472.

11 Hayashi, T., Yamamoto, K., and Kumada, M. (1974). *Tetrahedron Lett.* 15: 331–334.

12 Pioneering works on HPLC separation of enantiomers of chiral silicon molecules using chiral stationary phase, see: (a) Li, Y. and Kawakami, Y. (1998). *Macromolecules* 31: 5592. (b) Mori, A., Toriyama, F., Kajiro, H., Hirabayashi, K., Nishihara, Y., and Hiyama, T. (1999). *Chem. Lett.* 28: 549–550. (c) Schmid, T., Daiss, J.O., Surburg, H., and Tacke, R. (2003). *Organometallics* 22: 4343–4346.

13 Tacke, R., Reichel, D., Kropfgans, M., Jones, P. G., Mutschler, E., Gross, J., Hou, X., Waelbroeck, M., and Lambrecht, G. (1995). *Organometallics* 14: 251–262.

14 Jankowski, P., Schaumann, E., Wicha, J., Zarecki, A., and Adiwidjaja, G. (1999). *Tetrahedron: Asymmetry* 10: 519–526.

15 Huber, P., Bratovanov, S., Bienz, S., Syldatk, C., and Pietzsch, M. (1996). *Tetrahedron: Asymmetry* 7: 69–78.

16 Trzoss, M., Shao, J., and Bienz, S. (2002). *Tetrahedron* 58: 5885–5894.

17 Jankowski, P., Schaumann, E., Wicha, J., Zarecki, A., Adiwidjaja, G., and Asztemborska, M. (2000). *Chem. Commun.* 1029–1030.
18 Oestreich, M., Schmid, U.K., Auer, G., and Keller, M. (2003). *Synthesis* 2725–2739.
19 (a) Kawachi, A., Maeda, H., Mitsudo, K., and Tamao, K. (1999). *Organometallics* 18: 4530–4533.
20 Corriu, R.J.P. and Lanneau, G.F. (1974). *J. Organomet. Chem.* 64: 63–78.
21 Aouf, N.E., Djerourou, A.-H., and Blanco, L. (1994). *Phosphorus, Sulfur Silicon Relat. Elem.* 88: 207–215.
22 Tacke, R. and Heinrich, T. (2002). *Silicon Chem.* 1: 35–39.
23 Yamamoto, K., Kawanami, Y., and Miyazawa, M. (1993). *J. Chem. Soc., Chem. Commun.* 436–437.
24 Rendler, S., Auer, G., Keller, M., and Oestreich, M. (2006). *Adv. Synth. Catal.* 348: 1171–1182.
25 Oka, N., Nakamura, M., Soeda, N., and Wada, T. (2009). *J. Organomet. Chem.* 694: 2171–2178.
26 Bai, X.-F., Zou, J.-F., Chen, M.-U., Xu, Z., Li, L., Cui, Y.-M., Zheng, Z.-J., and Xu, L.-W. (2017). *Chem. Asian J.* 12: 1730–1735.
27 Kawakami, Y., Takeyama, K., Komuro, K., and Ooi, O. (1998). *Macromolecules* 31: 551–553.
28 (a) Kobayashi, K., Kato, T., and Masuda, S. (1987). *Chem. Lett.* 16: 101–104. (b) Kobayashi, K., Kato, T., Unno, M., and Masuda, S. (1997). *Bull. Chem. Soc. Jpn.* 70: 1393–1401.
29 Bauer, J.O. and Strohmann, C. (2014). *Angew. Chem. Int. Ed.* 53: 8167–8171.
30 Igawa, K., Takada, J., Shimono, T., and Tomooka, K. (2008). *J. Am. Chem. Soc.* 130: 16132–16133.
31 Li, N. and Guan, B.-T. (2017). *Adv. Synth. Catal.* 359: 3526–3531.
32 Schmidt, D.R., O'Malley, S.J., and Leighton, J.L. (2003). *J. Am. Chem. Soc.* 125: 1190–1191.
33 Yasutomi, Y., Suematsu, H., and Katsuki, T. (2010). *J. Am. Chem. Soc.* 132: 4510–4511.
34 (a) Kurihara, Y., Nishikawa, M., Yamanoi, Y., and Nishihara, H. (2012). *Chem. Commun.* 48: 11564–11566. (b) Koga, S., Ueki, S., Shimada, M., Ishii, R., Kurihara, Y., Yamanoi, Y., Yuasa, J., Kawai, T., Uchida, T., Iwamura, M., Nozaki, K., and Nishihara, H. (2017). *J. Org. Chem.* 82: 6108–6117.
35 (a) Igawa, K., Yoshihiro, D., Ichikawa, N., Kokan, N., and Tomooka, K. (2012). *Angew. Chem. Int. Ed.* 51: 12745–12748. (b) Wen, H., Wan, X., and Huang, Z. (2018). *Angew. Chem. Int. Ed.* 57: 6319–6323.
36 (a) Tomooka, K., Nakazaki, A., and Nakai, T. (2000). *J. Am. Chem. Soc.* 122: 408–409. (b) Nakazaki, A., Usui, J., and Tomooka, K. (2008). *Synlett* 2064–2068.
37 (a) Bertrand, G., Dubac, J., Mazerolles, P., and Ancelle, J. (1980). *J. Chem. Soc., Chem. Commun.* 382–383. (b) Bertrand, G., Dubac, J., Mazerolles, P., and Ancelle, J. (1982). *Nouv. J. Chim.* 6: 381–386.
38 (a) Shintani, R., Moriya, K., and Hayashi, T. (2011). *J. Am. Chem. Soc.* 133: 16440–16443. (b) Shintani, R., Moriya, K., and Hayashi, T. (2012). *Org. Lett.* 14: 2902–2905.

39 Zhang, Q.-W., An, K., Liu, L.-C., Zhang, Q., Guo, H., and He, W. (2017). *Angew. Chem. Int. Ed.* 56: 1125–1129.

40 Shintani, R., Maciver, E.E., Tamakuni, F., and Hayashi, T. (2012). *J. Am. Chem. Soc.* 134: 16955–16958.

41 Kumar, R., Hoshimoto, Y., Yabuki, H., Ohashi, M., and Ogoshi, S. (2015). *J. Am. Chem. Soc.* 137: 11838 11845.

42 Shintani, R., Kurata, H., and Nozaki, K. (2015). *Chem. Commun.* 51: 11378–11381.

43 Namba, T., Shibata, Y., Sugiyama, K., Teraoka, K., Uekusa, K., and Tanaka, K. (2018). *Chem. Lett.* 47: 787–790.

44 Onoe, M., Baba, K., Kim, Y., Kita, Y., Tobisu, M., and Chatani, N. (2012). *J. Am. Chem. Soc.* 134: 19477–19488.

45 (a) Djerourou, A.-H. and Blanco, L. (1991). *Tetrahedron Lett.* 32: 6325–6326. (b) Djerourou, A.-H. and Blanco, L. (1999). *Phosphorus, Sulfur Silicon Relat. Elem.* 149: 107–136.

46 Lu, X., Yang, W., Jiang, K., Yang, K.-F., Zheng, Z.-J., and Xu, L.-W. (2013). *Eur. J. Org. Chem.* 5814–5819.

47 Igawa, K., Yoshihiro, D., Abe, Y., and Tomooka, K. (2016). *Angew. Chem. Int. Ed.* 55: 5814–5818.

48 (a) Shintani, R., Takagi, C., Ito, T., Naito, M., and Nozaki, K. (2015). *Angew. Chem. Int. Ed.* 54: 1616–1620. (b) Shintani, R., Takano, R., and Nozaki, K. (2016). *Chem. Sci.* 7: 1205–1211.

49 Naganawa, Y., Namba, T., Kawagishi, M., and Nishiyama, H. (2015). *Chem. Eur. J.* 21: 9319–9322.

50 Shintani, R., Otomo, H., Ota, K., and Hayashi, T. (2012). *J. Am. Chem. Soc.* 134: 7305–7308.

51 Sato, Y., Takagi, C., Shintani, R., and Nozaki, K. (2017). *Angew. Chem. Int. Ed.* 56: 9211–9216.

52 (a) Sommer, L.H. and McLick, J. (1967). *J. Am. Chem. Soc.* 89: 5802–5806. (b) Sommer, L.H., McLick, J., and Golino, C.M. (1972). *J. Am. Chem. Soc.* 94: 669–670.

53 Koller, S.G., Bauer, J.O., and Strohmann, C. (2017). *Angew. Chem. Int. Ed.* 56: 7991–7994.

54 Suzuki, K. and Kawakami, Y. (2003). *Organometallics* 22: 2367–2369.

55 Nakazaki, A., Nakai, T., and Tomooka, K. (2006). *Angew. Chem. Int. Ed.* 45: 2235–2238.

56 (a) Lambert, J. and Urdaneta-Pérez, M. (1978). *J. Am. Chem. Soc.* 100: 157–162. (b) Lambert, J.B. and Schulz, W.J. Jr. (1989). *The Chemistry of Organic Silicon Compounds* (eds. S. Patai and Z. Rappoport), 1007–1014. Chichester: Wiley.

57 Pioneering works on preparation of enantioenriched chiral silyllithium, see: (a) Sommer, L.H. and Mason, R. (1965). *J. Am. Chem. Soc.* 87: 1619–1620. (b) Colomer, E. and Corriu, R.J.P. (1976). *J. Chem. Soc., Chem. Commun.* 176–177. (c) Colomer, E. and Corriu, R.J.P. (1977). *J. Organomet. Chem.* 133: 159–168.

58 Omote, M., Tokita, T., Shimizu, Y. et al. (2000). *J. Organomet. Chem.* 611: 20–25.

59 (a) Strohmann, C., Hörnig, J., and Auer, D. (2002). *Chem. Commun.* 766–767. (b) Strohmann, C., Bindl, M., Fraaß, V.C., and Hörnig, J. (2004). *Angew. Chem.*

Int. Ed. 43: 1011–1014. (c) Strohmann, C., Däschlein, C., Kellert, M., and Auer, D. (2007). *Angew. Chem. Int. Ed.* 46: 4780–4782.
60 Oestreich, M., Auer, G., and Keller, M. (2005). *Eur. J. Org. Chem.* 184–195.
61 Cavicchioli, M., Montanari, V., and Resnati, G. (1994). *Tetrahedron Lett.* 35: 6329–6330.
62 Adam, W., Mello, R., and Curci, R. (1990). *Angew. Chem. Int. Ed.* 29: 890–891.
63 Adam, W., Mitchell, C.M., Saha-Möller, C.R., and Weichold, O. (1999). *J. Am. Chem. Soc.* 121: 2097–2103.
64 Okada, Y., Oba, M., Arai, A., Tanaka, K., Nishiyama, K., and Ando, W. (2010). *Inorg. Chem.* 49: 383–385.
65 Ishimoto, R., Kamata, K., and Mizuno, N. (2009). *Angew. Chem. Int. Ed.* 48: 8900–8904.
66 Lee, M., Ko, S., and Chang, S. (2000). *J. Am. Chem. Soc.* 122: 12011–12012.
67 Sommer, L.H. and Lyons, J.E. (1969). *J. Am. Chem. Soc.* 91: 7061–7067.
68 Kamachi, T., Shimizu, K., Yoshihiro, D., Igawa, K., Tomooka, K., and Yoshizawa, K. (2013). *J. Phys. Chem. C* 117: 22967–22973.
69 Mori, K., Tano, M., Mizugaki, T., Ebitani, K., and Kaneda, K. (2002). *New J. Chem.* 26: 1536–1538.
70 Igawa, K., Kokan, N., and Tomooka, K. (2010). *Angew. Chem. Int. Ed.* 49: 728–731.
71 Igawa, K., Kuroo, A., Yoshihiro, D., and Tomooka, K. (2017). *Synlett* 28: 2445–2448.
72 Bratovanov, S. and Bienz, S. (1997). *Tetrahedron: Asymmetry* 8: 1587–1603.
73 Rendler, S., Auer, G., and Oestreich, M. (2005). *Angew. Chem. Int. Ed.* 44: 7620–7624.
74 (a) Oestreich, M. and Rendler, S. (2005). *Angew. Chem. Int. Ed.* 44: 1661–1664. (b) Rendler, S., Oestreich, M., Butts, C.P., and Lloyd-Jones, G.C. (2007). *J. Am. Chem. Soc.* 129: 502–503.
75 (a) Li, Y. and Kawakami, Y. (1999). *Macromolecules* 32: 548–553. (b) Uenishi, K., Imae, I., Shirakawa, E., and Kawakami, Y. (2002). *Macromolecules* 35: 2455–2460. (c) Kakihana, Y., Uenishi, K., Imae, I., and Kawakami, Y. (2005). *Macromolecules* 38: 6321–6326.For a review, see: (d) Kawakami, Y. and Li, Y. (2000). *J. Polym. Res.* 7: 63–72.
76 Tacke, R., Reichel, D., Jones, P.G., Hou, X., Waelbroeck, M., Gross, J., Mutschler, E., and Lambrecht, G., (1996). *J. Organomet. Chem.* 521: 305–323.

Index

Organosilicon Chemistry: Novel Approaches and Reactions, First Edition.
Edited by Tamejiro Hiyama and Martin Oestreich

b

e

i

k

l